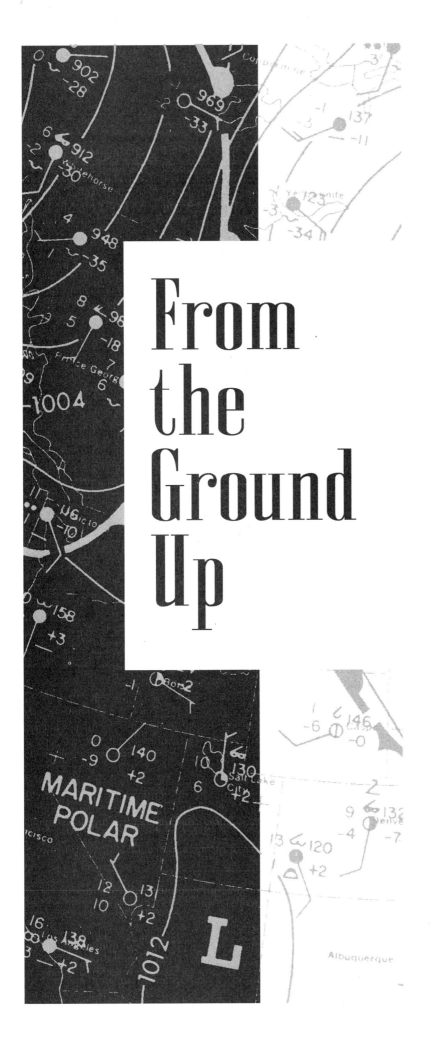

From the Ground Up

Published by
Aviation Publishers Co. Limited
Box 1361, Ottawa, Canada K1P 5R4
(613) 745-2943   Fax: (613) 745-9851
www.aviationpublishers.com

Acknowledgements

The publishers wish to express grateful appreciation for the assistance and co-operation which has been received in the preparation of this manual to officials of the Aviation Group, Transport Canada and to flying instructors and ground school lecturers from all across Canada who have submitted recommendations to improve this manual.

International Standard Book Number (ISBN) 09690054-9-0

International Standard Serial Number (ISSN) 0317-056X

Twenty-Seventh Revised Edition 1996

© Copyrighted

Photo Credit: Page 117. Fig. 6: Superstock/Four by Five
Photographer: Tom Rosenthal
Page 283 Fig. 36: Boeing Canada,
Dehavilland Division

Design and Production: WONG² + ASSOCIATES DESIGN GROUP INC.

So You Want To Be A Pilot

So you want to learn to fly. You are about to join a very select fraternity of those privileged individuals who have, in the words of the poet J. G. Magee, "slipped the surly bonds of earth and danced the skies on laughter-silvered wing".

Learning to fly is easy. The skill to handle an airplane in the air and during landing and take-off is acquired with a few hundred hours of patient application. Learning to be a proficient, professional and safe pilot is not so simple. It is a skill acquired through experience, judgment, persistent practice and ongoing training, It is, in fact, a challenge that never ends.

Professional pilots are, and always will be, highly sought, highly paid individual specialists because they are master craftsmen in their trade. So, what are the special skills that the professional pilot has?

Weather sense, for one thing: a knowledge of line squalls and thunderstorms and icing conditions, of stable and unstable air masses, of cold fronts, dewpoint and all the odds and ends that go to make up the science of meteorology.

They are expert navigators and understand how to plot headings and bearings. They have in depth knowledge of wind and drift problems. They are thoroughly conversant with such things as azimuth, isogonic lines and great circle tracks. They calculate altitude and airspeed corrections and understand the importance of determining density altitude before departing on a flight in hot weather conditions. They are proficient in the use of radio navigation aids, like ADF, Omni, DME, Loran and GPS.

Professional pilots know their airplanes and their engines. The lives of their passengers depend on the airworthiness of their equipment as much as on their own skills and knowledge — so they conscientiously superintend the service and maintenance of their airplanes. They understand fuel-air ratio and know how to get the last ounce of power and the most miles out of a given volume of fuel. They are familiar with all the invisible forces and couples that act on an airplane in flight and they know when the airplane has been subjected to any abnormal stresses that may lead to a structural strain.

In other words, professional pilots are individuals with whom one can fly with utmost confidence, based on the assurance that they not only rate officially as Grade A pilots and navigators but also as thorough technicians who are completely versed in every last-minute detail of their profession both on the ground and in the air.

"But," you may point out, "I have no ambition to become an airline captain. I am only interested in learning to fly as a private pilot. Is it necessary that I should learn all this technical stuff as well?"

Of course, it is necessary. Is your own life not every bit as precious to you as are the lives of its passengers to an airline company?

An airplane moves in a medium known as the atmosphere. This layer of air surrounding the earth for a depth of thousands of feet is a turbulent region of shifting winds, cross currents, storms, gusts and squalls. Invisible giants, the polar and equatorial air masses forever in conflict, make this atmosphere of ours a perpetual proving ground for the science of air navigation by frequent blanketing of entire areas with dense drop-curtains of cloud, fog, rain or snow.

An airplane moves in three-dimensional space which involves threefold problems in its control. It lacks buoyancy, is heavier than air and, hence, is dependent on the power from its engine to sustain it in flight. A forced landing, while not necessarily a hazardous undertaking, is nevertheless an undesirable course of action. Reliability is therefore a matter of vastly greater importance in the air than on land or sea. And reliability refers not only to the mechanical perfection of the airplane and its engine but also to the knowledge, judgment, and all-round proficiency that rides in the cockpit.

Time was when an older generation learned to fly by the seat of its pants. But time marches on and aviation has since swept ahead with giant strides. Many a private owner today will casually climb aboard his airplane to start off on a flight that to the pioneers of aviation would have seemed an epic undertaking. The pilot of today is equipped with information and knowledge that many decades of trial and error, of toil and effort and human sacrifice have placed at our disposal.

The pioneers in aviation had to get their experience the hard way and bore the scars of many a near thing. Today the ground school has become an international institution where those, who want to fly the scientifically sure way, may learn the things they should know the only sound and thorough way — FROM THE GROUND UP.

Contents

Part 1: Aircraft Operations — 7
The Airplane — 8
Theory of Flight — 19
    Flight Instruments — 37
Aero Engines — 49
    The Fuel System — 55
    The Carburetor — 57
    The Exhaust System — 65
    The Ignition System — 66
    The Electrical System — 68
    The Propeller — 69
    Engine Instruments — 72
    Operation of the Engine — 75
    Jet Propulsion — 81

Part II: Air Law — 85
Aeronautical Rules & Facilities — 86
    Aerodromes — 86
    The Canadian Airspace System — 94
    Rules of the Air — 100
    Air Traffic Rules and Procedures — 103

Part III: Meteorology — 113
Meteorology — 114
    Clouds — 115
    Pressure — 117
    Winds — 119
    Humidity, Temperature and Stability — 125
    Air Masses — 129
    Fronts — 130
    Clouds, Precipitation and Fog — 135
    Thunderstorms — 139
    Icing — 143
    Turbulence — 146
    High Level Weather — 147
    Weather Signs — 148
    Weather Information — 149

Part IV: Navigation — 163
Air Navigation — 164

Latitude and Longitude	164
The Earth's Magnetism	166
Units of Distance and Speed	172
Aeronautical Charts	173
Navigation Problems	184

Radio — **195**

Communication Equipment	199
Radio Communication Facilities	201
Radiotelephone Procedure	205

Radio Navigation — **215**

Omnirange Navigation System (VOR)	215
Radio Beacons	222
Instrument Landing System (ILS)	222
Microwave Landing System (MLS)	223
Automatic Direction Finder (ADF)	224
Distance Measuring Equipment (DME)	229
Flight Director	230
Area Navigation	231
Electronic Flight Instrument System (EFIS)	236
Radar and Radar Facilities	237
Emergency Locator Transmitter (ELT)	241

Part V: General Airmanship — 243

Airmanship — **244**

Care of the Airplane	244
Weight and Balance	249
Airplane Performance	253
Wake Turbulence	261
Airmanship	264
Emergency Procedures	277
Bush Sense	282
Ultra-Lights	285

Human Factors — **287**

Air Safety — **299**

Examination Guide — 305
Glossary — 315
Index — 329

For the sake of grammatical efficiency, the masculine pronoun is generally used in this text. Where used, it should be considered a generic term and read as he or she, him or her as required

PART 1

Aircraft Operations

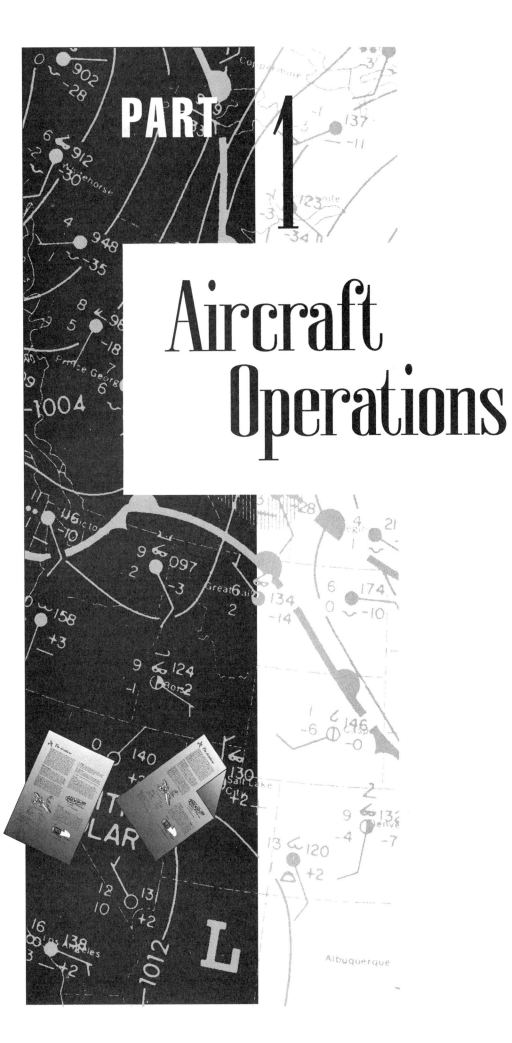

The Airplane

The reader may never experience the predicament of being down in the Arctic, 500 miles from the nearest outpost of civilization, with a damaged wing bracing strut or undercarriage. Canada's northcountry pilots, always noted for their outstanding resourcefulness, have been known, in many such emergencies, to successfully come home on hand whittled propellers, improvised struts and other ingenious make-shift repairs. The average pilot will probably never have occasion, in the course of his career, to rebuild or repair his own airplane, although there is these days a growing fraternity of aviation enthusiasts who are actually building their own airplanes. Whether one of these burgeoning "homebuilders" or simply a pilot who flies a production line model, a fundamental knowledge of the components of the airplane, their functions, structure, and particularly their limitations of strength and resistance to deterioration, is a very essential part of every pilot's qualifications.

The Airplane

The Canadian Aviation Regulations define an aeroplane as "a power-driven heavier-than-air aircraft deriving its lift in flight from aerodynamic reactions on surfaces that remain fixed under given conditions of flight". The terms "aeroplane" and "airplane" are for the most part interchangeable. The Canadian Air Regulations use the term "aeroplane" but common usage seems to favour "airplane". The latter is the form of choice of this manual.

There are numerous ways to classify a fixed wing airplane: by the position of its wings in relation to the fuselage, by the number of engines and by its undercarriage configuration. Airplanes may be of the high wing or low wing type. An airplane may be classed as single engine, if it has only one engine, or multi-engine, if it has two or more engines. An airplane is also classified according to the type of landing gear with which it is fitted for this determines the terrain from which it can operate: e.g. land, sea, ski. Landplanes can be further classified as having either conventional or retractable landing gear.

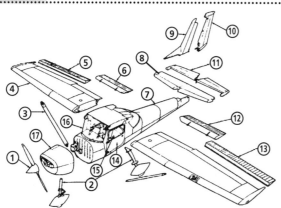

Fig. 1. Parts of an Airplane

1. Propeller
2. Landing Gear
3. Wing Strut
4. Wing
5. Right Wing Aileron
6. Right Wing Flap
7. Fuselage
8. Horizontal Stabilizer
9. Fin
10. Rudder
11. Elevator
12. Left Wing Flap
13. Left Wing Aileron
14. Door
15. Seat
16. Windshield
17. Engine Cowl

Parts of an Airplane

The essential components of an airplane are:
1. The fuselage or body
2. The wings or lifting surfaces
3. The tail section or empennage
4. The propulsion system, i.e. engine(s) with or without propeller(s)
5. Undercarriage or landing gear.

The **airframe** is the term used to describe the complete structure of an airplane, including the fuel tanks and lines, but without instruments and engine installed. It therefore includes the fuselage, wings, tail assembly and undercarriage.

The Fuselage

The fuselage is the central body of the airplane, designed to accommodate the crew, passengers and cargo. It is the structural body to which the wings, tail assembly, landing gear and engine are attached.

The fuselage is usually classed according to its type of construction.

Truss Type

In the early days, the fuselage was a frame made up of wooden members, wire braced. These materials are now obsolete, having been replaced by metal. The modern truss type fuselage is made up of steel tubes, usually welded or bolted together to form the frame. The **longerons** (three, four or more long tubes running lengthways) are the principle members and are braced, or held together, to form the frame by vertical or diagonal members, the whole assembly being in the form of a truss. Fig. 2 illustrates two types of steel tube truss type fuselage construction: N-Girder, and Warren Truss. The covering may be fabric, metal or composite.

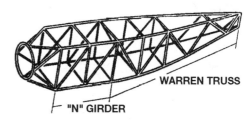

Fig. 2. Truss Type Fuselage.

Monocoque

The monocoque type of construction consists of a series of round or oval **formers** or **bulkheads** held together by **stringers** (long strips running lengthwise). The formers, or bulkheads, carry the loads, the stringers being merely superstructure. The early types of monocoque construction were of wood, plywood covered. Present monocoque construction is of metal, metal covered. Since the covering of the monocoque fuselage must be made stiff, the skin is capable of carrying some of the load. This is known as a **stressed skin** structure. A perfect stressed skin structure would be one in which the skin, in addition to providing the covering and forming the shape, would be capable of carrying all the load, without any internal bracing.

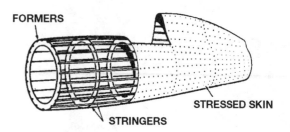

Fig. 3. Monocque (Stressed Skin) Fuselage.

The Wing

Most airplanes in use in general aviation today are monoplanes; that is, they have one pair of wings. Biplanes, those with two pairs of wings, are also to be found. They are usually restored antiques, agricultural spray planes, or sport and aerobatic airplanes. Wings come in a variety of shapes: rectangular, tapered from wing root to wing tip, elliptical, delta. They may be attached in different positions on the fuselage: at the top of the fuselage, known as high wing; at the bottom of the fuselage, low wing; or in the middle, mid wing. High wing airplanes may be externally braced with wing struts or may be fully cantilevered.

Five general systems of wing construction are now in use on modern airplanes. These are:

1. Metal frame, metal covered *(main strength in the covering, or skin, i.e. stressed skin)*.
2. Metal frame, metal covered *(main strength in the frame)*.
3. Metal frame, fabric covered.
4. Composite.
5. Wooden frame, fabric covered or plywood covered *(found mostly in classic and antique airplanes)*.

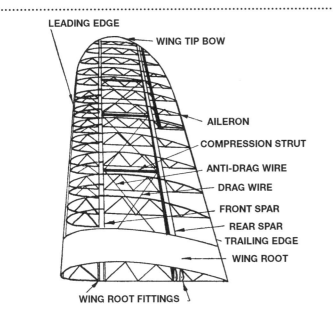

Fig. 4. Two-spar, Fabric-covered Wing.

The main members in a wing are the **spars**. These are beams, running the length of the wing from wing root to wing tip, which carry most of the load. The spars are intended to stiffen the wing against torsion or twisting. Some wings are constructed with two or more spars (**multispar**) and some with only one main spar (**monospar**). The latter type of single spar construction is found in certain models of modern airplanes which use a laminar flow airfoil wing design.

The **ribs** run from the leading to the trailing edge. They are cambered to form an airfoil section and their purpose is to give the wing its shape and to provide a framework to which the covering is fastened. To strengthen the leading edge, nose ribs are sometimes installed between the front spar and leading edge. These are generally known as **false ribs**.

Compression struts are spaced at regular intervals between the front and rear spars. They are usually steel tubes and are intended to take compression loads.

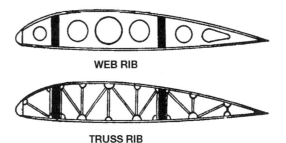

Fig. 5. Ribs

Further internal bracing is secured by **drag** and **anti-drag wires**. These are wires running diagonally from the front to the rear spars, the drag wires taking drag loads and the anti-drag wires anti-drag loads, as their names imply.

External bracing is secured in monoplane types by **wing bracing struts** which extend out from the fuselage to about the mid-section of the wing. In biplanes, **struts** are placed between the wings, well out towards the tips. These are braced by **incidence wires**, which run diagonally between the struts, and by **flying and landing wires**, which run diagonally between the struts and the fuselage. The flying wires transmit part of the load to the fuselage in flight and the landing wires support the weight of the wing on the ground.

Some wings are constructed with no external bracing at all. These are known as **cantilever** wings. Since there is no external support to such a wing, the spars must be made sufficiently strong to carry the load into the fuselage internally with no outside assistance.

Transmission of Loads — Internally. The load on a wing comes first on the skin. It is then transmitted to the ribs and from these to the spars and thence carried into the fuselage.

Externally. In an externally braced wing, part of the load is taken by the bracing struts or the flying or landing wires, as the case may be, and thence transmitted to the fuselage.

AILERONS. These are surfaces, usually of airfoil section, hinged to the trailing edge of the wing towards each wing tip for the purpose of **lateral control**. Their internal construction is much like that of the wing itself. They are usually hinged to the rear spar.

FLAPS. When fitted, these form a part of the wing structure. Like the ailerons, they are usually hinged to the rear spar. A full description of flaps and their function will be found in the Chapter **Theory of Flight**.

WING TIP BOW. This is generally a metal tube, curved to give the wing tip the particular shape required.

WING ROOT. The section of the wing nearest the fuselage. On low wing airplanes, it is reinforced to permit the passengers and crew to walk on it.

WING ROOT FITTINGS. The fittings which attach the wing, or the separate wing panels, to the fuselage.

WINGLET. A small nearly vertical winglike surface, usually of airfoil section, attached to the wing tip. The winglet is incorporated into the design of some modern airplanes. It is usually located rearward above the wing tip and is effective in reducing induced drag. (See Chapter **Theory of Flight**.)

CHORD. An imaginary straight line joining the leading and trailing edges of the wing. The **mean aerodynamic chord (MAC)** is the average chord of the wing.

SPAN. The maximum distance from wing tip to wing tip of an airfoil, wing or stabilizer.

The Tail Section or Empennage

The empennage is the tail section of the airplane and consists of a fixed vertical stabilizer or fin, the rudder, the stabilizer or tail plane, the elevators and all trimming and control devices associated with them. Instead of a fixed stabilizer and movable elevators, some airplanes have a one piece pivoting, horizontal stabilizer that is known as a stabilator.

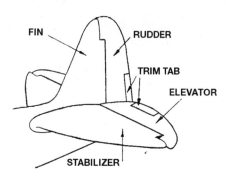

Fig. 6. The Tail Section.

THE TAIL PLANE or STABILIZER. An airfoil placed at the rear end of the fuselage to balance the airplane and provide **longitudinal stability.**

ELEVATORS. Surfaces hinged on the trailing edge of the stabilizer to give **longitudinal control.**

FIN. A fixed vertical surface placed ahead of the stern post to provide **directional stability.** The fin is usually offset from the centre to compensate for the corkscrew motion of the slipstream from the revolving propeller.

RUDDER. A movable surface hinged to the fin to give **directional control.**

TRIM TAB. An adjustable tab either fixed or hinged to a control surface (rudder, elevators and ailerons) that helps the pilot by eliminating the need to exert excessive pressure on the cockpit flight controls during the various phases of flight.

STABILATOR. A single airfoil section that replaces the combination of stabilizer and elevator. It is attached to the fuselage at a point around which it pivots.

Fig. 7. Stabilator.

CANARD. A few airplanes of modern design have replaced the familiar tail section with a canard that incorporates a horizontal stabilizer assembly at the front of the airplane. The arrangement, though much more streamlined, is reminiscent of the original Wright airplanes.

The airfoils comprising the tail unit assembly are similar to, but of lighter construction, than those of the main structure. The tail unit is positioned so that it is in the airflow and not blanketed by the main planes or other parts of the structure.

The Propulsion System

The propulsion system of the modern general aviation airplane is generally a gasoline powered, air cooled, internal combustion engine that drives a 2 or 3 bladed propeller. Many of the new business airplanes are jet powered as are most of the large transport type airplanes. The gas turbine, or jet, engine can also be used to drive a propeller and, in this configuration, is known as a turboprop engine.

Power plants and propellers are discussed in detail in the Chapter **Aero Engines.**

The Cowling

The cowling encloses the engine and streamlines the front of the airplane to reduce drag. The cowling provides cooling of the engine by ducting cooling air around the engine. On high performance airplanes, adjustable openings called **cowl flaps** are incorporated into the cowling to control the amount of cooling air circulating around the engine.

Engine Mountings

The engine is supported by a structure, usually of steel tubing welded together, called the **engine mount**, which is made flexible to absorb vibration from the engine and prevent it being transmitted to the fuselage. This is usually accomplished by **engine mount bushings** which are made springy in the direction of the engine rotation but rigid otherwise, in order to hold the engine steady fore and aft.

Fire Wall

Between the main structure and the engine is the **fire wall**. This is made of a heavy sheet of stainless steel or often a sandwich of asbestos between two sheets of dural. Openings for fuel and control lines are made small, with bushings to ensure a snug fit. The fuel tank must be behind the fire wall, whereas the oil tank may be ahead of it — oil being less flammable than gasoline.

Tank Installations

Fuel tanks may be carried in the wings or in the fuselage. See **Fuel System** in Chapter **Aero Engines.**

Landing Gear or Undercarriage

The function of the landing gear is to take the shock of landing and also to support the weight of the airplane and enable it to maneuver on the ground. The earliest type of main landing gear was a through axle, similar to the wheel and axle arrangement on a cart or wagon. This is now completely obsolete, having been replaced with more sophisticated, shock absorbing landing gear systems.

The landing gear on modern airplanes is either of the fixed gear type or retractable.

Fixed Undercarriage

On land airplanes, there are two basic classes of fixed gear undercarriage: main gear with a nose wheel, commonly called a tricycle gear, and main gear with a tail wheel. There are several types of undercarriage in use for the main gear. These are used with both the tail wheel and the tricycle gear configuration. They are **split axle, tripod, single spring leaf cantilever** and **single strut.**

The **split axle** type has the axle bent upwards and split in the centre to enable it to clear obstructions on the ground (Fig. 9). This type is used on airplanes such as the Piper PA-22. It is suspended on shock cords wound around a fuselage member which enables the whole assembly to spread when loads come on it. A strut or tie rod is usually incorporated to brace the structure against side loads.

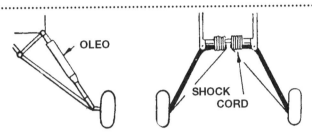

Fig. 8. Tripod Landing Gear.   Fig. 9. Split Axle

The **tripod** landing gear is illustrated in Fig. 8 and is typical of airplanes such as the Aeronca and the Champion. This gear consists of three members hinged so as to form a triangle. Two of these are rigid. The third is an oleo leg, designed to telescope and hence shorten its length when the load comes on the wheel. On landing, the whole assembly spreads outwards and upwards until springs, rubber discs, or other devices take the weight.

The **single leaf cantilever spring steel** type of main landing gear is used extensively on Cessna airplanes. The gear consists of a single strap of chrome vanadium steel bent to form the shape of the complete undercarriage structure. It is attached to the fuselage in a cradle bulkhead by bolts. It is capable of storing energy in initial impact, thereby producing quite low load factors. Low maintenance, simplicity and long service life characterize this gear type (Fig. 10).

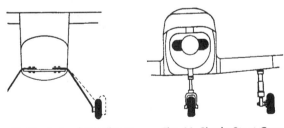

Fig. 10. Single Leaf Cantilever Spring Steel Gear.   Fig. 11. Single Strut Gear.

On the Cessna "Cardinal", a spring steel tubular gear replaces the more familiar single leaf gear described here. The spring steel tubular gear has the same characteristics as the single leaf type.

The **single strut** type is used on several modern, low wing, fixed gear airplanes such as the Piper Cherokee and the Beech Musketeer. This gear consists of a single leg or strut extending downward from its attachment point on the main spar. The strut usually incorporates a hydraulic cylinder or rubber biscuits for the purpose of absorbing the shock (Fig.11).

Retractable Gear

Retractable gears are made to retract or fold up into the wing or fuselage in flight. The mechanical means and methods for accomplishing this are many and varied. The wheel may fold sideways outwards towards the wing or inwards towards the fuselage. The latter is most common on high speed military airplanes when the wing camber is shallow. On some multi-engine airplanes the wheel folds straight back or forward into the nacelle and is left partly projecting in order to protect the belly of the ship in the case of a wheels-up landing. Some retractable undercarriages are made to turn through 90 degrees as they travel up and so fold into the side of the fuselage.

Most retractable undercarriage legs are cantilever, being a single oleo leg with no external bracing. They are hinged at the top to permit them to fold. The means of retraction may be a hand gear, electric motor, or motor-driven hydraulic pump. Where mechanical means are used, a hand gear is also provided to allow for lowering the gear in an emergency.

Making the undercarriage retractable is a common practice with both the tricycle and tail wheel configuration. In the case of tricycle gear, the nose wheel is also made retractable. In the case of a tail wheel, however, because it is small and causes little drag, it is fixed.

Nose Wheel Versus Tail Wheel

The practice of placing a steerable third wheel forward of the main gear has found universal acceptance in modern airplane design and is referred to as being a tricycle gear configuration. The landing gear configuration in which the third wheel is rearward of the main gear (i.e., at the stern of the airplane) is referred to as a tail wheel configuration (old timers fondly call such airplanes, tail-draggers). The modern trend to the tricycle gear configuration by most manufacturers is the result of certain advantages that this type of landing gear has over the tail wheel configuration. These advantages are: (1) Nose-over tendencies are reduced greatly. (2) Ground looping tendencies are reduced. (3) Visibility over the nose when taxiing, taking off or landing is superior due to the level flight position of the airplane while on the ground. (4) Greater maneuverability on the ground under high wind conditions due to the negative angle of attack of the wings. (5) Greater controllability on the ground in cross windconditions. Therefore, tricycle geared airplanes can use single runway airports (which are becoming more numerous) with greater safety in cross wind conditions than can tail wheel airplanes. (6) A novice can usually learn to maneuver a tricycle geared airplane on the ground in less time than he can master a tail wheel airplane.

Tail wheel airplanes have advantages too. These are: (1) The tail wheel has less parasite drag than a nose wheel due to its smaller size. (2) The tail wheel is cheaper and easier to build and maintain. (3) A broken tail wheel will not result in as much damage to an airplane as would a broken nose gear. (4) A tail wheel airplane can be more easily man handled on the ground and, because the tail is lower than that of a tricycle-geared airplane, it fits into some hangar space more easily. (5) When using rough sod, sand or gravel airports, the tail wheel airplane will sustain less propeller damage since the tips of the propeller are farther away from the ground and are less likely to pick up loose objects, such as stones and debris. (6) With constant use in rough fields, the tail wheel airplane is not as likely to sustain airframe damage since it is the main undercarriage which takes the bulk of the load and the shock when the airplane rides over depressions and irregularities on the ground. The main undercarriage (which hits the bumps first) is attached to a primary structure and is therefore stronger and more rigid than a nose gear (which in the tricycle gear configuration is the first to hit the bump) which is usually fastened to a weaker or non-primary part of the airframe. A tail wheel will easily absorb bumps that may be severe enough to damage a nose gear. (7) Tail wheel airplanes are more suitable for change-over to ski operations in winter.

On most modern airplanes, regardless of whether they have a fixed or retractable undercarriage, the nose wheel and the tail wheel are steerable by the pilot's controls.

Shock Absorbers
The purpose of the shock absorber is to prevent landing shock damage to the fuselage or body of the airplane. Pilots may accidentally impose heavy stresses due to faulty landings. If these stresses were not properly absorbed by the landing gear, they could easily cause failure in the airplane structure.

Shock absorbers generally are divided into four classes:

1. *Low Pressure Tires:* On some types of light airplanes, low pressure tires are the sole means provided for absorbing shocks. The principal difficulty with tires (and some of the other shock absorbing devices) is that they do not dissipate the shock but store it and kick the airplane back into the air after a rough landing.

2. *Oleo:* When the airplane on landing first contacts the ground, the momentum must be absorbed in the undercarriage. To absorb this energy on springs or rubber alone would result in the aircraft being bounced into the air again. On practically all modern airplanes, the energy produced on landing is dissipated by forcing oil (an incompressible fluid) from one side of a piston to the other through a **small orifice**. The displacement of the oil is thus **delayed**, cushioning the shock of landing for the reason that the bulk of the energy is absorbed in forcing the oil through the restricted orifice.

serves **only to absorb the shock of landing.** Further shocks experienced while taxiing or taking off are handled by devices such as the spring shown in Fig. 13 (Oleo-Aerol), or by compressed air, (Fig. 14. Oleo-Pneumatic).

3. *Rubber*: Two types of rubber shock absorbers are in use, usually in conjunction with the oleo, to cushion further shocks after landing. These take the form of rubber **discs** or **doughnuts** and **shock cord**, which is an elastic cord wound around two moving members.

4. *Spring Steel*: The spring steel type of landing gear, as described above, is in itself a shock absorber capable of storing energy.

Brakes
The advantage of the use of brakes on airplanes is two-fold:

1. They provide quick deceleration, or pull-up, after landing. For and high speed airplanes that land with faster initial, or "hotter", speeds, such quick deceleration is important, especially when landing on short runways.

2. Differential, or individually operated brakes, ensure better control after landing, to prevent ground loops, etc. They also provide better maneuverability on the ground. On some models of airplanes, steering while taxiing is accomplished only by the use of the brakes. They are needed to perform short radius turns.

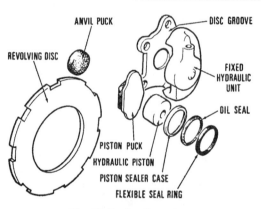

Fig. 15. Hydraulic Brake.

Due to the much higher landing speeds of modern airplanes, brakes have to be powerful, reliable and capable of dissipating heat very rapidly.

Nearly all airplanes use **disc brakes** operated by hydraulic pressure, sandwiching a rotating **disc** between two brake linings called **pucks.**

These pucks are located in a fixed cast unit, grooved to permit the disc to float freely. Attachment of the disc is attained by splitting the periphery into the wheel hub. This floating action allows the disc to move laterally during braking and permits the use of one moving puck. The fixed puck is called the **anvil**; the moving one is called the **piston puck**.

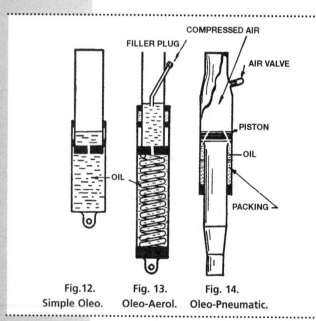

Fig.12. Simple Oleo. Fig. 13. Oleo-Aerol. Fig. 14. Oleo-Pneumatic.

The simple oleo (Fig. 12) consists of an **inner cylinder** which is attached to the fuselage and an **outer cylinder** fastened to the wheel. On landing, these will telescope and the oil will be displaced from the lower to the upper but is delayed in doing so by the restricted orifice. Since the oil, once displaced, will not return until the airplane again leaves the ground, the oleo leg

Pressure applied against the brakes that usually are part of the rudder pedal assembly is translated into hydraulic fluid pressure. The **hydraulic piston** responds to the increased pressure by pushing against the piston puck which in turn pushes the rotating disc against the anvil puck, allowing equal braking force (friction) on both sides of the disc. Special **flexible sealing rings** keep the puck-to-disc clearance automatically adjusted by returning the hydraulic piston to a neutral position after each braking action.

Disc brakes are so reliable that, normally, visual inspection is required only at 50 hour intervals. One precaution in their use is recommended. The parking brake should be left off and wheel chocks installed if the airplane is to be left unattended. Changes in the ambient temperature can cause the brakes to release or to exert excessive pressure.

A further problem can occur in airplanes that are flown infrequently (e.g. less than 100 to 200 hours per year). Since the discs are made of steel, they are subject to corrosion and rust, especially if exposed to unusual amounts of moisture, salt or industrial pollution. In an airplane that is used daily, the corrosion and rust are rubbed off by repeated use.

The prime element of the braking system is the hydraulic fluid. It transmits pressure and energy, lubricates the moving parts of the system and aids in cooling the working parts. It is important to check carefully the Owner's Manual to find out exactly what kind of brake fluid to use. Mixing different fluids negates the effectiveness of the hydraulic system. Some brake fluids can break down the rubber rings of incompatible systems. Brake fluid must be kept scrupulously free of contamination by dirt which can render the system effectively inoperative.

In some airplanes, the brakes are operated by pneumatic (air) pressure. A **pressure bag** is incorporated on the inside of the brake assembly. Air pressure admitted to this pressure bag causes it to expand, forcing the **brake shoes** to move radially outward against the surface of the **brake drum.**

The Control Systems
Ailerons
The ailerons are control surfaces attached to the trailing edge of the wing near the wing tip and are employed to bank the airplane. They move in opposite directions to each other and are controlled by movement of the control wheel or stick.

Three types of control systems are traditionally used to operate the ailerons. These are: (1) cables and pulleys, (2) push and pull rods and (3) torque tubes (Fig. 16). When stick control is used, any of these systems may be employed. With wheel control, cables and pulleys are generally used, although in some cases, push and pull rods may be utilized. In larger transport airplanes, the control systems are usually operated by a system of cables and pulleys aided by hydraulic systems. The new generation of transport airplanes have incorporated computerized control systems which allow operation of the aircraft controls (ailerons and also elevators and rudder) with electronic signalling. The controls are activated by electronic signals sent through wires from computers in the cockpit.

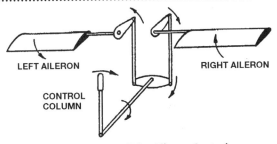

Fig. 16. Torque Tube Aileron Control.

When the control wheel is rotated to the right (or the control stick moved to the right), the left aileron moves down and the right aileron moves up. The lifting capability of the left wing is therefore increased at the same time as the lifting capability of the right wing is decreased. The left wing lifts and the right wing descends and the airplane rolls to the right. The airplane will continue to roll to the right, steepening the angle of bank, until the controls are neutralized establishing a particular angle of bank.

When the control wheel is rotated to the left, the left aileron moves up and the right one moves down and the airplane rolls to the left.

Elevators and Stabilators
A movable horizontal tail surface controls the movement of the airplane longitudinally and controls the angle of attack of the wings. The movable horizontal tail surface may be either elevators or stabilators.

The elevators or stabilators are operated by: (1) a system of cables and pulleys; (2) a rocking beam and cable; or (3) a push and pull rod system. These systems are connected to the pilot's control column.

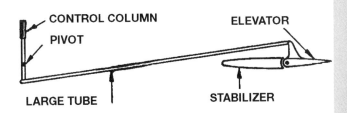

Fig. 17. Push and Pull Rod Elevator Control.

The elevators are hinged to the trailing edge of the horizontal stabilizer and are controlled by forward or aft movements of the control wheel. They move together. When the control wheel is pushed forward, the elevators move down, increasing the lifting capability of the tail. The tail rises and the nose of the airplane moves down. When the control wheel is pulled back, the elevators move up, the lift on the tail is decreased, the tail moves down and the nose of the airplane rises.

The stabilator is a one piece, horizontal tail surface that pivots up and down. It operates on the same principle as the elevators, moving up or down, changing its angle of attack and hence its lifting capabilities as the pilot pulls back or pushes forward on the control wheel.

Rudder
The rudder moves the airplane either left or right in a motion known as **yaw**. The rudder is attached to the trailing edge of the vertical stabilizer, or fin, and is connected to the rudder pedals by a cable system.

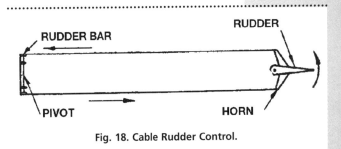

Fig. 18. Cable Rudder Control.

Pressure applied to the left rudder pedal displaces the rudder to the left into the airstream, increasing the pressure on the left side of the tail and forcing the tail to move to the right. The nose of the airplane moves to the left. Pressure applied to the right rudder pedal moves the nose of the airplane to the right. The rudder is used with the ailerons to achieve co-ordinated turns.

Trim Systems

Several types of trim devices are incorporated into the control system to help the pilot by eliminating the need to exert excessive pressure on the cockpit flight controls during the various phases of flight.

Trim Tabs

Trim tabs are adjustable devices located at the trailing edge of control surfaces such as elevators, rudders or ailerons. Their function is to permit the pilot to fly the airplane in a desired attitude, under various load and airspeed conditions, without the need to apply constant pressure in any particular direction on the flight controls.

A trim tab is, in effect, a control surface hinged to another control surface. It is designed to move above or below the chord line of the control surface to which it is attached and thereby create an aerodynamic force that assists the pilot in holding the control in the desired position. The trim tab, for example, is deflected downward in order to hold the control surface up. (See Fig. 19, in which a nose up attitude is achieved by use of trim tabs.)

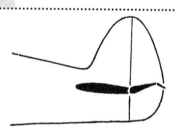

Fig. 19. Elevator Trim Tab.

The trim tab is operated from the cockpit by its own control which is located to be within easy reach of the pilot. A tab position indicator is incorporated in the control mechanism to show the nose up or nose down position of the tab setting. The mechanism that operates the trim tab is usually a system of wires and pulleys.

The trim tab is not the only device for affecting trim. Some trimming devices incorporate an adjustable spring tension as a means to exert pressure on the control surface to maintain the trimmed position. These are known as **bungees**.

Some form of inflight adjustable trim control is incorporated into the pitching plane of even the smallest airplane. Trim controls are also used on aileron controls, especially on multi-engine airplanes, and on rudder controls.

On airplanes in which in-flight adjustable trim is incorporated only on the elevator controls, ground adjustable trim tabs are often attached to ailerons or rudder. These have the effect of helping to correct any slight tendency of the airplane to roll or yaw as the result of less than perfect rigging.

Anti-servo tabs serve as trimming devices on stabilators.

Servo tabs are a device found most often on larger airplanes. They are connected directly to the control column. As in the case of the elevator, if the pilot moves the control column back, the servo tab is deflected downward. Air pressure on the tab deflects the elevator control upward to achieve a nose up attitude. The control column controls the servo tab, the elevator is free floating and moves in accordance with the tab deflection.

Adjustable Stabilizer

On some airplanes, longitudinal trim is achieved by adjusting the angle at which the stabilizer is attached to the fuselage. The leading edge of the stabilizer is moved up or down by means of a screw jack device that is controlled by a wheel or crank in the cabin. To effect a nose down attitude of the airplane, the leading edge of the stabilizer is rotated up, giving the stabilizer a higher angle of attack. The stabilizer, like trim tabs, can be set in any position between full up and full down. The adjustable stabilizer has the advantage of producing less drag than the conventional trim tab.

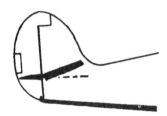

Fig. 20. Adjustable Stabilizer.

Movable Tail

On some airplanes, the entire empennage is hinged to pivot either forward or aft. A nose down attitude is achieved by rotating the tail aft. A nose up attitude results from rotating the tail forward.

Fig. 21. Movable Tail.

Cabin Pressurization

High altitude flight places the crew and passengers in a hostile environment, exposing them to the trauma of decreased oxygen in the air and decreased atmospheric pressure forces on the body. Cabin pressurization was devised as a method to maintain cabin pressure at a level lower than that of the actual flight altitude. This is done by increasing the air pressure in the cabin to provide sufficient oxygen to allow crew and passengers to breathe normally.

To have cabin pressurization, the cabin structure must be strong enough to withstand the normal twisting and flexing forces of flight. Doors, windows and control cable openings must be sealed. There must be a source of compressed air for pressurization and ventilation capable of large quantities of air flow. Regulation of temperature, pressure and rate of pressure change and a means of controlling the rate of outflow of air from the cabin all have to be addressed in designing a pressurized airplane.

Most general aviation airplanes with reciprocating engines utilize some air from their turbochargers (see **Turbochargers** in

Chapter **Aero Engines**) to achieve pressurization. The compressed air to serve the needs of the cabin passes through a venturi, called a flow control unit, that limits the flow from the turbocharger compressor so as not to starve the engine nor increase the volume of air flowing to the cabin beyond what is required. Each aircraft has a **design limited maximum allowable pressure differential (PSID)** between cabin and outside pressure. The greater the PSID, the higher the aircraft can fly while maintaining an acceptable cabin altitude. The PSID is the limiting factor in determining the aircraft's service ceiling.

Construction Materials

STEEL: Low carbon steels are tough, ductile and readily weldable but are incapable of being surface hardened except by case hardening.

Mild steels can be hardened, are strong but less ductile, less weldable. Used for fuselages and control surfaces.

High carbon steels exhibit increased strength and hardness but at the sacrifice of ductility and weldability.

Alloy steels, such as chrome moly, are very strong and resistant to impact and vibration. Used in the fabrication of fuselages. Alloy steels containing nickel (called stainless steel) are very corrosion resistant. Used for stressed skin structures, particularly in seaplane construction.

DURAL: An aluminum wrought alloy containing copper and magnesium. Has a very high tensile strength and fatigue endurance. Susceptible to corrosion but can be treated by anodizing. Used for ribs, tanks, bulkheads, propeller blades, fittings, etc.

ALCLAD: A sandwich of dural between two layers of pure aluminum (the aluminum layers constituting about 5-1/2% of the whole). The aluminum protects the dural and prevents corrosion. Very corrosion resistant. Used in seaplane construction. Needs no anodizing.

MAGNESIUM ALLOY: An alloy in which magnesium forms the principal constituent. Combines tensile strength with light weight (one-third lighter than aluminum). Used extensively in aircraft engine construction. Very corrodible in sea water and should always be anodized.

HONEYCOMB SANDWICH CONSTRUCTION: A metal honeycomb pattern between sheets of metal (Fig. 22). For cabin floor, door surfaces, etc., it offers the advantage of high strength/weight ratio, a smooth surface that does not buckle under load, and excellent bearing and bending properties in all directions. In supersonic airplanes, its ability to dissipate high temperatures makes it particularly suitable for wing skin structures.

COMPOSITE: Fibreglass cloth and epoxy resin molded over a foam form constitute the airplane structure. Composites have the advantage of weighing much less than aluminum and other conventional materials, while possessing great strength that is derived from their hidden fibres. A graphite-composite material is also being used in some new airplanes.

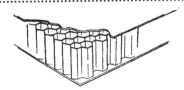

Fig. 22. Honeycomb Sandwich Construction

WOOD: Wood is still used in airplane construction for structural members. Plywood is used as a covering, giving a very smooth finish. Some airplanes are all wood; others are partly wood.

FABRIC: Airplanes whose fuselage structures are made of steel tubing may be covered with cotton, linen or synthetic fabric. The fabric is drawn taut either by the use of aircraft dope (cotton or linen) or by shrinking with a hot iron (synthetics such as dacron). Fabric is also used as the covering of wings whose spars and ribs are either all wood or all metal or a combination of both.

Corrosion

Corrosion must be treated as the enemy of all metal parts of an airplane. The attack may take place over an entire metal surface or it may be penetrating in nature, forming deep pits. There are a number of different types of corrosion.

1. *Oxidation:* This is produced by atmospheric conditions due to the moisture in the air. The effect is worse in the vicinity of salt water. The action consists of the dissolving of the surface by oxidation. Such oxidation is easy to detect. It may be removed and the surface treated with some preventative so further damage will not occur.

2. *Intercrystalline:* This type is more serious. It is caused by chemical or electrolytic action between the alloys in the metal itself. It may not become visible until considerable damage has been done. Surface protection aids very little in the prevention of this type of corrosion. The affected parts must be removed and replaced.

3. *Dissimilar Metals.* When metals of different chemical properties are in contact in the presence of moisture, the metal most easily oxidized will be subject to corrosion.

4. *Stress Corrosion.* When a metal part is overstressed over a long period of time under corrosive conditions, stress corrosion may result. Parts that are susceptible to stress corrosion are overtightened nuts in plumbing fittings, parts joined by taper pins that are overtorqued, fittings with pressed in bearings. Stress corrosion is not easy to detect until cracks begin to appear.

Corrosion fatigue is a type of stress corrosion that occurs where cyclic stresses are applied to a part or assembly. These stresses produce pores or cracks in the surface coating which allow moisture to penetrate.

5. *Fretting Corrosion* occurs when there is a slight movement between close fitting metal parts. The movement destroys any protective film on the metal surface and also produces fine particles of metal and oxide that tend to absorb and retain moisture.

A number of surface treatments have been developed to reduce or eliminate corrosion. Aluminum alloys are usually anodized, a process that provides a protective film. Steel parts are protected by cadmium plating, chrome plating or by phosphate processes that protect the surface from oxidation.

Stresses

A STRESS is the force, or combination of forces, exerting a strain. A pressure of your hand on the surface of, say, a small empty box is a stress.

A STRAIN is the distortion in form or bulk of a body due to stress. If the small box is crushed, it is said to be **strained**. A wire stretched is another example of **strain.**

There are five distinct types of stress:

1. *Compression* or "crushing", as in the case of the small box. Airplane wings are subjected to compression stresses.

2. *Tension* or "stretching", as in the case of the wire. Bracing wires in airplanes are usually in tension.

3. *Torsion* or "twisting". A screwdriver is subjected to severe torsional stress when forcing a screw into hardwood. Landing gear must be made to withstand torsional stresses.

4. *Shearing* or "cutting". The blades of scissors exert a shear stress on a piece of paper, which is "sheared" as a result.

5. *Bending*, as the name implies, means the bending of a long member due to a load or weight being imposed on it. Aircraft spars, or beams, must resist severe bending stresses.

An airplane structure in flight is subjected to many stresses due to the varying loads that may be imposed. The designer's problem is to try to anticipate the possible stresses that the structure will have to endure and to build it sufficiently strong to withstand these. This problem is complicated by the fact that an airplane structure must be **light** as well as **strong.**

Strength and lightness are essential in the structure of an airplane. Another factor almost as essential as strength is rigidity. Excessive deflection or bending under a load may lead to a loss of control with serious consequences.

Lack of rigidity may also lead to **flutter.** This is a rolling or weaving motion which arises when a deflection of a part of the structure causes the air forces on it to change in synchronism with its natural period of vibration. Flutter is most likely to occur in wings and control surfaces and may lead to structural failure. To prevent flutter, the wing and tail structures must be made stiff against both bending and twisting.

The narrow margin of safety permitted by weight limitations in airplanes makes it necessary that every member must bear its proper share of the load in every condition of flight. To attain uniform and adequate structural safety, it is essential to calculate what load each part may be called upon to carry. Such a determination of loads is called a **stress analysis**. This is a complicated mathematical process and is distinctly a job for only the trained engineer.

Loads and Load Factors

Airplane strength is measured by the total load the wings are capable of carrying. The load imposed on the wings depends on the type of flight. The wings must support not only the weight of the airplane but also additional loads imposed during maneuvers.

The WING LOADING of an airplane is the **gross weight** of the airplane divided by the **area** of the lifting surfaces and is expressed in lb. per sq. ft., i.e. the number of lb. that each sq. ft. of lifting surface must support. (The definition of wing loading may be slightly modified for particular applications, for example, with reference to ultra-light airplanes. See **Ultra-Lights** in Chapter **Airmanship**.)

The SPAN LOADING of an airplane is the gross weight divided by the span and is expressed in lb. per foot.

The POWER LOADING is the gross weight of the airplane divided by the hp of the engine(s) and is expressed in lb. per hp.

LOAD FACTOR: The weight of an airplane standing on the ground (or its weight due to gravity alone) is commonly referred to as a **dead load**. In flight, however, the weight of an airplane may be increased many times by **acceleration** (rate of change of speed) and/or by a change of direction. The additional loading imposed is called a **live load.**

The **load factor** is the ratio of the actual load acting on the wings to the gross weight of the airplane. In other words, it is the ratio of the live load to the dead load. When an airplane is in level flight, the lift of the wings is exactly equal to the weight of the airplane. The load factor is then said to be 1. In most maneuvers, such as a change in attitude, a banked turn, a pull out or any maneuver causing acceleration, centrifugal force enters the picture and brings about a change in the load factor. In a level turn at a bank angle of 30 degrees, for example, the load on the wings is increased to 1.15. In a 60 degree bank, the load factor goes up to 2. In a hard landing, the total load acting upward on the wheels may be as much as three times the weight of the airplane. The **landing load factor** in this case would be 3.

A load factor of 3 is often expressed as 3g. In this case, the letter g refers to gravity. Hence, 3g means a load on the wings equal to three times the weight of the airplane due to gravity alone.

There are, of course, maximum limits to which an aircraft is designed. These are usually referred to as **limit load factors**. The fact that it is sometimes possible to exceed these limits is evidence of the safety factor that is incorporated in all aircraft designs. Nevertheless, the limit load factor should not be exceeded intentionally because of the possibility of causing permanent set or distortion of the structure.

The flight maneuvers which impose high load factors are: steep turns, pull-outs, flick rolls, tail slides, and inverted loops. **These should be executed with due consideration on the pilot's part of the stresses which the particular airplane he is flying is designed to withstand.**

Airplanes which fly at several times their stalling speed are subject to excessive g loads in some circumstances. In an airplane that is flying at twice its stall speed, if the angle of attack is abruptly increased to obtain maximum lift, a load factor of 4 will be produced; at three times the stall speed, 9g would result; at four times the stall speed, 16g would result.

Weight also can result in high load factors. If an airplane is heavily loaded, the allowable load factors will be reduced accordingly and the pilot is likely to damage the structure in maneuvers that would normally be quite safe. **Therefore, when doing aerobatics, always be sure that the airplane is lightly loaded.**

In flying a heavily loaded airplane, a pilot should not trust his senses in determining the actual load on the wings. Because the heavy airplane is steady even in rough air, the pilot may get the false impression that the air turbulence is not excessive. However, the wings sense the actual load and may be about at their breaking point. On the other hand, a pilot flying an airplane that is lightly loaded may experience a good deal of buffeting and personal discomfort in rough air. Because of this, he may feel that the load factor is excessive whereas the wings which sense the actual load are not being over stressed. In any degree of turbulence, it is important to reduce the airspeed to prevent damage to the airplane structure. (See **Gust Load** below.)

Gust Load. Gusts are rapid and irregular fluctuations of varying intensity in the upward and downward movement of air currents. An airplane in a rising or descending current of air is not affected. When, however, the speed or direction of the air current changes abruptly (such as when flying at high speed through successive up-and-down gusts), load changes are imposed on the airplane structure. When an airplane flies out of a down-gust and immediately into an up-gust, for example, the effect on the wings is to suddenly increase the angle of attack. The **lift** is then in excess of the weight, and the airplane accelerates in an upward direction, just as it would if the pilot suddenly pulled back on the controls. If the total lift were to exceed the total weight by a factor of 2, the airplane would experience a 2g acceleration. This is known as a **gust load.**

Gusts, therefore, can impose very high load factors on the airplane. In fact, since gust loads can be sufficiently severe to be dangerous, it is wise to avoid, if at all possible, extremely rough air.

The faster the airplane is going, the more stress to which it is subjected when a vertical gust is encountered. For this reason, when flying in rough air conditions, it is safer to slow the airplane to a speed somewhat below the normal smooth air cruising speed.

On encountering any degree of turbulence, the airspeed should be reduced to the recommended maneuvering speed. Airplane manufacturers always specify in the Aircraft Owner's Manual a recommended **maneuvering speed (V$_A$)** for each model of airplane. This is the maximum speed at which full deflection of the controls can be made without exceeding the design limit load factor and damaging the airplane primary structure.

The airplane designer determines the maneuvering speed by a formula that multiplies the flaps-up, power-off stall speed at gross weight by the square root of the design limit load factor of the airplane. Most general aviation, normal category air-planes are certificated to withstand 3.8g. The maneuvering speed works out to be 1.9 times the stall speed at gross weight and it is this airspeed that is usually published in the Owner's Manual. This airspeed guarantees that the airplane will stall at the limit load factor.

However, this maneuvering speed is not always the best speed at which to penetrate turbulence. In the first place, an airplane flying into turbulence is flying with power. The power-on stall speed of an airplane is significantly less than its power-off stall speed. In the second place, turbulence is a form of wind shear which causes airspeed fluctuations. Rapid airspeed fluctuations of 5 to 15 knots in light turbulence and up to 25 knots in severe turbulence can be expected. Consequently, the best airspeed at which to penetrate turbulence should be at least 10 knots below the published maneuvering speed to compensate for the stall delaying effects of power and the effect of wind shear.

A further consideration is the fact that the published maneuvering speed is valid only when the airplane is at gross weight. Because stall speed decreases as weight decreases and because the maneuvering speed is a function of stall speed, a lightly loaded airplane should be flown at a slower airspeed in turbulence than one that is more heavily loaded. The lightly loaded airplane is accelerated more easily by gusts. A 20% decrease in weight requires a 10% decrease in maneuvering speed.

All of these factors demonstrate that the safest airspeed at which to penetrate turbulence is one that is somewhat less than the published maneuvering speed and, depending on the all-up weight, would range between 1.6 and 1.9 times the flaps-up, power-off stall speed. This speed is below V$_A$ but well above the stall.

While it is important to be concerned about subjecting the airplane to excessive structural loads when flying in turbulence, it is also essential to maintain control of the airplane at all times while in flight. Having adequate control to recover from the lateral and directional upsets that are the result of excessive turbulence requires flying at an airspeed at which the control surfaces are effective. The airspeed at which to fly in turbulence is therefore a compromise between structural and controllability margins.

The accepted procedure for flight in turbulence is to keep the wings level, maintain a normal pitch attitude and move the controls smoothly and slowly to recover from attitude displacement. Do not try to maintain altitude. Vertical air currents will probably cause significant altitude variations. Do try to maintain attitude and airspeed.

It is quite possible in turbulent conditions for the airplane to stall. In most instances, the wings stall and recover before the pilot even realizes what has happened. Safety is unlikely to be jeopardized unless the airplane has undesirable stall characteristics or is flying near to the ground as during the approach to landing when a stall can result in an accident. In these instances, a higher airspeed nearer to the published maneuvering speed to allow for a greater margin above stall is preferable.

These facts relating to loads are of critical importance and should be understood and intelligently applied so that you never impose loads on any airplane that you might be flying in excess of the limit load for which it was designed.

Log Books

The life of the airframe, engine(s) and propeller(s) is recorded in the Aircraft Technical Log which comprises an Airframe Log, a Record of Installations and Modifications, an Engine Log for each engine and a Propeller Log for each propeller. All maintenance, repairs, new installations, modifications, etc., must be completely recorded in the appropriate section of the Aircraft Technical Log.

A record of both flight time and air time and particulars of every flight is kept in a suitable Aircraft Journey Log. **Air time** is defined as the period of time commencing when the airplane leaves the supporting surface and terminating when it touches the supporting surface at the next point of landing. **Flight time** is defined as the total time from the moment an airplane first moves under its own power for the purpose of taking off until the moment it comes to rest at the end of the flight. Flight time is the time pilots should record in their log books. Air time and flight time should be recorded to the nearest 5 minutes (e.g. 1 hour 5 minutes, 1 hour 25 minutes, etc.) or to the nearest 6 minutes when using the decimal system (e.g. 1.1 hours, 1.5 hours, etc.).

In certain cases, log book information may be recorded in a computer data bank, rather than in a hard copy log book.

Although the Aviation Regulations do not require it, it is a recommended practice to keep log books for an ultra-light airplane as well, so that there is a record of maintenance, repairs, modifications, etc. and especially of time on the engine.

Inspection

An airplane must be inspected periodically by a qualified maintenance engineer and certified as airworthy in the Aircraft Log, as specified in Canadian Aviation Regulations and as detailed in the Airworthiness Manuals.

Theory Of Flight

Why learn about Theory of Flight?

The pilot today has a large variety of airplanes from which to choose. Some of these airplanes may fly at less than 100 knots top speed while others are capable of speeds well into the hundreds of knots. Some are single seaters carrying only the pilot, while others, even in the single engine light airplane classification, may carry 10 or more passengers. Some airplanes have laminar flow airfoil sections; others have airfoils of conventional design. A few light airplanes fly at 3 1/2 times their stalling speed; others do well to cruise at 1 1/3 times their stalling speed. Every one of these airplanes has different flight characteristics. If a pilot has a good grasp of the fundamentals of flight, he will understand what to expect of each different airplane that he may have the opportunity to fly. He will understand how best to handle each airplane as a result of his knowledge of the theory of design. He will comprehend the various loads to which an airplane of a particular design may be exposed while flying under abnormal or adverse conditions of flight. Not only to get the best performance but also to ensure the safety of each flight, an understanding of "Theory of Flight" is essential.

The study of theory of flight and aerodynamics can be a life time proposition. New theories are forever being put forward. Some questions have answers that are difficult to find. Others perhaps do not yet have adequate answers. The information that comprises this chapter can only be considered an introduction to a substantial but fascinating study.

Forces Acting on an Airplane in Flight

There are four forces acting on an airplane in flight. These are thrust, drag, lift and weight.

1. *Thrust*. The force exerted by the engine and its propeller(s) which pushes air backward with the object of causing a reaction, or thrust, in the forward direction.

2. *Drag*. The resistance to forward motion directly opposed to thrust.

3. *Lift*. The force upward which sustains the airplane in flight.

4. *Weight*. The downward force due to gravity, directly opposed to lift.

When thrust and drag are equal and opposite, the airplane is said to be in a state of **equilibrium**. That is to say, it will continue to move forward at the same uniform rate of speed. (Equilibrium refers to steady motion and not to a state of rest.)

If either of these forces becomes greater than the force opposing it, the state of equilibrium will be lost. If thrust is greater than drag, the airplane will **accelerate** or gain speed. If drag is greater than thrust, the airplane will **decelerate** or lose speed.

Similarly, when lift and weight are equal and opposite, the airplane will be in equilibrium. If lift, however, is greater than weight, the airplane will climb. If weight is greater than lift, the airplane will sink.

Let us first consider the force of lift.

Lift

If you consider the definitions cited by air authorities, a boy flying a kite could be construed to be a pilot in charge of an airplane! Ponder the idea a moment and it may not appear quite as absurd as it seems at first glance.

A kite is an inclined plane, the weight of which is supported in the air by the reaction of the wind flowing against it. If we substitute for the string, which holds the kite against the wind, the engine and propeller of an airplane, which move the wings forward against the airflow, we will see that the analogy of the kite is not without some validity.

The wings of an airplane are so designed that when moved through the air horizontally, the force exerted on them produces a **reaction** as nearly vertical as possible. It is this reaction that **lifts** the weight of the airplane.

Airfoils

An airfoil, or airfoil section, may be defined as any surface designed to obtain a reaction from the air through which it moves, that is, to obtain **lift**. It has been found that the most suitable shape for producing lift is a **curved** or **cambered** shape.

The **camber** of an airfoil is the curvature of the upper and lower surfaces. Usually the upper surface has a greater camber than the lower.

Fig. 1. An Airfoil Section

How Is Lift Created

What, then, causes this lift, you may ask.

Air flowing around an airfoil is subject to the **Laws of Motion** discovered by **Isaac Newton**. Air, being a gaseous fluid, possesses inertia and, according to Newton's First Law, when in motion tends to remain in motion. The introduction of an airfoil into the streamlined airflow alters the uniform flow of air. Newton's Second Law states that a force must be applied to alter the state of uniform motion of a body. The airfoil is the force that acts on the body (in this case, the air) to produce a change of direction. The application of such a force causes an equal and opposite reaction (Newton's Third Law) called, in this case, **lift**.

As the air passes over the wing towards the trailing edge, the air flows not only rearward but downward as well. This flow is called **downwash**. At the same time, the airflow passing under the wing is deflected downward by the bottom surface of the wing. Think of a water ski or surfboard planing over the water. In exerting a downward force upon the air, the wing receives an upward

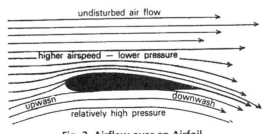

Fig. 2. Airflow over an Airfoil

counterforce. Remember Newton's Third Law — for every action there is an opposite and equal reaction. Therefore, the more air deflected downward, the more lift is created. Air is heavy; its weight exerts a pressure of 14.7 lbs per square inch at sea level. The reaction produced by the downwash is therefore significant.

The phenomenon defined by **Bernoulli's Principle** also has an effect in the production of lift by the wing moving though the air. Scientist Daniel Bernoulli discovered that the total energy in any system remains constant. In other words, if one element of an energy system is increased, another decreases to counter balance it. Take the example of water flowing through a venturi tube. Being incompressible, the water must speed up to pass through the constricted space of the venturi. The moving water has energy in the form of both pressure and speed. Within the venturi tube, pressure is sacrificed (decreased) to accelerate the speed of the flow.

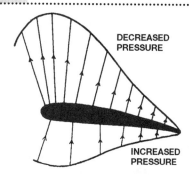

Fig. 3. Pressure Distribution Over An Airfoil

Air is a fluid, just like water, and can be assumed incompressible insofar as speed aerodynamics is concerned. As such, it acts exactly the same way as water in a venturi tube.

Picture the curved upper surface of a wing as the bottom half of a venturi tube. The upper half of this imaginary tube is the undisturbed airflow above the wing.

Air flowing over the wing's upper surface accelerates as it passes through the constricted area just as it does in the venturi tube. The result is a decrease in pressure on the upper surface of the wing that results in the phenomenon known as lift.

Relative Airflow
Relative airflow is a term used to describe the direction of the airflow with respect to the wing. (In other texts, it is sometimes called relative wind.) If a wing is moving forward and downward, the relative airflow is upward and backward. If the wing is moving forward horizontally, the relative airflow moves backward horizontally. The flight path and the relative airflow are, therefore, always parallel but travel in opposite directions.

Relative airflow is created by the motion of the airplane through the air. It is also created by the motion of air past a stationary body or by a combination of both. Therefore, on a take-off roll, an airplane is subject to the relative airflow created by its motion along the ground and also by the moving mass of air (wind). In flight, however, only the motion of the airplane produces a relative airflow. The direction and speed of the wind have no effect on relative airflow.

Angle of Attack and Centre of Pressure
The angle at which the airfoil meets the relative airflow is called the **angle of attack** (See Fig. 5).

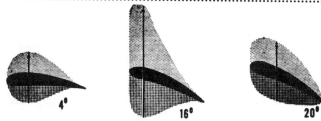

Fig. 4. Change of Pressure Distribution with Angle of Attack

Note: The envelopes indicated in Fig. 4 do not represent the actual areas of high and low pressure. These exist only in close proximity to the surfaces. They represent the comparative distribution of pressure as determined by pressure plotting. In Fig. 4, the darker shading represents increased pressure, the lighter shading, decreased pressure.

As the angle of attack is increased, the changes in pressure over the upper and lower surfaces and the amount of downwash (i.e. air deflected downward) increase up to a point (the stalling angle). Beyond this angle, they decrease (Fig. 4).

If we consider all the distributed pressure to be equivalent to a single force, this force will act through a straight line. The point where this line cuts the chord of an airfoil is called the **centre of pressure**.

Thus, it will be seen that as the angle of attack of an airfoil is increased up to the point of stall, the centre of pressure will move forward. Beyond this point, it will move back. The movement of the centre of pressure causes an airplane to be unstable.

Fig. 5. Angle of Attack

Weight
The weight of an airplane is the force which acts vertically downward toward the centre of the earth and is the result of gravity.

Just as the lift of an airplane acts through the centre of pressure, the weight of an airplane acts through the **centre of gravity (C.G.)**. This is the point through which the resultant of the weights of all the various parts of the airplane passes, in every attitude that it can assume.

Thrust
Thrust provides the forward motion of the airplane. There are several ways to produce this force — jets, propellers or rockets — but they all depend on the principle of pushing air backward with the object of causing a reaction, or thrust, in the forward direction. The effect is the same whether the thrust is produced by a propeller moving a large mass of air backward at a relatively slow speed or by a jet moving a small mass of air backward at a relatively high speed. A full description of engines and propellers will be found in the Chapter **Aero Engines**.

Drag
Drag is the resistance an airplane experiences in moving forward through the air.

For an airplane to maintain steady flight, there must be sufficient lift to balance the weight of the airplane, and there must be sufficient thrust to overcome drag.

Fig. 6 shows an airfoil moving forward through the air and depicts the principle known as **resolution of forces.** The vertical component (OL) is the lift and is used to support the weight of the airplane. The horizontal component (OD) is the drag. OR is the resultant reaction of these two components.

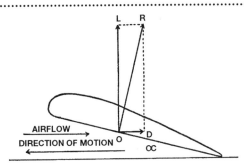

Fig. 6. Forces acting on an Airfoil

Since drag is a force directly opposed to the forward motion of the airfoil and, as the work of overcoming it is performed by the engine, it is desirable to have it as small as possible.

Drag is of two principal types.

1. *Parasite Drag* is the term given to the drag of all those parts of the airplane which do not contribute to lift, that is, the fuselage, landing gear, struts, antennas, wing tip fuel tanks, etc. In addition, any loss of momentum of the airstream caused by openings, such as those in the cowling and those between the wing and the ailerons and the flaps, add to parasite drag.

Parasite drag may be divided into two components: (1) **Form drag** refers to the drag created by the form or shape of a body as it resists motion through the air; (2) **Skin friction** refers to the tendency of air flowing over a body to cling to its surface.

Although parasite drag can never be completely eliminated, it can be substantially reduced. One method is to eliminate altogether those parts of the airplane that cause it. For this reason, retractable landing gear has been developed. Wing struts have been eliminated in favour of fully cantilevered wings. Another method is to streamline those parts that cannot be eliminated. Skin friction can be reduced substantially by the removal of dust, dirt, mud or ice that has collected on the airplane.

Even the most carefully designed individual parts must, however, be joined together to create the total airplane. Resistance caused by the effect of one part on another (i.e. where the wing is attached to the fuselage or the struts to the wings) is called **interference drag** and can be reduced by careful design in the fairing of one shape into another.

2. *Induced Drag* is caused by those parts of an airplane which are active in producing lift (i.e. the wing). It is the result of the wing's work in sustaining the airplane and is, therefore, a part of the lift and can never be eliminated. It increases as the angle of attack increases and decreases as the angle of attack decreases.

Induced drag can be reduced only during the initial designing of an airplane. A wing with a high aspect ratio (that is, with a very long span and a narrow chord) produces less induced drag than does a wing with a short span and a wide chord. Gliders and sail planes are therefore commonly designed with high aspect ratio wings.

The phenomenon, known as wing tip vortices, is testimony to the existence of induced drag.

As the decreased pressure over the top of the wing is less than the atmospheric pressure around it, the air flowing over the top surface tends to flow **inward.**

Fig. 7. Airflow over the Top Surface.

The air flowing over the lower surface, due to the lower pressure around it, tends to flow **outward** and curl upward over the wing tip.

Fig. 8. Airflow over the Bottom Surface.

When the two airflows unite at the trailing edge, they are flowing contra-wise. Eddies and vortices are formed which tend to unite into one large eddy at each wing tip. These are called **wing-tip vortices.** This disturbed air exerts a resistant force against the forward motion of the wing. This resistant force is known as induced drag.

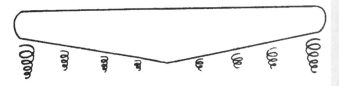

Fig. 9. Wing-tip Vortices.

In order to support the weight of an airplane, a large amount of air must be displaced **downward.** This displaced air must have somewhere to go and tends to flow **spanwise outwards,** as explained above. It is seeking to escape around the wing tips and to flow into the low pressure area over the upper surface of the wing where it spoils some of the lift potential. It will be obvious that the **heavier** the airplane and the **higher** the span loading on the wing, the more air it will displace downward, the greater will be the circulation of air, the greater the magnitude of the wing tip vortex created and the greater the induced drag.

Theory of Flight

The operational aspects of flight in areas where wing-tip vortices are present are discussed in the section **Wake Turbulence** in Chapter **Airmanship**.

Induced drag does not increase as the speed increases. On the contrary, it is greatest when the airplane is flying slowly, a few knots above the stalling speed when maximum lift is being realized at minimum speed.

The induced drag characteristics of a wing are not the same very near the ground as they are at altitude. During landing and take-off, the ground interferes with the formation of a large wing-tip vortex. Induced drag is, therefore, reduced when an airplane is flown very near the ground. This phenomenon is known as **ground effect**. (See **Ground Effect** in Chapter **Airmanship**.)

Although induced drag cannot be eliminated, it can be reduced by certain design features. For example, less induced drag is generated by a long, narrow wing than by a short, broad one. It has also been found that **winglets** are effective in reducing induced drag. Attached to the wing tip, the winglet, a small, vertical surface of airfoil section, is effective in producing side forces that diffuse the wing-tip vortex flow.

Lift and Drag Curves

As the amount of lift varies with the angle of attack, so too does the drag. Hence drag is the price we pay for lift. Although it is desirable to obtain as much lift as possible from a wing, this cannot be done without increasing the drag. It is therefore necessary to find the best compromise.

The lift and drag of an airfoil depend not only on the angle of attack, but also upon:

The shape of the airfoil
The plan area of the airfoil (or wing area) — S.
The square of the velocity (or true airspeed) — V^2.
The density of the air — p.
Hence the lift of an airfoil can be expressed as a formula by:
  $C_L \cdot 1/2 \, p \, V^2 \cdot S$
And the drag by: $C_D \cdot 1/2 \, p \, V^2 \cdot S$

The symbols C_L and C_D represent the **lift coefficient** and **drag coefficient** respectively. They depend on the shape of the airfoil and will alter with changes in the angle of attack.

The **lift-drag ratio** is used to express the relationship between lift and drag and is obtained by dividing the lift coefficient by the drag coefficient. C_L / C_D

The characteristics of any particular airfoil section can conveniently be represented by curves on a graph showing the amount of lift and drag obtained at various angles of attack, the lift-drag ratio, and the movement of the centre of pressure (Fig. 10).

Notice that the lift curve (C_L) reaches its maximum for this particular wing section at 18° angle of attack and then rapidly decreases. 18° is therefore the stalling angle.

The drag curve (C_D) increases very rapidly from 14° angle of attack and completely overcomes the lift at 22° angle of attack.

The lift-drag ratio (L/D) reaches its maximum at 0° angle of attack, meaning that at this angle we obtain the most lift for the least amount of drag.

The C.P. moves gradually forward till 12° angle of attack is reached and from 18° commences to move back.

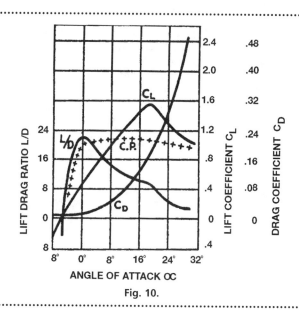

Fig. 10.

Note: The curves in Fig. 10 are for a particular airfoil section, the Clark Y. The values of C_L, C_D and L/D vs angle of attack will differ for every individual airfoil design. The general appearance of the co-ordinates in Fig. 10, however, is similar throughout a wide range of airfoil sections employed by aircraft designers today.

Thrust Available and Drag

The useful output of an aircraft engine is power. The power generated by a piston engine is used to turn a propeller which produces thrust by pushing a large volume of air backwards.

The greatest amount of thrust available from the propeller driven airplane engine occurs when the engine is running at full power and the airplane is standing stationary on the ground. As the airplane begines to move, the thrust available decreases with an increase in speed. The principal reason for this is that the drag curve rises rapidly as the airplane begins to move. Because they both have a relationship with drag, weight and airplane configuration affect the thrust/drag scenario; weight affecting induced drag and airplane configuration affecting parasite drag.

On take-off, full power is used for a few minutes, but within seconds after lift-off, the power is reduced to that required for climbing. After cruise altitude is reached, the power is reduced to cruise power which is between 65% and 75% of full rated power.

If the weight of the airplane is increased, the performance of the airplane is decreased. Every extra pound of weight carried by any airplane reduces its performance.

More power is required by the airplane at every increase in speed to produce the thrust required to keep the airplane flying at the desired speed and altitude. The power required curve is characteristically U shaped and similar to the drag curve. The power required to fly the airplane at a constant altitude varies with the drag existing at the different airspeeds. To increase the speed, weight and airplane configuration not changing, additional thrust/power is required. Testing and graphed results of such tests indicated that, for example, four times the thrust is required to double the speed.

Streamlining

In this age when everything from trains to children's toys is streamlined, it hardly seems necessary to explain what streamlining means. Basically, streamlining is a design device by which a body is so shaped that drag is minimized as the body moves forward through the air. A flat plate or a round ball moving through the air disturb the smooth flow of air and set up eddies behind them. A streamlined shape smooths out the airflow, eliminates the eddies, reduces the drag and minimizes the energy required to move that body through the air, as can be seen in Fig. 11.

Fig. 11. Effects of Streamlining.

Aileron Drag

In banking to make an airplane turn, one aileron is depressed and the other is raised. The downgoing aileron, being depressed into the compressed airflow on the underside of the wing, causes drag. The upgoing aileron, moving up into a more streamlined position, causes less drag. The drag on the downgoing aileron is known as **aileron drag** and, if not corrected for in the design of the aileron, tends to cause a yaw in the opposite direction to which the bank is applied.

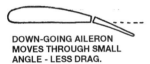

Fig. 12. Differential Ailerons.

Fig. 13. Frise Ailerons.

Both **frise** and **differential ailerons** have been designed to overcome aileron drag. On differential ailerons, the downgoing aileron moves through a smaller angle than the upgoing aileron. As a result, the downgoing aileron, although it still causes drag, causes less drag than an aileron not so designed. The upgoing aileron, moving through a larger angle, causes more drag. The combined effect is a balance that more or less eliminates aileron drag. In frise ailerons, the nose of the upgoing aileron projects into the airflow, while the downgoing aileron is streamlined.

The Boundary Layer

The boundary layer is a very thin sheet of air lying over the surface of the wing (and, for that matter, all other surfaces of the airplane). Because air has viscosity, this layer of air tends to adhere to the wing. As the wing moves forward through the air, the boundary layer at first flows smoothly over the streamlined shape of the airfoil. Here the flow is called the **laminar layer.**

As the boundary layer approaches the centre of the wing, it begins to lose speed due to skin friction and it becomes thicker and turbulent. Here it is called the **turbulent layer.** The point at which the boundary layer changes from laminar to turbulent is called the **transition point** (Fig. 14). Where the boundary layer becomes turbulent, drag due to skin friction is relatively high. As speed increases, the transition point tends to move forward. As the angle of attack increases, the transition point also tends to move forward.

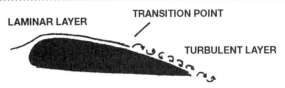

Fig. 14. Laminar and Turbulent Layer

Various methods have been developed to control the boundary layer in order to reduce skin friction drag.

Suction Method. One method uses a series of thin slots in the wing running out from the wing root towards the tip. A vacuum sucks the air down through the slots, preventing the airflow from breaking away from the wing and forcing it to follow the curvature of the wing surface. The air, which is sucked in, siphons through the ducts inside the wing and is exhausted backwards to provide extra thrust.

The **laminar flow airfoil** is itself a structural design intended to make possible better boundary layer control. The thickest part of a laminar flow wing occurs at 50% chord. The transition point at which the laminar flow of air breaks down into turbulence is at or near the thickest part. As can be seen in the accompanying illustration (Fig.15), the transition point at which the laminar flow of air becomes turbulent on a laminar flow airfoil is rearward of that same point on a conventional designed airfoil.

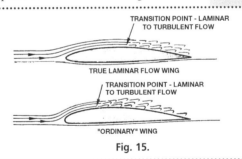

Fig. 15.

Vortex generators (Fig. 16) are small plates about an inch deep standing on edge in a row spanwise along the wing. They are placed at an angle of attack and (like a wing airfoil section) generate vortices. These tend to prevent or delay the breakaway of the boundary layer by re-energizing it. They are lighter and simpler than the suction boundary layer control system described above.

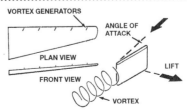

Fig. 16. Vortex Generators.

Couples

The principle of equilibrium has already been introduced at the beginning of this chapter in the discussion of the forces that act on an airplane in flight. When two forces, such as thrust and drag, are equal and opposite but parallel, rather than passing through the same point, they are said to form a **couple**.

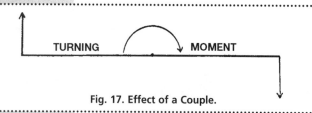

Fig. 17. Effect of a Couple.

A couple will cause a **turning moment** about a given axis as in Fig. 17.

If weight is ahead of lift, the couple created will turn the nose of the airplane down. Conversely, if lift is ahead of weight, the couple created will turn the nose of the airplane up.

If drag is above thrust, the couple formed will turn the nose of the airplane up. Conversely, if thrust is above drag, the couple formed will turn the nose of the airplane down.

Notice that in Fig. 18. Forces acting on an Airplane in Flight, weight is placed ahead of lift and drag is above thrust. As a result, when the engine is shut off and there is no thrust, the couple due to weight and lift will naturally tend to turn the nose down.

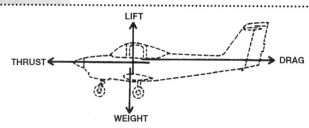

Fig. 18. Forces Acting on an Airplane in Flight.

In the case of flying boat design, it is practically impossible to have the drag above the thrust. Therefore, lift must be placed ahead of weight for normal flight. This leads to the unsatisfactory tendency, when the engine is shut off, to nose up with the consequent risk of stalling in the hands of an inexperienced pilot.

Design of the Wing

The type of operation for which an airplane is intended has a very important bearing on the selection of the shape and design of the wing for that airplane. If the airplane is designed for low speed, a thick airfoil is most efficient. A thin airfoil is best for high speed.

Conventional Airfoils

The following illustrations depict a selection of designs of airfoil sections. These are known as conventional airfoils.

Low camber — low drag — high speed — thin wing section
Suitable for race planes, fighters, interceptors, etc.

Deep camber — high lift — low speed — thick wing section
Suitable for transports, freighters, bombers, etc.

Deep camber — high lift — low speed — thin wing section
Suitable as above.

Low lift — high drag — reflex trailing edge wing section.
Very little movement of centre of pressure. Good stability.

Symmetrical (cambered top and bottom) wing sections.
Similar to above.

GA(W)-1 airfoil — thicker for better structure and lower weight — good stall characteristics — camber is maintained farther rearward which increases lifting capability over more of the airfoil and decreases drag.

Laminar Flow Airfoils

There is another type of airfoil in common use on modern airplanes. It is a fairly recent development and is known as the **laminar flow airfoil.**

Laminar flow airfoils were originally developed for the purpose of making an airplane fly faster. The laminar flow wing is usually thinner than the conventional airfoil, the leading edge is more pointed and its upper and lower surfaces are nearly symmetrical. The major and most important difference between the two types of airfoil is this: the thickest part of a laminar wing occurs at 50% chord while in the conventional design the thickest part is at 25% chord.

The effect achieved by this design of wing is to maintain the laminar flow of air throughout a greater percentage of the chord of the wing and to control the transition point. Drag is therefore considerably reduced since the laminar airfoil takes less energy to slide through the air. The pressure distribution on the laminar flow wing is much more even since the camber of the wing from the leading edge to the point of maximum camber is more gradual than on the conventional airfoil. However, at the point of stall, the transition point moves more rapidly forward.

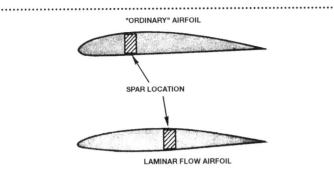

Fig. 19.

Planform

Planform refers to the shape of the wing as seen from directly above. Wings may be rectangular or elliptical or delta shaped. Some wings taper from wing root to wing tip, with the taper along the leading edge or along the trailing edge or, in some cases, with a taper along both edges.

The **aspect ratio** of a wing is the relationship between the length or span of the wing and its width or chord. It is computed by dividing the span by the average chord (the **Mean Aerodynamic Chord MAC**).

A wing, for example, that has a span of 24 feet and a chord of 6 feet has an aspect ratio of 4. A wing with a span of 36 feet and a chord of 4 feet has an aspect ratio of 9. The actual size, in area, of both wings is identical (144 sq. ft.) but their flight performance is quite different because of their differing aspect ratios.

A wing with a high aspect ratio will generate more lift and less induced drag than a wing with a low aspect ratio.

For this reason, gliders have wings with high aspect ratios.

Angle of Incidence

The **angle of incidence** is the angle at which the wing is permanently inclined to the longitudinal axis of the airplane.

Choosing the right angle of incidence can improve flight visibility, enhance take-off and landing characteristics and reduce drag in level flight.

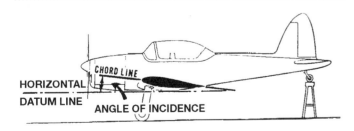

Fig. 20. Angle of Incidence.

The angle of incidence that is usually chosen is the angle of attack at which the lift-drag ratio is optimum (See Fig. 10). In most modern airplanes, there is a small positive angle of incidence so that the wing has a slight angle of attack when the airplane is in level cruising flight.

Wing Tip Design

Specially designed wing tips have been found to be effective in controlling induced drag and wing tip vortices. These vortices kill lift, create drag and cause instability at high angles of attack and at low airspeed. Any method that is effective in inhibiting their development increases the efficiency of the airplane.

The installation of **wing tip tanks** has several advantages. They increase the range of the airplane and distribute the weight over a greater portion of the wing but they also are effective in preventing the air from spilling over the wing tip, in reducing the intensity of the vortices and in reducing induced drag.

Small airplanes that do not use tip tanks may have **wing tip plates** installed on the wing tip. These plates have the same shape as the airfoil but are larger and prevent the air from spilling over the tips.

Another design feature is the **droop wing tip** that is also quite effective in decreasing vortex development.

Some mention has already been made of **winglets** (sometimes called **topsails**), small vertical winglike surfaces attached to the wing tip that are effective in inhibiting wing tip vortex development. They have the advantage of being able to develop sufficient lift to offset their own parasitic drag component.

Wash-out/Wash-in

To reduce the tendency of the wing to stall suddenly as the stalling angle is approached, designers incorporate in wing design a feature known as **wash-out and wash-in.** The wing is twisted so that the angle of incidence at the wing tip is less than that at the root of the wing. As a result, the wing has better stall characteristics because the section towards the root will stall before the outer section of the wing. The ailerons, located towards the wing tips, are still effective even though part of the wing is stalled. Increasing the angle of incidence is called wash-in and it increases the lift. Decreasing the angle of incidence is called wash-out and it decreases the lift.

The same improved stall characteristics are achieved by the device of changing the airfoil shape from the root to the tip. The manufacturer incorporates a wing shape at the tip which has the characteristic of stalling at a slightly higher angle of attack.

Wing Fences

Wing fences are fin-like vertical surfaces attached to the upper surface of the wing and are used to control the airflow. On swept wing airplanes, they are located about two-thirds of the way out towards the wing tip and prevent the drifting of air toward the tip of the wing at high angles of attack. On straight wing airplanes, they control the airflow in the flap area. In both cases, they give better slow speed handling and stall characteristics.

Fig. 21. Wing Fences.

Slots, Slats and Leading Edge Flaps

Slats are auxiliary airfoils fitted to the leading edge of the wing. At high angles of attack, they automatically move out ahead of the wing. At a high angle of attack, the low pressure just behind the leading edge on the top edge of the wing pulls the slat out of the wing. When the angle of attack is lower, the greater air pressure pushes the slat back into the wing. The angle of attack of the slat being less than that of the mainplane, there is a smooth airflow over the slat which tends to smooth out the eddies forming over the wing. Slats are usually fitted to the leading edge near the wing tips to improve lateral control.

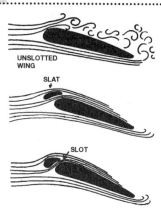

Fig. 22. Slotted Wings.

Slots are passageways built into the wing a short distance from the leading edge in such a way that, at high angles of attack, the air flows through the slot and over the wing, tending to smooth out the turbulence due to eddies. Sometimes the slots extend across the full span of the wing. Sometimes they are placed only ahead of the ailerons to keep the outer portion of the wing flying after the root has stalled. This installation design keeps the aileron effective and provides increased lateral control.

Some airplanes, particularly the large transport airplanes, have **flaps** on the leading edge of the wing. A jackscrew arrangement pushes the leading edge of the wing against a hinge on the lower surface causing the leading edge to droop. The **drooped leading edge flap** increases the camber of the wing and increases the lift coefficient at high angles of attack. The **Kreuger flap** performs much the same function. It is a special high lift shape that is extended out ahead of the leading edge by a hinged linkage.

Spoilers

Spoilers are devices fitted to the wing which increase drag and decrease lift. They usually consist of a long narrow strip of metal arranged spanwise along the top surface of the airfoil. In some airplanes, they are linked to the ailerons and work in unison with the ailerons for lateral control. As such, they open on the side of the upgoing aileron, spoil the lift on that wing and help drive the wing down and help the airplane to roll into a turn.

In some airplanes, spoilers have replaced ailerons as a means of roll control. The spoiler moves only upward in contrast to the aileron that moves upward to decrease lift and downward to increase lift. The spoiler moves only up, spoiling the wing lift. By using spoilers for roll control, full span flaps can be used to increase low speed lift.

Spoilers can also be connected to the brake controls and, when so fitted, work symmetrically across the airplane for producing drag and destroying lift after landing, thereby transferring all the weight of the airplane to the wheels and making braking action more effective.

Speed Brakes

Speed brakes are a feature on some high performance airplanes. They are a device designed to facilitate optimum descent without decreasing power enough to shock cool the engine and are especially advantageous in airplanes with high service ceilings. They are also of use in setting up the right approach speed and descent pattern in the landing configuration. The brakes, when extended, create drag without altering the curvature of the wing and are usually fitted far enough back along the chord so as not to disrupt too much lift and in a position laterally where they will not disturb the airflow over the tailplane. They are usually small metal blades housed in a fitting concealed in the wing that, when activated from the cockpit, pivot up to form a plate. On some types of aircraft, speed brakes are incorporated into the rear fuselage and consist of two hinged doors that open into the slipstream.

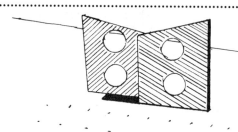

Fig. 23. Speed Brakes.

Flaps

Flaps are high lift devices which, in effect, increase the camber of the wing and, in some cases (as with the Fowler Flap) also increase the effective wing area. Their use gives better take-off performance and permits steeper approach angles and lower approach and landing speeds.

When deflected, flaps increase the upper camber of the wing, increasing the negative pressure on the top of the wing. At the same time, they allow a build up of pressure below the wing. During take-off, flap settings of 10° to 20° are used to give better take-off performance and a better angle of climb, especially valuable when climbing out over obstacles.

However, not all airplane manufacturers recommend the use of flaps during take-off. They can be used only on those airplanes which have sufficient take-off power to overcome the extra drag that extended flaps produce. The recommendations of the manufacturer should, therefore, always be followed.

Flaps do indeed increase drag. The greater the flap deflection, the greater the drag. At a point of about half of their full travel, the increased drag surpasses the increased lift and the flaps become air brakes. Most flaps can be extended to 40° from the chord of the wing. At settings between 20° and 40°, the essential function of the flaps is to improve the landing capabilities, by steepening the glide without increasing the glide speed. In an approach over obstacles, the use of flaps permits the pilot to touch down much nearer the threshold of the runway. Flaps also permit a slower landing speed and act as air brakes when the airplane is rolling to a stop after landing, thus reducing the need for excessive braking action. As a result, there is less wear on the undercarriage, wheels and tires. Lower landing speeds also reduce the possibility of ground looping during the landing roll.

Plain and split flaps increase the lift of a wing but, at the same time, they greatly increase the drag. For all practical purposes,

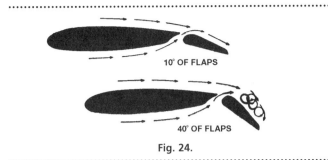

Fig. 24.

they are of value only in approach and landing. They should not normally be employed for take-off because the extra drag reduces acceleration.

Slotted flaps, on the other hand, including such types as Fowler and Zap, produce lift in excess of drag and their partial use is therefore recommended for take-off.

From the standpoint of aerodynamic efficiency, the Fowler Flap is generally considered to offer the most advantages and the fewest disadvantages, especially on larger airplanes, while double slotted flaps have won wide approval for smaller types.

On STOL airplanes, a combination of double slotted flaps and leading edge slats are common.

Changes in flap setting affect the trim of an airplane. As flaps are lowered, the centre of pressure moves rearward creating a nose down, pitching moment. However, in some airplanes, the change in airflow over the tailplane as flaps are lowered is such that the total moment created is nose up and it becomes necessary to trim the airplane "nose down".

The airplane is apt to lose considerable height when the flaps are raised. At low altitudes, therefore, the flaps should be raised cautiously.

Most airplanes are placarded to show a *maximum speed above which the flaps must not be lowered*. The flaps are not designed to withstand the loads imposed by high speeds. Structural failure may result from severe strain if the flaps are selected "down" at higher than the specified speed.

When the flaps have been lowered for a landing, they should not ordinarily be raised until the airplane is on the ground. If a landing has been missed, the flaps should not be raised until the power has been applied and the airplane has regained normal climbing speed. It is then advisable to raise the flaps in stages.

How much flap should be used in landing? Generally speaking, an airplane should be landed as slowly as is consistent with safety. This usually calls for the use of full flaps. The use of flaps affects the wing airfoil in two ways. Both lift and drag are increased. The increased lift results in a lower stalling speed and permits a lower touchdown speed. The increased drag permits a steeper approach angle without increasing airspeed. The extra drag of full flaps results in a shorter landing roll.

An airplane that lands at 50 knots with full flaps selected may have a landing speed as fast as 70 knots with flaps up. If a swerve occurs during the landing roll, the centrifugal force unleashed at 70 knots is twice what it would be at 50 knots, since centrifugal force increases as the square of the speed. It follows, then, that a slower landing speed reduces the potential for loss of control during the landing roll. It also means less strain on the tires, brakes and landing gear and reduces fatigue on the airframe structure.

There are, of course, factors which at times call for variance from the procedure of using full flaps on landing. These factors would include the airplane's all-up-weight, the position of the C.G., the approach path to landing, the desired rate of descent and any unfavourable wind conditions, such as a strong cross wind component, gusty winds and extreme turbulence. With experience, a pilot learns to assess these various factors as a guide to flap selection.

In some airplanes, in a cross wind condition, the use of full flaps may be inadvisable. Flaps present a greater surface for the wind to act upon when the airplane is rolling on the ground. The wing on the side from which the wind is blowing will tend to rise. In addition, a cross wind acting on full flaps increases the weather vaning tendency, although in an airplane with very effective rudder control even at slow speeds, the problem is not so severe. However, in many airplanes, the selection of full flaps deflects the airflow from passing over the empennage, making the elevator and rudder surfaces ineffective. Positive control of the airplane on the ground is greatly hampered. Since maintaining control of the airplane throughout the landing roll is of utmost importance, it may be advisable to use less flaps in cross wind conditions. In any case, it is very important to maintain the cross wind correction throughout the landing roll.

The Axes of an Airplane

There are three axes around which the airplane moves. These axes all pass through the airplane's centre of gravity, which is that point which is the centre of the airplane's total weight.

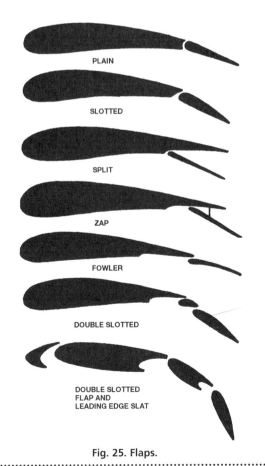

Fig. 25. Flaps.

Theory of Flight

The **longitudinal axis** extends lengthwise through the fuselage from the nose to the tail. Movement of the airplane around the longitudinal axis is known as **roll** and is controlled by movement of the ailerons. To move the ailerons, the pilot turns the control wheel either clockwise or counter clockwise (or moves the control stick either right or left). This action lowers the aileron on one wing and raises the aileron on the other wing. The downgoing aileron increases the camber of its wing, producing more lift and the wing rises. The upgoing aileron spoils the airflow on its wing, decreases the lift and the wing descends. The airplane rolls into a turn.

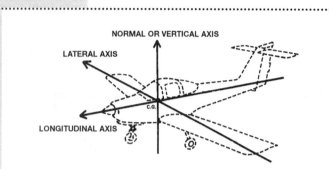

Fig. 26. The Axes of an Airplane.

The **lateral axis** extends crosswise from wing tip to wing tip. Movement of the airplane around the lateral axis is known as **pitch** and is controlled by movement of the elevators. To effect a nose down attitude, the pilot pushes forward on the control wheel or stick. The elevator deflects downward, increasing the camber of the horizontal tail surface and thereby increasing the lift on the tail. To effect a nose up attitude of the airplane, the pilot pulls the wheel toward him. The elevators are deflected upwards decreasing the lift on the tail, with a resultant downward movement of the tail.

The **vertical** or **normal axis** passes vertically through the centre of gravity. Movement of the airplane around the vertical axis is **yaw** and is controlled by movement of the rudder. Pressure applied to the left rudder pedal, for example, deflects the rudder to the left into the airflow. The pressure of the airflow against the rudder pushes the tail to the right. The nose of the airplane yaws to the left.

There is a distinct relationship between movement around the vertical and longitudinal axes of an airplane (i.e. yaw and roll). When rudder is applied to effect a yaw, to the right for example, the left wing (on the outside of the turn) moves faster than the inside wing, meets the relative airflow at a greater angle of attack and at greater speed and produces more lift. It is apparent, therefore, that the use of rudder along with aileron enhances the lifting capacity of the outside wing and produces a better co-ordinated turn.

In a roll, the airplane has a tendency to yaw away from the intended direction of the turn. This tendency is the result of aileron drag and is called **adverse yaw.** The upgoing wing, as well as gaining more lift from the increased camber of the downgoing aileron, also experiences more induced drag. The airplane, as a result, skids outward on the turn. The co-ordinated use of rudder and aileron corrects for adverse yaw.

Yaw comes in two varieties: static and dynamic. **Dynamic yaw** is movement or rotation about the normal axis of the airplane. **Static yaw** is a condition in which the airplane is flying with some angle of sideslip, in which the longitudinal axis of the airplane is not aligned with the airplane's flight path. In a **sideslip,** the nose of the airplane is yawed to the left or to the right. Any change in the sideslip angle would be a form of dynamic yaw.

Balanced Controls

Controls are sometimes dynamically balanced to assist the pilot to move them. Several of the various means by which an aerodynamic reaction is used to serve this purpose are illustrated in Fig. 27.

By having some of the control surface in front of the hinge, the air striking the forward portion helps to move the control surface in the required direction. The design also helps to counteract adverse yaw when used in aileron design.

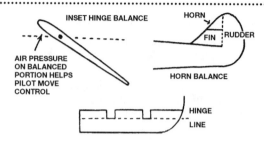

Fig. 27. Dynamic Balance.

Control surfaces are sometimes balanced by fitting a mass (usually of lead) of streamline shape in front of the hinge of the control surface. This is called **mass balance** and is incorporated to prevent flutter of the control surface which is liable to occur at high speeds. The mass may be attached as shown in Fig. 28 or it may be fitted inside that part of the control surface which lies in front of the hinge when this design feature has been used for the purpose of providing dynamic balance.

The exact distribution of weight on a control surface is very important. For this reason, when a control surface is repainted, repaired or component parts replaced, it is essential to check for proper balance and have it rebalanced if necessary. To do this, the control surface is removed, placed in a jig and the position of the centre of gravity checked against the manufacturer's specifications. Without any airflow over the control surface, it must balance about its specified C.G. This is known as **static balance.** For example, the aileron of the Bonanza is designed for a static nose heavy balance of 0.2 inch pounds. The C.G. of the aileron is forward of the hinge centre line causing the control surface to be nose heavy.

Fig. 28. Mass Balance.

Stability

An airplane in flight is constantly subjected to forces that disturb it from its normal horizontal flight path. Rising columns of hot air, downdrafts, gusty winds, etc., make the air "bumpy" and the airplane is thrown off its course. Its nose or tail drops or one wing dips. How the airplane reacts to such a disturbance from its flight attitude depends on its **stability** characteristics.

Stability is the tendency of an airplane in flight to remain in straight, level, upright flight and to return to this attitude, if displaced, without corrective action by the pilot.

Static stability is the *initial* tendency of an airplane, when disturbed, to return to the original position (the initial "wave" motion in Fig. 29).

Dynamic stability is the *overall* tendency of an airplane to return to its original position, following a series of damped out oscillations (the diminishing "wave" pattern in Fig. 29).

Fig. 29. Static and Dynamic Balance.

Stability may be (a) **positive,** meaning the airplane will develop forces or moments which tend to restore it to its original position; (b) **neutral,** meaning the restoring forces are absent and the airplane will neither return from its disturbed position nor move further away; (c) **negative,** meaning it will develop forces or moments which tend to move it further away. Negative stability is, in other words, the condition of **instability.**

A stable airplane is one that will fly "hands off " and is pleasant and easy to handle. An exceedingly stable airplane, on the other hand, may lack maneuverability.

An airplane which, following a disturbance, oscillates with increasing up and down movements until it eventually stalls or enters a dangerous dive would be said to be unstable, or to have negative dynamic stability.

An airplane that has positive *dynamic* stability does not automatically have positive *static* stability. The designers may have elected to build in, for example, negative static stability and positive dynamic stability in order to achieve their objective in maneuverability. In other words, negative and positive dynamic and static stability may be incorporated in any combination in any particular design of airplane.

An airplane may be *inherently stable* (that is, stable due to features incorporated in the design) but may become *unstable* due to changes in the position of the centre of gravity (caused by consumption of fuel, improper disposition of the disposable load, etc.).

Stability may be (a) **longitudinal,** (b) **lateral,** or (c) **directional,** depending on whether the disturbance has affected the airframe in the (a) pitching, (b) rolling, or (c) yawing plane.

Longitudinal Stability

Longitudinal stability is stability around the lateral axis of the airplane and is called pitch stability.

To obtain longitudinal stability, airplanes are designed to be nose heavy when correctly loaded. The centre of gravity is ahead of the centre of pressure. This design feature is incorporated so that, in the event of engine failure, the airplane will assume a normal glide. It is because of this nose heavy characteristic that the airplane requires a tailplane. Its function is to resist this diving tendency. The tailplane is set at an angle of incidence that produces a negative lift and thereby, in effect, holds the tail down. In level, trimmed flight, the nose heavy tendency and the negative lift of the tailplane exactly balance each other.

Two principal factors influence longitudinal stability: (1) size and position of the horizontal stabilizer, and (2) position of the centre of gravity.

The Horizontal Stabilizer

The tail plane, or stabilizer, is placed on the tail end of a lever arm (the fuselage) to provide longitudinal stability. It may be quite small. However, being situated at the end of the lever arm, it has great leverage. When the angle of attack on the wings is increased by a disturbance, the centre of pressure moves forward, tending to turn the nose of the airplane up and the tail down. The tailplane, moving down, meets the air at a greater angle of attack, obtains more lift and tends to restore the balance.

On most airplanes, the stabilizer appears to be set at an angle of incidence that would produce an upward lift on the tailplane. It must, however, be remembered that the tailplane is in a position to be in the downwash from the wings. The air that strikes the stabilizer has already passed over the wings and been deflected slightly downward. The angle of the downwash is about half the angle of attack of the main airfoils. The proper angle of incidence of the stabilizer therefore is very important in order for it to be effective in its function.

Centre of Gravity

The centre of gravity is, of course, very important in achieving longitudinal stability. If the airplane is loaded with the centre of gravity too far aft, the airplane may assume a nose up rather than a nose down attitude. The inherent stability will be lacking and, even though down elevator may correct the situation, control of the airplane in the longitudinal plane will be difficult and perhaps, in extreme cases, impossible.

Lateral Stability

Lateral stability is stability around the longitudinal axis and is called roll stability.

Lateral stability is achieved through (1) dihedral, (2) sweepback, (3) keel effect, and (4) proper distribution of weight.

Dihedral

The **dihedral** angle is the angle that each wing makes with the horizontal (Fig. 30). The purpose of dihedral is to improve lateral stability. If a disturbance causes one wing to drop, the unbalanced force (Fig. 30) produces a sideslip in the direction of the downgoing wing. This will, in effect, cause a flow of air in the opposite direction to the slip. This flow of air will strike the lower wing at a greater angle of attack than it strikes the upper wing. The lower wing will thus receive more lift and the airplane will roll back into its proper position.

Since dihedral inclines the wing to the horizontal, so too will the lift reaction of the wing be inclined from the vertical (Fig. 30B). Hence an excessive amount of dihedral will, in effect, reduce the lift force opposing weight.

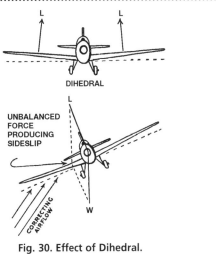

Fig. 30. Effect of Dihedral.

Theory of Flight

Some modern airplanes have a measure of negative dihedral, or **anhedral**, on the wings and/or stabilizer. The incorporation of this feature provides some advantages in overall design in certain types of airplanes. However, it does have an effect, probably adverse, on lateral stability.

Keel Effect
Dihedral is more usually a feature on low wing airplanes, although some dihedral may be incorporated in high wing airplanes as well.

Most high wing airplanes are laterally stable simply because the wings are attached in a high position on the fuselage and because the weight is therefore low. When the airplane is disturbed and one wing dips, the weight acts as a pendulum returning the airplane to its original attitude.

Sweepback
A sweptback wing is one in which the leading edge slopes backward.

When a disturbance causes an airplane with sweepback to slip or drop a wing, the low wing presents its leading edge at an angle that is perpendicular to the relative airflow. As a result, the low wing acquires more lift, rises and the airplane is restored to its original flight attitude.

Sweepback also contributes to directional stability. When turbulence or rudder application causes the airplane to yaw to one side (for example, the left as in Fig. 31), the right wing (B) presents a longer leading edge perpendicular to the relative airflow. The airspeed of the right wing increases and it acquires more drag than the left wing (A). The additional drag on the right wing pulls it back, yawing the airplane back to its original path.

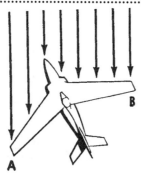

Fig. 31. Sweepback.

Directional Stability
Directional stability is stability around the vertical or normal axis.

The most important feature that affects directional stability is the vertical tail surface, that is, the fin and rudder. Keel effect and sweepback also contribute to directional stability to some degree.

The Fin
An airplane has the tendency always to fly head on into the relative airflow. This tendency which might be described as weather vaning is directly attributable to the vertical tail fin and to some extent also the vertical side areas of the fuselage. If the airplane yaws away from its course, the airflow strikes the vertical tail surface from the side and forces it back to its original line of flight. In order for the tail surfaces to function properly in this weather vaning capacity, the side area of the airplane aft of the centre of gravity must be greater than the side area of the airplane forward of the C.G. If it were otherwise, the airplane would tend to rotate about its vertical axis.

Flight Performance
Torque
The propeller usually rotates clockwise, as seen from the pilot's seat. The reaction to the spinning propeller causes the airplane to rotate counter clockwise to the left. This left turning tendency is called **torque**. The designer of the airplane compensates for torque in cruising flight by building a slight right turning tendency into the airplane. For example, the left wing may have a slightly greater angle of incidence than the right wing. Aileron trim tabs also are used to compensate for torque.

On take-off, torque affects directional control. Use of right rudder during the take-off roll corrects this condition.

Asymmetric Thrust
Another left turning tendency is the result of **asymmetric thrust**, or **P Factor**. At high angles of attack and high power settings, such as during take-off, the descending blade of the propeller (on the pilot's right) has a greater angle of attack than the ascending blade. This situation produces more lift from the right side of the propeller with a consequent yawing to the left. Right rudder pressure compensates for this tendency. Asymmetric thrust is significant only at high angles of attack. In level flight, both blades of the propeller meet the relative airflow equally and produce equal thrust.

Precession
The spinning propeller of an airplane acts like a gyroscope. One of the characteristics of a gyroscope is **rigidity in space;** that is, the rotating gyro tends to stay in the same plane of rotation and resists any change in that plane. If forced to change, precession results.

If an airplane changes suddenly from a nose up to a nose down position, as is the case during the take-off roll in a tail wheel airplane, the airplane will yaw sharply to the left as the pilot shoves the wheel forward to raise the tail. The application of right rudder compensates for the precession tendency.

Slipstream
The air pushed backward by a revolving propeller has a corkscrew motion. This causes an **increased** pressure on one side of the tail unit and a **decreased** pressure on the other side. The tail is consequently pushed sideways from the high pressure side towards the low, causing the airplane to yaw. The condition is corrected by offsetting the fin, or by offsetting the engine thrust line, or by fitting trim tabs on the rudder, or by a combination of two or all of these methods. In some airplanes, the rudder trim is adjustable by a control in the cockpit. In this way, the pilot is better able to compensate for the changes in pressure on the rudder as the airplane changes from climbing power, to cruise, to gliding.

The revolving slipstream from the propeller causes an airplane, especially tailwheel airplanes, as the throttle is opened to commence the take-off roll, to yaw to the left. As the airspeed increases, the tendency is less pronounced. Right rudder compensates.

It may be of interest at this point to mention the relative effects of the slipstreams of pusher and tractor type propellers. The tractor type of propeller located at the nose of the airplane pushes high speed turbulent air back over the airplane, thereby increasing considerably the drag of the fuselage and wing root sections. A pusher type of propeller, located at the rear of the airplane, allows better high speed performance due to the reduction of this drag. Because the tractor

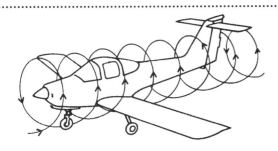

Fig. 32. Slipstream.

propeller bites into "clean" air, its efficiency is good whereas the pusher propeller bites into disturbed air. Nevertheless, from the standpoint of overall efficiency, the pusher propeller configuration is considered to have more to offer in performance benefits.

Climbing

The engine produces the energy that keeps an airplane flying. The throttle controls the output of this energy. **It is the function of the elevators to divide the energy, produced by the engine in the form of thrust, into speed and altitude.** The elevator does this by controlling the angle of attack of the wings. If, with no change in the thrust, the angle of attack is decreased, less energy is required to maintain lift and more of the total energy output is utilized to produce an increase in speed. If the angle of attack is increased, more energy is required to maintain lift and less energy is available for speed. If the pilot puts some back pressure on the control column (with no change in throttle setting), the airplane will climb and lose airspeed. Conversely, if the pilot puts some forward pressure on the control column (with no change in throttle setting), the airplane will descend and build up airspeed.

During level flight, the engine must produce a thrust equal to the drag of the airplane for the airplane to be in a state of equilibrium. If the power is increased, the pilot can maintain level flight at an increased speed by putting the nose down slightly (i.e. decreasing the angle of attack). If the pilot does not change the angle of attack, the airplane will begin to climb as a result of the increased thrust, since the increased speed of the relative airflow over the airfoils will produce more lift. By adjusting power (i.e. choosing any setting between that needed for normal, straight and level cruise and full power) and by varying the angle of attack, the pilot can flatten or steepen the angle of climb and the airspeed in the climb.

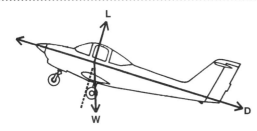

Fig. 33. Forces in a Climb.

Once established in a steady state of climb condition, the airplane is again in a state of equilibrium. In the climb attitude, the airplane is inclined away from the horizontal and, as a result, part of the weight acts rearward and combines with drag. Thrust, therefore, equals drag plus a component of weight and lift equals weight less that component of weight that is acting rearward.

The ability of an airplane to climb is dependent on the extra power that is available from the engine. At ever increasing altitudes, the density of the air decreases and the power of the engine drops off. The climb, therefore, becomes increasingly more shallow as greater altitudes are reached until further climbing is impossible. The airplane has then reached its absolute ceiling.

Every airplane has a BEST RATE OF CLIMB (V_Y). This is the *rate of climb* which will gain the most altitude in the *least time*. For every airplane there is an airspeed at a given power setting which will give the best rate of climb. The best rate of climb is normally used on take-off (after any obstacles are cleared) and is maintained until the airplane leaves the traffic circuit.

The BEST ANGLE OF CLIMB (V_X) is the *angle* which will gain the most altitude in a *given distance*. It is valuable in climbing out of restricted areas over obstacles. The airspeed for the steepest angle of climb is somewhat lower than the speed at which the best rate of climb is obtained. Because the airspeed for the best angle of climb is relatively slow, there is less air circulating around the engine to provide cooling and engine overheating is possible. The best angle of climb, therefore, should be maintained only until obstacles are cleared and then the nose of the airplane should be lowered to pick up the best rate of climb airspeed.

Every pilot should determine the airspeed for best rate of climb and for best angle of climb for the particular airplane he is flying. These airspeeds are usually given in the Airplane Flight Manual. However, it is necessary to bear in mind that these speeds will vary according to the all-up-weight of the airplane.

The rate of climb is not affected by the wind, since it is a vertical measurement of airplane performance and is not in any way related to groundspeed.

The angle of climb, on the other hand, is appreciably affected by the wind. When climbing into wind, the airplane moves over the ground at a lower speed and therefore takes longer to cover a given forward distance. The stronger the wind, the slower the ground speed, the steeper the angle of climb.

NORMAL CLIMB is a rate of climb that should be used in any prolonged cruise climb. The airspeed for normal climb is always indicated in the Airplane Flight Manual. It is a speed that is usually 5 to 10 knots faster than the airspeed for best rate of climb and as such provides better engine cooling, easier control and better visibility over the nose.

Gliding

In gliding, there is no power from the engine and the airplane is under the influence of gravity. Of the four forces, thrust is now absent and a state of equilibrium must be maintained by lift, drag and weight only.

In Fig. 34. Forces in a Glide, R represents the total reaction, i.e. resultant of lift and drag. This is equal and opposite to weight.

The angle at which the pilot chooses to glide determines the airspeed in the glide. The steeper the angle, the faster the airspeed; the shallower the angle, the slower the airspeed. At too fast an airspeed, structural damage to the airframe could result. At too slow an airspeed, the airplane could stall. The pilot must, therefore, choose a gliding angle that maintains an airspeed that is sufficient to maintain flight but not too fast to be unsafe.

When gliding with the power off, the airplane will tend to glide about 20% farther if the propeller is stopped than if it is windmilling.

Theory of Flight

The stopped propeller produces drag that is equal only to the parasite drag of its configuration. The propeller that is spinning acts as a windmill driving the engine, but without producing power. The power required to rotate the propeller and consequently the engine of the airplane is derived from the airflow and is about 10% of the rated power of the engine. The energy required to drive the

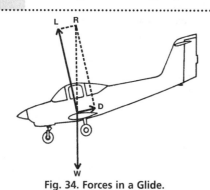

Fig. 34. Forces in a Glide.

propeller that is not producing positive thrust is therefore negative thrust or drag. Windmilling the propeller is like coasting an automobile in gear. The windmilling propeller in a tractor engine configuration directs disturbed air back over the lifting surfaces, inhibiting lift and creating drag. (Stopping a non-feathering propeller in flight should be done only in the event of an engine failure when there is no change of restarting the failed engine. The process of raising the nose and stopping the propeller takes skill and should be attempted only if you are confident of your ability to perform the procedure.)

Gliding Angle
BEST GLIDE SPEED FOR RANGE. An airplane will glide the farthest distance at the airspeed which results in an angle of attack that gives the maximum lift/drag ratio. This airspeed represents the optimum glide, a combination of airspeed and sink rate that allows an airplane to glide the farthest distance for altitude lost. This airspeed, alternately called optimum glide, maximum distance glide or gliding for range, is given in the Airplane Flight Manual.

If the pilot attempts to glide at an angle of attack either greater or less than that which gives the maximum L/D ratio, then in each case, the range will be decreased. If the angle of attack is decreased so that the airspeed increases, drag also increases and the path of descent will be steeper in still air. If the angle of attack is increased to flatten the glide, up to a certain point the rate of descent will decrease. However, the airspeed also decreases and the resultant decrease in speed over the ground means a steeper glide angle and reduced range.

Another factor which affects the glide path is, of course, the wind. A strong headwind or tailwind will tend to steepen or flatten the glide as the case may be.

When gliding into a fairly strong wind, greater distance may be covered over the ground if the nose of the airplane is kept somewhat lower than the attitude for best L/D ratio. For one thing, the increase in airspeed will yield an increase in ground speed which, in this case and contrary to gliding in still air or very light winds, will yield a shallower glide path. Secondly, by gliding at a slightly higher airspeed, the airplane will complete its glide in less time, having been subjected to the headwind for a shorter duration.

BEST GLIDE SPEED FOR ENDURANCE. An airspeed slightly less than that which gives the maximum L/D ratio is the airspeed to be used to achieve minimum sink. Sometimes, the object is to remain in the air for the longest period of time rather than to cover the greatest distance. Airplane Flight Manuals do not usually designate an airspeed for minimum sink since for most general aviation airplanes, it is not appreciably different from the airspeed for maximum L/D ratio. It is an airspeed of importance, however, under certain conditions and it is one with which glider pilots especially are very familiar. It can roughly be calculated as being approximately 1.1 times the power-off stalling speed.

Power Approach
The technique of gliding should be mastered by the student pilot in case at some time during his career he should have to make a forced landing with engine failure. The normal method of descent for landing, however, is the **power approach.** By applying power, the glide path may be flattened and the rate of descent more accurately controlled. Power approaches are advisable for light airplanes when landing in high winds or in gusty air conditions. Power approaches and power assisted landings are normally used when making landings on soft snow, sand or mud and by seaplanes landing on glassy water. To make a power approach: (1) Reduce power. (2) Allow the airplane to slow to approach speed. (3) Adjust to desired angle of descent. (4) Maintain constant airspeed and regulate rate of descent by manipulation of power. The normally recommended airspeed for power-on approaches to short field landings is 1.3 times the power-off stalling speed.

Turns
To make an airplane turn, the wings are rolled away from the normal horizontal position of level flight. The lift force, which always acts at 90° to the wing span, is, in a turn, inclined away from the vertical. Therefore, the vertical forces of lift and weight are no longer in balance, that is, are no longer in equilibrium. The airplane will descend unless the angle of attack is increased to produce more lift.

In a turn, the lift force has two components: one acting vertically and one acting horizontally. The vertical component opposes weight, while the horizontal component makes the airplane turn. This horizontal force is known as **centripetal force** and it counteracts the **centrifugal force** that, in a turn, tends to pull the airplane to the outside of the turn.

In Fig. 35, OW represents the weight of the airplane. The vertical component OA balances the weight of the airplane. OC represents the centripetal force necessary to counteract centrifugal force. OL is the resultant total lift factor.

The steeper the angle of bank, the more total lift is required to produce a level turn. The vertical component of the lift force must remain sufficient to compensate for weight. As the angle of bank increases, the total lift must be increased to provide sufficient centripetal force to overcome the increasing centrifugal force. This is accomplished by increasing the angle of attack by back pressure on the control column.

The steeper the angle of bank (for any given airspeed): (a) The greater the rate of turn. (b) The less the radius of turn. (c) The higher the stalling speed. (d) The greater the loading.

The higher the airspeed (for any given angle of bank): (a) The slower the rate of turn. (b) The larger the radius of turn.

Climbing and descending turns are executed like level turns. The factors acting on an airplane executing a climbing or descending turn are the same as those acting on the airplane in a level turn, except that, instead of maintaining a constant attitude, a constant climb or descent is maintained. In addition, there are considerations regarding power and attitude control.

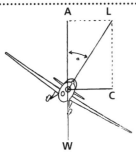

Fig. 35. Forces Acting in a Turn.

In a climbing turn, extra power over and above cruise power is required to achieve the climb configuration. In a descending turn, it is necessary to reduce power in varying degrees from cruise power to power fully off in order to achieve the descent speed and angle that is desired and that is safe and within limits.

The lateral stability of an airplane in a climbing or descending turn is affected by the angle at which the relative airflow meets each wing. In a descending turn, the inner wing, turning on a smaller radius, meets the relative airflow at a greater angle of attack and obtains more lift. The outer wing, however, because it is travelling faster also obtains more lift. The one compensates for the other and, therefore, in a descending turn, the angle of bank tends to remain constant.

In a climbing turn, the inner wing, describing an upward spiral, meets the relative airflow at the smaller angle of attack than the outer wing which is obtaining extra lift both from its extra speed and its greater angle of attack. Therefore, the angle of bank will tend to increase in a climbing turn.

Load Factors in Turns

In straight and level flight, an airplane has a load factor of 1, or 1 g. Turns increase the load factor. The steeper the angle of bank, the greater the load factor. A 60° turn produces a load factor of 2, making the effective weight of an airplane twice its normal weight. If an airplane weighs 2500 pounds in level flight, in a 60° turn, it will have an equivalent weight of 5000 pounds. A very steep turn may impose an increase in the loading as high as ten times the normal load. With some types of light airplanes, a bank of 80° or over can result in possible structural failure.

By reference to Fig. 36, it is readily apparent that the g load on the airplane structure increases with the angle of bank. However, other maneuvers than just turns impose high load factors on the airplane.

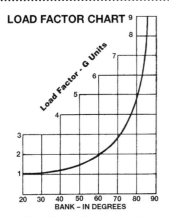

Fig. 36. Load Factors in Turn.

For example, when an airplane is pitched up more or less abruptly, the load factor is increased correspondingly. Therefore, while in a turn, should the pilot cause the airplane to abruptly pitch up, the g forces resulting from the pitching maneuver will be added to those caused by banking the airplane. The combined load factor may exceed the design limits of the airplane. Pilots should, therefore, be cautious about conducting maneuvers which require excessive movement of the controls in more than one plane.

Stall

A stall occurs when the wing is no longer capable of producing sufficient lift to counteract the weight of the airplane. A smooth laminar flow of air over the wing is necessary to produce lift. The stall occurs when the angle of attack is increased to the point where the steady streamlined flow of air is unable to follow the upper camber of the airfoil. The airflow separates from the wing, becomes turbulent or "burbles", the downwash and pressure differential are greatly reduced and loss of lift occurs. The airplane ceases to fly. This is called the **stall condition.**

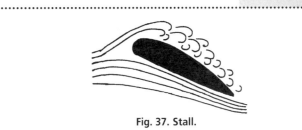

Fig. 37. Stall.

The point at which the airflow pulls away from the wing is called the **separation point.**

At high cruising speeds, the positive angle of attack is very low and the separation point is near the trailing edge of the wing. At slower speeds, the angle of attack must be increased to maintain a constant altitude and the separation point moves forward. When the angle of attack is increased to the point that the separation point moves forward far enough to exceed the design factor of the wing, the wing must stall. The stalling angle is commonly in the vicinity of 20° but varies with the shape of the airfoil.

Since few light airplanes have an angle of attack indicator, airspeed must be used as a guide in identifying the approach to a stall. On some airplanes, the approaching condition can be recognized by a light buffeting on the airframe as the air begins to burble over the wing. This buffeting may also be felt in the control wheel. As the angle of attack is further increased, the buffeting becomes more general and lateral control drops sharply as the ailerons lose their effectiveness in the separated airflow. When fully stalled, the wings lose all lift and the nose of the airplane drops.

On most airplanes, the stall occurs gradually. Because of the washout of the wings, the wing roots will stall before the wing tips. The alert pilot will recognize the symptoms and take corrective action while there is still aileron control and before all lift is lost.

The high lift airfoil, with a curved upper surface and a nearly flat bottom surface, generally stalls at a lower speed and a greater angle of attack than the more efficient high speed symmetrical or laminar flow airfoil used for modern general aviation airplanes.

An airplane properly loaded will stall at an **indicated airspeed** somewhere near the stalling speed published in the Airplane Flight Manual. This stalling speed, for all practical purposes, remains the same regardless of altitude.

It must be remembered that airplane attitude, airspeed and angle of attack are not consistently related.

> *An airplane will stall if the critical angle of attack is exceeded.*
> *It will stall at any airspeed if the critical angle of attack is exceeded*
> *It will stall at any attitude if the critical angle of attack is exceeded.*

Factors Affecting Stall

Weight: Weight affects the stalling speed of an airplane. Weight added to an airplane requires that it be operated at a higher angle of attack to produce the lift necessary to support that weight. Therefore the critical angle of attack will be reached at a higher airspeed.

Centre of Gravity: As the weight distribution moves within the allowable C.G. limits, the stalling speed as well as stability characteristics will be affected. As the airplane's centre of gravity moves from the most aft allowable position towards the most forward allowable position, the airplane's stalling speed will increase. The download on the horizontal tail surfaces increases as the airplane's centre of gravity moves forward. This download can be considered as part of the airplane's weight since it acts in the same direction as the weight force. Stalling speed, as we know, increases as the weight increases.

Of course, conversely, as the C.G. moves aft, the stalling speed decreases. This fact, however, should never be taken as license for a pilot, trying to reduce the stalling speed of his airplane, to load the airplane with the C.G. beyond the allowable aft limit. Decreased longitudinal stability, violent stall characteristics and poor stall recovery make improper loading a very poor policy.

Turbulence: Turbulence affects stall speed. An upward vertical gust causes an abrupt increase in angle of attack because of the change in direction of the air relative to the wing and could result in a stall if the airspeed of the airplane is at the same time relatively low.

Turns: As the angle of bank increases, the amount of lift required to sustain level flight also increases because of the increasing load factor that is integral to the action of banking an airplane. To increase lift, the pilot must increase the angle of attack of the airfoils. Therefore, in a turn, the stall angle is reached at a higher airspeed than in level flight. Most Airplane Flight Manuals have a chart similar to that in Fig. 38. that depicts the stalling speed at various angles of bank. There is, however, a fairly simple formula for determining stalling speed: normal stalling speed times the square root of the load factor being imposed. Typical load factor values and their square roots are shown in the following table:

Degree of Bank	Load Factor	Square Root
15°	1.04	1.02
30°	1.15	1.07
45°	1.41	1.19
60°	2.00	1.41
75°	3.86	1.96

The stalling speed of the airplane to which the chart in Fig. 38 applies is 64 knots. In a 30° banked turn, the stalling speed would be 64 kts x 1.07 (the square root of the load factor of 1.15), or 68.48 knots.

Flaps: The use of flaps, by increasing the lift potential of the wing, results in a reduction in stall speed, as indicated in the Stall Chart in Fig. 38.

Snow, Frost and Ice: An accumulation of frost, snow or ice on the wings will substantially alter the lifting characteristics of the airfoil and cause an increase in the stall speed and a decrease in the stall angle of attack. Even a very light layer of frost spoils the smooth flow of air over the airfoil by separating the vital boundary layer air. The airflow separates much farther forward than would normally be expected for the particular angle of attack. Lift is substantially reduced, the stall angle decreased and the stalling airspeed increased.

STALL SPEED, POWER OFF Gross Weight 2800 LBS. CONFIGURATION	ANGLE OF BANK		
	0°	30°	60°
FLAPS UP	64	69	91
FLAPS 20°	57	61	81
FLAPS 40°	55	59	78

Fig. 38. Stall Chart.

On the clean airfoil in Fig. 39, the separation point between the laminar and the turbulent airflow occurs near the trailing edge and the downwash angle is high. Downwash angle, it must be remembered, is a very important function of lift. On the frost covered airfoil in Fig. 40, the separation point has moved much farther forward, despite the fact that the angle of attack of the two airfoils is the same (about 12° which is about the angle for best angle of climb). The downwash angle is substantially reduced, as is lift.

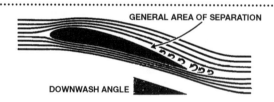

Fig. 39.

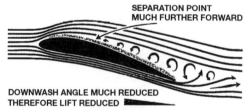

Fig. 40.

It is contrary to Air Regulations to take off in an airplane that has frost, snow or ice adhering to any of its critical surfaces. Critical surfaces are the wings, control surfaces, rotors, propellers, horizontal stabilizers, vertical stabilizers or any other stabilizing surfaces of the airplane and, in the case of an airplane that has rear mounted engines, the upper surface of the fuselage.

This is known as the **clean aircraft concept** and it is essential to the maintenance of flight safety. The critical surfaces must be inspected and determined to be free of contamination prior to take-off. Any accumulation of snow, frost or ice must be removed by placing the airplane in a heated hangar or by using approved de-icing solutions and/or methods. It is not acceptable to rely on the airplane's de-icing and anti-icing devices to do the job after take-off.

Frost, ice or snow formations that have a thickness and surface roughness comparable to medium or coarse sandpaper can reduce wing lift by as much as 30% and increase drag by as much as 40%. With a coating of frost such as this that appears relatively thin and insignificant, the airplane may manage to take off but as soon as the nose is raised to climb away, a stall may result at an angle of attack that would normally, on a clean wing, represent a safe climbing angle. The loss of lift and the decrease in the stalling angle of attack is not just the result of contamination of the wing surface but also and perhaps most significantly the contamination of the leading edge of the wing.

If a minimal coating of frost can have such a detrimental effect on lifting capacity, an accumulation of snow and ice will be even worse. It is possible for the stalling speed to be increased to such a degree that the airplane cannot reach a speed sufficient to achieve take-off, or if having achieved take-off to maintain flight. The weight of an accumulation of ice or snow, of course, adds to the total weight of the airplane but the weight factor is secondary to the effect the accumulation has on the airflow over the wing. Frost, snow, ice and slush should always be removed before take-off.

Dirt and bugs also disrupts the smooth surface and should be removed so that the surface is clean and uncontaminated.

It should never be assumed that loose snow or slush, or even water, will blow off during the take-off run. In fact, these may very well freeze to the wing and tail surfaces during take-off. There is evidence to show that there is a temperature decrease in the airflow over the wing as the pressure drops and lift is created. The temperature drop is only a few degrees, but, under the right conditions at ambient temperatures very near the freezing point, the airfoil surfaces could be cooled sufficiently during the take-off roll for any liquid to change to ice.

In airplanes where fuel tanks are located in the wings, the temperature of the fuel can significantly affect the surface temperature of the wing above and below these tanks. After a flight, the temperature of the fuel may be considerably colder than the ambient temperature and, depending on certain variables such as the water content of the precipitation and the wing surface temperature, clear ice may form on the wings above the fuel tanks. Such clear ice is hard to detect. This phenomenon is called **cold soaking**. Cold soaking can also cause frost to form on the wing under conditions of high relative humidity. This is a phenomenon that can occur in above freezing temperatures. In such instances, the frost tends to re-form quickly even when it has been removed.

On the other hand, warm fuel that is loaded into fuel tanks on a cold day may melt snow around the tanks. This liquid may then refreeze before take-off. (See also **Icing** in Chapter **Meteorology** and **Critical Surface Contamination** in Chapter **Airmanship**.)

Heavy Rain: Recent studies have indicated that an aircraft exposed to heavy rain experiences a loss of lift and an increase in drag as a result of the effect the rain has on the boundary layer and the surface of the airfoil. The rain drops that strike the leading edge are accelerated backward into the boundary layer and decrease air flow velocity. This causes premature separation of the boundary layer, an increase in drag and early stall.

The rain also causes a roughness of the airfoil surface. A thin water film forms on the surface. Raindrop impact craters and surface waves in the water film roughen the airfoil surface. The effect on lift and drag is not unlike that caused by frost.

It is possible that the aircraft may stall before the stall warning devices activate. These devices provide a warning signal just prior to the normal stall angle of attack. Airfoils contaminated with ice, snow, frost or heavy rain stall at an angle of attack much lower than this predetermined value.

Laminar airfoils are very sensitive to any surface roughness that changes the nature of the boundary layer and can be expected to be affected by heavy rain. Aircraft with a canard have also been shown to experience control difficulties in heavy rain.

The effect of heavy rain is more pronounced in high-lift configurations such as during take-off. The most significant and devastating decreases in aerodynamic performance are believed to be caused by torrential downpours associated with showers and thunderstorms.

Obviously the key to preventing the problems associated with heavy rain is to avoid the phenomenon. Do not penetrate heavy rain cells when landing, taking off or going around. Unfortunately, such heavy downpours cannot be predicted. If caught in one, expect a significant increase in descent rate and a decrease in airspeed.

Stall Warning Devices

Many light airplanes are fitted with a stall warning. This is a device which measures the angle of attack by means, usually, of a small vane extending forward of the leading edge of the wing. When the critical angle of attack is approached, it activates a warning device in the cockpit. This can be a red light, a bell or a buzzer of some sort.

The stall warning device is, however, calibrated to perform under clean wing conditions. It does not recognize the degraded performance of a contaminated wing and it cannot, therefore, be relied upon to give warning of an impending stall in icing conditions.

The best stall warning device is the pilot's training and experience in reaction to stalls. All pilots receive training in stalls but, after receiving their licence, many have such an aversion to them that they never go out and practice slow flight and stalls again. It is important to practice stalls sufficiently often that you develop an instinctive recognition of an incipient stall and an automatic reaction to avoid it.

Stall Recovery

Because insufficient lift is being generated by the wings to maintain flight, a stalled airplane starts to lose altitude. To recover from a stall, the pilot can

1. lower the nose to decrease the angle of attack, or
2. apply more power to accelerate the airplane. *If, however, the airplane is already under full power when the stall occurs, the only option for recovery from the stall is to lower the nose of the airplane.*

Spinning

Spinning may be defined as **autorotation** which develops after an aggravated stall.

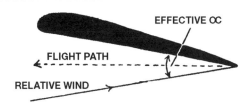

Fig. 41. Downgoing Wing.

When a wing is stalled, an increase in the angle of attack will decrease lift. If a disturbance causes a stalled airplane to drop one wing, or if rudder is applied to produce a yaw, the downgoing wing will have a greater angle of attack to the relative airflow, will receive less lift and will tend to drop more rapidly (Fig. 41). Drag on the downgoing wing increases sharply increasing the angle of attack of the downgoing wing still further and stalling it further. The nose of the airplane drops and autorotation sets in.

The upgoing wing will have a relative downward airflow, hence a **decreased** angle of attack and **increased** lift, and will rise more rapidly (Fig. 42).

The effect is to **accelerate** the rolling moment in the direction in which it first started. This is **spinning**.

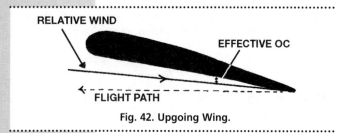

Fig. 42. Upgoing Wing.

Any attempt to correct a spin with aileron will only aggravate the roll. If the aileron on the downgoing wing is moved downward to bring the wing up, it will meet the airflow at a higher angle of attack and will therefore become more stalled than the wing itself. The upgoing aileron, on the other hand, will be unstalled and will have increased lift.

The spinning motion involves rolling, yawing and pitching. The airplane follows a helical or corkscrew path nose downward, rotating about a vertical axis. Pitch attitudes may vary from flat to steep. Forward and vertical speeds are both comparatively low.

The view from the cockpit is a steep, nose down attitude with a rolling motion about the spin axis. The airspeed is near stall. An angle of attack indicator (if installed) would show a fully stalled condition. The turn needle is fully deflected in the direction of the spin and the rate of descent is rapid. The g force acting on the airplane is 1.

The spin maneuver consists of **three stages**. The **incipient stage** occurs from the time the airplane stalls and rotation starts until the spin axis becomes vertical or nearly vertical. In the **developed stage,** the angles and motions of the airplane are stabilized and the flight path is nearly vertical. The third stage is **recovery**.

The aim in recovery from a spin is to upset the balance between the aerodynamic and inertia movements. Spin characteristics of different airplanes necessarily differ and the technique for spin recovery outlined in the Airplane Operating Manual must be followed. In the absence of manufacturer's recommendations, however, most light airplanes can be brought out of a spin by following these steps.

1. *Power to idle, ailerons neutralized.*
2. *Apply and hold full rudder opposite to the direction of rotation.*
3. *Control column positively forward far enough to unstall the airplane.*
4. *When rotation stops, neutralize the rudder, level the wings and recover smoothly from the resulting dive.*

Spiral Dive

A spiral dive is a steep descending turn in which the airplane is in an excessively nose down attitude. Excessive angle of bank, rapidly increasing airspeed and rapidly increasing rate of descent characterize it. It is, in fact, a hazardous maneuver. Structural damage can occur to the airplane if the airspeed is allowed to increase beyond normal limits. Excessive load factors may be produced in the pull up from the dive during recovery.

In some ways, a spiral dive resembles a spin. Don't confuse the two. In a spin, the airspeed is constant and low. In a spiral, the airspeed increases rapidly.

Recovery from a spiral dive must be taken promptly and in this sequence.

1. *Close the throttle and level the wings as nearly simultaneously as possible.*
2. *Keep straight.*
3. *Ease out of the dive.*
4. *Apply power as required to maintain height.*

Airspeed Limitations

Airspeed is the rate of movement of an airplane relative to the air mass through which it is flying. It is thrust impeded by drag. All airspeeds are a balance between thrust and drag, pitch attitude and power setting.

Loads greater than the weight of the airplane are produced by certain maneuvers and by gust conditions. In straight and level flight, the lift is equal to the weight. But an increase in speed will produce an increase in lift, which, in turn, will cause the airplane to climb. In order to maintain level flight when the airspeed is increased, the angle of attack must be reduced. An airplane flying in turbulent air may be subjected to severe vertical gusts which will, in effect, change the angle of attack on the wings. If the airplane is flying too slowly, the increased angle of attack may be great enough to cause the wings to stall. If, on the other hand, the airplane is flying too fast, the vertical acceleration may impose a load factor on the wings in excess of what they are strong enough to bear. For this reason, certain speeds are established by the manufacturer of each particular airplane and specified in the Flight Manual.

THE NEVER EXCEED, or MAXIMUM PERMISSABLE DIVE SPEED (V_{NE}). The maximum speed at which the airplane can be safely operated in smooth air. **A higher speed may result in structural failure, flutter, or loss of control.** (If you ever find yourself over the Never Exceed Speed inadvertently, throttle back and execute pull-outs with slow and firm pressure on the controls.)

MAXIMUM STRUCTURAL CRUISE, or NORMAL OPERATING LIMIT SPEED (V_{NO}). This is the cruise speed for which the airplane was designed and is the maximum safe speed at which the airplane should be operated in the normal category. **Warning: Do not EVER exceed this speed intentionally, even during descent, because of the possibility of unexpected gust loads.** The speed range between V_{NO} and V_{NE} (erroneously described as the Caution Range) should never be entered deliberately during normal operation.

MANEUVERING SPEED (V_A). The maximum speed at which the flight controls can be fully deflected without damage to the airplane structure. This speed is used for abrupt maneuvers or when flying in very rough air or in severe turbulence. At this speed, it is impossible for gusts to produce dangerous load factors. This is the speed least likely to permit structural damage to the airplane and yet allow a sufficient margin of safety above the stalling speed in gusty air. **Maneuvers which involve an approach to a stall, or full use of rudder or aileron control, should never be attempted above this speed.** (See also **Loads and Load Factors** in Chapter **The Airplane**.)

MAXIMUM GUST INTENSITY SPEED (V_B). The maximum speed for penetration of gusts of maximum intensity. In most light airplanes, the manufacturer does not recommend differing design speeds for V_B and V_A. However, for larger transport airplanes, there is a recommended design speed for maximum gust intensity as well as a maneuvering speed.

Note: Pilots flying high speed jets at high altitude may elect to maintain somewhat higher speeds when penetrating turbulent air. This is because the risk of an inadvertent stall (due to severe gust loads at high altitude) may present as great a hazard as the risk of structural strain. (Bear in mind when you get in the supersonic league that you can stall an airplane at almost any speed if you subject it to a sufficiently high g load.)

MAX. FLAPS DOWN SPEED (V_{FE}). The maximum speed at which the airplane may be flown with the flaps lowered. A speed in excess of this value may lead to a structural failure of the flaps.

*For other airspeed limitations, see **V Speeds** in the **Glossary**.*

Categories: Normal and Utility

Official aircraft specifications, which federal licensing authorities insist must be published before civil airplanes can be approved for certification, include operational information on speed and load limitations depending on how the airplane is to be operated. This information is also to be found in the manufacturer's manual. The information is usually published for two categories, **normal** and **utility**. A third category, **aerobatic**, is published for certain airplanes.

Normal Category

Maximum gross weight operations are permitted but certain maneuvers, such as spins, steep turns, etc., are prohibited.

Utility Category

Certain specified maneuvers may be performed but they must be carried out at reduced weight, narrower C.G. range and reduced airspeed. Baggage compartments and rear seats usually must be empty. When operated in the utility category, the following maneuvers are usually permitted: chandelles, lazy eights, steep turns, spins and stalls. The utility category is solely for the purpose of instructing and training pilots in certain flight maneuvers.

These operating limitations are set as a result of wing loading and C.G. considerations.

Positive limit load factors for the normal category are 3.8 times the gross weight. The reduced gross weight of the utility category permits a limit load factor of 4.4 times the gross weight.

When operating in the utility category, higher g forces can be imposed on the airframe. Because the weight is less than maximum gross weight, the ultimate strength factor of the airframe will not be exceeded.

The negative limit load factors for both normal and utility categories are approximately -1.52 and -1.76 respectively.

Aerobatic Category

In the aerobatic category, the positive limit load factor may be as much as 6.0 times the gross weight and the negative limit may be as much as -5.0 (depending on the particular make and model of airplane). When flying an airplane in the aerobatic category, it is essential to ensure that the weight and balance limits recommended by the manufacturer are respected. The weight and balance envelope is usually restricted when the airplane is to be flown in the aerobatic category.

Mach Number

The **mach number** (pronounced "mock") is the ratio of the speed of a body to the speed of sound in the air surrounding the body. The **speed of sound**, the rate at which sound travels in air, varies according to the temperature of the air. It is, by formula, proportional to the square root of the absolute temperature. At sea level at standard atmosphere conditions (15°C), the speed of sound is 660 knots. In the stratosphere where the temperature is about -60°C, the speed of sound will be only 575 knots.

The speed of sound is not dependent on temperature alone. Density also has a bearing but, temperature and density being so related, temperature is the controlling factor. The speed of sound is not, therefore, a function of height. In the Arctic regions of the world where temperatures commonly drop to -60°C, the speed of sound at sea level will drop to 575 knots. In tropical regions where the temperature even at considerable altitude may be well above 15°C, the speed of sound will be well over 660 knots.

The mach number is found by dividing the airspeed of the airplane by the speed of sound in the atmospheric temperature conditions existing at the time of the flight. An airplane flying at a mach number of .85 would be travelling at 85% of the speed of sound.

There is a critical mach number at which the airflow over the wing becomes sonic. At this airspeed, shock waves are formed on the wing and compressibility effects become apparent; drag increases, buffeting occurs and changes in lift and in the position of the centre of pressure cause changes in pitch. The shock wave on the upper surface upsets the lift distribution and causes a reduction in downwash over the tail. The airplane becomes increasingly unstable at increasing mach number. The swept wing design is somewhat effective in retarding this phenomenon. The shock wave tends to be experienced at the thick root first while the tips maintain lift. As a result, the critical mach number is raised. For this reason, high speed airplanes usually are designed with a swept wing. Some airplanes are designed so that, as there is an increase in mach number, stability tends to return.

Flight Instruments

An understanding of the various flight instruments and the way they work is essential to the pilot. With a basic knowledge of their characteristics and especially their limitations, a pilot is better able to interpret the information displayed on the instruments during various flight conditions.

Pitot Static Instruments

Instruments connected to the pitot static pressure system include the airspeed indicator, the altimeter and the vertical speed indicator. The system includes a pitot pressure source and a static pressure source.

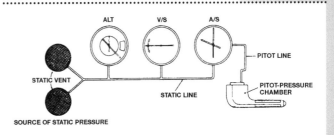

Fig. 43. Pitot Static System.

Theory of Flight

The **pitot pressure source** is usually located on the leading edge of the wing where it is clear of the slipstream, in a position to be as free as possible of air disturbance and facing the line of flight. When the airplane is in flight, the atmospheric pressure in the pitot pressure system is increased by dynamic pressure due to the forward motion of the airplane through the air. The airspeed indicator is the only instrument connected to the pitot pressure source.

The airspeed indicator, the altimeter and the vertical speed indicator are all connected to the **static pressure source** through which they are vented to allow air pressure inside their cases to equalize with the outside barometric pressure as the airplane gains or loses altitude. The ports, or vents (usually there are two), of the static pressure system are commonly located on opposite sides of the fuselage where they will not be affected by turbulence or by ram air pressures. Having two vents also compensates for any possible error in pressure that might occur on one of the vents when there are erratic changes in attitude, as in a steep turn.

One of the problems of the pitot pressure system is that the source is susceptible to the accumulation of dirt, water and ice. Blocked or partially blocked sources will cause inaccurate readings of the instruments. Complete blockage of the pitot pressure source, for example, (in combination with blocked drain holes) will trap pressure in the airspeed indicator and turn it into an altimeter. If the pitot pressure source is not completely blocked or if the system lacks integrity (has some air leaks), the airspeed indicator needle will rotate back to 0. On many modern airplanes, the pitot pressure source is electrically heated to prevent the build up of ice during flight. Failure to turn on pitot heat or failure of the electrical element does, however, happen. As ice builds up, gradually restricting the flow of dynamic pressure air through the pitot pressure system, the airspeed indication will gradually fall off. This situation could be particularly dangerous if the pilot, while flying in IFR conditions, tries to keep up his airspeed by continually reducing the angle of attack without referring to other flight and engine instruments that would alert him to the failed airspeed indicator. The airspeed of the aircraft could rapidly build up and exceed the red line, never exceed speed. Structural failure and in flight break up of the airplane would be the probable result.

A partially clogged static pressure system causes similar problems. Altimeter, airspeed indicator and vertical speed indicator will under-read in the climb, then slowly catch up when the airplane is levelled off. In the descent, the airspeed indicator will over-read, the VSI will indicate less than the true rate of descent and the altimeter will over-read.

For this reason, the pitot and static pressure sources should always be checked before a flight to see that they are clear. The pitot heat should always be turned on to prevent blockage of the dynamic pressure source. If ever, during a flight, you suspect blockage of the static pressure system because of contradictory readings of the instruments, open the alternate static source (most airplanes are so equipped) or, if necessary, remove the static source line entirely. If the needles move significantly, there is static pressure blockage.

Although the static pressure sources are located in a position where they ought not to be affected by ram air pressures, in certain maneuvers such as a sideslip, ram air will enter the static pressure source and cause erroneous readings on the pitot static instruments. On an airplane that has only one static pressure source, sideslipping in one direction (into the side where the port is located) will affect the instruments but sideslipping the other way will not.

The Altimeter

The altimeter is a special form of aneroid barometer (a barometer without liquid) which measures the pressure of the atmosphere. It is connected to the static pressure source through an outlet in the back of the case. This outlet serves as a vent to allow static atmospheric pressure to move into and out of the altimeter case as the airplane climbs or descends.

The atmospheric pressure at any point is due to the **weight** of the overlying air above, which decreases as the height above sea level increases. Hence, the instrument can be calibrated to read in terms of height. Under **standard air conditions** of 15°C, the weight of a column of air, one square inch in area, is 14.7 lb. at sea level. It exerts a pressure of 14.7 lb. per sq. in. This pressure is recorded on a barometer as 29.92 inches of mercury and by an altimeter as 0 feet. At 10,000 feet the weight of a one square inch air column has decreased to 10.11 lb., the corresponding barometric pressure to 20.58 inches, and the altimeter records 10,000 feet. The **decrease** in pressure is sensed by the altimeter and registered as an **increase** in height.

Fig. 44. Sensitive Altimeter.

The basic components of the altimeter are a stack of aneroid capsules located inside the case. These capsules are sealed and contain standard sea level pressure. Atmospheric pressure admitted to the case through the static pressure system causes these capsules to expand and contract. The expansion and contraction of these capsules transmits motion directly to gears and levers which rotate hands on the face of the altimeter. A large hand records altitude changes in hundreds of feet; a smaller hand records altitude in units of thousands of feet; and a third still smaller hand records altitude in units of ten thousand feet. The altimeter depicted in Fig. 44 is reading 10,400 feet.

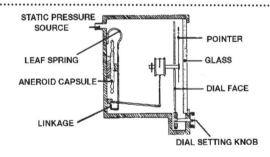

Fig. 45. Schematic of Simple Altimeter.

As the airplane climbs, the outside barometric pressure decreases and air moves out of the case through the static pressure system. As a result, the aneroid capsules expand, causing an increased altitude reading. As the airplane descends, air moves into the altimeter case and the capsules contract, causing a decreased altitude reading.

Pressure altimeters are calibrated during manufacture to indicate a true altitude in standard atmosphere conditions. The maximum allowable tolerance is plus or minus 20 feet at sea level. If, having set the current altimeter setting on the subscale and having

From the Ground Up

compared the altimeter reading to the known airport elevation, there is an error of more than plus or minus 50 feet, the instrument should be checked by maintenance.

Altimeter Errors
1. Pressure.
The heights at which airplanes are required to fly for air navigation purposes are **indicated heights above sea level.** Since the barometric pressure varies from place to place, an altimeter set to indicate height above sea level at the point of departure may give an erroneous indication after the airplane has flown some distance towards its destination. To correct for this, altimeters are fitted with a **barometric scale,** calibrated in inches of mercury, which allows the pilot to set the current altimeter setting in the altimeter setting window on the face of the instrument.

The altimeter setting, which is given in inches of mercury, may be obtained from towers and flight service stations in any particular locality over which the airplane is operating. In Fig. 44, the altimeter setting is 29.92 inches.

It is important, on a long cross country flight, for a pilot to get updated altimeter settings and regularly reset the altimeter. It is true that, if the airplane is operating well above the earth's surface, knowledge of the exact distance above the ground is of little immediate importance. However, if the airplane is operating at no great height above the ground or above the highest ground enroute, especially in conditions of poor visibility or when on instruments, knowledge of the actual separation between the airplane and the ground is of vital importance. Take, for example, a situation in which a flight altitude of 6000 feet has been selected to clear a 4800 foot mountain ridge that lies across the course near the destination. The altimeter setting at the airport of departure was 29.80". The altimeter setting at the destination airport, however, is only 29.20". If the pilot does not reset his altimeter, the airplane will clear the mountain ridge by only 600 feet instead of the 1200 feet expected.

An altimeter setting that is too high will give an altimeter reading that is too high. Each .10" Hg that is added to the altimeter setting will increase the indicated altitude recorded by the altimeter by approximately 100 feet. An altimeter setting that is too low will give an altimeter reading that is too low. One that is too high will give a reading that is too high. Therefore, when an airplane is flying from an area of high pressure into an area of low pressure (or from warm air into colder air) if a corrected altimeter setting has not been obtained by radio, the altimeter will read high.

"From high to low — watch out below".

In the northern hemisphere a drift to the right or starboard indicates that an airplane is flying towards an area of low pressure. A glance at the wind circulation around a low pressure area (see Fig. 9 in Chapter **Meteorology**) will indicate why this is so. Continued drift to starboard over a long period should act as a warning that an uncorrected altimeter may be reading high.

Abnormally High Pressure. Cold, dry air masses can produce barometric pressures higher than 31.00" Hg. On most standard altimeters, the barometric setting scale does not go higher than 31.00" Hg and the instrument will not indicate altitude accurately in conditions when the barometric pressure exceeds this figure. In such conditions, the aircraft's true altitude will be higher than the indicated altitude. It is possible to estimate the error in the indicated altitude for pressures above 31.00" Hg by adding 100 feet in aircraft altitude for each .10" Hg. Aircraft operating IFR and VFR, and especially those operating VFR at night, must exercise extra diligence both in flight planning and during the flight. When such conditions of abnormally high pressure exist, ATC will issue the actual altimeter setting and pilots are advised to set 31.00" Hg on their altimeters throughout the duration of the flight. To determine the suitability of an aerodrome from which or to which you intend to operate, increase the ceiling requirements by 100 feet and the visibility requirements by 1/4 SM for each .10" Hg over 31.00" Hg.

ALTIMETER SETTING REGION: When the altimeter setting is set on the barometric scale, the altimeter will register the **indicated height above sea level.** When you land at an airport from which the altimeter setting was obtained by radio, your altimeter will record the **altitude of the field above sea level.**

The altitude at which an airplane flies when using indicated height above mean sea level as its reference is known as **cruising altitude.**

In Canada, the area in which the indicated height above sea level is used is known as the Altimeter Setting Region. (See also **Altimeter Setting Region** in Chapter **Aeronautical Rules and Facilities**)

When taking off within the Altimeter Setting Region, or approaching to land, the altimeter should be set to the current altimeter setting of the field. If the altimeter setting is not available, prior to take-off, you can set the altimeter to read the elevation of the field above sea level. During flight, the altimeter should be progressively reset to the altimeter setting of the nearest station along the route of flight.

STANDARD PRESSURE REGION: Over trans-oceanic routes and certain continental areas, altimeter settings are not available. Flights over these areas are flown at pressure altitude (defined below). Because the altimeter is less reliable at high altitudes, flights above 18,000 ft. MSL are also flown at pressure altitude.

Pressure altitude is the height above sea level corresponding to a given barometric pressure under standard air conditions. When the barometric scale is set to read 29.92" Hg, the height recorded by the altimeter is pressure altitude.

In this case, the altitude at which the airplane is flying is referred to as the **flight level.** In reporting the flight level, the last two digits of the altitude are omitted. For example, an airplane flying at a pressure altitude of 15,000 ft. would report its height as "Flight Level 150 (FL 150)".

In Canada, the region in which the altimeter setting of 29.92" Hg. (1013.2 millibars) is used is known as the Standard Pressure Region. (See also **Standard Pressure Region** in Chapter **Aeronautical Rules and Facilities**)

When taking off within the Standard Pressure Region, the altimeter should be set to the current altimeter setting of the field; if this is not obtainable, to the elevation of the field above sea level. Immediately prior to reaching cruising altitude, the altimeter should be reset to 29.92" Hg. When approaching to land at an airport within the Standard Pressure Region, the altimeter should be re-set to the current altimeter setting of the field just prior to commencing the descent from cruising altitude or just prior to descending below the flight level at which a holding procedure has been conducted.

When flying out of the Standard Pressure Region into the Altimeter Setting Region, or vice versa, the altimeter should be adjusted to the setting appropriate to the area in which the airplane is being flown. See also **"Cruising Altitudes"**.

Field level pressure is the actual barometric pressure (not corrected to sea level) at any particular airport. If a pilot obtains the field level pressure by radio from an airport he is approaching and sets it on his barometric scale, his altimeter will register 0 feet when he lands.

Theory of Flight

2. Temperature.

The pressure altimeter is calibrated to indicate **true altitude** in standard atmosphere conditions (see **Standard Atmosphere** in Chapter **Meteorology**). Due to continual heating and cooling, the atmosphere at any given point is seldom at the temperature of standard air. In fact, the only time a pilot can be certain that the altimeter indicates true altitude is when the aircraft is on the ground at the airport for which the current altimeter setting is set on the sub-scale of an altimeter that has been precisely and accurately calibrated. When an aircraft is in flight, it can be assumed that the altitude indication of an altimeter is always in error as a result of temperature variations.

The amount of error depends on the degree to which the average temperature of the column of air between the aircraft and the ground differs from the average temperature of the standard atmosphere for the same column of air. If the actual temperature of the air column in which the airplane is flying is colder than standard air, the true altitude of the airplane above sea level will be lower than the indicated altitude. If the actual temperature is warmer than standard air, the true altitude will be higher. Extreme variations in temperature may cause the altimeter to register an indicated altitude as much as 2500 feet above or below the true altitude.

ALTITUDE CORRECTION CHART

A/D Temp °C	HEIGHT ABOVE THE ELEVATION OF THE ALTIMETER SOURCE (feet)												
	200	300	400	500	600	700	800	900	1000	1500	2000	3000	4000	5000
0°	0	20	20	20	20	40	40	40	40	60	80	140	180	220
-10°	20	20	40	40	40	60	80	80	80	120	160	260	340	420
-20°	20	40	60	60	80	80	100	120	120	180	240	380	500	620
-30°	40	40	60	80	100	120	140	140	160	240	320	500	660	820
-40°	40	60	80	100	120	140	160	180	200	300	400	620	820	1020
-50°	40	80	100	120	140	180	200	220	240	360	480	740	980	1220

NOTE: Values should be added to published altitudes.

Example: Aerodrome Elevation 2262   Aerodrome Temperature -50°

	ALTITUDE	HAA	CORRECTION	INDICATED ALTITUDE
Procedure Turn	4000 feet	1738 feet	+420 feet	4420 feet
FAF	3300 feet	1038 feet	+240 feet	3540 feet
MDA Straight-in	2840 feet	578 feet	+140 feet	2980 feet
Circling MDA	2840 feet	578 feet	+140 feet	2980 feet

Fig. 46. Altitude Correction Chart.

Since all altimeters in the same area are equally affected by temperature error, air traffic regulations require you to fly **only at indicated altitude**.

Abnormally Cold Temperature. It is important for a pilot to be able to calculate his true altitude above sea level. This is particularly essential in mountainous areas when the temperatures are very low. Pressure altimeters are calibrated to indicate true altitude under conditions of international standard atmosphere (ISA). Any deviation from ISA will result in an erroneous reading on the altimeter. In very cold temperatures, the true altitude will be much lower than the indicated altitude and could be critical in obstacle clearance. The very cold, dense air will cause an altimeter error of as much as 20%. The altimeter might read 8000 feet, for example, when the true altitude of the aircraft is only 6400 feet. In IFR operation, an aircraft maintaining the minimum obstacle clearance altitude (MOCA) while flying through the mountains may be well below a safe altitude if the pilot is trusting the altimeter indication and has not calculated the true altitude. Pilots experienced in winter operations in the mountains are conscientious about calculating true altitude before commencing any instrument approach to landing in order to ensure the safety of the flight.

The values derived from an altitude correction chart such as the one shown in Figure 46 should be added to the published procedure altitudes, minimum sector altitudes and DME arcs to ensure adequate obstacle clearance. The altimeter setting of the destination airport should be that used in the calculations.

To Calculate True Altitude. All computers currently in use are fitted with a sliding scale for correcting temperature error and converting indicated altitude to true altitude. However, the correction is based on the pressure altitude rather than the indicated altitude. To find the former at any time, simply set the barometric scale to 29.92 and the altimeter will record the pressure altitude. (See **Altitude Correction** in **Solving Problems with the Circular Slide Rule** in Chapter **Navigation**.)

Density Altitude. Barometric pressure and temperature both affect density. Density is an important factor in the take-off performance of modern airplanes. Low densities reduce engine thrust and aerodynamic lift. Density altitude, which is pressure altitude corrected for temperature, is important in calculating the safe fuel and payload permissible for take-off. Density altitude can be determined when the pressure altitude (at the airport) and the temperature are known. Most modern circular slide rules are calibrated to compute density altitude. However, if a computer is not available, the approximate density altitude may be found by the following simple formula:

Density altitude = pressure altitude + [100 **x** (actual temperature - standard temperature)].

To use this formula, there are three facts to remember.

1. *The temperature of standard air is 15°C at sea level.*
2. *The temperature of standard air decreases 2°C per 1000 feet of altitude.*
3. *For every 1°C difference from standard air, there is a 100 foot increase (or decrease, if the actual temperature is less than standard air) in the pressure altitude.*

Example: To find the density altitude:
Pressure altitude at airport .... 2000 feet
Actual temperature .... 25°C
The temperature of standard air at 2000 feet should be (15°C less 2° per 1000 feet) 11°C.
Therefore, the density altitude = 2000 + [100 x (25° - 11°)]= 3400 feet.
(See also **Density Altitude** in **Chapter Airmanship**.)

3. Mountain Effect.

Winds which are deflected around large single mountain peaks or through valleys of mountain ranges tend to increase speed which results in a local decrease in pressure (Bernoulli's Principle). A pressure altimeter in such an airflow would be subject to the decrease in pressure and would give an altitude reading that is too high. The error will be present until the airflow returns to its normal speed some distance downwind of the mountain or mountain range.

Winds blowing over a mountain range often create a phenomenon that is known as the mountain wave. The effect of a mountain wave often extends as far as 100 miles downwind of the mountains and up to altitudes much above the actual elevation of the mountain ridge. Within the mountain wave, downdrafts are usually very intense. They are normally more severe near the mountain and at the same height as the summit of the range. The increase in wind speeds that extend throughout the mountain wave causes a local drop in pressure within the wave. In addition, temperatures that vary from standard temperature complicate the situation, setting up the altimeter for a major error in altitude reading. In fact, altimeters may read as much as 3000 feet too high in severe mountain wave conditions. (See also **The Mountain Wave** in Chapter **Meteorology**.)

Although mountain waves usually generate severe turbulence, on occasion flight through a mountain wave may be very smooth. Such smooth conditions usually occur at night or when an overcast exists. As a result, there is no warning of the unusual flight conditions. An airplane, entering such a smooth and intense downdraft, will begin to descend but the altimeter will not register the descent until the airplane descends through an altitude level equal to the altimeter error caused by the mountain wave. Thus the pilot of the airplane in this intense downdraft may end up with much less ground separation than he expects without ever knowing that he was in an intense downdraft.

Altitude Definitions

Indicated altitude: the reading on the altimeter when it is set to the current barometric pressure.

Pressure altitude: the reading on the altimeter when it is set to standard barometric pressure (29.92" Hg).

Density altitude: pressure altitude corrected for temperature.

True altitude: exact height above mean sea level.

Absolute altitude: actual height above the earth's surface (altimeter set to field level pressure).

The Airspeed Indicator

The airspeed indicator tells the pilot the speed at which he is travelling through the air (not over the ground). The dial is calibrated in knots and miles per hour (Fig. 47).

The airspeed indicator is connected to both the pitot and static pressure sources. To give a reading of speed through the air, the instrument measures the difference between the pressure in the pitot pressure system and the pressure in the static system. When the airplane is standing on the ground, the pressure in the two systems is equal and the airspeed indicator registers 0. When the airplane is in motion, the pressure in the pitot pressure system is

Fig. 47. Airspeed Indicator.

increased by dynamic pressure due to the forward motion of the airplane through the air (pitot pressure is therefore the sum of atmospheric pressure and dynamic pressure). The airspeed indicator senses the total pressure in the pitot pressure system, subtracts the pressure in the static system and gives a reading of the dynamic pressure, the measure of the airplane's forward speed. This reading is displayed on a graduated scale on the face of the instrument and is called **indicated airspeed (IAS).**

The pitot pressure source is connected to the interior of a thin corrugated metal expansion box called the aneroid capsule and

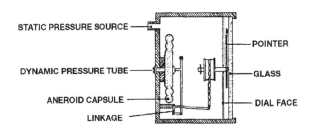

Fig. 48. Schematic of Simple Airspeed Indicator.

admits pitot pressure into this capsule. The static pressure source is connected to the inside of the instrument case and maintains the air in the case at the prevailing atmospheric pressure. Changes in dynamic pressure inside the aneroid capsule cause it to expand or contract. This movement is transmitted through a system of linkage to a hand which rotates around a dial calibrated in knots and/or miles per hour.

Airspeed Indicator Markings

Colour coded markings are used on the airspeed indicator to indicate safe operating ranges and operating limits.

1. Red. A red line is placed at the Never Exceed Speed (V_{NE}). This is the maximum speed at which the airplane can be operated.

2. Yellow. A yellow arc indicates the "caution speed" range. The lowest limit of the yellow arc is the Maximum Structural Cruising Speed (V_{NO}). The airplane should be operated in the "caution speed" range only in smooth air. However, since it is impossible to guarantee that turbulence will not be encountered and that the air will be smooth, the airplane should not be operated intentionally in this range.

3. Green. A green arc indicates the normal operating range. The lower limit of the green arc is the Power Off Stalling Speed with flaps and gear up (V_{SL}. The upper limit of the green arc is V_{NO} (the maximum cruising speed for normal operation).

4. White. A white arc depicts the speed range in which fully extended flaps may be used. The lower limit of the white arc is the Power Off Stalling Speed with flaps and gear down (V_{SO}). The upper limit on the white arc is the Maximum Flaps Down Speed (V_{FE}).

On the airspeed indicator depicted in Fig. 47, V_{SO} is 58 knots (67 mph), V_{FE} is 109 knots (125 mph), V_{SL} is 65 knots (75 mph), V_{NO} is 156 knots (180 mph) and V_{NE} is 197 knots (226 mph).

Maneuvering Speed is not marked on the face of the Indicator.

Airspeed Indicator Errors

The airspeed indicator is affected by several errors for which correction must be made. These are:

1. Density: The density of the air depends on atmospheric pressure and temperature. These are variable factors. Consequently a standard value for density has to be assumed in order that the airspeed indicator may be calibrated.

The standard for calibrating airspeed indicators is normal sea level pressure, 29.92 inches of mercury, at a temperature of 15°C. A rough correction may be made by adding 2% to the indicated airspeed for every 1000 ft. of pressure altitude.

For example: Indicated airspeed at 10,000 ft. — 130 knots.

Theory of Flight

Correction: 10 X 2% = 20%
20% of 130 kts. = 26 kts.
True airspeed = 130 + 26 = 156 kts.

More accurate corrections, allowing for the actual temperature and pressure, can be obtained with any suitable circular slide rule computer.

Some airspeed indicators, such as the one shown in Fig. 40, incorporate such a computer into the instrument. To determine TAS, pressure altitude and the outside air temperature must be known. The pressure altitude is set against the temperature in the window at the top of the instrument. TAS is read against the pointer on the window in the cruising speed range of the instrument (lower left.)

	AIRSPEED CORRECTION TABLE							
FLAPS UP	IAS	60	80	100	120	140	160	180	—
	CAS	68	83	100	118	137	156	175	—
*FLAPS DOWN 20°-40°	IAS	40	50	60	70	80	90	100	110
	CAS	58	63	68	75	84	92	101	110
*Maximum Flap Speed 110 knots, CAS									

Fig. 49. Airspeed Correction Table.

2. Position Error. Eddies that are formed as the air passes over the wings and struts are responsible for position error. The angle at which the pitot pressure source meets the airflow is also a cause. Position error is reduced by placing the pitot pressure source as far ahead of the leading edge of the wing as is practicable. The position error remaining is tabulated on a cockpit calibration card, or in tables in the Flight Manual. See Fig. 49.

3. Lag: This is a mechanical error due to the friction of the working parts of the instrument.

4. Icing: Ice formation, blocking the pitot and/or static pressure sources, has been a source of considerable trouble with airspeed indicators in the past. This has been largely eliminated in present day instruments by the adoption of electrically heated pitot pressure sources.

The question, of course, is when should pitot heat be used. Generally, in summer, it is not necessary to be concerned about moisture or ice in the pressure system. In spring, fall and winter, when outside temperatures drop and the moisture content of the air increases, the situation can be more critical. Any moisture, either in solid or liquid form, may cause an erroneous reading of the airspeed indicator. An accumulation of ice that completely blocks the pitot pressure source will cause a complete loss of airspeed indication. Even while the airplane is parked, if the pitot pressure source is not covered, rain, snow or frost may collect in the system and obstruct the free passage of air. For this reason, it is advisable to use pitot heat whenever the airplane is being operated in sensitive temperature or moisture conditions. As a rule of thumb, 10°C might be selected as the critical temperature. Turn the pitot heat on just prior to take-off and turn it off after landing since extended use of pitot heat on the ground may cause the heater element to burn out.

5. Water: Water in the system can cause very erratic airspeed indications. The errors may be high or low, depending on whether the water is in the dynamic or static pressure systems. Sufficient water in the pitot pressure system may block it off completely. The pitot pressure source should be covered when an airplane is standing in the open to prevent water getting into the system.

Airspeed Definitions

Indicated Airspeed (IAS) is the uncorrected speed read from the airspeed dial. It is the measurement of the difference between the total pressure (that is, the pitot pressure which is the sum of the atmospheric pressure and the dynamic pressure) and the atmospheric pressure in the pitot-static system.

Calibrated Airspeed (CAS) is indicated airspeed corrected for instrument error and installation error in the pitot-static pressure system. As the airplane flight attitude or configuration is changed, the airflow in the vicinity of the static pressure sources may introduce impact pressure into the static source, which results in erroneous airspeed indications. The pitot section of the system is subject to error at high angles of attack, since the impact pressure entering the system is reduced when the pitot pressure source is not parallel to the relative airflow. Performance data in Airplane Flight Manuals is normally based on calibrated airspeed.

Equivalent Airspeed (EAS) is calibrated airspeed corrected for compressibility factor. This value is very significant to pilots of high speed aircraft, but relatively unimportant to pilots operating at speeds below 250 knots at altitudes below 10,000 feet.

True Airspeed (TAS) is calibrated (or equivalent airspeed) corrected for the airspeed indicator error due to density and temperature. TAS is the actual speed of the airplane through the air mass. (See **Density** above.)

The Vertical Speed Indicator (VSI)

The vertical speed (vertical velocity, or rate of climb) indicator shows the rate, in feet per minute, at which the airplane is ascending or descending. The principle on which it operates is the change in barometric pressure which occurs with any change in height. This instrument is contained in a sealed case and is also connected to the static air pressure system.

Fig. 50. Vertical Speed Indicator.

Atmospheric pressure is led from the static pressure source directly into an aneroid capsule, or diaphragm, contained within the case of the instrument. Air is also permitted to leak at a relatively slow rate through a capillary tube into the case of the instrument. The difference between the quick change in pressure which occurs within the aneroid capsule and the relatively slow rate at which this pressure is equalized within the case causes the capsule to expand or contract. This movement is amplified and transmitted by linkage to the pointer on the dial of the instrument.

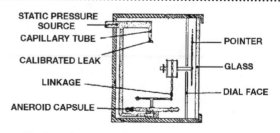

Fig. 51. Schematic of Simple Vertical Speed Indicator.

When the airplane loses altitude, pressure within the capsule increases almost immediately, while pressure within the case changes slowly. The capsule therefore expands and the pointer indicates DOWN in feet per minute.

When the airplane gains altitude, the process is reversed and the pointer indicates UP. When the airplane remains level, the pressures equalize and the pointer indicates 0.

Note that the vertical speed indicator registers the rate of climb or descent, not the attitude of the airplane. An airplane may gain height in a vertical up-current of air when it is flying perfectly level. The vertical speed indicator should be closely co-ordinated with the airspeed indicator. Corrections for altitude gained or lost in cruising flight should be made by nosing the airplane up or down (by use of the elevators). Intentional change in altitude should be made by increasing or decreasing power at a given airspeed (by use of the throttle).

Lag

A change in altitude must occur before the vertical speed indicator can indicate such a change. Although the indicator will show quite quickly a climbing or descending trend, there is a lag of from 6 to 9 seconds before it will indicate the correct rate of climb or descent. The instrument too will still indicate a vertical speed for a short time after the airplane has levelled off. If pitch changes are made slowly, the lag is minimal and the instrument will give a fairly accurate representation of the vertical speed at any given moment. If the pitch changes, however, are large and rapid, the amount of lag in the instrument will be sizeable.

Instantaneous vertical speed indicators are available that do not have lag and are therefore very accurate. They operate on a more complicated arrangement of pistons and cylinders that give an almost instantaneous indication of change in vertical speed.

The Radar Altimeter

The **radar altimeter** (also known as the absolute altimeter or the radio altimeter) indicates the actual height of the airplane above the earth, or above any object on the earth over which the airplane is passing. The principle is extremely simple. A radio transmitter in the airplane sends a signal towards the earth whose frequency changes at a definite rate with respect to time. This signal is reflected by the earth and returns as an **echo** after a time interval equal to twice the height divided by the velocity of the signal. During this interval, the frequency of the transmitter has changed and now differs from that of the echo by the rate of change of frequency times the time of transit. The reflected wave is combined with some of the outgoing wave in the airplane receiver and the difference, or "beat" frequency, is measured by a frequency meter. Since the reading of the meter is that of the "beat" frequency, it is proportional to the time delay of the echo hence to the **height** and can thus be calibrated in feet.

The radar altimeter depicted in Fig. 52 incorporates an AGL altitude scale with indications from 0 to 2500 feet and a DH select bug which the pilot, by rotating the DH knob, can select to the desired decision height on the altitude scale. The DH lamp lights when the decision height has been reached. A flag, when it drops into view, indicates that invalid altitude information is being displayed.

The radar altimeter is of immense value in an airplane flying on an IFR or night VFR flight. By displaying to the pilot exact information about height above the ground, it is an important instrument for use during landing approaches and in any other situation in which the pilot wants to maintain a safe altitude above the minimum enroute or minimum obstruction clearance altitude.

Radar altimeters are also used extensively by airplanes engaged in specialized types of flying, such as aerial survey, agriculture, etc., to maintain positive altitude above the ground regardless of rolling terrain, hills or valleys. The radar altimeter is also useful in determining the position of pressure systems. By comparing the readings of the radar altimeter and an altimeter set to pressure altitude, current changes in atmospheric pressure may be determined. In combination with his basic knowledge of the characteristics of pressure systems (i.e. highs and lows), a pilot is able to determine which direction to fly around a pressure system to take advantage of the best winds.

Gyro Instruments

Gyro instruments have made the art of piloting an airplane more precise. They are very useful in VFR flight. In instrument flight, they are invaluable. In most general aviation airplanes, there are three gyro instruments; the heading indicator, the attitude indicator, and the turn and slip indicator.

The Gyroscope

The gyroscope is a rotor, or spinning wheel, rotating at high speed in a universal mounting, called a gimbal, so its axle can be pointed in any direction (Fig. 53).

The peculiar actions of a gyroscope, though they may appear to defy physical laws, actually depend entirely upon Sir Isaac Newton's Laws of Motion.

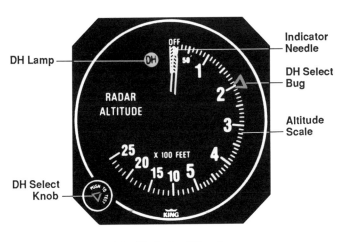

Fig. 52. Radar Altimeter.

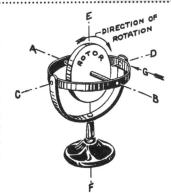

The Wheel, or rotor, is free to revolve in its supporting ring about the axis A-B. The supporting ring is free to revolve in the outer ring about the axis C-D at right angles to the rotation axis of the rotor. The outer ring is free to revolve about the vertical axis E-F on the pivot vearing at F.

Fig. 53. The Gyroscope.

Theory of Flight

All of the practical applications of the gyroscope are based upon two fundamental characteristics: gyroscopic inertia (or rigidity in space) and precession.

Gyroscopic Inertia is the tendency of any rotating body, if undisturbed, to maintain its plane of rotation. When the rotor in Fig. 53 is spinning about its axis A-B, the direction of this axis will remain fixed in space, regardless of how the base of the gyroscope is moved around it.

Precession is the tendency of a rotating body, when a force is applied perpendicular to its plane of rotation, to turn in the direction of its rotation 90 degrees to its axis and take up a new plane of rotation parallel to the applied force. The following is a simple illustration: Imagine a bicycle wheel spinning rapidly on a broom handle which is held horizontally in your hands. Attempt to push either end forward and one end will automatically raise in your hand while the other will drop. A push applied in the direction of the arrow at G (Fig. 53) would cause the rotor in its supporting ring to tend to rotate around the axis C-D.

Power Sources for Gyro Instruments

The gyro instruments require a source of power to drive their gyros. This power is supplied either by the electrical system of the airplane or by a vacuum system that functions by means of an engine-driven pump or a venturi.

Engine Driven Vacuum System

Air pressure differential is the principle on which the vacuum driven gyro instruments operate. A vacuum pump run by the airplane engine creates a partial vacuum in the system. A filtered air inlet upstream of the gyros allows air to rush into the system causing the gyro wheels to spin.

The gyro instruments require a suction of from 4 to 6 inches of mercury to operate. (Various manufacturers require differing vacuum settings for their particular instruments.) A vacuum pressure gauge, mounted on the instrument panel, indicates the amount of suction being generated in the system. The gauge indicates the relative difference between the outside air pressure and the air in the vacuum system. If there is not enough suction, the gyros will not spin fast enough for reliable operation. If there is too much suction, excessive wear of the gyro bearings will result from the gyros spinning too quickly.

The advantage of the engine driven vacuum system is that it starts operating as soon as the engine starts. On the other hand, in the event of engine failure, all power to the gyro instruments will be lost.

Some gyro instruments are driven by an engine powered pump that provides **positive pressure,** rather than a vacuum. High speed air is directed through the system causing the gyro wheels to spin.

Venturi Driven Vacuum System

A venturi is used in place of a vacuum pump on many light airplanes. It has the advantage of being inexpensive to install and simple to operate. However, its efficiency is dependent on the airspeed of the airplane and the venturi tubes themselves cause some aerodynamic drag.

The venturi tubes (there may be one or more) are usually mounted on the side of the airplane in a position to be in the airflow from the propeller. As the airflow enters the constriction of the venturi, the velocity of the airflow increases. A low pressure area is created within the venturi that results in a partial vacuum in the line leading from the venturi to the gyro instruments.

Electrically Driven Gyros

Electrically driven gyros were first developed for use in airplanes that flew at very high altitudes where atmospheric pressure was too low to operate a vacuum system. Alternating current from engine driven alternators or generators provides the power to drive the gyros.

Many of the newer electrically operated gyro instruments have the feature that they will not tumble in aerobatic maneuvers.

It is common practice to use a combination of electric and vacuum driven instruments for safety's sake, in case one system fails. Often the heading indicator and the attitude indicator operate on the vacuum system while the turn and slip is electrically operated. Large airplanes usually have two complete sets of instruments, one driven by a vacuum system, the other by electricity. A standby venturi powered system is also available for those airplanes whose gyro instruments rely only on an engine driven vacuum system. A three way valve ties the venturi into the system and automatically brings it on line if the engine driven vacuum pump fails.

Care of Gyro Instruments

The gyro is a precise instrument and requires special care. Gyro driven instruments should be caged before aerobatics to avoid tumbling the gyros which action can cause damage to the bearings. In addition, of course, the instrument is useless until the gyro is erected again.

Abrupt brake use should also be avoided, since it can impose acceleration loads on the gyro bearings.

Since airborne contaminants can damage the gyro bearings, it is good practice to periodically clean or replace the air filters of air driven units, especially if your airplane operates out of a dusty airport.

Gyro instruments that are driven by a venturi system will be disabled if ice forms in the venturi, blocking off the supply of air. Atmospheric conditions that are conducive to the formation of carburetor ice may produce ice in the venturi also.

Gyro instruments do regrettably sometimes fail but usually give some indication in advance that things are not well with them. Any abnormal operation, such as excessive precession, sluggish response or noise, should be viewed as an indication of impending failure. Gyro instruments, however, can fail without prior warning leaving the pilot without any indication that the instrument he is monitoring is giving incorrect information. Gyro instruments are being developed that incorporate a warning flag that warns of aberrant operation of the gyros.

Since shock damage is a sure way to turn a gyro instrument into junk, be sure that instruments are handled with care during removal for repair.

The Heading Indicator (HI)

The heading indicator (also known as the directional gyro) is an instrument designed to indicate the heading of the airplane and, because it is steady and accurate, to enable the pilot to steer that heading with the least effort.

The gyro wheel in the heading indicator is mounted vertically and spins about its horizontal axis at approximately 12,000 rpm. The spinning gyro wheel is mounted in an inner gimbal ring that is free to turn about the horizontal axis. The inner ring is, in turn, mounted inside an outer gimbal ring. The compass rose card on the face of the instrument is attached by a series of gears to the outer gimbal ring. As the airplane turns, the compass card rotates indicating a turn to the left or right.

Once the gyro starts spinning, it obeys the fundamental gyroscopic principle of rigidity in space. Thus the gyro wheel and the gimbal

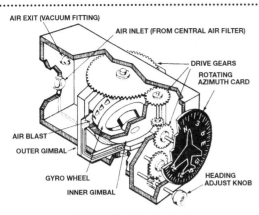

Fig. 54. Heading Indicator.

rings take up a fixed position in relation to the earth and it is the airplane that moves about them.

A heading indicator in common use today is shown in Fig. 54. The compass rose card turns as the airplane turns and the heading is read opposite the nose of the airplane pointer. As in the case of the compass, the figures are printed with the last 0 left off — 3 stands for 30°, 12 for 120°, etc.

The gyro is not, of its own accord, north seeking. It must, therefore, be synchronized with the magnetic compass at the beginning of a flight. For this purpose the instrument is fitted with a small knob (bottom left on the instrument in Fig 54). Pushing this knob in cages, or locks, the gyro system upright. By turning the knob, the compass rose card can be turned to any desired heading. Pulling the knob out releases the locking mechanism and leaves the gyro free to indicate changes in direction.

The magnetic compass is afflicted with many vagaries, including northerly turning error, acceleration and deceleration errors.

The heading indicator remains constant without swinging or oscillating and provides a means of accurate steering even in rough air. Precise turns can be made and stopped at any desired heading, as the instrument responds instantly without lag.

From all of which it might appear that the magnetic compass has had its day. Such would indeed be the case were it not for the fact that the heading indicator has a pet little gremlin of its own and must from time to time be reset, using the magnetic compass as a reference!

Precession Error
Frictional forces in the gyro system cause it to precess. This precession causes a creep or drift in the reading on the card, amounting to approximately 3° in 15 minutes.

The gyro is also subject to **apparent precession.** Gyroscopic inertia keeps the spinning gyro fixed in space so that, relative to space, it does not move. But the earth rotating underneath it gives it an apparent motion relative to the earth. This causes an apparent drift, or precession, which varies with latitude. At the equator, apparent precession is zero. At the poles, it is 15 degrees per hour.

Precession error, both mechanical and apparent, must be corrected for at regular intervals of about 15 minutes. It is important, while synchronizing the heading indicator with the magnetic compass, to hold the airplane straight and level to ensure that there is no compass error.

Limitations of the Heading Indicator
A gyro that operates on a vacuum system requires from 4 to 6 inches of mercury vacuum to operate. It should not be used for take-off until it has run for 5 minutes. This time is required to get up to operating speed. (If the gyro is operated by a venturi tube instead of a vacuum pump, it cannot be used for take-off at all.)

Heading indicators, such as that shown in Fig. 54, incorporate spill proof gyros. Because they have a self erecting mechanism, they remain functional in banks, climbs and glides at angles up to 85°. However, in maneuvers exceeding 85°, they give incorrect readings and must be reset.

The Attitude Indicator
The attitude indicator (also called the artificial horizon or the gyro horizon) provides the pilot with an artificial horizon as a means of reference when the natural horizon cannot be seen because of cloud, fog, rain or other obstructions to visibility. It shows the pilot the relationship between the wings and nose of the airplane and the horizon of the earth.

The natural horizon is represented by a horizon bar on the face of the instrument (Fig. 55). The attitude of the aircraft in relation to the horizon is indicated by a split bar or, on some instruments, by a miniature airplane. A pointer at the top indicates degrees of bank on an index scale graduated from 0° to 90° right or left.

In the attitude indicator, the gyro wheel is mounted horizontally and spins about its vertical axis. It is mounted in a universal gimbal ring system, free about both the pitching and rolling axes of the airplane and is therefore able to remain spinning in a horizontal plane parallel to the true horizon, regardless of the rolling or pitching movements of the airplane around it.

The horizon bar is attached by a pivoted arm to the gimbal ring and remains parallel to the natural horizon. The relationship of the split bar (miniature airplane) to the horizon bar is the same as the relationship of the actual airplane to the actual horizon. When the airplane is flying level, the split bar is lined up level with the horizon bar.

Fig. 55. Attitude Indicator.

When the airplane noses up, the gyro wheel remains horizontal. A relative down force is exerted on the pivoted arm to which the horizon bar is attached, causing the horizon bar to sink below the split bar.

In the case of a nose down condition the reverse action takes place.

The attitude indicator may be thought of and used in either of two ways:

1. Some pilots like to consider it as a window through the instrument panel and through the fog or clouds, in which case the horizon bar is where the natural horizon would be.

Theory of Flight

2. Others like to fly the miniature airplane with respect to the horizon bar.

When the airplane noses up, the miniature airplane rises above the horizon bar, indicating a nose high condition. When the airplane noses down, the miniature airplane sinks below the horizon bar, indicating a nose down condition. When the airplane banks, the miniature airplane banks on the horizon bar and the pointer indicates the degree of bank on the index scale.

When it is necessary to fly the airplane slightly nose up or down, according to altitude, power and load, the miniature airplane can be adjusted to match the horizon bar by means of a knob at the bottom of the case.

In the caged position, the gyro is locked with the miniature airplane showing level flight. If the instrument is uncaged during flight, the actual attitude of the airplane must be identical to that of the miniature airplane in the instrument (straight and level), otherwise the instrument will show false indications when uncaged.

Limitations of the Attitude Indicator

The gyro, of an attitude indicator that is driven by a vacuum system, requires 4 or more inches of mercury vacuum to operate and 5 minutes to get up to operational speed. An attitude indicator that is electrically driven is operational almost immediately.

Attitude indicators, such as that shown in Fig. 55 and especially those that are electrically driven, have virtually no limits of pitch and roll and will accurately indicate pitch attitudes up to 85° and will not tumble even in 360° roll maneuvers. The traditional style of attitude indicator that will frequently be found installed in airplanes of older manufacturer permits dives or climbs only up to 70° and banks up to 90° (vertical). If these limits are exceeded, the gyro will tumble. Therefore, during aerobatic maneuvers, the gyro should be caged.

Any force that disturbs the free rotation of the gyro will cause errors in the indications presented by the attitude indicator. Poorly balanced components, clogged filters, improperly adjusted valves, pump malfunctions, friction and worn parts will result in erratic and inaccurate indications.

Other errors are induced during normal operation. In a skidding turn, the gyro precesses toward the inside of the turn. After return of the aircraft to straight and level flight, the instrument shows a turn in the direction of the skid. In a normal turn, the gyro precesses towards the inside of the turn. The precession error is quickly corrected by the erecting mechanism.

Acceleration and deceleration also induce precession errors. During acceleration, the horizon bar moves down indicating a climb. During deceleration, the horizon bar moves up indicating descent.

The Turn and Slip Indicator

The turn and slip (or turn and bank) indicator combines two instruments in one and is sometimes called the needle and ball. The needle indicates the direction and approximate rate of turn of the airplane. The ball indicates the amount of bank in the turn; that is, whether there is any slipping or skidding in the turn (Fig. 56).

The ball is controlled by gravity and centrifugal force. It is simply an agate or steel ball in a liquid filled, curved glass tube. In a balanced turn, the ball will remain in the centre as centrifugal force offsets the pull of gravity. In a slip, there is not enough rate of turn for the amount of bank. The centrifugal force will be weak and this imbalance will be shown by the ball falling down toward the inside of the turn. In a skid, the rate of turn is too high for the amount of bank. The centrifugal force is too strong and this imbalance is indicated by the ball sliding toward the outside of the turn.

Fig. 56. Turn and Slip Indicator.

The turn needle is actuated by a gyro wheel operated either electrically or by a venturi tube or vacuum pump. Free air entering through a nozzle impinges on the paddles of the gyro wheel, causing it to rotate at approximately 9,000 rpm. The gyro wheel is mounted vertically and rotates about its horizontal axis.

The basic principle which governs the operation of the turn needle is gyroscopic precession. The spinning gyro wheel, or rotor, is mounted in a gimbal ring. When the airplane turns to the right or left, the gyro wheel precesses about its turning axis and rolls the gimbal ring. The rolling motion of the gimbal ring in turn rotates the turn needle on the face of the instrument. A spring returns the gyro to neutral when the airplane ceases to turn.

The turn indicator indicates the rate of the turn, not the amount of the turn. Thus a standard rate, or rate one, turn will give a rate of turn of 3° per second, or 360° in two minutes.

The instrument is usually calibrated to indicate a rate one turn when the turn needle is centred on one of the indexes seen either side of the centre index (Fig. 56).

Fig. 57. Straight & Level.   **Fig. 58. Left Turn. Correct Bank.**

Fig. 59. Straight. Rt. Wing Low.   **Fig. 60 Left Turn. Skid Outwards.**   **Fig. 61. Left Turn. Sideslip Inwards.**

- In a straight and level flight, the ball and needle are both centred (Fig. 57).
- In a correctly banked turn, the needle indicates the rate of the turn. The forces acting on the ball cause it to remain centred (Fig. 58).
- If one wing is permitted to drop, the ball will roll towards the side of the low wing. The needle in Fig. 59 shows the airplane to be flying straight, but the ball indicates it to be right wing low.
- If the airplane is not sufficiently banked in a turn, a skid towards the outside of the turn will occur. In Fig. 60, the needle indicates a left turn, the ball a right skid outwards.
- When the airplane is overbanked in a turn, it will sideslip inwards. The needle in Fig. 61 indicates a left turn, the ball a sideslip inwards.

46

From the Ground Up

Turn Co-Ordinator

Many modern airplane instrument panels use a turn co-ordinator to replace the more familiar turn and slip instrument. It works on the same principle as the turn and slip, although there are some differences in the construction of the instrument that enable it to react to roll as well as yaw. The gyro, instead of rotating about its horizontal axis as it does in the turn and slip, is canted at approximately 35°. As a result, during movement about both the yaw and roll axes, precession causes the gyro mechanism to roll slightly.

Fig. 62. Turn Co-Ordinator.

The gyro gimbal is connected to a miniature airplane on the face of the instrument. This little airplane tilts to inform the pilot of the rate of both yaw and roll. When the airplane is turning left or right, the miniature airplane banks in the direction of the turn. When the wing of the airplane is aligned with one of the marks, the airplane is in a standard rate, or rate one, turn.

The ball indicator gives the same indications as the ball indicator of the turn and slip indicator.

The Gyrosyn Compass

The gyrosyn compass combines the functions of both the directional gyro and the magnetic compass.

It provides stable compass headings in rough air, is north seeking like a magnetic compass, but is free from northerly turning error and oscillation. It does not precess and, therefore, does not require resetting as does the directional gyro.

The gyrosyn compass system incorporates a flux valve that senses the earth's magnetic lines of force through electromagnetic induction. They induce a voltage in the coils of a sensing unit in the flux valve. Direction can be determined from the induced voltages, which change with each change in the heading of the airplane. The flux valve is pendulously installed (to keep it horizontal) inside a case filled with damping fluid, usually in a wing tip, remote from local magnetic disturbances in the airframe.

Fig. 63. The Gyrosyn Compass.

The directional gyro unit is essentially the same as a standard directional gyro, except that it is slaved to the earth's magnetic lines of force by the signal received from the flux valve.

The dial of the gyrosyn compass system is shown in Fig. 63. The movable pointer points to the magnetic heading of the airplane. The course indicator (parallel lines) can be set to any desired heading for easy reference.

Some gyrosyn compasses are fitted with a dial switch. In one position, the complete system is in operation. In the other position, the flux valve is shut off and the instrument operates as a free directional gyro. This feature is for use in high polar latitudes or under any other circumstances where magnetic indications are unreliable.

Angle of Attack Indicator

The angle of attack indicator provides for the pilot a continuous readout of the margin above stall, regardless of aircraft weight, C.G. location, angle of bank or any of the other factors that affect indicated stall speed.

As a wing moves forward through the air, the airflow is divided into one section that flows over the wing and one section that flows under the wing. The point at which the airflow separates is known as the stagnation point. The stagnation point moves chordwise on the undersurface of the wing as the angle of attack changes. It moves forward as angle of attack decreases and aft as angle of attack increases to a maximum aft position at which the wing stalls.

The angle of attack indicator senses the changing position of the stagnation point by means of a sensing unit that is installed on the wing slightly beneath the leading edge. A spring loaded vane moves up and down as the stagnation point changes position and relays this information to the cockpit display. The indicator moves to the left towards the red or slow zone as the angle of attack increases and to the right towards the fast zone as the angle of attack decreases. The indicator displays both the current position of the stagnation point and also the position at which the airplane will stall.

An angle of attack indicator does not relate to airspeed and can therefore give a continuous, accurate readout of margin above stall, no matter what the flight attitude. It is an especially useful instrument during take-off, landing and steep banking maneuvers.

Mach Indicator

A mach indicator provides a continuous indication of the ratio of an airplane's airspeed to the local speed of sound. It expresses airspeed as a mach number by measuring and correlating dynamic and static pressures.

The instrument comprises two aneroid capsules enclosed in a sealed case that is connected to the airplane's static pressure system. One aneroid capsule is connected to the dynamic (pitot) pressure source, while the other is sealed and partly evacuated. This latter reacts only to static pressure and therefore measures altitude. The former reacts to both dynamic and static pressure and therefore measures airspeed. (Note: Mach number equals airspeed divided by the speed of sound).

A change in any sensed pressure causes appropriate expansion or contraction in one or both capsules. The capsules are geared to a pointer on the face of the instrument. This pointer reacting to the expanding and contracting capsules indicates the airspeed of the airplane in which it is installed.

Aero Engines

Present day aircraft engines are modern masterpieces of precision workmanship and quality material and represent the achievement of years of research and engineering experience in their design.

The fact is that the life and efficiency of an aero engine depends to an appreciable extent upon the use or abuse the motor suffers at the hands of the pilot. Sudden "gunning" of the throttle places undesirable stresses on the working parts of an engine. Switching off an overheated motor is a form of torture that should be punishable by a heavy fine. Operating an aero engine at continuously high power settings may progressively reduce the reliability of the engine and the safe period between overhauls.

Proper care and concern in the handling and running of an aero engine at the hands of an experienced pilot will keep that engine off the old age pension long beyond its normal span.

It is by no means necessary that a pilot should learn to be a mechanical engineer, but an elementary knowledge of the principles and construction of internal combustion engines is necessary if the proper procedure in running and handling is to be observed. (Note: An internal combustion engine is one in which the heat is created by the burning of the fuel within the engine itself, rather than in an external furnace, as in the case of steam engines.)

Very possibly some students who expect to fly as private pilots only will choose to skip over some of the details of engine design and construction that are described in this chapter. Those aspiring to be commercial pilots and who may some day fly as bush pilots in the more isolated areas of this country should have a more detailed knowledge of how an engine operates. Many a bush pilot has had to be his own mechanic. For this reason, a detailed description of an engine's operation is contained in the following pages.

Power

A piston driven aero engine is frequently referred to as a "power plant" and its product is power. First then, let us consider briefly the basic meaning of these terms. Power may be simply defined as the rate of doing work. Work, in turn, is heat transformed into energy. The amount of power an internal combustion engine can produce depends on the amount of heat which can be generated by the burning gases. This, in turn, is limited by the mechanical ability of the engine to transform as much of this heat as possible into useful work.

The standard unit in use for measuring the power produced by an engine is **one horse power.** This represents the amount of work done when 33,000 lbs. are raised 1 ft. in one minute.

The power developed within an internal combustion engine is called **indicated horse power.** Due to friction, etc., all this power is not available for useful work. The power which is available, after friction and other losses, is called **brake horse power (BHP).**

The total amount of indicated horse power an engine can produce may be represented by the formula:

Where
$$\frac{PLAN}{33,000} \quad \text{That is:} \quad \frac{P \times L \times A \times N}{33,000}$$

P = The mean pressure in lb. per sq. inch.
L = The length of the stroke in feet.
A = The area of the piston in sq. inches.
N = The number of impulses (or power strokes) per minute.

"P", the PRESSURE, in an internal combustion engine is obtained by admitting a mixture of vaporized gasoline and air into a cylinder, compressing this mixture, and burning it. As the gasses burn, they expand, exerting an enormous pressure on the piston head. The piston is driven down, pushing the connecting rod down, which in turn forces the crankshaft to turn at high speed. Note that the piston is driven down not by an "explosion" of gases, as is commonly supposed by beginners, but by their burning (or combustion) expansion and resultant pressure.

"L", the length of the STROKE, is the distance the piston travels up or down inside the cylinder.

"A", the AREA of the piston, means the area of the top or "head" of the piston.

"N", the number of IMPULSES, is the number of times the engine fires in one minute. In a 4-stroke engine this would be equivalent to half the number of revolutions of the crankshaft for each cylinder multiplied by the number of cylinders.

Aviation progress demands a continued search for greater and greater power output from aircraft motors. This involves constant improvement in the mechanical ability of engines to produce power from heat.

Since the horse power (hp) of an engine is the total product of the four factors referred to in the formula, it is obvious that an increase of all, or any one or more, of these will increase the power output of the engine.

"L" and "A", being dependent on the length of the stroke and area of the piston, are fixed and cannot be varied without altering the design of the engine.

"P" and "N", however, are variable and are the factors which can be stepped up in the search for greater power.

Some of the means adopted to boost "P" (mean effective pressure) are high compression ratios, higher octane fuels, supercharging, improved scavenging, etc.

A pilot may at will vary "N" by manipulation of the throttle (in the case of an airplane fitted with a fixed pitch propeller). Designers seek to improve this factor by designing high speed engines.

Power Definitions

M.E.P. — Mean Effective Pressure is the average pressure within the cylinder during the entire period of the power stroke, from the time the piston is at top dead centre until it reaches bottom dead centre.

METO POWER — Maximum Except Take Off Power.

B.M.E.P. — Brake Mean Effective Pressure is the M.E.P. available to produce the net power delivered at the propeller shaft (BHP).

Types of Combustion Engines

The three main types of piston engines in current use are **radial, in-line and horizontally opposed.** They derive their names from the arrangement of their cylinders around the central crankshaft.

The horizontally opposed type is most commonly used in general aviation airplanes.

Horizontally Opposed

The horizontally opposed engine, as the name implies, has two banks of cylinders working on the same crankshaft which lie directly opposite to each other in the horizontal plane. There may be four, six or eight cylinders.

Fig. 1. Horizontally Opposed Engine.

The outstanding advantage of this engine design is its flat, or "pancake" shape. Its small frontal area generates less drag.

Radial

The cylinders are arranged radially around a "barrel-shaped" crankcase. This type of engine always has an odd number of cylinders (five, seven, nine, etc.). This is necessary because the firing order is alternate (e.g. 1-3-5-2-4). The usual limitation is nine cylinders in a row. Larger engines of this type have two, three, or more rows working on a two or three throw crankshaft.

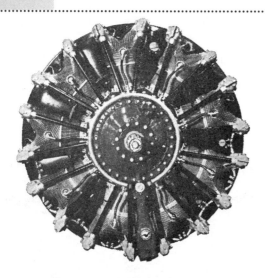

Fig. 2. Radial Engine.

Radial engines are dry sump or pressure lubricated.

The radial engine has an acceptable ratio of weight to horsepower developed and is easy to maintain. However, these advantages are outweighed by its poor shape that considerably increases parasite drag and reduces forward visibility from the cabin.

In Line

In this type of engine, the cylinders are arranged side by side in a row along the crankcase. Each separate piston works on an individual crank-throw.

The practical limit is six cylinders in one row, the reason being that a crankshaft long enough to accommodate a greater number of cylinders would be difficult to make stiff enough to avoid vibration. Where a larger number of cylinders is required, it is customary to arrange these in two or more banks. Hence, there are V-Type, X-Type and H-Type in-line engines which have two crankshafts side by side.

Fig. 3. In-Line Engine.

Some in-line engines are inverted, that is, the engine is installed upside down to provide better visibility for the pilot. These obviously must be of the dry sump or pressure lubricated type.

The in-line engine, with its small front area, generates little parasite drag.

The in-line engine is found today mostly in vintage airplanes.

Construction of a Reciprocating Engine

The basic parts of a reciprocating engine are the crankcase, cylinders, pistons, connecting rods, valves, valve operating mechanism, camshaft and crankshaft. In the head of each cylinder are located the valves and spark plugs. One of the valves is in a passage leading to the induction system; the other in a passage leading to the exhaust system. Inside each cylinder is a movable piston connected to a crankshaft by a connecting rod.

The cylinder and the cylinder head are finned to dissipate heat. The crankshaft converts the reciprocating motion of the piston into rotation and transmits power from the piston to drive the propeller.

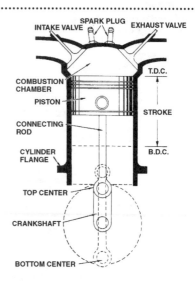

Fig. 4. Basic Parts of a Reciprocating Engine.

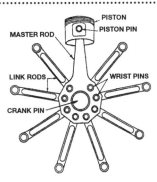

Fig. 5. Radial Connecting Rod.

The bore of a cylinder is its inside diameter. The stroke is the distance the piston moves from one end of the cylinder to the other, that is, from T.D.C. (top dead centre) to B.D.C. (bottom dead centre). See Fig. 4.

Reduction Gears

Many modern engines are geared. This means that the engine turns at a higher speed than the propeller which it drives. In this way, geared engines are made to develop greater power than direct drive engines with the same propeller speed.

The gearing consists of two sets of gear wheels. One of these is driven by the crankshaft and meshes with the other which, in turn, drives the propeller hub shaft. The reduction in speed is governed by the relative number of teeth in the two sets of gears.

Auxiliary Drives

The crankshaft drives the propeller. It is also made to drive various **auxiliary gears** which in turn drive oil pumps, magnetos, generators, dynamos, air compressors, and other essential auxiliaries.

The auxiliary gears are generally grouped in a **gear box** placed at the rear of the engine, to avoid increasing the frontal area.

In some cases a single flexible **half-time shaft,** driven by the crankshaft, is used to drive all the auxiliary gears.

The 4-Stroke Cycle

Almost all piston engines in use operate on what is known as the **four-stroke cycle.** This means that the piston travels four strokes (two up and two down) to complete one cycle. During this operation, the crankshaft revolves through 2 complete revolutions.

1. The Induction Stroke

The intake (sometimes called the inlet) valve open, the piston moves down from the top to the bottom of the cylinder creating a negative pressure. By this action, air, which surrounds the conventional suction carburetor and which is approximately at atmospheric pressure,

is drawn through the carburetor where it picks up a suitable amount of fuel (about one part gasoline to fifteen parts air) and the resulting fuel/air mixture rushes in past the open intake valve into the cylinder to fill the space above the piston. The amount of mixture which enters the cylinder will depend on the throttle opening. The exhaust valve remains closed.

2. The Compression Stroke

Both valves closed, the piston moves up from the bottom to the top of the cylinder, compressing the mixture.

The volume in the cylinder above the piston when it is at the bottom of the compression stroke compared to the volume when it has moved up to the top of the stroke is known as the **compression ratio**. e.g.3:1.

3. The Power Stroke

Both valves closed, the compressed mixture is fired by a spark plug. The burning gases, expanding under tremendous heat, create the pressure which drives the piston down with terrific force. This force is sufficient to complete the other three strokes in addition to providing the energy required for useful work.

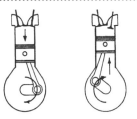

Fig. 8. Power   Fig. 9. Exhaust.

4. The Exhaust Stroke

Exhaust valve open, the piston moves up from the bottom to the top of the cylinder, pushing the burnt gases out past the open exhaust valve. The intake valve remains closed.

Timing

In discussing the four stroke cycle, we considered the various strokes as beginning and ending as the piston reached the top or the bottom of the cylinder. In actual practice, better performance is obtained from the engine by what is known as valve lead, lag, and overlap. Valves require time to open and close. They therefore are timed to open early and close late in order not to waste any of the induction or exhaust strokes.

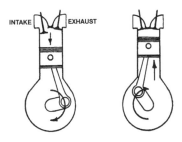

Fig. 6. Induction   Fig. 7. Compression.

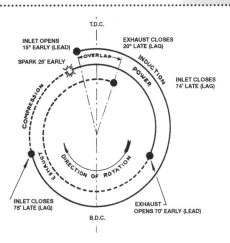

Fig. 10. Engine Timing Diagram.

Aero Engines

51

- Valve LEAD is timing the valve to open **early**.
- Valve LAG is timing the valve to close **late**.
- Valve OVERLAP is allowing both valves to remain open at the same time.

Fig. 10 shows a typical timing diagram for a modern high speed engine. T.D.C. refers to top dead centre, the point where the piston reaches the top of the cylinder, and B.D.C. to bottom dead centre. The circles represent the rotation of the crank-shaft, 180° between T.D.C. and B.D.C.

Note
1. The intake valve opens 15° early on the induction stroke.

2. The intake closes 75° late on the induction stroke in order to use the momentum of the incoming gases to force the maximum charge into the cylinder.

3. The spark occurs 25° early on the power stroke in order to allow complete combustion of the mixture so that the maximum pressure is reached when the piston passes T.D.C. and is travelling down on the power stroke.

4. The exhaust opens 70° early on the exhaust stroke because any pressure remaining in the cylinder after B.D.C. is passed opposes the motion of the piston coming up on the exhaust stroke.

5. The exhaust closes 20° late on the exhaust stroke. This means that both valves are open together, or **overlapped**, on the induction and exhaust strokes for 35°. The exhaust valve, by remaining open late, helps to scavenge the burnt gases. The momentum of the outgoing exhaust gases leaves a partial vacuum in the cylinder which accelerates the entry of the incoming charge.

The valve mechanism is operated by a camshaft, which is driven by a gear that mates with another gear attached to the crankshaft. The camshaft rotates at one-half the speed of the crankshaft.

Valve Clearances

A clearance is necessary between the valve stem and the rocker to prevent the valve being forced off its seat when it gets hot and expands. This is called the **valve clearance**, or sometimes **tappet clearance**, of the engine.

The clearances are set cold, allowance being made for the correct clearance to be attained when the valves reach their normal working temperature.

Clearances set too wide will cause a loss of power, vibration and excessive wear.

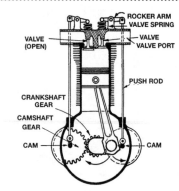

Fig. 11. Valve Mechanism.

Clearances set too close are apt to warp the valves and cause serious trouble.

The Two Stroke Cycle

Small engines that operate on the principle of a two stroke cycle are in common use on the ultra-light aircraft that are increasing in popularity among flying enthusiasts. They also power some airplanes registered in the new ARV category (Air Recreational Vehicle).

This engine takes only two strokes of the piston (one up and one down) to go through a complete power cycle. It differs from the four stroke cycle engine in that it uses the crankcase as a fuel mixture transfer pump. Charging the crankcase with fuel, compression of the fuel charge and ignition all take place on the upward stroke of the piston. Exhaust of the burned gases and transfer of the fuel to the cylinder take place on the downward stroke.

To accomplish all of this, there are three openings or ports in the cylinder (see Fig. 12). One port leads from the combustion chamber and carries away the exhaust gases (exhaust port). One port leads from the crankcase to the combustion chamber (inlet port.) The third opening is the inlet from the carburetor. These ports have no valve mechanism to open and close them as does the 4 stroke cycle engine. Instead, the piston acts as a slide valve as it moves up and down in the cylinder.

On the upward stroke of the piston, a vacuum is created in the crankcase and the fuel/air mixture is sucked into the crankcase through the inlet from the carburetor that is uncovered as the piston moves upward in the cylinder. At the same time, the piston seals off the exhaust and inlet ports and compresses the fuel/air mixture that is in the combustion chamber. Near the top of the stroke, the spark plug fires the compressed mixture. The hot, burning and expanding gases create the pressure that drives the piston down.

The descending piston seals off the inlet from the carburetor and begins to build up pressure in the crankcase. As the piston descends farther in the full travel of the stroke, it uncovers the exhaust port and the burned gases are allowed to escape. At the bottom of the stroke, the inlet port is uncovered and the compressed fuel/air mixture rushes through the inlet port from the crankcase into the combustion chamber. Some of the fuel/air mixture does escape through the exhaust port since both ports are uncovered at the same time. To reduce this loss to a minimum, the top of the piston is shaped to act as a barrier and direct the fuel mixture away from the exhaust port.

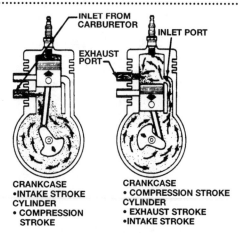

Fig. 12. Two Stroke Cycle.

The four stroke cycle engine, as will be explained in more detail in the section **Lubrication** which follows, uses the hollow crankshaft to distribute oil throughout the engine. The two stroke cycle engine, however, uses its hollow crankcase as a transfer pump for the fuel/air mixture and must therefore depend on a different lubricating system. In a two stroke cycle engine, fuel and oil are mixed in proportions specified by the manufacturer and this mixture is used to fill the fuel tank.

When the fuel/oil mixture passes through the carburetor, the fuel is vaporized and mixes with the air. The oil is broken up by the stream of air into tiny droplets which lubricate all the surfaces of the crank case, piston and cylinder. Some oil remains in the crankcase, but most passes into the combustion chamber where it is burned.

The engine manufacturer usually recommends the use of an oil that has no detergents as additives. Such additives do not burn completely and may leave undesirable carbon deposits on the pistons, spark plugs and cylinder surfaces.

Regular automobile gasoline is used in the two stroke cycle engine but the engine manufacturer usually recommends regular grade rather than premium. The lead in the latter causes lead deposits on the spark plugs.

The basic principles of carburetion and ignition are as applicable to the two stroke cycle engine as they are to the four stroke cycle engine, although the individual systems for achieving ignition and, more especially, carburetion will vary between the two types of engines. The explanations of carburetion and ignition which follow are, nevertheless, applicable to the two stroke cycle engine.

Cooling

We have previously considered the fact that the source of all power is heat. It will therefore be obvious that the modern aero engine, which is a unit designed to develop great power, must generate tremendous heat. The heat of combustion reaches temperatures inside the cylinder actually as high as 2500°C.

An appreciable portion of this heat is absorbed by the engine parts, the cylinder walls, piston heads, etc. This would cause excessive overheating, to the extent of actually fusing or melting the metal parts, if some means were not provided for dissipating it.

The engines of some airplanes use a liquid coolant, but, by far, the most common method of dissipating engine heat is by circulating cooling air around the engine cylinders. Both horizontally opposed and radial engines are air cooled. Some in line engines are air cooled; a few models were liquid cooled.

Air Cooling

In air cooled engines, **fins** are added to the cylinders to provide a greater area of metal to absorb the heat. Ram air passing over the fins absorbs this excess heat and carries it away. This cooling air enters the engine compartment through openings in the front of the engine cowl and is expelled through openings at the rear of the cowling.

Baffles are used to force the cooling air around the cylinders.

On low performance airplanes, the openings to expel the cooling air are of fixed size. In high performance airplanes, the size of these openings is usually controlled by **cowl flaps.** During high power settings, such as during take-off and climb, the forward speed of the airplane is low and engine power is high. More cooling air is necessary to dissipate the higher heat generated by the harder working engine. During cruise, less cooling air is required. Since the cooling air flowing over the cylinders causes drag, in a high performance airplane it is of benefit to reduce the amount of air entering the engine compartment when it is not needed. This is effected by the addition of cowl flaps that are controlled from the cabin. They are used fully open for taxi, take-off and climb and are closed or partially closed for cruise and descent.

Cooling fans are sometimes mounted on the front of the engine and are gear-driven from the engine crankshaft. They assist the flow of cooling air at high altitudes where the weight-flow of cooling air is becoming light.

On some airplanes, the exhaust gases are directed through **augmentor tubes**, or jet pumps, which produce a suction strong enough to increase the flow of cooling air past the cylinders. The augmentor tubes also produce additional thrust sufficient to off-set the drag of the cooling air.

Lubrication

Lubricating oil has four important functions to perform:

1. *Cooling* — Carries away excessive heat generated by the engine.

2. *Sealing* — Provides a seal between the piston rings and cylinder walls, preventing "blow-by" loss of power and excessive oil consumption.

3. *Lubrication* — Maintains an oil film between moving parts, preventing wear through metal to metal contact.

4. *Flushing* — Cleans and flushes the interior of the engine of contaminants which enter or are formed during combustion.

Requirements of Good Oil

1. *Viscosity* — Resistance to flow. Stickiness or body.

Good viscosity gives proper distribution of oil throughout the engine and prevents rupturing of the oil film which lubricates the engine parts over the wide range of temperatures in which aero engines work. An oil with a high **viscosity index** is one in which the changes in viscosity, due to widely varying operating temperatures, are small.

The use of oil of too high viscosity for existing climatic temperatures will cause **high oil pressure.**

The use of oil of too low viscosity for existing climatic temperatures will cause **low oil pressure.**

2. *High Flash Point* — The temperature beyond which a fluid will ignite. This should be in excess of the highest engine temperature.

3. *Low Carbon Content* — To leave as little carbon as possible should oil work past the scraper ring and burn. Good oil should also have a low wax content. Oils which have good resistance to deterioration and the formation of lacquer and carbon deposits are said to have good **oxidation stability**.

4. *Low Pour Point* — The temperature at which a fluid solidifies. Necessary for cold weather starting.

Additives

Some oils contain additives. These may be classified as follows:

1. *Detergents* — Those which improve engine cleanliness.
2. *Oxidation Inhibitors* — Those which improve oil stability.
3. *Anticorrosion Additives* — Those which deter corrosion.
4. *Pour Point Depressants* — Those which lower the pour point.

There are also various other "improvers" which increase film strength and viscosity and which decrease foaming characteristics.

It is important that oil which contains additives should be added ONLY to oil of the same type.

Grade of Oil

The grade of oil recommended for various seasonal aero engine requirements is designated by an **S.A.E. Number,** or by the **Saybolt Universal Viscosity**, which is expressed in "seconds". The engine oil grade is, therefore, frequently referred to as, say, S.A.E. 50, or as a "100 second oil". In the United States it is common practice to add 1000 to the Saybolt Viscosity, so that this particular grade could also be referred to as "Grade 1100". For engines of moderate horse-power, the following grades are generally used:

	Saybolt Viscosity	S.A.E. Number	U.S. Grade
Summer	120	60	1120
Fall or Spring	100	50	1100
Winter	80	40	1080
Arctic	65	30	1065

In order to eliminate the need to change oil seasonally, manufacturers of engine oil have developed a synthetic multi-viscosity oil. This oil remains thin in extremely cold weather but also, at high outside temperatures, maintains the same viscosity as the traditional summer grade of oil.

Oil Temperature

It is necessary to keep the oil temperature in an operating engine between certain well defined limits because the lubrication of the engine depends on the viscosity of the oil which, in turn, is governed by oil temperature.

If the oil gets too hot, its viscosity will be impaired and may not be enough to keep a good film of oil on the engine parts.

If it gets too cold, the oil will become thick and will not flow through the passageways, resulting in improper lubrication.

Oil temperature is monitored by means of an oil temperature gauge which is installed on the instrument panel of the airplane.

Engine manufacturers always specify operating limits which must be strictly observed, otherwise the life and reliabilIty of the engine may be seriously impaired.

Methods of Lubrication

1. *Force Feed:* In this method the oil is forced under pressure from a pressure pump through the hollow crankshaft where it lubricates the main and big end bearings. It is then sprayed through tiny holes to lubricate the remaining parts of the engine by a fine mist, or spray.

2. *Splash:* The oil is contained in a sump, or reservoir, at the bottom of the engine. It is churned by the revolving crankshaft into a heavy mist which splashes over the various engine parts. This type of lubrication is no longer used in engines that are manufactured today but it will be found in the engines of some of the older airplanes that are still flying.

An improvement of the splash method that incorporates a pressure delivery is, however, now in common use in modern engines. This is known as wet sump lubrication.

Force Feed: Dry Sump Lubrication

A dry sump engine is one in which the oil is contained in a separate tank and is forced under pressure from a **pressure pump** through the hollow crankshaft to lubricate the engine by the force feed method. The oil is then drained into a sump from which it is pumped by a **scavenging pump,** which has 20% greater capacity than the pressure pump, to ensure that the oil does not accumulate in the engine. Thence it passes through an **oil cooler** and is returned to the tank.

Fig. 13 shows the system of oil circulation in a dry sump engine. The **by-pass** around the **filter** is provided to prevent damage in case of failure (through neglect or carelessness) to clean the filter. In this event, dirty oil is considered definitely a lesser evil than no oil at all! The by-pass is also a safety device allowing the oil to return to the tank without damaging parts such as the filter in case the pressure should become excessive. The **pressure relief valve.** provides a means of regulating the oil pressure.

Oil pressure is monitored by means of an oil pressure gauge on the instrument panel of the airplane.

Wet Sump Lubrication

In the wet sump system of lubrication, the engine oil supply is contained in a sump, or pan, under the crankcase. The oil passes through a filtering screen in the bottom of the sump into the suction side of a gear type pressure pump. The pump moves the oil around the outside of its gears and imparts a predetermined pressure to it. A pressure relief valve is incorporated into the pump to ensure a steady pressure during all phases of engine operation.

The pressure pump which is engine driven delivers oil through passageways in the crankshaft and pushrods to all the bearings of the engine for lubrication and cooling. Some internal parts and surfaces of the engine may be splash lubricated or they may be sprayed by oil flung off rotating parts of the crankcase and connecting rods. After circulation, the oil drains through return tubes back to the sump. An oil cooler may be located in the return line to dissipate heat.

In some engines, the delivery of oil throughout the engine may be handled entirely by the pressure pump; in others, by a pressure-splash combination.

The wet sump method has the advantage of light weight and relative simplicity, being free of the exterior tank, tubing and linkages characteristic of the dry sump system. However, the capacity of the sump is limited by the size and design of the nacelle or cowling. Wet sump lubrication cannot be used on airplanes designed for aerobatic or inverted flight without special modifications because the oil would flood the crankcase.

Oil Dilution

Because of the difficulties experienced in starting an airplane in very cold weather, an oil dilution system is incorporated into the

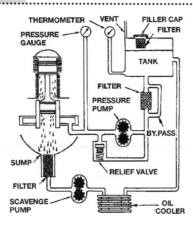

Fig. 13. Dry Sump Lubrication System.

From the Ground Up

lubrication system of some engines. Its purpose is to thin the oil with raw fuel immediately before the engine is stopped in order to lower the viscosity of the oil and thereby reduce the high cranking torque (resistance to turning) of a cold engine during the next start. It also ensures an immediate and adequate supply of oil to all moving parts after start up.

In such a system, a line connects the fuel pressure line to a special Y draincock in which a spring poppet valve is installed. This valve is operated manually from the cockpit.

In very cold weather, when the engine will be shut down for several hours or for overnight, a small amount of fuel is allowed to enter the oil line, just before the engine is stopped, by holding the dilution control open for a short time while the engine is running. Diluted oil flows through the entire engine, replacing the heavy oil, thereby facilitating the next start.

Before carrying out oil dilution, however, be sure to check the amount of oil in the oil tank and ensure that it is below capacity. Oil dilution will add an appreciable quantity of gasoline to the oil system and there must be room for the extra fluid. Otherwise the oil system will overflow. It may be necessary for the pilot to drain off some oil before starting oil dilution.

The amount of oil dilution necessary is dependent on the expected temperature against which it is necessary to protect. Oil dilution is recommended by most manufacturers at 5°C and lower. The elapsed time required to achieve the amount of oil dilution required depends on two factors: the rate of flow of gasoline into the oil and the amount of oil to be diluted. The engine manufacturer's instructions should be closely followed when carrying out oil dilution.

After starting the engine, the engine operating temperatures must be raised and maintained at a sufficiently high level to vaporize the gasoline. The vaporized gas is vented through the engine crankcase breather. This procedure is known as **boil-off.** Boil-off should be done prior to take-off and only in accordance with the manufacturer's instructions.

Over dilution can be caused by excess dilution time, by insufficient boil-off and, on occasion, by a leaking poppet valve. Excessive drop or fluctuation of oil pressure will be the result.

The introduction of gasoline into the lubrication oil will loosen carbon and sludge deposits within the oil system. This carbon and sludge is carried to the engine oil screen and collects there in sufficient quantity to collapse the screen. Consequently, the screen must be removed for inspection and cleaning regularly during periods when oil dilution is being carried out. Some operators use oil dilution regularly, even in warm weather, as a means to keep the oil system clear of carbon and sludge.

One problem associated with oil dilution that a pilot must watch for is oil venting. If the engine is warmed up too rapidly at too high power settings, the gasoline in the oil will tend to vaporize so rapidly that the pressure within the engine crankcase will result in oil as well as gasoline being blown out through the engine crankcase breather. Care in following recommended procedures for completing boil-off will prevent this happening.

The Fuel System

The aircraft fuel system stores and delivers the proper amount of fuel at the right pressure to meet the demands of the engine. It must deliver this fuel reliably throughout all phases of flight, including violent maneuvers and sudden acceleration and deceleration.

There are usually several tanks, even in a simple system, to store the quantity of fuel required to give the airplane reasonable range. These tanks are usually located in the wings, although extra tanks may be located in the cabin area. A line leads from each tank to a selector valve in the cockpit, by which the pilot is able to select the tank from which fuel is to be delivered.

Types of Fuel System
Fuel Pump Fuel System
With this type of fuel system, an engine driven fuel pump supplies the pressure that keeps the fuel flowing to the engine. This type of system is in use in all low wing airplanes and in any airplane with a higher performance engine. Such a system incorporates, as well as the basic pump, auxiliary electric pumps or booster pumps that serve in emergency in case the engine driven pump fails. The booster pump, operated by a switch on the instrument panel, is also used to start the fuel flowing under pressure before the engine is running.

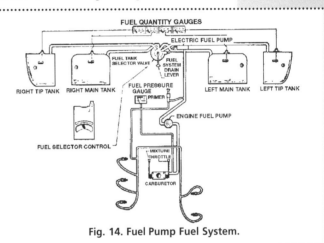

Fig. 14. Fuel Pump Fuel System.

A fuel pressure gauge, mounted on the cockpit panel, gives a visual indication that the fuel system is working by giving a reading of the pressure of fuel entering the carburetor.

Gravity Feed Fuel System
This is the simplest type of fuel system and is still in use on many high wing, low powered airplanes. The fuel tanks are mounted in the wings above the carburetor, with gravity causing the fuel to flow from the tanks, past the fuel selector valve to the carburetor. A drain allows removal of water and sediment trapped at the strainer. A primer sprays raw fuel into the intake manifold system or directly into the cylinders to aid engine starting, particularly in cold weather.

Fuel Tanks
The location, size and shape of fuel tanks vary with the type of airplane in which they are installed. They are most often located in the wings. Wing tip tanks are common.

The tanks are made of materials that will not react chemically with any aviation fuel. Aluminum alloy is most widely used. Synthetic rubber or nylon bladder type fuel cells are also in use. This latter type depends upon the structure of the cavity in which it sits to support the weight of fuel it holds.

Usually a drain is provided at the lowest point of the tank through which water and sediment, which are heavier than fuel and therefore settle to the lowest point of the fuel tank, can be drained.

Overflow drains are also incorporated to release fuel and prevent tanks from bursting when fuel expands in the tanks. Such expansion

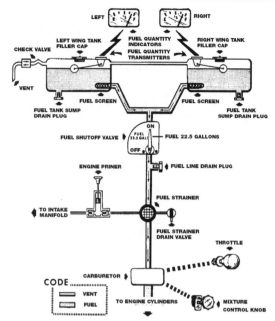

Fig. 15. Gravity Feed Fuel System.

takes place when an airplane is parked in direct sunlight and the fuel becomes heated.

The top of each tank is vented to the outside air to maintain atmospheric pressure within the tank. Tanks are also fitted with internal baffles to resist fuel surging caused by changes in the attitude of the airplane. The tanks are filled through a filler neck on the upper surface of the wing. Caps for the filler neck have locking devices to prevent accidental loss during flight.

Information about the capacity of each tank is usually posted near the fuel selector valve in the cockpit of the airplane.

Fuel Selector Valve

A fuel selector valve in the cockpit permits the pilot to select the tank from which he wishes to draw fuel. It also has an "Off" position which closes off the fuel flow entirely. When the airplane is parked, the "Off" position may be selected.

Fuel Lines and Filters

The various tanks and other components of the fuel system are joined together by fuel lines made of aluminum alloy metal tubing and flexible synthetic rubber or teflon hose.

Coarse mesh strainers are installed in the tank outlets and often in the tank filler necks. Other strainers, made of fine mesh, are installed in the carburetor fuel inlets and in the fuel lines.

The main strainer is located at the lowest point in the fuel system. Its purpose is to prevent any foreign matter from entering the carburetor and to trap any small amounts of water that may be present in the system.

It is good practice to drain the strainers prior to every flight to remove from the system any water that may be in the fuel. In cold weather, this water might freeze and stop the fuel flow. In warm weather, excessive water entering the carburetor could stop the engine.

Fuel Quantity Gauge

Fuel quantity gauges are mounted on the instrument panel and give a visual indication of the amount of fuel in each tank. A pilot should never assume that the instrument is correct but should make a visual check of each fuel tank before initiating a flight.

Fuel Primer

A small hand pump is located on the instrument panel. This is a primer that is used to pump fuel into the engine prior to starting. Ordinarily, it is needed only in cold weather. A pilot should be careful not to overprime since this action may flood the engine and make it hard to start and, in some cases, may even cause a fire.

Fuels

Fuels for modern high compression engines must burn slowly and expand evenly rather than explode quickly. The fuels which possess this quality are known as high octane fuels.

The **octane rating** of a fuel is arrived at as follows:
• **Octane** is a substance which possesses minimum detonating qualities.
• **Heptane** is a substance which possesses maximum detonating qualities.

The proportion of octane to heptane in a fuel is expressed as a percentage. Hence, 73 octane means 73% octane and 27% heptane characteristics in the fuel. The natural gas limit is 72 octane. Fuels of higher octane than this are treated with tetra ethyl lead or are "cracked" by a heat process which increases their volatility. They are also treated with sulphuric acid, lye, etc., to remove the gum, acid and other impurities.

Octane numbers go only as high as 100. Beyond this number, the anti knock value of the fuel is expressed as a **performance number.**

Fuel grades are usually indicated by two numbers, e.g.: Grade 80/87. The first number (80) indicates the octane rating at **lean mixture** conditions, and the second (87) at **rich mixture.** Grade 100/130 indicates a fuel with a lean mixture performance number (or octane rating) of 100, and a rich mixture performance number of 130.

The following are the approximate ratings and the colours for the most common fuels:

Low power output	Grade 80 (or 80/87)	Red
Medium power output	Grade 100 LL (low lead)	Blue
Medium power output	Grade 100 (high lead)	Green
Jet fuel	Kerosene	Clear or straw
MOGAS	P87-90	Green
MOGAS	P84-87	Undyed

It is the responsibility of the pilot to see that the proper grade of fuel is used in refuelling his airplane. The Airplane Flight Manual always specifies what grade of fuel should be used and usually this information is also marked on a placard in the cockpit and also next to the filler cap.

It is also essential to ensure that the correct type of fuel is used in refuelling the airplane. Occasionally, airplanes with reciprocating engines have been refuelled with jet fuel instead of with avgas. The engine may run briefly on jet fuel, but detonation and overheating will soon cause power failure (during take-off if you are really unlucky). Restrictor rings in the filler neck of the fuel tank will prevent the large jet fuel nozzle from fitting into the fuel tank. Pilots need to be wary of colourless fuel. Jet fuel, water and MOGAS are colourless. It is important to pay close attention when your airplane is being refuelled and be sure you are getting the type of fuel you require.

If the proper grade of fuel is not available and the engine must be operated, use the next higher grade. Never use a lower grade. The engine will overheat badly, detonation may occur and engine damage may result.

Unfortunately, the fuel manufacturers have, for the most part, ceased manufacture of 80/87 gasoline and operators of small airplanes fitted with engines that require 80/87 are having to use a higher grade of gasoline, such as l00 LL or 100/130. The higher octane fuels, however, have more lead content and may cause trouble, such as spark plug fouling. Even the l00 LL (low lead) gasoline that was developed to replace 80 octane, although it has less lead than regular 100, nevertheless has more lead than 80 octane fuel.

Use of a fuel of a higher octane than that recommended for your engine will probably lead to excess deposits in the combustion chamber. This is especially the case in cold weather. If you are regularly using 100 LL in an engine designed for 80/87, rotation of the plugs, good leaning techniques, oil changes every twenty-five hours and use of correct spark plugs will keep fouling problems to a minimum. In addition, it is advisable to avoid closed throttle operation (i.e. long idling descents) and practise power approaches. Don't let the engine cool too quickly. Heat helps to scavange possible lead deposits. Keep the cylinder head temperature in the normal operating range.

Automobile Gasoline/MOGAS. Although in recent years, the use of automobile gasoline has been conditionally approved for use in some types of aircraft, as a general rule automobile gasoline is not recommended as a replacement for aviation gasoline. The chemical specifications of automobile gasoline may vary widely on any continent due to local climatic conditions as well as state and provincial regulations. A pilot may, therefore, be unable to predict the performance of his engine under all operating conditions. Automobile gasoline is not as stable as aviation gasoline and may more easily come out of solution, especially if stored in the airplane fuel tanks for long periods. Some kinds of automobile gasoline will cause vapour lock and reduced fuel flow at high ambient temperatures and higher altitudes. Vapour lock may occur at temperatures above 24°C and at altitudes above 6000 feet. MOGAS is also more susceptible to carburetor icing. The continued use of unleaded automobile gasoline may cause accelerated valve guide wear, valve damage and valve seat recession in engines that were designed to operate on fuels with some lead content. The advantage of using aviation gasoline at all times is that it is manufactured to a common and internationally recognized standard. In addition, in order to minimize contamination, special care is taken in handling, transferring and delivering aviation gasoline. The engine and airframe manufacturers use aviation gasoline to obtain performance data on which the Operator's Manual is written. Thus the operator can be assured of consistent performance. Another point to remember is that some insurance companies will consider the policy invalid if the aircraft is operated with a gasoline that is not approved.

Additives. Water particles in suspension in the fuel may, in very cold weather, change into ice crystals and accumulate in fuel filters and fuel lines in sufficient quantity to block the fuel line and cause engine stoppage. Anti icing fuel additives inhibit the formation of these ice crystals. They should be used only if approved by the engine manufacturer and only in strict compliance with the manufacturer's instructions. Such additives should not be used without also consulting the airframe manufacturer because their chemical content may be incompatible with the aircraft fuel system cells, seals, etc. Lead scavenging additives have proved effective in reducing lead fouling of spark plugs but should also be used only if approved by both the engine and the airframe manufacturer.

Problems of the Fuel System
Detonation

Detonation is characterized by the inability of a fuel to burn slowly and is generally defined as an abnormally rapid combustion, replacing or occurring simultaneously with normal combustion. Detonation is also characterized by its almost instantaneous nature, as contrasted with the smooth progressive burning of normal combustion. Under conditions of detonation, cylinder pressures rise quickly and violently to peaks that are often beyond the structural limits of the combustion chamber. Detonation is dangerous and costly. It puts a high stress on engine parts and causes overheating, warped valves and piston damage. It is not always easy to detect. Cylinder head temperatures offer about the best appraisal of the unseen combustion process. A rapid increase in cylinder head temperature, unless explained by some other factor, often indicates detonation. Throttle reduction is the most immediate and surest remedy.

Cause: (1) Use of incorrect fuel. (2) Overheating, sometimes caused by too steep a climb that reduces the flow of air around the cylinders. (3) Too lean a mixture.

Temporary Remedy: In an emergency, use the mixture control full rich. This enriches the mixture, which in turn tends to cool the engine due to the evaporation of the raw gas in the cylinders. Also reduce the power as much as possible.

Permanent Remedy: Persistent detonation indicates that a fuel of too low octane is being used. **Use ONLY fuel of the octane rating specified by the engine manufacturer.** If the handbook is not available and there is any doubt in your mind, always use an octane rating on the high side, never one too low.

Pre-Ignition

Pre-ignition is another trouble-maker which is sometimes confused with detonation. Pre-ignition, however, is a premature ignition of the mixture due to glowing carbon particles, or "local hot spots". It is often experienced when attempting to start a hot engine and usually results in a backfire through the intake manifold.

Damage to an engine from pre-ignition can be disastrous, causing warped pistons, cracked cylinder heads and other serious damage.

Vapour Lock

Vapour lock in the fuel line can be caused by high atmospheric temperatures, causing the gas to vaporize and block the flow of liquid fuel in the line.

The Carburetor

The heat energy in an internal combustion engine is developed from the burning of a mixture of gasoline and air.

The combustion process in the cylinder relies on a proper mixture of this fuel and air to achieve optimum efficiency. The function of the carburetor is to measure the correct quantity of gasoline, vaporize this fuel, mix it with the air in the proper proportion and deliver the mixture to the cylinders.

The ratio of fuel to air is regulated by the pilot with the mixture control. The throttle regulates the flow of air into the engine and creates turbulence at the butterfly valve to assist in the mixing of fuel and air. The carburetor matches the flow of fuel with the air flow to achieve the ratio regulated by the mixture control.

In the fuel/air mixture delivered to the cylinders from the carburetor, the proportion of gasoline to air is governed by weight and not by volume. Mixture is a very precise measurement, as the fuel to air ratio for best power is only 2 to 4 pounds of fuel per 100 pounds of air different from the ratio which will cause a lean misfire condition.

The following fuel/air ratios (by weight) indicate the effect, on a typical engine, of varying the mixture ratio at full throttle, constant speed.

Running Mixture .... About 1 to 8
Best Power Mixture .... About 1 to 14
Chemically Correct Mixture .... About 1 to 15
Lowest Fuel Consumption Mixture .... About 1 to 18
Leanest Running Mixture .... About 1 to 20

Engine temperatures are substantially affected by the ratio of fuel to air in the mixture entering the combustion chambers. An engine will run hotter with a lean mixture than with a rich mixture because the lean mixture is slower burning and the cylinder walls are exposed to high temperatures for a longer period of time. A richer mixture, on the other hand, burns more quickly, exposing the cylinder walls to high temperatures for a shorter period. Also, the additional fuel in the fuel/air mix of a richer mixture helps to cool the engine.

Too rich a mixture (an excess of fuel), however, as well as lowering the combustion temperature, results in unburned wasted fuel being expelled through the exhaust system. It contributes to fouled spark plugs and combustion chamber deposits. Too rich a mixture may also cause rough engine operation, appreciable loss of power or actual engine failure.

Too lean a mixture may also cause rough engine operation, sudden "cutting out", "popping back" or back-firing, detonation, overheating or appreciable loss of power. Continual operation at too lean a mixture has also been responsible for engine failure.

The complete carburetor is a highly complicated unit incorporating many devices necessary to control the mixture ratio under widely varying conditions imposed on the modern airplane engine. As such, it is beyond the scope of this manual. However, an explanation of the basic working principles of a simple carburetor and some of its additional parts will be of interest to the student pilot.

How the Carburetor Works

Fuel flows through the fuel supply lines, past the fuel strainer and enters the carburetor at the **float (or needle) valve.** It flows into the **float chamber** where its level is controlled by a float which opens or closes the float valve as it rises or falls. When the float rises to a predetermined level, it shuts off the float valve. No additional fuel can then enter the carburetor until fuel is used by the engine.

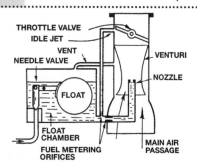

Fig. 16. Principle of a Simple Carburetor.

The float chamber is vented so that the pressure in the chamber equalizes with the atmospheric pressure as the airplane climbs and descends. The level of the gasoline in the float chamber governs the level of gasoline in the **nozzle,** as can be seen in Fig. 16. One of the problems of the carburetor is that sometimes the float may become punctured so that gasoline leaks into it increasing its weight. As a result, the level of gasoline in the chamber rises and gas overflows from the nozzle, thereby flooding the carburetor.

Outside air passes through the carburetor air filter which is located at the carburetor air intake in the front of the engine cowling. It is then drawn through the **venturi** (Fig. 16) where its speed is increased. A low pressure area is thereby created in the throat of the venturi. The reduced pressure around the nozzle draws the fuel, which is under atmospheric pressure, from the jet in the form of a fine spray.

The mixture of air and vaporized fuel, regulated in volume by the **throttle valve,** enters the intake manifold and is then distributed to the individual cylinders.

The throttle valve is connected directly to the throttle control on the instrument panel of the airplane. By means of the throttle, the pilot is able to control the amount of fuel/air mix that enters the engine thereby controlling the power output. Forward movement of the throttle opens the throttle valve, increasing the volume of fuel/air mixture and consequently the speed of the engine. Aft move-ment of the throttle closes the throttle valve and reduces the volume of fuel/air mixture entering the engine.

Some portion of the gasoline in the fuel/air mixture does not vaporize and is still in liquid form when it enters the engine. It is vaporized by the heat of the engine. It is, therefore, important to properly warm up the engine prior to take-off, so that there is sufficient engine heat to vaporize all the fuel and develop full power during take-off.

Idling

When an engine is idling, the throttle valve is closed and there is insufficient movement of air through the venturi to lower the pressure enough to draw fuel from the main nozzle. An **idle jet** is provided at the edge of the closed throttle valve where, owing to the narrow passage, the air accelerates and reduces the pressure enough to draw fuel from the idle jet.

Acceleration Pump

During rapid acceleration, the fuel in liquid form, which travels to the cylinders along with the vaporized fuel and air, is unable to increase its velocity as rapidly as does the air. This causes a temporary **leanness** at the cylinders which must be compensated for by temporarily adding more fuel at the carburetor. This is accomplished by the **acceleration pump.** The device consists of a piston controlled by the throttle which works up and down in a cylinder containing gasoline at the float level. The piston is drilled with two holes beneath which is a **check valve.** When the throttle is opened slowly, the check valve remains open and no fuel is pumped. When the throttle is opened quickly, however, the check valve closes and the fuel in the cylinder is pumped into the air stream through the **economizer discharge needle.**

Mixture Control

The need to have a mixture control system is occasioned by the fact that, as altitude increases, the density of the air decreases. Carburetors are normally calibrated for sea level operation, which means that the correct mixture of fuel and air will be obtained at sea level with the mixture control in the full rich position. As altitude increases, a given **volume** of air **weighs** less. It is obvious then that at higher altitudes, the proportion of air by **weight** to that of fuel will become less although the volume remains the same. (Regardless of altitude, the amount of fuel entering the carburetor remains approximately the same for any given throttle setting if the position of the mixture control is not changed.) The mixture, therefore, becomes over-rich, causing waste of fuel and loss of power.

To correct this condition, a mixture control is fitted to the carburetor. This device adjusts the amount of gas being drawn from the nozzle and thereby restores the proper fuel to air mix.

The mixture control, on some airplanes, is an automatic device. Most commonly, it is a manually operated control that is operated by the pilot.

The Manual Mixture Control

In a manually operated system, the amount of fuel that is drawn from the nozzle to mix with the air is governed by the mixture control that is located on the instrument panel. By adjusting this control, the fuel to air ratio can be adjusted either to enrich the mixture or to lean it.

There are many different forms of mixture control devices. The most common incorporates a **needle valve** which restricts the amount of fuel flowing from the float chamber to the nozzle. When the mixture control knob is moved fully forward, the needle valve retracts and allows fuel to flow freely to the nozzle. When the mixture control knob is moved aft, the needle moves into the fuel line to restrict the flow of fuel.

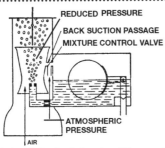

Fig. 17. Principle of the Back Suction Mixture Control.

Another type of control is the **back suction system,** which is depicted in Fig. 17. A back suction pipe or passage interconnects the air space in the float chamber to the reduced pressure area in the venturi. Closing the mixture control valve restricts the atmospheric pressure entering the float chamber and the back suction passage reduces the pressure in the chamber. This reduced pressure forces less fuel through the nozzle and **leans the mixture.**

Idle Cut-Off

Some engines are fitted with an idle cut-off control to stop the engine. This is simply an extreme "lean" position of the mixture control, which reduces the pressure within the float chamber to such a low value that no fuel at all flows through the jet. Idle cut-off will stop the engine immediately with no tendency towards pre-ignition or after-firing. It also makes the engine easy to start since the carburetor and fuel lines remain filled with gasoline.

Use of Manual Mixture Controls

Although specific instructions concerning mixture ratios are given for each type of airplane engine, a general rule to follow in using a manual mixture control is that the rich mixtures should be used at high power settings and the leaner mixtures at normal cruise power settings. Failure to observe the instructions pertaining to the use of the mixture control may easily result in engine overheating and detonation either of which will affect the reliability and useful life of the engine. In case of doubt, a mixture a little on the rich side is advisable as too lean a mixture can cause engine damage. The mixture control has the greatest effect, and all other engine controls a lesser effect, on engine operating temperatures.

Rich mixtures	-	*High power settings*
Leaner mixtures	-	*Cruise power settings*

A common method of adjusting a manually operated mixture control (for an engine with a fixed pitch propeller) is to observe the tachometer reading closely while moving the control from full rich towards lean. It will be observed that an increase in rpm occurs. The point at which this maximum rpm is first reached is called "rich best power". Further leaning will hold the rpm at maximum value for an appreciable movement of the control, whereupon still further leaning will cause a loss in rpm. The point where rpm begins to drop again from the maximum rpm value is called "lean best power".

This system of using the manually operated mixture control is, however, at best only approximate. The position of the mixture control should be rechecked with changes of operating condition, such as change in altitude, change in carburetor hot air setting, etc. (This method is not as effective with an engine having a constant speed propeller, since variation in mixture ratio will not appreciably alter the tachometer reading.)

More accurate methods of using the manual mixture control have made their appearance in recent years and are adaptable to most airplanes. These new systems are capable of providing a continuous instrument indication of the mixture ratio. These systems are of two main types: direct measurement and exhaust temperature measurement.

Fig. 18. Fuel Management System.

1. *Direct measurement.* A fuel flow gauge in the cockpit of the airplane tells the pilot the approximate volume per hour of fuel being metered to the engine. The gauge is used in conjunction with tables, supplied by the manufacturer and found in the Airplane Operating Manual, which chart the normal fuel flow settings to use with certain rpm, manifold pressure and TAS and when flying at various altitudes and at various outside or carburetor air temperatures. Fuel flow gauges of older manufacturer were somewhat unreliable as a guide to leaning an engine. They were very likely to go out of calibration and could not, therefore, be depended upon to give a correct reading of the exact amount of fuel being consumed per hour. Recent technology has improved the design of this gauge with solid state components that ensure a much more accurate and trustworthy indication of fuel flow while monitoring, at the same time, fuel remaining on board and flight time remaining at the current fuel flow. Even more advanced designs derive information from on board navigation equipment and are capable of displaying fuel flow information based on changing groundspeed.

2. *Exhaust temperature measurement.* The positive control of mixture is possible with the exhaust temperature systems. The EGT method uses a thermocouple in the exhaust stack near the exhaust valve of a cylinder to measure the temperature of the exhaust gases. This sensitive probe transmits its information to an indicator in the cockpit.

The exhaust gas temperature varies as the fuel/air mixture is adjusted by the mixture control. At full rich, a large amount of excess fuel is unburned; the exhaust gases are cooled resulting in a lower EGT reading. As the mixture is leaned, the amount of excess fuel is reduced and the temperature climbs. At the point where the ideal mixture

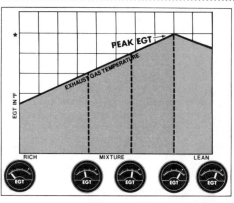

Fig. 19.

of fuel and air (i.e. complete burning of the fuel/air mixture) is reached, the EGT peaks. Leaning past this point results in a cooling of the exhaust gases by the excess air. The mixture is then said to be on the lean side of peak EGT.

At peak EGT, the fuel economy is greatest, but there is a slight loss in airspeed. Actually, airspeed will increase along with EGT up to a point approximately 100 degrees rich of peak EGT. After that, it drops off. A pilot, interested in a faster cruise, should enrich the mixture by about 100° from peak EGT to achieve best or maximum power. However, the slight loss of airspeed at peak EGT is more than compensated for by a 15% increase in fuel economy and range. The recommended mixture setting for maximum range at cruise power is peak EGT.

As in all aspects of engine operation, the recommendations of the engine manufacturer should be strictly followed in determining the temperature to which to lean when using the EGT system, if the manufacturer recommends a procedure that is different from that outlined above.

There are a number of different types of EGT gauges. A very basic system obtains an exhaust temperature reading from one cylinder only (usually that one which, according to the engine manufacturer, is the leanest cylinder). However, mixture distribution will vary from cylinder to cylinder and from engine to engine, even those of the same make and model. No aircraft engine is capable of perfect mixture distribution and the EGT will always vary from cylinder to cylinder. In fact, the leanest cylinder may even vary with variations in altitude, throttle and other conditions. To overcome this problem, the analyzer system (Fig.21) incorporates a probe in the exhaust stacks of each cylinder. The temperature gauge displays simultaneously the EGT of

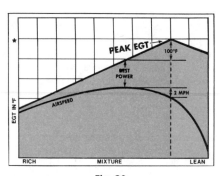

Fig. 20.

every cylinder. The pilot is, therefore, able to check the exhaust temperature in each cylinder, ensuring that no one has been leaned beyond peak EGT. In using the analyzer EGT system, the mixture should be leaned to peak EGT of the cylinder which reaches peak EGT first.

An added advantage of the analyzer system is the opportunity it affords to the pilot frequently to compare the running temperatures of each cylinder. If one is showing undue variance from the others, trouble may be indicated and a check and repair can be made before power loss or part failure occurs.

Fig. 21. Exhaust Gas Temperature Gauges.
Scanning Analyzer System   Analyzer System

A scanning analyzer system (Fig.21) digitally displays engine temperature information. It is programmed to continuously monitor EGT on all cylinders, as well as cylinder head temperature and oil temperatures and to activate a warning light if any temperature conditions varies from the normal operating range.

Automatic Mixture Control

Some engines are fitted with a mixture control which **automatically** compensates for changes in the pressure and temperature of the air entering the carburetor. The device consists of a sealed bellows containing gas which expands and contracts with changes in pressure and temperature. The movement of the bellows is used to operate the mixture control valve automatically. On some carburetors, a single automatic control is provided. On others, the pilot may select "Automatic Rich" or "Automatic Lean". The former is used for take-off and when higher power is required. The latter is used to give the most economical fuel consumption at normal cruise power. Provision is sometimes made for both manual and automatic mixture control. In this case, the pilot will be able to select "Full Rich" and "Idle Cut-off" in addition to "Auto-Rich" and "Auto-Lean".

Some automatic mixture control systems are calibrated to function properly only provided the pilot maintains the temperature of the fuel/air mixture in the carburetor venturi area at a constant reading of about 3°C(38°F). This is accomplished by the manual operation of the carburetor hot air control and by monitoring the carburetor temperature gauge. (The carburetor temperature gauge is described later in this chapter.) The temperature of 3°C(38°F) is selected as being outside the range in which carburetor ice is likely to form.

When to Lean the Engine

At cruise power, below approximately 75 per cent of the rated rpm of the engine, it is generally permissible and recommended to use, at any altitude above 5000 feet, the mixture control as an economy device. The mixture may be leaned to "lean best power" (or peak EGT). Care must be exercised, however, when using the mixture control as a fuel economy device, because selecting too lean a mixture will increase operating temperatures in the combustion chamber.

At high manifold pressures, the mixture control should be set at full rich. For taking off or landing at airports at any elevation up to 5000 feet density altitude and for climbing up to 5000 feet, the mixture control should be in the full rich position.

When taking off at high altitude airports where the density of the air is reduced, the mixture should be leaned in order to get the maximum power from the engine since an over-rich mixture results in loss of power. To lean the mixture prior to take-off, position the airplane at the end of the runway, lock the brakes and advance the throttle to full power. Adjust the mixture control by the method listed above to "rich best power".

For a given throttle opening and a given position of the mixture control, the mixture will become enriched as high altitude is attained. With loss in altitude, conversely, the mixture becomes leaner with unchanged mixture and throttle controls. Therefore, the mixture control should be readjusted for each 1000 feet of change in altitude both during climbs and descents. During a descent from altitude, it is not necessary to enrich the mixture at the beginning of the descent. The mixture should be adjusted gradually to avoid over-enriching the mixture. The enriched mixture has a cooling effect on the engine. So too does the reduced power to speed ratio during the descent.

Special consideration and limitations must be made in leaning turbocharged engines (see **Turbocharging**). In many turbocharged engines, a rich mixture is required to adequately cool the engine. Leaning during climb or on a hot day may result in excessive cylinder head and oil temperatures. The engine operating manual is the definitive source for leaning procedures and should always be followed.

When carburetor heat is being used, it may be necessary to lean the mixture slightly. Since the warmed air is less dense, the mixture will be richer.

Care must always be exercised in using the mixture control. Incorrect mixture ratios will cause certain variations in engine performance. At worst, improper leaning has led to shortened engine life, poor engine performance, and occasionally even engine failure. Even when it is not responsible for such serious problems, improper leaning, at least, has deprived airplane owners of a tremendous amount of fuel economy.

The most common misuse of the mixture control by pilots is selection of too lean a ratio. Many pilots, in an effort to conserve fuel and get the most air miles per gallon, commit this error. A "too rich" mixture will cause loss of power, but only seldom results in complete engine failure. "Too lean" a mixture, however, causes overheating of the cylinder parts and, in many cases, will lead to complete engine failure and substantial damage.

Engines that are operated at too lean a mixture for a considerable period of time will undoubtedly sustain more wear and will cost more to overhaul than will those engines operated with correct mixture selection. Pilots should realize that, in the long run, it is usually far cheaper to use a little more fuel during the running time between overhauls than to pay the extra overhaul costs.

Why Lean the Engine

Proper leaning of the engine is both practical and economical and will result in:

1. *Economy of fuel which means lower costs of operation.*
2. *A smoother running engine. Excessively rich mixtures make the engine run rough and cause vibrations which might cause damage to engine mounts and engine accessories.*
3. *A more efficient engine, giving higher indicated airspeeds and better airplane performance.*
4. *Extended range of the airplane at cruise.*
5. *Less spark plug fouling and longer life for spark plugs.*
6. *More desirable engine temperatures.*
7. *Cleaner combustion chambers and therefore less likelihood of pre-ignition from undesirable deposits.*

Carburetor Icing

Under certain **moist** atmospheric conditions, with air temperatures ranging anywhere from approximately -5°C to 30°C, it is possible for ice to form in the induction system. This can pose serious problems for the pilot if he does not take action to rid the carburetor of this ice that is gradually closing off the flow of fuel to the engine. The fact that carburetor ice is responsible for a great many aviation accidents is perhaps indicative that many pilots, especially inexperienced ones, do not have sufficient respect for or appreciation of the ever present dangers of this icing problem. Such accidents can be prevented only by greater pilot awareness and vigilance.

The chart depicted in Fig. 22 shows the range of temperature and relative humidity under which carburetor icing can occur.

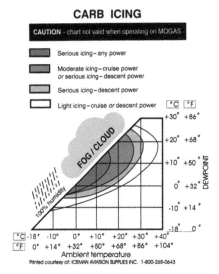

Fig. 22. Carburetor Icing Range.

The temperatures referred to in this chart are applicable only to aviation fuels. They are not valid if the airplane is being operated on MOGAS. Due to its higher volatility, MOGAS is more susceptible to the formation of carburetor ice.

Carburetor icing is usually indicated by a loss of power (a drop in manifold pressure with a constant speed propeller or a drop in rpm with a fixed pitch propeller). If severe enough, carburetor icing may cause complete engine failure, as the icing situation closes off entirely the induction flow.

Ice that forms in the carburetor is caused by two processes: the drop in temperature as heat is taken from the air in order to effect vapourization of the fuel and cooling due to the low pressure area in the carburetor.

Liquid gasoline must be changed into a vapour and mixed with air to become combustible. This process occurs in the carburetor. The heat that is required to change the liquid gasoline to a vapour is absorbed from the air that passes through the carburetor and into the manifold. The carburetor therefore is, in effect, a miniature refrigerator, the temperature in this mixing chamber dropping as much as 30 degrees Celsius below the temperature of the incoming air.

If this air contains large amounts of moisture (and most air contains some water vapour), the cooling process will cause the water vapour

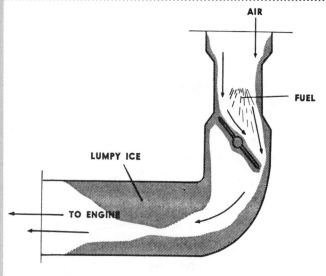

Fig. 23. Carburetor Ice.

to condense and possibly freeze on all surfaces of the carburetor but especially on the throttle butterfly. It is possible for this icing condition to occur, as can be seen, with outside air temperatures as high as 30°C, in bright sunshine with no sign of rain. A temperature of around 15°C should be regarded as the most suspect. The minimum relative humidity generally necessary for icing to occur is 50%, with the icing hazard increasing as the humidity increases. This type of icing is sometimes called **fuel vapourization ice.**

When atmospheric temperature is below -5°C, the danger of carburetor icing is not serious, since the possibility of water vapour being present in air at such low temperatures is very slight. Any small amounts of moisture that might be present in the air will be frozen and will pass through the system harmlessly.

Ice also forms as a result of the low pressure area that is created in the carburetor venturi. The rush of air across and around the throttle / butterfly valve, especially when it is in a partially closed position, causes a decrease in pressure which has a cooling effect on the fuel / air mixture. Moisture in this low pressure area will freeze and collect on the rear side of the throttle butterfly. This type of ice is sometimes called **throttle ice.** The presence of even a small amount of ice on the throttle valve may cause a relatively large reduction in airflow and therefore in engine power. A large collection will jam the throttle. The temperature drop resulting from this low pressure phenomenon is rarely more than 3 degrees Celsius.

In engine installations in which the fuel is introduced upstream from the throttle valve, as is the case with a float type carburetor, there is a fuel/air mixture at the throttle and any ice formation would be attributable to the cumulative effects of both the refrigeration process and the throttle ice phenomenon.

In a fuel injection system (see **Fuel Injection** below), only air passes the throttle and any ice formation would be attributable to throttle ice only. Icing is therefore not a problem in fuel injection engines when the temperature is above 4°C.

A third cause of ice in the carburetor system is the result of water present in the atmosphere as snow, sleet or supercooled liquid. Under these conditions, ice builds up on the air scoop, carburetor screen and carburetor metering elements. This type of ice is commonly called **impact ice.**

Prevention of Carburetor Icing

Modern airplane installations incorporate a means of directing heated air into the carburetor air intake. The carburetor hot air handle on the panel activates this process. Many airplanes are fitted also with a carburetor mixture temperature gauge which tells the pilot the temperature of the mixture. Since it requires much more heat to melt ice that has already formed than to prevent its formation, the carburetor air temperature should at all times be maintained at 4°C to 7°C in flight by using carburetor heat. Some airplanes may be fitted with a carburetor air inlet temperature gauge which measures the temperature of the air just before it enters the carburetor. In this case, the temperature should be maintained at 32°C.

For take-off, however, the carburetor heat control should always be in the cold air position, because the full power of the engine cannot be developed when the air is heated. Heated air has less density than colder air. Bring the carburetor air temperature within safe limits during run up, then select cold air for take-off. If icing conditions are suspected, hot air can be selected immediately after take-off. Since the full power of the engine is not required for cruising, moderate carburetor heat does not affect the performance of the engine in flight. Excessive carburetor heat, however, should be avoided when icing conditions are not present because detonation may result.

The possibility of ice formation may be detected (in airplanes that are without a carburetor air temperature gauge) by applying carburetor heat. If no ice is present, the application of heat will cause a loss of power indicated by a drop in manifold pressure (or in rpm, in the case of an engine fitted with a fixed pitch prop). When ice has begun to form in the carburetor, the application of full heat will remove the ice and permit the rpm in time to return to normal, but **initially** it will cause a further loss of power and a short period of engine roughness. Heated air directed into the induction system melts the ice, which goes into the engine as water, causing some roughness and the additional power loss. As soon as the ice is melted and the resulting water eliminated, the manifold pressure or rpm will return to the normal setting. This should occur in a minute or so. It is necessary for the pilot to understand what is happening so that he will not be frightened out of using heat by what he might consider a situation that has been worsened. Not to use heat in such a circumstance could result in losing the engine completely.

With experience, a pilot learns to recognize atmospheric conditions under which carburetor icing might occur. In an airplane fitted with a carburetor air temperature gauge, ice formation can be prevented by

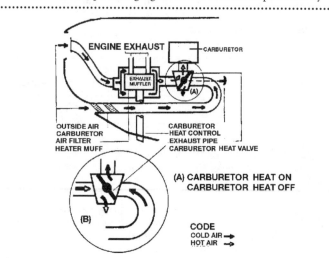

Fig. 24. Carburetor Heat System.

keeping the temperature of the air in the carburetor mixing chamber above the freezing level. A pilot, flying an airplane with no carburetor air temperature gauge of any sort, is well advised to check for carburetor ice by applying carburetor heat periodically. Certainly, it is preferable to anticipate the formation of carburetor ice and use heat as an anti-icer rather than to rely on it only as a de-icer. Any carburetor ice condition means loss of power; the engine exhaust therefore is not as hot; the carburetor hot air flow which is heated as it passes around the exhaust pipes is not as hot and is, in consequence, not as effective in melting a build up of ice, especially if it is forming rapidly.

How long should the use of heat be continued when icing conditions are proven to exist? If icing is severe, then heat should be used as long as the flight continues. Despite the moderate power loss, no amount of heat can damage an engine running at a cruise setting of 75% power or less. The possibility of detonation in the use of heat occurs only when the engine is running at high rpm, such as during take-off and climb. During icing conditions, the pilot's concern is keeping the engine running, not the possibility of detonation.

When using heat for extended periods, the mixture should be adjusted lean. Carburetor heat creates a richer mixture which may cause the engine to run rough. To correct this and also to compensate for the power loss when using heat, it is practical to adjust the mixture to a leaner setting.

Investigation of some forced landing incidents that were the result of carburetor icing have pointed up an interesting fact about this ever present hazard. One pilot using no carburetor heat but using a lean mixture experienced no carburetor icing. Another pilot, flying in the same air conditions, using no carburetor heat but flying at a full rich mixture developed ice and lost his engine. The refrigeration effect from full rich mixture created enough difference in induction air temperature to bring on icing.

To make use of the carburetor heat system effectively, it is necessary to understand that the hot air that is directed into the carburetor by activating the carburetor heat control is cold air that is heated as it passes over the engine's exhaust pipes before it enters the carburetor air intake. The degree to which this air can be heated is dependent on the heat of the exhaust. It can be readily appreciated that it is of little value to apply carburetor hot air **after** the engine has been throttled back. An idling engine produces little heat in the exhaust, comparatively speaking, and the resultant hot air may be inadequate to thaw any ice in the carburetor. A rule to remember, therefore, is:

> *Always apply carburetor heat before reducing power*

If ice is suspected while flying at a low power setting, it may be necessary to open the throttle to produce more heat so that the hot air entering the carburetor will be sufficiently hot to thaw the ice.

It should be realized also that partial carburetor heat can be worse than none at all under certain conditions. If the fuel/air mixture temperature is in the -5°C range, with no heat applied, the formation of ice is not as probable as if the temperature were brought up to -1°C by the use of partial heat. At temperatures around -10°C, any entrained moisture becomes ice crystals which pass through the system harmlessly. Application of partial heat might cause carburetor icing by melting these crystals and raising the temperature to the icing range. In an airplane without a carburetor air temperature gauge, partial heat should never be used.

In fact, on many light airplanes, there is no **effective** partial carburetor hot air position. Therefore, the carburetor heat control should be used either in the "full cold" or the "full hot" position unless there is a carburetor mixture temperature gauge installed which would give a positive indication of the result of partial heat. If the hot air control is moved only to the half way mark, a situation is created where the cold air entering the front air scoop meets "head on" the carburetor hot air which enters from the other direction. There is little or no mixing before it enters the carburetor and the turbulence created may cause rough running. When "full cold" is selected, all the hot air is prevented by a valve from entering the carburetor and is diverted elsewhere. When "full hot" is selected, the same valve repositions itself to shut off completely the cold air so that only hot air enters the carburetor. (The design features of the carburetor hot air control are such that a pilot should not assume that, by adjusting the control to a mark midway between the "full hot" and the "full cold" positions, a mixture of 50% hot air and 50% cold air will enter the carburetor.)

Awareness, therefore, of the possibility of icing conditions and a keen understanding of how ice affects engine operation are the pilot's best weapon against this hazard. Practising the following procedures may help prevent an accident due to carburetor icing:

1. *Be sure the carburetor heat system and controls are in proper working condition.*
2. *Always start the engine with carburetor heat in the COLD AIR position to avoid damage to the carburetor heat system.*
3. *Always, in the preflight check, include a check of carburetor heat availability and note the ON power drop.*
4. *Do not use carburetor heat while taxiing or during ground running. The cold air entering the carburetor by the regular intake is filtered but the hot air is not (see Fig. 24). When relative humidity is high, use carburetor heat before take-off to clear any possible ice from the system, but only if the airplane is standing on pavement or sod, not on sand. The propeller sets up a tiny vortex in the area of the air scoop that may suck abrasive particles of grit and dirt into the unfiltered air intake. (When carburetor icing conditions are suspected, many experienced pilots commence their take-off run with carburetor heat ON in order to be sure that ice will not form and choke off the engine during the take-off run and initial climb. The control is returned to the COLD position about half way down the take-off run so that full power is available for lift off and climb.)*
5. *In cold winter weather, hot air may be used for warm-up and taxiing prior to take-off. Extremely low temperatures cause the engine to run lean. Carburetor heat will enrich the mixture and help to vapourize the fuel. This procedure is recommended only in very cold weather, however, since the hot air that enters the carburetor is not filtered.*
6. *Use cold air selection for take-off. The less dense heated air that enters the induction system when carburetor heat is selected causes a reduction in power. Carburetor heat at high power settings can cause engine overheating and possibly detonation.*
7. *Remain alert for indications of icing, especially when relative humidity is 50% or more or when visible moisture is present.*
8. *Regard any drop in power as an indication of carburetor icing and apply carburetor heat without making an adjustment in the throttle setting.*
9. *When flying in icing conditions, use at least 75% of the full power available to obtain sufficient heat for the carburetor air heater. Maintain cylinder head temperatures as near maximum as possible.*
10. *Avoid clouds as much as possible.*
11. *Always apply full heat prior to closed throttle operations (e.g. descent) and leave on throughout throttled sequence. If the closed throttle operation is extended, open the throttle periodically so that enough engine heat will be produced to heat the carburetor*

hot air sufficiently to prevent icing.
12. *Remove carburetor heat in the event of a go-around. Always apply power first and then move the carburetor heat control to the cold position.*

Fuel Injection

As modern airplanes have become more sophisticated with a requirement for more powerful engines, the need for a more efficient system of distributing fuel to the cylinders became essential. Uneven fuel / air distribution, a common problem with conventional carburetors, results in uneven leaning of cylinders, an unacceptable situation for a high performance engine. The problem was solved through the introduction of the fuel injection system in which the fuel is individually metered to each cylinder. The engine therefore produces slightly more power and uses less fuel than a carbureted engine of equal displacement and compression ratio. The fuel injection system deposits a continuous flow of fuel into the induction system near the intake valve of each cylinder. The fuel is vapourized and sucked into the cylinder during the intake stroke.

Advantages of fuel injection are:

1. *More uniform distribution of fuel to all cylinders. The fuel is separately metered to each cylinder.*
2. *Better cooling, through the elimination of lean hot mixtures to some of the more distant cylinders.*
3. *Saving on fuel through more uniform distribution.*
4. *Increased power since the need to heat carburetor air is eliminated.*
5. *Elimination of the hazard of carburetor icing.*

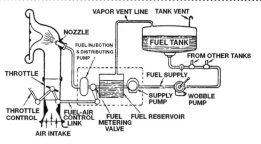

Fig. 25. Fuel Injection System.

Although carburetor ice (i.e. fuel vapourization ice) is not a problem in the fuel injection system, the possibility of ice buildup is not completely eliminated. Throttle ice can occur when the temperature is at 5°C or less. Impact ice can gather in bends in the system, in impact tubes and, especially, on the air filter.

Fig. 25 is a schematic diagram of a fuel injection system that is typical of smaller aero-engines. The fuel is pumped from the **fuel tank** by the **supply pump** to the **fuel reservoir** at about twice the rate required by the engine. The surplus fuel is by-passed through the return line back to the fuel tank, but at a restricted rate that maintains the fuel in the reservoir under pressure. The by-pass eliminates any tendency to vapour lock while in operation.

The fuel passes from the reservoir through the **metering valve** to the fuel **injection pump**. The metering valve controls the amount of fuel delivered to the fuel injection pump.

The fuel injection pump delivers the fuel to the **nozzles.** These atomize it into a fine spray which is discharged into the air stream entering the intake manifold. In some fuel injection systems, individual nozzles deliver the atomized fuel directly to the inlet ports of each cylinder.

The **throttle** which controls the air intake is linked to the fuel metering valve. The fuel and air control units, so linked, are calibrated so that they automatically supply the correct mixture for any particular position of the throttle.

The air required to achieve the proper fuel/air ratio enters the system through the air intake and passes through the air filter. Because impact ice can restrict or completely cut off the flow of air through the air filter, it is essential to have a source of alternate air that can be activated if the normal source of air is blocked or restricted. A manual control in the cockpit selects alternate air and bypasses the normal source.

Fuel injection engines have the reputation of being difficult to start when the engine is hot. When a hot engine is shut down, the heat inside the cowling may cause the fuel in the lines to vapourize. At the next start, the vapour in the lines prevents the engine from getting enough fuel to effect a start. Continued attempts to start will draw too much fuel for the amount of air available. The engine becomes flooded and cannot be started until it is cleared. To successfully start a hot engine, it is necessary to remove all the fuel vapours and get liquid fuel into the lines. This is done by activating the booster pump for a short period of time. Always refer to the Owner's Manual for the exact procedure for operating your engine under these conditions.

Supercharging

An engine designed to operate at normal sea level atmospheric pressure is called a **normally aspirated engine.** As we have already learned, as altitude increases, the density of the air decreases. As the air becomes less and less dense with altitude, the engine is capable of producing less and less power as indicated by the decreasing rate of climb and eventually the total inability of the airplane to climb any higher.

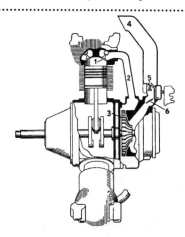

Fig. 26. The Supercharger.

The supercharger compresses the mixture of air and gasoline, thus maintaining the charge at a specified weight for full power output at high altitudes. 1. Intake Valve. 2. Intake Manifold. 3. Supercharger. 4. Airscoop. 5. Carburetor. 6. Throttle.

The **supercharger** is an internally driven compressor, powered directly from the engine. As much as 16% of the engine power can be required to drive the supercharger. It is installed downstream from the carburetor and compresses the fuel/air mixture after it leaves the carburetor. This is called **forced induction.**

Forced induction may be used to increase the power of an engine at low altitudes. In this case the pressure over and above sea level atmospheric pressure which is forced into the manifold is called **boost.**

From the Ground Up

When forced induction is used at high altitudes to make up the deficiency in pressure due to the lower density of the air (and hence maintain sea level power), it is called **supercharging.**

Turbocharging

In many modern airplanes, the job of supplying the engine with dense air when the airplane is operating in the thin air at altitude is accomplished by a turbocharger. A **turbocharger** is powered by the energy of the exhaust gases.

The hot exhaust gases, which are discharged as wasted energy in a normally aspirated engine, are directed through a turbine wheel, or impeller, and turn this wheel at high rpm. The turbine wheel is mounted on a shaft on which also is mounted a centrifugal air compressor. Each is enclosed in its own housing. The compressor, therefore, turns at the same speed as the turbine wheel. As more exhaust gases are directed over the turbine, the compressor will turn faster and the air supplied to the engine by the compressor will be denser allowing the engine to produce more power. The turbocharger is a particularly efficient system since it uses engine energy to maintain horsepower without using any engine horsepower as its source of power.

The turbocharger is installed between the air intake and the carburetor so that it compresses the air before it is mixed with the metered fuel in the carburetor.

The speed of the turbine depends on the difference in pressure between the exhaust gas and the outside air. The greater the difference, the less back pressure on the escaping exhaust gases and hence the higher the speed of the turbine and the greater the degree of compression provided by the turbocharger. As the plane climbs higher and higher into the area of less atmospheric pressure, the turbocharger, therefore, provides the engine with practically the same weight of air from sea level up to the critical altitude. At this altitude, the turbocharger is operating at its maximum capacity. If the airplane continues to climb above the critical altitude, engine power will diminish.

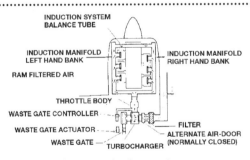

Fig. 27. Turbocharger System

Incorporated into the turbocharger system is a control that regulates the amount of exhaust gas which is directed to the turbine and that, therefore, regulates the speed of the turbocharger. When air of increased density is not needed from the turbocharger (at low altitudes or low power), a wastegate in the exhaust system is allowed to remain open and the exhaust gas is vented around the turbine wheel into the atmosphere. When dense air is required, the wastegate can be closed and the exhaust gases are forced into the turbine. Adjusting turbocharger speed to meet changing power requirements is a matter of controlling the flow of exhaust gases.

Control of the turbocharger is provided by either an automatic control or a manual control. There are two types of the latter.

The simplest form of manual control is the fixed bleed system which does not have a wastegate. Instead, some exhaust gas continuously escapes through an opening of predetermined size. The rest of the exhaust gas turns the turbocharger anytime the engine is running. Engine power is adjusted by the throttle and the amount of exhaust gas available to turn the turbocharger is directly related to the power developed at that particular throttle setting.

The more common manual control interconnects the throttle and wastegate with the cockpit throttle control. In this system, there is a programmed movement of the throttle plate in the carburetor or fuel injector and of the wastegate. As the throttle plate moves towards full open, the wastegate begins to close so that at full throttle, the throttle plate is fully open and the wastegate is fully closed. A pressure relief valve is incorporated into the system to protect the engine in the case of unintentional overactivation of the throttle.

With a manually controlled turbocharger system, the pilot must take care to limit throttle movement to keep the manifold pressure within the limit specified for the engine. During take-off, the throttle should be smoothly advanced until manifold pressure is about two inches below the maximum for which the engine is rated. As the turbocharger speed builds up, the manifold pressure will increase slightly to the maximum limit. As engine power deteriorates during climb, the throttle should be advanced slowly to maintain the desired manifold pressure until the full throttle position is reached at the critical altitude.

The automatic control system utilizes a pressure controller that senses differences in air pressure and adjusts the oil pressure which controls the position of the wastegate. No pilot action is required to activate its operation.

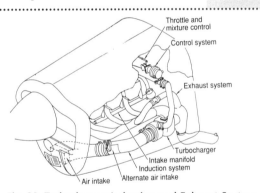

Fig. 28. Turbocharger Induction and Exhaust System.

In operating a turbocharger system, it is important to be aware of throttle sensitivity and the need for slow and smooth throttle movements. The turbocharger does not react instantly but needs time to follow throttle movements and then stabilize. The operation of any specific turbocharger system in any specific airplane should, of course, be carried out with close attention to the instructions of the manufacturer.

The great advantage of turbocharging is the increased performance at altitude. A turbocharged engine is able to deliver full power at altitudes much above the service ceiling of a normally aspirated engine. Better climb performance, faster cruise at altitude, better take-off performance at high density altitude airports are therefore possible.

Exhaust System

The exhaust system of a reciprocating engine is basically a scavenging system that collects and disposes of the high temperature, noxious gases, including the dangerous carbon monoxide, that are discharged by the engine. Its main function is to prevent the escape of these potentially destructive gases into the airframe and cabin. Modern exhaust systems must resist high temperatures, corrosion and vibration.

There are two main types of exhaust systems in use on reciprocating aircraft engines: the short stack system and the collector system. The former is used on non-turbocharged engines and on low powered engines. It is relatively simple, consisting of a downstack from each cylinder, an exhaust collector tube on each side of the engine and an exhaust ejector on each side of the cowling. Shrouds encircle each collector tube. Outside air passing through one of these is heated by the high exhaust temperatures and carried to the cabin to provide cabin heat. The other carries heated air for the carburetor hot air system.

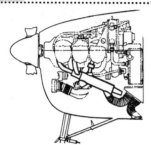

Fig. 29. Short Stack Exhaust System.

The collector system is used on most large engines and on all turbocharged engines. On the latter type of engine, the exhaust gases must be collected to drive the turbine compressor of the turbocharger (see **Turbocharging**). Individual exhaust headers empty into a welded, corrosion resistant collector ring that collects the exhaust from all the cylinders. One outlet from this ring routes the hot exhaust gas to the turbocharger. An exhaust tailpipe carries the exhaust gases away. The collector system raises the back pressure of the exhaust system. However, the gain in horse power from turbocharging more than offsets the loss in horse power that results from the increased back pressure.

On some aircraft engine installations, cooling air is drawn in and around the engine by means of a specially designed exhaust system called the **augmentor system.** The augmentors are designed to produce a venturi effect and work on the principle that the rush of exhaust gases being expelled from the cylinders creates a low pressure area that draws an increased airflow over the engine to augment engine cooling.

Most exhaust systems are made of stainless steel which is known chemically as 18-8.

The Ignition System

The function of the ignition system is to supply a spark to ignite the fuel/air mixture in the cylinders. The ignition system comprises two magnetos, two spark plugs in each cylinder, ignition leads and a magneto switch. The magneto, simply defined, is an engine driven generator that produces an alternating current. Its source of energy is a permanent magnet and it operates on the principle of "the polarity of a magnet".

The Polarity of a Magnet

When a bar of iron is magnetized, it acquires a **north** and a **south pole.**

Unlike poles attract. Like poles repel. Hence, the magnetic lines of force in a magnet pass from S. to N. inside the magnet and from N. to S. outside (Fig. 30).

The field in which these magnetic lines of force lie around a magnet is called its **magnetic field.**

If the bar is bent into a horseshoe, the lines of force will flow in the same manner from S. to N. inside and N. to S. outside the magnet (Fig. 31).

If a soft iron bar is placed between the poles of a horseshoe magnet, the lines of force will flow through the bar (Fig. 32) because the iron offers 280 times less resistance than does air.

Magnetism and electricity are inseparable. If an electric current is passing through a conductor, a magnetic field will be set up around that conductor. If, on the other hand, a conductor is made to cut the lines of a magnetic field, an electric current will be **induced** in the conductor.

Imagine that the soft iron bar (Fig. 32) is rotating, so that it is continually cutting the lines of the magnetic field. Now, if a coil of wire is wound around a bar (so that the coil forms a closed circuit), a current will be **induced** in the coil while the bar is rotating. The current will reverse every half revolution of the bar as the latter passes opposite poles of the magnet and is therefore known as an **alternating current.**

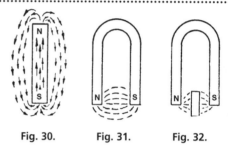

Fig. 30.   **Fig. 31.**   **Fig. 32.**

Here we have the primary element in what is known as the **rotating armature type magneto** — a horseshoe magnet with a rotating soft iron core, wound with a coil of wire, called the **armature,** generating a low tension current by induction. In other types of magnetos, the magnets rotate and the core is stationary. The principle, however, is the same.

The Magneto

The complete magneto combines all the elements of an entire ignition system; that is, it

1. *Generates a low tension current as above,*
2. *Transforms this to high tension,*
3. *Distributes the current to the individual spark plugs at the time it is desired to have them fire.*

The elementary principle of the rotating armature type magneto is illustrated in Fig. 33.

The **armature** revolves between the poles of the magnet, generating an **induced** low tension current in the **primary winding**.

The **contact breaker** is located in the primary circuit. With the breaker points of the contact breaker closed, the primary circuit is complete, and the primary low tension current flows through the breaker points to **ground.**

When the breaker points open, the flow of current in the primary winding is arrested. The magnetic field around the primary winding collapses, falling inward on itself and cutting the secondary winding. A **high tension** current is induced in the secondary winding, which consists of a coil of wire, lighter than the primary, and wound around the latter in a ratio of about 60 turns to 1.

The high tension current is then led through a collector ring to the **distributor.** The distributor has a rotating arm which aligns with separate segments as it rotates. Each segment is connected to an individual spark plug and distributes the current to the right plug at the exact time it is required to fire (No. 3 in Fig. 33).

The **magneto switch** is located on the instrument panel and operates in the primary circuit. When an engine is switched "off", the switch is closed. This means that the **primary circuit is directly grounded** and the breaker points cannot function. As long as the switch remains open (or on "contact"), the primary current will flow through the contact breaker which "makes" and "breaks" the primary (by the opening and closing of the points) and so continuously induces a high tension current in the secondary circuit at each exact instant that this is required. At the same instant, the distributor arm must align with the segment of the plug it is desired to fire.

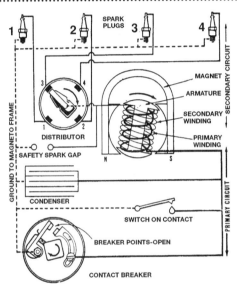

Fig. 33. Rotating Armature Type Magneto.

The **condenser** consists of alternate sheets of mica and tinfoil. It "soaks up" the induced current flowing in the primary when the latter is abruptly arrested by the opening of the breaker points. In this way, it prevents the current jumping across the gap and so burning the points.

The **safety spark gap** is located in the secondary circuit. An excessive voltage in the secondary circuit (usually due to a short circuit in the high tension lead) will jump the safety spark gap and go to ground, thus preventing damage to the armature.

Most light airplanes today are equipped with an **impulse coupling** which is fitted to one or both of the magnetos to facilitate starting. This is a mechanical device which permits the magneto to turn over quickly regardless of how slowly the propeller and crankshaft are turning. The impulse coupling consists of a set of spring loaded flyweights which wind up and release, thereby providing enough rotational speed to the magneto to produce a current that is sufficient to provide a start. The faster a magneto turns (up to a point), the hotter will be the spark it produces. An engine without such an impulse coupling may be difficult to start when it is cold because the slowly turning engine and the corresponding slowly turning magneto produce too weak a spark. The impulse magneto retards the spark somewhat to prevent propeller "kick back". For this reason, on those engines fitted with only one impulse magneto, it is the practice to start the engine on the impulse magneto only. The "impulse" part of the magneto automatically disengages after the engine has reached normal idling speed.

Solid State Ignition

The reader should bear in mind that there are, from time to time, new types of ignition devices being designed and introduced onto the industrial market. Solid state ignition systems, for example, are in use on some engines that are being installed in some amateur built airplanes. This type of system uses magnetic sensors to sense timing and solid state components to produce the spark, thus eliminating the distributor.

Dual Ignition

Modern aero engines are normally fitted with 2 spark plugs in each cylinder and 2 magnetos. One spark plug in each cylinder is fired by one magneto. The other magneto fires the other set of spark plugs. It is possible, therefore, to operate the engine on either magneto alone.

The purpose of dual ignition is two-fold.

1. *Safety. If one system fails, the engine will still operate.*
2. *Performance. Improved combustion of the mixture increases the power output and gives better engine performance.*

On some airplanes, the dual magnetos are driven by a single shaft. The single shaft means a reduction in weight and the need for only one drive at the accessory case but it does reduce the safety factor implicit in a system in which the two magnetos are driven by independent shafts.

The magneto switch, which is located on the instrument panel, allows the pilot to select either or both of the magneto systems. It usually has a position to select the left magneto system, a position to select the right magneto system, a position to select both systems at once, and an "off" position. The engine should always be operated on both magneto systems during take-off and normal flight operations. After landing, the magneto switch should be turned "off" since the engine of a parked airplane can fire if the ignition switch is on and someone moves the propeller.

The failure of one magneto in a dual system will cause a loss in power of approximately 75 rpm at normal engine cruising power.

In the event of failure or partial failure of a magneto or of part of the ignition system (i.e. spark plug, high tension leads, etc.), rough running of the engine is bound to occur. In such a case, an engine will often run more smoothly if the ignition switch is selected to the "good" magneto only. In the event that rough running or power loss occurs while in flight and if there is reason to suspect ignition malfunctioning, the pilot should try operating the engine on just one magneto at a time by switching from the "both" position to "left" and then to "right", allowing the engine to operate for a few seconds on each position to ascertain on which position optimum engine operation and smoothness result. (Smoothness is a very important consideration since a rough running engine can set up vibration which, on some airplanes, could eventually cause engine mount failure.) Operating an engine in flight at normal cruise power on one magneto will result in a power loss of about 3%. Provided the engine is running smoothly, no harm will be done to it if the pilot continues on to destination. However, it must be stressed that this is action to be taken if trouble develops in the magneto **during flight**. Certainly a single engine airplane would not attempt a take-off on one magneto after observing a rough magneto condition during ground runup.

Shielding

To prevent the ignition current interfering with the radio, the whole ignition system, magnetos, plugs, and wiring, are surrounded with a metal covering which is grounded. This is known as shielding.

Ignition Timing

The spark on modern high speed engines occurs early to allow complete combustion and maximum pressure to be developed when the piston passes top dead centre and is travelling down.

Ignition timing means timing the magneto to fire at the right time. Timing is, of course, critical to good engine performance. Advanced (that is, too early) firing of the spark plug results in loss of power and in overheating that can lead to detonation and pre-ignition, piston burning, scored cylinders and broken rings. The timing of the magnetos should be checked at least every 100 hours to ensure proper operation.

The Electrical System

It is beyond the scope of this manual to discuss in full detail the electrical system of an airplane. However, very few modern airplanes are without such a system and since the pilot is required to use it, he should be familiar with its basic operation and purpose.

An airplane's electrical system includes everything that operates electrically with the exception of the magnetos which are driven by the engine for the sole purpose of producing current to the spark plugs. This ignition system is not connected with the airplane's electrical system.

In many modern airplanes, the electrical system supplies power, not only to start the airplane, but also to operate a multitude of controls, such as the flaps, undercarriage, all radios, lights, heater fans, anti-icing and de-icing equipment, windshield wipers, etc.

The basic electrical system of an airplane is not unlike that used in an automobile. It consists of a storage battery, master switch and battery solenoid, starter motor and solenoid, generator (or alternator), voltage regulator, bus bar and circuit breakers (or other types of fuses). On some airplanes, a separate generator switch may be incorporated.

The electrical system is usually either a 12 or a 24 volt system and, of course, is direct current. Most light airplanes use the 12 volt system.

Electrical energy is supplied by the **storage battery.** To complete the circuit between the battery and the electrical system, a battery **solenoid** is incorporated which is activated by the **master switch.** (A solenoid is a switch of the artificial magnet type and is used because of its ability to operate as a remote controlled device.)

The purpose of the **starter motor** is to turn over the engine so that it will commence to operate. The starter switch activates the starter solenoid which, in turn, permits current to enter and drive the starter motor (provided the master switch is turned "on").

The electrical system includes either a **generator** or an **alternator** driven by the engine. The purpose of the generator or alternator is (1) to supply current to the electrical system, and (2) to recharge the battery. On most modern airplanes, alternators have replaced generators. The advantage of an alternator over a generator is that it produces sufficient current to operate the various electrical components at **low engine speeds** (i.e. taxiing) whereas a generator will not begin to supply an appreciable amount of current until the engine is turning at a somewhat faster speed.

The **voltage regulator** (1) prevents the generator (or alternator) from over-loading the system, and (2) prevents the battery from becoming overcharged.

The **bus bar** receives the current produced by the generator (or alternator) and battery. It is from the bus bar that the current passes through the various circuit breakers and branches out to the various electrical circuits which are connected to the components that require electrical current to operate. These are the electrical driven gyro instruments, stall warning device, pitot heat, landing and navigation lights, oil dilution system, fuel supply gauge, carburetor air temperature, exhaust temperature indicator, instrument and compass lights, cabin or dome lights, rotating beacon, radios, etc.

All electrical circuits are protected by **circuit breakers** or other types of fuses. These are incorporated into the system to protect the various components against damage caused by excess voltage or current, short-circuits, etc. Most circuit breakers are the "push to reset" type and every pilot should be familiar with their location in the cockpit and their operation in the airplane which he flies. It is not uncommon for an instrument or radio suddenly to go "dead" due to an electrical malfunction. A push on the "reset button", in such a case, may be all that is needed to make it operational again. However, if the reset buttom pops again, it is advisable not to attempt any further tries to reset. There may be a malfunction or short circuit in the component that could cause an electrical fire. The circuit breaker is a safety device designed to "fail" and prevent such a happening.

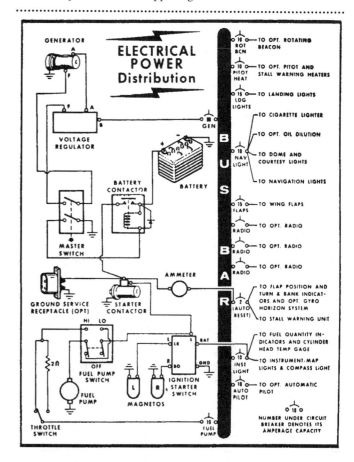

Fig. 34. Electrical System Typical of a Modern Light Aircraft

Most of an airplane's electrical components are automatically activated when the master switch is turned on. Those that are not (i.e., radio) should be left "off" until after the engine has been started. The needless use of electricity in the radio equipment means there is less current available for starting. Some types of radio equipment can be damaged by the voltage drop in the electrical system during

From the Ground Up

starting when all power comes directly from the battery before the generator or alternator are working to regulate voltage. **Failure to follow this rule may result in damage to the component and will cause unnecessary drain on the battery.**

In order for electrical components to function satisfactorily, the battery must be fully charged. In addition, the generator (or alternator) must be producing its rated current.

To ascertain that an airplane's electrical power source is functioning satisfactorily, the pilot may monitor an ammeter, voltmeter or generator warning light. These devices are usually mounted on or adjacent to the instrument panel. The **ammeter** measures in amperes the rate of flow of the electrical current being produced. It also indicates when power is being used from the battery (it registers in the "discharging" range) as happens when the electrical components are turned on when the engine is not running. The **voltmeter** indicates the voltage in the system. The **generator warning light** indicates whether or not the generator is functioning. (An airplane fitted with a voltmeter or ammeter will not have a generator warning light. This warning light is usually installed in the absence of either of the other two instruments.)

For the most part, a pilot need only know that, for satisfactory electrical operation, the ammeter should always show on the plus (+) side of the "0" on the gauge regardless of the amount of current being used by the various components. An ammeter registering discharge or minus (-) indicates that electrical energy is being drawn from the battery rather than from the generator and, if this situation continues for long, it will result in a discharged battery and, possibly, insufficient current to operate the various components satisfactorily.

In modern airplanes and especially in those used for instrument flying, a great deal of reliance is vested in electrical systems and, although it may not be necessary for a pilot to know in detail the internal working of each component, he should at least have a general knowledge of how the system works, how to operate the components properly, what can go wrong and what he can do to correct a minor malfunction in flight.

The most important action a pilot can take in ensuring the proper functioning of the electrical system of his airplane is to ensure that the battery is always fully charged. This is especially important in an airplane equipped with an alternator, for, unlike a generator which will charge a low battery during a flight, an alternator will not bring a dead battery back to life. As a result, if the battery is down on an alternator equipped airplane, electrically operated equipment may not function properly.

It is also important to ensure that all contacts between the battery, voltage regulator and the alternator or generator are clean and secure. The water level in the battery should be checked regularly. Most important, an aged battery that is no longer functioning properly should be immediately replaced.

The Propeller

The function of the propeller is to convert the torque, or turning moment, of the crankshaft into thrust, or forward, speed.

To do this, the propeller is so designed that, as it rotates, it moves forward along a corkscrew or helical path. In so doing, it pushes air backward with the object of causing a reaction, or thrust, in the forward direction. Unlike the jet engine which moves a small mass of air backward at a relatively high speed, the propeller moves a large mass of air backward at a relatively slow speed.

The propeller blade is an airfoil section, similar to the airfoil section of a wing. As such, it meets the air at an angle of attack as it rotates and thus produces lift and drag, in the same way that the airfoil section of a wing does. In the case of the propeller, however, these forces are designated as **thrust** and **torque** (Fig. 35).

Note: Propeller torque is entirely different from engine crankshaft torque. Propeller torque is drag. It is the resistance to the blades as they rotate and results in a tendency in the airplane to roll in a direction opposite to the rotation of the propeller. Engine crankshaft torque, on the other hand, is the **turning moment** *produced at the crankshaft. When the propeller is revolving at a constant rpm, propeller torque and engine torque will be exactly equal and opposite.*

A quick look at a typical propeller will remind the reader that it tapers towards the tip and appears to twist. In effect, the airfoil section alters from the hub of the propeller to the tip and the angle of attack decreases. The tip of a propeller rotates in a larger arc than does the hub and therefore travels at a greater speed. To produce the same amount of lift, or thrust, all along the length of the propeller, the angle of attack at the tip does not need to be so great as at the propeller hub. As a result, by means of this variation in airfoil section and angle of attack, uniform thrust is maintained throughout most of the diameter of the propeller.

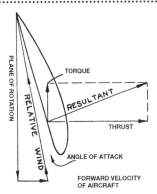

Fig. 35. Forces Acting on a Propeller Blade.

Propellers which are attached forward of the engine and which pull from the front of the airplane are called **tractors**. Those which are attached aft of the engine and push from behind are called **pushers**.

Pitch

The distance in feet a propeller travels forward in one revolution is called **pitch**.

If the propeller was working in a perfect fluid, the distance it would travel forward in one revolution would be a theoretical distance dependent on the blade angle and diameter of the propeller. This is called the **theoretical pitch** or **geometric pitch**.

In a medium such as air, however, the propeller encounters lost motion and the distance it travels forward is somewhat less than its theoretical pitch. This lesser distance is called the **practical pitch** or **effective pitch**.

The difference between the theoretical pitch and practical pitch is called **propeller slip**.

The angle at which the blade is set (like the angle of incidence of a wing) governs the pitch. Hence, there is **coarse pitch,** meaning that the blade is set at a large angle, and **fine pitch**, meaning that the blade is set at a small angle.

Coarse Pitch. A propeller set in coarse pitch will travel forward a greater distance with each revolution and hence the airplane will move forward at greater speed for a given rpm. In this way, it is much like the high gear in a motor car.

A coarse pitch propeller is best suited for high speed cruise and for high altitude flight.

Coarse pitch is also sometimes known as high pitch, or as decrease rpm, or as low rpm.

Fine Pitch. A propeller set in fine pitch, on the other hand, will have less torque, or drag, and will consequently revolve at higher speed around its own axis, thereby enabling the engine to develop greater power, much as does the low gear in a motor car.

A fine pitch propeller gives the best performance during take-off and climb.

Fine pitch is also sometimes known as low pitch, or as increase rpm, or as high rpm.

Types of Propellers

Fixed Pitch Propellers

Most training airplanes have fixed pitch propellers. The blade angles cannot be adjusted by the pilot. The angle of the blade is chosen by the manufacturer to give the best performance possible for all flight conditions, but at best it is a trade-off between take-off performance and cruise performance.

If a coarse pitch propeller is chosen, it will develop high cruising speed at comparatively low engine rpm with consequent good fuel economy. On the other hand, it will give poor take-off and climb performance.

If a fine pitch propeller is chosen, it will give good take-off and climb performance. However, a fine pitch propeller does not keep a good grip on the faster moving air in level flight, with the result that there is comparative inefficiency for cruising, both in the matter of speed and fuel economy.

Variable Pitch Propellers

To overcome the disadvantages of fixed pitch propellers, propellers whose blade angle, and consequent pitch, may be altered to meet varying conditions of flight have been developed.

Adjustable pitch propellers are those whose blade angle may be adjusted on the ground. They offer some advantage over the fixed pitch propeller in that the propeller pitch can be adjusted for varying flight situations. When operating at high altitude airports or when operating on floats when take-off and climb performance is critical, a fine pitch can be selected to give better performance for that particular type of operation. Like the fixed pitch propeller, however, the pitch cannot be altered during flight to select a better pitch angle for a changing flight condition.

Controllable pitch propellers are those whose blades can be adjusted by the pilot to various angles during flight.

Constant speed propellers are those whose blades automatically adjust themselves to maintain a constant rpm as set by the pilot.

The mechanism for varying the pitch of the propeller may be (1) mechanical, (2) hydraulic, or (3) electrical.

Mechanical Variable Pitch Propellers

The mechanism by which the pilot can adjust this type of propeller is operated by a control on the instrument panel that is directly linked to the propeller. Stops set on the propeller govern the blade angle and travel.

Hydraulic Variable Pitch Propellers

In hydraulic variable pitch propellers, the basic method used to change the propeller pitch is a hydraulically operated cylinder that pushes or pulls on a cam connected to gears on the propeller blades. This cam action is somewhat similar to an automatic screwdriver — push the handle in and the screwdriver rotates the screw.

There are two types of hydraulic mechanism: counterweight and hydromatic.

A **controllable pitch propeller** operated by the counterweight principle relies on oil pressure to move the cylinder that twists the blades toward fine pitch. This oil is drawn from the engine oil pressure system. (Occasionally a gear pump is used to boost the engine oil pressure to ensure quick and positive action on the larger type propellers.) The oil which changes the pitch is controlled by a pilot valve, operated by a lever and linkage from the cockpit. The force which moves the blades the opposite way, towards coarse pitch, is centrifugal force, generated by the counterweights. With the controllable pitch propeller, the pilot selects either coarse or fine pitch, depending on the flight attitude (cruise or climb), by means of the control in the cockpit that links to the pilot valve. The pitch of the propeller is then set in the selected pitch until the pilot elects to change it.

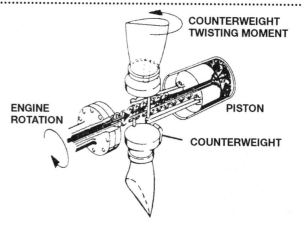

Fig. 36. Constant Speed Propeller Mechanism.

A **constant speed propeller** that operates by the counterweight mechanism uses the same oil pressure and counterweight principle to twist the blades towards the proper pitch angle to maintain a constant rpm. On the instrument panel, there are two power controls, a throttle and a propeller control. The throttle controls the power output of the engine which is registered on the manifold pressure gauge. The propeller control regulates the rpm of both the propeller and the engine. The desired rpm is set by the pilot. A propeller governor connected to the propeller control in the cockpit automatically changes the blade angle to counteract any tendency of the engine to vary from the setting. The governor is a device that consists of flyweights that are directly coupled to the engine and spin at engine speed. The governor operates the pilot valve which controls the flow of oil to the propeller. When the engine speed increases, centrifugal force causes the flyweights of the governor to fly outward, lifting the pilot valve and cutting off the supply of oil pressure to the propeller cylinder. This permits the counterweights to move the blades towards a coarser angle, thereby decreasing the engine rpm and automatically restoring the engine speed to the desired constant rpm. The reverse happens when an engine underspeed condition develops. Therefore, for any particular engine speed for any climb or cruise condition, the pilot merely selects the desired rpm. The constant speed propeller automatically maintains the proper blade angle required to keep the engine speed constant, regardless of the airspeed or altitude of the airplane.

If oil pressure is lost during flight, the propeller will automatically go into an extreme coarse pitch position where the blades are streamlined and cease to turn. This feature of the counterweight propeller is used when the propeller is installed on twin engine aircraft. The stopped propeller eliminates the asymmetric drag forces that would occur if the propeller was windmilling during single engine operation.

Propellers, whether controllable pitch or constant speed, that operate on the counterweight principle, should be shifted to fine pitch before the engine is stopped. The control should be left in this position in order to reduce the load on the engine during the start and warm up period.

There is one counterweight propeller, the Hamilton Standard, which operates differently and should be shut down in coarse pitch. This propeller was widely used on airplanes such as the deHavilland Beaver and the Norseman. Pilots who take an interest in classic airplanes should be aware of differing operating procedures for some equipment they may encounter. In all cases, the operating instructions of the manufacturer should be closely followed.

The **hydromatic constant speed propeller** makes use of a powerful force called centrifugal twisting moment which tends to turn the blades towards the fine pitch position. The use of this natural force eliminates the need for the counterweights. (Note: the centrifugal twisting moment moves the blades towards fine pitch, whereas the centrifugal force generated by the counterweights, as described above, is used to move the blades towards coarse pitch.)

The hydromatic constant speed propeller uses a piston to twist the blades. Oil entering the piston chamber, or dome, under high pressure moves the piston aft and the propeller blades are moved into coarse pitch. Oil entering the piston at engine pressure assists the centrifugal turning moment to move the blades towards fine pitch. The pilot valve, which is controlled by the governor, is a three way affair arranged (1) to admit the high pressure oil to the piston that results in moving the blade towards coarse pitch, (2) to close off the supply of oil thus holding the blade angle constant, and (3) to admit the engine pressure oil to the piston that results in moving the blades back towards fine pitch.

If oil pressure is lost during flight, the propeller will automatically go into fine pitch, enabling the engine (as long as it continues to run) to develop the most power that it can and to achieve the best performance under the circumstances. This type of propeller is used in single engine installations.

With the hydromatic constant speed propeller, the engine should be started with the propeller in fine pitch. This reduces the load, or drag, on the propeller during starting and warm up.

Electrical Variable Pitch Propellers
An **electrical controllable pitch propeller** is operated by an electrical motor which turns the blades through a gear speed reducer and bevel gears. The governor is similar in principle to that used with the hydraulic propeller, except that the movement of the flyweights in this case opens and closes electric circuits. One circuit contains a field which causes right-hand rotation of the motor. The other contains a field which causes left-hand rotation. The direction of rotation of the motor moves the blades towards fine pitch or coarse pitch, whichever is required.

A two way switch enables the pilot to select either "manual" or "automatic" operation. In the "manual" position, the propeller operates as a controllable pitch propeller. The pilot selects either coarse or fine pitch. The blades remain fixed in that position until the pilot chooses to adjust them. Using the control on "manual" reduces the current drain on the battery and reduces the wear on the control mechanism. In the "automatic" position, the propeller operates as a constant speed propeller, the governor automatically holding the engine at the rpm selected by the propeller control in the cockpit.

Feathering
In multi-engine airplanes, when one engine is stopped, it is desirable to **feather** the propeller on the dead engine. Feathering means turning the blades to the extreme coarse pitch position, where they are streamlined and cease to turn. Feathering reduces the drag on the blades. It stops the propeller from windmilling and possibly causing damage to the defective engine. It also stops excessive vibration.

For feathering or unfeathering, an auxiliary oil pressure supply is required, since the engine is no longer running. This pressure is supplied by an auxiliary, or feathering, pump operated by an electric motor. The auxiliary oil pressure is supplied to either face of the piston to move the blades towards "feather" or "unfeather", as the case may be, through the pilot valve system. A push-button **feathering switch,** or in some installations a **feathering lever,** is operated manually by the pilot to feather or unfeather the propeller.

Prop Reversing
It wasn't long before aviators wanted to use the power of their engines to stop their planes on the ground. This is accomplished by changing the blade angles of a controllable pitch propeller to a negative value during operation. The reverse pitch feature uses engine power to produce a high negative thrust at low speed. Reversing the pitch of a propeller is accomplished with the help of the auxiliary oil pump of the feathering system which assists the governor oil pump to twist the blades into reverse pitch.

To make propeller braking by reverse thrust as natural as possible to handle, an "instinctive" throttle quadrant arrangement is provided by the majority of airframe manufacturers. All throttle positions forward of the centre of the quadrant are for forward thrust operation (flight idle position to take-off power) and are referred to as the **alpha range.** The rear section of the quadrant is the reverse range and is referred to as the **beta range.** The farther back the throttle is retarded, the more engine power is developed for reverse thrust. The beta range control for ground handling is hydromechanical, using a cam and lever system to operate the pilot valve. Hydromechancial pitch locking devices or stops prevent inadvertent movement of the throttle levers into the beta range during flight or in the event of propeller malfunction.

In general, the instinctive action taken by a pilot to turn a plane to the right is to retard the right throttle and advance the left. Similarly, when the props are in reverse, appropriate reverse thrust braking power is achieved by retarding the right throttle and advancing the left.

Engine Instruments

The Oil Pressure Gauge

The oil pressure gauge (Fig. 37) is one of the principle engine instruments and is usually located on the instrument panel with the oil temperature and the fuel gauges. It indicates the oil pressure supplied by the oil pump to lubricate the engine. The instrument is calibrated in lb. per sq. in.

The oil pressure which you read on the gauge is generally obtained at a point between the oil pump and a pressure regulating valve whose function is to maintain any desired working pressure, for example 60 psi.

The oil pressure gauge should be checked immediately after the engine has been started. If the oil pressure does not register within 10 seconds, the engine should be shut down and checked.

In starting an engine with cold oil in the system, the oil pressure gauge will invariably read high, owing to the difficulty of forcing the sluggish oil through the small aperture in the pressure regulating valve. As the oil warms up and the flow through the pressure regulator improves, the pressure gauge will record the pressure accurately. High oil pressure should cause no concern until the oil is allowed to warm up and reach its normal viscosity, which might require 15 minutes. For this reason, in winter, a longer period of time is necessary to warm up the engine prior to flight.

If, on the other hand, it remains high, the engine is not getting proper lubrication. High oil pressure will force oil into the combustion chamber. Here it will burn, causing a smoky exhaust and badly carbonned piston heads, rings, valve seats, cylinder heads, etc.

Low pressure can cause more serious trouble still. If it is permitted to drop low enough, there will be no film of oil at all between the working surfaces of the engine and metal will be rubbing on metal, with such ruinous results as burned out main bearings, etc. Never let the oil pressure fall below the recommended minimum (which is roughly 40% of the maximum pressure at cruising power).

Fig. 37.
Oil Pressure Gauge.

Fig. 38.
Oil Temperature Gauge.

The Oil Temperature Gauge

The oil temperature gauge (Fig. 38) is normally located beside the oil pressure gauge. It gives a reading of the temperature of the oil in degrees Fahrenheit or Celsius.

There is an intimate relationship between the oil temperature and oil pressure, due to changes in the viscosity of oil which temperature changes affect.

In starting the engine with cold oil, when the pressure gauge reads high, the oil temperature gauge will read correspondingly low. As the oil warms up, both instruments will approach their normal readings at about the same rate.

An abnormal drop in oil pressure and coincident rise in oil temperature is a sure sign of trouble. However, even when the pressure shows no marked rise or fall, increasing oil temperature is a warning of excessive friction or overload in the engine.

Extremely low oil temperature is undesirable. Cold oil does not circulate freely and may cause scoring of the engine parts. Low temperature would be accompanied by a corresponding rise in pressure.

It should always be remembered that oil, in addition to lubrication, acts as a coolant.

The Cylinder Head Temperature Gauge

The cylinder head temperature gauge (Fig. 39) records the temperature of one (or more) of the engine cylinder heads. The instrument gives a reasonably good indication of the effectiveness of the engine cooling system. It should be monitored frequently during steep climbs to ensure sufficient cooling air is reaching the engine.

Extremely high cylinder head temperatures are an immediate sign of engine overloading. High head temperatures decrease the strength of metals and result in detonation, pre-ignition and eventual engine failure.

When operating the engine on lean mixture, the maximum cylinder head temperature permissible is lower than that permitted when operating on rich mixture. The engine manufacturer's recommended limits should be strictly observed.

Fig. 39. Cylinder Head Temperature Gauge.

Fig. 40. Carburetor Air Temperature Gauge.

The Carburetor Air Temperature Gauge

The carburetor air temperature gauge (Fig. 40) may be installed to indicate the temperature of the mixture entering the manifold, or it may record the temperature of the intake air entering the carburetor. Its purpose is to enable the pilot to maintain a temperature that will assure maximum operating efficiency and warn him of icing conditions in the carburetor that may lead to engine failure.

The carburetor air temperature gauge is, of course, the pilot's guide to the operation of the carburetor air heat control unit. Hot air is selected ON or OFF to keep the mixture temperature, or intake air temperature, within the recommended limits.

If the instrument is installed to record the mixture temperature, this should be maintained at around 4°C to 7°C (40°F to 45°F). If it is installed to record the intake air temperature, this should be maintained at about 29°C to 32°C (85°F to 90°F) when icing conditions exist.

The Outside Air Temperature Gauge

The outside air temperature (OAT) gauge records the ambient air temperature, that is, the temperature of the free air surrounding the airplane. To ensure that the temperature recorded is true, the element is shielded from the sun's radiation and located in a portion of the airflow that is relatively undisturbed. The temperature recorded by the gauge is not, however, entirely accurate. The dynamic pressure of the ram air causes a slight increase in temperature above that of the ambient air. The indicated temperature, therefore, must be corrected to get true air temperature (TAT). Knowledge of the true air temperature enables the pilot to select the proper manifold pressure, to calculate the true airspeed and altitude and warns of conditions that may cause ice formation.

At very high airspeeds, there is a very significant temperature rise on the body of the airplane. When an airplane moves through the air, it gets hot. Some of this heat is attributable to friction and some to the increase in pressure. At slow speeds, the increase in heat is insignificant, about 1°C at 90 knots. With increasing speed, the temperature rise reaches 4°C at 175 knots and 30°C at 500 knots. The reading on the OAT gauge will reflect this temperature rise of the body of the aircraft, rather than the true ambient air temperature.

The Tachometer

The tachometer, or rpm indicator (Fig. 41) is an instrument which shows the speed at which the engine crankshaft is turning in hundreds of revolutions per minute. The instrument usually incorporates a recording mechanism that keeps an accurate record of the engine hours. Tachometers are of many types, the more common being (1) mechanical, either centrifugal or magnetic, and (2) electrical, either direct or alternating current.

Fig. 41. The Tachometer.

On an airplane fitted with a fixed pitch propeller, the tachometer is the only instrument that displays information about the engine power that is being produced. It indicates the rpm at which the engine crankshaft and the propeller are turning. The power output absorbed by a fixed pitch propeller is proportional to the cube of its rpm. The throttle controls the rpm.

On an airplane fitted with a controllable pitch or a constant speed propeller, there are two instruments, the tachometer and the manifold pressure gauge, which display engine power information. In this case, the rpm settings, recorded on the tachometer, are controlled by the propeller control. Manifold pressure settings are controlled by the throttle.

Tachometer Markings

The tachometer is colour coded to give a ready indication of the proper range of engine operation. The green arc indicates the normal range of operation. Rpm settings within this range should be used for continuous operations. The yellow arc indicates the caution range in which there is a possibility of engine damage under certain conditions. The red line is the maximum limit.

Operation of the engine at greater speeds than those recommended may result in excessive mechanical stresses and may cause failure of major engine parts.

Green (Normal Operating Range)
Yellow (Caution Range)
Red (Maximum Allowable)

Fig. 42. Tachometer Markings.

The Manifold Pressure Gauge

On an airplane fitted with a controllable pitch or a constant speed propeller, there are two instruments which indicate information about engine power output, the tachometer and the manifold pressure gauge. Settings on the manifold pressure gauge are controlled by the throttle.

Colour range markings, similar to those on the tachometer, are incorporated on the dial of the manifold pressure gauge to indicate the normal operating range and operation limits.

The instrument is usually located on the instrument panel adjacent to the tachometer so that the pilot can refer to both instruments when making power settings.

The manifold pressure gauge indicates in inches of mercury the pressure of the fuel/air mixture in the engine intake manifold at a point between the carburetor and the cylinders. A manifold reading of 26" Hg indicates a pressure of about 13 pounds per square inch (psi) in the engine intake manifold (14.7 psi is equivalent to 29.92" Hg).

On an airplane fitted with a constant speed propeller, the rpm setting will remain constant since the propeller automatically changes its pitch to maintain the desired engine speed. The manifold pressure gauge therefore is the only instrument that records fluctuations in the engine power output. It is important for determining proper throttle settings and it is the instrument which will indicate power loss from occurrences such as the accretion of carburetor ice.

When the engine is not running, the reading on the manifold pressure gauge will be that of the existing atmospheric pressure. When the engine is running, the pressure inside the intake manifold is lower than that of the outside atmospheric pressure because the pistons create a partial vacuum.

The manifold pressure gauge is also an important instrument in an airplane with a turbocharged or supercharged engine. Turbocharging is accomplished by increasing the pressure of the air entering the carburetor; supercharging, by increasing the pressure of the fuel/air mixture after it leaves the carburetor. The pilot, in order to control turbocharging or supercharging, must know the pressure of the fuel/air mixture in the intake manifold. The manifold pressure gauge gives this information. On a turbocharged or supercharged engine, the pressure inside the intake manifold is higher than the outside atmospheric pressure and readings will normally be higher than those on normally aspirated engines.

Fig. 43. Manifold Pressure Gauge.

A drop in the reading of the manifold pressure gauge should be viewed as an indication of carburetor ice. If ice is forming in the carburetor, the airflow is restricted, causing a decrease in pressure in the intake manifold which will be indicated on the manifold pressure gauge.

Since the power of engines which are fitted with controllable or constant speed propellers and turbochargers or superchargers may be increased by either (a) increasing the rpm or (b) increasing the pressure, both of which are variables, it follows that there is an intimate relationship between the tachometer and the manifold pressure gauge.

If the rpm and manifold pressure of an engine at any given altitude are known, and the temperature correction for altitude applied, the brake horsepower of the engine can be determined from a chart. Re-versing the process, an engine may be operated up to its critical altitude at any power desired by maintaining a specified boost and rpm. Some engine manufacturers recommend higher rpm and lower boost while others recommend lower rpm and higher boost to obtain a given power. These factors are governed by the design of each particular engine. It is therefore most important that the recommended power settings for any particular engine, for each specified condition, be strictly adhered to. Tables of these recommended settings are published in the Airplane Flight Manual (see Fig. 44) and are also sometimes stated on a placard placed somewhere within the cockpit. The chart shown in Fig. 44 is to be used for flights at altitudes near 2500 feet. Similar charts will be prepared for other altitudes (e.g. 5000 ft, 7500 ft, 10,000 ft).

Excessive manifold pressure raises the compression pressure, resulting in high stresses on the pistons and cylinder assemblies. If the pressure exceeds the limits of the octane rating of the fuel that is in use, detonation will result. Excessive pressure also produces excessive temperature, which may cause scoring of pistons, sticking rings, and burned out valves.

When the propeller pitch control is moved, during flight, from the fine pitch (high rpm) to the coarse pitch (low rpm) position,

Standard Conditions • Zero Wind • Lean Mixture 2500 Feet 41.5 Gal. of Fuel (No Reserve) • Gross Weight-2500 Lbs.						
PROP. RPM	MP	% BHP	TAS MPH	GAL./ HOUR	ENDR. HOURS	RANGE MILES
2250	23	72.0	132.0	10.30	4.00	530
	22	68.0	128.0	9.80	4.30	545
	21	63.0	124.0	9.20	4.50	565
	20	59.0	120.0	8.60	4.80	580
2200	23	70.0	131.0	10.10	4.10	535
	22	66.0	127.0	9.50	4.40	555
	21	62.0	123.0	9.00	4.60	570
	20	57.0	118.0	8.40	5.00	585
2100	23	66.0	127.0	9.60	4.30	550
	22	62.0	123.0	9.00	4.60	565
	21	58.0	119.0	8.50	4.90	580
	20	53.0	114.0	8.00	5.20	595
2000	19	45.0	101.0	6.90	6.00	605
	18	41.0	92.0	6.40	6.50	595
	17	36.0	80.0	5.80	7.20	575
	16	32.0	66.0	5.20	8.00	530

Fig. 44. Airplane Performance Chart (Power Chart).

without an accompanying adjustment of the throttle, the manifold pressure gauge will register an increase, and, in the reverse situation (from coarse to fine pitch), it will register a decrease. The manifold pressure gauge registers a reading that is taken inside the intake manifold. When the rpm is decreased, the engine turns more slowly. The speed of flow of the fuel/air mixture through the manifold system is decreased. When the speed is decreased, the pressure increases (Bernoulli's Principle), hence, the increased reading on the manifold pressure gauge. The same situation in reverse occurs when the rpm is increased (from coarse to fine pitch). The engine speeds up, the fuel/air mixture speeds up passing through the manifold system, the pressure decreases and there is a decreased reading on the gauge.

> *When increasing power, increase the rpm first and then the manifold pressure.*
> *When decreasing power, decrease the manifold pressure first and then the rpm.*

Manifold Pressure as an Index to Power

In the era when there was nothing but the fixed pitch propeller, a pilot was able to understand the operation of his piston airplane engine much more readily than is the case today with the more complicated variable-pitch constant-speed propellers. With a fixed pitch propeller, opening the throttle increases the manifold pressure and, of necessity, also the rpm. The power that an engine with a fixed pitch prop is producing is determined by ref-erence to a tachometer. The manifold pressure cannot be developed independently of the rpm. For each throttle setting, or manifold pressure setting, there is a corresponding rpm or BHP of the engine.

However, with the variable-pitch constant-speed propeller, opening the throttle will not change the rpm, but it will increase the manifold pressure, propeller load and horsepower. It follows, therefore, that the power absorbed by this type of propeller is independent of the rpm, for by varying the pitch of the blades, the air resistance and the propeller load can be changed while the propeller speed remains constant.

The process by which the propeller load or torque, as it is more properly called, is absorbed by the propeller is by varying the pressure acting on the piston. Any factor which affects the pressure acting on the pistons changes the torque. All other factors being equal (i.e., mixture ratio, temperature, type of fuel used), the pressure developed in the cylinders will depend on the mass of the charge that can be forced into it. This, in turn, will depend on the pressure existing at the intake valve ports and is measured by manifold pressure (MP). Thus torque (propeller load) turns out to be a function of manifold pressure.

Consequently, with variable pitch propellers, a given propeller load may be absorbed at an infinite number of rpm and, since power is a function of both MP and rpm, a particular horsepower (BHP) may be developed at an unlimited number of combinations of MP and rpm.

There are other variable factors affecting engine power of which the pilot must be aware if he is to operate his airplane engine efficiently. (1) The most important factor affecting power output is the fuel/air ratio. Variations in fuel flow from standard carburetor settings may reasonably affect power by a factor as high as 8 per cent. (2) Inlet air temperatures affect the air density of the charge in the intake manifolds. As all conventional power charts are based on the assumption that "standard temperature conditions" prevail at the carburetor inlet, a correction must be applied to the airplane performance charts and tables. A conventional "rule of thumb" for making this correction is to add 1 per cent for each 6°C below standard to the chart BHP; or to subtract 1 per cent for each 6°C above standard.

In summing up, therefore, we see that in order to determine what percentage of BHP is being developed from an engine having a variable pitch propeller, we need to know the rpm, MP, mixture setting (rich or lean) and temperature of the air entering the carburetor. Knowing these variables and by reference to the airplane manufacturer's tables or graphs, the pilot is able to calculate BHP and fuel consumption and is able to anticipate air-plane performance.

Humidity and Power

Factors such as outside temperature and altitude have an important effect on the power output of an airplane engine. Does humidity have any effect? Yes, it does, but only to the extent of about a 2% change.

Water vapour in the air will reduce the performance of an engine to a slight degree. The water vapour in the fuel/air mixture that is metered into the engine cylinders is incombustible. Therefore, it follows that the fuel to air ratio is increased and the engine operates as on a somewhat richer mixture. The result is some power loss and incomplete combustion.

For the most part, this power loss is so negligible that it need not be considered. Testing has indicated a loss of less than 2% for 100% relative humidity at temperatures of 32°C. The only occasion when the effect of humidity might be considered is during take-off and landing in marginal conditions, as on a very hot day when the humidity is high and the airplane is heavily loaded. On such occasions, that 2% change might be worth remembering in calculating the take-off run the airplane will require and the climb performance it can achieve.

Operation of the Engine

Important Considerations About Engine Operation

The service life of the modern aircraft engine is dependent on good maintenance and operating procedures. This is especially true with the tightly cowled, high performance engines in today's sophisticated airplanes. It is all too easy to ruin a fine piece of equipment through improper use.

Never make abrupt movements of the throttle. Such action can lead to damage and eventual engine failure.

On take-off, open the throttle **slowly and steadily** to take-off power. In this way, the engine is able to accelerate in rpm at the same pace as the advancing throttle; the increase in rpm and manifold pressure keep pace and there is little possibility of overboost of the engine cylinders; the propeller governor/propeller pitch control mechanism has adequate time to respond to the increasing rpm without risking an overspeed condition; and, finally, temperature changes within the cylinder and piston assemblies take place more slowly reducing the possibility of overstress, cracking and breaking that are caused by very rapid temperature changes.

Once climbing speed is reached, reduce power **slowly** and always climb at the highest indicated airspeed that is consistent with safety and other considerations of the flight. At the higher climbing speed, the engine will receive better cooling which is essential at a time when higher power is being produced (climbing power). Also during climb, use a rich mixture to benefit from the cooling effect on the combustion chamber of the evaporating extra fuel. Where installed, cowl flaps should be in the full open position for the take-off and climb.

Avoid high speed dives with engine idling. This kind of action will cause sudden cooling and could cause engine damage, specifically, cracks in the cylinder heads around the spark plugs or valve ports. In the operation of any engine, cracked cylinders indicate that the pilot has been operating the engine at a fairly high temperature and then abruptly reducing power and thereby suddenly cooling the cylinder head. Gradual "let-down" without appreciably reducing engine power is best for the engine and is most efficient from the standpoint of obtaining maximum "air miles per gallon" performance.

Avoid long run ups that overheat the engine. The streamlined and close fitting cowlings of today's general aviation airplanes restrict the flow of cooling air over the engine during ground maneuvering. Over lengthy ground operation may overheat the cylinders which will then be suddenly cooled by the rush of cool air after take-off. The time to taxi from the ramp to the runway is usually sufficient to bring the engine up to its normal operating temperature. In the event that overheating does occur, turn the airplane into wind, ensure that all cowl flaps are open to their full extent and run the engine at about 1000 rpm to allow the propeller to direct cooling air back over the engine.

Avoid high power run ups, except when necessary for maintenance checks. They also cause overheating and, if carried out frequently, will produce a cumulative type of damage that shows up in the form of broken piston rings, scored pistons and cylinders. Turbocharged engines are very susceptible to overheating in high power run ups.

Be sure all cowl flaps and cooling devices are open during run ups and do not use carburetor heat.

Always use the power settings recommended by the manufacturer for cruise. Never over lean the mixture. Always double check that the method being used to lean the mixture is reliable.

The "normal rpm" stated by the engine manufacturer should never be exceeded for more than 5 minutes and then only when absolutely essential.

The "maximum permissible rpm" should never be exceeded more than momentarily in emergencies. Every engine is rated at a specified speed, or rpm, above which it may not be operated safely. To overspeed the engine, even momentarily, can cause excessive wear and, if the overspeed is high enough, serious damage and even engine failure.

The best cruise power setting is one at which the engine runs the most smoothly at approximately the rpm or MP recommended by the manufacturer (or at which the propeller load curve intersects the engine power curve, if these are available for the engine being operated). When selecting certain rpm/MP power combinations, a pilot should always "go by the book". Some engines are designed to cruise at 1600 rpm and 26" MP, while other engines are designed to be operated at 2600 rpm and 23" MP. Concerning rpm/MP combinations, there is no "rule of thumb" that applies to all engines. The manufacturer's performance charts or tables must always be referred to for every model of airplane. When given a choice of different power settings (various combinations of rpm and MP), always use the one that produces the least noise and least vibration. This action will prolong engine life, decrease the likelihood of damage or failure of parts or components and be less fatiguing to crew and passengers.

A further factor to consider when choosing between several rpm/MP combinations that produce the same percentage of power is fuel economy. Usually the best fuel economy is realized with a low rpm and high MP combination because internal friction is reduced and less fuel is, therefore, required to overcome the internal friction that is produced. This point is of particular importance to pilots undertaking long range flights.

In an airplane fitted with a fixed pitch, all metal propeller, it is important to avoid selecting an rpm that is not within the operating range recommended by the manufacturer. Both the engine and the propeller have a natural "resonance" and, at certain rpm, engine and propeller harmonics will coincide and cause sufficient propeller blade vibration to induce blade failure.

With turbocharged and supercharged engines, the "maximum permissible boost" must **never** be exceeded.

Boost pressures greater than "rated boost" are only permissible for 5 minutes at a time.

During a long descent in cold weather, apply sufficient power to keep the engine at normal operating temperature.

Above all, never switch off an overheated engine. After taxiing in, allow it to idle a sufficient time to cool off. A lot of residual heat has to dissipate.

If the engine is fitted with idle cut-off, this should always be used to stop the engine.

The life and dependability of an aero engine depend to an appreciable extent on the use or abuse the engine suffers at the hands of the pilot. The reason that some engines go their full time between overhauls while others do not can usually be credited to the care or lack of care by the pilot in handling the engine controls.

Maintenance and Care of the Engine

Airplanes should be flown frequently in order to prevent rust and corrosion from occurring inside the cylinders. Without frequent flights, water and acids collect in an engine. These are normally cooked out during flight. It is not sufficient, however, just to run up the engine on the ground in order to prevent this rusting process. If the oil temperature is not brought up to 75°C, water from combustion will be added to the engine oil. At 75°C water is still a part of the combustion process, but the oil is hot enough to vapourize the water. If the oil temperature is not raised to 75°C, this water will build up and slowly turn to acid. If it is impossible to fly the airplane even around the circuit, the best thing would be to pull the propeller through a full turn to recoat the cylinders with a film of oil. However, **never touch the propeller unless you have assured yourself that the magneto switches are OFF.**

Change oil at intervals recommended by the manufacturer to avoid excessive engine wear. As the oil flows throughout the engine, it becomes contaminated with carbon particles, water, acid and dirt. The old oil should be drained when the engine is hot as the oil then flows more freely and there is better assurance that all the contaminated oil will be cleared out of the engine.

Engine manufacturers now recommend that oil be changed every 25 hours or every 3 months, whichever comes first. Even if the aircraft has not been flown too often during the 3 months, the engine oil will become contaminated by acids as the airplane sits unused. Frequent oil changes will help prolong the life of the engine. If an airplane is to be put into storage for the winter (or for any prolonged period of time), it is wise to change the oil before putting it away and again when bringing it out of storage before flying it again.

An even better procedure for long term storage is to replace the regular oil with special inhibiting oil. After filling the oil tank with the inhibiting oil, run the engine for two or three minutes as per the instructions. Inhibiting oil is a special type of oil that is sticky and, therefore, clings to all metal parts inside the engine and prevents corrosion. When the engine is to be returned to service, it should be operated for a few minutes to bring the oil temperature up to a point that will permit the oil to flow. The inhibiting oil should then be drained and replaced with regular engine oil.

Installation of auxiliary oil filters are recommended to assist in preventing foreign matters from circulating through the engine and causing excessive wear. All oil filters should be regularly inspected and cleaned.

Service spark plugs regularly and replace them when they are worn to their limits. Be sure to use only the spark plugs recommended by the engine manufacturer. Swap top and bottom spark plugs every 25 to 50 hours. Top plugs scavenge better than bottom plugs. During operation, avoid over rich mixtures which contribute to spark plug fouling. Low operating temperatures coupled with rich fuel mixtures result in incomplete vapourization of the lead in the combustion chambers causing lead deposits on the plugs. Always, therefore, maintain engine operating temperatures in the normal range.

Inspect the ignition harness regularly to assure against worn insulation and seals that would cause misfiring.

Have the magneto points and timing checked as recommended by the manufacturer and have a compression check run regularly.

From the Ground Up

Check the intake manifold periodically to be sure there are no loose connections.

Air inlet filter screens should be cleaned regularly so they can perform their function of keeping sand and dust from entering the induction system. Ensure that the alternate air intake and carburetor heat valves operate correctly and seal properly. Any accumulation of dust or dirt near the entrance to the alternate air device should be removed to prevent it being drawn into the engine.

Care of the Propeller

Propellers should be checked regularly for damage. Failures of propellers almost always occur as the result of fatigue cracks which started at mechanically formed dents. Even a small defect such as a nick or dent if left unrepaired may develop into a crack. The crack, in turn, under great stress, will grow and blade failure is inevitable. Whenever defects are found, they should be repaired or the propeller replaced before further flight.

When performing an inspection on the propeller, inspect blades completely, not just the leading edge, for erosion, scratches, nicks and cracks. Regardless of how small the surface irregularity, it should be considered a source of fatigue failure. A visual inspection will sometimes miss small nicks and cuts that will be perceptible to the fingers. It is wise, therefore, to feel, with clean, dry hands, the entire surface of the blades.

Keep blades clean. A crack cannot be seen if it is covered with dirt or other foreign matter. As regularly as possible, wipe the blades with a clean cloth dampened with light oil. The oily cloth will remove substances that cause corrosion. Never hose down the propeller with water. Regular maintenance should include cleaning the propeller blades with a solvent and then waxing them with a paste wax. A propeller constructed of composite material will require different cleaning methods than those used on metal propellers. Always refer to the Owner's Manual when performing any kind of maintenance on a composite component.

Avoid engine run ups in areas containing loose rocks, gravel, etc.

Do not move the airplane by pushing or pulling on the propeller blades.

Do's and Don't's of Fuel Management

Do be thoroughly familiar with your plane's fuel system and its operation.

Do know the engine's hourly fuel consumption rate.

Do check your fuel supply visually if possible. Be sure you have the correct type of fuel. Observe the colour and odour of the fuel as you check the tank. Be sure all fuel caps (and oil tank caps) and covers are installed and secured properly.

Do check that each fuel tank vent is clear of obstructions (dirt, ice, snow, bent or pinched tubes, etc.). Fuel tanks must be vented to allow air to enter the tank as the fuel is used.

Do visually check the fuel selector valve when changing tanks.

Do make sure, in aircraft with a fuel gauge indicator selector, to change both the indicator selector and the fuel valve selector in switching tanks.

Do check the fuel selector valve carefully to be sure that it is operating properly. This is especially important in cold weather. It is possible for a fuel selector valve to freeze into position, making it impossible to select a fuel tank when needed. It is preferable to make this check before starting to taxi so that, when ready for take-off, the fuel flow from the take-off tank to the engine has been stabilized.

Do learn how to use the mixture control properly.

Do make a check during the preflight inspection of the fuselage and external wing surfaces for stains which may indicate fuel system leakage.

Do keep the tanks full when the aircraft is not in use. Water condenses on the walls of partially filled tanks and enters the fuel system.

Do periodically inspect and clean all fuel strainers.

Do practise good housekeeping in your routine maintenance.

Do calculate your usable fuel as 75% of your total capacity. To determine your **safe flight time limit**, divide your usable fuel by your fuel consumption rate. Include time from start up to shut down in total flight time. Fuel exhaustion is an all too common cause of accidents. Know your safe flight time limit and don't exceed it.

Do exercise care in flight planning, taking into account any situation which would cause an increase in fuel consumption. Do compute a reasonable time limit for your flight taking into consideration trip length, cruising altitude, wind (do not count on forecast tailwinds) and all up weight.

Do use only the fuel recommended by the engine manufacturer.

Do filter all fuel entering the tanks.

Do drain fuel sumps regularly on preflight checks. Draw a generous sample of fuel from each sump and screen drain it into a transparent container. Check that there is no water, rust or other contaminants in the sample.

Do assure that sump drains are fully closed after draining.

Do land somewhere short of your planned destination if the flight has been slower than anticipated and there is any doubt that you have enough fuel to complete the entire trip.

Don't fly beyond a refuelling stop unless the amount of fuel remaining is more than enough to get you to your destination.

Don't overlean the mixture to practice false economy.

Don't use additives that have not been approved by the manufacturer.

Don't change the fuel selector just prior to take-off. In fact, perform the engine run-up on the tank that will be used for take-off.

Don't change the fuel selector during the approach to landing. Change tanks sufficiently in advance to ensure that the fuel flow is stabilized.

Starting the Engine

1. Position the airplane so that dust will not be blown into hangars or on to other planes. In a strong wind, a light plane should be faced into wind for run up, but not where it will blow dust onto other planes.

2. Parking brakes full ON. Use chocks if parking brakes are inadequate to prevent the airplane from creeping forward during the warm-up.

3. Before starting, the propeller should be pulled through (i.e. rotated by hand) through several complete turns. This action releases any

accumulation of oil in the combustion chambers and also assists in partially priming the engine. (Be sure the switches are OFF before pulling through.) The propeller must not be motored over on "Contact" until the pilot is sure that there is no person or property that could be injured or damaged when the propeller starts.

When starting a seaplane at a buoy, a line should be placed loosely around one of the float undercarriage struts and not slipped until the pilot is sure that the engine will continue to run.

4. Carry out pre-starting procedures as per the instruction manual or as previous experience dictates (i.e. certain amount of prime depending on whether the engine is hot or cold, mixture FULL RICH, fuel selector ON, carburetor heat on COLD, etc.). In most airplanes, the throttle should be set 1/4" to 1/2" open from the fully closed position prior to engaging the starter switch. Immediately that the engine starts, the throttle should be moved towards the closed position until the tachometer registers about 600 to 700 rpm and the engine should be allowed to run at this rpm until oil pressure is obtained.

The oil pressure gauge should be checked immediately. If it does not register within 10 seconds, shut off the engine and investigate. However, in cold weather, when the oil is cold and sluggish, it may take as much as 30 seconds or more before any pressure registers on the gauge.

The action of throttling back immediately after starting the engine is very important to good engine handling. By doing so, engine wear is eliminated and friction reduced while the oil pressure is being built up within the engine. Secondly, in the event that chocks are not used and the brakes fail to hold, reducing the power immediately after the engine starts will prevent the airplane from rolling forward. On those airplanes that need to be started by swinging the propeller, reduction of throttle immediately after the engine starts is a safety precaution that prevents the airplane from rolling forward and seriously injuring the person who swung the propeller.

Selecting the exact throttle position, prior to starting, that will give 600 to 700 rpm is, in practice, difficult. Sometimes the setting will be accurate; sometimes it will produce an rpm of 1200 or more. Develop the habit of always closing the throttle slightly after the engine starts and then opening it until the rpm reaches the desired setting of about 700 rpm. Sometimes the engine may tend to "die" if the throttle has been closed too abruptly. Should this happen, the throttle should immediately be advanced sufficiently to keep the engine running.

After 1 minute of running in the 600 to 700 rpm range, the throttle should be adjusted as per the instruction manual to the recommended warm-up rpm which in most airplanes will be about 1000 rpm. At this rpm the engine is being warmed up gradually; there is some flow of cooling air being forced through the cowling by the propeller; and the engine oil pump is supplying adequate pressure to assure lubrication to all moving parts.

The airplane may be taxied to the take-off position during warm-up and, in fact, this procedure is recommended on all tightly cowled engines to eliminate excessive ground running of the engine. There is insufficient air flow through the engine while the airplane is on the ground to effect proper cooling. Ideally the engine warm-up should just be completed when the airplane reaches the take-off point and is in position for the pre-takeoff check and engine run up.

Procedure for Backfire During Starting

Due to overpriming, raw gasoline often collects in the lowest point of the carburetor induction system and poses a fire threat in the event that a "backfire" should occur. A backfire, as the name implies, is the burning of the combustible mixture back through the intake manifold. It commences in the cylinder when, due to abnormal conditions (that exist in starting), there is still burning within the cylinder at the moment the intake valve opens to permit a fresh charge of fuel to enter. The fresh charge ignites and burns back through the intake pipes to the carburetor area where a pool of raw gasoline, if present, could ignite and explode.

In most cases, any fire resulting from a backfire can be quickly quenched by continuing to crank the engine since this action will suck the flames and explosive vapours into the cylinder where they will do no harm. (The advantage of a well charged battery can be readily appreciated. It could be disastrous if the battery went "dead" just when it was needed to turn over the engine to eliminate an engine fire.)

Another form of "backfire" is the exhaust manifold fire. In actual fact, it is not technically a "backfire" and is therefore incorrectly named. This also is caused by overpriming and occurs when raw gasoline that was introduced into the intake manifold system by the primer passes through the cylinder and into the exhaust manifold without being burnt. In this case, a pool of raw gasoline could collect in the exhaust system and could be set afire by the hot exhaust gases which may follow starting. In most cases, starting the engine quickly will smother the flames before they can spread and do any damage.

The fact may now be appreciated that most engine fires which occur at the time of starting result from improper technique in use of the primer, throttle accelerator pump (which is the proper way to prime many engines) **or other priming devices.** Do not use the primer too far in advance of starting the engine. The fuel injected into the system gets more time to condense and run down the manifold to form a combustible pool somewhere in the system.

If an engine fire has started and it cannot be put out by getting the engine running, the fuel should be shut off and the primer locked and an external fire extinguisher or an extinguisher that is part of the engine installation should be used to put out the fire.

Running up the Engine

Always position the airplane into wind when doing an engine run up. The wind increases the flow of air through and around the engine, helping to keep the cylinders cool and preventing over-heating.

The throttle should always be opened and closed **slowly**.

1. Check oil pressure and temperature. Switch off the engine immediately if the pressure or temperature appear abnormal.

2. Check the rpm at full throttle. Throttle back to the engine speed recommended for a magneto check. Cut the switches one at a time to test each individual magneto, by moving the switch from BOTH to RIGHT, to BOTH, to LEFT, and back to BOTH. (You return to BOTH each time to allow the engine to regain its normal rpm and to clear the inoperative set of plugs in case they may have fouled with oil while their magneto switch was off.) The drop in rpm when the magnetos are operated singly as compared to the rpm when they are operated together should not be more than 75 rpm (unless the manufacturer specifies that a greater drop is allowable for that particular engine.) Rpm loss in excess of the allowable rate should be considered an indication of an ignition problem.

A check should also be made to ensure that both magnetos are properly grounded and that there is a complete loss of power when the ignition switch is turned to OFF. The grounding mechanism prevents the flow of electricity being generated in the magneto from reaching the spark plugs. A hot or live magneto is one which is not grounded. A hot magneto means that, even though the ignition switch is OFF, the engine could start if the propeller is moved even fractionally. Move the ignition switch from BOTH to OFF. If the engine continues to fire when the switch is OFF, there is a live or hot magneto. Shut off the fuel and check the magneto ground lead after the engine has stopped. The live magneto check should be conducted at **low power.** to prevent damage to the engine. To do this check at higher power means that the engine jumps from cruise rpm to zero rpm instantaneously (when the magnetos are properly grounded and do shut off when the ignition switch is moved to OFF), a procedure that can be very damaging to the crankshaft and other internal parts.

Check all instruments for proper indication while running up the engine, i.e. voltmeter or ammeter, manifold pressure gauge, fuel pressure, rpm indicator, vacuum, etc.

3. If the engine is fitted with a controllable or constant speed propeller, test its operation by moving the lever to coarse pitch. See that the required drop in rpm occurs. Return the lever to fine pitch after the test. Check the propeller feathering operation, if applicable.

4. Test the operation of the carburetor heat control by selecting "warm air". If the control is working satisfactorily, a drop in power will be noted. Check the operation of the carburetor air temperature gauge, if fitted.

5. See that the primer pump is switched off or that the manual primer is off and locked. Check for correct operation of the mixture control by moving it from full rich to full lean and back again. The engine should die when the mixture control is placed in the full lean position.

6. Idle the engine for a few moments to check the idling speed at proper working temperature.

During the running up, the pilot should check and listen intently for any sign of engine trouble. A minor adjustment on the ground may forestall a serious situation due to engine failure on take-off or in flight.

Never attempt to take off with a cold engine.

Maintenance Check of the Engine. Sometimes it is necessary to test an engine for maintenance purposes and to run it at full power. The brakes of the airplane will probably not be able to hold the airplane stationary during this procedure. It is essential, therefore, to chock the wheels and even to tie down the airplane. There are many recorded incidents of an airplane jumping the chocks when the brakes have not held, running into another aircraft or stationary installation and causing substantial property damage and personal injury. When carrying out "full power" testing of this sort, it is wise to position the airplane so that it is not in line with anything to which it could cause damage if the brakes should fail or if the aircraft does jump the chocks and move forward.

Engine Operation in Cold Weather

Operation in winter requires more care on the part of the pilot and an understanding of the effect of cold on the component parts of the power plant.

In cold weather, taking off with a cold engine and cold oil can put you in the embarrassing position of coping with a loss of power during a critical part of the take-off procedure. Aviation lubricants are heavier than the commonly used automotive engine oils and therefore require more time in warm up to obtain normal flow to function properly throughout an aircooled airplane engine.

At temperatures above freezing, a warm up of 2 or 3 minutes at 1000 rpm may be sufficient to bring engine temperatures within operating range. At temperatures below freezing, the warm up should be longer. With turbocharged powerplants, even more time is required to assure proper controller operation and to prevent manifold pressure overboost.

In cold weather, it is a good practice to rotate the propeller by hand several times before attempting to start the engine. This procedure helps to break the seal of cold congealed oil in the various parts of the engine. Be sure, however, that the magneto switches are OFF before touching the propeller. As an additional safety measure, turn the propeller backwards, rather than forwards.

Engines turn over rather sluggishly in cold weather and consequently there is a tendency to overprime to hasten the start. This practice may result in an excess of raw gasoline washing the oil from the cylinder walls, leaving them vulnerable to scoring by the piston. Excess fuel also increases the hazard of fire.

Oil. As soon as the temperature starts to dip in the fall, change to the weight of oil recommended by the manufacturer for cold weather operation. The heavier weight viscosity oil recommended for summer use simply will not flow in cold weather until it is warmed. It becomes thick and heavy and sluggish and will not flow through the very small clearances in the engine. As a result, not only will it fail to lubricate the moving engine parts during a cold start, but, in some cases, it will actually prevent the engine from turning at sufficient speed to achieve a start at all. The lighter viscosity oil recommended for cold weather operation will flow through the small clearances and will provide the essential lubrication to prevent damage during the cold start. Adequate warm-up, of course, is still necessary before take-off to ensure that the oil is flowing normally throughout the engine.

Very recently, several oil companies have developed a multi-viscosity oil which eliminates the need to change oil seasonally. This oil remains thin in extremely cold weather, but also maintains, at high outside temperatures, the same viscosity as the traditional summer grade of oil.

The crankcase oil breather should be checked regularly to assure that it is open. Moisture is a natural by-product of the heating and cooling and operation of a gasoline engine. This moisture is vented to the atmosphere through the oil breather line. In the winter time when temperatures are below freezing, it is possible for the breather line, in that portion which protrudes from the lower cowling of the aircraft, to freeze shut. Should this happen, pressure will build up in the oil system causing the oil filler cap to blow off or a seal to rupture with resultant loss of the oil supply.

To prevent such an occurrence, most aircraft require an oil breather line modification: a V shaped notch that is cut into the tubing several inches upward from the end, in that portion of the breather line that is inside the cowling where the heat of the engine prevents the line from freezing. If the end of the breather line freezes, the notch permits venting of the accumulating moisture. The notch is cut in such a way that, as long as the breather line is open, the moisture laden oil vapours are vented through the line protruding below the cowling. The moisture is vented through the notch only when the end of the breather line is blocked.

Winter Kit. As well as changing to winter grade oil, the manufacturer will have other recommendations for winterizing that should be followed. This usually involves installation of a winter kit. Among other things, a winter kit will probably include baffle plates for installation in the air inlets in order to cut down the amount of abnormally cold air directed around the engine. With baffle plates installed, the engine will be able to reach and maintain its normal operating temperatures. Intake pipes are lagged, that is, wrapped with insulating material such as felt or asbestos. This is done to ensure that the vapourized fuel in the fuel/air mixture does not condense back into its liquid form in abnormally cold intake pipes. The oil tank may also be lagged in order to keep the oil warm.

All hoses, flexible tubing and seals should be checked for signs of deterioration, cracks or hardening. They get brittle with age and vibration. Primer lines especially should be inspected for leaks around fittings to reduce fire hazard.

Ignition leads should be inspected and spark plugs cleaned and reset, or replaced if necessary. Magneto timing should also be checked for accuracy. Magneto timing will change with wear. Cold weather starts require an ignition system in optimum condition.

Fuel. Fuel tank sumps sometimes freeze during cold weather. If nothing will drain from the sump, it is highly probable that water in the fuel has turned to ice and blocked the sump. The only solution is to put the airplane into a warm hangar until the ice thaws and the water can be drained out of the sumps. The best way to prevent water condensing and turning to ice in the fuel tanks is to fill the tanks immediately after flight.

At temperatures above freezing, any water in the fuel sinks to the bottom of the tank where it can be removed with normal sump draining. However, at temperatures below freezing, minute water droplets in the fuel sometimes freeze into tiny ice crystals and remain suspended in the fuel. Normal draining of the sumps will not remove them. These ice particles then may congregate in narrow passages, such as the fuel filters, fuel selector valve, etc., and restrict fuel flow. An anti-icing additive cures the problem, but it must be used with caution and exactly according to the manufacturer's instructions. Incorrectly used, anti-icing additives may damage fuel tanks and certain carburetor parts.

Always check the fuel selector valve for proper movement since it can easily become frozen into position and impossible to move. For this same reason, do not run a tank dry before switching tanks during flight. In case the fuel selector valve has become frozen and you are unable to switch tanks, there will still be enough fuel in the tank in use to divert to a nearby airport or to have engine power for a precautionary landing.

Turbocharged Engines. Avoid overboosting turbocharged engines in cold weather take-offs. Cold air is heavier than air at a more moderate temperature and therefore power output of the engines is increased. For every 5 degrees Celsius below standard air temperature, engine power increases by 1%.

Spark Plugs. The engine doesn't start at sub zero temperatures? A common cause of this problem is moisture that has frozen on the spark plug electrodes. Remove at least one plug from each cylinder and warm them.

Battery. A battery that is not fully charged is apt to freeze at below freezing temperatures. Any battery that is allowed to sit unused will tend to discharge. In cold weather, the electrolyte (the sulphuric acid/water mix) in the battery will break down and change to water. The water may then freeze and possibly burst its case. If the airplane will not be used until the next day or for an even longer period, remove the battery and keep it in a warm place. An unused battery (in storage) should be given a slow charge (about 3 amps per hour) every 5 or 6 weeks.

Preheat. In very cold weather, preheating the engine from an auxiliary heating source is essential. In fact, engine manufacturers usually recommend preheating at any temperature below -8°C and require it at any temperature below -12°C. The heavy oils used in air cooled aviation engines (even those recommended for cold weather operation) do not flow at such low temperatures. Inadequate lubrication would result from a cold start and permanent damage, such as broken piston rings, scored cylinders and pistons, could result. Preheating the engine, as well as warming the metal parts, warms the oil and ensures that it will flow readily and lubricate all engine parts as soon as the engine starts.

When heating an engine by means of an external source, it is most important that a well insulated cowling cover be used to keep the heat within the engine thus assuring that it will warm all engine parts evenly.

Carburetor Heat. Use carburetor heat as required. In some cases, it may be necessary to use heat to vapourize the fuel, as gasoline does not vapourize readily in cold weather.

However, do not use carburetor heat in such a way that the mixture temperature is raised just to freezing level. Such action may induce carburetor ice. (See Section **Carburetor Icing**.) A carburetor air temperature gauge is a very valuable instrument during cold weather operations.

Propeller. If the airplane is equipped with a constant speed propeller, change rpm settings with the prop pitch control about every half hour to help prevent the oil from congealing in the prop dome. Congealed oil may make changing pitch impossible.

Descents. Try to avoid power off descents by starting to lose altitude far enough back from your destination that the entire descent can be conducted with power. If a rapid descent is unavoidable, use gear and flaps to induce drag so that some engine power (heat) may be maintained.

Engine Fault Finding Table

1. Engine will not start.
- **No Spark at Plugs** (plugs fouled or incorrectly set — H.T. leads crossed — switch "off" or defective switch — mag wires grounded — defective mag — wet plugs, leads or mag).
- **Weak Spark** (breaker points dirty or incorrectly set — faulty condenser).
- **Insufficient Fuel** (tanks empty — fuel cocks "off" — insufficient priming — air vents in tanks clogged — leaks or stoppages in fuel system — defective fuel pump — float needle valve sticking — altitude control open — carburetor tuned too weak — vapourization of fuel in lines of fuel injection engine caused by hot engine).
- **Flooding** (excessive priming — fuel pressure too high).
- **Cold Engine** (engine stiff to turn).
- **Water or Dirt in Fuel.**
- **Mechanical Defects** (plugs loose — cylinders scored — piston rings stuck, broken or scored — loose intake connections — poor compression and leaky valves).
- **Insufficient Cranking Speed.**
- **Incorrect Throttle Setting.**

2. Engine Misses
- **Defective Plugs** (oiled up, incorrectly set, or cracked).
- **Defective Magneto** (timing incorrect).
- **Water or Dirt in Fuel Line.**
- **Mechanical Defects** (sticking valve — crack in cylinder).
- **Low Compression in One or More Cylinders.**
- **Cold Engine or Overheated Engine.**
- **Lack of Fuel** (fuel tank vents clogged — fuel cock not fully open — lean mixture).

3. Backfire.
- **Incorrect Timing** (H.T. leads crossed — spark too far advanced).

- **Defective Carburetor** (Sticking float, needle or toggles).
- **Lean Mixture** (dirty filters — altitude control open — fuel pressure low — leaks in fuel system).
- **Water, Dirt or Air Lock** (in carburetor or fuel line).
- **Sticking Valve.**
- **Hot Engine** (carbon in cylinders — overheated valves, plugs, etc.).

4. *Loss of Power.*
- **Defective Plugs** (oiled up — incorrectly set).
- **Incorrect Timing** (magnetos defective — not properly synchronized — faulty wiring — breaker points burned or pitted).
- **Insufficient Fuel** (leaks or stoppages in the fuel system — defective fuel pump — float needle valve sticking — weak mixture — throttle or carburetor throttle valve not fully open — altitude control open — carburetor iced up).
- **Excessive Fuel** (rich mixture — fuel pressure too high — air intake restricted).
- **Improper Lubrication** (wrong grade of oil — incorrect oil temperature — leaks — low oil pressure — dirty filters).
- **Carburetor Heat Used Unnecessarily.**
- **Carburetor Icing.**
- **Excessive Cylinder Temperatures** (engine tends to seize).
- **Loss of Compression** (plugs loose — piston rings worn — valve seats worn — clearances out — springs weak — valves warped, stuck, or seating improperly — incorrect timing).
- **Poor Fuel** (or water or dirt in fuel).
- **Engine Too Cold, or Overheating.**
- **Engine Needs Overhaul.**
- **Pre-ignition.**
- **Internal Friction.**
- **Detonation.**

5. *Rough Running.*
- **Propeller or Engine Bearers Loose or Out of Balance.**
- **Bent Crankshaft.**
- **Defective Magnetos** (timing, wiring, etc.)
- **Incorrect Mixture** (too rich or too lean).
- **Wrong Valve Clearance** (or sticking valve).
- **Detonation, or Pre-ignition.**
- **Defective Plugs.**
- **Cold Engine.**

6. *Engine Stops.*
- **Lack of Fuel** (tanks exhausted — gas cocks accidentally knocked closed — failure to switch over to full tank when the one in use runs dry — water or dirt in fuel — fuel tank air vents clogged — filters clogged — air lock in fuel system — mixture control accidentally moved to idle cut-off).

7. *Engine continues to run with switches off.*
- **Defective Switch or Wiring.**
- **Pre-ignition.**
- **Overheating.**

8. *Engine will not take throttle.*
- **Lean Mixture.**
- **Cold Engine.**

9. *Blue smoke from exhaust.*
- **Worn or Stuck Piston Rings.**

10. *Black smoke or long red flame from exhaust.*
- **Excessive Rich Mixture.**

11. *Engine overheating.*
- **Detonation.**
- **Insufficient Oil Supply.**
- **Stuck Rings.**
- **Lean Mixture.**

12. *Excessive oil consumption.*
- **Blocked or Frozen Breather** (causing excessive pressure).
- **Sticking Valves.**
- **Worn Piston Rings.**

13. *Long yellow flame from exhaust..*
- **Lean Mixture.**

Jet Propulsion

It is **not** the blast of hot gases out the exhaust of a jet engine that provides the driving force, but rather the **reaction** to the exhaust jet. Jet propulsion is based on Sir Isaac Newton's Third Law of Motion, namely that every action will have an equal and opposite reaction. (Like the "kick" you get from a shotgun.)

Imagine a sphere (Fig. 45) suspended by a cord and filled with air under pressure. The sphere will remain motionless, because the pressure is exerted evenly against all sides. The pressure at A is balanced by the equal and opposite pressure at B. The pressure at C is balanced by the equal and opposite pressure at D.

Now punch a hole in the sphere allowing air to escape at A. The pressures at C and D are still equal and opposite. The pressure at B pushes the sphere forward. The force exerted at B is equal and opposite to the momentum of the air escaping at A. A is the Jet. B is the reaction to the jet, which is known as thrust.

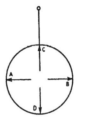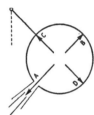

Fig. 45. Principle of Jet Propulsion.

The Ram Jet

The simplest form of jet engine is the **ram jet**. This is little more than a streamlined "stove" in which fuel mixed with air is burned and exhausted out the "stovepipe" to produce a thrust reaction. The air which enters the intake is "rammed" into the combustion chamber by the forward speed of the airplane (Fig. 46). As this high speed ram air passes through the diffuser section, as indicated in Fig. 46, the speed of the air decreases while the pressure increases. This is brought about by the increase in cross section of the diffuser (Bernoulli's Principle for Incompressible Flow). The difference in pressure is known, in a ram jet engine, as the **pressure ratio**. (When we get to the turbojet engine, pressure ratio has a slightly different meaning.)

The slowed down ram air is mixed with fuel introduced by the fuel injectors at C. This volatile mixture of fuel and air, once ignited, burns, expands and roars out of the exhaust nozzle at tremendous velocity at E. This creates a **thrust reaction** similar to that which we observed in the case of the punctured balloon. The difference between the high pressure in the forward part of the engine and the low pressure towards the rear, imparts momentum to the jet stream rushing out of the nozzle. This momentum has its equal and opposite reaction in the form of thrust, the force which drives the engine ahead and propels the airplane forward.

Initially, the engine is started by a spark, or igniter plug, the same as a piston engine, but as soon as the fuel is ignited, the ignition is switched off and the fire burns continuously.

The ram jet is extremely simple and particularly suited to high altitude and high speed. First, however, it must be carried aloft by some other means of power and impelled at sufficient forward speed to enable it to get started. Its practical use appears limited to missiles, supersonic airplanes, etc.

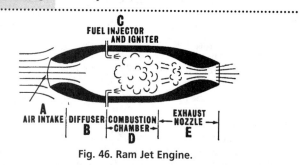

Fig. 46. Ram Jet Engine.

The Turbojet Engine

In the turbojet engine, the intake air is compressed and rammed into the combustion chamber by a compressor, which is driven by a turbine wheel. The turbine in turn is driven by the momentum of the hot exhaust gases.

The basic principle of the turbojet engine, together with some of its construction features, is illustrated in Fig. 47, the cutaway drawing of the General Electric J-85 engine.

The engine is started by a starter motor. Air enters the **air inlet duct** (1) past the **inlet guide vanes** which guide the flow of air at the right angle to the first stage compressor rotor. The air is compressed by the **compressor** (2). The compressor illustrated has 8 rows of **rotor blades** one behind the other, axially, which rotate around a shaft, and is therefore referred to as an **eight-stage axial flow compressor**. Between the rows of rotor blades are rows of stationary **stator vanes** (2a). They are fastened to the engine case and form a complete ring around the inside of the engine casing. Their purpose is to convert the velocity imparted by the rotor blades into pressure and to direct the air at the proper angle to the next set of blades where the process is repeated.

The **mainframe section** (3) serves as the main structural component. In many turbojet type engines, the section behind the compressor functions as a **diffuser**.

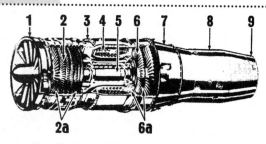

Fig. 47. General Electric J-85 Turbojet Engine.

The air from the compressor enters the **burner section** (4) through holes in the burner skin. (These also help to reduce the intense heat of the burning gases.) **The igniter plug(s)** and **fuel nozzles**, located within the burner section, ignite the gases initially and admit the fuel which burns with the compressed air to produce terrific heat. The heated gases are allowed to drop in pressure and accelerate through rows of **nozzle guide vanes,** the whole assembly of vanes being generally called the **turbine nozzle(s).** Whereas in the compressor the vanes convert velocity to pressure, in the turbine they convert pressure to velocity. The rotational velocity is arrested by the blades or buckets of the **turbine wheel** (6), thereby producing shaft horsepower to drive the compressor.

Because of the very high temperature of the gases entering the turbine, only a part of the pressure built up in the compressor is needed to make the turbine drive the compressor. That is, there is pressure left over after the turbines to make the gas do more work downstream.

The **exhaust cone** (7) collects and straightens the gas flow. The length of the **exhaust duct**, or **tail pipe** (8) determines the position of the **exhaust nozzle** (9), an orifice at the rear of the duct, whose size determines the speed of the gases as they emerge from the engine. Increasing the speed of the gases increases their momentum and hence the amount of thrust produced by the engine.

The **gas generator** consists of the parts of the engine which produce gas at a higher pressure and temperature than at the engine inlet. This includes the compressor, the diffuser, the burner and the turbine. The term "gas" has nothing to do with gasoline. A gas turbine engine is one which is operated by a gas, as distinct from one operated by steam vapour or water.

The engine is controlled by the amount of fuel fed to it. Cutting off the fuel stops the engine.

The fuel used is kerosene. Any hydrocarbon can be used in a jet engine as long as it can be blown through a nozzle and burned.

Why do some jet engines leave black smoke trails? Usually, the reason is poor combustion. The fuels used in gas turbine engines consist almost entirely of hydrocarbons, that is, compounds of hydrogen and carbon. When the fuel is burned, the hydrogen and carbon combine to form, ideally, water and carbon dioxide. If, however, the combustion is not complete, particles of carbon are formed which are visible as black smoke.

The engine is lubricated by a pressure spray or flow of oil which is fed to the three bearing assemblies. Synthetic oil is the type most widely used on turbojet engines.

Temperatures in the combustion chambers and around the turbine in a jet engine run as high as 700°C. Because of the heat of the jet stream, a person should not stand nearer than 150 to 200 feet directly back of the tailpipe.

The jet engine weighs roughly one quarter the weight of a piston engine and propeller of similar power output. The frontal area of the axial flow jet engine is approximately one tenth that of the frontal area of a radial piston engine.

The jet engine's power is rated in **pounds of static thrust** instead of horsepower. At a speed of 325 knots, one lb. of thrust equals roughly one horsepower. At 525 knots at sea level, the jet engine's 5000 lb. of static thrust is equivalent to 8000 true thrust horsepower. Assuming 66% propeller efficiency for the piston engine, this would be equivalent to 12,000 hp.

Thrust is simply force measured in pounds (or kilograms).

Like the piston engine, the jet engine's power depends upon the weight of air it can consume. As altitude increases, the density of the air (i.e. its weight) decreases. It is for this reason that turbochargers are used to maintain the power of piston engines at altitude. The turbine engine also suffers from loss of power at altitude, but this is compensated for by the lower drag (resistance to forward speed) and the "ram" effect of the jet's higher speed, which rams more air into the engine to produce a "turbocharging" effect.

The **pressure ratio** of a gas generator usually means the ratio of pressure at the compressor exit to that at the compressor inlet.

Where the expression "engine pressure ratio (EPR)" is used, it means the pumping pressure ratio of the whole gas generator, i.e. ratio of pressure at exit of the compressor drive turbine to the compressor inlet.

Pressure ratios (of the compressor) at maximum rpm are mostly in the range 6 to 1 to 15 to 1. The EPR is usually 1.5 to 2 to 1.

Turboshaft and Turboprop Engines

In addition to jet propulsion, the gas turbine engine can also be used (in lieu of the reciprocating piston engine) to drive a propeller, in which case it is known as a **turboprop** engine, or to drive a helicopter rotor, in which case it is known as a **turboshaft** engine. One of the simplest of several different methods presently in use is the **fixed shaft** method. In this arrangement, the turbine drives both the compressor and the shaft directly.

In some gas turbine engines two turbines are used, one to drive the compressor and the other to drive the shaft. When these turbines are mechanically free of one another, the engine is called a **free turbine** engine. The Pratt and Whitney Aircraft of Canada PT6 engine illustrated in Fig. 48 is an example of a free turbine turboprop engine.

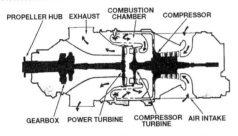

Fig. 48. PT6 Free Turbine Turboprop Engine.

The arrows indicate the gas flow through the PT6 engine. Air enters the air intake and is compressed by a three axial stage and single centrifugal stage compressor before it enters the combustion chamber. The hot gases from the combustion chamber drive the compressor turbine. They then drive the separate power turbine, which in turn drives the propeller shaft. The gear box reduces the propeller speed to a fraction of the speed at which the turbine drives the drive shaft.

The By-Pass Engine

The propeller generates thrust by imparting a relatively small acceleration to a relatively large mass of air. The jet generates thrust by imparting a relatively large acceleration to a relatively small mass of air. For low air speeds, the propeller is most efficient. For high airspeed, the jet excels because the jet engine produces its greatest propulsive efficiency at high airspeeds. At low airspeeds, its thrust efficiency falls off and its fuel consumption becomes excessive. The by-pass engine was designed to improve the efficiency of the jet engine at low airspeeds. Part of the air from the low-pressure compressor by-passes the high-pressure compressor, the combustion chambers and the turbine. This relatively low speed air is mixed with the high speed hot gases from the combustion chambers in the tail pipe, thus reducing the overall velocity of the jet stream. Pure by-pass engines are no longer in production, but the principle is used in the turbofan engine.

The Turbofan Engine

The turbofan engine (Fig. 49) is the practical application of the principle of the by-pass engine. It was developed to incorporate the best features of the turbojet, which has excellent cruise speed capability, and the turboprop, which has short field take-off capability.

The turbofan is a turbojet with a fan attached at the forward end ahead of the compressors. The fan diameter is much less than that of the propeller of a turboprop engine but it contains many more blades and moves the air at a greater velocity. The diameter of the fan is, however, greater than that of the engine proper, so that some of the air passing through the blades is accelerated backward on the outside of the engine, providing extra thrust. This extra thrust is available for increased take-off, climb and cruise performance, reduced fuel consumption and hence better payload/range.

Axial flow, multiple compressor turbofan engines are widely used to power large aircraft, providing efficient, quiet and economical operation. They are also used as the powerplant on some of the small general aviation jet airplanes.

In recent years, turbofan engines have been improved by changes in the design of turbine and compressor blades and vanes, by use of improved alloys, by improving internal engine cooling systems and by modification of fan blade design. The engines are, as a result, more fuel efficient and develop more thrust for their weight.

The fan of the turbofan engine includes fixed inlet guide vanes and fixed outlet guide vanes. Air from the fan is divided into two airflows by the two stage fan: the primary air is directed to the main engine core, the secondary air provides by-pass thrust. The fan turns at the same rate and is driven by the same turbine section as the low pressure compressor.

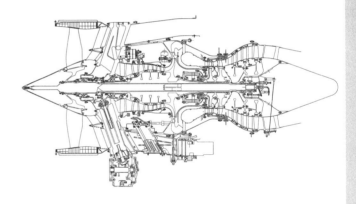

Fig. 49. Pratt & Whitney 305 Turbofan Engine.

The low pressure compressor (N_1) usually consists of eight stages, including the two fan stages, and is driven by the last three stages of the turbine. The high pressure compressor (N_2) consists of seven stages and is driven by the first stage of the turbine. The casings that house the compressors include automatic bleed valves which open to prevent compressor surge and stall. A diffuser changes the direction of the high velocity, high pressure air and directs it to the combustion chamber. Fuel is supplied through a fuel manifold inside the forward end of the main combustion chamber. The air for the turbofan core engine is compressed, ignited and discharged in the same manner as in a turbojet engine. There are three combustion chambers. The hot gases from the individual combustion chambers are directed to the turbine rotor which drives the compressors.

The turbofan engine is equipped with thrust reversers. The effect of the reverser is to reverse the direction of the exhaust and fan discharges and thus cause the engine to deliver reverse thrust.

There are three classifications of turbofan engines; low by-pass, medium by-pass and high by-pass. In the low by-pass engine, the fan and compressor sections are utilizing approximately the same mass airflow and have a by-pass ratio of 1 to 1. In the medium by-pass turbofan engine, the fan is slightly larger in diameter and provides a by-pass ratio of 2 or 3 to 1. The high by-pass turbofan engine with fan ratios of 4 to 1 and up has an even larger diameter fan with approximately 80% of the thrust produced by the fan and 20% by the core engine.

Thrust Reverser

Propeller driven airplanes, as we already have learned, are capable of reverse thrust by changing the pitch of the propeller blades in order to reverse the airflow and use the power of the engine to stop the airplane on the ground. Turbine airplanes are also capable of reverse thrust, using engine power as a deceleration force.

There sre several methods of obtaining reverse thrust in jet engines.

1. *Clamshell type doors,* operated pneumatically, reverse the exhaust gas stream. The ducts through which the exhaust gases are deflected remain closed during normal operation. When the pilot selects reverse thrust, the doors rotate to uncover the ducts and close the normal gas stream exit. Baffles, called cascades, direct the exhaust gas stream forward to achieve reverse thrust. When the reverse is not in use, the clamshell doors retract and nest neatly around the engine nacelle.

2. *The bucket target system* uses bucket type doors to reverse the hot gas stream. Thrust reverser doors are actuated by a conventional pushrod system, hydraulically powered.

3. *The cold stream reverser system* is activated by an air motor or by hydraulic rams. When the engine is operating in forward thrust, the cold stream nozzle is covered by blocker doors (flaps) and by the movable cowling. On selection of reverse thrust, the movable cowling moves rearward, the blocker doors fold and blank off the cold stream nozzle, diverting the airflow through the cascade vanes.

For aerodynamic reasons, the gas cannot be directed in a completely forward direction. A discharge angle of approximatley 45 degrees means that the effective power in reverse thrust is proportionately less than the power in forward thrust for the same throttle setting.

The safety features incorporated into these systems ensure that the reverse thrust lever used to select reverse thrust cannot be moved to the reverse thrust position unless the engine is running at a low power setting. Once reverse thrust is selected at low power, the pilot will then open up the engine to high power to use the full power of the engine to decelerate. However, if the reverser is not in the full and correct reverse thrust position, high engine power cannot be selected. Operation of the thrust reverse system is indicated to the crew by a series of lights. Baffles, called cascades, deflect the jet blast forward to achieve the reverse thrust.

A thrust reverser must not affect engine operation either when the reverser is operating or when it is not. It must withstand high temperatures and be mechanically strong but light in weight. When not in use, it should retract into the engine nacelle in a streamlined design. A thrust reverser should be able to produce in reverse at least 50% of the full forward thrust of which the engine is capable.

For N speeds and other terminology applicable to turbine engines, see the Glossary.

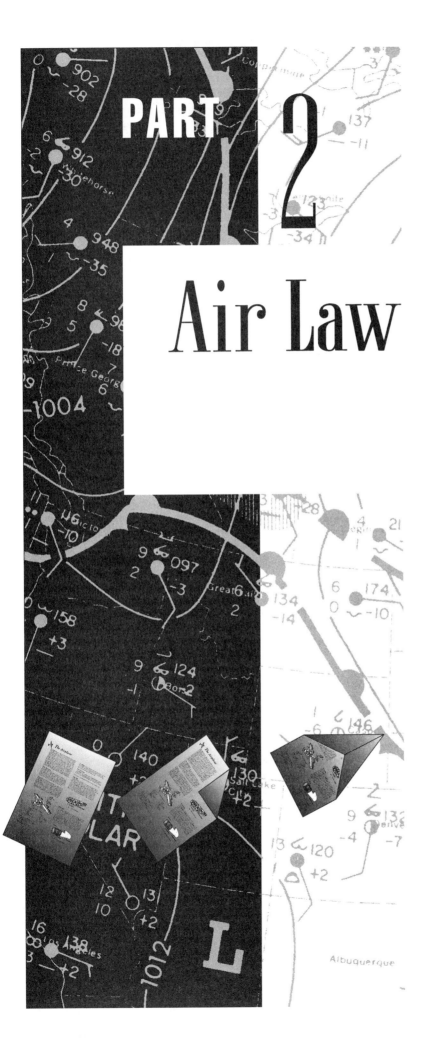

PART 2

Air Law

Aeronautical Rules & Facilities

Across the continent, there is an intricate system of aeronautical facilities designed to facilitate the efficient movement of air traffic. It won't be long before a new pilot with his new license will be anxious to take off from his home and very familiar airport to visit some of the hundreds of other airports throughout the country. In order not to be intimidated by new, different and bigger airports, an understanding of the basics of airport design is an essential part of a pilot's training. To reach those other airports, he will have to cope with the intricacies of controlled and uncontrolled airspace, airways, control zones, cruising altitudes, etc., etc. This chapter is designed to introduce the pilot to aerodromes, to the airspace system and to the regulations and procedures of flight in the Canadian airspace.

Aerodromes

The many aerodromes across the continent vary widely in the facilities they offer to the itinerant pilot. The large metropolitan airports with several runways, complicated patterns of taxiways and sophisticated lighting systems seem very complex in comparison to the single strip sod runway that constitutes the aerodrome of many small communities throughout the country. There are, however, certain standard features that apply to every aerodrome, no matter how large or how small.

The term, **aerodrome,** is defined as any area of land or water designed for the arrival, departure, movement and servicing of aircraft and includes buildings, installations and equipment there situated.

The term, **airport,** is defined as any aerodrome in respect of which a certificate is in force. Some airports are designated "international airports" to support international commercial air transport. An airport certificate testifies that the airport meets airport certification safety standards.

Aerodromes are classified as (1) certified for public use, (2) certified for private use, (3) registered or (4) military. The majority of aerodromes in Canada are listed in the **Canada Flight Supplement** or the **Water Aerodrome Supplement.** The classification of the aerodrome is listed with the information on the aerodrome operator in these publications.

An airport, that has been issued a **Public Use Certificate (Pub)** is open to all aircraft. An airport that has a **Private Use Certificate (Pvt)** is private property and is not open, except in emergency, to itinerant aircraft. The permission of the owner is required prior to use. Aerodromes that are not certified may be **registered (Reg)** for the purpose of publishing aeronautical information in the **Canada Flight Supplement.** Access to a registered aerodrome may be public or limited to private use. Since the information on the aerodrome is provided by the owner/operator, it is advisable to verify the status and condition of the aerodrome prior to use. A **military (Mil)** aerodrome may be used by civil aircraft only if prior permission is obtained or in an emergency. In the event of an in-flight emergency, any aerodrome in Canada may be used at the discretion of the pilot.

The landing of any aircraft should be reported to the airport operator by the pilot of the aircraft.

Those parts of an airport or aerodrome used for the surface movement of aircraft (**movement area**) include the maneuvering areas and aprons. The **maneuvering area** is comprised of those parts of the airport intended for the taking off and landing of aircraft and the movement of aircraft, in other words, the runways and taxiways. The **apron** is the area intended for the loading and unloading of passengers and cargo, the refuelling, servicing, maintenance and parking of aircraft and the move-ment of aircraft, vehicles and pedestrians necessary for such purposes.

Runway Numbering

Airport runways are numbered for the purpose and convenience of identification. The number a runway is assigned corresponds to its magnetic bearing, rounded off to the nearest 10°. For convenience, the last zero is omitted. Thus Runway 09 would be the runway which runs from west to east (090° magnetic). Since runway numbers are taken to the nearest 10° magnetic, a runway which lies 018° magnetic would be considered as being 020° magnetic and would be numbered 02.

The runway number is displayed at the approach end of each runway. A single runway would, therefore, have different numbers at each of its two ends. These numbers would be 180° apart, or reciprocals of each other. For example, runway 09 would be numbered 27 at the other end; runway 02 would be 20.

While airports throughout most of Canada are assigned numbers based on their magnetic bearings, those in the Northern Domestic Airspace are assigned numbers based on true bearings. (The Northern Domestic Airspace, defined in Section **The Canadian Airspace System,** comprises the northern and arctic areas of Canada.) In this northern area, compass indications are unreliable because of the nearness of the north magnetic pole and therefore true bearings are considered more reliable for runway numbering purposes.

At larger aerodromes, there are sometimes two or more runways parallel to each other. The letter R below the number of a runway indicates that it is the right of a pair of parallel runways. An L indicates the left runway. The letters are always assigned as viewed from the direction of approach. Triple runways are designated L, C or R and would be called, for example, three six left, three six centre and three six right.

Runway Markings

The perimeters of unpaved runways are required to be delineated with frangible, weatherproof markers that are clearly visible both on the ground and from the air. These markers, that may be either pyramid shaped or cone shaped, are evenly spaced at intervals of not more than 300 feet along the sides of the runway. Additional markers are placed at the four corners of the runway at right angles to the centre line.

At certified aerodromes, these markers are painted in alternate stripes of international orange and white; at other aerodromes, they are solid international orange.

Evergreen trees about 4 or 5 feet high are sometimes used to mark the perimeters of snow covered landing strips.

Paved Runways

In addition to the number that is painted in large white figures at the approach end of the runway, lines are usually painted down the centre of the runway to help the pilot align the airplane during landings and take-offs.

Sometimes the entire paved portion of the runway is not usable as the landing area. In this case, the **threshold** of the usable portion is displaced and its position is marked by a line across the runway with arrowheads pointing to it. Centre line arrows also point to the threshold line. A **displaced threshold** usually is made necessary because of obstacles at the end of the runway that require additional clearance during the approach. The paved area behind the displaced threshold can be used for taxiing, the landing rollout and the initial take-off roll. The displaced threshold portion of the runway may be used for landing but it is the responsibility of the pilot to ensure that the descent path safely clears all obstacles. A **relocated threshold** is made necessary if a section of the runway is closed either temporarily or permanently. The closed portion is marked with closed markings and is not to be used for taxiing, for the initial take-off roll or for the landing rollout. A **turnaround bay** may be provided at the threshold end of a runway that is not directly served by a taxiway. It is a widened area which can be used for turnaround but it does not give sufficient clearance from the runway edge for holding while other airplanes use the runway.

Sometimes runways are constructed with lengthy overrun and undershoot areas. Such **pre-threshold areas** are paved but are non-load bearing and are marked for their entire length with yellow chevrons **Stopways** at the end of runways in the direction of take-off are prepared as suitable areas in which an airplane can be stopped in the case of an abandoned take-off. A stopway is marked over the entire length with yellow chevrons.

Sometimes, special threshold markings are included as part of the runway marking scheme. These consist of 2 sets of four or more bars parallel to the sides of the runway. They serve to indicate that a non-precision approach facility such as a VOR serves the runway and that the runway can be used for landing under instrument conditions as well as VFR.

Runways for which there are precision approach facilities, such as an ILS, have additional landing zone markings at 500 foot intervals down the length of the runway. Touchdown zone stripes, consisting of 2 sets of 3 parallel white stripes are painted at a distance of 500 feet from the threshold of the runway; fixed distance markers, either 2 thick parallel bars or 2 sets of 5 parallel stripes, indicate 1000 feet; further sets of stripes indicate 1500, 2000, 2500 and 3000 feet.

Runway marking varies depending on runway length and width. A description of the full range of runway markings is published in the Transport Canada publication **Aerodrome Standards and Recommended Practices.**

Taxiways

Taxiways are an important part of the aerodrome facilities. They enable airplanes to move to and from runways without interfering with traffic taking off and landing.

Yellow lines are usually painted down the centre of taxiways. A broad yellow line across it indicates the end of the taxiway where an airplane must "hold short" until ready for take-off. At a controlled airport, the airplane must hold at this position until receiving clearance from the tower to enter the runway itself for take-off. If no taxi holding position is established, an airplane should hold at sufficient distance from the edge of the active runway so as not to create a hazard to arriving and departing traffic (200 feet is the recommended distance.)

An airplane may not cross an active runway at a controlled airport without first obtaining clearance to do so. The broad yellow line indicates the position at which it must hold until clearance is received.

At the large airports where there are quite a number of taxiways, they are assigned letters for identification (e.g. A) and would be referred to as taxiway alfa.

Airside Guidance Signs

Airside guidance signs are intended to provide direction and information to taxiing aircraft and to assist in the safe and expeditious movement of aircraft. Airside guidance signs are illuminated at airports that are used at night.

Operational guidance signs provide direction or facilities information and have black inscriptions on a yellow background. The inscriptions incorporate arrows, numbers, letters or pictographs. A **location sign** is used to identify a taxiway and never contains arrows. A **direction sign** is used to identify intersecting runways and will contain arrows to identify the angle of intercept. Direction signs are normally used in conjunction with location signs. A **runway exit sign** identifies a taxiway exiting a runway. Some airports provide rapid exit taxiways which are angled at 30° to the runway to facilitate the rapid clearance of airplanes from the active runway. **Destination signs** provide directions to certain facilities on the airport. Other signs may include information on VOR and DME checkpoints, parking areas, etc.

Fig. 1. Airside Direction Sign.

Mandatory instruction signs are used to identity **holding** positions beyond which pilots must have ATC clearance to proceed. **A holding position sign** is installed at all taxiway to runway intersections and shows the designation (i.e. number) of the departing runway. At uncontrolled airports, pilots are required to hold at points marked by these signs until they have ascertained that there is no air traffic conflict. Mandatory instruction signs are red with reflective white lettering and symbols.

Aerodrome Markings

Certain ground markings indicate the status of some aerodromes and pilots are required to comply with these markings.

A large cross, either white or yellow and at least 20 feet in length, displayed at each end of a runway or taxiway indicates that that runway or taxiway is unserviceable (Fig. 2). For night operations any unserviceable portion of a runway is closed off by placing red lights at right angles to the centreline across both ends of the unserviceable area. In addition, the runway lights for the unserviceable area should be turned off.

Fig. 2. Aerodrome Markings.

If an unserviceable portion of any maneuvering area or taxiway is small enough that it can be bypassed by an airplane with safety, red flags are used to outline the area of unserviceability. At night, the area is marked by steady red lights. Sometimes, in the interest of safety. one or more flashing red lights may be used as well.

Wind Indicators

The runway in use is determined by the wind direction since airplanes, during landing and take-off, operate most efficiently and safely by flying as directly into the wind as possible. All airports are therefore required to display a device that indicates which way the wind is blowing. There are several types of wind indicators, as depicted in Fig. 3.

The **windsock** or **wind cone**, (on the left in Fig.3), the old reliable good samaritan to homecoming airmen since World War I, is the traditional symbol of flying fields the world over. The wind always blows in the big end and out the small end. The velocity of the wind is indicated by the amount of extension of the windsock. A standard Transport Canada windsock will fly perfectly horizontal in a wind of 15 knots or more but will hang at about 30 degrees below the horizontal in a wind of only 6 knots. A fluttering windsock indicates gusty conditions. At night, the windsock is illuminated.

The **tetrahedron,** or wind T, is designed like an arrow whose small end points into the wind.

Aerodromes with runways greater than 4000 feet in length will have a wind direction indicator located at each end of the runway. Where the runways are less than 4000 feet in length, a single wind indicator will be centrally located about mid way along the runway.

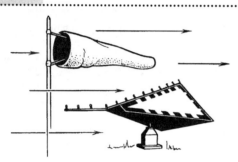

Fig. 3. Wind Direction Indicators.

Aerodrome Lighting

For night operations, **landing and take off areas** (i.e. runways) are indicated by two parallel lines of white lights that are visible at least 2 miles in all directions. Low intensity white lights are sometimes installed down the centre line of runways. Some registered aerodromes may use reflective markers in place of lights to mark the edges of the runway. They must be capable of reflecting aircraft landing lights from a distance of 2 NM.

The **runway threshold** is indicated by green lights. These lights when viewed from the back are red and indicate the end of the runway for traffic landing and taking off when the other end of the runway is active.

The edges of **taxiways** are marked by blue lights. Sometimes green lights are installed down the centre line of taxiways. Clearance bars, which consist of at least 3 flush mounted, unidirectional yellow lights showing in the direction of approach, are sometimes installed in taxiways to indicate a specific holding position.

At public use aerodromes, runway, taxiway and approach lights (see below) are normally operated on an "as required" basis and are turned on about 5 minutes before the ETA of the approaching airplane. At private use aerodromes, lights may be available only during limited hours or by prior request. In some cases, the aerodrome lights can be activated by radio control from the aircraft. **Aircraft radio control of aerodrome lighting(ARCAL)** is effected via the VHF transmitter in the aircraft by keying the microphone a given number of times within a specified number of seconds. Each activation will illuminate the lights for a given period (usually 15 minutes). Pilots are advised to key the activating sequence even if the lights are on, in order to have the full 15 minutes available for their approach. Specific information about the ARCAL installation at a particular aerodrome is listed in the **Canada Flight Supplement.**

An **airport beacon** helps a pilot to locate an airport amidst all the other confusing ground lights of a community. The beacon is a white light, visible for about 10 n. miles on a clear night, that rotates at a constant speed producing highly visible light flashes at regular intervals of about 2 or 3 seconds. The airport beacon usually operates continuously during night time hours. However, if the airport is located near the centre of an urban area where other rotating types of lights would detract from the effectiveness of the beacon on the airport or if the airport is easily recognizable because of other visual aids, the operation of the beacon may be waived. The airport beacon is not operated during the day. Some rotating beacons have been replaced with strobe type beacons which flash about 30 times per minute.

Obstruction lights are used to mark tall buildings and towers that might be flight hazards. These may be red lights that are either steady or flashing or they may be flashing white strobe lights.

Approach Lights

Many airports have an installation of approach lights that extends from the centreline of the runway back along the approach path. They help the pilot to align his airplane with the runway. There are various types of approach light systems. Most are medium or high intensity lights. Some are steady; some flash in sequence towards the threshold.

The installation of approach lights for runways for which there are precision instrument approach systems may, for example, include, in addition to the green threshold lights, a bar of red lights (at about 200 feet) that indicates the imminence of the threshold. Rows of 5 white lights are spaced at 100 foot intervals commencing at 300 feet from the threshold and extending back for a distance of 3000 feet. A distance marker (a long bar of white lights) indicates 1000 feet from the threshold.

Visual Approach Slope Indicator System (VASIS)

The visual approach slope indicator system is a lighting system designed to help a pilot maintain a correct glide path on the approach to the runway. VASIS installations are being phased out and will eventually be replaced by the PAPI system.

The system consists of a set of upwind and a set of downwind lights. Each light unit projects a beam of white light in its upper part and a red light in its lower part. The light units are so arranged that when an airplane is approaching the runway on the proper approach slope, the downwind lights will appear white and the upwind lights will appear red (Fig. 4).

If the airplane is below the proper approach slope, both sets of lights will be red. If it is above the approach slope, both sets of lights will be white (Fig. 4).

Two Bar Vasis V1, V2

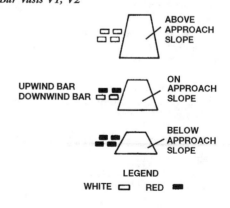

Fig. 4. VASIS Approach Lights.

The lights are so arranged that the approach slope angle is about 2.5° to 3° and if the pilot maintains this approach slope, the airplane is assured of safe obstacle clearance and will land between the two sets of lights. The VASIS provides safe wheel clearance over the runway threshold and also ensures that, if the airplane touches down between the two sets of lights, the available runway length is used to full advantage.

In certain atmospheric conditions, the white lights sometimes appear yellowish, orange or brown. The red lights, however, are not affected and the principle of colour differentiation is still applicable.

At some airports used by the very large jet airplanes, a 3 bar VASIS has been adopted to accommodate the greater eye to wheel heights of these airplanes. The third bar is positioned farther down the runway and is, in effect, an extra upwind set of lights. Pilots of small airplanes may ignore the most distant set of lights during an approach to a runway equipped with the 3 bar VASIS.

A variation of the VASIS called the T-VASIS is installed at a few airports. It eliminates the requirement to discriminate colour except in a dangerously low approach when the lights will appear red. The T-VASIS installation consists of 10 lights on each side of the runway; 3 lights in a row parallel to the runway, 4 lights in a bar at right angles to the runway and another 3 lights in a line beyond the bar. On the proper approach slope, only the bar of 4 lights is visible. On a too high approach, the 3 upwind lights also are visible producing an inverted T. On a too low approach, the 3 lights downwind of the bar combine with the bar lights to show an upright T.

Precision Approach Path Indicator (PAPI)
The PAPI system has been developed to replace VASIS which is due to be phased out. PAPI consists of four light units installed on the left side of the runway in the form of a wing bar.

If the airplane is flying the proper approach slope, the two units nearest the runway show red and the two units farthest from the runway show white. If the airplane is above the approach slope, 3 or all 4 units will show white (depending on the degree of displacement from the approach slope). If the airplane is below the approach slope, 3 or all 4 will show red (Fig. 5).

To accommodate the greater eye to wheel heights of large jet airplanes and to ensure safe wheel clearance over the runway threshold, the position of the PAPI light bar in relation to the threshold can be varied.

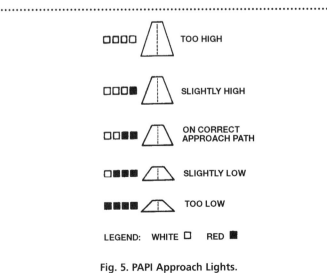

Fig. 5. PAPI Approach Lights.

Aerodrome Traffic Procedures
A pattern for traffic movement has been established for use at all aerodromes. It is called a **traffic circuit** (in the U.S., it is called a traffic pattern) and it expedites and separates airplanes using the same aerodrome.

It is the responsibility of every pilot, for safety and efficiency, to learn and follow the proper traffic procedures when coming in to land at an aerodrome.

The following definitions apply to portions of the traffic circuit:

The **upwind side** is the area on the opposite side of the landing runway from the downwind leg. Approach should be made into this area at or above circuit height.

The **circuit joining crosswind** is a corridor, lying within the airspace between the centre of the landing runway and its upwind end, linking the upwind side and the downwind leg.

The **downwind leg** is a flight path, opposite to the direction of landing, which is parallel to and at a sufficient distance from the landing runway to permit a standard rate-one turn to the base leg.

The **base leg** is a flight path at right angles to the direction of landing and sufficiently downwind of the approach end of the landing runway to permit at least a 1/4 mile final approach leg after completion of a standard rate-one turn to final approach.

The **final approach leg** is a flight path in the direction of landing, commencing at least 1/4 mile from the runway threshold, wherein an airplane is in line with the landing runway and descending towards the runway threshold.

The Traffic Circuit at Uncontrolled Airports
An uncontrolled airport is one at which no tower is in operation. (In the United States, it is called a non-tower airport.) Some airports have towers which operate only during specified hours and these airports are considered to be uncontrolled during those times when the tower is shut down.

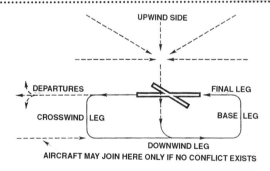

Fig. 6. The Traffic Circuit at Uncontrolled Airports.

Note: Standard Left Hand Circuit. If a right hand circuit is designated, the opposite of this diagram is applicable. Circuits are normally flown at 1000 ft. AAE.

Where no mandatory frequency procedures are in effect, you should approach the traffic circuit (Fig. 6) from the upwind side of the runway and enter crosswind at circuit height. Taking due account of other traffic, join the circuit on the downwind leg. You may join the circuit directly on the downwind leg provided you have ascertained without doubt that there is no conflict with other traffic. Where mandatory frequency procedures are in effect and traffic advisories available, you may approach straight in or at 45° to the downwind leg and join the circuit at circuit altitude or you may approach straight in to the base or final approach legs. However, you must be alert for other VFR traffic entering the circuit at these positions and for IFR traffic doing straight in or circling approaches. (In the United States, there is only one recommended entry to the traffic pattern at a non-tower airport, a 45° entry to the downwind leg at the midpoint of the runway at pattern altitude.)

When in the vicinity of an uncontrolled airport, it is imperative that you be especially alert for other traffic and exchange information by communicating on the MF or ATF (see below). All radio equipped aircraft should monitor the designated frequency and follow reporting procedures.

When approaching a strange airport, you should enter the airport zone throttled back to slow speed and letting down in a gradual descent. Never dive into the area from a height or out of clouds.

Circuit height should be reached before entering the traffic circuit.

In joining the circuit, always take care to avoid cutting off other airplanes and overtaking the airplane ahead of you when approaching to land.

If it is necessary to cross the airport prior to joining the circuit (to study an unfamiliar airport, to determine the runway in use, or for any other reason), the cross over should be done at least 500 feet above circuit height and descent to circuit altitude should be made on the upwind side or well clear of the traffic circuit.

Circuit altitude is established as 1000 feet above the elevation of the aerodrome unless otherwise specified because of particular circumstances at a certain location. Except when the cloud and ceiling situation prevents it, all aircraft in the traffic circuit must maintain the 1000 foot circuit altitude on the crosswind and downwind legs.

Landing should be made on or parallel to the runway most nearly aligned into wind. However, you as pilot have the final authority and responsibility for the safe operation of your airplane and may select another runway in the interest of safety if you choose.

After landing, clear the runway as quickly as possible by turning off at the nearest taxiway. If it is necessary to taxi back along the runway in use, turn left 90° and watch for airplanes landing before proceeding to taxi. All turns made to clear the runway should be made to the left. It is standard practice for a pilot landing behind an airplane that has just touched down to pass to the right as he overtakes the other airplane.

Left hand circuits are normally in use at most airports in Canada. At a few, right hand circuits have been designated and in that case, the reverse of the traffic flow pictured in Fig. 6. would apply.

If you are flying circuits, you should, after each take-off, reach circuit altitude before joining the downwind leg.

If you are taking off from an uncontrolled airport, always check the final approach for traffic coming in to land before moving out onto the runway. It is also wise to check the final approach paths to other intersecting runways in case an airplane is landing from another direction.

In taking-off and departing the circuit, you should climb straight ahead on the runway heading until clear of the traffic circuit. Any turn while operating in the traffic circuit should be made to the left (except of course when a right hand circuit is in effect). However, you may make a right hand turn to depart from the vicinity of the aerodrome when the aircraft is well beyond the circuit area.

Mandatory Frequency (MF)

At certain selected uncontrolled airports, mandatory frequencies have been designated. Ordinarily these are airports at which there is an instrument approach and at which there is a flight service station (FSS) or a community aerodrome radio station (CARS), both of which provide airport and vehicle advisory service. An MF may also be designated at an airport with an RCO (remote communication outlet) if it provides remote airport and remote vehicle advisory service or at any other airport where an MF is considered necessary because of special traffic conditions. An MF may also be designated at a controlled airport for use during the period of the day when the control tower is not in operation.

(Flight service stations, community aerodrome radio stations and remote communication outlets are discussed in the Chapter **Radio**.)

Mandatory frequencies have been established in the interest of safety, on the premise that a full exchange of information concerning the movement of traffic will prevent unsafe conditions from developing. Aircraft operating within the specified area of an MF are required to be radio equipped. Pilots must report their positions and indicate their intentions and must monitor the mandatory frequency while operating within the specified area surrounding the airport at which the MF is designated.

If you are intending to land at or take off from one of these airports, you shall establish communication with the ground station on the published MF frequency. En route VFR traffic operating within the specified area shall also use the MF to advise of their position and intentions and to obtain traffic information.

In most cases, the **specified area** associated with the MF is 5 nautical miles in radius from the airport on which it is centred and extends up to 3000 feet AAE.

The mandatory frequency, call sign, and distance are listed with other communication information about the airport in the **Canada Flight Supplement.** For example, the listing might read: MF — radio 118.7 04-12Z 5NM 3050 ASL. In this instance, the mandatory frequency is 118.7 MHz and is monitored by a flight service station. The call sign is the name of the FSS/airport and the word RADIO (i.e. Dawson Radio). The MF is in force from 0400 to 1200 UTC. (This example refers to an airport at which the MF is in effect during the hours when the control tower is closed. If the MF is in force 24 hours a day, no hours of operation are mentioned in the **Canada Flight Supplement** listing.) The specified area within which communication with the station must be established lies in a 5 nautical mile radius from the airport and extends up to 3050 feet ASL which is 3000 feet above the elevation of the airport.

You should make your first transmission on the MF at least 5 minutes before entering the specified area and you should report your position, altitude, arrival procedure intentions and estimated time of landing. You must maintain a listening watch on the mandatory frequency while in the specified area. Upon joining the traffic circuit, you should report your position in the pattern. You should report again when you are established on the final approach. A final report should be transmitted when you are clear of the runway after landing.

If the ground station fails to respond to your initial transmission, all further reports that you would normally make shall be broadcast. To **broadcast** means to make a radio transmission that is not directed to any particular receiving station. In this case, the transmission is broadcast for the purpose of advising any other traffic in the specified area of your intentions.

If you are intending to take off from an airport for which an MF has been designated, you should request airport advisory service on the MF prior to taxiing in order to have a good idea of what traffic is operating in the area of the airport. You should report your intentions and maintain a listening watch on the MF frequency. When you are ready to take off, you should report your departure procedure intentions and ascertain both by radio and by visual observation that there is no traffic before moving onto the runway. You should report when you are clear of the circuit and monitor the MF until you are well clear of the specified area.

En route VFR traffic intending to fly through the specified area must report position, altitude and intentions prior to entering the area, maintain a listening watch on the MF while in the area and report clear of the specified area when leaving it. If at all possible, it is advisable to avoid passing through the specified area of an MF in order to minimize conflict with local traffic and to reduce radio congestion on the MF. Monitoring the MF frequency while in proximity to the area is nevertheless a wise practice.

Where an MF is operated by an RCO, it is important to remember that the RCO is operated remotely by an FSS at another location. The operator is some distance away and cannot see what is going on at the airport. He is able to pass on only information that he has been given on the radio.

Only aircraft equipped with two-way radio capable of communication with a ground station are permitted to operate on the maneuvering area of an airport for which a mandatory frequency has been designated or under VFR within the specified area. However, special provision has been made to permit NORDO aircraft to operate if prior notice of the intentions of the pilot has been given to the FSS or CARS. NORDO aircraft must, of course, be especially observant while operating within the specified area and must enter the traffic circuit in a position which will require the aircraft to complete at least the downwind leg and the base leg of the circuit before turning onto the final approach path. Information on the presence of NORDO aircraft operating either in the air or on the ground is provided to other traffic operating within the MF area. Advice on the arrival of a NORDO aircraft is included in the traffic advisory for 5 minutes before and for 10 minutes after its ETA; on the departure, just prior to take-off and for 10 minutes after departure.

(The FAA in the U.S. has established common traffic advisory frequencies (CTAF) to be used when operating in the vicinity of non-tower airports.)

Aerodrome Traffic Frequency (ATF)

At uncontrolled airports that are considered to be relatively active, aerodrome traffic frequencies are designated instead of mandatory frequencies. The designated ATF will normally be the ground station where one exists. Usually this would be the UNICOM frequency 122.8 MHz. Where no ground station exists, the designated ATF would be 123.2 MHz used in the broadcast mode. Vehicle operators, who are trained and authorized to do so, can communicate with pilots on the ATF and provide information about vehicles and aircraft on the maneuvering area and about runway conditions.

The specified area associated with an ATF is normally 5 nautical miles in radius from the airport on which it is centred and extends up to 3000 feet AAE.

The ATF, call sign, hours of operation if applicable and distance are listed with other communication information about the airport in the **Canada Flight Supplement.** For example, the listing might read: ATF — unicom ltd hrs O/T tfc 122.8 5NM 4300 ASL. In this instance, the ATF is monitored by the ground based UNICOM station during limited hours and during this time, you should use the call sign UNICOM (i.e. Otter Lake UNICOM). At other times, you should use the call sign "traffic" (i.e. Otter Lake Traffic) to relay your intentions in the broadcast mode. The specified area lies in a 5 nautical mile radius from the airport and is capped at 4300 feet ASL which is 3000 feet above the airport elevation.

If you are intending to land at or take off from an airport for which an ATF has been designated, you would follow the same procedures as those outlined for an airport at which an MF has been designated. In other words, if you are intending to land at the airport, you would make your first transmission before entering the specified area advising of your position, altitude, intentions and expected time of landing, and would make further reports as required. If you are intending to take off from the airport, you would follow the same procedures in making reports of your intentions as you would do if operating where an MF is in operation.

NORDO aircraft may operate without restriction at an airport where there is an ATF. However, if you are the pilot of a NORDO aircraft, you must be especially observant since neither the other aircraft nor the airport service vehicles can be alerted to your presence. You should fly a full circuit so that runway availability and traffic can be visually ascertained.

If you are intending to land at an uncontrolled airport for which neither an MF or an ATF has been designated, in the interest of safety, you should transmit your intentions in the broadcast mode on the frequency of the ground station, if one exists, or on the general frequency 123.2 MHz, if there is no ground station. You should switch from 126.7 MHz (the frequency you should be monitoring during flight in uncontrolled airspace) to 123.2 MHz (or the frequency of the ground station) when you are about 5 to 10 nautical miles from the airport and broadcast a position report and your landing intentions. You should monitor the frequency throughout the period you are in the vicinity of the airport and report again when you are established in the traffic pattern (as you would do if there was an MF or ATF designated for that airport).

The Traffic Circuit at Controlled Airports

The traffic circuit at a controlled airport is not dissimilar from that at an uncontrolled airport. It consists also of a crosswind leg, a downwind leg, a base leg and a final approach leg (Fig. 7) The principal difference is that you must establish communication with the control tower.

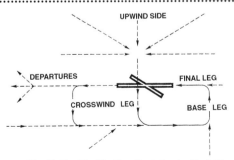

Fig. 7. The Traffic Circuit at Controlled Airports.

Note: Standard Left Hand Circuit. If a right hand circuit is designated, the opposite of this diagram is applicable. Circuits are normally flown at 1000 ft. AAE.

You must establish and maintain radio communications with the control tower prior to operating within the control zone served by an operational control tower. You must, therefore, call the control tower on the appropriate frequency prior to entering the control zone, give your identification and position and request landing instructions. It is advisable to make this initial call about 5 minutes prior to entering the zone. If the control zone is Class B or Class C airspace, the appropriate clearance must be received from the controlling agency prior to entry into the classified airspace.

The tower controller will advise the runway in use, wind direction and speed, altimeter setting and any other pertinent information and then will clear you to enter the circuit. **"Cleared to the Circuit"** authorizes you to join the circuit on the downwind leg at circuit height. If, because of your position in relation to the runway in use, it is necessary to proceed crosswind prior to joining the circuit on the downwind leg, do so as indicated in Fig. 7, approaching the active runway from the upwind side at a point midway between each end of the runway staying clear of the approach and departure paths of the active runway. When joining the circuit, you must conform as closely as possible to the altitude, speed and size of the circuit being flown by other traffic.

The airport controller may clear you to a straight in approach and, in this instance, you may join the traffic circuit on the final approach leg without having executed any other portion of the circuit.

Once established in the traffic circuit, you should advise the tower of your position (e.g. "Foxtrot Romeo Lima Tango is downwind"). The tower will then give you your landing instructions. For example:

Tower: **Piper Foxtrot Romeo Lima Tango you are number one.**
Or,
Tower: **Piper Foxtrot Romeo Lima Tango you are number two. Follow Cessna 185 now on base leg.**

You must have landing clearance prior to landing. Normally, the controller will clear you to land as you turn onto final. If this does not happen, it is your responsibility as pilot to request landing clearance in sufficient time to accommodate the operating characteristics of your airplane. If you do not receive landing clearance, you must pull up and make another circuit. Even after landing clearance is given, the tower may advise you to pull up and go around again if the situation on the runway becomes unsafe for landing.

If, after landing clearance is accepted, the situation is such that you, as pilot, feel that there is a hazard to the safe operation of your flight, you should advise ATC of your intentions and go around again. If, for example, the cross wind component is too much for the capabilities of your airplane, you may request another runway that is more into wind if one exists. Always advise ATC of your intentions.

After landing you should clear the runway without delay by continuing forward to the nearest available taxi strip or turn off point. Continue to taxi until you have crossed the taxi position hold line, or until you are at least 200 feet from the runway. You must not exit a runway onto another runway unless authorized by ATC to do so. If you have landed beyond the last turn off point, proceed to the end of the runway, turn off and wait for permission to taxi back to an intersection. Do not turn and taxi back against the direction of landing traffic unless instructed to do so by the tower. When clear of the active runway, the tower will advise you to switch to ground control who will give you instructions and authorization to taxi to the parking areas.

At some of the larger controlled airports, more facilities than just the tower are available. Runway and weather information is broadcast on the automatic terminal information service (ATIS) (see Chapter **Radio**). Always listen to the ATIS before contacting the tower and then advise the tower that you have the ATIS information. Extended areas of Class C airspace surround some control zones and it is necessary to contact the area controller before contacting the tower (see **The Canadian Airspace System,** following). Always check the **Canada Flight Supplement** and/or a VTA chart if applicable for special procedures that are in force at any airport at which you intend to land.

If you are intending to take off from a controlled airport, you must contact ground control for taxi instructions before starting off towards the active runway. At some of the larger airports, you must contact clearance delivery even before contacting ground control to advise of your intentions. Ground control will give you instructions on how to proceed to the active runway and will then advise you to switch to the tower frequency for take-off instructions. When cleared for take-off, you shall acknowledge and take off without delay. Once airborne, remain tuned to the tower frequency during the time you are operating within the control zone and preferably until you are at least 10 miles outside it. You do not require permission to change from the tower frequency once you are clear of the control zone and should not request release or report clear when there is considerable frequency congestion.

(See also **Radio Communication Facilities** in Chapter **Radio** and **Take-Off/Landing Procedures** in Chapter **Airmanship.**)

Sequential and Simultaneous Operations

ATC procedures allow for sequential and/or simultaneous operations on intersecting runways at some controlled airports in the interest of increasing and expediting airport traffic.

During sequential operations, controllers may not allow an arriving airplane to cross the arrival threshold or a departing airplane to commence its take-off roll until (1) the preceding airplane has passed the intersection, or (2) in the case of an arriving airplane, it has competed its landing roll and has turned off the runway, or (3) in the case of a departing airplane, it is airborne.

Simultaneous intersecting runway operations (SIRO) are permitted on intersecting runways only if the two airplanes involved are both arriving or one is arriving and the other is departing. Certain conditions are required, the main one being that there is sufficient runway length before the intersection for the arriving airplane to come to a stop and hold short of the intersection.

A **land and hold short operation (LAHSO)** clearance should be accepted by the pilot only if he is confident that he can bring the aircraft to a full stop before the intersection or can exit the runway at the convenient taxiway before reaching the hold short point. A LAHSO clearance once accepted must be adhered to. Controllers require a full read back of a LAHSO clearance.

In addition, SIRO/LAHSO may be carried out only if (1) the weather minima of a 1000 foot ceiling and 3 n.m. visibility are met, (2) the braking action is good and the runways are bare, (3) there is a tailwind of not more than 5 kts or a crosswind on a wet runway of not more than 15 kts, and (4) the pilot of the landing aircraft can accept a clearance to "hold short" of the intersecting runway. SIRO is not authorized if thunderstorms, turbulence, wind shear or other conditions exist that may adversely affect the aircraft's ability to "hold short".

Nordo (without radio) at a Controlled Airport

Aircraft without radio (NORDO) are not permitted to operate at most large controlled airports served by the scheduled air carriers.

Where they are permitted to operate (less busy controlled airports), they are directed by visual signals. You, as pilot, must be alert for the light signals from the tower letting you know what to do.

Before initiating a NORDO flight, you should contact the control tower to inform the controllers of your intentions and to secure a clearance for operation within the control zone. The tower will then be expecting you and will be prepared to give you light signals. To operate in Class C airspace, you **must** secure such a prior clearance.

Departure. You should taxi with caution to the runway in use but must stay at least 200 feet from the edge of the runway until clearance is received to take off. If stopped at any time by a red light, you must wait for further clearance before proceeding. When ready for take-off, turn the aircraft towards the tower to attract the attention of the controller. Acknowledge signals by full movement of rudder or ailerons or by taxiing the aircraft to the authorized position.

The following are authorized light signals to aircraft on the ground:

Flashing Green Light — Cleared to taxi.
Steady Green Light — Cleared for take-off.
Flashing Red Light — Taxi clear of landing area in use.
Steady Red Light — Stop.
Flashing White Light — Return to starting point on airport.
Blinking Runway Lights — Vehicles and pedestrians are to vacate the runway immediately.

Arrival. To join the traffic circuit, you must do so from the upwind side of the runway, join crosswind at circuit height as in Fig. 7 and turn onto the downwind leg. You must conform to the size and speed of the circuit and maintain adequate separation from airplanes ahead of you.

If it is necessary to cross the airport, before joining the circuit, for the purpose of determining the runway in use or to obtain other landing information, you must do this at least 500 feet above circuit height. Descent to circuit height then should be made in the upwind area of the active runway, prior to joining the crosswind leg of the circuit.

Before turning on final, you must check for any airplanes on a straight in approach.

The tower will give you landing clearance on final approach by means of a light signal. If landing clearance is not given, you must pull up and go around again.

The following are authorized light signals to aircraft in the air:

Steady Green Light — Clear to land.
Steady Red Light or Red Flare — Do not land. Continue in circuit. Avoid making sharp turns, climbing or diving after you receive this signal.
Flashing Green Light — Recall signal. Return for landing (usually to recall an airplane which has taken off or has been previously waved off with a red light). This will be followed by a steady green light when the approach path and landing area is clear.
Alternating Red and Green Light (U.S.) — Danger. Be on alert. This signal may be used to warn you of such hazards as danger of collision, obstructions, soft field, ice on runways, mechanical failure of your undercarriage, etc. The danger signal is not a prohibitive signal and will be followed by a red or green light as circumstances warrant.
Flashing Red Light — Airport unsafe. Do not land.
Red Pyrotechnical Light — The firing of a red pyrotechnical light, whether by day or night and notwithstanding any previous instruction, means "Do not land for the time being".

By day, acknowledge all light signals from the tower by rocking the wings of the airplane; at night, by a single flash of the landing light.

Ground Control Signals to Aircraft

STOP — START ENGINE(S) — CUT ENGINES

FOR TAIL TO PORT: — FOR TAIL TO STARBOARD — SLOW DOWN
TURNS WHILE BACKING

CHOCK'S REMOVED: — CHOCKS INSERTED: — SLOW DOWN ENGINE(S) ON INDICATED SIDE
CHOCKS

TO PROCEED UNDER FURTHER GUIDANCE BY SIGNALMAN — THIS BAY — PROCEED TO NEXT SIGNALMAN

TURN TO YOUR LEFT: — TURN TO YOUR RIGHT: — MOVE AHEAD
TURN

ENGAGE BRAKES: — RELEASE BRAKES:
BRAKES

ENGAGE BRAKES; RAISE ARM AND HAND, WITH FINGERS EXTENDED, HORIZONTALLY IN FRONT OF BODY, THEN CLENCH FIST
RELEASE BRAKES; RAISE ARM, WITH FIST CLENCHED, HORIZONTALLY IN FRONT OF BODY, THEN EXTEND FINGERS.

MOVE BACK

DAY: MAKE RAPID HORIZONTAL FIGURE-OF-EIGHT MOTION AT WAIST LEVEL WITH EITHER ARM, POINTING AT SOURCE OF FIRE WITH THE OTHER.
NIGHT: SAME AS FOR DAY BUT USING WANDS.

FIRE — ALL CLEAR

When cleared to land by a green light signal, come in on a straight glide or power approach for a distance of not less than 3000 feet. Do not S-turn. Continue to watch the tower on final approach for further signals.

No taxi clearance is required after landing, except to cross any runway or to taxi back to a turn off point.

If your airplane is equipped with receiver only (RONLY), it is your responsibility to advise the airport controller of this fact, preferably by filing a flight plan. Otherwise, he will instruct you by means of visual signals.

The arrival and departure procedures for RONLY aircraft are the same as those for NORDO aircraft, except that the controller may request acknowledgement of his transmissions in a specific manner (i.e. rocking the wings of the airplane in flight).

The Canadian Airspace System

For the purposes of air traffic control, separation and control of airplanes flying by reference to instruments and by visual reference, national defence, search and rescue, etc., the Canadian airspace has been divided in a number of different ways. It is necessary for a pilot to have an understanding of these divisions.

Domestic Airspace

The Canadian Domestic Airspace (CDA) includes all airspace over the Canadian land mass, the Canadian Arctic and Archipelago and certain areas over the high seas. It is divided into two areas, the Northern Domestic Airspace and the Southern Domestic Airspace. The boundaries of these two divisions are illustrated in Fig. 8.

Northern Domestic Airspace (NDA)

In close proximity to the north magnetic pole, the earth's lines of force dip vertically towards the pole, and the compass, which lies in a horizontal plane, loses its ability to point the way. There is, therefore, a large area of Canada in which magnetic compass readings are unreliable. Certain navigation procedures are recommended for operation in the area.

All aircraft operating in the Northern Domestic Airspace must fly at an altitude or flight level that is appropriate to their direction of flight as determined by **true** track calculations. Runway numbering is oriented to and surface winds are reported in degrees true. Aircraft operating at night or under IFR must be equipped with a gyroscopic direction indicator.

Southern Domestic Airspace (SDA)

All aircraft in level cruising flight within the Southern Domestic Airspace must maintain an altitude or flight level that is appropriate to their direction of flight as determined by the **magnetic** track.

Airways and air routes are based on magnetic tracks. Airport runways are assigned numbers based on their magnetic bearing. Surface winds are reported in magnetic degrees.

Altimeter Regions

Canadian airspace is divided, for purposes of altimetry, into an Altimeter Setting Region and a Standard Pressure Region. The geographic limits of these two regions are illustrated in Fig. 9.

The Altimeter Setting Region

Prior to take-off from an airport located in the Altimeter Setting Region, a pilot must set the aircraft altimeter to the current altimeter setting of that airport or, if that altimeter setting is not available, to the elevation of the airport.

While cruising within the Altimeter Setting Region, pilots must adjust their altimeters to the reported current altimeter setting of the nearest station along their route of flight, or where such stations are separated by more than 150 nautical miles, to the nearest station to the route of flight. Altimeter setting (QNH) is the setting made to an altimeter so it will indicate altitude ASL.

When approaching the airport of intended landing, the altimeter must be set to the current altimeter setting of that airport.

The Altimeter Setting Region includes the airspace up to 18,000 feet only.

The Standard Pressure Region

The Standard Pressure Region includes that area of Northern Canada depicted and so designated in Fig. 9 and also all airspace above 18,000 feet anywhere in Canada.

In the Standard Pressure Region, the altimeter is set to standard pressure (29.92" Hg). For take off and climb, from an airport in this region, the altimeter should be set to the current altimeter setting (or the airport elevation) and reset to the standard pressure setting immediately prior to reaching the cruising altitude. For descent and landing, the current altimeter setting of the airport of intended landing should be set on the altimeter.

Altimeters must be reset after entering and before leaving the Standard Pressure Region, unless otherwise authorized by ATC. A transition level exists between the Altimeter Setting Region and the Standard Pressure Region. This is FL 180. When climbing (immediately after ascending through FL 180), change your altimeter from the current altimeter setting (QNH) to standard pressure. When descending (immediately before passing through FL 180), change from standard pressure to the QNH altimeter setting.

FL 180 is not used for cruising in Canada.

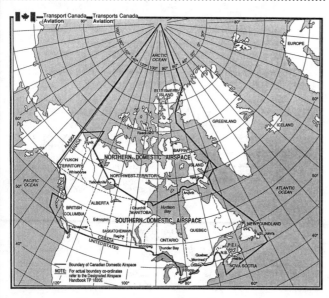

Fig. 8. Canadian Domestic Airspace.

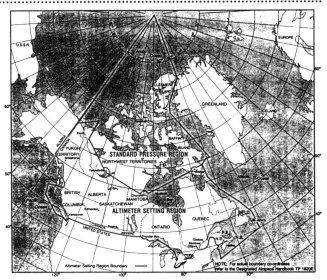
Fig. 9. Altimeter Regions.

Identification Zones

The defence authorities of both Canada and the United States have established a number of Air Defence Identification Zones for the security control of air traffic. It is imperative that pilots acquaint themselves with the locations and boundaries of these zones. They are denoted on aeronautical charts, radio facility charts and the **Canada Air Pilot.**

The Canadian ADIZ lies off each coast of Canada and extends across the roof of the North American continent from Baffin Island on the east to Alaska on the west.

The procedures and regulations for flight within an ADIZ are more fully discussed in the Section **Air Traffic Rules and Procedures,** following.

Sparsely Settled Areas

Much of the geographical area of Canada is virgin land with very few settlements. Flight in this Sparsely Settled Area, the boundaries of which are defined in Fig. 10, requires special precautions and procedures because of limited navigation facilities, severe weather conditions, limited weather information, limited fuel supplies and servicing facilities. These procedures are described in the Section **Air Traffic Rules and Procedures,** following.

High Level Airspace

Canadian Domestic Airspace is also divided vertically. All airspace 18,000 feet ASL and above is considered high level airspace. All airspace below 18,000 feet ASL is considered low level airspace.

The high level airspace is divided into three regions.

The **Southern Control Area (SCA).** The boundaries of the Southern Control Area are the same as those of the Southern Domestic Airspace (see Fig. 8 and Fig. 11). Within the Southern Control Area (i.e. all airspace above 18,000 feet ASL), all air traffic is controlled.

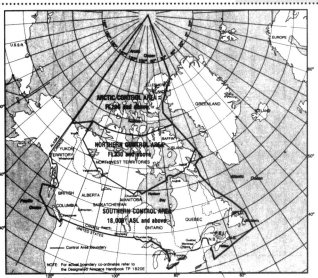
Fig. 11. Southern, Northern and Arctic Control Areas.

The **Northern Control Area (NCA)** extends from the northern limits of the Southern Control Area to a line approximately following the 72°N parallel of latitude. It comprises all airspace at and above FL 230. Within the Northern Control Area, all air traffic is controlled.

The **Arctic Control Area (ACA)** extends north from the northern boundary of the Northern Control Area to the north pole. It comprises all the airspace at and above FL 280. Within the Arctic Control Area, all traffic is controlled.

In the Northern Domestic Airspace, there is airspace between 18,000 feet ASL and the floors of the Northern Control Area (FL 230) and the Arctic Control Area (FL 280) which is high level airspace but which is uncontrolled (and therefore Class G airspace in the airspace classification system — see below). All high level airspace in the Southern Domestic Airspace (i.e. 18,000 feet ASL and above — Southern Control Area), however, is controlled airspace.

Encompassed within the high level airspace are the following types of **controlled high level airspace:** high level airways and that portion of any terminal control area, control area extension, transition area, Class F restricted airspace, Class F advisory airspace, military operations area or Class F danger area that may extend above 18,000 feet. Also in the high level airspace are high level air routes which are uncontrolled airspace.

High level airways are prescribed tracks between specified radio aids to navigation in the high level airspace. Air traffic control is provided.

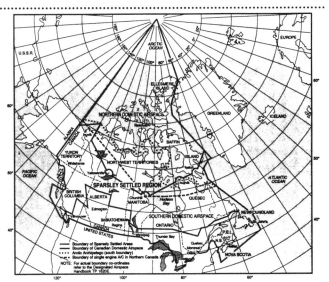
Fig. 10. Sparsely Settled Areas of Canada.

Aeronautical Rules & Facilities

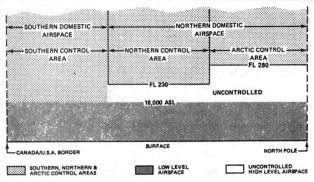

Fig. 12. Vertical Dimensions of Southern, Northern and Arctic Control Areas..

High level air routes are prescribed tracks between specified radio aids to navigation but along which air traffic control is not provided. High level air routes exist only in the Northern Domestic Airspace between 18,000 feet ASL and the bases of the Northern and Arctic Control Areas.

Low Level Airspace

The low level airspace comprises all airspace within the Canadian Domestic Airspace below 18,000 feet ASL. Not all of the low level airspace is controlled. Boundaries of low level controlled airspace are indicated on pilotage charts by shaded lines, the solid edge of which defines the outer limits of the controlled area. Types of **controlled low level airspace** include low level airways, control area extensions, control zones, transition areas, terminal control areas, military terminal control areas, Class F restricted airspace, Class F advisory airspace and Class F danger areas.

Low Level Airways

Airways are routes between points along which aircraft can navigate by following NDB or VOR signals. **VHF/UHF airways** (also known as **VOR or Victor airways**) are followed by reference to the radials projected by chains of VHF omnirange stations located approximately 100 miles apart along the airways. They are designated by the letter V and numbers (e.g. V52).

LF/MF airways are followed by reference to signals from low frequency transmitters such as non-directional beacons.

The basic width of a VHF/UHF airway is 4 nautical miles on each side of the centre line; of an LF/MF airway, 4.34 nautical miles on each side of the centre line. Unless is it otherwise published on VFR Navigation Charts, an airway has its base at 2200 feet above ground (AGL). It extends up to the base of the overlying high level airspace.

That portion of a low level airway that extends upwards between 12,500 feet and 18,000 feet ASL is designated Class B airspace and the rules and procedures that apply to flight in Class B airspace must be followed. Below 12,500 feet ASL, a low level airway is Class E airspace. VFR aircraft following a low level airway in Class E airspace are not subject to ATC control. ATC control is, however, provided for IFR flight.

When flying VFR along airways, fly down the centre line of the airway.

Exercise caution when crossing an airway. IFR traffic is using the airway at the 1000 foot levels and VFR traffic is using the airway at the 500 foot levels.

Control Areas Extension (CAE)

Control area extensions are established at some airports to provide additional controlled airspace to handle IFR traffic. At some busy airports, the controlled airspace contained within the associated control zone and converging airways is not sufficient for the maneuvering required to separate arriving and departing IFR traffic. A control area extension provides the required additional controlled airspace. The CAE surrounds and overlies the core control zone and IFR traffic is controlled by the area control centre (ACC). Such control area extensions are usually circular with a defined nautical mile radius. They extend upwards from 2200 feet AGL to 18,000 feet ASL. A control area extension is usually classified as Class E airspace. That portion of a control area extension that lies between 12,500 feet and 18,000 feet ASL is designated as Class B airspace. Below the base of a control area extension, a transition area is established between 2,200 feet AGL and 700 feet AGL to accommodate IFR traffic doing instrument approaches. In vertical cross section, therefore, a control area extension may have the shape of an inverted wedding cake.

Control area extensions have also been established to connect areas of controlled airspace, such as those that connect the domestic airways structure with the oceanic control areas. A CAE, such as this, may be based at other altitudes, such as 2000 feet, 5500 feet or 6000 feet ASL, etc.

Control Zones

Control zones are designated around certain aerodromes to keep IFR aircraft within controlled airspace and to facilitate the control of VFR and IFR traffic.

These control zones are established around airports where there are operating control towers. The area of control begins at the surface of the earth. The upper limit is usually 3000 feet AGL but may vary at specific airports. Control zones associated with a terminal control area normally have a 7 nautical mile radius. Others have a 5 nautical mile radius, although there are a few that are only 3 nautical miles in radius. Military control zones usually have a 10 nautical mile radius and are capped at 6000 feet AAE.

Control zones will be classified as B, C, D or E depending on the classification of the surrounding airspace.

Certain airports which are very busy with a great deal of traffic are restricted to aircraft with prior reservations. Aircraft wishing to land at or take off from these airports are required to arrange for a slot time for their intended operation. Such control zones may be classified as Class F.

VFR traffic may operate within a control zone only if the weather minima are met. (See **Weather Minima for VFR Flight** below.) When weather conditions are below VFR minima, a pilot operating VFR may request special VFR (SVFR) authorization to enter the control zone. (See **Special VFR** below.)

Transition Area

Transition areas are established between the outer perimeter of control zones (which are based at ground level) and control area extensions and terminal control areas (which are based at 2200 feet AGL). The transition area extends upward from 700 feet AGL to the base of the overlying controlled airspace. The area provided will normally be circular with a 15 nautical mile radius measured from the centre of the airport. The transition area is established to accommodate IFR traffic doing instrument approaches.

Terminal Control Area (TCA)

Terminal control areas are established at airports with a high volume of traffic to provide IFR control service to arriving, departing and en route aircraft.

Terminal control areas may be Class A, B, C, D or E airspace. TCAs usually extend into the high level airspace. That portion of the TCA that extends above 12,500 feet is Class B airspace regardless of the classification of the rest of the TCA. Any portion that may extend above 18,000 feet would be Class A airspace. The TCA operating rules are established according to the classification of the airspace within the TCA. IFR traffic is normally controlled by a terminal control unit (TCU). During periods when the TCU is not in operation, the area control centre (ACC) will provide service.

A TCA will normally have a circular configuration centred on the geographic co-ordinates of the primary aerodrome. The control zone is 7 nautical miles in radius and extends from the ground to 3000 feet AGL. The first level of the TCA is 12 nautical miles in radius and is based at 1200 feet above the airport elevation. The second level of the TCA is 35 nautical miles in radius and is based at 2200 feet AAE. The third level of the TCA is 45 nautical miles in radius and is based at 9500 feet AAE. For convenience, on maps and charts, the base is converted to the equivalent height above sea level. For example, if the airport elevation is 900 feet ASL, the base of the first level of the TCA would be 2100 feet ASL, the second level would be 3100 feet ASL and the third level would be 10,400 feet ASL.

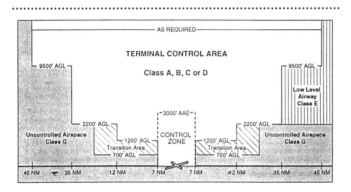

Fig. 13. Terminal Control Area.

As is the case with control area extensions, a TCA takes the shape of an inverted wedding cake.

The airspace between the floor of the terminal control area (and the floor of a control area extension) and the ground is uncontrolled airspace. VFR traffic may operate there without contacting the ATC unit controlling the overlying controlled airspace. However, the pilot of a VFR aircraft must contact the appropriate ATC unit before entering the controlled airspace.

Pilots operating in this uncontrolled airspace must exercise extra caution. Large jet transport airplanes may be operating as low as the floor of the controlled airspace on their approach and departure paths and the wake turbulence resulting from their passage will settle down into the airspace below the controlled area where it may pose a serious hazard to small airplanes. Airplanes operating in the uncontrolled area below the floor of any controlled airspace should, therefore, maintain sufficient vertical separation from the floor altitude so as not to create a hazardous situation either for themselves or for the jet transport airplanes which may be operating there.

Military Terminal Control Area (MTCA)
A military terminal control area is a terminal control area in which air traffic control is provided by a military terminal control unit. Special provisions for military aircraft prevail, such as VFR above FL 180.

Classification of Canadian Airspace
The Canadian Domestic Airspace is divided into seven classifications, identified by a single letter: A, B, C, D, E, F or G.

Flight within each class of airspace is governed by specific rules which are defined in the **Airspace Classification Structure and Use Regulations.** These rules depend on the classification of the particular airspace and not on the name by which the airspace is commonly known. For example, control zones and terminal control areas may be classified as B, C, D or E. Weather minima, however, are still related to the common name of the controlled or uncontrolled airspace.

Because of variations in dimension and classification of terminal airspace, the actual organization of the airspace is charted for VFR pilots on the VFR maps. The information on individual airports in the **Canada Flight Supplement** indicates the classification of the airspace surrounding each airport.

Class A Airspace
Class A airspace is that airspace within which only IFR flight is permitted.

It includes all controlled high level airspace between 18,000 feet ASL and Flight Level 600 inclusive. It includes, therefore, the Southern Control Area, the Northern Control Area and the Arctic Control Area (see above). It may also include any other airspace so designated by the Minister on either a permanent or temporary basis.

All flights require an ATC clearance to enter Class A airspace, must operate in accordance with instrument flight rules and are subject to ATC clearances and instructions. ATC separation is provided to all aircraft. The pilot must have an instrument rating and the airplane must be IFR equipped and have an approved and functioning transponder and approved and functioning automatic pressure altitude reporting equipment (i.e. Mode C).

All flights operating in Class A airspace above 18,000 feet ASL must use Standard Pressure Region procedures in setting their altimeters.

Class B Airspace
Class B airspace is that airspace where an operational need exists to provide air traffic control to IFR aircraft and to control VFR aircraft. Operations may be conducted under either IFR or VFR but all aircraft are subject to ATC clearances and instructions and ATC separation is provided to all aircraft.

It includes all controlled low level airspace above 12,500 feet and up to but not including 18,000 feet ASL. Such controlled airspace comprises the designated airways and any other control area that may extend upwards to this level. Control zones and associated terminal control areas may also be classified as Class B airspace. It may also include any other airspace so designated by the Minister.

A VFR flight must file a flight plan indicating altitude and route requested and receive an ATC clearance prior to entering Class B airspace. No special licence, rating or endorsement is required by the VFR pilot to operate in Class B airspace. The airplane must be capable of two way radio communication with the appropriate ATC facility and a listening watch must be maintained at all times. The aircraft must also be equipped with radio navigation equipment capable of utilizing navigation facilities to maintain the flight planned route. The airplane must be operated in VFR weather conditions at all times. When it becomes evident that it is not possible to operate the aircraft in VMC (VFR meteorological conditions) at the altitude or along the route specified in the ATC clearance, the pilot shall request an amended clearance that will enable the aircraft to be operated in VMC to the destination or to an alternate aerodrome. If the airspace is a control zone, the pilot shall request authorization to operate in special VFR flight.

The aircraft must be equipped with a functioning sensitive pressure altimeter that has been tested, inspected and certified within the past 24 months and with an approved and functioning transponder with Mode C capability.

The procedures for flight within Class B airspace are more fully discussed in the Section **Air Traffic Rules and Procedures,** following.

Class C Airspace

Class C airspace is a controlled airspace within which both IFR and VFR flights are permitted. However, VFR flights require a clearance from ATC to enter. ATC separation is provided between IFR traffic and, as necessary, to resolve conflicts between VFR and IFR aircraft and, upon request, between VFR aircraft that are radar identified and in communication with ATC. Traffic information is issued to aircraft to warn of other air traffic that may be in close proximity.

Terminal control areas and associated control zones may be classified as Class C airspace. Class C airspace becomes Class E airspace when the appropriate ATC unit is not in operation.

To enter and fly in Class C airspace, under VFR conditions, a pilot must hold a valid pilot's licence or a student pilot permit. No special licence, rating or endorsement is required.

The airplane must be equipped with a functioning two-way radio capable of communication with the appropriate ATC unit and, at all times while within Class C airspace, a listening watch must be maintained on the frequency assigned by the ATC unit. Airplanes must be equipped with a serviceable and functioning transponder with Mode C capability.

In the event of a communications failure while operating VFR in Class C (or Class B) airspace, the aircraft must maintain VFR at all times, divert to and land at, complying with the established NORDO arrival procedures, any suitable aerodrome and inform the appropriate ATC unit of the circumstances of the failure. If the airplane is transponder equipped, the pilot should select code 7600. If the communication failure is only in the transmitting capability, he should maintain a listening watch on appropriate frequencies for control messages or further clearances.

If the communication failure occurs while operating outside of Class C (or Class B) airspace precluding the pilot from obtaining the appropriate clearance to enter the airspace and if no suitable nearby aerodrome is available, the pilot is permitted to enter the classified airspace, continue VFR and comply with NORDO arrival procedures.

NORDO aircraft are permitted in Class C airspace during daylight hours and in VFR weather conditions if a clearance to enter has been obtained from the appropriate ATC unit prior to operation within the controlled area. This type of clearance is usually obtained by telephone prior to initiating the flight.

Authorization from the appropriate ATC unit must be obtained prior to entering Class C airspace or before taking off from an airport within Class C airspace. While airborne, authorization to enter this airspace is required whether the aircraft is intending to land at the airport or merely proceeding through the airspace.

Before entering Class C airspace, a pilot must contact the appropriate ATC unit and relay the following information: type and registration (in phonetic letters) of the aircraft; position (either by an accurate fix using a map or by a VOR radial and distance) to enable the controller to make an identification on the radar screen; altitude and if climbing or descending; destination; and transponder status. A VFR flight should select code 1200 on the transponder. If ATIS is available, the pilot should obtain this information prior to contacting the ATC controller and so inform him on the initial contact.

Authorization for flight in Class C airspace does not relieve the pilot of the responsibility of avoiding other aircraft, of maintaining terrain and obstruction clearance and of remaining in VFR weather conditions.

Class D Airspace

Class D airspace is a controlled airspace within which both IFR and VFR flights are permitted. VFR flights must establish two way communication with the appropriate ATC facility prior to entering Class D airspace. ATC separation is provided only to IFR traffic. Other traffic will be provided with traffic information. Upon request and equipment and workload permitting, conflict resolution instructions will be issued. ATC may instruct VFR traffic to remain clear of the Class D airspace.

Terminal control areas and associated control zones may be classified as Class D airspace. Class D airspace becomes Class E when the appropriate ATC unit is not in operation.

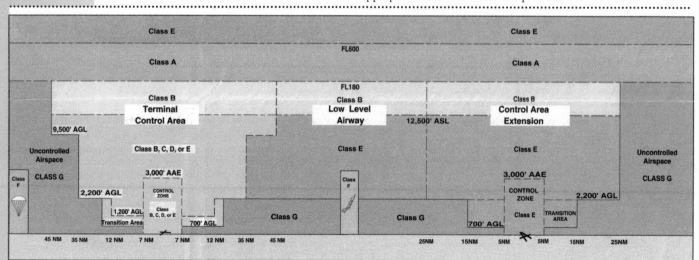

Fig. 14. Canadian Airspace Classification.

An airplane operating VFR in Class D airspace must be equipped with radio equipment capable of two way communication with the appropriate ATC unit and, at all times while in Class D airspace, the flight crew must maintain a listening watch on the radio frequency assigned by ATC. Certain Class D airspace may be specified as a transponder area in which case the airplane must be equipped with a transponder with Mode C capability.

NORDO aircraft are permitted in Class D airspace during daylight hours and in VFR weather conditions, provided permission to enter is obtained from the appropriate ATC unit prior to operating within the airspace.

Class E Airspace
Class E airspace is designated where an operational need exists for controlled airspace but the requirements for Class A, B, C or D are not met. Both IFR and VFR flights are permitted to enter Class E airspace but ATC separation is provided only to IFR traffic. There are no special requirements for VFR traffic.

Low level airways, control area extensions, transition zones and control zones at airports without an operating control tower may be classified as Class E airspace. Certain Class E airspace may be specified as a transponder zone in which case the aircraft must be equipped with a transponder with Mode C capability.

Class F Airspace
Class F airspace is airspace of defined dimensions within which activities must be confined because of their nature and within which limitations may be imposed on aircraft which are not part of those activities. The restrictions may be on either a permanent or a temporary basis.

Class F airspace may be classified as either **advisory** or **restricted** and can be controlled or uncontrolled airspace or a combination of both.

When areas of Class F airspace are inactive, they assume the rules of the appropriate surrounding airspace.

All designated Class F airspace is published in HE and LO Charts and on VFR aeronautical charts. Each restricted and advisory area within Canada has been assigned an identification code group which consists of (1) the nationality letters CY, (2) the letter A for advisory or the letter R for restricted (or the letter D for danger area if the restricted area is over international waters), (3) a three digit number which identifies the area, the first digit of the number identifying the geographic area of Canada in which the restricted or advisory area is situated (e.g. 1 for B.C., 2 for Alta., 3 for Sask., etc.) and (4) in the case of advisory areas, a letter in brackets which indicates the type of activity within the area: (A) aerobatics, (F) aircraft test area, (H) hang gliding, (M) military operations, (P) parachuting, (S) soaring and (T) training. Thus an area coded CYA 511(T) denotes training activity in advisory area 511 in Ontario, Canada.

Advisory Airspace
Airspace may be classified as Class F advisory airspace if there is activity there which, for flight safety purposes, non-participating pilots should be aware of. There are no specific restrictions which apply to the use of advisory airspace but VFR traffic is encouraged to avoid flight in advisory airspace unless participating in the activity taking place. Non- participating flights may enter advisory areas at their own discretion but extra vigilance is essential and pilots of both participating and non-participating aircraft are responsible for collision avoidance.

Pilots intending to fly in a Class F advisory area should monitor an appropriate frequency, broadcast their intentions when entering and leaving the area and communicate, when necessary, with other users to ensure flight safety in the area. In Class F advisory uncontrolled airspace, 126.7 MHz would be an appropriate frequency. In Class F advisory controlled airspace, ATC will assign a frequency.

Restricted Airspace
A restricted airspace is one of defined dimensions above the land areas or territorial waters of Canada within which flight is restricted in accordance with certain specified conditions. A restricted airspace may be designated because of aerial activity, surface activity or to protect a ground installation.

No one may conduct aerial activities within active Class F restricted airspace unless permission has been obtained from the user agency. The user agency is the civil or military agency or organization responsible for the activity for which the Class F airspace has been provided. It has the jurisdiction to authorize access to the restricted airspace.

Any restricted airspace which may be established over international waters will be designated as a **danger area.**

Temporary restricted areas may be designated to restrict flight around and over forest fires areas, well fires, disaster areas, etc. for the purpose of ensuring the safety of flight operations in support of the occurrence. Because the restriction is temporary in nature, the restricted airspace is not considered to be classified Class F airspace in accordance with Airspace Regulations. Such temporary restricted areas are authorized in the Canadian Aviation Regulations and by NOTAM. The NOTAM sets out the location and dimensions of the fire area, the airspace involved in the restriction. Notwithstanding whatever restrictions are imposed on flight near a forest fire by a NOTAM, no one may operate an aircraft below 3000 feet AGL within 5 nautical miles of the limits of the forest fire area.

Altitude Reservation
An altitude reservation is a block of controlled airspace reserved for the use of a specific agency during a specified time. Information about an altitude reservation is published in a Class I NOTAM. Pilots should avoid altitude reservations. ATC will not clear flights into an active reservation.

An altitude reservation may be confined to a fixed area (stationary) or moving in relation to the aircraft that operate within it (moving). Civil altitude reservations are normally for a single aircraft, while those for military use are normally for more than one aircraft.

Class G Airspace
Class G airspace is airspace that has no been designated Class A, B, C, D, E or F and within which ATC has neither the authority nor the responsibility for exercising control over air traffic.

It is, therefore, uncontrolled airspace. Included in this category are low level and high level air routes and aerodrome traffic zones as well as all other airspace that is not classified as controlled. Although ATC units do not exercise control over air traffic, ATC units and flight service stations do provide flight information and alerting services.

Flight in instrument conditions is permitted in Class G airspace only if the pilot in command is the holder of a valid pilot licence and a valid instrument rating.

Low Level and High Level Air Routes
Air routes are similar to airways in most respects except that along air routes air traffic control services are not available. Air routes are not, therefore, within controlled airspace.

Air routes extend between L/MF or VOR navigation aids. They extend from the surface of the earth up to Class A airspace. They are 9 nautical miles wide and are designated by colours and numbers (e.g. Amber Route 5, Blue Route 7).

Air routes exist mainly in the northern areas of Canada.

Aerodrome Traffic Zone (ATZ)
At the present time, there is one aerodrome traffic zone in Canada, at Fort Frances, Ontario. The airport there does not have a control tower facilities but it does have an approved instrument approach procedure. The zone has a radius of 5 nautical miles and extends from the surface of the ground up to 3000 feet AGL. Control zone weather minima applies within this zone.

Pilots entering or operating within an ATZ should use the procedures outlined for operation at uncontrolled airports.

Flight Information Region (FIR)
The Canadian Domestic Airspace is also divided into seven flight information regions within which flight information service and alerting service are provided. These regions are centred on Vancouver, Edmonton, Winnipeg, Toronto, Montreal, Moncton and Gander.

Rules of the Air

Air Regulations are undergoing constant change and revision. A limited amount of information regarding regulations is included in this manual. Unfortunately, during the lifetime of this edition, some of this information may become outdated because of the ongoing revision process in the regulatory system. We do, therefore, urge you to refer to the most recent amendments to the aviation regulations to ensure that your knowledge in this area is absolutely current.

In addition, regulations may differ from country to country. Although there is, for the most part, a uniformity in the regulations pertaining to flight procedures, there may be subtle differences that would impact on your flight patterns. If you are flying in a country other than Canada, you must acquaint yourself with the regulations pertaining to flight in that country.

Indeed, it is your responsibility as a pilot to make yourself familiar with and to conform to all the regulations and requirements which pertain to the airplane which you fly and the conditions under which it is being flown.

The government published **Canadian Aviation Regulations** (which replace the Air Regulations and the Air Navigation Orders) and the revisions to this legislation cover all the rules pertaining to operation of aircraft in Canada. They should be carefully studied. The regulations that were included in the Air Regulations, Air Navigation Orders and other bodies of legislations are consolidated into the CAR (Canadian Aviation Regulations). Each part of the CAR deals with a particular topic, such as registration, certification and marking, or licensing, or visual flight rules, etc. In format, the new CAR somewhat resemble the ANOs.

The **A.I.P. Canada (Aeronautical Information Publication)** is especially useful for it is updated on a regular schedule and incorporates Canadian Aviation Regulations and other aeronautical information that is pertinent to the basic requirements to fly within Canadian airspace. A.I.P. amendments are published about 4 or 5 times a year and are distributed to all licensed pilots.

Information Circulars and A.I.P. Canada Supplements are also distributed to all pilots and should be carefully scrutinized. Class I NOTAMS are distributed to all flight service stations and are available to pilots on a continuing basis in those locations. Information Circulars cover practically every subject to do with aviation in Canada. They are informative in nature, although some may be mandatory. NOTAMS deal mostly with matters of an urgent nature, such as danger and restricted areas, obstructions, airport construction, changes in navigation and control procedures.

Certificate of Airworthiness
No airplane may be flown unless it is registered and has a flight permit or C of A (certificate of airworthiness) issued by the aviation authority in the country of registration. The airplane must conform with respect to equipment, maintenance and operation to the conditions specified in the certificate. The keeping in force of the certificate is dependent upon the aircraft being maintained in accordance with the standards set out in the Airworthiness Manual. The C of A is continuous but may be out of force if the owner or operator fails to take action on airworthiness directives and manufacturer's service bulletins, scheduled inspections or maintenance and the rectification of defects which adversely affect flight safety.

An Annual Airworthiness Information Report must be submitted on the anniversary date of the issue of the certificate. The report requires the owner/operator of the aircraft to submit information on the most recent inspection, recent significant aircraft damage, flying hours in the reporting period, recent modifications, and aircraft, engine and propeller data. If this report is not filed, the C of A will automatically expire. It is the responsibility of the owner or operator to ensure that there is proper documentary evidence in the aircraft journey log to prove that all required standards have been met.

Airplane Logs and Licences
An airplane may not be flown unless there is carried on board the licences of all members of the flight crew. The licences must be valid for the airplane being flown (i.e. single engine, multi-engine, etc.) and for the type of operation (land, sea, night, etc.). Also on board must be the certificate of registration and the certificate of airworthiness of the airplane, proof of insurance coverage, the licence for the radio equipment installed in the airplane and the radio operator's licence of the pilot or other crew members, and the journey log of the airplane. An ultra-light airplane is exempted from the requirement to carry on board the journey log and a certificate of airworthiness.

A journey log and a technical log must be maintained for every airplane, other than an ultra-light. It is the responsibility of the pilot to keep the log books up to date by recording the facts of every flight.

Pilot's Licence
No person may act as pilot-in-command or co-pilot of an airplane unless he is the holder of a valid pilot's licence, endorsed for the particular class and type of aircraft being flown (i.e. airplane, balloon, glider, ultra-light, helicopter, gyroplane).

The holder of a student permit may fly an airplane only under the supervision of an authorized instructor and may not carry passengers. The flight must be conducted within Canada, by day and under VFR.

The holder of a pilot licence may not exercise the privileges of his licence unless he has acted as pilot-in-command or co-pilot of an aircraft within the five years preceding the flight or has met the written examination requirements for the licence within the 12 months preceding the flight. If passengers are carried, he must have, within the 6 months preceding the flight, completed at least 5 take-offs and

From the Ground Up

landings in an aircraft of the same category and class. These may be conducted either by day or night, if the flight is to be conducted by day, but must be conducted by night, if the flight is to be conducted in whole or in part by night.

Every holder of a pilot licence shall maintain a personal log book and record the particulars of every flight: the date of the flight, the type and registration of the aircraft, the capacity in which he acted and the names of other flight crew members, the place of departure, the place of arrival, any intermediate stops and any instrument approaches, the amount of flight time, the conditions under which the flight was conducted, and, in the case of a flight in a glider, the method of launch used for the flight. Log book information may be kept in a computer as an alternative to entries on paper.

The holder of a pilot licence shall not exercise its privileges under instrument flight rules (IFR) unless the licence is endorsed with an instrument rating.

A private pilot may carry passengers by day in any airplane which has been endorsed on his licence, but not for remuneration. A commercial pilot may carry passengers for hire or reward in any airplane endorsed on his licence, that does not exceed 12,500 lbs. Pilots carrying passengers by night must have their licences endorsed for night flying.

Dual Instruction

No person is authorized to give instruction unless he holds a flight instructor rating. Commercial or airline pilots may, however, give a check out to a licensed pilot on a new type of airplane. Flying time in the log book is entered as dual, pilot-in-command, or co-pilot in accordance with the duties the pilot performed.

Right of Way and Rules of the Air

The few basic rules outlined below do not by any means attempt to cover the entire subject of Air Traffic Regulations. It is vitally important that pilots keep themselves fully informed on current air traffic matters by reference to the circulars issued regularly by Transport Canada and to amendments to the Canadian Aviation Regulations and the A.I.P. Canada.

It is the responsibility of the pilot not to fly his aircraft in such close proximity to another aircraft as to create a collision hazard.

When two aircraft are approaching head-on or approximately so, each should alter course to the right in order to avoid any danger of collision.

An aircraft overtaking another, whether climbing, descending or in horizontal flight, shall alter its heading to the right to pass on the right. (This is the only air traffic rule which differs from motor traffic regulations.) It shall not resume its original heading until it has entirely passed and is clear of the other aircraft.

When two airplanes are on converging courses at approximately the same altitude, the airplane that has the other on its right must give way. The airplane that has the right of way should maintain its course and speed but has, at the same time, a responsibility to take any such action that is necessary to avoid collision.

The aircraft that is required to give way shall not pass over or under or cross ahead of the other aircraft unless it does so at a distance sufficient not to create any risk of collision.

Based on their ability to maneuver, aircraft have priority for the right-of-way in the following order:

1. Fixed or free balloons.
2. Gliders.
3. Airships.
4. Power driven fixed wing or rotary wing airplanes.

An airplane towing another has priority over other aircraft having mechanical power.

If two airplanes are approaching to land at the same time, the airplane flying at the greater height must give way to the airplane at the lower altitude. However, the pilot of the latter must not take advantage of this requirement to maneuver in front of another aircraft that is about to land. The pilot of a power driven airplane must give way to a motorless airplane, if both are approaching to land at the same time at the same airport.

No person shall take off or attempt to take off until there is no apparent risk of collision with any other aircraft.

An aircraft that is in flight or maneuvering on the ground or water must give way to another aircraft that is about to land.

Over the built up area of any city, town or other settlement or over any open air assembly, the minimum altitude at which an aircraft may be flown is an altitude that will permit, in the event of an emergency, the landing of the aircraft without creating a hazard to persons or property on the ground. In any case, this minimum altitude may not be less than 1000 feet above the highest obstacle within a radius of 2000 feet from the aircraft.

Elsewhere than over the areas mentioned above, the minimum altitude is 500 feet above the highest obstacle within a radius of 500 feet from the aircraft.

Except when taking off or landing, an airplane may not be flown at a height of less than 2000 feet over an aerodrome.

Single engine airplanes in Canada may be flown over open water beyond gliding distance from the shore only if they are not engaged in a commercial air service.

Night Requirements

NIGHT is any period of time during which the centre of the sun is more than 6° below the horizon. That is, for any place where the sun rises and sets daily, it is the period between the end of evening civil twilight and the beginning of morning civil twilight. It may be taken as commencing not less than one half hour after sunset and ending one half hour before sunrise.

SUNRISE occurs when the upper limb (edge) of the sun appears to be on the horizon. SUNSET occurs when it is about to disappear. Times of sunrise, sunset and twilight at any place on any date may be found in the Air Almanac.

Night Equipment – Airplanes

Aircraft operating at night must be equipped with the following approved, serviceable and functioning flight instruments: an airspeed indicator, a sensitive pressure altimeter, a direct reading magnetic compass, a turn and bank indicator, a gyro magnetic compass or heading indicator (if the flight is to be conducted beyond the immediate vicinity of the airport), and a means to illuminate the flight instruments. An aircraft that is to be flown within the Northern Domestic Airspace must have on board a means of establishing direction that is not dependent on a magnetic source. Each crew member must have access to a reliable time piece and a functioning flashlight. Any aircraft operating in controlled airspace must be equipped with a functioning two-way radio communication system.

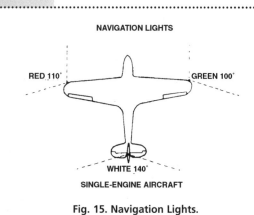

Fig. 15. Navigation Lights.

Night Lighting – Airplanes

Airplanes operating at night, in the air and while maneuvering on the ground, must also be equipped with a functioning navigation light system.

On the left wing tip, a steady **red** light and, on the right wing tip, a steady **green** light, each visible for 2 miles from dead ahead through an unobstructed angle of 110°. On the tail, in a position as far aft as possible, a single steady **white** light, visible for a distance of 2 miles through an angle of 140°. (If installation of a single light is impractical, two rear facing steady white lights may be displayed.)

An anti-collision light must also be installed on any airplane that will be operated at night. This high intensity flashing light may be either white or red or red/white segmented. It must be visible through 360° and project 30° above and below the horizontal plane of the aircraft.

Navigation lights must be displayed by all aircraft operating at night in Canada, and between sunset and sunrise in the U.S.

Between sunset and sunrise, an aircraft moored on the water and anchored to a fixed object either on land or in the water must display a white light visible in all directions for a distance of 2 miles unless it is moored at a place where aircraft are customarily moored (i.e. a seaplane base).

Anti-collision beacons should not be on while flying in dense cloud because of the possible flicker vertigo effect on the pilots. In any event, they should be switched off at the first indication of dizziness.

Fuselage lights, when installed, are white lights mounted on the top and bottom of the fuselage in line, spanwise, with the running lights on the wing tips.

Over Water Flights

Seaplanes must be equipped with an approved life jacket for every person on board the airplane. An approved life jacket for an adult is one that is capable of providing a total buoyancy of at least 15.9 kg. It must also be equipped with a means of illumination for the purpose of facilitating the location of persons in the water.

All persons must be informed of the location and method of use of the life saving equipment carried for their use.

A single engine land airplane may make an overwater flight if it remains within gliding distance of land. Beyond gliding distance from land, it must be equipped with approved life jackets for all persons on board. If it is operating more than 50 nautical miles from shore, it must be equipped with a life raft(s) with sufficient capacity to accommodate all persons on board the airplane. The life raft must be equipped with emergency stores such as water purification tablets, food, flares, signal mirror, first aid kit, etc.

Aerobatics

Any maneuver intentionally performed by an aircraft, involving an abrupt change in its attitude, an abnormal attitude, or an abrupt variation in speed is considered to be aerobatic flight.

Aerobatics are not permitted over any urban or populous area. They are not permitted in controlled airspace or in any air route or over any assembly of people except with written authorization from Transport Canada. Aerobatics are permitted in uncontrolled airspace. Certain advisory areas, classified as Class F airspace, have been especially designated as areas in which aerobatics may be carried out.

The performance of aerobatics must not in any way constitute a hazard to air traffic.

Passengers may not be carried in an aircraft performing aerobatics. The pilot must be the sole occupant of the aircraft except when accompanied by a qualified instructor who is giving aerobatic instruction.

Aerobatics should always be conducted at a safe altitude. The recovery from a practice spin should be made at an altitude not lower than 2000 feet above the ground.

Although there is no requirement that a pilot conducting aerobatic flight must wear a parachute, it is a recommended safety precaution.

Aircraft Occurrences

It is the responsibility of the pilot, operator, owner or any crew member involved to report an aircraft occurrence (accident or incident) as soon as possible to the Transportation Safety Board of Canada (TSB). Such a report may be made to the Regional TSB office or to the nearest flight service station or air traffic control unit who will pass on the information to the TSB.

A reportable aviation accident is defined as an occurrence associated with the operation of an aircraft that takes place between the time the first person boards the aircraft for the purpose of flight and the time the last person disembarks in which (1) any person suffers death or serious injury, or (2) the aircraft sustains substantial damage or is destroyed.

A reportable aviation incident is defined as an occurrence involving an aircraft with a gross weight of over 5700 kg where an engine fails; smoke or fire occurs; a malfunction of an aircraft system causes difficulties in controlling the aircraft; fuel shortage or depressurization occurs; the aircraft is refuelled with incorrect fuel; there is loss of separation or risk of collision during the flight; or, during landing or take-off, there is failure of the aircraft to execute a safe and effective procedure.

In the case of an accident, the information to be reported includes the type, model, nationality and registration marks of the aircraft, the name of the owner or operator of the aircraft, the name of the pilot-in-command, the date and time of the accident, the last point of departure and the point of intended landing of the aircraft, the position of the aircraft with reference to some easily identified geographical point, the number of crew members and passengers and whether they were killed or seriously injured, the nature of the accident and the extent of damage to the aircraft, a description of any dangerous goods aboard the aircraft, the name and address of the person making the report.

If an aircraft is missing, a report with similar information should be filed with an FSS or ATC unit.

The police should be notified as soon as possible to secure and guard the accident site until the arrival of the investigating team. Except to rescue or remove survivors, extinguish a fire or to prevent danger to any person or property, nothing at the site should be touched or removed. The preservation of any marks on the ground or objects along the accident trail is also of critical importance.

The purpose of accident or incident investigation is to further aviation safety and to attempt to prevent recurrence of that type of occurrence through the dissemination of safety information. Therefore, assist accident investigators as much as possible as they try to determine the cause of the occurrence.

Confidential Aviation Safety Reporting Program (CASRP)

A confidential aviation safety reporting system has been established by the TSB whose purpose is to enable aviation occurrences, deficiencies and discrepancies in the Canadian aviation system to be reported to the Board on a confidential basis. The TSB encourages pilots who are aware of such occurrences and deficiencies to report them using the form that was especially prepared for this purpose. Analysts use the reports to identify and make recommendations to correct safety deficiencies.

The legislation establishing the system prohibits the identification of the person submitting the report. In any legal or disciplinary proceedings arising from the report, the source is guaranteed anonymity.

Explosives and Dangerous Goods

Explosives or other dangerous articles or substances, including magnetized materials, may not be transported in an airplane except with permission from the Minister of Transport (Can.).

Air Traffic Rules and Procedures

For flight within Canadian airspace, there are many procedures and regulations which a pilot must know and which he must follow to ensure separation from other traffic and the safe and efficient operation of his own flight. Procedures and regulations are constantly undergoing change and it is a pilot's responsibility to keep himself informed of such changes by consulting the Canadian Aviation Regulations and the AIP Canada.

For definitions of terms and services in Air Traffic Control, see the **Glossary.**

Air Traffic Services

The following air traffic control and information services are provided by ATC through area control centres (ACCs), terminal control units (TCUs) and control towers:

Airport Control Service — is provided by airport control towers to aircraft and vehicles on the maneuvering area of an airport and to aircraft operating in the vicinity of an airport.

Area Control Service — is provided by ACCs to IFR and VFR flights operating within specified control areas.

Terminal Control Service — is provided by ACCs or TCUs to IFR and VFR flights operating within specified control areas.

Terminal Radar Service — is provided by ATC units to IFR aircraft operating within the radar service area.

Alerting Service — notifies appropriate organizations regarding aircraft in need of search and rescue services.

Airspace Reservation Service — provides reserved airspace for specified air operations in controlled airspace and disseminates information about these reservations.

Aircraft Movement Information Service — collects, processes and disseminates information about aircraft flights operating into or within the Air Defence Identification Zones.

Customs Notification Service (ADCUS) — provides advance notification to customs officials for transborder flights at specified ports of entry.

Flight Information Service — is provided by ATC units to assist pilots by supplying information about known hazardous flight conditions.

Clearances and Instructions (ATC)

An **air traffic control clearance** is an authorization from an ATC unit for an aircraft to proceed within controlled airspace under specific conditions. If you, as pilot, are unsure of the exact meaning of any part of an ATC clearance, you must ask for clarification before accepting it. Once you accept it, you are required to comply with an ATC clearance. You also must read back the text of the clearance if you are on an IFR flight. If VFR, you must read back the text of the clearance only if requested by ATC to do so.

A clearance is identified by use of some form of the word "clear" in its text.

Acceptance of an air traffic control clearance or instruction (below) does not relieve you, the pilot, of the responsibility of avoiding other traffic. If all or part of an air traffic control clearance is unacceptable to you, because, for example, of the operational capabilities of yourself or your airplane, you must so inform ATC and ask for other instructions. Even after accepting a clearance, if you subsequently find that you cannot comply, you should so inform ATC and ask for new directions. If you find it necessary to change speed, altitude or course while proceeding on an ATC clearance, you must inform ATC. Separation from other flights is predicated by ATC on the assumption that the clearance is being exactly followed.

An **air traffic control instruction** is a directive issued by an ATC unit for air traffic control purposes. You are required to comply with and acknowledge receipt of an ATC instruction which is directed to you provided the safety of the aircraft is not jeopardized. An instruction is always worded in such a way as to be readily identified, although the word "instruct" may not be used.

Provision is made for pilots to deviate from an ATC clearance or instruction to follow ACAS/TCAS resolution advisories. ATC must be informed as soon as possible of the deviation and the pilot should return quickly to the last ATC clearance or instruction received and accepted. ATC is not responsible for separation for an aircraft responding to an ACAS resolution advisory until the aircraft has returned to the last ATC clearance received and accepted or an alternate ATC clearance or instruction has been issued.

Position Reports

A **position report** is a report from an airplane en route to ATC made upon passing a reporting point. A position report must include identification of the aircraft, position, time of passing the reporting point, altitude, type of flight plan and destination. The estimated time of arrival at the destination or at the next reporting point may be included. That reporting point should be named (to identify the route).

Pilots on VFR flights are not required to make position reports but are encouraged to do so. These reports should be made to the nearest flight service station (FSS).

A **reporting point** is a geographical location in relation to which the position of an airplane is reported. Compulsory reporting points are shown on Radio Facility Charts as solid triangles ▲, non compulsory reporting points as outline triangles △. Reporting points are usually navigational aids such as a VOR. However, they may be non-directional radio beacons or the intersection of an omni radial with the radial which lies along the airway being flown. (In the latter case, an arrow indicates the direction of the intersection position line.)

Flight Rules

VISUAL FLIGHT RULES (VFR). The rules which apply when flying by means of visual reference to the ground.

INSTRUMENT FLIGHT RULES (IFR). The rules which apply when flying by means of reference to the instruments in the cockpit.

It is the responsibility of the pilot in command to determine whether a flight will be conducted in accordance with visual or instrument flight rules.

IFR flight is only permissible when the pilot has a valid instrument rating and the airplane is suitably instrument and radio equipped for IFR flight.

Visual Flight Rules only are dealt with in this manual.

Flight Plans and Itineraries

For any VFR flight that will be conducted beyond a radius of 25 nautical miles from the airport of departure, the pilot in command is required to file a VFR flight plan or a VFR flight itinerary with an ATS unit such as an air traffic control unit, a flight service station or a community aerodrome radio station. The pilot also has the option of leaving a flight itinerary with a responsible person who will notify Search and Rescue in the event that the flight is overdue. The pilot also has the option of filing with ATS a VFR flight notification. Flight notifications are scheduled to be phased out in the middle of 1996.

Notwithstanding the above, a VFR flight plan must be filed for any flight to or from a military aerodrome. (Prior permission from the military authorities is also required before landing at a military field.)

A VFR flight plan must also be filed for any transborder flight between Canada and the United States, and for any flight that will penetrate an Air Defence Identification Zone.

In the sparsely settled areas of Canada where communication facilities are inadequate to permit submitting a flight plan to ATS or closing it after arrival, the pilot has the option of notifying a responsible person of his flight itinerary.

VFR Flight Plan

The completion of a flight plan is simply a matter of inserting the requested information in the appropriate boxes on the form. A VFR flight plan shall contain the following information:

a) type of flight plan (VFR)
b) aircraft identification (i.e. registration). If the flight is a MEDEVAC, the suffix /M should be entered after the call sign.
c) the type of aircraft and the type of NAV/COMM/SSR equipment expressed by means of the equipment code.

NAV/COMM EQUIPMENT
/N — No NAV/COMM/Approach equipment
/S — Standard (VHF, ADF, VOR and ILS)
/A — Loran A   /P — Doppler
/C — Loran C   /R — RNav
/D — DME   /T — Tacan
/E — Decca   /U — UHF
/F — ADF   /V — VHF
/H — HF   /W — RNPC certification
/I — INS   /X — MNPS certification
/L — ILS   /Y — CMNP certification
/M — Omega   /Z — Other equipment
/O — VOR

If the letter Z is used, specify in "Other Information" what other equipment is carried.

SSR EQUIPMENT
/N — Nil
/A — Transponder. Mode A
/C — Transponder, Mode A and Mode C
/X — Transponder, Mode S without either aircraft identification or pressure altitude transmission
/P — Transponder, Mode S with pressure altitude transmission but no aircraft identification transmission.
/I — Transponder, Mode S with aircraft identification transmission but no pressure altitude transmission
/S — Transponder, Mode S with both pressure altitude and aircraft identification transmission

For example: C172/O/V/N — a Cessna 172 with VOR and VHF only and no transponder

PA66/SDH/C — a Piper Malibu with VHF, ADF, VOR, ILS, DME, HF and transponder with Mode A and C

d) true airspeed in knots or mach number
e) point of departure, using the 3 or 4 letter identifier of the aerodrome
f) flight altitude/level and route (on the VFR flight plan, it is necessary only to indicate VFR))
g) destination airport
h) time of departure (UTC) — proposed/actual
i) estimated elapsed time en route expressed in hours and minutes
j) alternate airports (not required on a VFR flight plan)
k) fuel on board expressed in hours and minutes. The fuel on board at the beginning of a VFR flight must be sufficient to fly to the destination, taking into account anticipated winds and other weather conditions, with a reserve that would allow a further 45 minutes of flight at cruising speed.
l) type of emergency locator transmitter
Type A or AD — automatic ejectable or automatic deployable
Type F or AF — fixed or automatic fixed
Type AP — automatic portable
Type P — personnel
Type WS — water activated or survival

From the Ground Up

Fig. 16. Canadian Flight Plan.

An IFR flight plan has been chosen as an example because it meets all the complete requirements for flight planning. VFR flight plans are somewhat less detailed. In the example above: the aircraft is an Airco with standard NAV/COM equipment (VHF, VOR and ILS) and a DME and a transponder with Mode A and Mode C. (a) indicates that the point of departure is CYXD (Edmonton). (b) states the initial cruising altitude will be 70 (7000 feet) via V301W (Victor airway 301W) to CYYC (Calgary); then via V304 (Victor airway 304) to CYEA (Empress, Alta). From this point, the cruising altitude will be 50 (5000 feet) via R18 (red airway 18) to CYYN (Swift Current); then via V300 (Victor airway 300) to CYXB (Broadview, Sask.); and then via V304 (Victor airway 304) to (c) the point of first intended landing, CYWG (Winnipeg). (d) names the alternate airports CYQK (Kenora) and CYYI (Rivers). This latter information is not required in a VFR flight plan. Other portions of the sample flight plan should be completely self-explanatory.

m) the type of communication equipment to be used en route. If standard frequencies are used, specify VHF, UHF or HF as appropriate. Frequencies other than standard should be listed. The word RONLY means receiver only or the word NORDO means no radio

n) navigation and approach aids. Indicate the type of navigation equipment installed in the aircraft by checking off the appropriate boxes.

o) the total number of persons on board at the point of first departure

p) pilot's name

q) name and address of aircraft owner

r) other information such as whether the aircraft is on wheels or or skis, is a seaplane or amphibian, the colour of the aircraft, duration of stops, the nature of any hazardous or dangerous cargo on board. When the point of departure is from a location without air/ground communication with ATS, it is required to include in the "Other Information" section a telephone number that may be used by ATS to conduct communication searches if the flight is overdue.

s) SAR time. This block is used when the pilot requires a SAR time which is different than the SAR time that is automatically provided (1 hour for a VFR flight plan and 24 hours for a flight itinerary).

t) where the arrival report is to be filed. If the flight is to a remote location, it is necessary to include specific information on how the arrival report will be filed. For example: XY tower — telephone — MacKay farm — Anytown. When the pilot has filed a flight plan, search and rescue services will be alerted if the aircraft is overdue by 1 hour. The pilot may specify his preferred SAR initiation time if he wishes it to be other than the automatic 1 hour.

u) customs notification if the flight is a transborder flight (ADCUS),

number of U.S. and non U.S. citizens on board

v) pilot's licence number

A U.S. VFR flight plan requests basically the same information in a somewhat different order.

Flight plans should be filed at any flight service station, community aerodrome radio station or air traffic control unit either directly or through a communication base such as an airport office, an operations office such as a flying club or a fixed base operator. A flight plan may be telephoned to an FSS. If telephoning is impossible, a VFR flight plan may be filed by radio with the nearest FSS as soon as practical after take-off. However, it is not advisable, because of possible overloading, to use the air/ground communication frequencies for filing flight plans when any alternate method is available.

Flight plans may also be filed via a **direct user access terminal system (DUATS)**. A DUATS service, that meets the standards of and is approved by Transport Canada, ensures that flight plans sent to ATS are complete and correctly formatted.

A single flight plan may be filed for a flight that will include one or more stopovers as long as the stopovers are of short duration (e.g. boarding passengers, refuelling, etc.). The stops must be indicated by name on the route section of the flight plan form; the duration of each stop must be indicated in the other information section. When intermediate stopovers are planned, the estimated elapsed time must be calculated as the total time to the destination, including the duration of each stopover.

Composite Flight Plan. A composite flight plan may be filed for a flight, the first portion of which will be conducted VFR and the last portion IFR. In filing such a flight plan, both the VFR and the IFR field of the flight plan are to be checked. The rules that govern VFR shall apply to the VFR portion of the flight and the rules that govern IFR shall apply to the IFR portion of the flight. Upon approaching the point where the IFR portion of the flight shall commence, the pilot must contact the appropriate ATC unit for clearance. The request may be made through a flight service station or directly to an ATC unit. Until ATC clearance for IFR is received and acknowledged by the pilot, the flight must be flown in weather conditions (VMC) that permit VFR flight.

Arrival Report. Closing of a flight plan, (i.e. reporting your arrival) must be submitted to an ATC unit or to an FSS or CARS within one hour after landing or at the SAR initiation time specified in your flight plan. If this is not done, the alerting service of the flight service station or the ATC unit will contact stations along your intended route and if they are unable to obtain information as to your whereabouts, they will contact the search and rescue service. In the event that you have had to terminate your VFR flight at an aerodrome other than the one to which the flight was planned or if you have had to make any changes in the flight route, ETA, etc., you must notify ATC as soon as possible of the change in the flight plan. Alerting service and subsequent search and rescue action is based on the estimated elapsed time indicated in the flight plan.

The pilot is at all times responsible for the closing of his flight plan and should never assume that the tower personnel will automatically file an arrival report. An **arrival report** should include (1) the aircraft registration, (2) the type of flight plan, (3) the aerodrome of departure, the date and time of arrival and (4) the aerodrome of arrival. Flight plans may not be closed in the air.

Pilots who plan to file an arrival report from a remote location, such as a farm, ranch, lake, etc., must be sure that adequate communication facilities are available at destination to permit the filing of the report.

Once you have filed a flight plan or a flight itinerary, **follow it.** If a deviation from your intended route or altitude becomes necessary, advise ATC or a flight service station as soon as practicable of the intended change. If the flight is being conducted in controlled airspace, you must receive an air traffic control clearance before making the intended change. Search and rescue operations are conducted based on the information in the flight plan. If you have deviated from your flight plan and are involved in a crash or forced landing, rescue planes may lose valuable time searching for you in the wrong place.

Flight Itinerary

With the exception of VFR flight to and from a military aerodrome or flight within an identification zone, pilots have the option of filing a VFR flight itinerary in lieu of a VFR flight plan. A flight itinerary must be filed with a responsible person, an air traffic control unit, a flight service station or a community aerodrome radio station. A responsible person is an individual who has agreed to notify the proper authorities in the event that the flight is overdue. (The proper authorities are an air traffic control unit, an FSS, a CARS or a Rescue Co-ordination Centre.) The pilot filing a flight itinerary has the option of specifying an SAR initiation time. If no such specific time is specified, SAR services will be alerted automatically if no arrival report has been filed by 24 hours after its expected time of arrival.

A flight itinerary shall contain the following information:

a) aircraft registration and type
b) estimated duration of flight and ETA at destination
c) route of flight or the specific boundaries of the area of flight operations
d) location of overnight stops, if applicable.

Upon the termination of a flight for which a flight itinerary was filed, the pilot must file an arrival report with ATS, if the itinerary was filed with ATS, or with the responsible person with whom the itinerary was filed. The arrival report must be filed within 24 hours of the ETA specified on the flight itinerary or by the SAR initiation time, if that option was exercised by the pilot. The pilot on a flight itinerary who realizes that his ETA will be later than that specified on the flight itinerary should make every possible attempt to contact ATS about the delay so that an unnecessary SAR is not undertaken.

Flight Notification

As well as flight plans and flight itineraries, there is also provision in the legislation for the filing of a flight notification. However, Transport Canada has advised that flight notifications will be phased out sometime in 1996.

A flight notification was originally intended to be used whenever the pilot could not meet the arrival report requirements of a flight plan but wished the flight information and alerting services provided by ATC. Usually it was used when it was not practical to determine the exact length of the flight beforehand or when the destination would not have the facilities to file an arrival report with ATC. Flight notifications were mostly used by pilots flying in the sparsely settled areas of Canada.

A flight notification required the same information as a flight plan as well as the following: the location of overnight stops, if any were to be made, whether the aircraft was equipped with wheels, skis, floats or a combination thereof and the colours of the aircraft. An arrival report for a flight notification is required to be filed within 24 hours from the time indicated on the flight notification form of the expected arrival at destination.

Trans Border Flights

A flight plan must be filed for any flight that crosses the border into the United States. Strictly speaking, the Canadian customs regulations are such that all flights, both of a commercial and private nature can be required to clear outward from Canada when proceeding to points in the U.S.A. or beyond. However, in actual practice, private flights (but not commercial flights) are permitted to depart for points in the U.S.A. without first obtaining an outward clearance from Canadian customs authorities. Having crossed the border, the first landing must be made at a U.S. customs port of entry to clear customs in-bound. Upon returning to Canada, private flights are not required to clear outbound from the U.S.A. (Commercial flights of any kind, i.e. charter, air taxi, cargo, etc., must often meet special requirements to comply with the laws of the United States.) The first landing in Canada after taking off from a point in the U.S.A. must be made at a Canadian customs port of entry. Regulations require that you notify the customs authorities of your expected time of arrival in advance. Upon landing, you **must not leave your airplane** until customs and immigration officers arrive at the airport to clear you. On transborder flights, you must have your pilot's licence in your possession and the airplane registration and airworthiness certificates must be on board.

A trans border flight plan should be filed to the customs port of entry only. The stop for customs should not be considered as a stopover on a flight plan to a farther destination. Delays created in the process of clearing customs could result in unnecessary alerting service procedures being initiated.

If you add ADCUS (Advise Customs) to your flight plan, ATC will automatically notify the customs authorities of your estimated time of arrival. This saves you the inconvenience of making a long distance phone call. However, at least an hour is needed for the ADCUS information to be relayed to the customs officials. If your transborder flight is less than one hour's duration, a telephone call should be made directly to the customs office. On flights from Canada to the United States, the number of U.S. and non-U.S. citizens on board must be included in the flight plan (for example: ADCUS — 2 US — 2 Non). On flights from the United States to Canada, this information is not required.

A new program **CANPASS - Private Aircraft** simplifies customs clearance for low risk travellers at a number of small airports in Canada. A CANPASS - Private Aircraft permit allows the permitholder to land at one of the designated airports anytime the airport is open (not just during customs' hours), report to customs using a 1-800 number and make a declaration by telephone.

Cruising Altitudes

Cruising altitudes appropriate to the direction of flight must be maintained at all times regardless of whether a flight plan has, or has not, been filed.

In the Southern Domestic Airspace, cruising altitudes are based on magnetic tracks. In the Northern Domestic Airspace, cruising altitudes are based on true tracks.

Altitudes below 18,000 feet are stated in thousands of feet. Altitudes at or above this height are referred to as flight levels (i.e. FL 180).

On airways, IFR cruising altitudes are assigned by ATC. In uncontrolled airspace, a cruising altitude appropriate to the direction of flight should be chosen by a pilot and maintained.

Cruising Altitudes – Canada
VFR
At altitudes below 12,000 feet ASL — On or Off Airways

000° to 179° — ODD thousands plus 500 feet (1500', 3500', 5500' MSL, etc.)

180° to 359° — EVEN thousands plus 500 feet (2500', 4500', 6500' MSL, etc.)

In level cruising flight, at or below 3000 feet AGL, the specified cruising altitudes are not compulsory but should be flown if practicable.

At altitudes above 12,000 feet but below 18,000 feet ASL — In Uncontrolled Airspace

000° to 179° — ODD thousands plus 500 feet (13,500', 15,500', 17,500' MSL)

180° to 359° — EVEN thousands plus 500 feet (12,500'. 14,500', 16,500' MSL)

At altitudes above 18,000 feet ASL but below FL 280 — In Uncontrolled Airspace

000° to 179° — ODD flight levels (FL 190, FL 210, FL 230, etc.)

180° to 359° — EVEN flight levels (FL 200, FL 220, FL 240, etc.)

000° to 179°		180° to 359°	
IFR	VFR	IFR	VFR
1000		2000	
3000	3500	4000	4500
5000	5500	6000	6500
7000	7500	8000	8500
9000	9500	10000	10500
11000	11500	12000	12500
13000	13500	14000	14500
15000	15500	16000	16500
17000	17500	FL 180	FL 180
FL 190	FL 190	FL 200	FL 200
FL 210	FL 210	FL 220	FL 220
FL 230	FL 230	FL 240	FL 240
FL 250	FL 250	FL 260	FL 260
FL 270	FL 270	FL 280	FL 280
FL 290	—	FL 310	—
FL 330	—	FL 350	—
FL 370	—	FL 390	—
etc	—	etc	—

Fig. 17. Cruising Altitudes - Canada.

VFR in Class B Airspace (12,500 to 18,000 feet ASL — On Airways)

In Class B airspace between 12,500 and 18,000 feet AGL, VFR flights are controlled and are assigned cruising altitudes.

000° to 179° — ODD thousands (13,000', 15,000', 17,000' MSL)

180° to 359° — EVEN thousands (14,000', 16,000' MSL)

IFR
In controlled airspace, the altitude or flight level to be flown is the one assigned by air traffic control. In uncontrolled airspace, when operating IFR in level cruising flight, the altitude or flight level must be appropriate to the direction of flight.

At altitudes below FL 290

000° to 179° — ODD thousands or flight levels (3000', 5000' MSL, etc. or FL 190, FL 210, etc.)

180° to 359° — EVEN thousands or flight levels (2000', 4000' MSL, etc. or FL 200, FL 220, etc.)

At altitudes above FL 290

000° to 179° — At 4000 foot intervals beginning at 29,000' pressure altitude (FL 290, FL 330, FL 370, etc.)

180° to 359°— At 4000 foot intervals beginning at 31,000' pressure altitude (FL 310, FL 350, FL 390. etc.)

Cruising Altitudes–U.S.

Because most Canadian pilots sooner or later will embark on a cross country flight that will take them into the United States, the U.S. cruising altitudes, which are slightly different than the Canadian ones, are explained here.

VFR — On or Off Airways
At altitudes below FL 290 (But above 3000')

000° to 179° — Fly ODD thousands plus 500 ft. (3500', 5500', etc. MSL)

180° to 359° — Fly EVEN thousands plus 500 ft. (4500', 6500', etc. MSL)

At altitudes above FL 290

000° to 179° — Fly at 4000' intervals beginning at 30,000' pressure altitude. (FL 300, 340, etc.)

180° to 359° — Fly at 4000' intervals beginning at 32,000' pressure altitude. (FL 320, 360, etc.)

Note: Altitudes for VFR apply to "VFR on top"

IFR — On or Off Airways
At altitudes below FL 290

000° to 179° — Fly at ODD thousands (1000', 3000', etc. MSL or FL 190, 210 etc.)

180° to 359° — Fly at EVEN thousands (2000', 4000', etc. MSL or FL 180, 200 etc.)

At altitudes above FL 290
000° to 179° — Fly at 4000' intervals beginning at 29,000' pressure altitude. (FL 290, 330, etc.)

180° to 359° — Fly at 4000' intervals beginning at 31,000' pressure altitude. (FL 310, 350, etc.)

Minimum Cruising Altitudes

Minimum Obstruction Clearance Altitude (MOCA) — the minimum MSL altitude along an airway segment which will permit the airplane to clear obstructions by 1000 ft.

Minimum En Route Altitude (MEA) — the minimum MSL altitude which will clear obstructions and also permit satisfactory radio reception. It is the minimum altitude to be considered for IFR flight plans. (See "Table of VHF Reception Distance")

Both MOCA and MEA are indicated on Enroute Charts.

Cruising Speeds

All Canadian controlled airspace below 10,000 feet ASL is considered as a "speed limit area", and all aircraft must conform to the regulations pertaining to operation within this area.

Within 10 nautical miles of a controlled airport and at an altitude of less than 3000 feet AGL, airplanes may be flown at an indicated airspeed no greater than 200 knots.

In controlled airspace but beyond the 10 nautical mile limit around controlled airports and below 10,000 feet ASL, airplanes may be flown in a climb, descent or in level flight at an indicated airspeed no greater than 250 knots, unless otherwise authorized by an ATC clearance or instruction.

Only if the minimum safe speed for a particular aircraft is greater than the above mentioned speeds, may the aircraft be operated at a higher speed.

Weather Minima for VFR Flight

Within Control Zones and Aerodrome Traffic Zones
Visibility minima for flight in control zones and ATZs are
Flight Visibility and Ground   Not less than 3 miles
Visibility (when reported)

Aircraft shall be operated at a height above the surface and at a distance from any cloud that is not less than the following:
Distance from Cloud   500' vertically
                             1 mile horizontally
Height above surface   500'

The aircraft must be operated at all times with visual reference to the surface.

Other Controlled Airspace
Weather minima for flight in controlled airspace are
Flight Visibility   Not less than 3 miles
Distance of Aircraft from Cloud 500' vertically
                                 1 mile horizontally

Uncontrolled Airspace
Weather minima for flight in uncontrolled airspace are

At or above 1000' above surface
Flight Visibility By Day   Not less than 1 mile
            At Night   Not less than 3 miles
Distance from clouds   500' vertically
                                 2000' horizontally

Below 1000' above surface
Flight Visibility By Day   Not less than 2 miles
            At Night   Not less than 3 miles
Distance from clouds   Clear of cloud

(With publication of the Canadian Aviation Regulations in 1996, the elevation AGL that determines the criteria for distance from clouds in uncontrolled airspace will be changed to 1000 feet. Prior to the publication of the CARs, the determining elevation is 700 feet AGL.)

Within Controlled Airspace (U.S.)
In the U.S., weather minima for VFR are slightly different. VFR on top is permitted.

1200 feet or less above the surface
Visibility   1 mile
Distance from clouds   Clear of cloud

More than 1200 feet above the surface
Visibility   3 miles
Distance from clouds   500' below
                          1000' above
                          2000' horizontally

More than 10,000 feet above the surface
Visibility   3 miles
Distance from clouds   1000' above
                          1000' below
                          1 mile horizontally

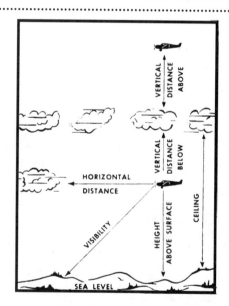

Fig. 18. Weather Minima References.

Special VFR (SVFR)

Within control zones, air traffic control may authorize a pilot to fly **special VFR** under weather conditions that are below VFR minima. The rules for special VFR are designed to allow VFR rated pilots to arrive at and depart from controlled airports when the weather is below VFR limits. The pilot must request special VFR. It is not offered automatically by the ATC unit working the control zone. The controller cannot authorize SVFR on his own. It must be approved by the ACC or TCU who will authorize it only if the IFR traffic flow permits. If there is actual or anticipated IFR traffic, special VFR may be denied or delayed.

A pilot who has received a clearance from an ATC unit to proceed in the control zone under special VFR may ask for a radar vector (at an airport where surveillance radar is available) in order to have assistance, under restricted visibility, to locate the airport. This service must be specifically requested. It is not offered automatically. It is the pilot's responsibility, even when being directed by radar vectors, to remain clear of cloud and to refuse a vector that would prevent him from maintaining contact with the surface.

Special VFR may be authorized only if the flight visibility and the ground visibility (when reported) are each not less than 1 mile. In the case of a helicopter, the flight and ground visibility must be not less than 1/2 mile. Because of the limited visibility, it is advisable that the helicopter be operated at a reduced airspeed that will give the pilot-in-command adequate opportunity to see other air traffic or obstructions in time to avoid them.

The aircraft requesting SVFR must be equipped with a functioning two way radio capable of maintaining communications with the ATC unit. The pilot-in-command shall keep a listening watch on a frequency that permits the receipt of ATC clearances and instructions.

The aircraft must be operated clear of cloud and within sight of the surface of the earth at all times. It must be operated at not less than 500 feet above the ground except during take-off and landing.

Special VFR at night is authorized only for the purpose of allowing the aircraft to enter the control zone in order to land.

VFR Over the top (VFR OTT)

With the introduction of the **VFR over the top** rating, pilots who do not hold an IFR rating have a new option for dealing with poor weather conditions along part or all of their intended route. The VFR OTT rating allows a pilot to conduct a flight in VFR conditions above the en route cloud layer providing the following conditions exist:

1) the flight is conducted during the day (VFR OTT is not permitted at night).
2) departure and climb above the en route cloud layer can be done under VFR.
3) descent and arrival at the destination can be accomplished under VFR.
4) during the cruise portion of the flight conducted above the cloud layer, the aircraft is operated at a vertical distance from cloud of at least 1000 feet.
5) where the aircraft is operated between two cloud layers, the vertical distance between the layers is at least 5000 feet.
6) the weather at the destination aerodrome is forecast to have a sky condition of scattered cloud or to be clear; ground visibility of five miles or more with no forecast of precipi-tation, fog, thunderstorms or blowing snow. The forecast conditions shall be valid, in the case of an aerodrome forecast (TAF), for one hour before and for two hours after the estimated time of arrival. If no aerodrome forecast is available and the forecast used is an area forecast (FA), the forecast conditions shall be valid for one hour before and for three hours after the estimated time of arrival.

Acquisition of the rating requires a minimum of 15 hours dual instrument time of which a maximum of 5 hours may be instrument ground time.

Minimum VFR Flight Altitudes

The regulations impose minimum flight altitudes for VFR flight based on the type of area over which the flight is being conducted.

Built Up Areas. Except when landing or taking off, the minimum altitude at which an aircraft in VFR flight may be flown over the built up area of any community or over any open air assembly of people is an altitude that, in an emergency, would permit the landing of that aircraft without creating a hazard to persons or property on the ground. In no case can the altitude be less than 1000 feet above the highest obstacle within a radius of 2000 feet from the aircraft.

Areas Which Are Not Built Up. Except when landing or taking off, the minimum altitude at which an aircraft may be flown over areas that are not built up is 500 feet above the highest obstacle within a radius of 500 feet from the aircraft.

Non Populous Areas or Open Water. Provided no hazard is created for people or property on the ground, an aircraft may be flown over non populous areas or over open water at lower altitudes so long as it is at least 500 feet from any person, vessel, vehicle or structure.

Special exemptions are made to these minimums only for aircraft engaged in recognized special purpose operations, such as fire fighting, crop spraying, pipeline and power line surveying, etc.

VFR Flight in Class B Airspace

Class B airspace is part of the controlled airspace system. VFR flight is permitted, but it is subject to the control services provided by ATC. All aircraft are subject to ATC clearances and instructions and ATC separation is provided. No special licence, rating or endorsement is required to act as pilot-in-command of an aircraft in VFR flight in Class B airspace.

VFR flights must be conducted in accordance with procedures designed for use by IFR flights. However, the flight must be conducted in VFR weather conditions (i.e. weather minima for control areas) with visual reference to the ground.

Before entering the Class B airspace, the pilot in command of the airplane must file a VFR flight plan stating the altitude at which the flight is to be conducted and the route that is to be followed. (e.g. Flight Altitude 14,000'/VFR, Route V71).

Clearance for VFR flight within Class B airspace is given only if the altitude requested is available and if traffic conditions are such that the flight can be accommodated. Clearance is not usually given prior to take off but rather upon receipt of a position report that the flight has reached the last 1000 foot altitude below the base of the Class B airspace.

A functioning two way radio is, of course, essential and, throughout the flight, the pilot must maintain a listening watch on the appropriate radio frequency and must make position reports as required by the ATC unit. Installed equipment in the aircraft must also include radio navigation equipment that is capable of utilizing navigation facilities to enable the aircraft to be operated in accordance with the flight plan. A transponder capable of Mode C altitude reporting is also required.

The pilot is responsible at all times for maintaining VFR flight. If the weather observed ahead falls below VFR minima, he may request an amended ATC clearance which will allow him to maintain VFR. If this is not possible, he must leave Class B airspace by the shortest route exiting either horizontally or by descending. When clear of the Class B airspace, he must take whatever action is necessary to continue the flight under VFR conditions and, as soon as possible, inform ATC of the action taken.

If the aircraft is operating VFR in a control zone that is designated Class B airspace and if the weather falls below VFR minima, the pilot must land at the aerodrome on which the control zone is based.

It is essential that the assigned altitude be maintained with precision and that no deviation be made from the clearance without advising ATC.

It may seem that the Class B airspace is just another unnecessary complication for the VFR pilot. However, it is common practice for airlines to fly instrument flight rules most of the time regardless of the weather; they probably are not looking out of the window watching for unexpected traffic. With modern high speed airplanes, the rate of closing between two approaching airplanes can be in excess of 500 knots, or 8 miles a minute. This means that with 3 miles visibility, a 200 knot airplane approaching a 300 knot airplane would have 22 seconds to alter course. In practice, the time is not so long. There is a lag between the time the pilot sights another airplane and the time that he takes corrective action. There is a further lag in the response of the airplane. These lags take up an average of 15 seconds, leaving 7 seconds to avoid a collision. The purpose in providing control services to and separation of all traffic within control areas is to prevent collision potential.

To prepare for a VFR flight in Class B airspace, check the appropriate Enroute Low Altitude Charts for distances between reporting points along the route. Take along a computer.

You are required to make position reports en route when passing all compulsory reporting points or when requested by ATC. On filing a position report, it is necessary to state your estimated time of arrival at the next reporting point. This means that you have to compute groundspeeds and ETA's as you go.

ATC will clear you to a specific point. This may be your destination airport, an omni or other radio station, or a reporting point en route. The location to which you are cleared is known as the **clearance limit.**

Should you arrive at the clearance limit without clearance beyond, or holding instructions, you must immediately request further clearance and meanwhile hold at the assigned altitude. The reason for this is that another airplane may be flying at your altitude beyond this point. (If you do not wish to hold, however, you may request a clearance to exit the Class B airspace.)

A flight plan may not be changed en route without authority from an ATC centre. If you should desire to make a change in altitude, route or destination, obtain an amended clearance from ATC.

Emergency While VFR in Class B Airspace

If instrument weather conditions are encountered, leave the Class B airspace immediately either horizontally or by descending. Advise ATC as soon as possible of your action.

In the event of radio failure, leave the Class B airspace immediately and, when clear, proceed with the flight maintaining VFR. Report your action as soon as possible after landing.

In the event of engine failure or any other cause necessitating an immediate descent, advise ATC immediately (using the emergency frequency 121.5 MHz if necessary).

Holding Pattern

There are several situations in flying in which a pilot may be asked to hold over a particular fix. This can happen if clearance to proceed past a clearance limit cannot be given. Sometimes on arrival at an airport, a pilot is asked to hold over a beacon until landing clearance can be given (IFR flights).

For the pilot who is maintaining visual reference to the surface, adhering to a precise holding pattern is not absolutely necessary. It is, however, good practice to learn to fly a holding pattern precisely.

The standard form of holding pattern is the elliptical, or race track pattern. It is illustrated and explained in Fig. 19. The holding fix may be a marker beacon, VOR, non-directional radio beacon or any facility which can be definitely identified by a radio signal when you are directly overhead.

Do not leave the holding pattern until cleared by ATC to do so.

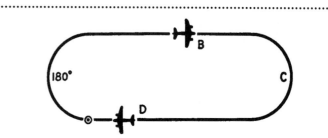

Fig. 19. Standard Holding Pattern.

Over the radio fix at A, start a rate one turn to the right (180°—1 minute.) Fly 1 minute outbound, if at or below 14,000 ft ASL, or 1-1/2 minutes, if above 14,000 ft ASL (B). Execute a rate one turn to the right at C. Fly the inbound course to the radio fix (D) — parallel to B. (Note: The wind should be compensated for.) The holding pattern is flown in a specified direction in relation to the holding fix. Assuming the top of the page to be north, the holding pattern illustrated would be east and the aircraft would be said to be "Holding East of (name of Radio Fix)". The track outbound (B) would be 090° and the track inbound (D) 270°. Note: The outbound time should be increased or decreased, according to the wind conditions, to effect 1 minute (or 1-1/2 minutes) inbound to the fix.

VFR Visual Holding Pattern

Because of traffic congestion, VFR flights may be asked to **orbit visually** over a geographic location, VFR checkpoint or call up point until they can be cleared to the airport. The pilot is expected to proceed to the specified geographic location, orbit within visual contact of the checkpoint and be prepared to proceed to the airport immediately upon receipt of a clearance to do so. While orbiting the checkpoint, the pilot should make left hand turns. Terrain and collision avoidance are his responsibility.

Identification Zones

Special procedures are in effect for airplanes operating in the Air Defence Identification Zones (ADIZ). These rules are applicable to all aircraft. Unidentified aircraft which enter an ADIZ will be directed by NORAD aircraft to land at a published aerodrome port of entry where they will be cleared by Canada Customs and the RCMP.

To enter or fly within an ADIZ, it is required that the pilot-in-command file an IFR flight plan, a DVFR flight plan or a flight itinerary with an air traffic control unit, a flight service station or a community aerodrome radio station. No deviation from the flight plan is permitted without prior notification to the ATC unit. When prior notification is impossible, ATC must be advised of the deviation as soon as possible.

The flight plan or flight itinerary should be filed prior to take-off. However, in the event there is no facility for the transmission of flight plan information, the flight plan or flight itinerary may be filed in the air as soon as possible after take-off. An aircraft, making an air file, may be asked to fly at a speed of less than 150 knots for 5 minutes or more to enable the ATC unit to make a positive identification.

For both aircraft whose point of departure is an airport outside the ADIZ and who intends to penetrate the ADIZ and for aircraft intending to take off from an airport within the ADIZ, the flight plan or flight itinerary shall include the estimated time and place of ADIZ penetration. As soon as possible after take-off, the pilot shall establish radio communication with the ATC unit, a flight service station or community aerodrome radio station and make a position report, advising ATC of any variance in the time or place of ADIZ penetration from that indicated in the flight plan or flight itinerary. Revised estimates shall be reported if the airplane will not be within 5 minutes of its forecast time of arrival at a reporting point, the point of penetration of the ADIZ or the destination within the ADIZ. Deviations of more than 20 nautical miles from the centre line of the intended route or from the estimated point of penetration of the ADIZ must be reported.

In the case of an aircraft taking off from a location within the ADIZ, it will be required to establish radio contact as soon as possible and make a position report.

A functioning two way radio is required equipment and the pilot must maintain a listening watch on the appropriate frequency throughout the time that the aircraft is within the boundaries of the ADIZ. In the event of communication failure, a VFR flight is required to proceed in accordance with the flight plan or to land at the nearest available airport and advise ATS.

An arrival report must be filed when closing a flight plan or flight itinerary.

The rules of the Security Control of Air Traffic and Air Navigation Aids Plan (SCATANA) will be implemented only in time of war. The SCATANA rules are, however, periodically tested. Pilots must, therefore, while operating within Canadian Domestic Airspace or the ADIZ, maintain a listening watch on an ATC frequency at all times. When notified that SCATANA rules are in effect, pilots must comply with all instructions from ATC. VFR and IFR flights may be ordered to change course, or altitude, or even to land at the nearest facility. On such occasions, a pilot planning a flight within an ADIZ shall obtain approval for the flight prior to take-off and shall provide position reports, if operating in uncontrolled airspace, at least every 30 minutes during the flight or, if operating in controlled airspace, as requested by the ATC unit.

Radar Assistance

The Canadian Armed Forces can provide assistance in an emergency to civil aircraft operating within the ADIZ. Assistance consists of track and groundspeed checks and the position of the aircraft in geographic reference or by bearing and distance from the station. Position of heavy cloud in relation to the aircraft can also be indicated.

To obtain assistance in the NWS area, call "Radar Assistance" on 126.7 MHz. In ADIZ areas, contact should be made on 121.5 MHz or on the UHF frequencies 243.0 MHz or 364.2 MHz.

Interception of Civil Aircraft

Interceptions are made only where the possibility is considered to exist that an unidentified aircraft may be truly hostile in intent. Intercepted aircraft should maintain a steady course, avoid attempting evasive action and follow these procedures.

a) Follow the radio and visual instructions given by the intercepting aircraft, interpreting and responding to the prescribed visual signals and procedures that have been established for use by both the intercepting aircraft and by the intercepted aircraft. A copy of these procedures and signals is required to be carried on board any aircraft operating in Canadian airspace. They are outlined in the **Canadian Aviation Regulations** and are also included in th **Canada Flight Supplement** and the **A.I.P. Canada.**

b) If possible, advise the appropriate ATC unit of the interception.

c) Attempt to establish radio communication with the intercepting aircraft on the emergency frequencies 121.5 MHz or 243.0 MHz giving the identity and position and the nature of the flight.

d) Unless instructed to do otherwise by the appropriate ATC unit, select Mode A Code 7700 if you have a transponder.

Should you receive instructions by radio from any source which conflict with those received from the intercepting aircraft, you should request immediate clarification while continuing to comply with the instructions received from the intercepting aircraft.

Sparsely Settled Areas

To fly in the sparsely settled areas of Canada, an airplane must be capable of two way radio communication with a ground station in the area. An emergency locator transmitter of an approved type that transmits on the distress frequency is mandatory. A portable manually activated ELT is a recommended extra.

When operating within the sparsely settled areas or when operating over water at a distance of more than 50 miles from shore, a pilot is expected to continuously monitor the emergency frequency 121.5 MHz, unless he is carrying out communications on other VHF frequencies or if cockpit duties or aircraft electronic equipment limitations do not permit simultaneous monitoring of two VHF frequencies.

Survival Equipment

The Canadian Aviation Regulations require certain survival equipment to be carried on board any aircraft operated over land, especially in the sparsely settled areas and anywhere where rescue is more difficult because of inaccessibility. This equipment must be sufficient for the survival on the ground, for a minimum of 72 hours, of each person carried on board. In selecting the equipment to be carried, special consideration needs to be given to the geographic area, the season of the year and the anticipated seasonal climatic variations of the area over which the aircraft will be operating.

Locating and saving people in aeronautical emergencies has been greatly improved by the introduction and widespread use of ELTs and the detection capability of the SARSAT/COSPAS system. Detection and location of crash sites and the rescue of survivors is commonly effected very quickly and rarely takes more than 72 hours.

In a survival situation, the first rule is to provide shelter that will keep you dry and protected from the wind in order to prevent hypothermia. The following items are therefore suggested: a tent; an 8 foot by 8 foot tarpaulin, ideally bright orange in colour (it can be used to make a shelter and can also be laid out to attract the attention of searching aircraft); a saw to cut branches to make a shelter; personal rain protection; space blankets; an air inflated mattress or unicell foam pad to provide insulation from wet and cold ground; one sleeping bag for each two persons; a sewing kit to repair clothing; mosquito head nets and insect repellent (in spring and summer) and tape to tape jacket sleeves and trouser pant bottoms for protection from insects. A good survival manual is useful.

Since rescue can be expected reasonably quickly, food is no longer considered a critical item in survival kits. If you choose to carry food, food with a calorific value of about 1500 calories for each person on board is sufficient for 72 hours. Water is more important than food. Water purification tablets will ensure you have safe drinkable water. If you prefer to boil your water, you will need a billy kettle or other suitable container. You might also include, in your survival kit, 500 ml of drinking water for each 4 persons.

In order to make a fire, you will need waterproof matches carried in a waterproof container, a candle to helping in lighting a fire, fuel tablets and an axe.

In order to attract the attention of searching aircraft, a holographic mirror is the best item you can carry. It is effective over 20 miles. A 2 sided signalling mirror with a hole in the centre is second best, but it requires some training and skill to use it effectively. For night signalling, a good strobe light can be seen on a clear night up to 8 miles away. A flashlight is effective for about one-half mile.

Every aircraft is required to have a first aid kit on board. A basic kit should contain antiseptic wound cleaner and disposable applicators, a variety of bandages, burn dressings, wound dressings, gauze dressings, adhesive dressings and adhesive tape, hand cleaner and cleansing towelettes, tweezers, scissors, splint set with padding, eye pad/shield and a first aid manual.

Survival gear should be stowed securely in a rear baggage compartment where it is less likely to be damaged or become inaccessible in the event of a crash. Survival gear stowed in a forward baggage compartment (such as one in the nose) is likely to be damaged in a crash.

Dress for Survival

All crew members and passengers should be dressed in, or have on board the aircraft, clothing that will be adequate for survival in the coldest conditions for the season of the year in which the flight is conducted. Even in summer, loose fitting cotton pants and a long sleeve cotton shirt should be the minimum that you wear. The long pants and long sleeve shirt will help to protect you from sun burn and from insect bites as well as chilly temperatures at night. Sturdy shoes and cotton socks are essential.

Layering materials and wearing outer garments that are loose fitting will minimize heat loss.

One of the greatest problems in any emergency or crash landing is the possibility of fire. Getting out of the aircraft as quickly as possible and avoiding inhaling the hot gases should be your first concern. Inhaling the hot air can burn the lung tissues and result, in the worst case scenario, in pneumonia. So hold your breath as you escape from a burning aircraft.

Your chances of surviving a fire with minimum burns and injury are greatly enhanced by wearing clothing made of proper materials. Wool, cotton and a fire retardant fabric called Nomex are the best choice. Most synthetics, such as polyester, nylon and rayon, will melt and shrink. The fabric in contact with the skin will melt into the skin and solidify making the burn injury worse. Cotton, wool and especially Nomex will decompose and char instead of melting.

Exposed skin has no protection against burn injury from fire. So choose your flying apparel carefully and wear it.

Be sure also that what you choose to wear is clean. The bacteria on dirty clothing (or skin) will invade your skin through a burn and cause severe infection.

Uncontrolled Airspace

Within the uncontrolled airspace in Canada, aircraft may operate free of the control of an ATC unit. However, all aircraft are required at all times to conduct their activities with regard to the Canadian Aviation Regulations.

When cruising in uncontrolled airspace, the lack of information of other aircraft in close proximity may constitute a potential hazard. Pilots, therefore, are advised to monitor continuously the frequency 126.7 MHz whenever practicable and to make periodic en route position reports and, especially, to report an intended change in flight altitude to the nearest FSS on 126.7 MHz. Such a report should include the last known position, the estimated next position, the present altitude and intention (climb/descent) and the planned altitude.

At certain uncontrolled airports, mandatory frequencies or aerodrome traffic frequencies have been established. Where an MF or ATF is designated, all pilots are required to call on the published frequency and provide details of their intentions to land at the airport. (See **Mandatory Frequency** and **Aerodrome Traffic Frequency** in Section **Aerodromes**.)

PART 3

Meteorology

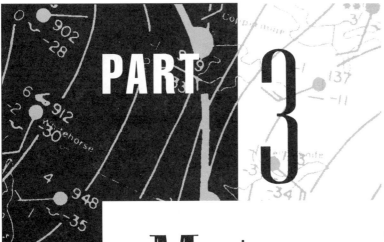
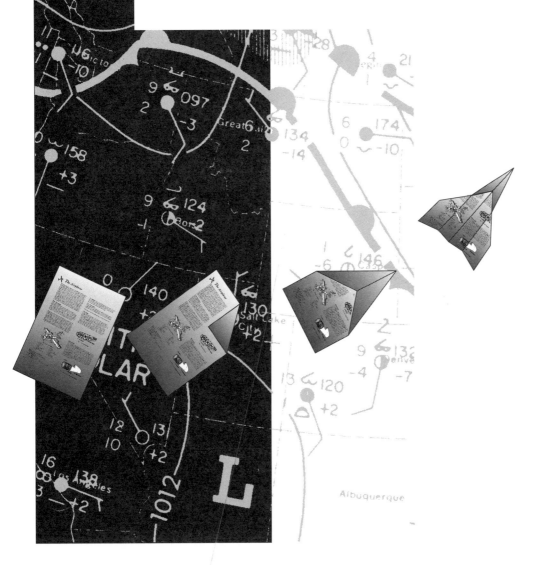

Meteorology

"Man lives at the base of an invisible ocean of air termed the atmosphere." Pick.

This atmosphere is forever in a state of commotion and physical change, giving rise to weather conditions which vary throughout the range of an extremely vast scale. The airman not only lives at the base of this sea of air but navigates through it. The weather, therefore, is a matter of vital concern to him, particularly conditions such as fog, ice formation, thunderstorms and line squalls, which present unusual hazards to flying.

To minimize the hazards to air navigation that are constantly being manufactured in the "weather factory", a vast world-wide meterological organization has been built up, to collect, analyze and broadcast information relative to the ever changing phenomena of the upper air.

The pilot can today avail himself of last minute weather reports and forecasts along all the regularly established air routes. In addition, he can secure much valuable weather data with reference to areas located off the organized airways. He must, however, possess sufficient "weather sense" to be able to size up and deal with sudden changing conditions which may be encountered at any stage along his route. The few brief notes which follow are intended to cover the highlights of the subject only. The student of aviation will be well advised to include meteorology among the subjects marked for further detailed study and read some of the excellent manuals which are available on the subject.

Beyond the atmosphere lies space and the most challenging adventure man has ever set his mind to embark upon. Some of you who read these lines will be destined to play an active role in the conquest of space. But the approach to that mysterious realm of the stars lies through the atmosphere. Navigation through this atmosphere will be the primary concern of this chapter on meteorology.

The Atmosphere

Air, which is the material of which the atmosphere is composed, is a mixture of invisible gases. At altitudes up to 250,000 feet, the atmosphere consists of approximately 78% nitrogen and 21% oxygen. The remainder is made up of argon, carbon dioxide, several other gases and water vapour. Water vapour acts as an independent gas mixed with air.

Water vapour is found only in the lower levels of the atmosphere. From the standpoint of weather, however, it is the most important component of the air. Because it can change into water droplets or ice crystals under atmospheric conditions of temperature and pressure, water vapour is responsible for the formation of clouds and fog. However, just when and under what conditions water vapour will change to a visible form is difficult to predict and is made more difficult by the fact that the amount of water vapour in the air is never constant but varies from day to day and even from hour to hour.

The lower layer of the atmosphere contains an enormous number of microscopic impurities such as salt, dust and smoke particles. They are important to aviation for they are often present in sufficient quantities to reduce visibility. They also have a function in the condensation process.

The upper layers of the atmosphere do not contain dust particles or impurities off which the sun's light can reflect and, for this reason, appear deep cobalt blue to black in colour.

The atmosphere has weight. Although the weight of the atmosphere is only about one millionth the weight of the earth, it does exert a force or pressure on the surface of the earth. A square inch column of air weighs approximately 14.7 lbs. at sea level. This weight diminishes with height. At 20,000 feet, a square inch column weighs 6.75 lbs.

The characteristics of the atmosphere vary with time of day, season of the year and latitude. Consequently only average values are referred to in this manual.

Properties of the Atmosphere

Mobility, capacity for expansion and capacity for compression are the principal properties of the atmosphere. These characteristics are of the utmost importance in a study of weather for they in combination are the cause of almost all atmospheric weather phenomena. The capacity for expansion is especially important. Air is forced to rise by various lifting agents (thermal, frontal or mechanical means). Rising in areas of decreasing pressure, the parcel of air expands. In expanding, it cools. The cooling process may bring the temperature of the parcel of air to the degree where condensation occurs. Thus clouds form and precipitation may take place. Conversely, sinking air, as the external pressure increases, decreases in volume with an attendant rise in temperature.

Divisions of the Atmosphere

The atmosphere consists of four distinct layers surrounding the earth for a depth of many hundreds of miles. They are, in ascending order, the troposphere, the stratosphere, the mesosphere and the thermosphere.

THE TROPOSPHERE. This is the lowest layer of the atmosphere and varies in height in different parts of the world from roughly 28,000 feet above sea level at the poles to 54,000 feet at the equator. Within the troposphere, the pressure, density and temperature all decrease rapidly with height. Most of the "weather" occurs in the troposphere because of the presence of water vapour and strong vertical currents. In the upper regions of the troposphere, winds are strong and the fast moving jet streams occur.

The top layer of the troposphere is known as the **tropopause**. Here the temperature ceases to drop and remains substantially constant at about -56°C. The height of the tropopause varies, as already stated, from the poles to the equator, but also from summer to winter.

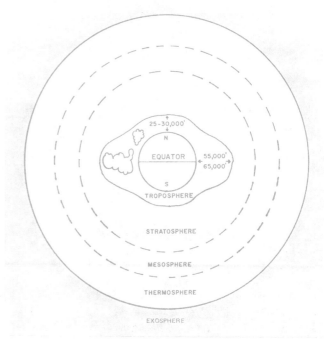

Divisions of the Atmosphere.

THE STRATOSPHERE. For a distance of about 50,000 feet above the tropopause, there is a layer known as the stratosphere in which the pressure continues to decrease but in which the temperature remains relatively constant in the vicinity of -56°C. This layer also varies in thickness, being quite deep over the poles and thinner over the equator. Water vapour is almost non-existent and air currents are minimal. The top layer of the stratosphere is called the **stratopause.**

THE MESOSPHERE. . The mesosphere is characterized by a marked increase in temperature. At a height of about 150,000 feet, the temperature reaches 10°C. The rise in temperature is due to the presence of a layer of ozone which absorbs more of the sun's radiation. In the top layer of the mesosphere, called the **mesopause,** the temperature again drops rapidly reaching a level of about -100°C at 250,000 feet above the earth.

THE THERMOSPHERE. Temperature again begins to rise in the thermosphere and increases for an indefinite distance into space, rising as high as 3000°C at 400 miles. This does not mean that a space ship, if it was cruising at this height, would experience a temperature of 3000°C by contact with the atmosphere. The temperature in these rarefied layers is based on the kinetic theory of gases. The only heat the space ship would experience would be what it would receive from the radiation of the sun. The spectacular auroras form in the upper regions of the thermosphere.

Within the thermosphere is a region known as the **ionosphere,** which extends from about 50 miles to 250 miles above the earth's surface. Within this region, free electron density is very great and affects radio communication. The ionosphere is the layer which reflects low and medium frequency radio waves back to the earth. Very high frequency radio waves, however, penetrate this layer.

Beyond the ionosphere lies a layer so thin that the pressure drops to little more than a vacuum. Under these conditions, the concept of temperature has little meaning and is usually replaced by definitions of the energy states of individual molecules. This thin upper layer is known as the **exosphere.**

SPACE. Since air becomes gradually thinner with increasing altitude, the upper limit of the atmosphere is, for all practical purposes, difficult to define. Now that man has invaded the cosmic realm of outer space, the question arises as to just where space actually begins. At an altitude of 90 to 100 miles, one is entering the realm of satellites and aerodynamic lift is no longer a prerequisite for maintaining height above the earth. This region has been accepted by some authorities as the boundary of outer space. Ninety miles is recognized as the limit of national sovereignty.

Standard Atmosphere

The decrease with height of pressure, density and temperature which occurs in the lower layers of the atmosphere is not constant but varies with local conditions. However, for aeronautical purposes, it is necessary to have a standard atmosphere. There are several different standard atmospheres in use but they vary only slightly.

The **ICAO standard atmosphere** for the continent of North America, based on summer and winter averages at latitude 40° assumes the following conditions:

1. *the air is a perfectly dry gas.*
2. *a mean sea level pressure of 29.92 inches of mercury.*
3. *a mean sea level temperature of 15°C.*
4. *the rate of decrease of temperature with height is 1.98°C per 1000 feet.*

The chart in Fig. 1 plots standard atmosphere through various altitude levels.

Altitude	Pressure in inches of Mercury	Temperature in degrees Celsius	Relative Density
Sea Level	29.92	+15	.1
5,000ft.	24.89	+ 5.1	0.86
10,000ft.	20.58	- 4.8	0.74
15,000ft.	16.88	-14.7	0.63
20,000ft.	13.75	-24.6	0.53
30,000ft.	8.88	-44.4	0.37
40,000ft.	5.54	-56.5	0.24
50,000ft.	3.44	-56.5	0.15

Fig. 1. Standard Atmosphere.

Clouds

To a pilot knowledgeable in the science of meteorology, clouds are an indication of what is happening in the atmosphere. The location and type of cloud are evidence of such weather phenomena as fronts, turbulence, thunderstorms and tell the pilot what type of conditions may be expected during flight.

Classifications of Clouds

Clouds are classified into four families: high clouds, middle clouds, low clouds and clouds of vertical development.

As well, clouds are identified by the way in which they form. There are two basic types: cumulus and stratus.

Cumulus clouds form in rising air currents and are evidence of unstable air conditions.

Stratus clouds form in horizontal layers and usually form as a layer of moist air is cooled below its saturation point.

Clouds from which precipitation falls are designated **nimbus clouds.**

The cloud heights referred to below are for the temperate regions. In the polar regions, clouds tend to occur at lower heights and in the tropics at greater heights.

Fig. 2. Cirrus.

High Clouds
The bases of high clouds range from 16,500 feet to 45,000 feet and average about 25,000 feet in the temperate regions. They are composed of ice crystals.

Cirrus (Ci). Very high. Thin, wavy sprays of white cloud made up of slender, delicate curling wisps or fibres. Sometimes takes the form of feathers or ribbons or delicate fibrous bands. Often called cats' whiskers or mares' tails.

Cirrocumulus (Cc). Thin clouds, cotton or flake-like. Often called mackerel sky. Gives little indication of future weather conditions.

Cirrostratus (Cs). Very thin high sheet cloud through which the sun or moon is visible, producing a halo effect. Cirrostratus is frequently an indication of an approaching warm front or occlusion and therefore of deteriorating weather.

High clouds have little effect on flying. Some moderate turbulence may be encountered.

Middle Clouds
The bases of middle clouds range from 6500 feet to 23,000 feet. They are composed of ice crystals or water droplets, which may be at temperatures above freezing or may be supercooled.

Altocumulus (Ac). A layer or series of patches of rounded masses of cloud that may lie in groups or lines. Sometimes they indicate the approach of a front but usually they have little value as an indication of future weather developments.

Altocumulus Castellanus (Acc). Altocumulus with a turreted appearance. Instability, turbulence and shower activity are characteristic. Altocumulus castellanus may develop into cumulonimbus.

Fig. 3. Altocumulus.

Altostratus (As). A thick veil of grey cloud that generally covers the whole sky. At first, the sun or moon may be seen through the cloud, but they disappear as the cloud gets thicker. The presence of altostratus indicates the near approach of a warm front. Some light rain or snow may fall from thick altostratus. Icing may occur in this cloud.

There is usually little turbulence associated with middle clouds unless cumulus clouds are embedded in them or unless altocumulus is developing.

Low Clouds
The bases of low clouds range from surface height to about 6500 feet. They are composed of water droplets which may be supercooled and sometimes of ice crystals.

Fig. 4. Altostratus.

Stratus (St). A uniform layer of cloud resembling fog but not resting on the ground. Drizzle often falls from stratus. When stratus cloud is broken up by wind, it is called **stratus fractus**.

Stratocumulus (Sc). A layer or series of patches of rounded masses or rolls of cloud. It is very often thin with blue sky showing through the breaks. It is common in high pressure areas in winter and sometimes gives a little precipitation.

Nimbostratus (Ns). A low layer of uniform, dark grey cloud. When it gives precipitation, it is in the form of continuous rain or snow. The cloud may be more than 15,000 feet thick. It is generally associated with warm fronts.

Little turbulence occurs in stratus. The low cloud bases and poor visibility make VFR operations difficult to impossible.

Clouds of Vertical Development
The bases of this type of cloud may form as low as 1500 feet. They are composed of water droplets when the temperature is above freezing and of ice crystals and supercooled water droplets when the temperature is below freezing. They may appear as isolated clouds or may be embedded in layer clouds.

Cumulus (Cu). Dense clouds of vertical development. They are thick, rounded and lumpy and resemble cotton balls. They usually have flat bases and the tops are rounded. They cast dense shadows

Fig. 5. Cumulus.

From the Ground Up

and appear in great abundance during the warm part of the day and dissipate at night. When these clouds are composed of ragged fragments, they are called **cumulus fractus (Cf)**. Flight under the bases of cumulus clouds is usually bumpy.

Towering Cumulus (TCu). Cumulus clouds that build up into high towering masses. They are likely to develop into cumulonimbus. Rough air will be encountered underneath this cloud. Heavy icing may occur in this cloud type.

Cumulonimbus (Cb). Heavy masses of cumulus clouds that extend well above the freezing level. The summits often spread out to form an anvil shaped top that is characteristic of thunderstorm and showery conditions. Violent vertical currents exist within the cloud. Hail is frequently present within the cloud and may occasionally fall from it. A line of cumulonimbus is often an indication of a cold front. The cloud should be avoided because of its turbulence, the danger of heavy icing and violent electrical activity. Cumulonimbus clouds may be embedded in stratiform clouds.

Fig. 6. Cumulonimbus.

Sky Condition

There are five types of sky condition, relating to cloud cover. In the new METAR weather reporting code, the sky is divided in 8 sements(called **oktas**), in contrast to the SA code in which the sky was divided into 10 segments

1. *Clear: No clouds.*
2. *Few: 2 oktas (8ths) or less of the celestial dome is covered by clouds.*
3. *Scattered: 3 to 4 oktas of the celestial dome is covered by clouds.*
4. *Broken: 5 to 7 oktas of the celestial dome is covered by a layer of clouds based aloft.*
5. *Overcast: 8 oktas of the celestial dome is covered.*

Pressure

Atmospheric pressure is important to the pilot for two reasons. Firstly, airplane altimeters are operated by atmospheric pressure and must be properly set to give correct readings of altitude. Secondly, pressure distribution in the atmosphere controls the winds which are of great importance to the pilot in planning cross country flights.

Atmospheric Pressure

The pressure of the atmosphere at any point is due to the weight of overlying air. Pressure at the surface of the earth is usually measured by the **mercury barometer** and is expressed in inches of mercury (written "Hg). The mercury barometer consists of an open dish of mercury into which the open end of an evacuated glass tube is placed. Atmospheric pressure forces mercury to rise in the tube. The greater the pressure, the higher the column rises. A measurement expressed in inches is, in effect, the length of the column of mercury, the weight of which will balance a column of air extending from the ground to the top of the atmosphere (Fig. 7).

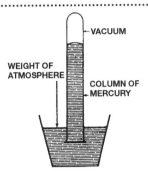

Fig. 7. Principle of the Mercury Barometer.

Pressure, however, is a force and in meteorological work it is common to employ a unit, the hectopascal (hPa), to measure it. A hectopascal is a pressure exerted on an area of 1 square centimeter by a force of 1000 dynes.

One hectopascal is equal to one millibar, the term in common usage prior to the adoption of metric terminology.

A pressure expressed as 29.92 inches of mercury is equivalent to 1013.2 hectopascals.

In the weather information given on public radio and television broadcasts in Canada, atmospheric pressure is expressed in kilopascals (kPa). One kilopascal equals 10 hectopascals. A pressure expressed as 103.32 kPa is equivalent to 1033.2 hectopascals.

Station Pressure, Sea Level Pressure and Altimeter Setting

Station pressure is the actual atmospheric pressure at the elevation of the observing station. It is, in other words, the actual weight of a column of air extending up from the station level to the outer limit of the atmosphere. The value is determined directly from the mercury barometer at the observing station. Since the weight of the atmosphere decreases with altitude, it follows that the atmospheric pressure reading at a station at 5000 feet elevation will be less than that at a station at 1000 feet and still less than that at a station at sea level.

To have a consistent record of the distribution of atmospheric pressure, it is necessary to reduce the station pressures to a common level. This standard is called **mean sea level (MSL) pressure.** The reduction to MSL pressure involves adding to the station pressure the weight of an imaginary column of air extending from the station level down to mean sea level. In determining MSL pressure, local temperature must be taken into account. The temperature value is based on the average of the surface temperatures at the time of observation and for 12 hours before the time of observation. Mean sea level pressure is expressed in hectopascals.

To make an altimeter in an airplane correctly read the true height above mean sea level, it must be set to a standard atmospheric pressure. This pressure reading is called the **altimeter setting.** It differs slightly from the mean sea level pressure. In reducing station pressure to sea level pressure for altimeter setting purposes, the standard sea level temperature of 15°C and the standard lapse rate of 1.98°C per 1000 feet is used in computing the equivalents. When correctly set, the altimeter will then read the true elevation of the airport at which the airplane is parked. (Conversely, if a pilot does not know the altimeter setting but does know the elevation of the airport at which the airplane is parked, he can dial in the correct elevation and get also the correct altimeter setting.) Altimeter setting is reported in inches of mercury.

Pressure Systems

The pressure readings that are taken at various weather reporting stations all over North America are transmitted to forecast offices and are plotted on specially prepared maps. Areas of like pressure are joined by lines, wind direction arrows are entered and the result is a **weather map** that gives the weather man a symbolic picture of the weather over the whole continent.

The lines that join, on a weather map, areas of equal barometric pressure are called **isobars.** These lines are drawn on the map at intervals of four hectopascals, above and below the value of 1000 hectopascals. When the isobars are drawn in, they form definite patterns. They never cross, but form roughly concentric circles and form themselves into distinct areas of high and low pressure.

The various types of pressure systems are: lows, secondary lows, troughs, highs, ridges and cols. They are illustrated in Fig. 8 as they would appear on a typical weather map.

The pressure patterns on the weather map are very like contour lines on a topographic map; the high pressure areas correspond to hills and the low pressure areas to valleys. It is important to recognize that the pressure systems are relative to the pressure around them. A high pressure area, in the centre of which the pressure reading is, for example, 1000 hectopascals, is classed a high because the surrounding pressure is less than 1000 hectopascals. A pressure area with the same pressure reading of 1000 hectopascals at the centre would be classed as a low if the surrounding pressure is higher than 1000 hectopascals.

These pressure systems are constantly moving or changing in appearance. Lows may deepen or fill and highs may build or weaken. Most systems move in a general west to east direction.

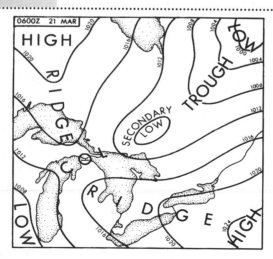

Fig. 8. Pressure Systems.

If we put down on a map beside the barometer readings and wind arrows the state of the weather, we shall see that, in the high pressure areas, it is usually fine and clear, probably cooler, while in the low pressure areas, it is generally rainy and cloudy on the east side and probably fine on the west side. The weather, in fact, is intimately connected with the shape of the isobars.

Low Pressure Areas

Areas of low pressure are called cyclones, depressions or simply lows. A low is a region of relatively low pressure with the lowest pressure at the centre.

A low may cover a small region such as a county, or it may extend across half a continent. Some are much deeper than others. A tornado, for example, is a very deep, small but concentrated low. A deep low is one where the pressure is very low in the centre and the isobars are rather close together. A shallow depression is low in the centre, but not much lower than the surrounding areas.

Depressions seldom stay long in one place but generally move in an easterly direction. Their average rate of movement is 500 miles (about 800 kilometers) a day in summer, and 700 miles (about 1100 kilometers) a day in winter. Their drift is generally to the north-east or south-east. Only rarely is there an exception to this pattern of easterly drift of low pressure areas.

Secondary Low

A secondary low is a smaller disturbance of a cyclonic nature which forms within the area dominated by the main depression. The secondary centre revolves around the main centre in an anti-clockwise direction. Secondaries are frequently associated with thunderstorms in summer and gales or heavy precipitation in winter.

Trough of Low Pressure

A trough is an elongated U-shaped area of low pressure with higher pressure on either side, which may bring about a gradual windshift.

The term "trough" is also applied to the V-shape formed by the sharp bending or kinking of the isobars along a frontal surface, sometimes referred to as a V-shaped depression. (see Fronts). Sudden windshifts may be expected, accompanied by the type of weather generally associated with fronts.

Col

A col is a neutral region between two highs and two lows. Weather conditions are apt to be unsettled. In winter, the mixing of air of dissimilar air masses frequently produces fog. In summer, showers or thunderstorms may occur. While it is quite possible for weather conditions to be fair, generally speaking, cols may be regarded as regions of undependable weather.

High Pressure Areas

A high, or anti-cyclone, is an area of relatively high pressure, the pressure being higher than that of the surrounding regions. The pressure is highest at the centre and decreases towards the outside. The accompanying weather is usually fine to fair, clear and bright, with light, moderately cool breezes. The temperature becomes cooler at night with possibly a little frost. In winter, a high very often brings clear, cold weather.

Occasionally, however, an anti-cyclone occurs in which a persistent cloud sheet develops, causing dull, gloomy weather.

The winds associated with high pressure are usually light and rather variable. They circulate in a clockwise direction around the centre, blowing outwards into the lows.

From the Ground Up

Highs move much more slowly across the country than depressions and occasionally remain almost stationary for days at a time.

Ridge of High Pressure
An anti-cyclone ridge is a neck or ridge of high pressure with lower pressure lying on either side. The weather in a ridge is generally fine to fair.

Pressure Changes
Pressure readings are taken at regular intervals (usually hourly) at weather stations. Weather maps are prepared four times a day at six hour intervals. From these readings and maps, the changes in pressure can be observed and approaching weather forecast. If a low, for example, is approaching a station, the pressure will steadily fall. Once the centre of the low has passed by, the pressure will begin to rise. This pattern of changing pressure is called **pressure tendency.**

Pressure Gradient
If some of the air were removed from a room, the pressure would be reduced and the pressure outside would force air in through the doors and windows until the room was again filled with the normal amount of air. Similarly, in the case of low and high pressure areas, there is a tendency for the higher pressure air in a high to flow towards the area of lower pressure.

The speed at which this movement of air occurs depends on the **pressure gradient.** The pressure gradient is defined as the rate of change of pressure over a given distance measured at right angles to the isobars.

The steepness of the pressure gradient is measured by the nearness of the isobars. Where the isobars are spaced widely apart, the pressure gradient is shallow and the movement of air (wind) is slow or light. Where the isobars are very close together, the gradient is steep and the wind is strong.

Coriolis Force
The air moving from a high pressure area to a low pressure area does not flow directly from the one to the other. It is deflected to the right in the Northern Hemisphere by a force called the coriolis force and as a result flows parallel to the isobars.

Anything moving above the surface of the earth will continue to move in a straight line if no force acts on it, but the earth in its rotation moves under the moving body. The moving body is, therefore, apparently deflected to the right in the Northern Hemisphere. This is known as **Fennell's Law.** The apparent deflecting force is called **coriolis force** (Fig. 9).

Hence, the wind does not blow straight into an area of low pressure from all sides but is deflected to the right and blows around

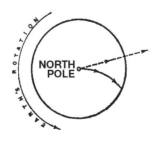

Cut out a paper disk to represent the earth. Place a pin through the centre to represent the north pole. Rotating the disk around the pin (anti-clockwise) to represent the earth's rotation, attempt to draw a straight line (the dotted line in the figure) to represent a wind blowing from north to south. The line will curve as shown, or in other words, be deflected to the right.

Fig. 9. Coriolis Force.

the area of low pressure **anti-clockwise** in the Northern Hemisphere. Coriolis force causes water to swirl **anti-clockwise** in a wash basin when the plug has been removed.

In the case of a high pressure area, the air flows out from the area of high pressure, but is deflected to the right because of the coriolis force. As a result, the wind blows clockwise round an area of high pressure in the Northern Hemisphere.

In either case, if you stand with your back to the wind, the low pressure area will be on your left side. This is known as **Buys Ballot's Law.** In the Southern Hemisphere, the reverse is the case.

Surface Friction
In the lower levels of the atmosphere, a third force acts on the direction and speed of the air moving from areas of high pressure to areas of low pressure. This force is **surface friction.** Friction between the air masses and the surface of the earth tends to slow down the movement of the air, thereby reducing the wind speed. This in turn retards the coriolis force. As a result, air tends to move across the isobars at a slight angle inwards towards the centre of the lows and outwards from the centres of the highs.

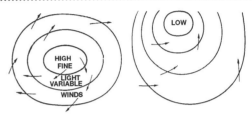

Fig. 10. Surface Friction Effect on Air Movement Across isobars.

The more surface friction, the greater the angle at which the flow of air is deflected from a flow parallel to the isobars. Over water, it is less than over land where the roughness of the terrain may cause a deflection of as much as 40°. Over water, the deflection is rarely more than 10°. The friction effect is greatest near the surface of the earth but may be carried aloft by turbulence to heights as great as 2000 feet. Above this altitude, it is practically negligible and the winds tend to flow parallel to the isobars as a result of the coriolis force.

Centrifugal Force
Centrifugal force, acting on the circulating flow of air around the high and low pressure areas, tends to increase wind speed in the high pressure areas and decrease it in the low pressure areas.

Convergence and Divergence
Convergence. The inflow of air **into** an area is called convergence. It is usually accompanied by an upward movement of air to permit the excess accumulation of air to escape. Areas of convergent winds are favourable to the occurrence of precipitation in the form of thunderstorms, rain, hail or snow.

Divergence. When there is a flow of air **outwards** from a region, the condition is known as divergence. The outflow is compensated by a downward movement of air from aloft. Areas of divergent winds are not favourable to the occurrence of precipitation.

Winds
The horizontal movement of air, called **wind**, is a factor of great importance to a pilot as he plans a flight. Upper winds encountered en route will affect groundspeed either favourably or detrimentally and thus have a bearing on time en route and fuel consumption. Surface winds are important in landing and take-off.

Hemispheric Prevailing Winds

Since the atmosphere is fixed to the earth by gravity and rotates with the earth, there would be no circulation if some force did not upset the atmosphere's equilibrium. The heating of the earth's surface by the sun is the force responsible for creating the circulation that does exist.

Because of the curvature of the earth, the most direct rays of the sun strike the earth in the vicinity of the equator resulting in the greatest concentration of heat, the largest possible amount of reradiation and the maximum heating of the atmosphere in this area of the earth. At the same time, the sun's rays strike the earth at the poles at a very oblique angle, resulting in a much lower concentration of heat and much less reradiation so that there is, in fact, very little heating of the atmosphere over the poles and consequently very cold temperatures.

Cold air, being more dense, sinks and hot air, being less dense, rises. Consequently, the rising warm air at the equator becomes even less dense as it rises and its pressure decreases. An area of low pressure, therefore, exists over the equator.

Warm air rises until it reaches a certain height at which it starts to spill over into surrounding areas. At the poles, the cold dense air is sinking. Air from the upper levels of the atmosphere flows in on top of it increasing the weight and creating an area of high pressure at the poles.

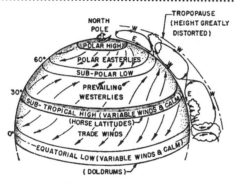

Fig. 11. Wind Circulation.

The air that rises at the equator does not flow directly to the poles. Due to the rotation of the earth, there is a build up of air at about 30° north latitude. (The same phenomenon occurs in the Southern Hemisphere.) Some of the air sinks, causing a belt of high pressure at this latitude.

The sinking air reaches the surface and flows north and south. The air that flows south completes one cell of the earth's circulation pattern, as shown in Fig. 11. The air that flows north becomes part of another cell of circulation between 30° and 60° north latitude. At the same time, the sinking air at the north pole flows south and collides with the air moving north from the 30° high pressure area. The colliding air is forced upward and an area of low pressure is created near 60° north. The third cell circulation pattern is created between the north pole and 60° north.

Because of the rotation of the earth and the coriolis force, air is deflected to the right in the Northern Hemisphere. As a result, the movement of air in the polar cell circulation produces the **polar easterlies.** In the circulation cell that exists between 60° and 30° north, the movement of air produces the **prevailing westerlies.** In the tropic circulation cell, the **northeast trade winds** are produced. These are the so-called permanent wind systems of the earth. They are illustrated in Fig. 11.

Upper Level Winds

There are two main forces which affect the movement of air in the upper levels. The pressure gradient causes the air to move horizontally, forcing the air directly from a region of high pressure to a region of low pressure. The coriolis force, however, deflects the direction of the flow of the air (to the right in the Northern Hemisphere) and causes the air to flow parallel to the isobars (or **contours** as the lines joining areas of equal pressure are called on Upper Level Weather Charts.)

Winds in the upper levels will blow clockwise around areas of high pressure and counterclockwise around areas of low pressure.

The speed of the wind is determined by the pressure gradient. The winds are strongest in regions where the isobars are close together.

Surface Winds

Surface friction plays an important role in the speed and direction of surface winds. As a result of the slowing down of the air as it moves over the ground, wind speeds are less than would be expected from the pressure gradient on the weather map and the direction is changed so that the wind blows across the isobars into a centre of low pressure and out of a centre of high pressure.

The effect of friction usually does not extend more than a couple of thousand feet into the air. At 3000 feet above the ground, the wind blows parallel to the isobars with a speed proportional to the pressure gradient.

Even allowing for the effects of surface friction, the winds, locally, do not always show the speed and direction that would be expected from the isobars on the surface weather map. These variations are usually due to geographical features such as hills, mountains and large bodies of water. Except in mountainous regions, the effect of terrain features that cause local variations in wind extends usually no higher than about 2000 feet above the ground.

Land and Sea Breezes

Land and sea breezes are caused by the differences in temperature over land and water. The sea breeze occurs during the day when the land area heats more rapidly than the water surface. This results in the pressure over the land being lower than that over the water. The pressure gradient is often strong enough for a wind to blow from the water to the land.

The land breeze blows at night when the land becomes cooler. Then the wind blows towards the warm, low pressure area over the water.

Land and sea breezes are very local and affect only a narrow area along the coast.

Mountain Winds

Hills and valleys substantially distort the airflow associated with the prevailing pressure system and the pressure gradient. Strong up and down drafts and eddies develop as the air flows up over hills and down into valleys. Wind direction changes as the air flows around hills. Sometimes lines of hills and mountain ranges will act as a barrier, holding back the wind and deflecting it so that it flows parallel to the range. If there is a pass in the mountain range, the wind will rush through this pass as through a funnel with considerable speed. The airflow can be expected to remain turbulent and erratic for some distance as it flows out of the hilly area and into the flatter countryside.

Day time heating and night time cooling of the hilly slopes cause day to night variations in the airflow. At night, the sides of the hills cool by radiation. The air in contact with them becomes cooler and therefore denser and it blows down the slope into the valley. This is a **katabatic wind** (sometimes also called a **mountain breeze**). If the slopes are covered with ice and snow, the katabatic wind will blow, not only at night but also during the day, carrying the cold dense air into the warmer valleys. The slopes of hills not covered by snow will be warmed during the day. The air in contact with them becomes warmer and less dense and, therefore, flows up the slope. This is an **anabatic wind** (or **valley breeze**).

In mountainous areas, local distortion of the airflow is even more severe. Rocky surfaces, high ridges, sheer cliffs, steep valleys, all combine to produce unpredictable flow patterns and turbulence. The most treacherous of the local wind systems associated with mountains, however, is the **mountain wave**.

The Mountain Wave

Air flowing across a mountain range usually rises relatively smoothly up the slope of the range but, once over the top, it pours down the other side with considerable force, bouncing up and down, creating eddies and turbulence and also creating powerful vertical waves that may extend for great distances downwind of the mountain range. This phenomenon is known as a mountain wave. Fig. 12 illustrates the type of wave pattern which may be activated by a very strong wind. Note the up and down drafts and the rotating eddies formed downstream.

If the air mass has a high moisture content, clouds of very distinctive appearance will develop.

Cap Cloud. Orographic lift causes a cloud to form along the top of the ridge. The wind carries this cloud down along the leeward slope where it dissipates through adiabatic heating. The base of this cloud lies near or below the peaks of the ridge; the top may reach a few thousand feet above the peaks.

Lenticular (Lens Shaped) Clouds form in the wave crests aloft and lie in bands that may extend to well above 40,000 feet.

Rotor Clouds form in the rolling eddies downstream. They resemble a long line of stratocumulus clouds, the bases of which lie below the mountain peaks and the tops of which may reach to a considerable height above the peaks. Occasionally these clouds develop into thunderstorms.

The clouds, being very distinctive, can be seen from a great distance and provide a visible warning of the mountain wave condition. Unfortunately, sometimes they are embedded in other cloud systems and are hidden from sight. Sometimes the air mass is very dry and the clouds do not develop.

The severity of the mountain wave and the height to which the disturbance of the air is affected is dependent on the strength of the wind, its angle to the range and the stability or instability of the air. The most severe mountain wave conditions are created in strong airflows that are blowing at right angles to the range and in very unstable air. A jet stream blowing nearly perpendicular to the mountain range increases the severity of the wave condition.

The mountain wave phenomenon is not limited only to high mountain ranges, such as the Rockies, but is also present to a lesser degree in smaller mountain systems and even in lines of small hills.

Mountain waves present problems to pilots for several reasons.

Vertical Currents. Downdrafts of 2000 feet per minute are common and downdrafts as great as 5000 feet per minute have been reported. They occur along the downward slope and are most severe at a height equal to that of the summit. An airplane, caught in a downdraft, could be forced to the ground.

Turbulence is usually extremely severe in the air layer between the ground and the tops of the rotor clouds.

Wind Shear. The wind speed varies dramatically between the crests and troughs of the waves. The wind speed variation is usually most dramatic near the mountain range and is responsible for severe wind shear.

Altimeter Error. The increase in wind speed results in an accompanying decrease in pressure which in turn affects the accuracy of the pressure altimeter. (See also **Mountain Effect** in the Section **The Altimeter** in Chapter **Theory of Flight**.)

Icing. The freezing level varies considerably from crest to trough. Severe icing can occur because of the large supercooled droplets sustained in the strong vertical currents.

When flying over a mountain ridge where wave conditions exist: (1) Avoid ragged and irregular shaped clouds — the irregular shape indicates turbulence. (2) Approach the mountain at a 45 degree angle. If you should suddenly decide to turn back, a quick turn can be made away from the high ground. (3) Avoid flying in cloud on the mountain crest (cap cloud) because of strong downdrafts and turbulence. (4) Allow sufficient height to clear the highest ridges with altitude to spare to avoid the downdrafts and eddies on the downwind slopes. (5) **Always remember that your altimeter can read over 3000 ft. in error on the high side in mountain wave conditions.**

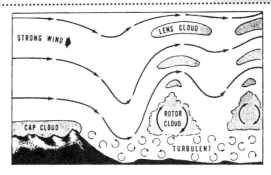

Fig. 12. The Mountain Wave.

Gustiness

A **gust** is a rapid and irregular fluctuation of varying intensity in the upward and downward movement of air currents. It may be associated with a rapid change in wind direction. Gusts are caused by mechanical turbulence that results from friction between the air and the ground and by the unequal heating of the earth's surface, particularly on hot summer afternoons.

Squalls

A **squall** is a sudden increase in the strength of the wind of longer duration than a gust and may be caused by the passage of a fast moving cold front or thunderstorm. Like a gust, it may be accompanied by a rapid change of wind direction.

Diurnal Variations

Given an isobaric pattern and a pressure gradient that are fundamentally identical, surface winds can usually be expected to be stronger and gustier during the day than during the night. In unstable air, there are ascending and descending currents that provide a vertical link between winds at different levels in the convective layer. In stable air, especially in an inversion that typically occurs at night, this link does not exist and winds can be very different above and below the inversion.

When the sun sets and night falls, the nocturnal inversion develops (the surface air cools by radiation and there is warm air above the cooler surface layer). With no surface heating, turbulence diminishes and the convective current link between the low level air affected by the earth's surface and the free flowing air above disappears. The surface layer of air becomes more strongly influenced by the frictional effects of the earth's surface. The surface winds decrease in speed and back in direction (see **Veering and Backing** below).

When the sun rises, warming breaks down the nocturnal inversion and vertical currents develop. The resulting turbulence causes the faster air aloft to be brought down to the surface and the slower surface air to be carried aloft. Since the wind direction at the higher level is parallel to the isobars and its speed is greater than the surface wind which is slowed by friction, the mixing causes the surface wind to increase in speed, veer in direction and become gusty. It reaches its greatest speed during the afternoon when the temperature rises to a maximum value.

Diurnal variations do not occur over large lakes and oceans since nocturnal inversions do not develop over water.

Eddies–Mechanical Turbulence

Friction between the moving air mass and surface features of the earth (hills, mountains, valleys, trees, buildings, etc.) is responsible for the swirling vortices of air commonly called eddies. They vary considerably in size and intensity depending on the size and roughness of the surface obstruction, the speed of the wind and the degree of stability of the air. They can spin in either a horizontal or vertical plane. Unstable air and strong winds produce more vigorous eddies. In stable air, eddies tend to quickly dissipate. Eddies produced in mountainous areas are especially powerful.

The bumpy or choppy up and down motion that signifies the presence of eddies makes it difficult to keep an airplane in level flight. Because they are present predominantly in the lower levels, pilots encounter these eddies during landing and take-off when the bumpiness and sudden drops and increases in wind speed and direction may be critical.

Dust Devils

Dust devils are phenomena that occur quite frequently on the hot dry plains of mid-western North America. They can be of sufficient force to present a hazard to pilots of light airplanes flying at low speeds.

They are small heat lows that form on clear hot days. Given a steep lapse rate caused by cool air aloft over a hot surface, little horizontal air movement, few or no clouds, and the noonday sun heating flat arid soil surfaces to high temperatures, the air in contact with the ground becomes super-heated and highly unstable. This surface layer of air builds until something triggers an upward movement. Once started, the hot air rises in a column and draws more hot air into the base of the column. Circulation begins around this heat low and increases in velocity until a small vigorous whirlwind is created. Dust devils are usually of short duration and are so named because they are made visible by the dust, sand and debris that they pick up from the ground.

Dust devils pose the greatest hazard near the ground where they are most violent. Pilots proposing to land on superheated runways in areas of the mid-west where this phenomenon is common should scan the airport for dust swirls or grass spirals that would indicate the existence of this hazard.

Tornadoes

Tornadoes are violent, circular whirlpools of air associated with severe thunderstorms and are, in fact, very deep, concentrated low pressure areas. They are shaped like a funnel hanging out of the cumulonimbus cloud and are dark in appearance due to the dust and debris sucked into their whirlpools. They range in diameter from about 100 feet to one half mile and move over the ground at speeds of 25 to 50 knots. Their path over the ground is usually only a few miles long although tornadoes have been reported to cut destructive swaths as long as 100 miles. The great destructiveness of tornadoes is caused by the very low pressure in their centres and the high wind speeds which are reputed to be as great as 300 knots.

Wind Speed and Direction

Wind speeds for aviation purposes are expressed in knots (nautical miles per hour). In the weather reports on Canadian public radio and television, however, wind speeds are given in kilometers per hour.

In a discussion of wind direction, the compass point from which the wind is blowing is considered to be its direction. Therefore, a north wind is one that is blowing from the north towards the south. In aviation weather reports, area and aerodrome forecasts, the wind is always reported in **degrees true.** In ATIS broadcasts and in the information given by the tower for landing and take-off, the wind is reported in **degrees magnetic.**

Veering and Backing

The wind **veers** when it changes direction clockwise. Example: The surface wind is blowing from 270°T. At 2000 feet it is blowing from 280°T. It has changed in a right-hand, or clockwise, direction and is called a veering wind.

The wind **backs** when it changes direction anti-clockwise. Example: The wind direction at 2000 feet is 090°T, at 3000 feet 085°T. It is changing in a left-hand, or anti-clockwise, direction and is called a backing wind.

In a descent from several thousand feet above the ground to ground level, the wind will usually be found to back and also decrease in velocity, as the effect of surface friction becomes apparent. Above the friction/surface layer (2 to 4 thousand feet), the pressure gradient and the coriolis force are in balance and the air is moving parallel to the isobars. In the friction layer near the surface, wind speed starts to decrease due to friction. Because of the reduced wind speed, the coriolis force is also reduced and the air is not deflected to the right as much. Conversely, in a climb from the surface to several thousand feet AGL, the wind will veer and increase.

Wind Speed

There are certain tell tale signs that help a pilot to determine the speed of the wind.

A wind of 1 to 2 knots: smoke drifts and water ripples.

A wind of 4 to 6 knots: leaves rustle; small wavelets break the surface of water bodies but there are no white caps.

A wind of 7 to 10 knots: leaves and small twigs move as do flags; scattered whitecaps break the surface of water bodies.

A wind of 11 to 16 knots: small branches move; long waves and frequent whitecaps can be seen on water surfaces.

A wind of 17 to 21 knots: small trees sway; many long whitecaps break the water surface.

A strong wind of 25 to 31 knots: large branches move; large waves, spray and white foam crests break the surface of water bodies.

From the Ground Up

Wind Shear

Wind shear is the sudden "tearing" or "shearing" effect encountered along the edge of a zone in which there is a violent change in wind speed or direction. It can occur at high or low altitude and is most often associated with strong temperature inversions or density gradient. It produces churning motions and consequently turbulence. Under some conditions, wind direction changes of as much as 180 degrees and speed changes of as much as 80 knots have been measured.

The effect on airplane performance of encountering wind shear derives from the fact that the wind can change much faster than the airplane mass can be accelerated or decelerated. Severe wind shears can impose penalties on an airplane's performance that are beyond its capabilities to compensate, especially during the critical landing and take-off phase of flight.

In Cruising Flight

In cruising flight, wind shear will likely be encountered in the transition zone between the pressure gradient wind and the distorted local winds at the lower levels. It will also be encountered when climbing or descending through a temperature inversion and when passing through a frontal surface. Wind shear is also associated with the jet stream (see below). Airplanes encountering wind shear may experience a succession of updrafts and downdrafts, reductions or gains in headwind, or wind shifts that disrupt the established flight path. It is not usually a major problem because altitude and airspeed margins will be adequate to counteract the shear's adverse effects. On occasion, however, the wind shear may be severe enough to cause an abrupt increase in load factor which might stall the airplane or inflict structural damage.

Near the Ground

Wind shear, encountered near the ground, is more serious and potentially very dangerous. There are four common sources of wind shear encountered near the ground: frontal activity, low level wind shear associated with thunderstorms, temperature inversions and strong surface winds passing around natural or man-made obstacles.

Frontal Wind Shear. Wind shear is usually a problem only in fronts with steep wind gradients. If the temperature difference across the front at the surface is 5°C or more and if the front is moving at a speed of about 30 knots or more, wind shear is likely to be present. Frontal wind shear is a phenomenon associated with fast moving cold fronts but can be present in warm fronts as well.

Low Level Wind Shear. Low level wind shear, associated with thunderstorms, occurs as the result of two phenomena, the gust front and downbursts. As the thunderstorm matures, strong downdrafts develop, strike the ground and spread out horizontally along the surface well in advance of the thunderstorm itself. This is the **gust front.** Winds can change direction by as much as 180° and reach speeds as great as 100 knots as far as 10 miles ahead of the storm. The **downburst** is an extremely intense localized downdraft flowing out of a thunderstorm. The power of the downburst can exceed aircraft climb capabilities. The downburst (there are two types of downbursts: **macrobursts** and **microbursts**) usually is much closer to the thunderstorm than the gust front. Dust clouds, roll clouds, intense rainfall or virga (rain that evaporates before it reaches the ground) are clues to the possibility of downburst activity but there is no way to accurately predict its occurrence. (See also **Thunderstorms.**) The two diagrams in Fig. 13 depict the possible accident sequence when an aircraft encounters severe wind shear during the take-off or landing approach.

Temperature Inversions. Overnight cooling creates a temperature inversion a few hundred feet above the ground that can produce significant wind shear, especially if the inversion is coupled with the low level jet stream.

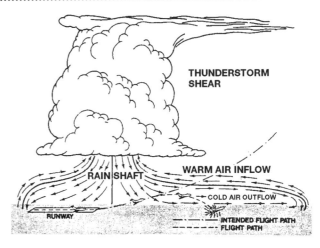
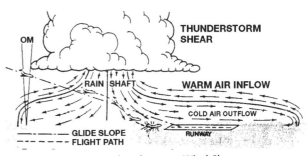

Fig. 13. Thunderstorm Wind Shear.

As a nocturnal inversion develops, the wind shear near the top of the inversion increases. It usually reaches its maximum speed shortly after midnight and decreases in the morning as daytime heating dissipates the inversion. This phenomenon is known as the **low level nocturnal jet stream.** The low level jet stream is a sheet of strong winds, thousands of miles long, hundreds of miles wide and hundreds of feet thick that forms over flat terrain such as the prairies. Wind speeds of 40 knots are common, but greater speeds have been measured. Low level jet streams are responsible for hazardous low level shear.

As the inversion dissipates in the morning, the shear plane and gusty winds move closer to the ground, causing wind shifts and increases in wind speed near the surface.

Surface Obstructions. The irregular and turbulent flow of air around mountains and hills and through mountain passes causes serious wind shear problems for aircraft approaching to land at airports near mountain ridges. Wind shear is a phenomenon associated with the mountain wave (see **Mountain Wave** above). This type of wind shear is almost totally unpredictable but should be expected whenever surface winds are strong.

Wind shear is also associated with hangars and large buildings at airports. As the air flows around such large structures, wind direction changes and wind speed increases causing shear.

Wind shear occurs both horizontally and vertically. Vertical shear is most common near the ground and can pose a serious hazard to airplanes during take-off and landing. The airplane is flying at lower speeds and in a relatively high drag configuration. There is little altitude available for recovering and stall and maneuver margins are at their lowest. An airplane encountering the wind shear phenomenon may experience a large loss of airspeed because of the sudden change in the relative airflow as the airplane flies into a new, moving air mass. The abrupt drop in airspeed may result in a stall, creating a dangerous situation when the airplane is only a few hundred feet off the ground and very vulnerable.

The operational aspects of flight in wind shear zones is discussed in the Chapter **Airmanship.**

The Jet Stream

Narrow bands of exceedingly high speed winds are known to exist in the higher levels of the atmosphere at altitudes ranging from 20,000 to 40,000 feet or more. They are known as **jet streams.** As many as three major jet streams may traverse the North American continent at any given time. One lies across Northern Canada and one across the U.S. A third jet stream may be as far south as the northern tropics but it is somewhat rare. A jet stream in the mid latitudes is generally the strongest.

The jet stream appears to be closely associated with the tropopause and with the polar front. It typically forms in the break between the polar and the tropical tropopause where the temperature gradients are intensified (See Fig. 14). The mean position of the jet stream shifts south in winter and north in summer with the seasonal migration of the polar front. Because the troposphere is deeper in summer than in winter, the tropopause and the jets will normally be at higher altitudes in the summer.

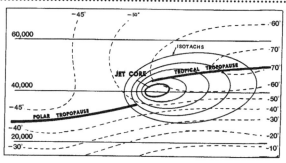

Fig. 14. The Jet Stream and the Tropopause.

Long, strong jet streams are usually also associated with well developed surface lows beneath deep upper troughs and lows. A low developing in the wave along the frontal surface lies south of the jet. As it deepens, the low moves near the jet. As it occludes, the low moves north of the jet which crosses the frontal system near the point of occlusion. The jet flows roughly parallel to the front (See Fig. 15).

The subtropical jet stream is not associated with fronts but forms because of strong solar heating in the equatorial regions. The ascending air turns poleward at very high levels but is deflected by the coriolis force into a strong westerly jet. The subtropical jet predominates in winter.

The jet streams flow from west to east and may encircle the entire hemisphere. More often, because they are stronger in some places than in others, they break up into segments some 1000 to 3000 nautical miles long. They are usually about 300 nautical miles wide and may be 3000 to 7000 feet thick. These jet stream segments move in an easterly direction following the movement of pressure ridges and troughs in the upper atmosphere.

Winds in the central core of the jet stream are the strongest and may reach speeds as great as 250 knots, although they are generally between 100 and 150 knots. Wind speeds decrease toward the outer edges of the jet stream and may be blowing at only 25 knots there. The rate of decrease of wind speed is considerably greater on the northern edge than on the southern edge. (Note the closeness of the isotachs, lines joining points where winds of equal speed have been recorded, on the polar edge of the jet in Fig. 14.) Wind speeds in the jet stream are, on average, considerably stronger in winter than in summer.

The air in a jet rotates slowly around the core with the upward motion on the equatorial edge. If the air is moist, cirrus clouds will form in the ascending air. Their ragged, windswept appearance is always an indication of very strong winds. Sometimes quite dense cirrostratus clouds form in broad bands over the ridge of the jet stream, following the flow of the jet. They may be 10 to 12 thousand feet thick. The clouds taper off, sometimes quite sharply, at the polar edge of the jet. Dense, jet stream cirriform cloudiness is most prevalent along mid latitude and polar jets.

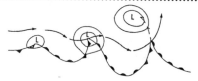

Fig. 15. Jet Stream Position Relative to a Frontal System.

Like other weather phenomena, jet streams grow, shift, intensify, decline and die. Since they do shift location rapidly from day to day and with the seasons, their position can best be found with reference to specially constructed charts prepared by weathermen. Knowing the location of the jet stream is important in planning long range flights at high altitudes. For obvious reasons, on a flight eastbound, a pilot would want to take advantage of the excellent tail winds the jet stream would provide, whereas on the flight westbound, he would want to avoid it.

Clear Air Turbulence (CAT)

Closely associated with the jet stream is clear air turbulence (CAT), a bumpy, turbulent condition that occurs in cloudless sky and can be severe enough to be a hazard to modern high performance airplanes. It occurs at high altitudes, usually above 15,000 feet and is more severe near 30,000 feet. CAT develops in the turbulent energy exchange along the boundary between contrasting cold and warm air masses. Such cold and warm advection along with strong wind shear develops near the jet stream especially where curvature of the jet stream increases sharply.

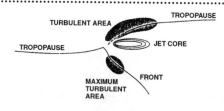

Fig. 16. Turbulence Associated with The Jet Stream.

The most probable place to expect CAT is just above the central core of the jet stream near the polar tropopause and just below the core. Clear air turbulence does not occur in the core. CAT is encountered more frequently in winter when the jet stream winds are strongest. Nevertheless, CAT is not always present in the jet stream and, because it is random and transient in nature, it is almost impossible to forecast.

Clear air turbulence may be associated with other weather patterns, especially in wind shear associated with the sharply curved contours of strong lows, troughs and ridges lying aloft, at or below the tropopause, and in areas of strong cold or warm air advection. Mountain waves create severe CAT that may extend from the mountain crests to as high as 5000 feet above the tropopause. Another form of CAT which may be encountered at lower levels is wing tip vortices, which are discussed in the Chapter **Airmanship.**

Since severe CAT does pose a hazard to airplanes, pilots should try to avoid or minimize encounters with it. These rules of thumb may help.

Avoid jet streams with strong winds (150 knots) at the core. Strong wind shears are likely above and below the core.

CAT within the jet stream is more intense above and to the lee of mountain ranges.

If the 20 knot isotachs (lines joining areas of equal wind speeds) are closer than 60 nautical miles on the charts showing the locations of the jet stream, wind shear and CAT are possible.

Turbulence is also related to vertical shear. If the wind speed is increasing by more than 5 knots per thousand feet, turbulence is likely. Vertical shear is also related to temperature gradient. If the 5°C isotherms on the upper level charts are closer than 120 nautical miles (2 degrees of latitude), there is usually turbulence.

Wind shift areas associated with pressure troughs are frequently turbulent. The sharpness of the wind shift affects the degree of turbulence.

Curving jet streams are likely to have turbulent edges, especially those that curve around a deep pressure trough.

When moderate or severe CAT has been reported or is forecast, adjust speed to rough air speed immediately on encountering the first bumpiness or even before encountering it to avoid structural damage to the airplane.

The areas of CAT are usually shallow and narrow and elongated with the wind. If jet stream turbulence is encountered with a tail wind or head wind, a turn to the right will find smoother air and more favourable winds. If the CAT is encountered in a cross wind, it is not so important to change course as the rough area will be narrow.

Establish a course across a sharp pressure trough line rather than parallel to it if turbulence is encountered in an abrupt wind shift associated with the trough.

Mountain waves produce intense vertical and horizontal wind shear. If the flight path crosses a mountain range where a mountain wave condition may exist, select turbulence penetration speed and avoid areas where the terrain drops sharply.

Humidity, Temperature and Stability

Humidity

Of all the elements which compose the lower atmosphere, water vapour is the most variable. Although it forms but a small proportion of the total mass of air at any time, its effects from a flying point of view are of great importance. It is the only gas which can change into a liquid or a solid under ordinary atmospheric conditions and it is because of this characteristic that most of the" weather" develops.

The moisture in the atmosphere originates principally from evaporation from the earth's water bodies (oceans, lakes, rivers, etc.) and from transpiration from the earth's vegetation. It exists in the atmosphere in two forms. In its invisible form, it is water vapour. In its visible form, it is either water droplets or ice crystals.

By a process called **condensation,** water vapour changes into water droplets. By a process called **sublimation,** water vapour changes into ice crystals. In its visible form, as either water droplets or ice crystals, moisture forms clouds and fog. Further developments within a cloud may lead to precipitation.

The amount of water vapour that a given volume of air can contain is governed by its temperature. Warm air can hold more moisture than cold air. When a mass of air contains the maximum amount of water vapour it can hold **at a given temperature,** it is said to be **saturated**. If the temperature falls any lower after the air is saturated, some of the invisible water vapour will condense out in the form of visible water droplets. By this process of **condensation**, clouds, fog and dew are formed.

If the temperature is below freezing when the saturation occurs, the water vapour changes directly into ice crystals without passing through the visible water droplet stage. This process is known as **sublimation.**

The process is reversible. If the visible water droplets or ice crystals are heated, they will turn back into the invisible gas, water vapour. Therefore, if fog or clouds are heated, they will disappear.

Any change of state, even though there is no change of temperature, involves a heat transaction. During the process of condensation, as water vapour changes to visible water droplets while remaining at the same temperature, energy is released to the atmosphere in the form of heat. This heat is known as the **latent heat of vaporization.** In the reverse process, heat is absorbed when liquid evaporates to its invisible state at the same temperature.

Melting and freezing involve the exchange of the **latent heat of fusion.** In the process of sublimation, the change directly from water vapour to ice crystals, the release of energy in the form of heat is equal to the latent heat of vaporization plus the latent heat of fusion.

Proof of the energy inherent in latent heat is seen in the thunderstorm and in the hurricane. The tremendous power of these two weather phenomena is derived from the energy released in the form of the latent heat of vaporization and of fusion as the water vapour changes to water droplets and to ice crystals. These violent forms of weather customarily occur in very warm, moist air since more energy is released during condensation and sublimation at warmer temperatures than at cooler ones.

The process of condensation or sublimation cannot take place unless there are microscopic particles present in the air on which the water vapour can condense. The atmosphere contains a vast quantity of impurities, such as fine dust from deserts, smoke from industrial regions, salts from the oceans, and seeds, pollen, etc. Since these originate from the earth, they exist only in the lower layers of the atmosphere. They act as **condensation nuclei** on which the condensation of water vapour takes place when air is cooled below its saturation temperature (dewpoint).

Supercooled Water Droplets

One would expect that visible water droplets in the air would freeze as soon as the ambient temperature drops to 0°C. In fact, liquid water droplets often persist at temperatures well below 0°C and are then referred to as supercooled.

The reason for the existence of liquid water droplets at temperatures well below freezing is complicated. Briefly, it might be said that the freezing process is initiated by the nuclei on which the water droplets form. These nuclei are of different composition and because of their varying chemical characteristics, some of them do not initiate the freezing process until quite low temperatures. The temperature at which a supercooled water droplet freezes also depends on its size. Large droplets freeze at temperatures only slightly below freezing whereas very minute droplets may remain in liquid form until the temperature nears -40°C. Below -40°C, very few, if any, supercooled droplets exist. Supercooled water droplets are often found in abundance in clouds at temperatures between 0°C and -15°C.

Dewpoint

The temperature to which unsaturated air must be cooled at **constant pressure** to become saturated (without the addition or removal of any water vapour) is called the **dewpoint.**

When the spread between the temperature and dewpoint is very small, the air can be said to be nearly saturated and a slight drop in temperature may cause condensation in the form of clouds, fog, or precipitation.

Relative Humidity

The **relative humidity** is the ratio of the actual water vapour present in the air to the amount which the same volume of air would hold if it were saturated (at the same atmospheric pressure and temperature). Saturated air has 100% relative humidity. Completely dry air has 0% relative humidity.

When a given mass of air is heated and no new water vapour is added, the relative humidity of air **decreases.** If the mass of air is cooled, the relative humidity **increases.** If cooling continues long enough, the relative humidity will reach 100% and the air will be **saturated.**

Hence, the **smaller** the spread between temperature and dewpoint, the **higher** will be the relative humidity.

Fog or low clouds are likely to form when the temperature is within 2°C. of the dewpoint.

Absolute humidity expresses the **weight** of water vapour per unit volume of air. It is usually stated in grains of water vapour per cubic foot of air. (A grain is 1/7000th part of a lb.)

Dew and Frost

On clear, still nights, vegetation often cools by radiation to a temperature below the dewpoint of the adjacent air. Moisture, known as **dew,** then collects on the leaves.

Frost forms in the same way, when the dewpoint is colder than freezing. Water vapour sublimates directly as ice crystals and adheres to any object (such as a metal airplane) which has lost sufficient heat by radiation cooling to be cooler than the dewpoint. Sometimes dew forms and later freezes. Frozen dew is hard and transparent. Frost, however, is white and opaque.

Temperature

The source of energy which warms the earth's surface and its atmosphere is the **sun.** The method by which the heat is transferred from the sun to the earth is known as **solar radiation.** Radiation itself is not heat. The temperature of a body is affected only if it can absorb radiation.

Some of the solar radiation that reaches earth is absorbed in the stratosphere and the ionosphere but the rest passes through the lower portions of the troposphere and is absorbed by the earth. The earth, in turn, radiates energy back into the atmosphere. This outgoing radiation is known as **terrestrial radiation.** On a world wide basis, the average heat gained through incoming solar radiation is equal to the heat lost through terrestrial radiation in order to keep the earth from getting progressively hotter or cooler. However, regional and local imbalances between solar and terrestrial radiation cause temperature variations that have great significance in weather formation.

Some of the outgoing terrestrial radiation is absorbed by the lower levels of the atmosphere. The rest passes out into space. The lower levels of the atmosphere are therefore not heated directly by the sun. The sun heats the earth and the earth heats the atmosphere. This fact is of the greatest importance in an understanding of weather. **The atmosphere is heated from below and not from above.**

The amount of solar energy received by any region varies with time of day, with seasons, with latitude and with surface topography. Temperatures can, therefore, vary widely.

Diurnal Variation. During the day, solar radiation exceeds terrestrial radiation and the surface of the earth becomes warmer. At night, solar radiation ceases but terrestrial radiation continues and cools the surface. Warming and cooling of the atmosphere occur as a result of this diurnal imbalance.

Seasonal Variation. Because the axis of the earth is tilted to the plane of its orbit, the angle at which the solar radiation strikes the earth varies from season to season. The Northern Hemisphere receives more solar energy in June, July and August and is therefore warmer and receives less in December, January and February and is therefore cooler.

Latitude. The sun is more directly overhead in equatorial regions than at higher latitudes. The tropics consequently receive the most radiant energy and are warmer than the polar regions where the slanting rays of the sun deliver less energy over a given area.

Topography. Land surfaces absorb more solar radiation than do water surfaces and radiate it more readily. Land surfaces therefore warm up more rapidly during the day and cool more rapidly at night. All land surfaces do not, however, absorb radiation at a uniform rate. There is great variation in radiation absorption by varying types of land surface. Wet soil (as found in swamps and marshes) is almost as effective as water in suppressing temperature changes. Heavy vegetation acts as an insulation against heat transfer. The greatest temperature changes occur over arid, barren surfaces such as deserts and rocky plains.

Some of the solar radiation is reflected back out to space by the earth's surface and is not absorbed at all. Some of this reflection is due to the angle at which the radiation strikes the surface but the principal cause of reflection is the type of surface. A snow surface, for example, can reflect 90% of the radiation.

Clouds greatly affect the rate of radiation. A layer of clouds will reflect a high percentage of the incoming solar radiation back out to space, drastically reducing the amount of energy reaching the earth to warm it. On a cloudy night, the clouds absorb the outgoing terrestrial radiation and radiate a considerable part of it back to earth, hindering the escape of heat.

How the Atmosphere is Heated

The lower layers of the atmosphere are heated by radiation, as we have learned, and to some small degree by conduction. Air is, however, a poor conductor of heat and thus conduction plays only a small role. The heat is distributed to the higher layers of the troposphere by several processes.

Convection. The air over a very warm surface becomes buoyant and rises rapidly through the atmosphere. A compensating flow of cold air descends to take its place. A vertical circulation is thereby created that distributes the heating through the upper layers.

Advection. A flow of air that moves from a cold area over a warm area will be heated in its lowest layers by the warm earth over which it is flowing. Warming of the air in this manner is known as advection heating.

Turbulence. Mechanical turbulence which is the result of friction between the air and the ground causes a mixing process that spreads the surface heat into the air aloft.

Compression. There are some weather systems which are favourable for the development of sinking air. This occurs in anti-cyclonic pressure systems or in air flowing down the side of a mountain range. As the air descends, it reaches regions of increased atmospheric pressure and is compressed and its temperature rises. This phenomenon is called **subsidence.**

Temperature Scales

The scales commonly used in meteorology to measure the temperature are the Celsius, Fahrenheit, and Absolute Scales. The Celsius Scale is used in Canada for all meteorological temperature readings.

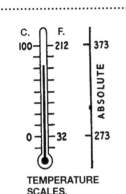

The CELSIUS SCALE has a freezing point of 0° and a boiling point of 100°.

The FAHRENHEIT SCALE has a freezing point of 32° and a boiling point of 212°.

The ABSOLUTE SCALE assumes an absolute zero of -273°C. The absolute temperature can be found by adding 273 to the Celsius Temperature, For example, +10°C = 283° Absolute. -10°C = 263° Absolute.

To convert Celsius to Fahrenheit, or Fahrenheit to Celsius, use the following formula:

Degrees F. = 9/5 C. + 32
Degrees C. = 5/9 (F. - 32)

For example: 20°F equals 5 / 9 (20 - 32) equals -6.7°C.
20°C equals (9 / 5 x 20) + 32 equals 68°F.

Nine degrees on the Fahrenheit Scale equal five degrees on the Celsius Scale. Do not, however, interpret this to mean that 9°F equals 5°C (9°F equals -12.8°C)

Isotherms

On meteorological maps, lines joining places of equal temperature are known as **isotherms**.

Density and Temperature

The **density** of air means its **mass** per unit of volume. Cold air is dense because the molecules which compose it are moving relatively slowly and are packed closely together. Warm air is **less dense** because the molecules which compose it move about rapidly. Hence, they take up more space and there are fewer molecules in a given volume. Since cold air is denser, it is therefore **heavier** and tends to sink due to the force of gravity. Warm air, being lighter, is pushed up by the denser cold air and tends to rise.

How the Atmosphere is Cooled

Since the atmosphere is heated from below, the temperature usually decreases with height through the troposphere. The rate of decrease with height is called the **lapse rate.** The average lapse rate is about 2°C per thousand feet. This figure is based on the lapse rate of I.C.A.O. Standard Air which is defined as 1.98°C per thousand feet. It is this figure which is universally used in the calibration of altimeters.

This lapse rate, however, can really be only a theoretical figure. In practice, it seldom exists since there are such wide variations in air masses and the cooling process. Sometimes, the temperature remains unchanged through several thousand feet. Sometimes, the temperature will rise with height. (See **Inversions and Isothermal Layers** below.)

The atmosphere is cooled by several processes. At night, solar radiation ceases but terrestrial radiation continues. The temperature of the earth, therefore, gradually decreases. Air in contact with the cooling earth will in turn be cooled. This cooling process is known as **radiation cooling.** It rarely affects more than the lower few thousand feet of the atmosphere. Radiation cooling will be reduced if a blanket of clouds is present. The clouds absorb the terrestrial radiation and reflect it back to earth, slowing down the rate at which the earth cools.

If the circulation is such that air from a warm region moves over a colder region, the air will be cooled. Cooling due to this process is known as **advection cooling.**

The most important cooling process in the atmosphere is cooling resulting from expansion as air is forced to rise. As a parcel of air rises, it encounters lower pressure and expands. As it expands, the temperature of the air decreases. In a rising current of air, the temperature decreases at a rate that is entirely independent of the lapse rate in the surrounding non-rising air. Such a temperature change results from the **adiabatic process,** the word adiabatic meaning that the temperature change takes place without adding or taking away heat from outside the parcel of air. Conversely, if a parcel of air should sink, it will be compressed by the increasing pressure and its temperature will rise (adiabatic heating).

Adiabatic Lapse Rates

Saturated air cools by expansion at a different rate than does unsaturated air. The **dry adiabatic lapse rate** is considered to be 3°C per 1000 feet.

If a parcel of air rises and cools until the temperature reaches the dewpoint, condensation will occur, since the air has now become saturated. The change of water vapour to water droplets involves a heat transaction, the latent heat of vaporization, which causes the air to cool at a slower rate than that at which unsaturated air cools. The **saturated adiabatic lapse rate** is about 1.5°C per 1000 feet. The saturated adiabatic lapse rate, however, shows considerable variation because it is dependent on the rate at which the water vapour is condensing and the figure of 1.5°C must be regarded as an average value only. In actuality, the saturated adiabatic lapse rate can vary from 1°C to almost 3°C.

If saturated air aloft begins to sink, it will be compressed and its temperature will rise. As it warms, it becomes unsaturated and will heat at the dry adiabatic lapse rate. Adiabatic heating of air flowing down the eastern slope of the Rocky Mountains produces the warm, dry Chinook wind.

Understanding and applying the adiabatic lapse rate is important to you as a pilot. Knowing surface temperature and dewpoint, you are able to determine at what height you might expect the bases of the clouds to be. You are also able to determine at what altitude you might expect to encounter icing conditions. You are able to calculate the temperature at various altitudes.

Let us consider the following situation. The surface temperature is 15°C and the dewpoint is 5°C. The surface elevation is 1500 feet. At what height might the bases of the convective type clouds be expected to be? We know that the temperature in a rising column of unsaturated air decreases 3°C per 1000 feet. The calculation to find the answer to this problem is not a simple case of determining the spread between the surface temperature and the dewpoint and dividing that figure by 3. The fact is that the dewpoint also falls as the column of unsaturated air rises. It decreases at a rate of about .5°C per 1000 feet. Therefore, in a rising column of unsaturated air, temperature and dewpoint converge at a rate of about (3° less .5°) 2.5° per 1000 feet. In this problem, the spread between temperature and dewpoint on the surface is 10°. Ten divided by 2.5 is 4. The bases of the clouds can be expected to be at a height of approximately 4000 feet AGL or 5500 feet ASL. The dewpoint temperature at this altitude would be 3°C.

A quicker method by which to calculate the bases of cumulus clouds is to determine the spread between temperature and dewpoint and multiply by 400.

At what height might icing conditions be encountered in the cloud?

In saturated rising air, the lapse rate averages about 1.5°C per 1000 feet. The base of the cloud is at 5500 feet ASL at a temperature of 3°C. The spread between the dewpoint and freezing is 3°. Three divided by 1.5 is 2. The freezing level, therefore, could be expected at about 2000 feet above the bases of the clouds, or at 7500 feet ASL.

What is the temperature at 10,000 feet ASL? The temperature at 5500 feet is 3°C. For the remaining 4500 feet, the temperature decreases 1.5°C per 1000 feet, giving a cooling of 4.5 x 1.5 or 6.6°. The temperature at 10,000 feet would, therefore, be 3° less approximately 7°, or -4°C.

It must be remembered that the temperature within a rising column of air is quite different than the temperature in the surrounding non-rising air.

The foregoing discussion of lapse rates may leave the impression that lapse rates are constant. In fact, they are not but vary considerably around the stated averages. Many factors, such as moisture content of the air, stability, decrease in lapse rate with height, decrease in dewpoint with height will affect the rate at which temperature decreases with height.

Inversions and Isothermal Layers

Normally, the temperature of the atmosphere decreases with height. However, this is not invariably the case. Sometimes, warmer air may be found at higher altitude. Such a reversal of normal conditions is known as an **inversion** if the temperature is actually increasing with height. In an **isothermal layer,** the temperature remains constant throughout a layer of some depth.

Inversions and isothermal layers can occur on a clear, still night when the cold ground cools the air above it in the lower levels. The temperature at the top of an inversion so formed may be 15° to 20°C warmer than the temperature at the surface. If there is a wind, the turbulence will mix the air in the lower few thousand feet of the atmosphere and distribute the cooling effect. In this case, the inversion will be much weaker or the temperature may be so uniform as to produce an isothermal layer.

Inversions may also occur as cold air, which is denser than warm air, flows into a low lying area such as a valley and becomes trapped there. Warm air, lifted above colder air over a frontal surface (see **Fronts** below), is another cause of an inversion.

The intensity of an inversion is weak during windy conditions, under a cloud cover and in maritime areas and it is strong under calm conditions and in valleys. A ground based inversion favours poor visibility by trapping fog, smoke and other obstructions in the lower levels of the atmosphere.

Stability

The normal flow of air tends to be horizontal. But disturbances may cause vertical updrafts or downdrafts to develop. Air that will resist upward or downward displacement and tends to return to its original horizontal level is said to be **stable**. Air which tends to move further away when disturbed is **unstable.** The vertical currents associated with an unstable condition may cause turbulence and, if intensive enough, thunder shower activity.

When air rises, it expands and cools adiabatically. If a mass of rising air (cooling by expansion) is still **warmer** than the air surrounding it, it is considered to be **unstable**. If disturbed, it will tend to rise further.

If a mass of rising air is **cooler** than the air around it, it is considered to be **stable**. If disturbed, it will tend to sink back to its original level.

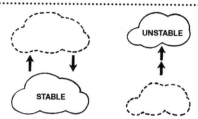

Fig. 17. Stable and Unstable Air.

The lapse rate in the part of the air which is not rising is one of the most important factors which determines the stability of air. If the lapse rate is steep, there is colder air aloft which will tend to sink to the surface if the air is disturbed and the warm air of the lower levels will rise. If the lapse rate is shallow so that there is only a slight temperature decrease with height or if there is warmer air on top, there will be no tendency for vertical motion to develop.

The relationship between lapse rate and stability, therefore, depends on the steepness of the lapse rate. Unstable air is indicated by a steep lapse rate; stable air by a shallow lapse rate, an isothermal layer or an inversion. It follows, also, that any modification of the lapse rate results in modification of stability. If the lapse rate becomes steeper, the air becomes more unstable. If the lapse rate becomes less steep, the air becomes more stable.

A lapse rate can be steepened by increasing the lower level temperature (as happens with day time heating of the earth) or by decreasing the temperature aloft. It can be made more shallow by decreasing the temperature in the lower levels (as happens with night time cooling of the earth) or by increasing the temperature aloft (as in an inversion).

Since the lapse rate is seldom uniform with height, there is considerable variation in the degree of stability or instability. In some situations, the rising currents may reach 20,000 feet or more; in other situations, the rising air will rise only a few thousand feet. It is possible for air to be stable near the surface and unstable aloft, but more commonly, it is unstable in the lower few thousand feet and stable aloft.

So far, we have considered masses, or parcels, of air which have, for any reason, been displaced upwards. If a **layer** of air becomes warmer in the lower levels or cooler in the upper levels, it will have a steep temperature lapse rate. It will therefore become **unstable.** A layer of air which has a small temperature lapse rate will be **stable.** There will be little tendency for vertical currents to develop.

Flight characteristics of stable air
- *poor low level visibility. Fog may occur.*
- *stratus type cloud*
- *steady precipitation*
- *steady winds which can change markedly with height*
- *smooth flying conditions*

Flight characteristics of unstable air
- *good visibility (except in precipitation)*
- *heap type cloud (cumulus or cumulonimbus)*
- *showery precipitation*
- *gusty winds*
- *turbulence may be moderate to severe*

Lifting Agents

There are five principal conditions that provide the lift to initiate rising currents of air.

1. *Convection.* The air is heated through contact with the earth's

From the Ground Up

surface. The rising columns are usually local and separated by areas of sinking air. They result from unequal heating of different types of land surface and especially from the different surface temperatures of land and water areas.

2. *Orographic Lift.* Air moving up a sloping terrain, such as a mountainside, will continue its upward movement, especially if it is unstable.

3. *Frontal Lift.* When different air masses meet, warm air is forced aloft by the advancing or receding wedge of cold air. (See **Fronts** below).

4. *Mechanical Turbulence.* Friction between the air and the ground disrupts the lower levels of the air into a series of eddies. These eddies are usually confined to the lower few thousand feet of the atmosphere but may extend higher if the air is unstable and surface winds are strong.

5. *Convergence.* In a low pressure area, the winds blow across the isobars into the centre of the low. Air accumulates in the centre of the low and the excess air is forced to rise.

Vapour Trails (Contrails)

The white vapour, or condensation trails (contrails), you see high up in the blue in this age of jet airplanes owe their origin to two different causes.

1. EXHAUST TRAILS. When, in the combustion process, the hydrogen and carbon in aviation fuel are burned, the exhaust is composed of a colourless gas, which is a product of the burned carbon, and water vapour, which is a product of the burned hydrogen. Both of these by-products are invisible. The water vapour will remain invisible as long as the humidity of the surrounding air has not reached its saturation point. Warm air can contain much more invisible water vapour than can cold air before it becomes saturated. In the extreme low temperatures encountered at very high altitudes, the cold air is incapable of absorbing the excess water vapour coming out of the exhaust. The water vapour therefore condenses into a visible cloud of water droplets or ice crystals. This is known as an **exhaust trail.**

2. WING TIP TRAILS. As we learned in **Theory of Flight (Drag)**, vortices, in the form of eddies rotating with a corkscrew motion, are formed off the tips of an airplane wing in flight. These rapidly rotating vortices have considerable centrifugal force acting outwards, which causes a rarefaction and therefore an expansion of the air in the middle of the vortex. Air, which expands, cools. If the vortex is strong enough and the humidity of the air high enough, this cooling will cause condensation. The white cloud-like trails which form off the wing tips are known as **wing tip trails.**

Air Masses

In the past, the weatherman based his predictions of the weather mainly upon the existence and movement of high and low pressure areas and the wind and weather systems which are associated with them. Today, the whole system of weather forecasting is based upon the properties of air masses (of which pressure is only one factor), the changes which occur as an air mass moves away from its source, and the weather phenomena which can be predicted along the front where two air masses of different properties come in contact.

An **air mass** may be defined as a large section of the troposphere with uniform properties of temperature and moisture in the horizontal. An air mass may be several thousands of miles across.

It takes on its original properties from the surface over which it has formed. An air mass which has formed over the ice and snow surfaces of the Arctic would be cold and dry. An air mass that has formed over the South Pacific would be warm and moist.

An air mass which has formed over a large body of water and is therefore moist is referred to as **maritime air.** One which originates over a large land area and is therefore dry is referred to as **continental air.**

The three main sources of the air masses of North America are: (1) The Arctic Region, which extends from the north pole south to the permafrost line. (2) The Polar Region, which extends south from the permafrost line to where the mean temperature is 10°C. (3) The Tropical Region, which lies below Latitude 30°N.

The principal air masses of North America are as follows:

Continental Arctic (cA) and **Continental Polar (cP):** Cold dry air masses which originate over the intensely cold ice and snow covered surface of the far north and bring a cold wave when they move south.

Maritime Arctic (mA) and **Maritime Polar (mP):** These are cold air masses which form over the Arctic and acquire moisture as they move south over the cold waters of the North Atlantic and North Pacific oceans. The Maritime Polar air mass is air which has moved farther out and over the ocean and, through contact with the ocean surface, has become more warm and moist than Maritime Arctic air.

Maritime Tropical (mT): Because the continent of North America narrows down towards its southern extremity, most of the tropical air in the southern latitudes is in contact with the warm ocean surface and becomes hot and moist. Source regions for Maritime Tropical air are: The Gulf of Mexico, the Caribbean Sea, and the Tropic of Cancer regions of the North Atlantic and North Pacific oceans.

In winter, the most common air masses that form over North America are Maritime Polar, Maritime Arctic and Continental Arctic. Continental Polar seldom appears over this continent. In most cases, Maritime Arctic and Maritime Polar air masses found over the continent have entered from the west. Maritime air masses found over the Atlantic Ocean affect only the east coast. The air mass in the southern portion of the continent is Maritime Tropical.

In summer, snow and ice melt, leaving numerous small lakes in the northern reaches of Canada and Alaska. These lakes provide moisture that affects the forming air masses. In summer, therefore, Maritime Arctic is the principal air mass and Continental Arctic rarely appears. Maritime Polar which enters from the Pacific Ocean and Maritime Tropical are the other common summer air masses.

Weather in an Air Mass

There are three main factors that determine the weather in an air mass — moisture content, the cooling process and the stability of the air.

Some air masses are very dry and little cloud develops. Maritime air has a high moisture content and cloud, precipitation and fog can be expected.

Even if the air is moist, condensation and cloud formation occur only if the temperature is lowered. The cooling processes that contribute to condensation and the formation of clouds are (1) contact with a surface cooling by radiation, (2) advection over a colder surface and (3) expansion brought about by lifting. It follows, therefore, that cloud formation within an air mass is not uniform. Clouds may form, for example, in an area where the air is undergoing orographic lift even though the rest of the air mass is clear.

Stability of the air is of prime importance. In stable air, layer cloud and poor visibility are common. Good visibility and cumulus cloud are common in unstable air.

Modification of Air Masses

Although the characteristics of an air mass are determined by the area over which it forms, as an air mass moves away from its source area it is modified by conditions over which it passes. If the modification is extensive both horizontally and vertically, the air mass is given a new name. Continental Arctic air, for example, by moving out over the oceans, becomes moist and may be renamed Maritime Polar or Maritime Arctic.

Temperatures may vary considerably within an air mass as the air is warmed while passing over relatively warm surfaces and cooled while passing over relatively cool surfaces.

Moisture content may also vary. The air mass will pick up moisture as it passes over lakes, wet ground or melting snow. Moist air passing over a mountain range may lose a great deal of moisture as precipitation as it ascends the western slope of the range. The air will be much drier when it reaches the prairies.

Stability characteristics also may vary considerably. Air warming from below by radiation becomes unstable as convective currents develop. Air that is cooled from below becomes more stable as vertical motion is blocked. Changes in stability alter weather conditions quite considerably.

These variations in the properties of a particular air mass during its life are not unusual. As a result, it is impossible to say that a certain weather pattern will always occur in an air mass that is classified, for example, as Continental Arctic.

Nevertheless, it is possible to define in broad terms some typical characteristics of the particular air masses.

Continental Arctic (cA)
- *forms over a region that is covered with ice and snow.*
- *moisture content is low.*
- *as it moves south, it is gradually heated. Strong winds set up turbulence and, if it acquires moisture, stratocumulus clouds with light snow will develop.*
- *if it moves over open water, such as the Great Lakes, it will be heated and acquire moisture. A steep lapse rate and accompanying instability give rise to cumulus clouds and snow showers. Eddying, gusty winds cause restricted visibility in blowing snow.*
- *usually follows a path from the polar regions across the prairies and into the eastern part of the continent. It rarely affects the British Columbia coast.*

Maritime Arctic (mA)
- *forms over Siberia or Alaska and travels across the North Pacific where it becomes moist and unstable.*
- *stratocumulus clouds.*
- *when the air is lifted by coastal mountains, thunder storms develop. Precipitation occurs in the form of snow showers, snow and rain.*
- *loses most of its moisture as precipitation on the western slopes of the Rockies and is dry when it reaches the prairies.*
- *mA that develops over the North Atlantic affects only the east coast of North America.*
- *in summer, mA develops in polar regions and moves south over the lakes of the northern part of the continent. Daytime heating makes the air unstable.*

Maritime Polar (mP)
- *reaches the Pacific Coast after a long journey over the ocean. It is more extensively modified than mA and the air mass is somewhat more stable.*
- *orographic lift along the mountains produces extensive cloud formation and considerable rain. It is, therefore, drier when it reaches the prairies.*

Maritime Tropical (mT)
- *forms over oceans and water bodies of the Tropics.*
- *very warm and moist.*
- *rarely penetrates north of the Great Lakes in winter, although it does frequently appear aloft.*
- *unstable when it is lifted at a front, giving snow, freezing rain, rain, severe icing and turbulence.*
- *extensive fog often occurs, especially on the east coast.*

Cold Air Mass
Instability
Turbulence
Good Visibility
Cumuliform Clouds
Precipitation: Showers
Hail, Thunderstorms

Warm Air Mass
Stability
Smooth Air
Poor Visibility
Stratiform Clouds and Fog
Precipitation: Drizzle

Fronts

The troposphere is made up of air masses. The transition zone between two air masses is called a **front.** The interaction of these air masses, along their frontal zones, is responsible for weather changes.

Polar Front

The present day theory, which explains the formation of depressions or lows, was developed by the Norwegians and is known as the **Polar Front Theory.** It is based on the fact that the polar regions are covered by a mass of cold air and the equatorial regions by a mass of warm air. In the temperate zone, the two air masses meet. Air masses do not usually mix. The transition zone between the two masses is therefore narrow and is called a **polar front**.

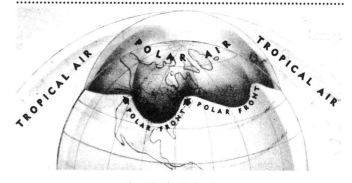

Fig. 18. The Polar Front.

From the discussion of air masses, it is evident that the cold air dome covering the polar region, in fact, encloses two other domes and that frontal systems will be found along the zones separating them. Continental Arctic air makes up the smallest and most northerly dome and its edge, the Continental Arctic front, is the most northerly frontal system to be found on the weather map. A dome of Maritime Arctic air overlies the cA air and its edge lies somewhere between the Continental Arctic front and the polar front. The fronts are named for the colder air mass involved in the system. The sloping surface of each dome acts as a lifting surface for the warmer air mass associated with it. Fig. 19 illustrates in vertical cross section the four air masses and the three frontal surfaces separating them.

A **front,** correctly defined, is the transition zone between two air masses as it appears on the surface. The sloping side of the cold air is called a **frontal surface.**

The discussion below deals with the development of frontal waves and depressions along the polar front. Similar frontal waves and depressions occur along the Continental Arctic and the Maritime Arctic fronts.

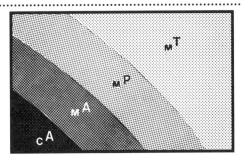

Fig. 19. Air Masses.

Development of a Frontal Depression

The air on the northern side (considering the northern hemisphere) of this surface of separation is termed **arctic and polar air.** It is normally cold and dry. The air on the southern side is termed **tropical air.** It is normally warm and moist.

Due to the difference in the properties of these two air masses, the polar front is known as a surface of discontinuity. Depressions form along this **surface of discontinuity** (the polar front) and are the means whereby interchange takes place between the warm and cold air masses.

Along the polar front, the cold polar air flows from the north-east towards the south-west on the north side (Fig. 20) while the warm air flows from south-west towards the north-east on the south side.

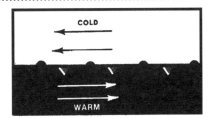

Fig. 20.

The arrangement is not a stable one but is subject to continual disturbances due to the warm air bulging north and cold air bulging southwad (Fig. 21). This northward bulge, once having started,

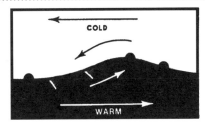

Fig. 21.

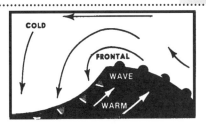

Fig. 22.

continues to develop, the cold air on the northward side swinging round at the back, setting up a counterclockwise circulation and emphasizing the bulge. This bulge is the new depression (low pressure area) just born. It normally travels northeastward along the polar front. The polar front has now been bent or broken into two sections, the warm front (⬤⬤⬤⬤⬤) and the cold front (▲▲▲▲▲). This deformation is called a frontal wave (Fig. 22).

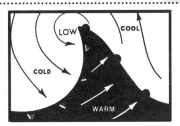

Fig. 23.

The area contained within the bulge of the polar front is the warm sector and contains the warm or tropical air, while the rest of the area is composed of the polar air.

The pressure at the peak of the frontal wave falls and the low pressure area deepens. The surface winds become stronger and the fronts begin to move. Both fronts are curved in the direction toward which they are moving. The peak of the wave is called the crest. At the cold front, the cold flow of air from the northwest is undercutting the warm southwesterly flow. At the warm front, the southwesterly flow of warm

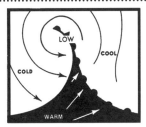

Fig. 24.

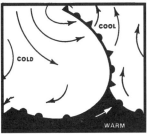

Fig. 25.

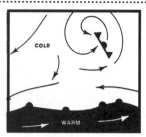

Fig. 26.

air is overrunning the retreating flow of cold air. At the same time, the entire system in the immediate future is moving over the ground in a direction parallel to the isobars in the warm sector (See Fig. 32, isobars in the warm sector). The wave is rotating about the centre.

The cold front moves faster than the warm front (Fig. 23) and soon catches up with it. The two fronts merge causing an occluded front, or occlusion (▲▲▲) (Fig. 24). As the occlusion increases in dimension, the low pressure area weakens and the movement of the fronts slows down. Sometimes a new frontal wave may begin to form on the westward portion of the polar front (Fig. 25). In the final stage, the two fronts have become a single stationary front again (▲▼▲). The low pressure area has virtually disappeared (Fig. 26).

A COLD FRONT is that part (or parts) of a frontal system along which cold air is **advancing** and is coloured blue on the weather map.

A WARM FRONT is that part (or parts) of a frontal system along which cold air is **retreating** and is coloured red on the weather map.

Types of Front
The Warm Front
As a mass of warm air advances on a retreating mass of cold air, the warm air, being lighter, ascends over the cold air in a **long gentle slope.** As a result, the cloud formation associated with the warm frontal system may extend for 500 or more nautical miles in advance of it. Warm fronts usually move at relatively slow speeds and therefore affect a vast area for a considerable length of time.

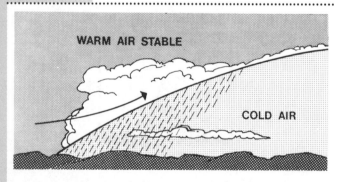

Fig. 27. A Warm Front with Overrunning Warm, Stable Air.

If the warm air is moist and stable, stratiform clouds develop in a distinctive sequence. The first signs of an approaching warm front are high cirrus clouds which thicken to cirrostratus and altostratus as the warm front approaches. The ceiling gradually falls and there follows a long belt of steady rain falling from heavy nimbostratus cloud. Precipitation may lead the frontal surface by as much as 250 nautical miles.

If the warm air is moist and somewhat unstable, cumulonimbus and thunderstorms may be embedded in the stratiform layers.

Heavy showers in advance of the surface front can then be expected.

Very low stratus and fog throughout the frontal zone are typical characteristics of warm fronts.

The passing of the warm front is marked by a rise of temperature, due to the entry of the warm air, and the sky becomes relatively clear.

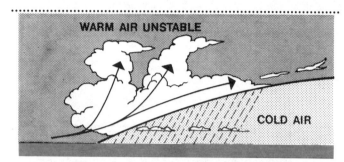

Fig. 28. A Warm Front with Overrunning Warm, Unstable Air.

The Cold Front
When a mass of cold air overtakes a mass of warm air, the cold air being denser stays on the surface and undercuts the warm air violently. Surface friction tends to slow down the surface air while the upper air, not being subject to the friction, retains its speed and tends to catch up. As a result, the slope of the advancing cold front is quite steep. The rapid ascent of the warm air gives rise to cumuliform cloud. The width of cloud cover is usually relatively narrow, only about 50 nautical miles (Fig. 29). Precipitation in the form of rain, snow or hail will be of the showery type and quite heavy. The cumuliform clouds that form at the frontal surface frequently build up into thunderstorms creating some of the most violent weather a pilot may encounter. Sometimes, a continuous line of thunderstorms (a squall line) will develop ahead of a particularly fast moving cold front.

Fig. 29. A Cold Front Undercutting Warm, Moist, Unstable Air.

A sharp fall in temperature, a rise in pressure and rapid clearing usually occur with the passage of the cold front.

Sometimes, an advancing cold front will be relatively slow moving. Because it does not undercut the warm air so violently, a rather broad band of clouds develops extending a fair distance behind the frontal surface. If the warm air is stable, these clouds will be stratiform; if the warm air is unstable, they are cumuliform and possibly thunderstorms. With passage of the frontal surface, clearing is more gradual.

The Stationary Front
There is generally some part of a front along which the colder air is neither advancing nor retreating. There is no motion to cause the front to move because the opposing air masses are of equal

pressure. The surface wind tends to blow parallel to the front and the weather conditions are similar to those associated with a warm front although generally less intense and not so extensive. Usually a stationary front will weaken and eventually dissipate. Sometimes, however, after several days, it will begin to move and then it becomes either a warm front or a cold front.

Occluded Fronts

With the progress of time, as a depression advances, the cold front gradually overtakes the warm front and lifts the warm sector entirely from the ground. It is simply a case of the cold air catching up with itself as it flows around the depression. Thus only one front remains which is called an **occluded front** or **occlusion.** An occluded depression soon commences to fill up and die away.

The cold air, in the distance it has travelled, may have undergone considerable change. Therefore it may not be as cold as the air it is overtaking. In this case (**cool** air advancing on **colder** air), the front is known as an **occluded warm front** or a **warm occlusion** and has the characteristics of a warm front, with low cloud and continuous rain and drizzle. If the warm air is unstable, heavy cumulus or cumulonimbus cloud may be embedded in the stratiform cloud bank (Fig. 30).

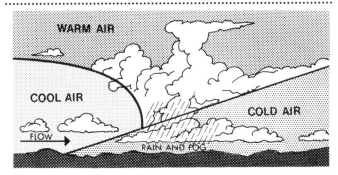

Fig. 30. A Warm Front Occlusion Lifting Warm, Moist, Unstable Air.

If the cold air is colder than the air it is overtaking (**cold** air advancing on **cool** air), the front is known as an **occluded cold front** or a **cold occlusion.** A cold occlusion has much the same characteristics as a warm front, with low cloud and continuous rain (Fig. 31). If the warm air is unstable, cumulonimbus and thunderstorms are likely to occur, with the violent turbulence, lightning and icing conditions associated with these clouds.

It will be noted that, in the case of either a warm or cold occlusion, **three** air masses are present, a cool air mass advancing on a cold air mass, or a cold air mass advancing on a cool air mass, with, in either case, a warm air mass lying wedge shaped over the colder air. This wedge shaped mass of warm air is known as a **trowal** in Canada. (In some other countries, it is called an **upper front.**)

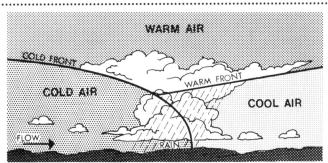

Fig. 31. A Cold Front Occlusion Lifting Warm, Moist, Stable Air.

Upper Fronts

In Canada, the term upper front refers to a non-occlusion situation. Sometimes, cold air advancing across the country may encounter a shallow layer of colder air resting on the surface or trapped in a topographical depression. The advancing cold air rides up over the colder, heavier air. The cold front which is the leading edge of the advancing cold air, therefore, leaves the ground and moves along the top of the colder air. It is then known as an **upper cold front.**

Sometimes, the structure of the advancing cold front is such that the cold air forms a shallow layer for some distance along the ground in advance of the main body of cold air. The frontal surface of the main mass of cold air, in this situation, will usually be very steep. The line along which the frontal surface steepens is also known as an upper cold front.

On occasion, an advancing warm front rides up over a pool or layer of cold air trapped on the ground. A station on the ground does not experience a change of air mass because the front passes overhead. This is known as an **upper warm front.**

Sometimes, the surface of the cold air that is retreating ahead of an advancing warm front is almost flat for some distance ahead of the surface front and then steepens abruptly. The line along which the surface of the retreating cold air steepens sharply is also called an upper warm front.

Frontal Weather

The theory of the polar front, which for the sake of simplicity has been described in the form of its original conception, might leave the impression that depressions form only along some well defined line lying somewhere midway between the poles and the equator. Air masses are in a constant state of formation over all the land and water areas of the world. Once formed, they tend to move away from the source regions over which they form. The same frontal processes and phenomena occur whenever a mass of warm air and a mass of cold air come in contact.

There is a widespread impression among pilots that fronts always bring bad weather and that all bad weather is frontal. Actually some fronts have little or no weather associated with them. A slight change of temperature and a windshift may be the only evidence that the front has gone through. And, of course, bad weather can develop without the passage of a front. Fog, for example, generally occurs when no fronts are present and severe thunderstorms may develop in an air mass which has no frontal characteristics.

Another common misconception is that the front is a thin wall of weather. This false idea is perhaps occasioned by the line that indicates a front on a weather map. The line on the map only shows the surface location at which the pressure change, windshift and temperature change occur. The actual weather associated with the front may extend over an area many miles in width, both well ahead and also for many miles behind the actual line on the weather map.

A front itself is actually a transition zone between two large air masses with different properties of temperature and moisture. Each individual air mass may extend over hundreds of thousands of square miles. Everywhere along the boundary of an air mass, where it overrides or undercuts the air mass upon which it is advancing, and for a considerable height upward from the surface as well, there is a frontal zone. The frontal zone aloft is called a **frontal surface.** The frontal zone on the ground is called the **front.** The frontal weather associated with the front, therefore, can be expected to extend for hundreds of miles along the boundary of the air mass.

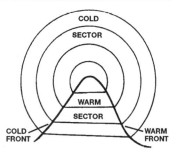

Fig. 32. Development of a Depression Showing Isobars.

Note: The isobars are bent into the form of a "V" or "trough" at both the warm and cold fronts.

Frontogenesis means a front which is increasing in intensity.

Frontolysis means a front which is decreasing in intensity.

If you examine the diagrams showing fronts on a weather map, you will notice that all fronts lie in regions of lower pressure. The isobars are bent sharply at a front. These two factors are characteristic of all fronts.

Weather at the Cold Front

Cold fronts are not all the same. The weather associated with a cold front may vary from a minor windshift to severe thunderstorms, low ceilings, restricted visibility and violent gusty winds. The severity of the weather is determined by the moisture content and stability of the warm air mass that the cold air mass is undercutting and the speed of the advancing cold front.

Fast moving cold fronts may travel across the country with a speed of 30 knots or more. If the warm air that is being undercut by the cold air mass is very moist and unstable, towering cumulus clouds and thunderstorms are likely to develop. Heavy rain or hail may be associated with the front. A slower moving cold front advancing on more stable and drier air in the warm sector will produce less severe weather conditions, stratus or altocumulus clouds with light or no precipitation.

A long line of cumulus clouds on the western horizon is usually an indication of an approaching cold front. Sometimes a deck of altocumulus cloud or decks of stratus and stratocumulus extending ahead of the front will mask the main frontal cloud from the view of the high flying or low flying pilot respectively.

Weather Changes
Surface Wind: The wind direction will always veer as the front passes. Gustiness may be associated with the windshift.

In flying through a cold front, the windshift may be quite abrupt and occurs at the frontal surface rather than at the front. The windshift is always such that an alteration in course to starboard is required, no matter which way you are flying through the front.

Temperature: On the ground, the temperature may drop sharply as the front passes, but usually it drops gradually. The air immediately behind the front has been warmed in passing over the warm ground. Therefore, it may be several hours before the temperature drops to the true value of the cold air mass.

In flying through a cold front, there will be a noticeable temperature change when passing through the frontal surface.

Visibility: Visibility usually improves after passage of a cold front. If the front is moving fairly rapidly, the width of frontal weather generally is less than 50 miles. If the front is moving slowly, however, flight operations may be affected for many hours.

Pressure: The approach of a cold front is accompanied by a decrease in pressure. A marked rise will be noticed when the front has passed.

Turbulence: Turbulence may be associated with the cold front if it is active, although thunderstorms are not always present. Even in cases where there are no clouds, turbulence may be a problem. As a rule, flight through an active cold front can be expected to be rough.

Precipitation: The frontal rain or snow is usually narrow, especially if it is showery in character.

Icing in the turbulent cumulus clouds can be severe.

Line Squalls
A long line of squalls and thunderstorms which sometimes accompanies the passage of a cold front is called a line squall (or squall line). It is usually associated with a fast moving cold front that is undercutting an unstable warm air mass. It may form anywhere from 50 to 300 nautical miles in advance of the front itself. The line squall is a long line of low, black, roller like cloud, which often stretches in a straight line for several hundred miles and from which heavy rain or hail falls for a short time. Thunder and lightning frequently occur. The squall is also accompanied by a sudden wind change from southerly or south-westerly to north or north-westerly, together with a sudden drop in temperature and a rise in barometric pressure. The actual wind squall lasts only for a few minutes but is often extremely violent, constituting a serious menace both to shipping and to airplanes. The signs indicating the approach of a line squall are unmistakable. Airplanes on the ground should be immediately hangared. Those in the air should at all costs avoid this violent weather phenomenon.

Weather at the Warm Front
Warm front changes are usually less pronounced than cold front changes. The change is also generally very gradual. However, the weather at a warm front is usually more extensive and may cover thousands of square miles.

A wide variety of weather characterizes warm fronts. The weather may even vary along a given front.

The degree of overrunning and the moisture content and stability of the overrunning warm air determine the severity of the weather. If the warm air is very moist, the cloud deck forming in the overrunning air may extend for hundreds of miles up the slope of the retreating cold air. If the warm air is unstable, thunderstorms may be embedded in the cloud deck.

High cirrus cloud is the first sign of the approach of an active warm front. Cirrostratus soon follows (the high thin cloud which causes a halo around the sun or moon). The cloud gradually thickens and the base lowers until a solid deck of altostratus/altocumulus covers the area. Low nimbostratus moves in, merging with the altostratus, with the result that a solid deck of cloud extending from near the surface to 25,000 feet or more covers the whole area. Precipitation is usually heavy.

Weather Changes
Windshift: With the passage of a warm front, the wind will veer but the change will be much more gradual than in the case of a cold front.

When flying through a warm front, the windshift will occur at the frontal surface and will be more noticeable at lower levels. When flying through a warm front, the windshift is such that a course alteration to starboard is necessary.

Temperature: The warm front brings a gradual rise in temperature.

A pilot flying through the frontal surface will notice a more abrupt temperature rise.

Visibility: Low ceilings and restricted visibility are associated with warm fronts and, because warm fronts usually move quite slowly, these conditions persist for considerable time.

When rain falls from the overrunning warm air, masses of irregular cloud with very low bases form in the cold air. Fog is frequently a condition 50 nautical miles ahead of an advancing warm front.

Turbulence: Cumulonimbus clouds are frequently embedded in the main cloud deck and these storms are responsible for the most severe turbulence associated with a warm front. However, these storms and the turbulence they occasion are less severe than those associated with cold fronts. The principal problem with these storms is that they cannot be located by sight since they are embedded in the main cloud cover.

Precipitation: The first precipitation begins in the region where the altostratus layer of cloud is from 8000 to 12,000 feet above the ground. As the front approaches, the precipitation becomes heavier. Occasional very heavy precipitation is an indication of the presence of thunderstorms.

Winter Warm Front

In winter, when temperatures in the cold air are below freezing and temperatures in the lower levels of the warm air are above freezing, snow and freezing rain can be expected.

Snow falls from that part of the warm air cloud that is high and therefore below freezing in temperature. From the lower cloud, where temperatures are above freezing, rain falls. However, as the rain falls through the cold air (of the cold air mass that the warm air is overrunning), it becomes supercooled and will freeze on contact with any cold object. This is known as freezing rain (ZR).

In the area ahead of the freezing rain, there is a region where the rain falling through the cold air becomes sufficiently supercooled to freeze and falls to the ground as ice pellets (IP).

A pilot approaching the frontal surface at higher altitudes may not encounter the ice pellets, but the pilot flying at quite low altitudes can expect to encounter snow, ice pellets and then freezing rain.

Icing is a problem associated with warm fronts in winter. Snow is not responsible for icing, unless it is very wet when it can stick to an airplane and form ice. Freezing rain, however, causes a rapid build up of ice. Icing will also be a problem in the cloud layers.

Weather at Trowals and Upper Fronts

The weather that occurs with a trowal is a combination of cold and warm front conditions. The cloud pattern ahead of the approaching trowal is similar to that of a warm front. Cold front cloud formations will exist behind it. Cumulus buildups and thunderstorms are likely to be interspersed with stratiform clouds, continuous precipitation and widespread low ceilings. In winter months, freezing rain and severe icing conditions are likely hazards as the rain aloft in the occluded warm air falls through the freezing temperatures of the ground based cold sectors. The maximum precipitation, convective activity and icing conditions usually occur in the northeast sector of the low and extend some 50 to 100 miles ahead of the occluded front.

Clouds, Precipitation and Fog

Clouds

Clouds form when the invisible water vapour that is present in the air changes into its visible form as water droplets or ice crystals.

The process by which water vapour changes into water droplets is called **condensation** and occurs when the relative humidity is high, when condensation nuclei are present in the air and when there is cooling of the air.

The level at which water vapour condenses and becomes visible is known as the **condensation level.** This level is, in practice, the base of the clouds. If the cloud forms at ground level, it is called fog rather than cloud.

Except at temperatures well below freezing, clouds are composed of very small droplets of water which collect on microscopic water absorbent particles of solid matter in the air (such as salt from evaporating sea spray, dust, and smoke particles). The abundance of these particles, called **condensation nuclei,** on which the droplets form, permits condensation to occur generally as soon as the air becomes saturated. If the condensation nuclei are particularly abundant, condensation may occur at less than 100% relative humidity.

Clouds which form at temperatures well below freezing are usually composed of small particles of ice known as ice crystals which form directly from water vapour through the process of **sublimation.** When the temperature is between freezing and about -15°C, clouds are composed largely of supercooled water droplets with some ice crystals as well. (See **Supercooled Water Droplets** in section **Humidity.**)

Saturated warm air holds much more water vapour than does saturated cold air. Cooling saturated warm air will result in more water vapour condensing into visible water droplets than is the case when cooling saturated cold air. Denser, thicker cloud formations occur when condensation occurs in a warm air mass.

Clouds are formed in two ways. (1) Air, in which water vapour is present, is cooled to its saturation point and condensation occurs. The cooling process will occur as warm air comes in contact with a cold surface or with a surface that is cooling by radiation or as air is affected by adiabatic expansion. (2) Air, without a change in temperature taking place, may absorb additional water vapour

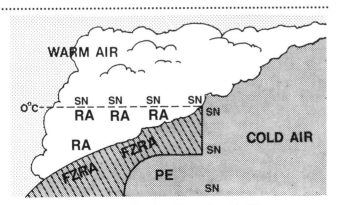

Fig. 33. Precipitation in a Warm Front in Winter.

until its saturation point is reached with the result that clouds are formed.

Of these, the most common cause of cloud formation is adiabatic expansion, that is, cooling due to expansion brought about by lifting.

Stability of the air, of course, is one of the major factors which determines the strength and extent of vertical motion and therefore cloud formation. In stable air, the cloud forms in horizontal sheets of stratus cloud. In unstable air, cumulus clouds develop.

The lifting process is initiated by a number of different phenomena.

Orographic Lift

Air blowing against a range of hills or mountains is forced upward, reaches a region of lower pressure, expands and cools. Condensation will occur when the dewpoint is reached. The type of cloud formed will depend on the moisture content and on the stability of the air. The slope and height of the terrain and the strength of the wind component that produces the upslope flow also have an effect.

If the air is dry, very little cloud will form. Stratus cloud will form if the air is moist and stable, cumulus and cumulonimbus if the air is moist and unstable.

A long bank of cloud from which rain may fall forms on the windward side and on the upper parts of the hills or mountains. Due to the fact that the rising ground is always in the same place, orographic clouds and rain are typically persistent and usually widespread. The descending air on the leeward side of the mountains will be compressed and heated, causing dissipation of the clouds.

Convection

Warm air rises. Owing to the heating of the ground by the sun, rising currents of air occur. The upward movement of air is known as **convection.** (The downward movement of air is known as **subsidence.**) As currents of air rise due to convection, they expand. The expansion is accompanied by cooling. The cooling produces condensation; a cumuliform cloud forms at the top of each rising column of air. The cloud will grow in height as long as the rising air within it remains warmer than the air surrounding it. The height of the cloud, however, is also dependent on the stability of the air in the mid levels of the troposphere. If the air is stable, fair weather cumulus, only a few thousand feet thick, will form. If the air is unstable through the troposphere right up to the tropopause, towering cumulus and cumulonimbus will develop. Heavy rainshowers and thunderstorms are typically associated with convection clouds.

Convection also occurs when air moves over a surface that is warmer than itself. The air is heated by advection and convective currents develop. Warming of air by advection does not depend on daytime heating. Convection will, therefore, continue day or night so long as the airflow remains the same.

Frontal Lift

When a mass of warm air is advancing on a colder mass, the warm air rises over the cold air on a long gradual slope. This slope is called a warm frontal surface. The ascent of the warm air causes it to cool and clouds are formed, ranging from high cirrus through altostratus down to thick nimbostratus from which continuous steady rain may fall over a wide area.

When a mass of cold air is advancing on a mass of warm air, the cold air undercuts the warm air and forces the latter to rise. The slope of the advancing wedge of cold air is called a cold frontal surface. The clouds which form are heavy cumulus or cumulonimbus.

Heavy rain, thunderstorms, turbulence and icing are associated with the latter.

Turbulence

When a strong wind blows over a rough surface, the friction between the ground and the air produces **mechanical turbulence,** or **eddy motion.** The intensity of the turbulence is dependent on the roughness of the underlying surface, the strength of the wind and the instability of the air.

The eddy motion consists of irregular up and down currents. The air in the upward current cools and, if sufficient moisture is present and if the turbulence is vigorous, condensation may take place in the upper part of the turbulent layer. The cloud layer has an undulating base which is lower in the rising eddies than in the sinking eddies. The top of the cloud marks the top of the turbulent layer and tends to be very flat. Very often an inversion exists above the turbulent layer and it is this which blocks further vertical motion. The cloud layers tend to be stratocumulus in form.

Sometimes convection occurs in combination with the mechanical turbulence and then cumulus clouds will develop and will be embedded in the stratocumulus layer.

If the air is very stable, the mechanical turbulence is dampened and confined to very low levels. Only if the air is very moist will clouds form in this slight lifting action and then they will be stratus in form. If the wind is strong, stratus factus may be associated with the stratus layer.

Convergence

When air piles up over a region as at the centre of a low pressure area, convergence is said to be occurring. The excess air is forced to rise; it expands and cools and, when the condensation level is reached, clouds form. Since all fronts lie in regions of low pressure, convergence is often a contributing factor to frontal weather.

Precipitation

Precipitation occurs when the water droplets (visible as a cloud) grow sufficiently in size and weight to fall due to gravity. In clouds with temperatures above freezing, vertical air currents cause the droplets to move about. As a result, they collide with other drops and gradually grow in size as they absorb those drops with which they collide. They gain momentum until they fall through the air as rain. A single water droplet must grow enormously in order for precipitation to take place. The average raindrop is about one million times larger than a cloud water droplet. This process is known as **coalescence.** Precipitation due to coalescence alone generally occurs only in warm climates.

In a stable cloud such as stratus, there is very little vertical motion, not even enough to sustain small water droplets. They frequently escape and drift slowly to the earth. This form of precipitation is called drizzle.

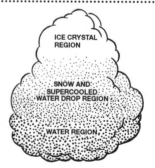

Fig. 34. Regions of a Cloud.

A second mechanism by which precipitation occurs requires that ice crystals and water droplets exist side by side in a cloud at temperatures below freezing. The ice crystals grow at the expense of the water droplets. The droplets tend to evaporate and the resulting water vapour sublimates on the ice crystals. The ice crystals grow in size and weight. They are sustained in the cloud until they grow large enough that their terminal velocity exceeds the updraft velocity in the cloud. They then fall as precipitation. If the temperature below the region of formation is above freezing, the crystals will melt, coalesce with other drops and will arrive at the earth as rain. If the temperatures are cold all the way to the ground, the ice crystals will aggregate into snow flakes. In Canada, heavy rainfall usually occurs as a result of a combination of sublimation on ice crystals and coalescence.

Two facts are therefore significant. If the ice crystals are necessary for the occurrence of heavy precipitation, the cloud from which the rain is falling must have built up well above the freezing level. Since the size of a raindrop is a function of the turbulence in the parent cloud, large drops and heavy precipitation are an indication of strong vertical motion.

Steady precipitation falls from a layer of stratus cloud. A shower, a sudden heavy burst of precipitation, falls from a well developed cumulus or cumulonimbus cloud, which may be embedded in a stratus layer.

Precipitation may take many forms.

Drizzle
Precipitation in the form of very small drops of water which appear to float is called **drizzle** At temperatures at or below the freezing level, drizzle will freeze on impact with objects and is known as **freezing drizzle.**

Rain
Precipitation in the form of large water droplets is called **rain.** **Freezing rain** is composed of supercooled water droplets that freeze immediately on striking an object which is itself at a temperature below freezing.

Hail
Observations have revealed the fact that water drops, certainly in the liquid form, can exist with temperatures as low as -40°C. It is clear, then, that small drops can be **supercooled** a long way without freezing.

Most big clouds formed as a result of an upward current of air are divisible into three well defined regions.

First there is the lowest layer where the cloud particles are in the form of water drops.

Next there is a region where some of the water droplets are frozen into ice crystals (snow) but some are still liquid but supercooled.

Third there is the highest region of the cloud where the water vapour sublimates into minute ice crystals.

There is no sharp dividing line between the snow and supercooled water regions. For some distance, the ice crystals and supercooled water drops are co-existing. When a supercooled water drop collides with an ice crystal, it at once freezes on the latter, imprisoning a little air which causes it to freeze in the form of soft ice. As it falls through the supercooled region, more soft ice is deposited on it, increasing its size. The ball of soft ice so formed then falls through the water region. Water freezes on it in the form of hard, transparent ice. Finally the ball falls out of the base of the cloud as a hailstone, a hard, transparent layer of ice covering a soft, white core.

Sometimes gusts carry hailstones back up to the top of the cloud, in which case the whole process is repeated, perhaps several times. In this way, very large hailstones are formed.

The vertical gusts which produce very large hailstones may have speeds in excess of 85 knots.

The conditions which produce hail are very similar to those in which thunderstorms originate. Hence hail is often encountered in a cumulonimbus thundercloud.

Snow Pellets (Soft Hail)
If the water region lying below the supercooled region of the cloud is not of great depth, the hailstone does not acquire the hard, transparent covering and arrives at the ground as the original soft, white ice. It is then known as a **snow pellet** or **soft hail.**

Snow
In the formation of **snow,** the invisible water vapour in the air sublimates directly into ice crystals, without passing through any intermediate water stage. Snow flakes are formed of an agglomeration of ice crystals and are usually of a hexagonal or star like shape. **Snow grains** are tiny snow crystals that have acquired a coating of rime. They fall from non-turbulent clouds.

Ice Prisms
Ice prisms are tiny ice crystals in the form of needles. They may fall from cloud or from a cloudless sky. They exist in stable air masses and at very low temperatures.

Ice Pellets
Ice pellets are formed by the freezing of raindrops. They are hard transparent globular grains of ice about the size of raindrops. They generally rebound when striking the ground.

Precipitation and Cloud Type
Each of the various forms of precipitation is associated with a particular type of cloud. While the table below is generally typical, exceptions do occur.

Drizzle, freezing drizzle, snow grains	*Stratus and stratocumulus*
Snow or rain (continuous)	*Thick altostratus and nimbostratus*
Snow or rain (intermittent)	*Thick altostratus and stratocumulus*
Snow showers, rain showers	*Altocumulus, heavy cumulus and cumulonimbus*
Snow pellets, hail, ice pellet showers	*Cumulonimbus*
Ice pellets (continuous)	*Any cloud that gives rain — the temperature is below freezing*
Ice prisms	*No cloud necessary*

Fog
Fog is, in fact, a cloud, usually stratus, in contact with the ground. It forms when the air is cooled below its dewpoint, or when the dewpoint is raised to the air temperature through the addition of water vapour.

To form a water drop in the atmosphere (the basis of fog formation), there must be present some nucleus on which the water may form. Dust, salt, sulphur trioxide, smoke, etc. provide this function.

Given a sufficient number of condensation nuclei, the ideal conditions for the formation of fog are high relative humidity and a small temperature dewpoint spread and some cooling process to initiate condensation. Light surface winds set up a mixing action which spreads and increases the thickness of the fog. In very still air, fog is unlikely to form. Instead dew will collect.

Fog is most likely to occur in coastal areas where moisture is abundant. Because of the high concentration of condensation nuclei, it is also common in industrial areas.

Smoke and dust in the air over large cities produce the "pea soup" fogs characteristic of large industrial centres. The carbon and dust particles cause such fogs to be dark. Otherwise, when composed of water drops only, fog is white in colour.

Fog is usually dissipated by sunlight filtering down through the fog or stratus layer. This results in heating from below.

Types of Fog

RADIATION FOG is formed on clear nights with light winds. The ground cools losing heat through radiation. The air in direct contact with the earth's surface is cooled. If this air is moist and the temperature is lowered below the dew point, fog will form. The ideal conditions for the formation of radiation fog are a light wind which spreads the cooling effect through the lower levels of the air, clear skies that permit maximum cooling and an abundance of condensation nuclei. This type of fog is commonly called **ground fog**, since it forms only over land. Radiation fog normally forms at night but sometimes it thickens or even forms at sunrise as the initial slight heating from the sun causes a weak turbulence. Radiation fog tends to settle into low areas, such as valleys, and it is usually patchy and only a few hundred feet thick. It normally dissipates within a few hours after sunrise as the sun warms the earth and radiation heating causes the temperature to rise.

ADVECTION FOG is caused by the drifting of warm damp air over a colder land or sea surface. This type of fog may persist for days and cover a wide area. It occurs most frequently in coastal regions. Widespread fog forms when moist air from a warm region of the ocean moves over colder waters. It will persist for lengthy periods since the water surface is not affected by daytime heating. Advection fog will spread over land if the circulation is from the sea to a colder land surface and will persist until the direction of the wind changes. Although it may dissipate or thin during the day from daytime heating, it will reform at night. The warm sector of a frontal depression is also favourable for the formation of advection fog.

UPSLOPE FOG is caused by the cooling of air due to expansion as it moves up a slope. A light upslope wind is necessary for its formation.

STEAM FOG forms when cold air passes over a warm water surface. Evaporation of the water into the cold air occurs until the cold air becomes saturated. The excess water vapour condenses as fog. Steam fog occurs over rivers and lakes, especially during the autumn.

PRECIPITATION-INDUCED FOG is caused by the addition of moisture to the air through evaporation of rain or drizzle. This type of fog is associated mostly with warm fronts and is sometimes known as **frontal fog.** The rain, falling from the warm air, evaporates and saturates the cooler air below.

ICE FOG forms in moist air during extremely cold calm conditions. The tiny ice crystals composing it are formed by sublimation and are often called needles. Ice fog is caused by the addition of water vapour to the air through fuel combustion. The very cold air cannot hold any additional water vapour and the excess sublimates into visible ice crystals. Ice fog may appear suddenly when an aircraft engine is started.

Haze

Haze is composed of very small water droplets, dust or salt particles so minute that they cannot be felt or individually seen with the unaided eye. Haze produces a uniform veil that restricts visibility. Against a dark background, it has a bluish tinge. Against a bright background, it has a dirty yellow or orange hue.

Smoke, industrial pollutants and smog from vehicular exhausts are responsible for the thick blanket of haze that severely restricts visibility in some urban and industrial areas. When flying in such conditions, visibility is very poor, especially when flying into the sun.

Haze is a problem only in very stable air. In unstable conditions, the particles scatter.

Sky Condition and Visibility

The amount of cloud cover aloft is reported as FEW (a trace of cloud), SCT (scattered), BKN (broken) or OVC (overcast). A cloud ceiling is said to exist at the height of the first layer for which the coverage symbol of BKN or OVC is reported. (See also **Aviation Weather Reports**).

VISIBILITY is one of the most important elements of weather from the standpoint of aircraft operations. It, in conjunction with the height of cloud layers aloft, determines whether an airport is open to traffic. VFR operations generally require a minimum of three miles ground and flight visibility and a height of cloud layers aloft that ensures that the aircraft can maintain 500 feet distance from the cloud base and 500 feet above the surface of the ground.

Restrictions to visibility would include cloud, precipitation, fog, haze, smoke, blowing dust or snow. Blowing snow, dust and sand can produce very poor visibility conditions. Blowing snow can be responsible for optical illusions. Precipitation, in the form of rain, snow and drizzle, appreciably reduces visibility. Drizzle which occurs in stable air is often accompanied by fog or smog.

Reduced visibility is a function of the stability of the air. If the air is stable, impurities that contribute to haze are trapped in the lower levels. Stable air is also favourable to drizzle and fog. If the air is unstable, vertical currents scatter the haze particles but cause blowing snow and dust which also contribute to reduced visibility.

VISIBILITY means the distance at which prominent objects may be seen and identified by day and prominent lighted objects by night.

FLIGHT VISIBILITY is the average range of visibility forward from the cockpit of an airplane in flight. It is sometimes called "air to air visibility".

SLANT RANGE VISIBILITY is the distance a pilot can see over the nose of the airplane towards the ground. It is sometimes called "approach visibility".

GROUND VISIBILITY is the visibility at an airport, as reported by an accredited observer.

PREVAILING VISIBILITY is the distance at which objects of known distance are visible over at least half the horizon. It is reported in miles and fractions of miles.

RUNWAY VISUAL RANGE (RVR) represents the distance a pilot will be able to see the lights or other delineating markers along the runway from a specified point above the centreline that corresponds to eye level at the moment of touchdown. RVR is reported in hundreds of feet. A device, called a transmissometer, that is installed adjacent to the runway samples a specified portion of the atmosphere and converts the sample into an estimate of the runway visual range. The reading derived from a transmissometer located adjacent to the runway threshold is reported as RVR "A"; that from a transmissometer located adjacent to the runway midpoint is reported as RVR "B". RVR information is available from ATC, the control tower and the flight service station. The actual RVR reading is provided to pilots if the RVR is less than 6000 feet.

VMC and IMC

Visual meteorological conditions (VMC) is a term used by meteorologists to indicate that visibility, distance from cloud and ceiling are equal to or better than the minima under which flight according to the visual flight rules (VFR) may be conducted.

Instrument meteorological conditions (IMC) is a term that indicates that visibility, distance from cloud and ceiling are below minima and flight can be conducted only under instrument flight rules (IFR).

Thunderstorms

A thunderstorm is a weather phenomenon whose passage creates extremely serious hazards to flying. It has aptly been described as a cumulus cloud gone wild. It is always accompanied by thunder and lightning, strong vertical drafts, severe gusts and turbulence, heavy rain and sometimes hail. It is a weather condition of which a pilot should be enormously respectful. It has been estimated that, at any one time, there are about 1800 thunderstorms in progress over the earth. With this kind of frequency, every pilot is sure occasionally to come in contact with one.

The basic requirements for the formation of a thunderstorm are unstable air, some form of lifting action and a high moisture content. Since these are also the requirements for the formation of a harmless cumulus cloud, it follows that the intensity of the conditions is the key to development of a thunderstorm. These violent weather factories occur when an air mass becomes unstable to the point of violent overturning. Such unstable atmospheric conditions may be brought about when air is heated from below (convection), or forced to ascend the side of a mountain (orographic lift) or lifted up over a frontal surface (frontal lift). The resulting buoyancy causes air which is warmer than its environment to push up in the form of convection currents, like drafts up a chimney flu.

If a mass of superheated moist air rises rapidly, an equal amount of cooler air rushes down to replace it.

When these conditions lead to the development of a thunderstorm, the area in which the rising and descending currents are active is called a **thunderstorm cell.** A thunderstorm may be composed of a number of such cells. As a storm develops, each successive cell grows to a greater height than did the previous one.

There are three distinct stages in the life cycle of a thunderstorm. Every thunderstorm begins life as a **cumulus** cloud. The cloud starts growing upward, driven by the latent heat of vaporization and/or fusion as water vapour condenses. Strong updrafts prevail throughout the cell and it rapidly builds up into a towering cumulonimbus cloud. Temperatures within the cell are higher than temperatures at the same level in the surrounding air, intensifying still more the convective currents within the cell. There is usually no precipitation from the storm at this stage of its development since the water droplets and ice crystals are being carried upwards or are kept suspended by the strong updrafts.

In its **mature** stage, the buildup of a towering cumulonimbus thunderstorm may reach heights as great as 60,000 feet. The updrafts may attain speeds of 6000 feet per minute. As the water droplets grow large enough to fall, they drag air down with them, starting a downdraft in the middle region of the cell that accelerates downward. The speed of the downdraft, although not as great as the updrafts, may nevertheless be as high as 2000 feet per minute. Violent turbulence is associated with the up and down drafts. The appearance of precipitation on the ground is evidence that the thunderstorm cell is in its mature stage. The mature stage lasts for about 15 to 20 minutes, although some thunderstorms have been known to last as long as an hour. Lightning, microbursts, gust front wind shear, hail and tornados are all phenomena associated with a thunderstorm in its mature stage.

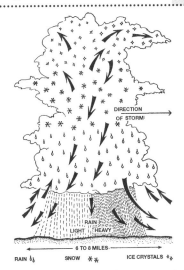

Fig. 35. Thunderstorm Cell in Mature Stage.

Then the cell begins to **dissipate.** The cool precipitation tends to cool the lower region of the cloud and the cell loses its energy. The downdraft spreads throughout the whole area of the cell with the exception of a small portion at the top where updrafts still occur. The rainfall gradually ceases. The top of the cell spreads out into the familiar anvil structure.

Individual thunderstorms are usually no more than 10 nautical miles in diameter. However, they do tend to develop in clusters of two or more. Such clusters, with individual thunderstorms in various stages of development, may cover vast areas and last for many hours, travelling great distances across the country during their life cycle.

There are two main types of thunderstorm: **air mass thunderstorms** and **frontal thunderstorms.**

Air mass thunderstorms usually form either singly or in clusters on hot summer days in warm moist air. Being scattered, there is usually VFR weather around them and they are therefore easy to avoid. They form either as a result of convection or orographic lift.

Frontal thunderstorms are associated most commonly with an advancing cold front but do develop in warm fronts as well. They usually form a line that may extend for hundreds of miles. They pose a special hazard to pilots because they are often embedded in other cloud decks and are therefore impossible to see. An advancing line of frontal thunderstorms should be avoided. Squall line thunderstorms rolling along in advance of a cold front are particularly violent.

Thunderstorm Weather

Thunderstorms produce very complex wind patterns in their vicinity. Wind shear can be found on all sides of a thunderstorm cell, in the downdraft directly beneath the cell and especially at the gust front that can precede the actual storm by 15 nautical miles or more.

In the cumulus stage of a thunderstorm's development, there is an inflow of air (updraft) into the base of the cloud. As the thunderstorm matures, strong downdrafts develop and the cold air rushing down out of the cloud spreads out along the surface of the ground well in advance of the thunderstorm itself undercutting the warm air in such a manner as to resemble a cold front. This is the **gust front** and turbulence within this lower level of spreading air is severe. The gust front generates strong, gusty winds near the surface which can change direction by 180° and gust up to 50 knots in a matter of seconds upon gust front passage.

The severe downward rush of air and its outburst of damaging winds on or near the ground is commonly called a **downburst.** Downbursts can be classified as either macrobursts or microbursts. A **macroburst** is a large downburst with a diameter of 2 nautical miles or more when it reaches the earth's surface and with damaging winds which last from 5 to 20 minutes. The most intense of these cause tornado like damage.

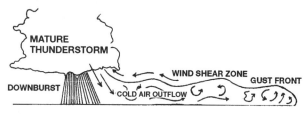
Fig. 36. Gust Front.

Smaller downbursts with a surface outflow diameter of less than 2 n. miles and peak winds that last less than 5 minutes are called **microbursts.** Wet microbursts occur in the presence of storm clouds with precipitation reaching the ground. Dry microbursts originate in moisture laden towering cumulus clouds. The downward flowing column of air contains precipitation (called **virga**) which evaporates before reaching the ground. The evaporation of the water appears to further cool the air, increasing the intensity of the microburst.

Downdrafts in microbursts can have vertical speeds as great as 6000 feet per minute. As they near the ground, these downdrafts spread out in all directions to become horizontal winds with speeds as high as 80 knots. They rapidly change direction and even reverse, causing very severe and dangerous wind shear. Because of the rapidly changing winds, both in speed and direction, an airplane encountering a microburst may lose so much lift that it cannot remain airborne. Depending upon the intensity of a thunderstorm, microbursts may occur anywhere up to 10 miles from the storm cell itself. Sometimes microbursts are concentrated into a line structure and, under these conditions, activity may continue for as long as an hour. Once microburst activity starts, multiple microbursts in the same general area are common.

Lightning flashes are the visible manifestation of the discharge of electricity produced in a thunderstorm. Areas of positive and negative charge accumulate in different parts of the cloud until the difference in electrical potential reaches a critical value and the air breaks down electrically. A lightning flash, a large electrical spark in the atmosphere, then follows. The mechanism by which electrification occurs in the thunderstorm is complex and varied. Ice particles and supercooled water droplets impacting on pellets of soft hail, the upward transport of rain in the strong vertical currents, coalescence of droplets, mass movement of charge by air flow are all factors.

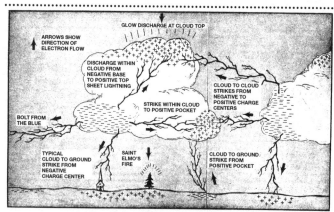
Fig. 37. Electrical Activity in a Thunderstorm.

In an active thunderstorm, the positive charge usually collects in the upper portion of the cloud and the negative charge in the cloud base. Flashing in a single thunderstorm cell may reach a peak of five discharges in a minute. Besides the activity within the individual cell, lightning may travel from one cloud to another or from the cloud to the ground. Occasionally, it will travel from the ground to the cloud, if a negative charge has accumulated in the ground.

Thunder is the noise which accompanies a lightning flash. It is attributed to the vibration set up by the sudden heating and expansion of the air along the path of the lightning flash.

Hail may be regarded as one of the worst hazards of thunderstorm flying. It usually occurs during the mature stage of cells having updrafts of more than average intensity. The formation of hail has been covered earlier in this section.

Icing in a thunderstorm is encountered at or above the freezing level in the areas of heaviest turbulence during the mature stage of the storm. The altitudes within a few thousand feet of the freezing level, above or below, are especially dangerous.

The **barometric pressure** ahead of a thunderstorm falls abruptly as the storm approaches, rises quickly when the rain comes, and returns to normal when the storm subsides. Occasionally after a storm, the pressure falls below normal, then rises to near normal again. All this can happen in a matter of 10 to 15 minutes.

The dangers of flying in or close to a thunderstorm are:

1. *Turbulence.* Turbulence, associated with thunderstorms, can be extremely hazardous, having the potential to cause overstressing of the aircraft or loss of control. Thunderstorm vertical currents may be strong enough to displace an aircraft up or down vertically as much as 2000 to 6000 feet. The greatest turbulence occurs in the vicinity of adjacent rising and descending drafts. Gust loads can be severe enough to stall an aircraft flying at rough air (maneuvering) speed or to cripple it at design cruising speed. Maximum turbulence usually occurs near the mid-level of the storm, between 12,000 and 20,000 feet and is most severe in clouds of the greatest vertical development.

Severe turbulence is present not just within the cloud. It can be expected up to 20 miles from severe thunderstorms and will be greater downwind than into wind. Severe turbulence and strong out-flowing winds may also be present beneath a thunderstorm. Microbursts can be especially hazardous because of the severe wind shear associated with them.

2. *Lightning.* Static electricity may build up in the airframe, interfering with operation of the radio and affecting the behaviour of the compass. Trailing antennas should be wound in. "Lightning blindness" may affect the crew's vision for 30 to 50 seconds at a time, making instrument reading impossible during that brief period. Lightning strikes of aircraft are not uncommon. The probability of a lightning strike is greatest when the temperature is between -5°C and 5°C. If the airplane is in close proximity to a thunderstorm, a lightning strike can happen even though the aircraft is flying in clear air. Lightning strikes pose special hazards. Structural damage is possible. The solid state circuitry of modern avionics is particularly vulnerable to lightning strikes. Electrical circuits may be disrupted. The possibility of lightning igniting the fuel vapour in the fuel cells is also considered a potential hazard.

3. Hail. Hailstones are capable of inflicting serious damage to an airplane. Hail is encountered at levels between 10 and 30 thousand feet. It is, on occasion, also encountered in clear air outside the cloud as it is thrown upward and outward by especially active cells.

4. Icing. Heaviest icing conditions occur above the freezing level where the water droplets are supercooled. Icing is most severe during the mature stage of the thunderstorm.

5. Pressure. Rapid changes in barometric pressure associated with the storm cause unreliable altimeter readings.

6. Wind. Abrupt changes in wind speed and direction in advance of a thunderstorm present a hazard during take-off and landing. Gusts in excess of 80 knots have been observed.

Very violent thunderstorms draw air into their cloud bases with great intensity. Sometimes the rising air forms an extremely concentrated vortex from the surface of the ground well into the cloud, with vortex speeds of 200 knots or more and very low pressure in its centre. Such a vortex is known as a **tornado.**

7. Rain. The thunderstorm contains vast amounts of liquid water droplets suspended or carried aloft by the updrafts. This water can be as damaging as hail to an aircraft penetrating the thunderstorm at high speed. The heavy rain showers associated with thunderstorms can cause contamination of the wing surface that results in early stall. If encountered during approach and landing, heavy rain can reduce visibility and cause refraction on the windscreen of the aircraft, producing an illusion that the runway threshold is lower than it actually is. Water lying on the runway can cause hydroplaning which destroys the braking action needed to bring the aircraft to a stop within the confines of the airport runway. Hydroplaning can also lead to loss of control during take-off.

St. Elmo's Fire

If an airplane flies through clouds in which positive charges have been separated from negative charges, it may pick up some of the cloud's overload of positive charges. Weird flames may appear along the wings and around the propeller tips. These are called **St. Elmo's Fire.** They are awe-inspiring but harmless. If the airplane flies in the vicinity of a cloud where negative charges are concentrated, its positive overload may discharge into the cloud. In this case, it is the airplane which strikes the cloud with lightning! The electricity discharges cause a noisy disturbance in the lower frequency radio bands but do not interfere with the very high frequencies. This **precipitation static,** as it is called, tends to be most severe near the freezing level and where turbulence and up and down drafts occur.

Thunderstorm Avoidance

Because of the severe hazards enumerated above, attempting to penetrate a thunderstorm is asking for trouble. In the case of light airplane pilots, the best advice on how to fly through a thunderstorm is summed up in one word — DON'T.

Detour around storms as early as possible when encountering them en route. Stay at least 5 miles away from a thunderstorm with large overhanging areas because of the danger of encountering hail. Stay even farther away from a thunderstorm identified as very severe as turbulence may be encountered as much as 15 or more nautical miles away. Vivid and frequent lightning indicates the probability of a severe thunderstorm. Any thunderstorm with tops at 35,000 feet or higher should be regarded as extremely hazardous.

Avoid landing or taking off at any airport in close proximity to an approaching thunderstorm or squall line.

Microbursts occur from cell activity and are especially hazardous if encountered during landing or take-off since severe wind shear is associated with microburst activity. Dry microbursts can sometimes be detected by a ring of dust on the surface. Virga falling and evaporating from high based storms can cause violent downdrafts.

The gust front, another zone of hazardous wind shear, can be identified by a line of dust and debris blowing along the earth's surface.

Swirls of dust or ragged clouds hanging from the base of the storm can indicate tornado activity. If one tornado is seen, expect others since they tend to occur in groups.

Do not fly under a thunderstorm even if you can see through to the other side, since turbulence may be severe. Especially, do not attempt to fly underneath a thunderstorm formed by orographic lift. The wind flow that is responsible for the formation of the thunderstorm is likely to create dangerous up and down drafts and turbulence between the mountain peaks.

Reduce airspeed to maneuvering speed when in the vicinity of a thunderstorm or at the first indication of turbulence.

Do not fly into a cloud mass containing scattered embedded thunderstorms unless you have airborne radar.

Do not attempt to go through a narrow clear space between two thunderstorms. The turbulence there may be more severe than through the storms themselves. If the clear space is several miles in width, however, it may be safe to fly through the centre, but always go through at the highest possible altitude.

When flying around a thunderstorm, it is better to fly around the **right** side of it. The wind circulates anti-clockwise and you will get more favourable winds.

If circumstances are such that you must penetrate a thunderstorm, the following few simple rules may help you to survive the ordeal:

Go straight through a front, not across it, so that you will get through the storm in the minimum amount of time.

Hold a reasonably constant heading that will get you through the storm cell in the shortest possible time.

Before entering the storm, reduce the airspeed to the airplane's maneuvering airspeed to minimize structural stresses.

Turn the cockpit lights full bright. (This helps to minimize the risk of lightning blindness.) Turn on the pitot heat and check the carburetor heat. Fasten the seat belts. Secure loose objects in the cabin.

Try to maintain a constant attitude and power setting. (Vertical drafts past the pitot pressure source and clogging by rain cause erratic airspeed readings.)

Avoid unnecessary maneuvering (to prevent adding maneuver loads to those already imposed by turbulence).

Determine the freezing level and avoid the icing zone. Avoid dark areas of the cell and those areas of heavy lightning.

Do not use the autopilot. It is a constant altitude device and will dive the airplane to compensate for updrafts, causing excessive airspeed, or climb it in a downdraft, creating the risk of a stall.

Lightning Detection Equipment

Lightning detection equipment, installed in the airplane, can help a pilot avoid thunderstorms.

This equipment detects the electromagnetic discharges associated with vertical air currents. All thunderstorms contain strong updrafts and downdrafts. These opposing ascending and descending air currents rub against each other, generating static electricity. The electrons tend to accumulate in positive and negative charges and when they have built up sufficiently, the potential difference will cause a current discharge. This discharge manifests itself not only as lightning but also in the radio spectrum. The lightning detection equipment picks up the radio frequencies from these discharges, a computer processes the signals, plots them by range and azimuth and presents them on a small, circular, radar like screen. The static electrical discharges picked up by the equipment may or may not be associated with lightning. It receives these signals through 360 degrees around the airplane and from as far away as 200 nautical miles.

Fig. 38. Lightning Detection Equipment.

Each static discharge is represented by a bright green dot on the cathode ray tube display. Clusters of dots indicate areas of thunderstorm activity. The display can be programmed to 3 different range settings, 40, 100 and 200 miles. It is most accurate on the 40 mile range. On the 200 mile range, the equipment sees everything but range is not so accurate. Generally, the display is more accurate and easier to read as the storm intensifies. In heavy electrical activity, the system has a problem called **radial spread.** The dots tend to spread over the display screening the areas between major clumps of storm caused dots.

Lightning detection equipment has some advantages over weather radar. Radar measures rainfall intensity. Lightning detection equipment is capable of detecting turbulence in clouds that have little or no precipitation. It is also able to see through areas of heavy precipitation to detect turbulent areas beyond. It does not, however, see rain.

There is an advantage in having both lightning detection equipment and weather radar installed in the aircraft, since they each provide valuable information in locating thunderstorms. Models of this equipment are available that interface with radar, displaying information from both systems on the same screen.

Lightning detection equipment is not dependent on line of sight. It will see, for example, the weather behind mountains. The system can, therefore, be used on the ground to determine weather for a 200 mile radius and is a useful flight planning tool.

Weather Radar

Airborne weather radar is one of the best instrument aids that a pilot can have in locating and avoiding thunderstorms. It is able to detect and display on the cockpit radar screen any significant weather that lies ahead on the flight route. The radar equipment does this by measuring precisely rainfall density of targets under observation. The antenna of the weather radar radiates a very narrow and highly directional beam, in the X band of the radio spectrum, straight ahead of the aircraft. The beam is cone-shaped and from 3 to 10 degrees in diameter. (Beam width is a function of antenna size and type.) The antenna scans left and right to cover a sector of about 120 degrees.

Although weather radar is not able to detect turbulence itself, the intensity of precipitation within a storm is a reliable indication of the amount of turbulence within a storm since strong drafts and gusts are necessary to produce water drops of significant size and quantity. Radar sees only water drops that are large enough to be affected by gravity and fall as rain. Because of the characteristics of X band radio waves and water, raindrops reflect the radiated beam back to the radar receiver. The sum of the reflections from all the raindrops appears on the screen as a target.

The computerized receiver measures the rainfall rate and grades the targets into levels which are represented on the screen by colours, green for level 1 which is light rain, yellow for level 2 which is medium rain and red for level 3 which is heavy rain. Areas of steep rain gradients are easy to see because of the colour coding. A precipitation rate that changes from minimum to maximum over a short distance is known as a **steep rain gradient** and usually is associated with a shear zone. Any target that is showing red is said to be **contouring,** is considered to be a storm and must be avoided and detoured. Areas of no precipitation between targets remain black and are called **corridors.**

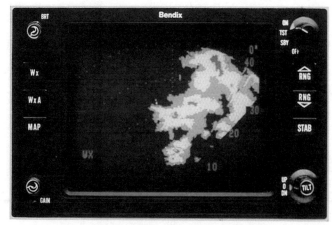

Fig. 39. Airborne Weather Radar.

The airborne display is gradated into mileage rings. Distance to the storm as well as its bearing with respect to the airplane's heading are therefore displayed. As a result, the pilot is able to select a safe and smooth flight path through thunderstorm areas. It is wise to give all contouring targets at least a ten mile clearance. A corridor between 2 targets should be at least 20 miles wide before considering it a safe passage.

Most weather radar systems manufactured today present a digital display that does not fade between sweeps. Some equipment incorporates automatic tilt to compensate for the attitude of the airplane. On some models, tilt is handled manually. Since the weather radar only can display targets illuminated by the radar beam, tilt management of the radar antenna is essential. The tilt feature controls the up and down angle of the antenna and consequently the plane of scan of the antenna. This feature is important in evaluating weather. The antenna beam does not see the whole storm, but only a 3 to 10 degree slice of it. By setting the tilt higher, the beam scans the upper region of the cell. By setting it lower, the beam scans the lower region of the cell. Since rain is most concentrated in the middle and lower regions of a thunderstorm, the best part of a storm to scan is this middle/lower area that gives the best indication of size and intensity.

It does take skill and training to use airborne weather radar most effectively. Interpreting the display is not an exact science but depends on the pilot's general knowledge of thunderstorms, the quality of the pre-flight briefing that he has received, and his familiarity with the limitations of the radar equipment in his airplane.

There are some limitations to weather radar. Moisture in relatively close proximity to the airplane can scatter the radar beams. This problem, called **attenuation,** means that heavy rain areas can block out a radar return from significant weather that lies beyond. Moisture and ice on the radar dome **(the radome)** installation on the nose of the airplane can diminish the radar signal. Useful range of the weather radar is only about 90 to 100 miles.

Icing

When an airplane, **while flying at a level where the temperature is at or below freezing,** "strikes" a supercooled water droplet, the droplet will freeze and adhere to the airplane. Dangerous icing can occur in cloud, freezing rain, or freezing drizzle.

The cloud in which icing most frequently occurs in winter is stratocumulus, but the heaviest deposits are encountered in cumulus and cumulonimbus. Clouds composed of ice crystals (such as cirrus) do not present an icing hazard. The ice crystals do not adhere to the wing.

The more dangerous types of icing are encountered in dense clouds, composed of heavy accumulations of large supercooled drops, and in freezing rain.

The seriousness of icing depends on the air temperature, the temperature of the aircraft skin and the amount of water striking the aircraft.

A supercooled water droplet freezes if disturbed. When struck by an aircraft, the drop begins to freeze immediately but, as it freezes, it releases heat to raise its temperature to 0°C. Freezing by impact then ceases and the remaining liquid in the drop begins to freeze more slowly as a result of cold surroundings. At very low temperatures, a large part of the drop freezes by impact. At higher temperatures, a smaller part of the drop freezes by impact leaving a greater amount to freeze more slowly.

How fast this liquid part of the drop freezes depends on the temperature of the aircraft skin. The higher the temperature, the more the drop will spread from the point of impact before freezing is complete.

Whether or not a drop freezes completely before another drop strikes the same spot is another factor affecting the character of icing. The amount of water intercepted by an aircraft in a given time is called its **rate of catch.** This rate varies with the liquid water content of the cloud, the size of the water droplets, the airspeed and the type of wing of the aircraft.

The liquid water content varies from level to level within the cloud. Generally, the amount of supercooled water in a cloud increases with height when the temperature is just a little below 0°C but decreases with height when the temperature is well below freezing, since at such low temperatures, more drops will freeze into ice crystals reducing the liquid content.

The size of droplets also affects the rate of catch. Small drops tend to follow the airflow and are carried around the wing. Large, heavy drops tend to strike the wing. When a small drop does hit, it will spread back over the wing only a small distance. The large drop spreads farther.

As for airspeed, the number of droplets struck by the aircraft in a certain time increases as the airspeed increases.

The curvature of the leading edge of the wing also has an effect on the rate of catch. Thin wings catch more droplets than do thick wings.

The rate of catch is, therefore, greatest for an aircraft with thin wings flying at high speed through a cloud with large droplets and a high liquid water content.

How Icing Affects the Airplane

Ice collects on and seriously hampers the function of not only wings, control surfaces and propellers, but also windscreens and canopies, radio antennas, pitot and static pressure sources, carburetors and air intakes. Turbine engines are especially vulnerable. Ice forming on the intake cowling constricts the air intake. Ice on the rotor and stator blades affects their performance and efficiency and may result in flame out. Chunks of ice breaking off may be sucked into the engine and cause structural damage.

The first structures to accumulate ice are the surfaces with thin leading edges: antennas, propeller blades, horizontal stabilizers, rudder, landing gear struts. Usually the pencil-thin outside air temperature gauge is the first place where ice forms on an airplane. The wings are normally the last structural component to collect ice.

Sometimes, a thin coating of ice will form on the windshield, preceded in some instances by frosting. This can occur on take-off and landing and with sufficient rapidity to obscure the runway and other landmarks during a critical time in flight.

Icing of the propeller generally makes itself known by a slow loss of power and a gradual onset of engine roughness. The ice first forms on the spinner or propeller dome and then spreads to the blades themselves. Ice customarily accumulates unevenly on the blades, throwing them out of balance. The resulting vibration places undue stress on the blades and on the engine mounts, leading to their possible failure.

If the propeller is building up ice, it is almost certain that the same thing is happening on the wings, tail surfaces and other projections. The weight of the accumulated ice is less serious than the disruption of the airflow around the wings and tail surfaces. The ice changes the airfoil cross section and destroys lift, increases drag and raises the stalling speed. At the same time, thrust is degraded because of ice on the propeller blades and the pilot finds himself having to use full power and a high angle of attack just to maintain altitude. With the high angle of attack, ice will start to form on the underside of the wing adding still more weight and drag. Landing approaches and landing itself can be particularly hazardous under icing conditions. Pilots should use more power and speed than usual when landing an ice-laden airplane.

If ice builds up on the pitot and static pressure sources, flight instruments may cease operating. The altimeter, airspeed and rate of climb would be affected. Gyroscopic instruments powered by a venturi would be affected by ice building up on the venturi throat.

Ice on radio antennas can impede VOR reception and destroy all communications with the ground. Whip antennas may break off under the weight of the accumulating ice.

Note: Carburetor icing may occur in clear air at higher than freezing temperatures. This subject is examined in the Chapter Aero Engines.

Tail Plane Stall

Because of the narrow camber of the airfoil section and the small radius of curvature of the leading edge, tail planes are very susceptible to ice accretion. The tail plane has an ice collection efficiency that is twice or three times that of the wings. If there is a half inch of ice on the wing, the accumulation on the tail plane may be an inch or more. In certain circumstances, ice may be accumulating on the tail plane when there is no visible ice on the wing.

The principal danger attributable to tail plane ice occurs during approach and landing when the flaps are extended or after nose down pitch and speed adjustments are made following flap extension. Most tail planes are set at an angle of incidence to provide negative lift to balance the lift of the wings. During approach and landing, the tail is already at a high angle of attack. Extension of the flaps causes a major change in the downwash on the tail and a further increase in the angle of attack of the tailplane. If, at the same time, there is substantial ice contamination of the tail plane, tail plane stall is likely to occur.

Buffeting, reduction in the elevator effectiveness, unusual nose down trim changes and nose down pitch warn of impending tail plane stall. The corrective action is **not** to increase the airspeed. Increasing the airspeed increases the margin of safety when the wing is approaching the stall condition, but it decreases the margin of safety when the tail plane is near stall because the increased airspeed increases the downwash on the tail plane. In a tail plane stall, it is necessary to correct the condition by retracting the flaps and applying nose up elevator. Then increase the power, commensurate with the new flap setting. Land with a reduced flap setting.

Types of Icing

The three main types of ice accretion, in order of their hazard to flying, are as follows:

Clear Ice

A heavy coating of glassy ice which forms when flying in dense cloud or freezing rain is known as clear ice or glaze ice. It spreads, often unevenly, over wing and tail surfaces, propeller blades, antennas, etc. Clear ice forms when only a small part of the supercooled water droplet freezes on impact. The temperature of the aircraft skin rises to 0°C with the heat released during that initial freezing by impact of part of the droplet. A large portion of the droplet is left to spread out, mingle with other droplets before slowly and finally freezing. A solid sheet of clear ice thus forms with no embedded air bubbles to weaken its structure. As more ice accumulates, the ice builds up into **a single or double horn shape** that projects ahead of the wing, tail surface, antenna, etc. on which it is collecting. This unique ice formation severely disrupts the airflow and is responsible for an increase in drag that may be as much as 300 to 500%.

The danger of clear ice is great owing to (1) the loss of lift, because of the altered wing camber and the disruption of the smooth flow of air over the wing and tail surfaces, (2) the increase in drag on account of the enlarged profile area of the wings, (3) the weight of the large mass of ice which may accumulate in a short time, and finally (4) the vibration caused by the unequal loading on the wings and on the blades of the propeller(s). When large blocks break off, the vibration may become severe enough to seriously impair the structure of the airplane.

When mixed with snow or sleet, clear ice may have a whitish appearance. (This was once classified as rime-glazed but it is now considered to be a form of clear ice.)

Rime Ice

An opaque, or milky white, deposit of ice is known as rime. It accumulates on the leading edges of wings and on antennas, pitot pressure sources, etc. For rime to form, the aircraft skin must be at a temperature below 0°C. The drop will then freeze completely and quickly without spreading from the point of impact. It is also dependent on a low rate of catch of small supercooled water droplets.

Rime forms when the airplane is flying though filmy clouds. The deposit has no great weight. Its danger lies in the aerodynamic alteration of the wing camber and in the choking of the orifices of the carburetor and instruments. Rime is usually brittle and can easily be dislodged by de-icing equipment.

Occasionally, both rime and clear ice will form concurrently. This is called **mixed icing** and has the bad features of both types.

Frost

A white semi-crystalline frost which covers the surface of the airplane forms in clear air by the process of sublimation. This has little or no effect on flying but may obscure vision by coating the windshield. It may also interfere with radio transmission and reception by coating the antenna with ice. It generally forms in clear air when a cold aircraft enters warmer and damper air during a steep descent.

Aircraft parked outside on clear cold nights are likely to be coated with frost by morning. The upper surfaces of the aircraft cool by radiation to a temperature below that of the surrounding air.

Frost which forms on wings, tail and control surfaces must be removed before take-off. Frost alters the aerodynamic characteristics of the wing sufficiently to interfere with take-off by increasing stall speed and reducing lift.

Frozen dew may also form on aircraft parked outside on a night when temperatures are just below freezing. Dew first condenses on the aircraft skin and then freezes as the surface of the aircraft cools. Frozen dew is usually clear and somewhat crystalline, whereas frost is white and feathery. Frozen dew, like frost, must be removed before take-off.

In fact, any snow or moisture of any kind should be removed since these may freeze to the surface while the airplane is taxiing out for take-off. The heat loss due to the forward speed of the airplane may be sufficient to cause congelation.

Intensity of Icing

Icing may be described as light, moderate and severe (or heavy). In severe icing conditions, the rate of accretion is such that anti-icing and de-icing may fail to reduce or control the hazard. A change in heading and altitude is considered essential. In moderate icing, a diversion may be essential since the rate of accretion is such that there is potential for a hazardous situation. Light icing is usually not a problem unless the aircraft is exposed for a lengthy period.

Clear ice is considered more serious than rime ice since the rate of catch must be high to precipitate the formation of clear ice.

The seriousness of an icing situation is, of course, dependent on the type of aircraft and the type of de-icing or anti-icing equipment with which the aircraft is equipped or the lack of such equipment.

Icing in Clouds and in Precipitation

Cumulus. Severe icing is likely to occur in the upper half of heavy cumulus clouds approaching the mature cumulonimbus stage especially when the temperatures are between -25°C and 0°C. The horizontal extent of such cloud is, however, limited so that the aircraft is exposed for only a short time.

Stratus. Icing is usually less severe in layer cloud than in cumulus type clouds but it can be serious if the cloud has a high water content. Since stratus cloud is widespread in the horizontal, exposure to the icing condition can be prolonged. Icing is more severe if cumulus clouds are embedded in the stratus layer.

Freezing Rain is common ahead of warm fronts in winter. Serious icing occurs when the aircraft is flying near the top of the cold air mass beneath a deep layer of warm air. Rain drops are much larger than cloud droplets and therefore give a very high rate of catch. In freezing temperatures, they form clear ice.

Freezing Drizzle. Drizzle falls from stratus cloud with a high water content. As the droplets fall through the clear air under the cloud, their size decreases due to evaporation. Therefore, icing in freezing drizzle is usually maximum just below the cloud base where the drops are largest. Icing is of the clear ice type.

Snow and Ice Crystals do not adhere to cold aircraft and do not usually constitute an icing problem. However, if the aircraft is warm, the snow may melt as it strikes the warm surface and ice accretion may result. If supercooled water droplets are also present with the snow, a rapid build up of rough ice can occur.

Protection From Icing

Many modern airplanes that are designed for personal and corporate use, as well as the larger transport type airplanes, are fitted with various systems designed to prevent ice from forming (anti-icers) or to remove ice after it has formed (de-icers).

1. *Fluids.* There are fluids which are released through slinger rings or porous leading edge members to flow over the blades of the propellers and the surfaces of the wings. A fluid is an anti-icing device since it makes it difficult for ice to form.

2. *Rubber Boots.* Membranes of rubber are attached to the leading edges. They can be made to pulsate in such a way that ice is cracked and broken off after it has already formed. This is a de-icing device.

3. *Heating Devices.* Heating vulnerable areas is a method for preventing the buildup of ice. Hot air from the engine or special heaters is ducted to the leading edges of wings, empennages, etc. Electrically heated coils protect pitot pressure sources, propellers, etc.

Icing Avoidance

Few single engine airplanes, or even light twin-engine types, incorporate any means of ice prevention and none are certified to fly in icing conditions. Even for those aircraft that are equipped with de-icing and anti-icing systems, certification testing probably did not produce quantitative documentation on how effective the de-icing equipment would be in coping with severe icing or, in fact, anything heavier than light icing.

A few tips are therefore in order to help pilots avoid or get out of icing situations.

When ice formation is observed in flight, there is only one certain method of avoiding its hazards and that is to get out of the ice-forming layer as quickly as possible. This may be done by quickly performing a 180° turn to fly back out of the icing situation. If the accumulation of ice has already become serious, it may be necessary to make a precautionary landing. The second alternative would be to descend and fly "contact" below the ice forming zone. The advisability of this course would depend on the ceiling and visibility along the route at the lower level. The third alternative is to climb above the ice forming zone. This alternative would obviously require an airplane having good performance and fitted with radio and proper instruments for flying "over the top".

Prompt action on the radio is important when icing starts. Information about the latest weather for altitudes above and below will help the pilot to make the decision on what action to take.

In any event, the decision must be made rapidly since, once ice has started to form, the condition may become critical in a matter of approximately six minutes.

Freezing rain and freezing drizzle account for the most rapid and most hazardous buildup of ice on an aircraft. Freezing precipitation is commonly accepted as forming from precipitation aloft falling through an above zero warm layer into a sub zero zone at the surface. The warm precipitation then either supercools in the cold air at low level or freezes on contact with a cold surface. Large supercooled droplets in this situation develop through a **condensation-coalescence process.** However, recent studies by the Cloud Research Physics Division of the Canadian Atmospheric Environment Service have found that freezing precipitation (especially freezing drizzle) can occur without a region of warm air aloft. Supercooled warm rain falls from cloud that is entirely below 0°C (called **supercooled warm rain process**). This phenomenon is particularly common on the east coast of Canada where 60 to 75% of freezing drizzle events occur with no warm air aloft. They occur when winds are from the north and east (i.e from over the ocean). Since the classic course of action to get out of an icing condition is to climb into the warm air aloft, pilots are cautioned to understand that in the situation described above, there may be no warm layer aloft. The higher you go, the smaller will be the freezing droplets and the smaller will be the rate of catch. However, there will be no warm air to melt the accumulated buildup.

Pilots flying in light airplanes which are not fitted with an outside air temperature gauge will be well advised to have one installed as this instrument will warn of temperatures that are conducive to icing conditions.

Avoid flight into an area where icing conditions are known to exist. Do not fly through rain showers or wet snow when the temperature is near 0°C. Do not fly into cumulus clouds when the temperature is low.

Always consult a weather office or flight service station to obtain a forecast about expected icing conditions before taking off on any flight in fall or winter.

Do not remain in icing conditions any longer than necessary. For that reason, during climbs or descents through a layer in which icing conditions exist, plan your ascent or descent to be in the layer for as short a time as possible. However, keep your speed as slow as possible consistent with safety. Speed of an airplane affects accretion of ice. The faster an airplane moves through an area of supercooled water drops, the more moisture it encounters and the faster will be the accumulation of ice.

If ice has started to build up on the airplane, do not make steep turns or climb too fast since stalling speed is affected by ice accumulation. Fuel consumption is greater due to increased drag and the additional power required. Land with more speed and power than usual. Do not land with power off.

With the advent of the jet age, the problem of icing has taken on some surprising new aspects. At one time, the pilots of airplanes flying through high cirrus clouds did not worry about ice forming on the airplane as cirrus clouds are composed of ice crystals rather than water droplets. With the increased speeds of which jet airplanes are capable, the heat of friction is sufficient to turn the ice crystals in the cloud to liquid droplets which subsequently freeze to the airplane.

Turbulence

Turbulence is one of the most unpredictable of all the weather phenomena that are of significance to pilots. Turbulence is an irregular motion of the air resulting from eddies and vertical currents. It may be as insignificant as a few annoying bumps or severe enough to momentarily throw an airplane out of control or to cause structural damage.

Most of the causes of turbulence have been mentioned in other sections of this chapter since turbulence is associated with fronts, wind shear, thunderstorms, etc.

In reporting turbulence, it is usually classed as light, moderate, severe or extreme. The intensity of the turbulence is determined by the nature of the initiating agency and by the degree of stability of the air.

Light turbulence momentarily causes slight changes in altitude and/or attitude or a slight bumpiness. Occupants of the airplane may feel a slight strain against their seat belts.

Moderate turbulence is similar to light turbulence but somewhat more intense. There is, however, no loss of control of the airplane. Occupants will feel a definite strain against their seat belts and unsecured objects will be dislodged.

Severe turbulence causes large and abrupt changes in altitude and/or attitude and, usually, large variations in indicated airspeed. The airplane may momentarily be out of control. Occupants of the airplane will be forced violently against their seat belts.

In extreme turbulence, the airplane is tossed violently about and is impossible to control. It may cause structural damage.

Whether turbulence will be light or more severe is determined by the nature of the initiating agency and by the degree of stability of the air.

There are four causes of turbulence.

1. Mechanical Turbulence. Friction between the air and the ground, especially irregular terrain and man-made obstacles, causes eddies and therefore turbulence in the lower levels. The intensity of this eddy motion depends on the strength of the surface wind, the nature of the surface and the stability of the air. The stronger the wind speed, the rougher the terrain and the more unstable the air, the greater will be the turbulence. Of these factors that affect the formation of turbulence, stability is the most important. If the air is being heated from below, the vertical motion will be more vigorous and extensive and the choppiness more pronounced. In unstable air, eddies tend to grow in size; in stable air, they tend not to grow in size but do dissipate more slowly.

Fig. 40. Mechanical Turbulence.

Turbulence can be expected on the windward side and over the crests of mountains and hills if the air is unstable. There is less turbulence on the leeward side since subsidence stabilizes the air. Mountain waves produce some of the most severe turbulence associated with mechanical agencies. In strong winds, even hangars and large buildings cause eddies that can be carried some distance downwind.

Strong winds are usually quite gusty; that is, they fluctuate rapidly in speed. Sudden increases in speed that last several minutes are known as squalls and they are responsible for quite severe turbulence.

2. Thermal Turbulence. Turbulence can also be expected on warm summer days when the sun heats the earth's surface unevenly. Certain surfaces, such as barren ground, rocky and sandy areas, are heated more rapidly than are grass covered fields and much more rapidly than is water. Isolated convective currents are therefore set in motion which are responsible for bumpy conditions as an airplane flies in and out of them. This kind of turbulence is uncomfortable for pilot and passengers. In weather conditions when thermal activity can be expected, many pilots prefer to fly in the early morning or in the evening when the thermal activity is not as severe.

Convective currents are often strong enough to produce air mass thunderstorms with which severe turbulence is associated.

Turbulence can also be expected in the lower levels of a cold air mass that is moving over a warm surface. Heating from below creates unstable conditions, gusty winds and bumpy flying conditions.

Thermal turbulence will have a pronounced effect on the flight path of an airplane approaching a landing area. The airplane is subject to convective currents of varying intensity set in motion over the ground along the approach path. These thermals may displace the airplane from its normal glide path with the result that it will either overshoot or undershoot the runway (See Fig. 41).

3. Frontal Turbulence. The lifting of the warm air by the sloping frontal surface and friction between the two opposing air masses produce turbulence in the frontal zone. This turbulence is most marked when the warm air is moist and unstable and will be extremely severe if thunderstorms develop. Turbulence is more commonly associated with cold fronts but can be present, to a lesser degree, in a warm front as well.

4. Wind Shear. Any marked changes in wind with height produce local areas of turbulence. When the change in wind speed

From the Ground Up

and direction is pronounced, quite severe turbulence can be expected. Clear air turbulence is associated at high altitudes with the jet stream. (See also **Wind Shear** and **The Jet Stream**.)

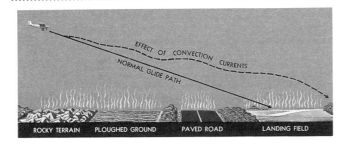

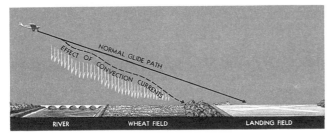

Fig. 41. The Effect of Thermal Turbulence.

High Level Weather

In recent times, there has been a rapid increase in the number of general aviation airplanes capable of flight at high levels. As a result, some comment about weather conditions at these heights, between 20 and 60 thousand feet, would seem to be in order. There are certain aspects of weather at high levels that differ from the weather that predominates at the lower levels. For the most part, the weather at the lower levels does not extend to these great heights.

The troposphere, the lowest layer of the atmosphere, extends up to a level between 20 and 60 thousand feet. The principal feature of the troposphere is a general decrease of temperature with height.

The stratosphere extends beyond the troposphere. The principal feature of the stratosphere is that temperature remains constant with height (at about -50°C). The boundary between the two layers is called the tropopause. The tropopause rises to greater heights over the equator than over the poles. Generally the tropopause lies at about 20 thousand feet over the poles and rises to a height of about 50 thousand feet over the equator.

In the polar air, the temperature in the troposphere decreases with height up to the tropopause which lies on the average at about 30,000 feet. At this point, the temperature has about reached the mean average temperature of the stratosphere. Therefore, there is no further decrease in temperature with height into the stratosphere. In the tropical air, the temperature in the troposphere also continues to decrease with height up to the tropopause but because the tropopause lies at a greater height (about 50,000 feet) that temperature will be lower than the mean average of the stratosphere. Therefore, the temperature in the stratosphere rises slightly with height until the mean average temperature is reached (Fig. 42).

The troposphere is composed of a number of different air masses with different characteristics and temperatures. The height of the tropopause over warm air masses originating in hot regions is quite high, over cold air masses originating over arctic regions quite low. As these air masses move away from their source regions, they carry their tropopauses with them. As a result, at the polar front that separates the warm tropical air mass from the cold polar air mass, the height change of the tropopause occurs rather abruptly, producing a distinct break in it. Sometimes the two parts seem to overlap and sometimes there is quite a gap between them (Fig. 43).

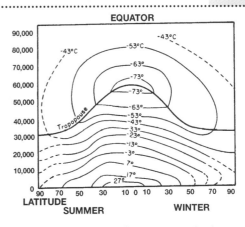

Fig. 42. Normal Distribution of Temperature in the Atmosphere.

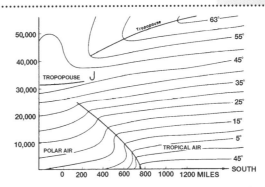

Fig. 43. Temperature Cross Section through a Typical Front.

The position of the tropopause can sometimes be visually recognized quite clearly. There is frequently a haze layer with a definite top at the tropopause. Anvil tops of thunderstorms spread out at the tropopause even though the convective core of a particularly active cell may extend up into the stratosphere. Winds are normally strongest at the tropopause, decreasing above and below it. Wind shear and turbulence occur near the tropopause.

High level charts are prepared by the weather service indicating pressure distribution at heights of 18,000 feet (500 hectopascals) and 34,000 feet (250hPa). The contour lines represent the approximate height of the pressure level indicated by the map. The wind direction is parallel to the contour lines and flows from west to east. The winds are strongest where the contour lines are closest, where the gradient is the steepest (Fig. 44). These charts are of immense value to pilots planning high level flights.

High Level Winds

In the temperate latitudes, the prevailing winds blow from the west to the east. The sub polar low and the subtropical high created by the circulation cells at 60° north latitude and 30° north latitude are responsible for this general wind pattern. (See **The Hemispheric Prevailing Winds**.)

The most important wind pattern in the high levels, however, is the jet stream, a narrow band of extremely high speed winds, that flows from west to east and takes a snake like pattern in its path across the hemisphere. The jet streams are closely associated with the tropopause, especially with the break in the tropopause that occurs over the polar front. (See also **The Jet Stream** in the Section **Winds**.)

Turbulence at High Levels

Cumulonimbus clouds very often extend to very great heights even several thousand feet into the stratosphere. Severe turbulence is associated with such towering buildups even at high levels.

The most severe turbulence at high levels is clear air turbulence, a bumpy condition that occurs in cloudless sky and, for that reason, is encountered without warning. It can be severe enough to be hazardous to high performance airplanes. Clear air turbulence is associated with wind shear within the jet stream. (See also **Clear Air Turbulence** in the Section **Winds**.)

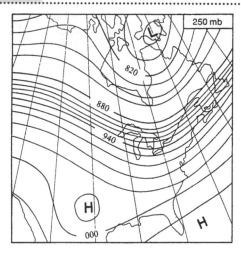

Fig. 44. 250 hPa Chart.

Visibility at High Levels

Usually there is little smoke or dust at altitudes above 20,000 feet. In some cases, smoke from large forest fires and smoke and dust from volcanic eruptions may rise to the tropopause. However, such occurrences are not common.

Haze layers which are not visible to a ground observer are frequently present in the upper troposphere. They are less dense than cirrus cloud but nevertheless do restrict visibility to zero when the airplane is flying into the sun. Such haze layers occur in stagnant air masses.

Most cloud above 20,000 feet is composed of ice crystals. The cirrus layers may vary in thickness from less than 100 feet to more than 20,000 feet. Cirrus cloud is frequently found above 30,000 feet and sometimes above 40,000 feet.

Substances, such as dust, smoke and haze, that are obstructions to visibility originate on the earth's surface and must be carried aloft by strong vertical currents. In the stratosphere, temperature is constant and therefore the air is stable. Obscuring materials therefore cannot be transported above the tropopause. Visibility in the stratosphere is excellent. However, it is very difficult to spot other airplanes in the stratosphere, possibly because there is lack of detail in the surroundings to provide contrast or to provide a focal point on which the eyes can focus. The condensation trail formed by the passage of an airplane is about the only way to spot that airplane at distances greater than 5 nautical miles. (See also **Vapour Trails**.)

Icing at High Levels

Icing above 20,000 feet is rare since most clouds are composed of ice crystals.

However, very high speed airplanes may experience icing problems in cirrus clouds. The heat due to friction is sufficient to melt the ice crystals in the cloud to liquid droplets which subsequently freeze to the cold exterior surfaces of the airplane.

Canopy Static

Canopy static is sometimes a problem at high levels. It is caused by the discharge of static electricity generated by the friction between solid particles in the atmosphere, especially the ice crystals in the cirrus clouds, and the plastic covered airplane surfaces. It reduces radio reception and can occur in such rapid succession that it seems to be a continuous disturbance. Changing altitude will generally eliminate the problem.

Weather Signs

Look for cloudy unsettled weather when
- The barometer is falling.
- The temperature at night is higher than usual.
- The clouds move in different directions at different levels.
- High thin white clouds (cirrus) increase. A large ring appears around the sun or moon and stays there until the overcast clouds thicken and obscure the sun or moon.
- Summer afternoon clouds darken.

Look for steady precipitation when
- There have been signs of unsettled weather.
- The wind is south or southeast, the pressure falling. Rain (or snow) within a day if the pressure is falling slowly. If it is falling rapidly, rain soon with winds increasing.
- The wind is southeast to northeast, the pressure falling. Rain (or snow) soon.
- Thunderclouds developing against a south or southeast wind.

Look for showers when
- Thunderclouds develop in a westerly wind.
- Cumulus clouds develop rapidly in the early afternoon in the spring or summer.

Look for clearing weather when
- The barometer rises.
- The wind shifts into the west or northwest.
- The temperature falls.

Look for continued bright weather when
- You can look directly at the sun when it sets like a ball of fire.
- The barometer is steady or slowly rising.
- Cloudiness decreases after 3 p.m. or 4 p.m.
- Morning fog breaks within two hours after sunrise.
- There is a light breeze from the west or northwest.
- There is a red sunset.

Look for higher temperatures when
- The barometer falls. In summer a falling barometer may indicate cloudy weather which will be cooler than in clear weather.
- The wind swings away from the north or the west.
- The morning sky is clear, except when the barometer is high or is rising in wintertime, or if the wind is strong from the north or west.

Look for lower temperatures when
- The wind swings from the southwest into the west, or from the

west into the northwest or north.
- When skies are clearing (although clearing skies in the morning will likely mean warmer weather by afternoon, particularly in the summer).
- In the winter, the barometer rises.
- Snow flurries occur with a west or north wind.
- Pressure is low and falling rapidly, wind east or northeast and backing slowly into north. (The fall in temperature will be gradual.)

Weather Information

The format and method by which aviation weather information is distributed is subject to frequent revision. The reader is advised to check recent NOTAMS and the AIP for any changes that might have been incorporated since publication of this manual.

Weather Charts

Weather maps are prepared on a regular schedule by the Atmospheric Environment Service (AES) and disseminated to all weather offices and flight service stations. They are used by meteorologists and briefers to give weather briefings to pilots. Along with weather reports and forecasts, they provide a comprehensive picture of the weather.

Surface Weather Charts

Surface weather charts are issued in all countries to show the state of the weather on the date and hour of publication of the chart.

The construction of the weather chart (map), so far as the representation of high and low pressure areas, with their accompanying wind systems, temperatures and frontal conditions will be more readily understood by reference to the sections **Isobars, High and Low Pressure Areas, Air Masses and Fronts** which were discussed in the preceding pages of this Chapter.

Surface pressure patterns can be considered as representative of the atmosphere up to 3000 feet AGL. Weather up to any level that is visible from the surface is also included.

A typical surface weather chart prepared by the Canadian Meteorological Service is reproduced as an insert in the **Examination Guide** at the end of this book. This chart depicts the surface weather at a given time for all of Canada and the United States. The information used to compile such a map is received from weather stations all over the continent who make hourly reports. The charts themselves are issued four times a day at six hour intervals. However, it is several hours after the information is received before the charts can be prepared and distributed, so that they are always several hours out of date when received at the weather office. The time of the observations on which the chart is based is indicated in the title box on the corner of the chart. Reporting stations are marked by circles and weather information is placed around the station circle in a standard pattern called a **station model.** All of the symbols used on a weather chart are depicted and explained on the reverse side of the weather map insert, in the Chapter **Examination Guide.**

Upper Level Charts

Upper level charts are similar in construction to surface charts, except that they show flow patterns aloft. Actual weather conditions in the upper air are measured twice a day at 0000Z and at 1200Z. The data is plotted on constant pressure level charts called **analyzed charts** (**ANAL**). These charts always depict conditions as they were at a specific time **in the past**.

Upper level charts are prepared for 850, 700, 500 and 250 hectopascals (representing on the average the 5000 ft/1000 m, 10,000 ft/ 3000 m, 18,000 ft/5500 m and 34,000 ft/10,400 m levels). These charts show reported atmospheric conditions at the pressure levels, such as wind speed and direction, temperatures, moisture content and frontal and pressure systems. Used in conjunction with the surface weather charts, they give a picture of the extent of the weather systems, cloud levels, thunderstorms, rain, etc.

The solid lines on an upper level ANAL chart represent the approximate height of the pressure level indicated by the map and are called **contour lines** rather than isobars as they are called on surface charts. The contours are labelled in decameters (10's of metres). On a 500 hPa chart, for example, 540 means 5400 metres. On a 250 hPa chart, 020 means 10,200 metres. Contours are spaced 60 metres apart on 850, 700 and 500 hPa charts and 120 metres apart on 250 hPa charts.

Dashed lines represent isotherms connecting areas of equal temperature and are labelled in 5 degree increments. They appear only on the 850 and 700 hPa charts.

Fronts are indicated on upper level charts only when they extend up to the pressure level of the chart and appear usually only on the 850 and 700 hPa charts. Information from a reporting station is indicated as on a surface chart but is called a **plotting model** rather than a station model. Information about wind direction and speed is included in the plotting model.

Prognostic Charts

Prognostic surface charts are issued to show what the weather map is expected to look like, a specified number of hours hence. They are actually graphic forecasts of the expected movement of high and low pressure areas, with their accompanying weather systems, and hence are valuable to a pilot in predicting changes which may occur before his actual time of departure or during the progress of his flight. The charts are issued 48 hours before the time for which they are valid and a second revised chart is issued 12 hours later (36 hours before the time for which it is valid). As with surface charts, the pressure patterns depicted can be considered representative of the atmospheric weather up to 3000 feet AGL.

Upper level prognostic charts (PROG), like their surface counterparts, are useful in anticipating weather conditions which may be expected to exist at some later specified time. They are prepared for FL 240, 340 and 450. They are issued 12 hours before the time for which they are valid and depict forecast winds and temperatures for the chart level. They are issued four times a day.

Winds are depicted using arrow shafts with pennants (50 knots each), full feathers (10 knots each) and half feathers (5 knots each). The orientation of the arrow indicates wind direction in degrees true and a small number at the pennant end of the arrow shaft gives the 10s digit of the wind direction.

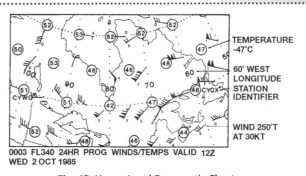

Fig. 45. Upper Level Prognostic Chart.

Temperatures in degrees Celsius are presented in circles at fixed grid points for the flight level. All temperatures are negative unless otherwise noted.

Wind and temperature information from these charts, in conjunction with the upper winds and temperature forecast (FD) and the significant weather charts (see following) can be used to determine wind shear and clear air turbulence (CAT).

Significant weather prognostic charts are prepared for FL 100 — 250 (the lower levels of the atmosphere) and for FL 250 — 600 (the mid and high levels). They show occurring and forecast conditions considered to be of concern to aircraft operations. Like PROG's, they are issued 4 times a day, 12 hours before the time for which they are valid.

The charts for FL 250 — 600 forecast any significant weather, such as thunderstorms, cloud heights when they extend into the heights depicted by the chart, tropopause heights, jet streams, turbulence, severe squall lines, icing and hail, widespread sand and dust storms, tropical cyclones and hurricanes, and the surface positions of frontal systems associated with significant weather phenomena.

The charts for the lower levels (FL 100 — 250) depict any of the above conditions which are applicable and also include information on moderate to severe icing, cloud layers of significance, marked mountain waves, the freezing level line at 10,000 feet and the surface positions and direction of movement of frontal and pressure systems. The recorded pressure at the centre of high and low pressure areas is indicated in hectopascals; for example, H x 1020 or L x 996. The symbols used in significant weather charts are depicted and explained on the reverse of the weather map insert in the **Examination Guide** at the end of the book.

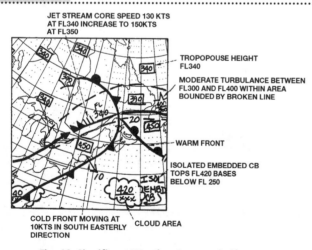

Fig. 46. Significant Weather Prognostic Chart.

Interpreting the Weather Map
In the insert reproduction of the surface weather chart referred to above, note the frontal pattern, the distribution of pressure, and the general circulation of the wind. The following observations are a few of the many and varied deductions that a pilot might make by studying the weather map.

A cold front, now lying off the Atlantic coast, brought sunny skies to the Maritime Provinces. The high pressure area has also brought cool temperatures. A quasi-stationary front, lying in an east west line through Maryland and Virginia will probably move towards Nova Scotia, bringing cloud and precipitation to that area. An arctic cold front is moving southward out of the prairies towards southern Ontario.

Temperatures in the cold air are near freezing. A warm front has moved across British Columbia and has now reached central Alberta. A widespread area of rain and low cloud is affecting the weather there.

Weather Observing Systems
Weather observations are taken every hour at selected aerodromes and at other locations throughout Canada. These observations are taken by specially trained observers or by an automated system.

The **automated weather observations system (AWOS)** is designed to electronically collect weather data and to disseminate partial or full surface weather observations.

At major airports, the **operational information display system (OIDS),** a fully automatic, computer controlled system, provides continuous updated information. It calculates both the now and the mean wind. The **now wind** is averaged over the last 5 seconds; the **mean wind** is averaged over the last 2 minutes. The mean wind is given in ATIS broadcasts, landing and taxiing information and is imputted into the AWOS data. The now wind is given for landing and take-off clearance and on pilot request. OIDS also determines the current altimeter setting and reads the runway visual range (RVR) as an average over the last 1 minute.

Awos Reports
Automated weather observation systems are generating weather observation data at hundreds of sites in Canada. The installation of these automated systems at airports that were not served previously by a manned weather observation station has meant that approved IFR approaches can be carried out at these airports. Data from AWOS stations is identified in weather reports by the word AUTO. The reports from these autostations are given a code number which appears directly after the word AUTO. Only AUTO 2 and AUTO 5 reports are considered usable for aircraft operations. Because pilots have access to weather information through DUATS and through electronic bulletin boards, it is important that they identify and use only those AWOS reports that are for aviation use.

AUTO 2 reports give cloud height and sky cover, visibility, temperature, dewpoint, mean sea level pressure, wind velocity and precipitation. A station issuing an AUTO 2 provides only hourly weather observation and does not issue specials. As a result, the report may be almost an hour old when it is used by the pilot. An AUTO 2 report does not include an altimeter setting. AUTO 2 reports should not be used by pilots who are self briefing for flight but should be used only in consultation with a qualified aviation weather briefer.

AUTO 5 reports are generated by a newer type of equipment and contain data on cloud base up to 10,000 feet AGL, sky cover, visibility, temperature, dewpoint, wind velocity, altimeter setting, precipitation occurrence, type and density. Human observers can add remarks to an AUTO 5. AUTO 5 special reports are issued for significant changes in cloud heights, visibility and wind criteria.

A voice generated message of the AWOS report has been developed which will, in the future, give aviation users access to the information via VHF radio or by telephone.

AWOS reports are sometimes unreliable due to computer software technical problems. Pilots should be cautious when planning a flight based only on AWOS generated weather reports. If possible, verify the AWOS report by comparing it with manned observations.

Aviation Weather Reports
An **aviation weather report** is a statement of weather conditions actually existing at a particular place at a given time. They are taken hourly by trained observers or by automatic weather observing stations (AWOS) at hundreds of stations throughout Canada and the United States.

Information about current weather is also available via the **weather radar network** which issues reports in graphic form. Observations of actual conditions observed by pilots during flight are available as **PIREPs**. **Runway surface condition reports** and **James Brake Index reports** detail runway conditions at aerodromes all across the country.

Aviation Routine Weather Report (Metar)

Weather data has for many years been transmitted in a special code called the SA code. About mid 1996, a new code will be adopted and weather data will be transmitted in the code that is called **METAR**. METAR is the name given to the international meteorological code for an aviation routine weather report. METAR observations are normally taken and disseminated on the hour. A SPECI, a special aviation weather report, is reported when weather changes of significance to aviation are observed. METAR and SPECI reports are not encoded by the observer but are generated by computer software. METAR reports replaced the SA aviation weather reports with which pilots have long been familiar. The weather data is transmitted in code in the following sequence. The code is composed of groups which are always in the same relative position to one another. When a weather element or phenomenon does not occur, the group is omitted.

Report name; station identifier; date and time of observation; wind and gust information; variations in wind direction; prevailing visibility; runway visual range and variations in RVR; present weather; sky condition; obscured sky; temperature and dewpoint; altimeter setting; recent weather; wind shear; remarks.

The aviation weather reports describe the existing weather conditions at specific times and at specific stations. They do not, therefore, give the complete weather picture. Between stations that are a hundred miles apart, the weather can be drastically different than that reported at either station. In addition, weather conditions can change very quickly and be quite different from the condition reported at a particular airport in the time it takes you to fly there.

For the full weather picture, consult the reports over several hours to determine the either improving or deteriorating trend and also consult the forecasts.

Symbols Used in Metar Weather Reports

Report Name
The code name METAR or SPECI is given in the first line of text. A METAR report is the regular hourly report. A SPECI report is issued when significant changes in weather conditions occur off the hour.

Station Indentifier
Weather reporting stations are assigned four letter identifiers in both Canada and the United States. In the U.S., the first letter of the identifier is K to denote a U.S. station. The following three letters are an abbreviation of the airport name, as, for example, "PIT" for Pittsburgh (KPIT). In Canada, the first letter of the identifier is C to denote a Canadian station. The following three letters identify the station. The first letter of the three letter group indicates the type of weather reporting station.

- Y or Z — reporting station co-located with an airport. Z is used only if the Y could cause a conflict with an identifier in use in the U.S.A.
- W — reporting station not co-located with an airport
- U — reporting station co-located with a radio beacon.

For example: CYOW — Ottawa International Airport — weather reporting station is co-located with the airport. WMN — Mount Forest, Ontario — weather reporting station is not located at an airport.

Date and Time of Observation
A 6 digit grouping giving the date and the time of observation in UTC (universal coordinated time) is included in all reports. The official time of observation (on the hour) is used for all METAR reports that do not deviate from the official time by more than 10 minutes. In SPECI reports, the time is reported in hours and minutes and refers to the time at which the weather changes occurred that required the issue of the report. The word AUTO follows the date and time grouping when the report is from an automatic weather observation station (AWOS). A correction to a METAR or SPECI is denoted by the letters CCA (for the first correction, CCB (for the second correction), etc.

Therefore, a date and time grouping 201200Z AUTO indicates a report taken on the 20th day of the month at 1200 UTC by an automated weather station. A date and time grouping 170500Z CCA indicates the first correction of a report taken on the 17th day of the month at 0500 UTC.

Wind and Gust Information
The 2 minute mean wind direction and speed are reported in a 5 digit grouping. Information on gust activity is preceded by the letter G and follows the wind direction and speed grouping.

Wind direction is that from which the wind is blowing. It is always reported in a 3 digit grouping given in degrees true and rounded off to the nearest 10 degrees. The third digit is therefore always a zero. A wind blowing directly from the north is reported as 360. Wind speeds are reported in knots in a 2 digit grouping.

Fig. 47.

In the case of mean wind speeds of 3 knots or less and a variable direction, the speed grouping is encoded as VRB.

Calm is encoded as 00000KT

Gust information is included if gust wind speeds exceed the mean wind speed by 10 knots or more in the 10 minute period preceding the observation. The letter G indicates the existence of gust conditions and the 2 or 3 digit group that follows the G reports the peak gust. If the gust criteria are not met, the G grouping is omitted.

Wind units are reported as KMH, KT and MPS, the standard ICAO abbreviations for kilometers per hour, knots and meters per second, respectively. In Canada and the U.S., wind speed is reported in knots.

Therefore, a wind grouping 04010G25KT indicates a wind blowing from 040° true at 10 knots, gusting to 25 knots. A wind grouping 270VRB indicates a light and variable wind blowing from 270° true.

Variations in Wind Direction
A grouping composed of 3 digits, the letter V and 3 more digits is used to report variations in wind direction. It is included only if, during the 10 minute period preceding the observation, the wind direction varies 60 degrees or more and the mean speed exceeds 3 knots. The two extreme directions are encoded in clockwise order.

Therefore, 060V130 indicates that the wind is varying from 060° to 130°.

Prevailing Visibility

A grouping composed of no more than 3 digits and the letters SM reports the prevailing visibility in statute miles and fractions thereof. If the visibility in any one sector of the runway is less than the prevailing visibility, that information is reported as remarks at the end of the METAR report.

In Canada and the U.S., visibility is reported in statute miles. In other countries, it is reported in metres.

Therefore, 1/2SM indicates a prevailing visibility of 1/2 of a statute mile.

Runway Visual Range and Variations in RVR

Runway visual range information is included whenever the prevailing visibility is 1 mile or less and/or the runway visual range is 6000 feet or less.

The grouping is identified by the letter R which precedes a 2 digit group that designates the runway (e.g. 08). The letters L, C or R (left, centre or right) are added if there are 2 or more parallel runways. A slash followed by a 4 digit grouping reports the visual range in hundreds of feet. The letters FT indicate that the units for RVR are feet. A slash followed by the letter D, U or N indicates the trend in the RVR. A distinct upward or downward change of 300 feet or more is encoded as U or D (upward or downward). A N indicates no distinct change has been observed. When it is not possible to observe the tendency, the tendency indicator is omitted.

Therefore, R29L/4000FT/U indicates a runway visual range for runway 29 left of 4000 feet. The trend indicates an upward, improving change.

Present Weather

The grouping representing present weather may consist of up to three groups. Each group may contain from two to nine characters.

Present weather comprises **weather phenomena** which may be one or more forms of precipitation, obscuration or other phenomena. Weather phenomena are preceded by one or two **qualifiers** which describe the intensity of or proximity to the station of the phenomenon, the other of which describes the phenomenon.

Intensity is reported as (-) light, (no sign) moderate or (+) heavy.

The **proximity qualifier VC** (for vicinity) is used in conjunction with the following phenomena:

DS	Duststorm		SS	Sandstorm
FG	Fog		SH	Shower
PO	Dust or Sand Swirls		BLDU	Blowing Dust
BLSA	Blowing Sand		BLSN	Blowing Snow
FC	Tornado, Waterspout or Funnel Clouds			

VC is used when these phenomena are observed within 8 KM (5 SM) but are not at the station. VCFC (tornadoes in the vicinity) shall be reported without regard to any distance restriction. When VC is associated with SH (showers) in a report, the type and intensity of precipitation is not specified.

The **descriptor**. No present weather group has more than one descriptor.

MI	Shallow		BC	Patches
DR	Drifting		BL	Blowing
SH	Shower(s)		TS	Thunderstorm
PR	Partial		FZ	Freezing

The descriptors MI (shallow), PR (partial) and BC (patches) are used only in conjunction with fog (FG). For example, BCFG indicates fog patches.

The descriptors DR (drifting) and BL (blowing) are used only in combination with snow (SN), dust (DU) and sand (SA). Drifting is used when the phenomenon is below 2 metres; two metres or more is defined as blowing. If snow and blowing snow are occurring together, both are reported but in separate groups (for example SN BLSN).

SH (shower) is used only in combination with one or more of rain (RA), snow (SN), ice pellets (PE), snow pellets (GS) and hail (GR). For example, SHPE indicates ice pellet showers; -SHRAGR indicates light showers composed of rain and hail.

TS (thunderstorm) will be reported either alone or in combination with one or more of the precipitation types. The end of a thunderstorm is the time at which the last thunder is heard, followed by a 15 minute period with no further thunder.

FZ (freezing) is used only in combination with fog (FG), drizzle (DZ) and rain (RA).

Weather phenomena. Different forms of precipitation are combined in one group, the dominant form being reported first. The intensity qualifier represents the overall intensity of the entire group, not just one component of the group.

Precipitation forms are reported as follows:

DZ	Drizzle		RA	Rain
SN	Snow		SG	Snow Grains
PE	Ice Pellets		GR	Hail
GS	Snow Pellets		IC	Ice crystals
UP	Unknown Precipitation (AUTO)			

Obscuration. Obstructions to vision are reported whenever the prevailing visibility is 6 miles or less. An obscuration occurring simultaneously with one or more forms of precipitation is reported in a separate group.

BR	Mist (VSBY equal to or greater than 5/8)			
FG	Fog (VSBY less than 5/8)			
HZ	Haze		FU	Smoke
SA	Sand		DU	Dust
VA	Volcanic Ash			

Other Phenomena are reported as follows:

PO	Dust/Sand Whirls		SS	Sandstorm
DS	Duststorm		SQ	Squalls
+FC	Tornado/Waterspout		FC	Funnel Cloud

Therefore, -RADZ BR indicates light rain and drizzle with mist, visibility less than 5/8 statute miles. +BLSN indicates heaving blowing snow.

Sky Condition

Sky condition for **layers aloft** are reported in a group that symbolizes layer coverage and the height of the layers above station elevation. Cloud amount is encoded as follows:

SKC — *Sky clear. No cloud present.*
FEW — *Trace to 2 oktas (2/8 of the sky is covered).*
SCT — *3 to 4 oktas (3/8 to 4/8 of the sky is covered).*
BKN — *5 to less than 8 oktas (5/8 to less than 8/8 of the sky is covered).*
OVC — *8 oktas (8/8 of the sky is covered).*

In METAR reporting, the celestial dome is divided into 8 segments, each called an **okta**.

There is no provision in METAR for reporting thin layers or partially obscured conditions. The symbol -X, common in SA reports, is reported in METAR as SKC.

A cloud ceiling is said to exist at the height of the first layer for which the coverage symbol BKN or OVC is reported.

Significant convective clouds, when observed, are identified by adding the letters CB (cumulonimbus) or TCU (towering cumulus) to the cloud group without a space.

The **height of the layer** above station elevation is reported in increments of 100 feet to a height of 10,000 feet, thereafter in increments of 1000 feet.

100 feet	*001*	*5000 feet*	*050*
400 feet	*004*	*10,000 ft.*	*100*
500 feet	*005*	*11,500 ft.*	*110*
1000 feet	*010*	*25,000 ft.*	*250*
3000 feet	*030*	*30,000 ft.*	*300*

The term CAVOK will **not** be used in Canada in METAR reports. Reports from other countries may use the term. When it is used, the term implies that the following conditions are occurring: visibility of 10 kilometers or more; no cloud below 5000 feet; no precipitation, thunderstorm, sandstorm, duststorm, shallow fog or low drifting dust, sand or snow. The term CAVOK is still used in Canada by ATC and is used in ATIS messages.

Therefore, SCT 050 indicates scattered clouds at 5000 feet.

BKN 030 BKN100 indicates a broken sky condition at 3000 feet and at 10,000 feet.

SCT 030 SCT 040TCU indicates scattered clouds at 3000 feet and scattered towering cumulus at 4000 feet.

FEW 004 BKN 080 indicates a few clouds at 400 feet and a broken condition at 8000 feet.

Obscured Sky (Vertical Visibility)
When the sky is obscured, a three digit group following the letters VV is included in the METAR report. VV is the group identifier and the vertical visibility is reported in units of hundreds of feet. The inclusion of a vertical visibility in the METAR report indicates an obscured sky condition.

Therefore, VV010 indicates that the sky is obscured and vertical visibility is 1000 feet.

Temperature and Dewpoint
The air temperature and the dewpoint are reported in 2 two digit groups separated by a slash (/). They are rounded to the nearest whole degree Celsius. Negative values are preceded by the letter M.

Therefore, 15/10 indicates an air temperature of 15°C and a dewpoint temperature of 10°C.

M05/M10 indicates an air temperature of minus 5°C and a dewpoint temperature of minus 10°C.

Altimeter Setting
The altimeter setting is given in hundredths of inches using four digits preceded by the letter A. (Internationally, the altimeter setting is reported to the nearest hectopascal and is preceded by the letter Q.)

Therefore, A2992 indicates an altimeter setting of 29.92 inches of mercury.

Q1013, in an international METAR report, indicates an altimeter setting of 1013 hectopascals.

Recent Weather
The METAR report includes information on recent weather. The group is preceded by the letters RE followed, without a space, by the appropriate abbreviation for the applicable weather when the phenomena were observed during the hour since the last routine report but not at the time of observation. The RE group may also be included in SPECI reports.

The following may be reported as recent phenomena: freezing precipitation; moderate or heavy rain, drizzle or snow; moderate or heavy ice pellets, hail or snow pellets; moderate or heavy blowing snow; sandstorm or duststorm; thunderstorm; volcanic ash.

The same weather phenomenon will be reported in both the present weather and in the recent weather groups only if the phenomenon was of greater intensity during the period since the last METAR report.

Therefore, a moderate rainshower at the time of observation with a heavy rainshower 20 minutes before the time of observation would be coded SHRA in the present weather grouping and as RERA in the recent weather grouping.

REBLSN indicates moderate blowing snow has occurred since the last routine report.

Wind Shear
The wind shear group, identified by the indicator WS, is used to report low level wind shear (within 1600 feet AGL) on the take-off and approach path. The runway designator RWY is normally followed by a 2 digit group to indicate the runway number but the letters L, C or R may be included if there are parallel runways.

Therefore, WS RWY18 indicates that windshear was encountered on runway 18.

WS RWY20R indicates that windshear was encountered on runway 20 right.

WS ALL RWY indicates that windshear was encountered on all runways.

Remarks
The remarks grouping is identified by the indicator RMK and includes information on cloud type and layer opacity (in oktas), sea level pressure (SLP) in hectopascals and any other information significant to aviation.

Therefore, RMK SC3 CI0 SLP134 indicates 3/8 opacity of stratocumulus, 0 opacity of cirrus and sea level pressure of 1013.4 hPa.

RMK ST8 ACSL OVR RDG NW SLP992 indicates 8/8 opacity of stratus, Mountain wave clouds (altocumulus standing lenticular) were reported over the ridge northwest of the station. Sea level pressure of 999.2 hPa.

The following abbreviations are used for the various cloud types:

CI	*cirrus*	*ST*	*stratus*
CS	*cirrostratus*	*SF*	*stratus fractus*
CC	*cirrocumulus*	*SC*	*stratocumulus*
AS	*altostratus*	*CU*	*cumulus*
AC	*altocumulus*	*CF*	*cumulus fractus*
ACC	*altocumulus castellanus*	*TCU*	*heavy cumulus*
NS	*nimbostratus*	*CB*	*cumulonimbus*

Aviation Weather Report (Metar)
Example:

METAR CYWG 172000Z 30015G25KT 3/4SM R36/ 4000

FT/D -SN BLSN BKN008 OVC040 M05/M08 A2992 REFZRA
WS RWY36 RMK SF5NS3 SLP134

METAR: METAR report.

CYWG: Code letters of reporting station (CYWG means Winnipeg).

172000Z: Date and time of observation (17 means 17th day of the month; 2000Z means 2000 UTC).

30015G25KT: Wind and gust information (300 means that the wind is from 300° true; 15 means windspeed is 15 knots; G25KT means gusts to 25 knots have been reported).

3/4SM: Prevailing visibility (3/4SM means 3/4 statute mile).

R36/4000FT/D: Runway visual range (R36 means runway 36; /4000FT/D means visual range is 4000 feet and is decreasing).

-SN BLSN: Present weather (-SN means light snow; BLSN means blowing snow).

BKN008 OVC040: Sky condition (BKN008 means broken cloud cover at 800 feet; OVC040 means overcast at 4000 feet).

M05/M08: Temperature and dewpoint (MO5 means air temperature of minus 5°C; MO8 means dewpoint temperature of minus 8°C).

A2992: Altimeter setting (A2992 means altimeter setting of 29.92 inches of mercury).

REFZRA: Recent weather (REFZRA means freezing rain was reported since the last scheduled report but is not present at the time of this observation).

WS RWY36: Windshear (WS means windshear was reported on runway 36).

RMK SF5NS3 SLP134: Remarks (RMK means remarks; SF5 means 5 oktas of stratus fractus cloud; NS3 means 3 oktas of nimbostratus cloud; SLP134 means sea level pressure of 1013.4 hPa).

SA Coded Weather Report

Weather reports transmitted in the SA code took the following format:

Station Identification: Type of Report and Time: Sky Condition and Ceiling: Visibility: Weather and Obstructions to Vision: Barometric Pressure: Temperature and Dewpoint: Wind: Altimeter Setting: Clouds and Obscuring Phenomena: Remarks: Pressure Tendency. An example of an SA weather report would be as follows:

YWG   SA   0900   3 SCT   M7 OVC   1   1/2   R-FK 926/5/3/0404/925/SF3ST7 VSBY SW 1/2F 304

Decoded, this reads: Station is Winnipeg. Time of the report is 0900Z. Sky condition is scattered at 300 feet with an overcast measured ceiling at 700 feet. Visibility is 1 1/2 statute miles in light rain, fog and smoke. Barometric pressure is 992.6 hectopascals. Temperature is 5°C. Dewpoint is 3°C. Wind is from 040° true at 4 knots. Altimeter setting is 29.25 inches of mercury. Stratus fractus clouds cover 3/10th of the sky and stratus clouds cover 7/10th of the sky. The pressure is rising.

Weather Radar Network (EC/DND)

Specially designated weather surveillance radars are used at a number of sites in Canada. Observations are made of precipitation areas and convection type storms. The computer generated composite reports are transmitted in graphic form and give a visual representation of

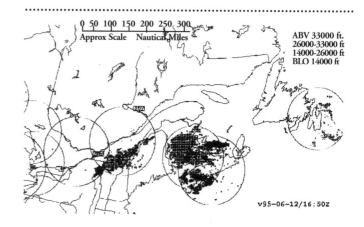

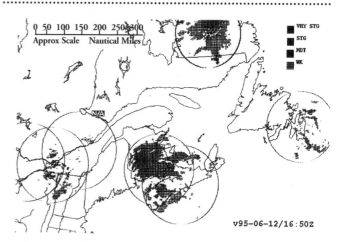

Fig. 48. Weather Radar Maps Showing Intensity and Height.

the radar report. They are transmitted on the AES communication circuits and are also available on computer hook-ups and on Internet.

The radar presents a display of weather in a 130 nautical mile (200 kilometer) radius. The rate of movement (velocity) of the precipitation areas is determined by the doppler system (see **Doppler** in Chapter **Radio Navigation**) and is accurate for a 65 nautical mile (120 kilometer) radius. Radar does not show turbulence. Turbulence is usually associated with heavy rainfall and can be presumed, therefore, to be present in any area showing heavy precipitation.

Pilot Reports (Pireps UA)

Observations of actual conditions taken by pilots during flight are also available on the communication circuits.

Example:

UA/OV YXU 04835 1537 FL090 / TP BE80 / SK OVC 070 / TA -20 / IC LGT RIME BLO 070 / RM WIND COMP HEAD 010 MH 048 TAS 210

Decoded, the above reads:

Pilot report. Location: London VORTAC, 048 degree radial, 35 nautical miles. Time: 1537Z at 9000 feet. Aircraft type: Beech Queen Air. Sky condition: overcast with tops at 7000 feet. Air temperature: minus 20 degrees Celsius. Icing: light rime below 7000 feet. Remarks: head wind component 10 knots, magnetic heading 048 degrees, true airspeed 210 knots.

In PIREPS, cloud heights are always given as a height above sea level (ASL).

RSC JBI Report

The Flight Manuals for all airplanes specify performance data relating to landing distance, cross wind limitations, etc. These figures presuppose bare, dry runways. Any contaminant, such as water, snow or ice will degrade braking capability, substantially increase the landing roll and make directional control difficult if a crosswind is also involved. Because the condition of the runway at the airport of destination is, therefore, of as much concern as the state of the weather, it is important to have some source for this information.

Contaminated runways are also of concern to the aircraft that is taking off. If the runway is wet or icy, the accelerate-stop distance or the required take-off distance could exceed the available runway length. Studies have shown that the accelerate-stop distance is increased by approximately 15% on wet runways, 50% on snow, 75% on water deeper than 3 mm and 100% on ice covered runways. Take-off distance is increased 10% when 15 mm of loose snow or 8 mm of slush covers the runway.

Runways in Canada are, on average, wet or contaminated one third of the time during the five coldest months of the year. Runway surface condition reports are therefore issued to advise pilots of surface contamination. Included in an RSC report is a report on the braking action that is derived from a James Brake Decelerometer.

The James Brake Decelerometer is an instrument which measures the decelerating forces acting on the test vehicle (on which it is installed) when the brakes are applied. The resulting figures are called the James Brake Index (JBI).

When available, a JBI reading is included as part of the runway surface condition (RSC) report. RSC/JBI reports are transmitted on the AES communication circuits as a special type of NOTAM. They are available as a voice advisory from the control tower at controlled airports and from flight service stations at uncontrolled airports.

The RSC section of the report provides runway surface information describing in plain language the runway condition. The JBI section describes braking action quantitatively.

A RSC/JBI NOTAM is distributed when there is slush or wet snow on the runway; when there is loose snow on the runway that exceeds 1/4 inch in depth; when there is packed snow, ice or frost on the runway; when the JBI reading is .4 or less; and when the runway is not cleared to the full width. When only a portion of the runway width has been cleared, the report will also include a description of the remaining portion of the runway (e.g. depth of snow, windrows, etc.)

The JBI readings are on a scale from 0 (non-existent braking) to 1 (excellent braking). Low numbers, such as .2, .3, represent low braking coefficients. JBI reports are issued at such times as the JBI reading for any runway falls to .4 or below. When the JBI reading exceeds .4, a "Normal Winter Conditions" text is transmitted, cancelling previous reports.

A JBI report takes the following format: station identifier, title (JBI), runway number, temperature celsius, runway average JBI reading and the time UTC (year, month, day, hour and minute)

CYOW JBI 07/25 -4 .28 9502 251200Z.

In this case, the report concerns runway 07/25 at Ottawa International Airport. The temperature is -4°C. The braking action on the runway is rated at an average of .28 for the full length of the runway. The readings were taken in February 1995 on the 25th day at 1200UTC.

An RSC/JBI report typically appears as follows:

CYOW RSC ALL RWYS COVERED 4 INCHES WET SNOW 9502 0600 CLEARING EXPECTED TO COMMENCE 9502 251000
or
CYOW RSC 07/25 SNOW DRIFTS 3-4 INCHES 9501 071000

CYOW JBI 07/25 -15 .30 9501 071000
or
CYOW RSC 07/25 WET WITH .25/.75 SLUSH CENTRE THIRD 9502 281100

Decoded, this report reads: A runway surface condition report for runway 07/25 indicates that the runway is wet with slush on the centre section varying from a depth of 1/4 inch to 3/4 inch. Time of the report is February 1995, 28th day at 1100 UTC

Aviation Weather Forecasts

A **weather forecast** is a statement of **anticipated** conditions expected at a particular place, over an area, or along a route during a given period, based on available information. The forecaster will usually state the degree of confidence with which the forecast is made.

In planning a flight, the pilot must have reasonable knowledge of the following facts:

1. Whether the ceiling and visibility at destination (or at alternate landing places) will be sufficiently good to make a safe landing. If the flight is to be made by visual contact, the ceilings and visibilities must be VFR along the entire route.

2. The wind conditions at various flight levels to estimate the time and fuel required, and to choose a route and altitude to fly that will give the most favourable winds.

3. At what levels to expect icing conditions and how to avoid severe icing zones.

4. Where thunderstorms, turbulence or hail may be encountered, so that he may delay the start of the flight or land en route to avoid dangerous weather in the air.

Since the weather is constantly changing, the conditions shown in a weather report may change during a flight and a forecast of the changes must be obtained before departure.

The following code identifies the various types of forecasts:

TAF — Aerodrome Forecast
FACN — Area Forecast
FDCN — Winds and Temperatures Aloft Forecast
WSCN — AIRMET Message

Area Forecast (FACN)

An **area forecast** is a general statement of weather conditions to be expected over a particular region. They are prepared every 6 hours and are issued approximately one half hour before the beginning of their valid period (at 0000Z, 0600Z, 1200Z and 1800Z). They are valid for 12 hours and each new forecast replaces the last 6 hours of the preceding one. Forecasts also include a 12 hour outlook that applies to the period beyond the main forecast. The area forecast includes **regional forecasts** for each region in the area.

Cloud heights in area forecasts are always stated as a height above sea level (ASL) unless otherwise noted. A forecast of cloud above

20,000 feet ASL is indicated by the slash symbol (/) placed directly before the amount (e.g. /BKN).

Example:

FACN01 CYEG 171730

18-06

OTLK 06-18

HGTS ASL UNLESS NOTED

PROG

FLAT RDG FORT SMITH—MCMURRAY—LETHBRIDGE AT 1800Z MOVG TO MEDICINE HAT—NORTH BATTLEFORD 250 MI E OF FORT SMITH BY 0600. SLOLY DPNG TROF PRINCE GEORGE—FORT NELSON. AIR UNSTBL AND FAIRLY MOIST. UPSLP CLDS OVR SRN ALTA.

YEG — 1

LETHBRIDGE RGN

CLDS AND WX. 50 OVC 80 110 BKN 140 BCMG 50 BKN 80 110 BKN 140 AFT 0400Z.

ICG. LGT RIME ICGIC. FRLVL 40.

OTLK VFR

YEG — 2 — 3 — 4

CALGARY CORONATION EDSON REGIONS

CLDS AND WX. 60 BKN CU100 /-SCT BCMG /-SCT AFT 0100Z.

ICG. LGT RIME ICG IN CU. FRLVL 40.

OTLK VFR

YEG — 5

EDMONTON RGN

CLDS AND WX. 60 BKN CU100 OCNLY 60 BKN TCU 150 6RW- BCMG 100 BKN 130 BY 0200Z.

ICG. MDT CLR ICG IN TCU 60-150 OTRW LGT ICGIC. FRLVL 40.

TURBC MDT VCNTY TCU TIL 2400Z.

OTLK VFR

Decoded, the above reads:

Area forecast (FACN) issued by the weather office at Edmonton International at 1730Z on the 17th of the month and valid from 1800Z to 0600Z on the 18th. Outlook valid from 0600Z to 1800Z on the 18th.

All heights above mean sea level unless noted.

Prognosis: Flat ridge overlying Fort Smith, McMurray and Lethbridge at 1800Z is moving towards Medicine Hat and North Battleford. Will be 250 nautical miles east of Fort Smith by 0600Z. Slowly deepening trough between Prince George and Fort Nelson. Air unstable and fairly moist. Upslope clouds over Southern Alberta.

YEG 1: Lethbridge Region

Clouds and Weather: 5000 feet overcast with tops at 8000' and 11000' broken with tops at 14000'. Becoming 5000' broken with tops at 8000' and 11000' broken with tops at 14000' after 0400Z.

Icing: Light rime icing in cloud. Freezing level 4000 feet.

Outlook: VFR.

YEG 2 — 3 — 4: Calgary, Coronation, Edson Regions.

Clouds and weather: 6000' broken. Cumulus clouds with tops at 10000'. High thin scattered cirrus. Becoming high, thin scattered cirrus after 0100Z.

Icing: Light rime icing in cumulus clouds. Freezing level 4000'.

Outlook: VFR

YEG 5: Edmonton Region.

Clouds and Weather: 6000' broken, cumulus clouds with tops at 10000'. Occasionally 6000' broken, towering cumulus clouds with tops at 15000'. Visibility 6 statute miles in light rain showers. Becoming 10000' broken, tops at 13000' by 0200Z.

Icing: Moderate clear icing in towering cumulus from 6000' to 15000'. Otherwise light icing in cloud. Freezing level 4000'.

Turbulence: Moderate turbulence in vicinity of towering cumulus until 2400Z.

Outlook: VFR

The code for forecasts of weather outlook is as follows:

IFR — Ceilings less than 1000 feet AGL and/or visibility less than 3 statute miles.

MVFR — Marginal VFR. Ceilings between 1000 and 3000 feet AGL and/or visibility between 3 and 5 statute miles.

VFR — Ceilings more than 3000 feet AGL and visibility more than 5 statute miles.

The cause of IFR and MVFR conditions is given in the outlook by indicating the ceiling or visibility restrictions. For example: IFR CIG F would indicate that the outlook is for IFR weather due to ceilings and fog. If winds are forecast to be greater than 20 knots, the word WND is included. For example: VFR WND would indicate that the outlook is for VFR weather with winds or gusts in excess of 25 knots.

In area forecasts, the forecast visibility is included if it is expected to be 6 statute miles or less. The forecast wind is included if it is expected to be 25 knots or more. The term LYRS indicates that the cloud is not a solid mass of cloud, but that random layers exist in the space between the bottom and top layers.

Aerodrome Forecast (TAF)

For many years, aerodrome forecasts have been transmitted in a code called FTCN. About mid 1996, aerodrome forecasts will be reported in a new code known as TAF, the international meterological code for an aerodrome forecast. TAF forecasts replace the familiar FTCN forecasts.

An **aerodrome forecast** states in specific terms the expected weather conditions that will affect landing and take-off at the aerodrome. Wind, low level wind shear, prevailing visibility, significant weather and sky condition are given for an area within 5 nautical miles of the aerodrome. Aerodrome forecasts are issued only for those aerodromes for which routine hourly and special reports are available.

Aerodrome forecasts are transmitted on the Environment Canada telecommunications network and on the ADIS circuit (automated data interchange system) as soon as possible but not later than 20 minutes prior to their period of coverage. They are issued at least four times daily at 0440, 1040, 1640 and 2240 UTC. The normal validity period is 24 hours, beginning at 0500Z, 1100Z, 1700Z and 2300Z. An aerodrome forecast is valid from the time of issue until it is amended or until the next scheduled TAF is issued. When a new TAF is issued, it automatically cancels the previous one.

Aerodrome forecasts shall always contain, as a minimum, information about the following elements: wind, visibility, weather and cloud. They are transmitted in the following format:

Report Name
The code name TAF may be followed by the letters AMD to indicate an amended or corrected forecast.

Location Identifier — Date & Time — Period of Coverage — Automatic Station Indicator
The four letter ICAO station identifier (as in METAR) is followed by the seven digit group that indicates the date and time of origin of the forecast. Following that is a six digit group that indicates the period of coverage of the forecast. This period is normally 24 hours, except in the case of amended TAFs. The word AUTO is included if the forecast is based on observations from an automatic weather station (AWOS).

Therefore, CYUL 090440Z 090505 AUTO indicates an aerodrome forecast for Montreal issued on the 9th day of the month at 0440 UTC. It is valid from 0500 UTC on the 9th to 0500 UTC on the 10th. The forecast is based on observations from an automatic weather station.

Wind
The forecast wind is encoded as in a METAR report giving direction in degrees true and speed in knots with an indication of gust activity. The code VRB is used when the wind speed is 3 knots or less but may be used for higher speeds when it is impossible to forecast a single direction (as, for example, when a thunderstorm passes).

Therefore, 35020G45KT indicates that the wind direction is forecast to be from 350° true at 20 knots, gusting to 45 knots.

Low Level Wind Shear
When strong, non-convective, low level wind shear is forecast to occur over the aerodrome, this optional group will be used. The coded grouping begins with the letters WS followed by a 3 digit grouping indicating the height in hundreds of feet AGL of the shear zone. A slash followed by a 5 digit group indicates the wind speed and direction of the shear zone.

Therefore, WS 015/20015KT indicates that wind shear is forecast at 1500 feet AGL over the aerodrome. The wind is from 200° true at 15 knots.

Prevailing Visibility
The prevailing visibility is encoded as in a METAR report, except that visibility greater than 6 statute miles will be indicated by the code P6SM.

Therefore, 3/4SM indicates that the visibility is forecast to be 3/4 statute mile.

Significant Forecast Weather
Significant forecast weather is encoded with the same codes as used for **weather phenomena** in METAR reports. Intensity and proximity qualifiers, descriptors, precipitation and obscuration are included as required. The abbreviation NSW, for no significant weather, may be used.

The codes TEMPO (for temporary fluctuation) or BECMG (for becoming) are used to forecast the end of significant weather. In this case, the code NSW (no significant weather) is included.

The code TEMPO or BECMG is also used to forecast a significant change in weather or visibility. In this case, all weather and visibility groups are repeated, including those which are unchanged.

If a different weather condition is **not** indicated after TEMP or BECMG, then the previously given condition continues to apply.

Therefore, -RA BR TEMPO RA BR indicates a forecast of light rain and mist and a temporary fluctuation that will bring moderate rain and mist.

3SM -DZ BR BECMG 1SM -DZ BR indicates a forecast visibility of 3 statute miles in light drizzle and mist, becoming visibility of 1 statute mile in light drizzle and mist.

24015KT 4SM BLSN BECMG 30015KT indicates a forecast wind from 240°T at 15 knots, visibility of 4 statute miles in blowing snow, becoming wind from 300°T at 15 knots (visibility of 4 statute miles in blowing snow continues to apply).

Sky Condition
Sky condition is encoded as in a METAR report, using the symbols SKC, FEW, BKN, OVC or VV. Forecast cloud type is not identified except in the case of CB layers. The letters CB then follow the coded height. The specific height of cirriform cloud is given in TAF forecasts.

When TEMPO or BECMG is used to forecast that clouds and/or obscuration are no longer expected, SKC (sky clear) will be indicated.

When TEMPO or BECMG is used to forecast a significant change in sky condition, all layers will be repeated including those which are unchanged.

Therefore, SCT008 BKN015CB BKN250 indicates scattered cloud at 800 feet, broken cumulonimbus at 1500 feet, broken cloud at 25,000 feet.

SCT100 BECMG SKC indicates scattered cloud at 10,000 feet becoming sky clear.

Probability and Time of Occurrence
In order to indicate the probability of changing conditions that would constitute a hazard to aviation, the indicator PROB is included with digits indicating the percentage of probability and the probable time period in UTC.

Therefore, PROB 30 20Z23Z 1/2SM +TSRAGR indicates a 30% probability between 2000 and 2300 UTC of visibility of 1/2 statute mile and thunderstorms, heavy rain and hail.

Forecast Change and Time
When a permanent change is forecast to occur rapidly, a part-period forecast comes into effect at the time of the change. It is indicated by the letters FM and it supersedes all prior conditions.

Therefore, 24010KT P6SM SKC FM 1640Z 27015KT P6SM SKC indicates that the wind is forecast to be from 240° true at 10 knots, visibility greater than 6 statute miles and sky clear. After 1640 UTC, the wind is forecast to change to 270° true at 15 knots. Visibility will continue to be greater than 6 statute miles and the sky will be clear.

A **temporary fluctuation** of some weather element is indicated by the code TEMPO and a six digit group indicating the time of the specified period of fluctuation. The changing conditions are indicated.

Therefore, 09004KT 1 1/2SM BR VV004 TEMPO 12Z15Z 3/4SM BR VV001 indicates a forecast wind from 090° true at 4 knots, visibility of 1 1/2 statute miles, mist (sky obscured), vertical visibility of 400 feet. A temporary fluctuation is forecast for between 1200 and 1500 UTC of visibility of 3/4 statute mile, mist (sky obscured) and vertical visibility of 100 feet.

A **permanent change** in some of the forecasted weather elements is indicated by the code BECMG with a time indication.

Therefore, FM 1200Z VRB02KT 2SM BR SKC BECMG 13Z15Z P6SM SCT020 indicates that from 1200 UTC the wind direction is forecast to be variable at 2 knots, visibility will be 2 statute miles in mist, sky will be clear. Between 1300 and 15000 UTC, conditions are forecast to become visibility greater than 6 statute miles with scattered cloud at 2000 feet.

Example:

TAF

CYOG 011640Z 011717 28015KT P6SM -SNRA FEW015 OVC040 TEMPO 17Z24Z 2SM -SNRA BR OVC015 FM0000Z 28015KT P6SM BKN030 BKN250 TEMPO 00Z03Z -SHRA FM 1000Z 30015KT P6SM SKC.

Decoded, the above reads:

Aerodrome forecast for Windsor issued on the 1st day of the month at 1640Z and valid from 1700Z (on the 1st) to 1700Z (on the 2nd).

Wind 280° true at 15 knots, visibility greater than 6 statute miles, light snow and rain, few clouds at 1500 feet, overcast at 4000 feet.

A temporary fluctuation between 1700Z and 0000Z will bring visibility of 2 statute miles, light snow and rain, mist, overcast at 1500 feet.

From 0000Z, the wind will be from 280° true at 15 knots, visibility greater than 6 statute miles, broken sky cover at 3000 feet, broken at 25,000 feet.

A temporary fluctuation between 0000Z and 0300Z will bring light rainshowers.

From 1000Z, the wind will be from 300° true at 15 knots, visibility greater than 6 statute miles, sky clear.

FTCN Coded Aerodrome Forecast

An FTCN aerodrome forecast was transmitted in the following format:

FTCN 241645

YYC 241705 30 SCT C90 OVC 1412 2100Z C20 OVC 6RW- 1415. 0200Z C30 BKN 1415.

Decoded, this reads: A 12 hour aerodrome forecast issued at 1645Z on the 24th day of the month for Calgary and valid from 1700Z on the 24th to 0500Z on the 25th. Scattered clouds at 3000 feet AGL Ceiling 9000 feet AGL overcast. Surface wind will be from 140° at 12 knots. At 2100Z the ceiling will become 2000 feet AGL overcast. Visibility of 6 statute miles in light rain showers. Surface wind will be from 140° at 15 knots. At 0200Z, the ceiling will rise to 3000 feet AGL with broken clouds. Surface winds will be from 140° at 15 knots.

Upper Winds and Temperature Forecast (FDCN)

An **upper winds and temperature forecast** provides an estimate of upper wind conditions and temperatures at selected levels. Wind direction is given in degrees true to the nearest 10° and wind speeds in knots. Data for the production of FDs is derived from a variety of atmospheric data sources taken at 32 sites twice daily at 0000Z and 1200Z. This data is used to issue periodic forecasts during the 24 hour period. The time of issue and the validity period is indicated in the body of the forecast.

Some lower level groups are omitted. No winds are forecast within 1500 feet of the station elevation and no temperatures are forecast for the 3000 foot level or for a level within 1500 feet of the station elevation.

Example:

FDCN01 CWAO 171530Z

BASED ON 1200Z

DATA VALID 171800Z FOR USE 1500-2100Z

	3000	6000	9000	12000	18000
YVR	2021	2425-07	2430-10	2434-18	2542-26
YYF	2523	2432-04	2338-08	2342-13	2448-24
YXC		2431-02	2330-06	2344-11	2352-22
YYC		2426-03	2435-06	2430-12	2342-22
YQL		2527-01	2437-05	2442-10	2450-21

Decoded, the above reads:

Winds and temperature forecast, issued by the Montreal Meteorological Centre at 1530Z on the 17th day of the month.

Based on data fed into the computer at 1200Z. Forecast is valid for 1800Z on the 17th day of the month and for use for 3 hours either side of 1800Z, i.e. from 1500Z to 2100Z.

Vancouver: (YVR): At 3000 feet, wind from 200° at 21 knots. At 6000 feet, wind from 240° at 25 knots and temperature -7°C. At 9000 feet, wind from 240° at 30 knots and temperature -10°C. At 12,000 feet, winds from 240° at 34 knots and temperature -18°C. At 18,000 feet, winds from 250° at 42 knots and temperature -26°C.

FDCN1 KWBC 081750Z

DATA BASED ON 081200Z

VALID 090000Z FOR USE 1800Z-0300Z TEMPS NEG ABV 24000

	24000	30000	34000	39000
YVR	2973-24	293040	283450	273763
YYF	3031-24	314041	304551	304763
YQL	2955-28	306845	307455	791159

Decoded, the above reads:

Winds and temperature forecast, issued by the U.S. Meteorological Service in Washington at 1750Z on the 8th day of the month.

Based on data fed into the computer at 1200Z on the 8th. The forecast is valid for use from 1800Z on the 8th to 0300Z on the 9th. Temperatures above 24,000 feet should be read as negative values.

YQL: At 24,000 feet, the wind is from 290° at 55 knots and the temperature is -28°C. At 30,000 feet, wind from 300° at 68 knots and temperature -45°C. At 34,000 feet, wind from 300° at 74 knots and temperature -55°C. At 39,000 feet, wind from 290° at 111 knots and temperature -59°C.

When the wind speed is greater than 100 knots (100 to 199 knots), 50 is added to the direction code and 100 is subtracted from the speed code. Therefore, in the YQL forecast for 39,000 feet, subtract 50 from 79 to determine that the wind direction is from 290° and add 100 to 11 to determine that the wind speed is 111 knots.

If the windspeed is forecast to be greater than 200 knots, the wind group is coded as 199 knots. Thus, 6699 is decoded as the wind from 160° at 199 knots or greater.

The code 9900 is used to indicate light and variable winds or wind speeds of less than 5 knots.

Winds aloft are given for a number of representative levels from 3000 feet ASL to 53,000 feet ASL. Winds at heights not reported can be obtained by interpolation.

What is the wind at Penticton (YYF) at 8000 feet?

The wind at 6000 feet is from 240° at 32 knots.

The wind at 9000 feet is from 230° at 38 knots.

Hence, the wind is backing approximately 3° per 1000 feet. At 8000 feet, it would be from approximately 233°.

The velocity of the wind is increasing 2 knots per 1000 feet. At 8000 feet, it would be blowing at approximately 36 knots.

The same kind of interpolation can be made for the temperature.

"Weather Sense"

The pilot gets information about the weather along his intended route from the aviation weather reports, from the area and aerodrome forecasts and from a study of the weather map. One of the most important attributes that a good pilot possesses is the ability to use this information to form in his own mind a three dimensional picture not only of the current weather but also of how that weather may develop and change, deteriorate or improve during his flight.

A pilot needs to understand when and how a particular weather condition could develop to become hazardous and to plan ahead on alternative action. It is, therefore, always advisable to look at a weather map with a view to looking for weather, wind circulation, ceilings and visibility not only along the route but also to either side of the route to be flown in order to know which way to turn in case the weather "closes in" at your terminal. Always have an "out" in case things turn bad.

This is a skill acquired through knowledge and experience and one which is invaluable to adopting a true defensive attitude towards weather.

Weather is constantly in a state of change. Weather systems travel across the country, sometimes quite rapidly, sometimes intensifying as they move. Local geographical conditions cause local peculiarities in the weather over a particular area that can develop very suddenly.

Sometimes in a matter of minutes the weather at a particular landing strip may change from clear to below VFR. Often aerodromes within a few miles of each other have drastically different conditions of visibility and ceiling.

Some weather conditions are invisible and therefore are encountered without warning. Windshifts, severe vertical currents and turbulence are hazards that occur in clear air.

A defensive attitude towards weather is therefore an important concept for a pilot to cultivate. It is founded in a thorough understanding of how, when and why weather phenomena develop and what sequence of events and combinations of circumstances can change good weather into bad. It is strengthened by a resolve never to take off on a flight without obtaining a thorough weather briefing and to keep a constant check on the weather by means of radio during a flight.

It is an ability to read weather signs, to recognize situations in which, for example, turbulence may be present and to be prepared to avoid or compensate for them. It is an ability to recognize little signs that give warning of a developing situation that could affect the safe outcome of a flight, smoke trails, for example, or ripples on water, etc., that indicate wind direction and turbulence that might be encountered during landing. It involves alertness in noticing and recognizing a deteriorating weather situation, such as loss of visual reference in worsening visibility and to turn back out of it before you get boxed in with no room to maneuver.

Above all, it involves the wisdom to cancel a planned flight or to turn back when deteriorating conditions present a situation with which you cannot cope. It is knowing and respecting your own limitations.

Weather Information Sources

Information about prevailing weather conditions and forecasts of expected weather developments is available in Canada from Atmospheric Environment Service (AES) weather offices, flight service stations and weather stations. In the U.S., the same type of information is available from the flight service stations at the airport and at the weather bureau office in the town.

Trained weather experts will discuss your proposed flight and will review with you the latest weather maps, aviation weather reports at your point of departure and destination and for points along your intended route, area and aerodrome forecasts, radar precipitation patterns, sigmets, PIREPS, etc. If you are not familiar with the format of the coded reports, they will explain them to you in plain language and will advise you as regards the existing and anticipated weather conditions along your route. This service is known as a **weather briefing.**

When requesting weather information either in person or by telephone, always identify yourself as a pilot and give your pilot licence number. Indicate the type of aircraft, its registration, the type of flight plan (VFR or IFR), your intended route, altitude, destination, departure time and time of return. When filing VFR, advise if you are equipped and prepared for IFR flight. This information helps the flight service or weather station personnel to select the information that is pertinent to your needs.

Tell the briefer to speak slowly as you are going to copy everything he says. Special weather briefing forms are available on which to notate this information so that it is available to you later for reference.

Your chance of totally understanding a weather briefing and coming away from it with a clear three dimensional picture of what is happening in the skies over your area of the country is greatly enhanced if you do some preparation before contacting anyone in the weather service. Make a habit of watching the TV weather channels. Then check the weather information in the local paper. By getting as much information as possible from the publicly available weather services, you will know quite a bit about the day's weather before you even approach the weather briefer.

Aviation Weather Information Service (AWIS)

Flight service stations are especially equipped to provide weather information to pilots both at the preflight and inflight stages.

At the preflight stage, the flight service station personnel will provide a weather briefing to help you in planning your flight. They are authorized to provide, in plain language, a description of the weather for your point of departure, destination and alternates along the route. Included in the offered information are regular and special weather reports, aerodrome and area forecasts, PIREPs, SIGMETs,

AIRMETs, and wind and temperature forecasts. The information available at the flight service station covers an area within a 500 nautical mile radius of that station. If you are planning a flight beyond 500 nautical miles, a more complete weather picture is available from the nearest AES weather office. Telephone service from the flight service station to the weather office is usually available.

At the inflight stage, the flight service station personnel provide en route aircraft with updated information. You should not request an "initial" weather briefing while airborne because this practice ties up the FSS radio frequency. However, if you need information to help you make a decision about terminating a flight or altering course, contact any FSS that is within communication range. An important aspect of this VHF inflight weather service is the solicitation of pilot reports on upper winds, cloud tops, icing and turbulence and the relaying of this information to other pilots.

If inflight information is required, VHF radio contact should be established with any in-range FSS on the frequency 126.7 MHz or, where appropriate, on the HF frequency 5680 KHz. The flight service station has the facility to reply on the non directional beacon frequency, the VOR frequency or the frequency 126.7 MHz. You should specify what frequency you are monitoring when calling for weather information from an FSS.

Aviation Weather Briefing Service (AWBS)

The aviation weather briefing service is a fully interpretive weather briefing service which is available at some flight service stations. Access is also available by a toll free 1-800 number and by a local telephone number. Briefers, who are trained and equipped to the AWBS standard, are authorized to provide, in addition to the weather information included under AWIS, an interpretation and adaptation of meteorological information to fit the changing weather situation and the special needs of the user. They consult and advise on special weather problems and, on request, will provide flight documentation for long range flights.

Transcribed Weather Broadcasts (TWB)

In order to provide a means of mass dissemination of aviation weather information, recorded weather data is continuously broadcast on the primary low frequency navaid (NDB). Similar recorded weather information is continuously broadcast by many FSS in the U.S. The TWBs are also available on multiple access telephone lines (see The **Canada Flight Supplement** for listings). The broadcast data includes flight precautions, weather synopses, significant en route weather, upper winds, selected pilot reports, radar reports, hourly and special aerodrome weather reports, terminal forecasts and selected NOTAMs. These broadcasts are updated as new reports or forecasts are received.

Duats

Aviation weather is also available from Direct User Access Terminal System (DUATS) service via computer terminals. The DUATS service is provided by vendors or owned by users and provides not only weather but NOTAM information. Flight plans may be filed via an approved DUATS installation.

Pilots Automatic Telephone Weather Answering Service (PATWAS)

At selected locations, a continuous recording of meteorological and NOTAM information is available by public telephone. The information and locations of this service are listed in the **Canada Flight Supplement.**

PATWAS recordings normally include general weather conditions derived from the appropriate area forecasts, hourly and special reports for selected stations, aerodrome forecasts for selected stations, forecast winds and temperatures aloft, icing, freezing level and turbulence, selected PIREPs and daily sunrise and sunset times.

Volmet Broadcasts

The VOLMET service is provided by Gander Radio on a number of HF frequencies to broadcast weather information, aerodrome forecasts, special reports, SIGMET messages, etc. to flights operating in the Gander Oceanic and the Gander, Moncton, Montreal and Toronto domestic FIRs.

Weather Radar (RAREPS)

The AES operates a number of radar units located all across Canada that report, at least once every hour and more often if necessary, when precipitation is detected. These radar observations are disseminated through the communication circuits and are available at flight service and weather stations to assist in preflight and inflight planning. The radar reports are included in the transcribed weather broadcasts.

Sigmets (WACN)

Significant meteorological advisories (SIGMETS) are messages to aircraft in flight of severe and hazardous weather conditions. These would include severe turbulence, severe icing, thunderstorms, squall lines, hurricanes, heavy hail, widespread dust and sand storms, marked mountain waves, volcanic ash and low level wind shear. They are broadcast immediately upon receipt on the navigational aids voice channels; they are included in the scheduled broadcasts for the duration of the period for which they are valid; they are transmitted directly to specific aircraft when such action is considered appropriate by the FSS personnel; and they are issued on request to aircrews over normal air-ground communications channels. They are also transmitted through the communication circuits.

Airmets (WSCN)

AIRMETS (airmen's meteorological advisories) are intended for aircraft in flight to notify pilots of potentially hazardous weather conditions not described in the area forecast but not requiring a SIGMET.

They advise of freezing precipitation, thunderstorms, moderate icing, moderate turbulence, extensive areas of visibilities of less than 3 statute miles, broken or overcast cloud conditions of less than 1000 feet AGL (i.e. IFR conditions), surface winds of 20 knots or more and gusts of 30 knots or more when no winds were forecast and wind direction when the difference between the forecast and observed wind direction is greater than 60°. An AIRMET is issued when the area forecast has not anticipated the developing poor weather situation. AIRMETs are worded in abbreviated plain English.

Environment Canada Bulletin Board Service

Environment Canada offers to the aviation industry a national and international package of aviation weather products on their bulletin board service. It provides up to date weather information through a menu driven system that allows computer users to directly access Environment Canada databases. The aviation products and services available include hourly weather reports, aerodrome and area forecasts, upper wind forecasts, AIRMETs, SIGMETs, aviation charts and satellite pictures.

Internet

Nav Canada, Canada's air navigation system provider, has launched a weather information web site on the Internet. The site provides a selection of current aviation weather information. To access information on any airport in Canada, enter the four letter identification (such as CYOW). The weather is displayed with graphics and text. It is updated every hour or more frequently when significant weather conditions could affect aviation. Area and aerodrome forecasts, significant weather advisories and surface weather observations are included. The Nav Canada web site is: www.navcanada.ca

PIREPS

Pilot reports (PIREPS) are reports by pilots in flight of any unusual weather conditions encountered (such as unpredicted thunderstorms, tornadoes, hail, severe icing or turbulence, mountain waves, low ceilings or visibility, etc.). They are especially valuable as they provide up-to-date information that is not so promptly available from forecasts or ground reporting stations. They may also be the only information available for areas between reporting stations, especially mountain and large water areas where weather commonly develops. Pilot reports should be made to any FSS or ATS (air traffic service) facility over normal en route navaid voice frequencies. They will be broadcast immediately by the station receiving them if the information concerns a condition considered to be a hazard to aviation. They will also be included in subsequent scheduled weather broadcasts.

Flight service station personnel are required to solicit PIREPs when any of the following weather conditions are known to exist: ceilings below 2000 feet, visibility less than 3 statute miles, the presence of moderate or heavy precipitation, turbulence, icing, thunderstorms, winds in excess of 50 knots or when conditions differ substantially from those forecast. FSS specialists are also required to obtain PIREPs during the climb-out and approach phases of flight when less than VMC conditions exist.

PIREPs should be passed directly to an FSS specialist via 126.7 MHz, if airborne, or by a toll-free or collect telephone call, after landing.

In transmitting a PIREP, you should include time of observation, type of aircraft, altitude, position and meteorological conditions.

It is good practice to make a habit of filing pilot reports on weather conditions you encounter when flying cross country. Report good weather as well as bad, especially if the conditions are better than forecast and/or improving. That information may be very useful to a fellow pilot facing a GO/NO GO decision back at home base.

Clear air turbulence (CAT) PIREPS provide the best information on this phenomenon and any pilot encountering CAT should report it to the radio facility with which he is in contact.

Wind shear PIREPS are also valuable, since strong wind shear close to the ground is a hazard to aircraft during landing and take-off. Since no instruments are available to detect and measure this phenomenon, any pilot encountering it should report it immediately. Pilots using INS should report the wind and altitude both above and below the wind shear layer. Pilots without this equipment should report loss or gain of airspeed and the altitude at which it was encountered and the general effect on the aircraft of the wind shear encounter.

Airframe icing should also be reported. Since the accumulation of ice on an airplane can be hazardous, knowledge of the existence of icing conditions will assist other pilots to avoid them. Icing conditions are reported as to intensity (trace, light, moderate or severe) and as to type (rime or clear).

Volcanic ash is potentially dangerous to aircraft and any encounter with it should be reported immediately. Pilot reports on this phenomenon can provide valuable information on the spread of volcanic ash from an eruption. Volcanic ash can rise rapidly to altitudes above 60,000 feet and exist in hazardous concentrations up to 1000 NM from the source. Volcanic ash is not detectable on radar.

Cavok

In air-ground communications when transmitting meteorological information to arriving aircraft, the term CAVOK (kav-oh-kay) is used to indicate the weather condition in which there is no cloud below 5000 feet, or below the highest minimum sector altitude, which- ever is higher, in which there is no cumulonimbus, no precipitation or thunderstorm activity, fog or drifting snow and in which the visibility is 6 statute miles or more. The term is also used in ATIS messages. It is not used in aviation weather reports or forecasts.

PART 4

Navigation

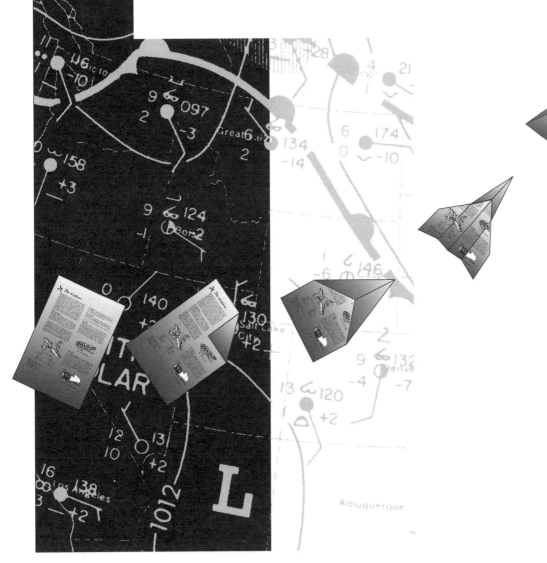

Air Navigation

The earth on which we live is a ball, or sphere, technically an oblate spheroid. Not knowing any better, the human race has for centuries lived on the outside of the sphere, enduring heat and cold, clouds and rain, snow and sleet, fog, mist and mortal uncertainty. In due course, cities of the future will be located underground in a world of perfectly regulated conditions of heat, light and fresh air. Meanwhile, we are here today and gone tomorrow and must make the best of our cabbage-patch world, such as it is.

Man is a restless soul; travel and curiosity have always been two of his most cultivated and persevering habits. To make it possible to move about on this great terrestrial sphere, he must have some master plan to enable him to define position, direction and distance.

There are several methods of navigation used by pilots to find their way from place to place on this earth of ours.

Pilotage — navigation by reference only to landmarks.

Dead reckoning — navigation by use of predetermined vectors of wind and true airspeed and precalculated heading, groundspeed and estimated time of arrival.

Radio navigation — navigation by use of radio aids, that is, navigation signals broadcast by radio stations on the ground or from satellites.

Celestial navigation — navigation by measuring angles to heavenly bodies (sun, moon and stars) to determine position on the earth.

Inertial navigation — navigation by self-contained airborne gyroscopic equipment or electronic computers that provide a continuous display of position.

Most pilots use these various methods of navigation in combination. The use of pilotage by itself is limited by visibility (you have to be able to see the landmarks) and by familiarity with the area over which the flight is being conducted. Therefore, pilotage in combination with dead reckoning is a more effective method of navigation and is the subject of this chapter.

Celestial navigation is touched upon briefly but being a specialized art used mostly in the Arctic areas where the compass is unreliable, it is beyond the scope of this manual to describe it at length.

Radio navigation and inertial navigation are discussed in a later chapter.

Latitude and Longitude

As a well-planned modern city is an orderly gridwork of intersecting streets and avenues, so ancient master minds have divided the surface of our sphere into a geometrical pattern of intersecting circles called the **graticule.** Those running north and south are **meridians of longitude.** Those running east and west are **parallels of latitude.** In exactly the same way that Wun Wing Lo's Cafe may be located at the corner of 49th Street and 11th Avenue, so the town of Riverton may be located at the intersection of the 75th meridian west and the 40th parallel north.

Meridians of longitude are semi great circles joining the true or geographic poles of the earth. They are also called true meridians since they join the geographic poles, as opposed to magnetic meridians which join the magnetic poles.

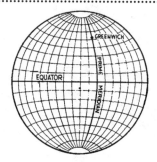

Fig. 1. Latitude and Longitude.

Longitude is measured from 0° to 180° east and west of the **prime meridian** which is the meridian which passes through Greenwich, England. The prime meridian is numbered 0°. The meridian on the opposite side of the earth to the prime meridian is the 180th and is called the **international date line** for here the time changes a day.

Longitude is measured in degrees (°), minutes (') and seconds ("). There are 60 minutes in a degree and 60 seconds in a minute.

Parallels of latitude are circles on the earth's surface whose planes lie parallel to the equator. The **equator** is a great circle on the surface of the earth lying equidistant from the poles.

Latitude is measured from 0° to 90° north or south of the equator, which is numbered 0°. Latitude is expressed in degrees (°), minutes (') and seconds (").

Geographical Co-Ordinates

The location of cities, towns, airports, etc. may be designated by their geographical co-ordinates, that is, the intersection of the lines of latitude and longitude marking their position on a map.

In Fig. 2, the co-ordinates of the airport at Matagami are 49° 46'N, 77°48'W. In other words, the airport is located 49 degrees, 46 minutes north of the equator and 77 degrees 48 minutes west of the prime meridian.

Fig. 2. Geographical Co-ordinates..

On a chart, the lines representing the meridians and parallels are numbered (in Fig. 2, 78°00'W, 77°30'W and 50°N). Each small mark along the lines represents one minute.

One position on the earth's surface is related to another by the **change of latitude** (written Ch.Lat.) and the **change of longitude** (Ch.Long.) between the two places.

If an airplane is to proceed from a place, Dunnville, to a place, Harrisburg, the Ch.Lat. is named north or south according to whether Harrisburg is north or south of Dunnville. Similarly, the Ch.Long. is named east or west, depending on whether Harrisburg is east or west of Dunnville.

Example:
Harrisburg	Lat.	60°27'N.	Long.	40°20'W.
Dunnville	Lat.	45°30'N.	Long.	15°30'E.
	Ch. Lat.	14°57'N.	Ch. Long.	55°50'W.

In the above example, Harrisburg is obviously north of Dunnville by 14°57' of latitude, hence its Ch.Lat. is 14°57'N. Harrisburg is west of Dunnville, but as Dunnville is east of the prime meridian, the Ch.Long. in this case is the sum of the longitude of the two places. Not that any community 15° east of the prime meridian is likely to have a name such as Dunnville, but the example is merely chosen to help you to get the general idea.

Time and Longitude

The earth rotates about its own axis. It also revolves in an elliptical orbit around the sun. As a result of these revolutions of the earth, it appears to us as though the sun were revolving around the earth instead.

The time between one apparent passage, or transit, of the sun over a meridian and the next passage over that same meridian is called an **apparent solar day** and varies throughout the year.

To provide a convenient method of measuring time, an imaginary sun called the **mean sun,** is assumed to travel at a uniform rate of speed throughout the year. The interval between two successive transits of the mean sun is called a **mean solar day.**

The mean solar day is divided into 24 equal hours. The mean sun is assumed to travel once around the earth every mean solar day, and therefore travels through an angle of 360° of longitude in that time. Hence, **mean time** may be expressed in terms of longitude, and vice versa.

For example:
24 hrs.=	360° Long.		360° Long.=	24 hrs.
1 hr.=	15° Long.		1° Long.=	4 min.
1 min.=	15' Long.		1' Long.=	4 sec.
1 sec.=	15" Long.		1" Long.=	1/15 sec.

Time will vary on different meridians on the earth at any particular instant from 0 hours to 24 hours. The mean time on any particular meridian is called its **local mean time (LMT).**

In order to have a universal standard time for reference at any point on the earth, a mean value, based upon measurements of time in a number of places on the earth, has been established and is known as **coordinated universal time (UTC).** UTC is also referred to as Z time. Coordinated universal time replaces **Greenwich mean time (GMT)** which was the universally accepted standard for the measurement of time until December 1985.

UTC is the local mean time for the prime meridian.

The LMT of any place east of the prime meridian is **ahead** of UTC. For example: Cairo 14:00 = UTC 12:00.

The LMT of any place west of the prime meridian is **behind** UTC. For example: UTC 12:00 = Toronto 07:00.

If every place kept its own local mean time, confusion would result. For this reason the world is divided into 24 time zones (Fig. 3). Each zone is 15° of longitude wide and keeps the time of the mid meridian of the zone. The zones are numbered from 1 to 12 east of the prime meridian (minus) and 1 to 12 west of the prime meridian (plus). If either the zone number or the longitude of the mid meridian of the zone is known, zone time can readily be converted to UTC (Fig. 3). It will be seen that when it is noon at the prime meridian, it is midnight on the 180th meridian. Here the time changes a day. Travelling from east longitude to west longitude (i.e. from Tokyo to San Francisco), the time goes back a day (i.e. from Wednesday to Tuesday). Travelling from west longitude to east longitude (i.e. San Francisco to Toyko), time goes forward a day (i.e. from Tuesday to Wednesday).

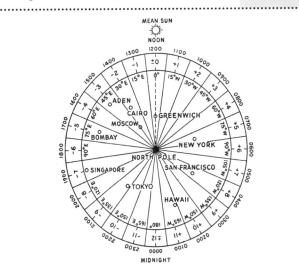

Fig. 3. Zone Time and Longitude.

Air traffic control and meteorological facilities in both Canada and the United States operate on UTC, or Z time, not local zone time. A pilot flying on a long cross country flight would be wise to adjust his personal wrist watch (or airplane clock) to UTC and refer to it when radioing position reports. Such a precaution will eliminate any chance of error in reporting time since it is often most difficult to determine exactly when one zone (which, to add to the confusion, might be on local daylight saving time) ends and another begins. To be one hour in error when reporting to an air traffic control unit could have serious consequences.

When travelling westward into a new time zone, watches are turned back one hour from the previous zone. When travelling eastward, watches are advanced one hour each time a new time zone is reached.

The relationship of UTC to standard zone time in North America is indicated in the following example:

12:00 (noon) UTC equals
08:30 NST (Newfoundland Standard Time)
  (UTC is NST plus 3 1/2 hours.)
08:00 AST (Atlantic Standard Time)
  (UTC is AST plus 4 hours.)

07:00 EST (Eastern Standard Time)
(UTC is EST plus 5 hours.)
06:00 CST (Central Standard Time)
(UTC is CST plus 6 hours.)
05:00 MST (Mountain Standard Time)
(UTC is MST plus 7 hours.)
04:00 PST (Pacific Standard Time)
(UTC is PST plus 8 hours.)

Note: the above abbreviations are frequently contracted to "Z" (Coordinated Universal Time), "A" (Atlantic Standard Time), "E" (Eastern Standard Time), etc.

Great Circles and Rhumb Lines

A **great circle** is a circle on the surface of a sphere whose plane passes through the centre of the sphere and which, therefore, cuts the sphere into two equal parts (Fig. 4). Only one great circle may be drawn through two places on the surface of the sphere that are not diametrically opposite each other. The shortest distance between these two points (A and B in Fig. 4) is the shorter arc of the great circle joining them.

The equator is a great circle. The meridians are only semi great circles since they do not completely encircle the earth but run only half way around it from pole to pole. With the exception of the equator, the parallels of latitude are not great circles as their planes do not pass through the centre of the earth.

Since a great circle route represents the **shortest distance** between two points on the surface of the earth, its advantage to the air navigator will be obvious. Most long-distance flights are flown over great circle routes or modified great circle routes.

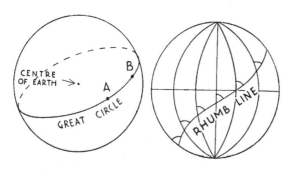

Fig. 4. Great Circle   **Fig. 5. Rhumb Line**

A great circle, however, does not cross the meridians it meets at the same angle. For this reason, the heading must be changed at frequent intervals to enable the airplane to maintain a great circle route.

A **rhumb line** is a curved line on the surface of the earth, cutting all the meridians it meets at the same angle (Fig. 5). All parallels of latitude are rhumb lines. The meridians and the equator are also rhumb lines while also being great circles. Any rhumb line that is not at the same time a parallel of latitude, will spiral into the poles (Fig. 5).

When two places are not situated on the equator, or on the same meridian, the distance measured along the rhumb lines joining them will not be the shortest distance between them. However, the direction of the rhumb line is constant. A rhumb line route offers to the navigator, therefore, the advantage of following a constant heading.

Headings and Bearings

Direction is measured in degrees clockwise from north which is 0° (or 360°). East is 90°, south is 180° and west is 270°.

The direction of any point on the surface of the earth from an observer is known, if the angle at the observer between a meridian passing through the observer and a great circle joining him to the object is known. This angle is the **azimuth** or the **bearing** of the point. The angle is measured clockwise from the meridian through 360° (Fig.7).

The angle between the meridian over which an airplane is flying and the line representing the direction in which the airplane's nose is pointing (the fore and aft axis), measured clockwise, is the airplane's **true heading** (Fig.6).

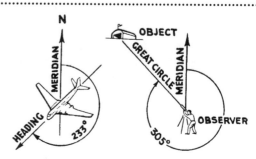

Fig. 6. Heading (233°).   **Fig. 7. Bearing (305°).**

Since the meridians are **imaginary lines,** the angle between a meridian and a particular heading or bearing cannot be measured directly. By the aid of a compass, however, the direction of the meridian may be determined and the angle ascertained.

The Earth's Magnetism

The earth is a magnet and like any other magnet has a north and south magnetic pole. Lines of force flow between these two poles creating a magnetic field that surrounds the earth. A compass needle will be influenced by the earth's magnetic field and will lie parallel to one of the magnetic lines of force with its north seeking pole pointing to magnetic north. (This does not necessarily mean that it is pointing directly towards the magnetic north pole).

The magnetic line in which the compass needle lies is called a **magnetic meridian.**

By referring to Fig. 8, you will note that the earth's magnetic lines of force are horizontal at the equator, but become gradually vertical towards the poles. There are two forces acting on a compass needle. One, the horizontal H force, acting horizontally, tends to hold the needle level (B). The other, the vertical Z force, acting vertically, tends to pull the seeking end of the needle down (A and C). A freely suspended magnet under the influence of a large directional (H) force and a small vertical (Z) force would lie parallel to the earth's surface. This occurs near the magnetic equator. As the magnet is moved north or south from the equator, H force decreases and Z force increases; the pole seeking end of the magnet would be pulled down from the horizontal. The angle between the horizontal and the plane the magnet lies in under the influence of Z force is called **dip** and from latitudes north or south of 45 degrees is quite large, increasing to near 90 degrees at the poles. Dip must be corrected or compensated for in a compass during construction.

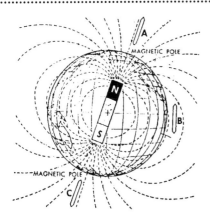

Fig. 8. The Earth's Magnetism.

Variation

The magnetic meridians do not coincide with the true meridians because the north magnetic pole does not coincide with the true, or geographic, pole. Averaged over thousands of years, the position of the north magnetic pole will roughly correspond with the position of the geographic pole, but, at any given time, the two poles can be very far apart. In 1990, the north magnetic pole was located at 77.6°N, 103°W, near King Christian Island in the Canadian Arctic, approximately 1400 km south of the true north pole.

The angle between the true meridian and the magnetic meridian in which the compass needle lies is called **magnetic variation** (also known as **magnetic declination**).

Since the magnetic north pole is not stationary but moving in an erratic circle, the variation is not constant at any one place, but changes slowly from year to year. This is called **annual change**. The Canadian Department of Energy, Mines and Resources about once a decade, replots the position of the north magnetic pole and updates its charts, repositioning magnetic meridians and assigning corrected magnetic variation where necessary.

Isogonic Lines

The direction of the earth's magnetic field is measured, periodically, over most of the earth's surface. The results of such a survey are plotted on a chart. Lines are drawn on the chart joining places having the same variation and these lines are called **isogonic lines** or **isogonals.** Isogonals are not straight lines but bend and twist due to the influence on the magnetic field of local magnetic bodies below the earth's surface.

The isogonals are numbered east and west according to whether the compass variation is to the east or west of true north.

Isogonic lines are represented on aeronautical charts by dashed lines with the amount of variation indicated. In Fig. 2, the 15°W isogonic line can be seen to pass near the Matagami Airport.

Agonic Lines

In each hemisphere there will be places where the north pole and north magnetic pole will be in transit, that is, where they will lie in the same straight line. These places will, therefore, have no magnetic variation. Lines drawn through places of no variation are called **agonic lines.** Like the isogonic lines, they twist and curve, due to the local attraction of magnetic bodies in the earth.

The agonic line for the western hemisphere passes just to the west of Thunder Bay, Ontario, and just to the west of Chicago, Illinois.

East of this line the variation is westerly and west of it the variation is easterly.

The Effect of Variation

It can, therefore, be seen that if the magnetic pole lies west of the true pole from a given point, the compass needle will point west of true north. Hence, the magnetic meridian will lie west of the true meridian.

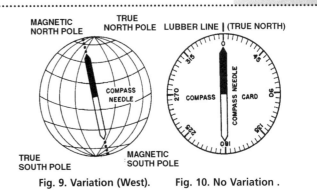

Fig. 9. Variation (West).   Fig. 10. No Variation.

In this case, the variation is named west (Fig. 9).

Similarly, if the compass needle points east of true north, the magnetic meridian will lie east of the true meridian, and the variation will be named east.

Figures 10, 11, and 12 are intended to represent a compass installed in an airplane whose nose is pointed towards true north.

The lubber line of a compass is always permanently fixed in the fore and aft line of the airplane and therefore always indicates the direction in which the airplane is heading. The compass card is always read against the lubber line.

The compass needle pivots and is free to point always towards magnetic north, regardless of what direction the airplane may be heading.

In Fig. 10, there is no variation. The magnetic meridian, in which the compass needle lies, coincides with the true meridian in which the longitudinal axis of the airplane lies.

Fig. 11 illustrates westerly variation. The compass needle is deflected 30° west of the true meridian. The compass card is always attached to the compass needle and, therefore, when the compass needle is deflected towards the west, it drags the whole compass card around with it. As a result, the reading at the lubber line is more than it should be. Although the airplane is actually pointed towards true north, the compass reads 30°.

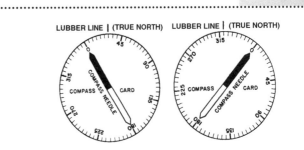

Fig. 11. Variation 30° West.   Fig. 12. Variation 40° East.

Fig. 12 illustrates easterly variation. When the needle is deflected east of the true meridian, the compass card is rotated so that a smaller reading is opposite the lubber line. Although the airplane is in fact pointed towards true north, the compass reads 320°.

Conversion of True and Magnetic Headings

It is necessary for a pilot to be able to convert true headings and bearings to magnetic headings and bearings, and vice versa, rapidly in the air.

To convert Magnetic to True:
    Subtract .... *Westerly Variation*
    Add .... *Easterly Variation*

To convert True to Magnetic:
    Add .... *Westerly Variation*
    Subtract .... *Easterly Variation*

An easy way to remember how to apply variation is to remember the rhyme:
    Variation West, Magnetic Best (Better than True)
    Variation East, Magnetic Least (Less than True)

Problem 1. To convert a magnetic bearing of 80° to a true bearing when the variation is 13°W.
    Magnetic Bearing .... *80°*
    Variation .... *13°W*
    Hence, True Bearing .... *67°*

Problem 2. To convert a true heading of 136° to a magnetic heading when the variation is 8°E.
    True Heading .... *136°*
    Variation .... *8°E*
    Hence, Magnetic Heading .... *128°*

The Magnetic Compass

In some general aviation airplanes, the magnetic compass is the only direction finding instrument installed in the airplane. Even in airplanes that have more complicated direction finding devices, such as a heading indicator, these instruments must be set with reference to the magnetic compass.

A compass is a reliable, self contained instrument that requires no external power source. It will continue to perform satisfactorily in a steady climb or glide, provided that neither is greater than 20°. In turns or rapid changes of speed, however, the compass will not give an accurate reading of direction. To use the instrument properly, a pilot must understand how a compass works and to what errors it is subject.

Fig. 13. Magnetic Compass.

Construction of the Magnetic Compass

The magnetic compass consists of north seeking magnets (usually two) attached to a float to which is also attached a **compass card**. This complete magnet system is mounted on a pivot and is free to rotate. The whole assembly is mounted within the **compass bowl**.

To damp out oscillations of the magnetic system which may be caused by turbulence, the bowl is filled with alcohol or white kerosene. This liquid also helps to reduce the weight of the compass card and the magnets that is carried by the pivot. The liquid also helps to lubricate the pivot point. An expansion chamber is provided to allow for expansion of the liquid due to temperature changes. The bowl and container are of brass, which is non-magnetic.

The compass card is usually graduated in 5° divisions (the compass headings). The last 0 is omitted from the numbers, which are shown as 3 for 30, 33 for 330, etc. The card is read through a window on which is painted a white vertical line called the lubber line.

The **lubber line** indicates the direction the airplane is headed and should be exactly in line with, or parallel to, the fore and aft (longitudinal) axis of the airplane.

Since the headings are painted on the reciprocal side of the card, which the pilot sees looking forward, the card appears to be going the wrong way when the airplane is turning.

The magnetic compass is affected by anything metal that is placed too near to it. Photographic exposure meters, screwdrivers and other tools, earphones, manuals bound with steel covers can affect the compass and cause distorted readings. For example, in one reported incident, an exposure meter lying within four inches of the compass caused a 23° error; in another incident, a pair of pliers lying six inches from the compass caused a 10° error.

Compass Errors
Deviation

When magnets mounted in simple direct reading compasses are installed in airplanes, they usually do not point directly in the direction of magnetic north but, due to the influence of magnetic fields associated with the metal in the airplane's frame and engine, are deflected slightly to what is called **compass north.**

The angle through which the compass needle is deflected from the magnetic meridian is called **deviation.**

Deviation is named east if the north seeking end of the needle is deflected to the east of magnetic north. It is named west if the needle is deflected west of magnetic north.

If deviation causes the compass needle to be deflected west of the magnetic meridian, the compass card will be dragged around anti-clockwise and the reading opposite the lubber line will be more than it should be.

Similarly, if the compass needle is pulled over east of the magnetic meridian, the compass card will be rotated clockwise and the reading opposite the lubber line will be less than it should be.

To convert Compass to Magnetic:
    Subtract .... *Westerly Deviation*
    Add .... *Easterly Deviation*

To convert Magnetic to Compass:
    Add .... *Westerly Deviation*
    Subtract .... *Easterly Deviation*

An easy way to remember how to apply deviation is to remember the rhyme:
> *Deviation West, Compass Best (Better than Magnetic)*
> *Deviation East, Compass Least (Less than Magnetic)*

Problem 1. To convert a magnetic heading of 270° to a compass heading when the deviation is 3° East.

Magnetic Heading	*270°*
Deviation	*3° East*
Compass Heading	*267°*

Problem 2. To convert a compass heading of 002° to a magnetic heading when the deviation is 4° West.

Compass Heading	*002°*
Deviation	*4° West*
Magnetic Heading	*358°*

Because a compass can be affected by anything metal, it is essential that pilots avoid placing metallic objects near the compass during flight.

Correction for Deviation by Swinging the Compass: The effect of deviation is neutralized in airplanes as far as possible by a procedure called **swinging the compass.** The compass is fitted with corrector magnets, which are permanently installed and connected by gears. They are located in the top and bottom of the compass case and are adjusted by screws. As the compass is swung, the screws are adjusted to eliminate deviation. It is rarely possible to eliminate all the deviation. What remains after the compass has been swung is tabulated on a **compass deviation card** to enable the pilot to make allowance for it. A typical compass deviation card is illustrated in Fig. 14.

The compass is swung on a compass swinging base located remote from any magnetic influences, such as metal hangars, gas pumps, etc. There should be no metal objects in the vicinity of the compass or on the person performing the swing. Magnetic bearings, every 30°, are laid out on the base. The airplane is lined up with each bearing and the compass heading compared with each in turn, The method of correcting deviation is detailed below. The compass should be swung with the engine running and all radio and electrical equipment functioning.

Deviation can also be determined by the use of a **landing** (or master) **compass.** As the aircraft is lined up on the various compass headings, the readings of the aircraft compass are compared to the readings of the landing compass to determine deviation.

The simplest method of swinging the compass is as follows:
1. *Place airplane on N.* Take out deviation.
2. *Place airplane on E.* Take out deviation.
3. *Place airplane on S.* Take out **half** the deviation.
4. *Place airplane on W.* Take out **half** the deviation.
5. Commencing with the corrected reading on W, place the airplane successively on magnetic headings every 30°; that is, W, 300°, 330°, N, 030°, 060°, E, 120°, 150°, S, 210°, 240°. Take compass readings on all 12 points. Enter the readings on the deviation card (Fig. 14) which is then placed near the compass in the airplane. Enter details in the Aircraft Log.

FOR.......	N	030	060	E	120	150
STEER.....	003	031	061	090	122	147

FOR.......	S	210	240	W	300	330
STEER.....	179	209	241	270	298	334

Fig. 14. Compass Deviation Card.

The deviation card illustrated reads: "For N Steer 003". This means that on a heading of north magnetic, the compass has a 3° deviation west (Deviation West, Compass Best). It is therefore necessary to stee 003° by the compass to make good a magnetic heading of north. And so on for the other points of the compass.

Some types of retractable undercarriages will be found to be magnetized to such an extent that they will cause considerable deviation of the compass when retracted. In such cases, the compass must be swung in the air with the undercarriage retracted.

This can be done by the following simple method:

Select a road or railroad on the map whose bearing is as nearly as possible magnetic north and south and another as nearly as possible magnetic east and west. Smooth air is essential, preferably with no wind. Align the airplane heading on magnetic north by sighting up the north-south road or railroad. Set the heading indicator. Hold the heading constant by the heading indicator and correct the magnetic compass to match the heading on the heading indicator. Repeat the process on east.

The heading indicator can now be used to align the airplane on the other headings and make the necessary corrections by the same method as that detailed above for swinging the compass on the ground.

Note: The heading indicator is subject to precession error. The airplane should not be flown more than 5 minutes without realigning the heading indicator on one of the roads selected.

If a driftmeter is available, the compass may be swung in the air with prevailing wind conditions. This and other methods which require the use of special navigation equipment are beyond the scope of this pilot-navigator text but may be found in the many handbooks which deal more extensively with navigation.

If roads or railroads are scarce in the area, the airport runways may be utilized. Airport runway bearings are numbered to the nearest 10° magnetic and accurate bearings to the nearest degree should be obtained.

Note: The use of the omnirange to establish magnetic headings for compass swinging is not recommended. The needle is not sufficiently sensitive for this purpose, except when close in to the station. In addition, the VOR is subject to certain course errors.

Magnetic Dip

The earth's lines of force are horizontal at the equator but become vertical towards the poles, This causes the compass to tend to **dip** in higher latitudes. (See Fig. 8)

To compensate for dip, the magnet system is (1) balanced pendulously on a pivot bearing with the centre of gravity of the system below or lower than the pivot point, or (2) the top part of the pivot is so constructed that the centre of buoyancy of the system is well above its centre of gravity, or (3) a combination of both methods. As one end of the magnet system begins to dip below the horizon, the C of G is moved out of line with the pivot so that the weight of the system counteracts Z Force or, as one end of the magnet system tends to dip, the centre of buoyancy moves out of line with the pivot and C of G, counteracting Z Force. The resultant dip, in "normal" latitudes, is about 2 or 3 degrees (Fig. 15). Note the relative positions of the pivot and centre of gravity and keep them in mind in the discussion of northerly turning error, acceleration and deceleration errors that follows.

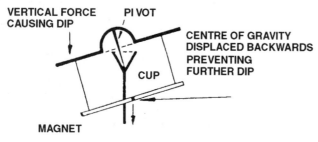

Fig. 15. Magnetic Dip.

Northerly Turning Error

When an airplane in flight executes a normal turn, it banks and the wing on the inside is lowest. The plane of gyration of the magnetic system of the compass tends to maintain a position parallel to the transverse plane of the airplane due to the fact that the centre of gravity of the magnetic system is placed below the point at which the system pivots. As a result of the turn, the forces acting on the magnet system undergo certain changes which have a marked effect upon the direction indicating properties of the compass and, under certain conditions, large apparent deviations will be observed.

When an airplane is in straight and level flight, H, the horizontal component of the earth's total force, is exerting its maximum influence while Z, the vertical component, is exerting a force in the vertical plane of the magnet system but can cause no deflection of the system. If the magnet system is tilted sideways through 90° (i.e. to the east or to the west), the system will take up a position under the combined influence of H and Z, parallel to the direction of the lines of force of the earth's magnetic field at the position. But if the system is now turned in azimuth until its normal axis of rotation is parallel to the meridian, H is no longer able to affect the magnet system, which will come entirely under the effect of Z, and will take up a vertical position, north-seeking end downwards in the northern hemisphere and upwards in the southern. Consequently, the north-seeking end of the magnet system will rotate towards the inside of the turn in the northern hemisphere (centripetal force which acts through the pivot towards the centre of the turn) and towards the outside in the southern (centrifugal force which acts through the centre of gravity towards the outside of the turn). This is the magnetic factor responsible for northerly turning error.

There is also a mechanical factor which has a slight effect on the compass during turns. When the airplane commences a turn, the liquid in the compass bowl tends to turn with the bowl. This liquid swirl drags the magnet system with it and causes a rotation of the system towards the centre of the turn.

The error is most apparent on headings of north and south. When making a turn from a heading of north, the compass may briefly give an indication of a turn in the opposite direction but it then corrects itself to turn with the airplane. It does, however, under register the turn (or lags). The compass is said to be "sluggish" in the northerly quadrant. When making a turn from south, it gives an indication of a turn in the correct direction but at a much faster rate than is actually occurring. It over registers (or leads). In the southern quadrant, the compass is said to be "lively". The error is less pronounced in turns from south than from north.

The effect of northerly turning error is greatest over the poles and gradually decreases towards the magnetic equator, where it is zero. The deflection of the compass is maximum on north and south headings and nil on east and west.

An understanding of the vagaries of the magnetic compass caused by northerly turning error has been of invaluable assistance to pilots who have experienced loss of gyros, or who are flying an airplane that is not equipped with gyros, at a time when they have to fly by reference to instruments. In fact, the method was first used by World War I pilots. The method is to establish a south heading. The compass on south is very sensitive and leads any turn. Therefore, if the compass indicates that the airplane is turning, control movements should be applied to return it to its original heading. The tilt of the compass card will indicate when one wing is down.

Accurate turns onto any heading are quite simple if use is made of a heading indicator. However, if there is no heading indicator and turns must be made from north or south headings by reference to the magnetic compass, the following simple rules may prove helpful:

On turns from north, northerly turning error causes the compass to lag.

On turns from south, northerly turning error causes the compass to lead.

The greatest effect will be experienced when the turn commences near to the north and south headings and will be less noticeable when the commencement of the turn is nearer to east and west headings.

The fact should be realized that, often in a turn, an airplane may be subject to gust loads which will further aggravate the constant behaviour of a compass. Under certain conditions, an airplane may be turning

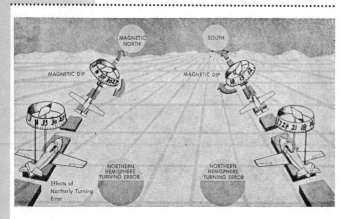

Fig. 16. Northerly Turning Error.

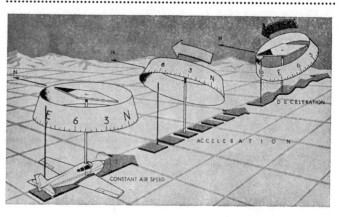

Fig. 17. Acceleration Error.

but the compass, being subject to outside forces, does not indicate a turn at all but maintains a constant heading. When the airplane levels out for straight and level flight, the compass will spin rapidly to indicate the new heading.

In an airplane with a turn and bank indicator, a pilot may make reasonably accurate turns to a particular heading by banking the airplane into a standard rate 1 turn and allowing 1 second for every 3 degrees of the compass (i.e. 30 seconds to turn 90 degrees).

While reading the compass when holding north or south headings, always make sure the wings are level and the turn needle is centred.

Acceleration and Deceleration Errors
When an airplane increases its speed, a force is set up in the direction in which the airplane is moving. This is **acceleration** and will act through the pivot of the compass. The magnet system, however, tends to lag behind due to **inertia** acting through the centre of gravity. On east and west headings, this causes a **turning moment,** tending to rotate the magnet system and so causing deflection of the compass card.

When the airplane decreases its speed, **deceleration** causes a similar effect but the deflection will be in the opposite direction.

On E. and W. headings, **acceleration** causes the compass to register a turn towards **north.**

On E. and W. headings, **deceleration** causes the compass to register a turn towards **south.**

The deflection of the compass will occur only while the airplane is accelerating or decelerating. Immediately the airplane is again in equilibrium, the compass will return to its normal magnetic heading.

When reading the magnetic compass on east or west headings, always make sure that the airspeed is kept constant.

The effect of acceleration and deceleration compass error is nil on north and south headings.

Dry Vertical Card Compass
The **dry vertical card compass** looks like a heading indicator (directional gyro) but is in fact a compass. The display features a miniature airplane that points to the headings on a vertical card that rotates. All headings are simultaneously displayed, so that alternate headings (180°, 90°, 45°, etc.) are easily determined.

The vertical card compass utilizes a sensing magnet which is not pendulous and which does not float in liquid. Instead it rotates in a fixed plane in relation to the compass housing. It is heavily damped and limited to only 20 degrees of pitch and roll. As a result, although the vertical card compass is still subject to lead and lag as a result of northerly turning error and to acceleration and deceleration error, the errors are less noticeable than in the standard compass. The vertical card compass is generally more stable and consistent in its readings.

The vertical card compass is more sensitive to magnetic and electrical disturbances in the airplane and requires careful swinging.

The Astro Compass
Close to the magnetic pole, the earth's magnetic lines of force dip vertically towards the pole. The compass needle lies in a horizontal plane and, as the strength of the horizontal component of the earth's magnetic field tends to disappear in the close vicinity of the pole, the magnetic compass loses its ability to point the way, and becomes anything but a dependable companion. In a large area of Northern Canada the magnetic compass is therefore unreliable. Within this area, called the Northern Domestic Airspace, aircraft require some means of determining direction that is not dependent on a magnetic source. Navigation systems based on geosynchronous satellites are invaluable in the remote areas where the compass is unreliable. Until satellite navigation made its appearance, the astro compass was the method by which aviators navigated in the north. The astro compass is not influenced by the earth's magnetic field, but indicates direction with reference to the sun, moon, planets or stars. These heavenly bodies have been man's guiding beacons in the skies since time immemorial.

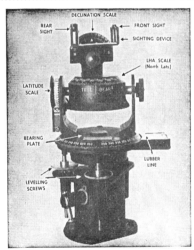

Fig. 19. The Astro Compass.

The astro compass is an instrument designed to enable a pilot or navigator to determine the true heading of the airplane. (It can also be used to take a bearing on an object.)

In the simplest possible terms, the astro compass is a bearing plate with a movable sighting device mounted above it. The bearing of a heavenly body can be set on the bearing plate and when the movable sights are properly aligned on the heavenly body, the north mark on the bearing plate will coincide with true north. The bearing plate is then read off against a lubber line (aligned in the fore and aft axis of the airplane) to indicate the true heading. The true heading is set on the heading indicator and the airplane steered with reference to the heading indicator in lieu of the magnetic compass.

The astro compass will automatically find the true bearing (or azimuth) of a heavenly body if the following information is known:

Fig. 18. Dry Vertical Card Compass.

1. The latitude.
2. The local hour angle of the body.
3. The declination of the body.

The position in latitude and longitude must be derived through dead reckoning navigation.

The position of the sun, stars and other astronomical bodies can be accurately predicted for any given moment of time. They are recorded in a handy volume called the **Air Almanac.** For any particular date, hour and minute in UTC, the Almanac lists the Greenwich hour angle, the declination of the sun, moon, Aries, or the planets, the sidereal hour angle and declination of the stars normally used for navigation purposes.

The GREENWICH HOUR ANGLE (GHA) of a heavenly body is the angular distance between the Greenwich meridian and the meridian of the body. It is measured westward through 360°.

The DECLINATION of a heavenly body is its angular distance north or south of the celestial equator (hence it corresponds to **latitude**). It is measured in degrees and minutes north or south of the celestial equator, the same as is latitude.

The SIDEREAL HOUR ANGLE (SHA) of a Star is the angular distance between the First Point of Aries and the meridian of the star measured westward from Aries through 360°.

The FIRST POINT OF ARIES (γ) sometimes referred to as the vernal equinox, is a point on the celestial equator arbitrarily chosen as a reference point from which to measure the hour angles of the stars.

To find the Greenwich hour angle (GHA) of a star, look up the sidereal hour angle of the star in the Air Almanac and also the Greenwich hour angle of Aries. The GHA of the star is the algebraic sum of both.

GHA ☆ = GHA γ + SHA ☆

The LOCAL HOUR ANGLE (LHA) of a heavenly body is the angular distance between your meridian and the meridian of the body, measured westward. It is the algebraic sum of the GHA of the body and your longitude.

LHA = GHA + E (or - W) longitude

*Note: In measuring the hour angles of heavenly bodies, it is simpler to think of the position of the body as a point on the surface of the earth, rather than one in the sky. In other words, bring the body down on the ground for convenience in measuring. Such a point is called the **sub stellar point.** It is defined as the point on the surface of the earth directly beneath the heavenly body at any given instant of time and is where the line joining the body to the centre of the earth cuts the surface.*

From the foregoing, it will be clear that to find true heading by the astro compass, it is necessary to have a reasonable knowledge of position, to know the correct time (UTC) and to be able to get a sight on a heavenly body.

The astro compass, in addition to the bearing plate and sighting device mentioned previously, has a number of movable drums and scales (Fig. 19).

To find true heading by astro compass:
1. Level the instrument.
2. Set the latitude (nearest degree) on the latitude scale.
3. Set the local hour angle of the body on the LHA scale.
4. Set the declination of the body on the declination scale.
5. Rotate the bearing plate until the body on which a sight is being taken is lined up in the sighting device.
6. Read the true heading against the lubber line.

Finding the Sun's True Bearing

In a booklet called **Finding The Sun's True Bearing,** Transport Canada has published tables that give the sun's true bearing for locations from 40 degrees north latitude to 85 degrees north latitude, based on 10 day intervals throughout the year and eight minute intervals throughout the day.

The use of these tables eliminates the need for an astro compass and the somewhat cumbersome procedure for determining position by this instrument. With the book, an accurate clock, a heading indicator and, of course, the sun in the sky, it is relatively simple to keep an accurate check on true bearing.

By dead reckoning, it is necessary to determine the geographical co-ordinates of the position on the earth over which the airplane is flying. Let us suppose that the airplane is over a landmark whose co-ordinates are 55°15'N and 120°10'W. The time is 1809 UTC and the date is June 9th.

Turn to the page in **Finding The Sun's True Bearing** for 55 Degrees north latitude.

55 DEGREES NORTH LATITUDE				
HOURS	1000	1008	1016	1024
DATE				
APR 30	139.0	141.5	144.0	146.6
MAY 10	137.7	140.2	142.8	145.5
MAY 20	136.2	138.8	141.5	144.2
MAY 30	134.7	137.3	140.0	142.8
JUN 9	133.3	136.0	138.7	141.5
JUN 19	132.3	134.9	137.6	140.4
JUN 29	131.7	134.4	137.0	139.8

Fig. 20. Sample Table.

Multiply the number of whole degrees of longitude by 4(120 x 4 is 480) to obtain the correction in minutes for the local civil time (LCT). Convert this value to hours and minutes (480 minutes is 8 hours 00 minutes). Since you are west of the prime meridian, subtract this correction from the time in UTC (1809 minus 0800) to determine the LCT 1009.

On the 55 degree north latitude page, find on the tables the date June 9th and the time LCT 1008 (closest to 1009). According to the tables, the sun's true bearing for that time, date and latitude is 136 degrees.

Turn the airplane so that it is facing directly into the sun. Set the heading indicator to 136 degrees.

Units of Distance and Speed

• A STATUTE MILE is a distance of 5280 feet.
• A NAUTICAL MILE (6080 feet) is the average length of one

minute of latitude. For all practical purposes, it may be taken as the length of one minute of arc along any great circle.
- A KILOMETER is a distance of 1000 meters.
- A KNOT is a speed of one nautical mile per hour.

Conversions
Speed
- 66 nautical miles = 76 statute miles.
- To convert knots to mph, multiply knots by 1.15.
- To convert mph to knots, divide mph by 1.15.
- To convert kilometers per hour to knots, multiply by .54.
- To convert kilometers per hour to mph, multiply by .62.

A scale of nautical miles (based on the scale of the chart at mid latitude) is printed on all I.C.A.O. aeronautical maps.

Practically all circular slide rule computers have statute mile/nautical mile conversion indexes printed on the outer scale.

The abbreviation, officially adopted for nautical miles is n.miles or n.m. or NM; for statute miles, s.miles or s.m. or SM.

Hours and Minutes
To convert minutes to hours, divide by 60 (60 min. = 1 hr.). For example, 30 min. equals 30 ÷ 60 = .5 hrs.

To convert hours to minutes, multiply by 60. For example, .75 hrs. equals .75 x 60 = 45 min.

Time in Flight.
To find the time in flight, divide the distance by the groundspeed. For example, the time to fly 120 n. miles at a groundspeed of 80 knots is 120 ÷ 80 = 1.5 hrs (1.5 hrs. x 60 = 90 min). Answer: 1 hr. 30 min.

Distance.
To find the distance flown in a given time, multiply groundspeed by time. For example, the distance flown in 1 hr. 45 min. at a groundspeed of 120 knots is 120 x 1.75 = 210 n. miles.

Groundspeed.
To find the groundspeed, divide the distance flown by the time. For example, an airplane flies 270 n. miles in 3 hrs. The groundspeed is 270 ÷ 3 = 90 knots.

Aeronautical Charts

The earth is a sphere and therefore its surface cannot be represented accurately on a flat plane. The map maker's problem can be readily appreciated by cutting a rubber ball in half and then attempting to flatten out one of the halves in a flat plane.

Since the surface of a sphere cannot be accurately projected onto a map, a map must show the portion of the earth's surface it represents with some distortion.

There are four basic elements in map construction. These are:
   1. Areas.   *3. Bearings.*
   2. Shapes.   *4. Distances.*

According to the particular purpose of a map, one or more of these elements is preserved as nearly correct as possible, with consequent unavoidable distortion in the remaining elements.

The mathematical bases on which maps are constructed are termed **projections.**

There are two principal types of chart projections used in air navigation charts, the **Lambert Conformal Conic Projection** and the **Transverse Mercator Projection.**

The Lambert Conformal Conic Projection

The basic idea upon which the Lambert Conic Projection is developed is that of supposing a cone to be superimposed over the surface of a sphere (Fig. 21). If the cone were opened and unrolled, the meridians and parallels would appear as shown in the sketch (Fig. 22).

On the Lambert Conformal Conic Projection, the angles between meridians and parallels will be the same on the map as they are on the ground. The term conformal refers to this characteristic. Scale must be the same along both meridians and parallels.

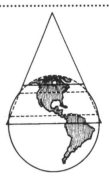

Fig. 21. Lambert Conformal Conic Projection.

Fig. 22. Lambert Conformal Conic Development.

The properties of the Lambert Projection are:
1. Meridians are curves or straight lines converging towards the nearer pole. If they are curves, the curvature is often so small as to be inappreciable. The angle which one meridian makes with another on the earth is called **convergency.** It varies with latitude. At the equator, there is no convergency between meridians. At the poles, the meridians converge at angles equal to the change of longitude between them.

2. Parallels of latitude are curves which are concave towards the nearer pole. On any but large scale maps, the curvature is considerable.

3. The scale of distance is practically uniform throughout the entire map sheet, the maximum distortion being not more than 1/2 of 1%.

4. A straight line drawn between any two points on the map may be assumed, for all practical purposes, to represent **an arc of a great circle.**

Since a great circle does not cross every meridian at the same angle, a straight line on the map will not have the same bearing when measured on two or more different meridians. Hence a straight line, or great circle route, cannot be flown without changing heading at regular intervals.

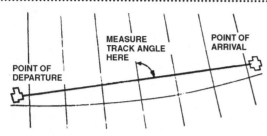

Fig. 23. Measuring the Track.

To make good a given track, or straight line, on this type of map, it is necessary to change heading 2° for every 3° of longitude. Flying **east,** the 2° is added. Flying **west,** it is subtracted.

For flights up to roughly 300 miles, the heading change referred to above may be averaged by measuring the course or track on the meridian nearest the centre (Fig. 23).

The VFR Navigation Charts (VNC Series) and the World Aeronautical Charts (WAC Series) are based on Lambert Conformal Conic Projections.

The Mercator Projection

The principle on which the Mercator Projection is based is that of a cylinder which has its point of tangency at the equator. Its approximate form may be visualized by imagining a light at the centre of a globe that casts a shadow of the meridians and parallels on a cylinder of infinite length enclosing it. As will be seen in Fig. 24, the shadows of the more northerly parallels of latitude will be cast wider apart on the cylinder than they actually are on the sphere.

The shadows of the meridians on the cylinder will be straight and parallel lines to infinity, whereas on the sphere they converge to meet in a point at the poles. It is, in other words, as though the meridians were stretched apart at the poles to the same distance they are apart at the equator. This causes extreme exaggeration of longitude in northerly areas.

The principal characteristics of the Mercator Projection are:

1. Meridians are straight and parallel lines.

2. Parallels of latitude are straight and parallel lines.

3. A straight line drawn between any two points on the chart will represent a **rhumb line**.

4. Owing to the method of projection, there is no constant scale of distance on a Mercator chart and areas are greatly exaggerated in high latitudes.

Mercator charts are, as a rule, graduated on the right and left hand sides for latitude, and at the top and bottom for longitude. The divisions of the longitude scale are only to be used for laying down and taking off the longitude of a place. **The longitude scale must never be used for measuring distance.** Since 1 minute of latitude is always equal to one nautical mile, the latitude scale is used for measuring distance.

The Mercator Projection is relatively precise in depiction of distances in the equatorial regions. However, distortion, as has been stated, becomes more pronounced with distance from the equator.

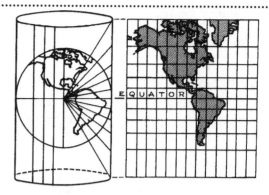

Fig. 24. Mercator Projection and Development.

The Transverse Mercator Projection

To overcome this problem, the Mercator technique can be applied by rotating the cylinder 90 degrees so that the point of tangency is a meridian of longitude rather than the equator. In this case, the chart is accurate along the selected meridian.

Such a projection is called a **Transverse Mercator Projection**.

The Transverse Mercator Projection, because of its method of production, is quite accurate in depicting scale, especially on charts covering a relatively small geographical area. Depending on the area to be depicted, any one of the 360 meridians of longitude can be selected as the point of tangency for the chart projection.

The VFR Terminal Area Charts (VTA Series) are based on Transverse Mercator Projections.

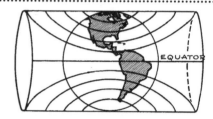

Fig. 25. Transverse Mercator Projection.

Types of Aeronautical Charts
VFR Navigation Charts (VNC Series)

VFR Navigation Charts are designed primarily for visual navigation and are most useful during flight at lower altitudes and at slower speeds. They were introduced some years ago to replace the Canadian Pilotage Charts which are now no longer published. Each of the VFR Navigation Charts replaces four of the old CPC Series. Each chart therefore covers a fairly extensive portion of Canada, reducing quite considerably the number of charts required to cover the whole of the Canadian land mass and the Artic islands.

The chart is printed on both sides, the northern half of the area to be depicted on one side and the southern half on the reverse. The title of the chart and other aeronautical information is printed on two sides of a white border on one edge of the map.

Each chart is identified by the name of a principal landmark on the chart (e.g. Toronto, Winnipeg, Vancouver, Yellowknife, Cambridge Bay).

The scale of the chart is 1: 500,000 or about one inch to eight miles.

VFR Navigation Charts are based on the Lambert Conformal Conic

Projection and conform to the characteristics of that projection.

To keep pace with rapidly changing facilities, the charts are reprinted at regular intervals with updated information. The publication date is always shown with the chart identification.

World Aeronautical Charts (WAC SERIES)

World Aeronautical Charts are designed also for visual navigation and are most useful during flight at the higher altitudes and at greater speeds. Each chart depicts a sizeable portion of the geographical area of Canada. Nineteen charts cover the whole country.

Each chart is identified by a letter and a number. For example, E17 covers the area from Thunder Bay west to Regina, and from the 48th parallel north to Thompson, Manitoba.

WAC Charts are based on the Lambert Conformal Conic Projection and conform to the characteristics of that projection.

The scale of the chart is 1: 1,000,000 or about one inch to sixteen miles.

The chart is printed on both sides of the sheet, the northern half of the area to be represented on one side and the southern half on the reverse. The title of the chart and other aeronautical information is printed on two sides of a white side border on one edge of the map.

VFR Terminal Area Charts (VTA SERIES)

VFR Terminal Area Charts are large scale charts (1: 250,000) published for those airports in Canada which have been designated as classified airspace for control purposes. Before entering such classified airspace, a VFR flight must contact the appropriate ATC unit and receive clearance to enter the area. VTA charts depict VFR call up points outside the area and VFR check points within the area. Radio communication information and other information that is necessary for conducting flight through the area is given on the chart. The edge of the area is marked by a solid blue line. The edge of the control zone is marked by a dashed blue line.

VFR Terminal Area Charts are based on the Transverse Mercator Projection and conform to the characteristics of that projection.

A more detailed discussion of classifed airspace is given in the Chapter **Aeronautical Rules and Facilities.**

Enroute Charts

Enroute Charts provide information for radio navigation over designated airways systems. **Enroute Low Altitude Charts (LO)** are intended for use up to, but not including, 18,000 feet. **Enroute High Altitude Charts (HE)** are intended for use at 18,000 feet and above. **Terminal Area Charts** depict aeronautical radio navigation information in congested areas at a larger scale and are intended for use from the surface up to 18,000 feet ASL.

Enroute charts do not portray any cities, towns, or topographical features, but depict all radio navigation aids, including airways, beacons, reporting points, communication frequencies, etc.

Uncontrolled areas are tinted a basic light green and controlled areas (such as airways, control zones, terminal control areas, etc.) are shown in solid white. VHF facilities are overprinted in black (omnirange stations, victor airways, etc.). The boundaries of control zones are depicted by dashed green lines.

The directions, or bearings, of all airways (magnetic), distances between check points, minimum enroute altitudes, VOR change-over points, station identification letters, radio frequencies, and other pertinent radio navigation information is clearly shown in bold, easy to read characters.

The scale is not constant, but varies to suit the requirements of each individual chart, or series of charts.

Each chart contains two separate map sheets printed back to back. A table of map symbols and a list of all radio communication facilities available within the area covered by the charts is printed on the front sheet.

Canada Flight Supplement

Published as a supplement to the Aeronautical Information Publication (AIP Canada), the **Canada Flight Supplement** lists all the aerodromes shown on VFR Terminal Area Charts, VFR Navigation Charts and World Aeronautical Charts for both the southern and northern regions of Canada. The booklet is designed to be used in conjunction with the aeronautical charts and should be carried by any pilot departing on a long cross country flight.

The Canada Flight Supplement is made up of six sections: (1) General, (2) Aerodrome and Facility Directory, (3) Planning, (4) Radio Navigation and Communications, (5) Military Flight Data and Procedures and (6) Emergency.

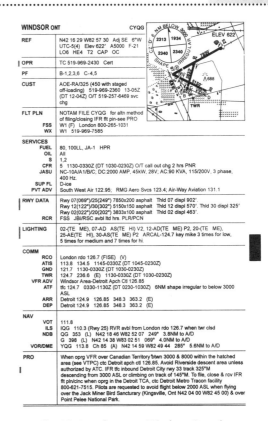

Fig. 26. Aerodrome at Windsor, Ontario.

The GENERAL section has information of a general nature: abbreviations, a cross reference of aerodrome names and abandoned aerodromes, conversions tables, the phonetic alphabet, a legend of chart symbols and a comprehensive section explaining the annotations and codes used in the Aerodrome and Facility Directory.

The AERODROME AND FACILITY DIRECTORY lists all aerodromes alphabetically by the community in which they are located. For example, Toronto/Buttonville, Toronto/Downsview,

Toronto City Centre, Toronto/Lester B. Pearson, etc. The word heliport identifies a location that is for helicopter use only.

A sketch of the airport is included showing runway layout, taxiways, locations of buildings and tower (if present). Also included in the sketch is an obstacle clearance circle (OCC). The circle is divided into four quadrants and printed within each quadrant is the height above sea level, plus 1000 feet, of the highest obstacle that is located within the geographical area that the circle describes. The quadrants are aligned with magnetic cardinal points in the Southern Domestic Airspace and with true cardinal points in the Northern Domestic Airspace. The centre of the circle describes the centre of the aerodrome. The radius of the circle is noted on the circumference. If a control zone exists at that airport, the class of airspace and the ceiling of the control zone is indicated. If the zone does not conform to the standard shape, the words "shape irregular" are included.

Also given is the 3 or 4 alpha/numeric identifier of the aerodrome, its geographical co-ordinates, its distance and direction from the community, the magnetic variation, the time conversion from UTC, its elevation and a list of the aeronautical charts on which the aerodrome is depicted.

Information is given about the operator of the aerodrome and the facilities available under the categories: Public Facilities (telephone, food, taxi, accommodation, car rental, etc. available in the terminal or near the aerodrome), Customs, Flight Planning (FSS, weather, TWB, etc.), Services (fuel, oil, maintenance services, etc.), Runway Data, Lighting, Communications (radio, ATIS, ground, tower, mandatory frequency, aerodrome traffic frequency, arrival and departure control, etc.), Navigation (ILS, NDB, Tacan, Vortac, DME, PAR, etc.), Procedures (recommended routes, altitudes, special circuit procedures, etc.).

The PLANNING section gives information on such subjects as flight plans, notifications and itineraries, position reports, PIREPs, equipment prefixes and suffixes, transponder codes, SCATANA, classification of the airspace, weather minima, cruising altitudes, flight restrictions, preferred IFR and RNAV routes.

The RADIO NAVIGATION AND COMMUNICATIONS section lists pertinent information about navigation aids, north warning system beacons, commercial broadcasting stations and volmet stations.

The MILITARY FLIGHT DATA AND PROCEDURES section lists information that is mainly pertinent to military operations in Canada and the North Atlantic.

The EMERGENCY section gives information on transponder operation, unlawful interference, traffic control light signals, fuel dumping, search and rescue, emergency radar assistance, communication failure procedures, military visual signals, interception of civil aircraft, interception signals.

The booklet is published jointly by Transport Canada and the Department of National Defence. A new edition of the Canada Flight Supplement is published every 56 days to ensure that the information it contains is as current as possible.

Water Aerodrome Supplement

Information about all water aerodromes that are shown on Canadian aeronautical charts is contained in the **Water Aerodrome Supplement.**

The format of this booklet is much like that of the **Canada Flight Supplement.** An introductory section contains explanatory information about annotation and codes used in the Directory.

LYNN LAKE (ELDON LAKE) MAN KD3

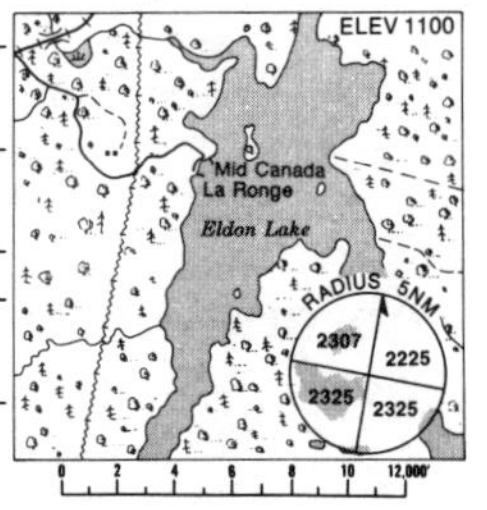

REF	N56 49 W101 01 3NNW 11°E UTC-6(5) Elev 1100' A5024 D-13 E-16
OPR	Mid Canada Sky Svcs 204-356-2532 Reg; La Ronge Avn Svc Ltd 204-356-2457 Reg
PF	B-1 D-2,3,4,5,6
FLT PLN FSS WX	NOTAM FILE CYYL 204-356-2242 W2
SERVICES FUEL OIL S	 100LL, JB All 5
A/D DATA	Open water Jun-mid Oct. Depth 3', Slipway & gin pole. Skiplanes in win.
COMM RADIO MF	 5680 122.3 126.7 (V) rdo 122.3 5NM 4100 ASL
NAV NDB VOR/DME	 YL 395 (M) N56 49 51 W101 04 13 105° 2NM to A/D YYL 112.6 Ch 73 N56 51 51 W101 04 31 (1219') 135° 3.4NM to A/D
CAUTION	Land A/D 3NM NNW.

Fig. 27. Water Aerodrome at Lynn Lake, Manitoba.

In the AERODROME AND FACILITY DIRECTORY, water aerodromes are listed alphabetically and a sketch of the water body on which the aerodrome is located is included. Also included is an obstacle clearance circle similar to that included with aerodrome layouts in the **Canada Flight Supplement.** The 3 or 4 letter alpha/numeric identifier, the geographical co-ordinates, magnetic variation, time conversion from UTC, elevation and a list of the aeronautical charts on which the aerodrome is depicted are given. Also included is information about the operator of the aerodrome, public facilities, flight planning services, services such as fuel, oil, radio and navaids and a final remarks section with pertinent information about special procedures, hazards, available mooring, etc.

The sections entitled PLANNING, RADIO NAVIGATION AND COMMUNICATIONS and EMERGENCY contain information exactly similar that included in the same sections in the **Canada Flight Supplement.**

Basic Chart Information

Scale

The scale of the map is the relationship between a unit of distance (i.e. one inch) on the map to the distance on the earth that the unit represents. On most aeronautical charts, the scale is expressed as a representative fraction.

VFR Navigation Charts have a scale of 1: 500,000, that is, one inch on the map is equal to 500,000 inches on the ground (or approximately one inch to 8 miles).

World Aeronautical Charts have a scale of 1: 1,000,000, that is, one inch on the map is equal to 1,000,000 inches on the ground (or approximately one inch to 16 miles).

VFR Terminal Area Charts have a scale of 1: 250,000, that is, one inch on the map is equal to 250,000 inches on the ground (or approximately one inch to 4 miles).

A graduated scale line also gives scale information. On aeronautical charts, three such scale lines are printed on the border of the chart. One line represents kilometers, one statute miles and the other nautical miles. A scale line based on 1: 500,000 is represented in Fig. 28.

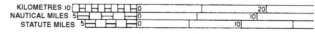

SCALE 1:500,000

Fig. 28. A Graduated Scale Line

A conversion scale for converting feet to meters is also printed on the margin of aeronautical charts.

Latitude and Longitude

A graticule (grid-like pattern of meridians and parallels) is depicted on aeronautical charts. The meridians are graduated in minutes of latitude and the parallels in minutes of longitude. This makes it possible for latitude and longitude to be measured on the chart. In addition, the subdivided meridians provide a scale of nautical miles, since a minute of latitude may be considered as a nautical mile.

On VFR Navigation Charts (VNC), the meridians of longitude and the parallels of latitude represented on the chart are numbered at the edge of the chart. Parallels and meridians are also conveniently numbered within the body of the chart. Meridians and parallels, at 30' intervals, are divided into 1 minute divisions.

On WAC Charts, each meridian of longitude and parallel of latitude printed on the body of the map is marked off in minutes of latitude and longitude. The numbers of each parallel and meridian appear frequently on the graticule.

On VFR Terminal Area Charts (VTA), latitude and longitude are indicated on the edge of the chart but not throughout the body of the chart.

Relief

The representation of relief, or ground elevation above sea level, on aeronautical maps is of primary importance. On maps used for air navigation, it is essential that height and position of the highest parts of hills and mountains should be clearly shown and that sufficient detail should be given to make easy the recognition of prominent features. Relief is shown by contour lines, layer tinting and spot heights.

Layer Tinting

The map is coloured to represent different levels of elevation. On the white border of every map, an elevation legend is printed to show what colours are used for different elevations.

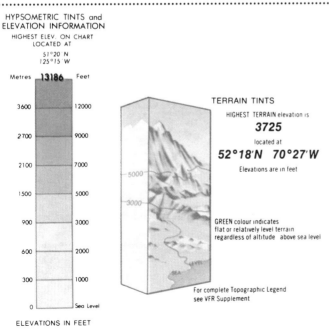

Fig. 29. Layer Tinting Legends.

On VFR Navigation Charts and VFR Terminal Area Charts, the legend is called a hypsometric tint scale (See Fig. 29). The colour coding ranges from green through shades of yellow to dark bronze to represent increasing elevations in thousand foot (300 meter) increments.

On WAC Charts, a coloured pictorial scene gives terrain characteristic tints. The colour coding ranges from green through shades of grey to yellow for the highest elevations.

All bodies of water including rivers and streams are coloured blue.

Where differences in elevation are considerable, a layer tinted map is very easily read, since the relief stands out as on a model, but it is not possible to show minor variations of height and the impression may be that the ground within the tinted areas is level.

Contours

Contour lines are drawn on a chart joining points of equal elevation above mean sea level. The height of a contour line (in feet or meters) is indicated on the map by figures. The gradient (steepness) of a slope is indicated by the horizontal distance between the contour lines. The closer the contours, the steeper the slope of a hill or valley.

Spot Heights

Figures printed on a map show the height above sea level at any particular point. Especially high elevations are marked by a dot with the spot height beside the dot. The higher the elevation of the hill represented by a spot height, the larger in size are the numerals indicating its elevation. The highest spot on a particular map is represented by the largest size of numerals and the numeral is boxed with the layer tinting removed within the box.

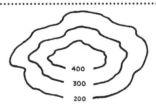

Fig. 30. A Hill Represented by Contour Lines.

The highest elevation on a particular chart is also indicated on the white border of the chart. On VFR Navigation Charts and VFR Terminal Area Charts, it is shown, together with its geographical co-ordinates, at the top of the hypsometric tint legend (See Fig. 29). On WAC Charts, the highest elevation with its geographical co-ordinates is printed as shown in Fig. 31.

Relief Portrayal

Elevations are in feet
Highest Terrain elevation is 12972   located at 53°07'N 119°09'W

Fig. 31. Elevation Information.

Isogonic Lines

Isogonic lines (lines joining places of equal magnetic variation) are depicted by dashed lines. The degree of variation is printed at regular intervals along the line.

Communities, Roads, Railways

Yellow squares represent towns and small villages. Hamlets are represented by small circles. A city is depicted by a yellow area outlined in black that corresponds to the actual shape and size of the community.

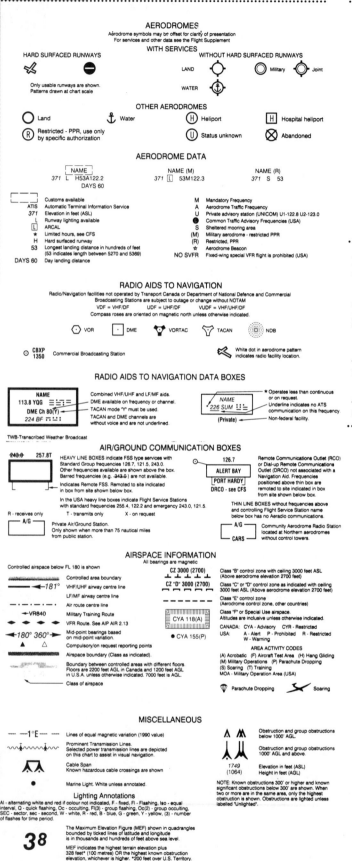

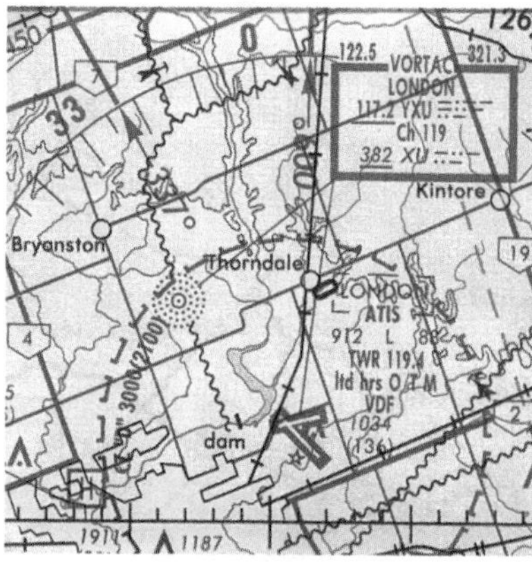

Fig. 33. Aerodrome at London, Ontario

Highways are indicated by red or brown lines, double lane highways by double lines. Railways are shown as black lines.

Aerodromes

Small aerodrome are shown by a circle as depicted in Fig. 32. The code letter or symbol included in the circle indicates its status.

Aerodromes with hard surface runways are depicted at their location on the chart by a diagram of the landing areas.

Airport information is also given in a box adjacent to the schematic depiction of the runway configuration (Fig. 33, top).

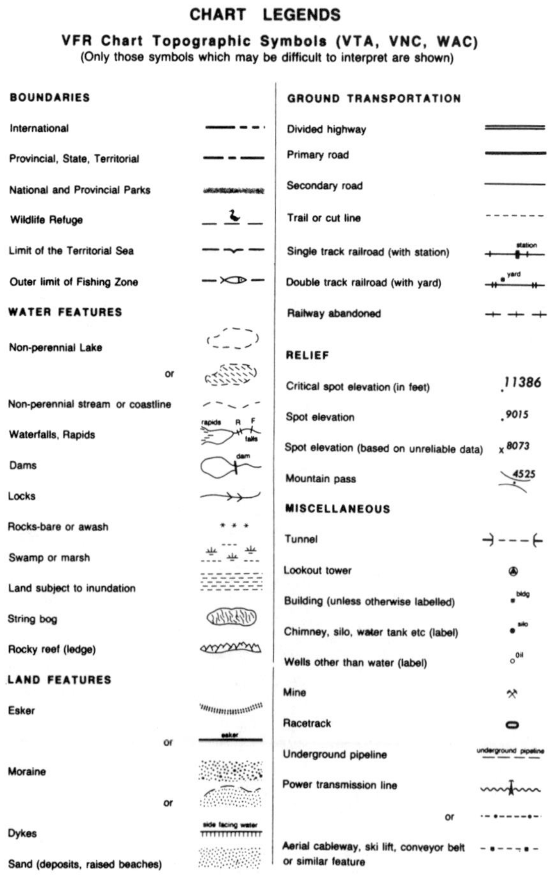

Fig. 32. Aeronautical Information Legend.

Fig. 34. Topographical Symbols

From the Ground Up

Class F Airspace
Areas that have been designated as Class F Airspace are marked on aeronautical charts and information about them printed on the chart. These include restricted and advisory areas and altitude reservations.

Compass Rose
A circle overprinted on a chart divided into 360°, from which directions may be measured, is called a compass rose. The centre of the compass rose is a transmitting navigational facility such as a VOR or a TACAN station. The compass rose is oriented on the chart on magnetic north.

It has a diameter of approximately 19 nautical miles on VFR Navigation Charts (1: 500,000).

A more detailed study of radio navigation facilities appears in the Chapter **Radio Navigation.**

Aeronautical Information
The date of issue of a particular chart is always included with the information printed with the title of the chart. Always use up to date charts in planning a cross country flight. Radio frequencies change, high rising obstructions are built and airport facilities change and it is essential to have the latest information in order to conduct a safe and efficient flight.

A legend of aeronautical information explaining the symbols and data appearing on the chart is always printed somewhere on the sheet. On VTA, VFR Navigation and WAC Charts, it appears on the white side border. This legend includes information on Aerodromes, Radio Facilities and Airspace Information, etc. (Fig. 33). A legend of topographical symbols used on aeronautical charts is shown in Fig. 34.

Plotting Instuments
The Navigation Plotter
Of great assistance to a pilot in plotting and planning flights is an instrument such as the navigation plotter. It combines a protractor and a straightedge in one device which also incorporates a mileage scale for both 1: 500,000 and 1: 1,000,000 charts.

The plotter is made of clear plastic so that details of the chart can be seen through it.

With the straightedge, the pilot can draw a track from the airport of departure to the planned destination.

The direction of a track is determined by using the protractor portion of the plotter. It is numbered from 000° to 180° on the outside scale and from 190° to 360° on the inside scale. The outside scale is used for easterly tracks and the inside scale for westerly tracks.

To use the plotter, place the hole in the centre of the plotter over an intersection of the track line and one of the longitude lines on the chart. A point somewhere near the mid point of the track is best chosen to obtain greater accuracy.

Place a pencil point through the hole and rotate the plotter until the top edge of the straightedge is aligned with the track line.

Read the direction of the track in degrees true where the longitude line of the chart intersects the scales. (Use the outer scale for easterly tracks and the inner scale for westerly tracks.)

For measuring tracks that are almost directly north and south, a latitude line may be used as a line of reference and the small scale at the centre of the protractor used to determine direction.

In using the straightedge to determine the distance from the airport of departure to the destination, be sure to use the correct side of the straightedge for the type of chart in use. The mileage scale on one side of the straightedge is 1: 500,000 for VNC Charts and is marked off in both statute and nautical miles. The reverse side of the straightedge has a scale of 1: 1,000,000 for WAC Charts and is also marked off in both statute and nautical miles.

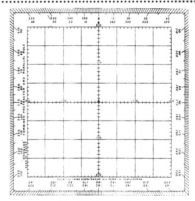

Fig. 36. Protractor

The Protractor
A navigation plotter, such as that described above, is not the only instrument that can be used to plot tracks. A simple protractor and a ruler will serve the purpose just as well.

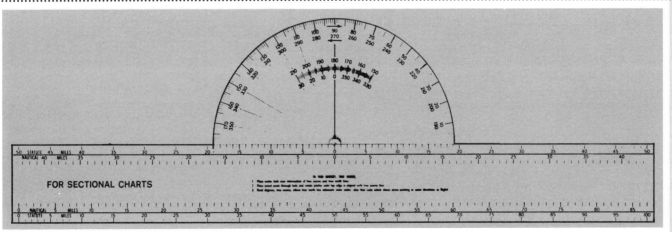

Fig. 35. Navigation Plotter.

Air Navigation

A protractor, being square, can be used both for determining direction and as a straightedge. The instrument has a compass rose graduated in 360° marked around the outer edges. It is transparent so that, when placed on a map, the map is visible through it.

Place the protractor on the map with the hole in the centre lying on the track at a poin **where the north-south line on the protractor lies along the meridian.** If this is not convenient, one of the parallel lines may be lined up parallel with the nearest meridian. The track is read off where it cuts the edge of the protractor.

Ruler

In addition to a protractor, a pilot will require a ruler to measure distance. A mileage scale is printed on every aeronautical chart and it is a simple matter to measure the distance from the airport of departure to the destination and lay this distance off against the chart scale to determine the mileage. The distance scales on ICAO charts in a given series are nearly, but not exactly, constant and are also affected by humidity. A ruler which is constructed mathematically to scale may not exactly correspond to the map sheet you are using. For practical air pilotage purposes, the difference is inappreciable. For extreme accuracy, note the difference between the ruler and the map scale at the 100 mile mark and apply it.

Preparing the Map for a Flight

Map reading, either on the ground or in the air, calls for a clear understanding of the scale of the map and a "sense" of that scale (in other words, a sense of proportion). It also requires an understanding of the direction of true or magnetic north and of the symbols used on the map.

The direction of north should never be in doubt. Remember that the right and left hand edges of the map sheet are not always parallel to the meridians.

In preparing for a cross country flight, draw a line on your chart from your point of departure to your destination. Using a plotter, as described above, determine the direction of the track in degrees true. Having determined the direction, it is then necessary to apply the magnetic variation to find the direction of the track in degrees magnetic (See **Magnetism**).

The magnetic variation is generally clearly stated on aeronautical charts. On flights up to 300 miles, the variation should be averaged for the entire route. This is done by using the isogonic line which intersects the track (as nearly as possible) midway between the point of departure and the destination. On flights of longer duration, the heading should be altered at regular intervals to allow for accumulated changes in variation.

Deviation will also have to be applied to determine the compass heading to fly to make good the desired track over the ground. This information can be found on the compass deviation card in the airplane.

Now measure the track and divide it into equal intervals of 10 to 20 miles each. During the flight, these divisions will help you to estimate your groundspeed, quickly to measure distance to landmarks and airports and to monitor fuel and time en route.

Study the map in the vicinity of the track line for landmarks that you will easily be able to identify during the flight. Check the ground elevation along the route and select a flight altitude that will clear any high ground at sufficient altitude to comply with regulations and safety considerations. Check for any high rising obstructions such as broadcast towers and plan to fly at an altitude well above them. Be sure that your track does not proceed through a restricted area.

Do as much preparation in advance as possible. Once in the air, you are kept busy flying the airplane.

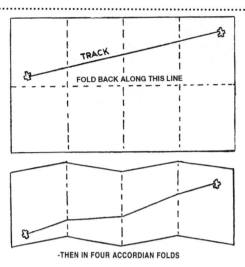

Fig. 37. Folding a Map for Use in Flight.

Folding a Map or Chart

A map should be folded into a strip about eleven or twelve inches wide with the track lying somewhere about the centre of the strip.

It should then be folded "accordion" fashion, so that successive portions of the track can be read by turning the folds over one by one (Fig. 37).

If more than one map sheet is used, the sheets should be numbered and arranged in the order in which they will be required. This is important.

To Plot a Track Between Points On Different Charts

1. Lay a separate piece of paper over the north chart so that the side edge is aligned with A, one end of the required track (i.e. the point of departure or the destination). Mark A on the edge of the paper. On the bottom edge, mark D and E, two points common to both charts. There is an overlap of about 3 miles between the north and south portions of the chart.

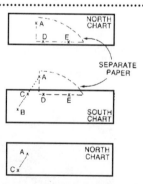

Fig. 38. Plotting a Track.

2. Position the paper on the common points of the south chart. Using a ruler to connect A and B (the other end of the required track), draw a line from B to C near the edge of the chart, a common point on both charts.

3. Locate point C on the north chart and draw a line connecting C and A to complete the required track.

Landmarks

Water — Can be seen at great distance and offers the best landmark of all. During wet seasons, it must be remembered that flood areas may easily be mistaken for lakes. Some rivers and streams may be obscured by trees in summer and therefore be hard to identify.

Heights of Land — Can be seen for very great distances but are not as numerous as other landmarks.

Roads — It is sometimes difficult to distinguish main roads from unimportant ones since even secondary roads are paved. Double lane throughways are very noticeable and are especially good landmarks. The intersection of two main roads is also a good landmark.

Railway Lines — These do not show up as clearly as roads, but are less numerous and therefore serve as distinctive and useful landmarks. Railway crossings or junctions are especially prominent and present distinctive patterns.

Towns — Towns can usually be readily identified by shape, and particularly by the pattern of roads or railway lines entering or leaving them. Smoke haze over larger cities makes them sometimes very difficult to recognize, particularly in dull weather and when approaching them towards the sun.

Golf Courses and Race Tracks — These are both fairly good landmarks and are possible emergency landing grounds for airplanes having reasonably low landing speeds.

Flying by Map Reading

Skill in map reading does not begin and end in the ability to pick out conspicuous landmarks and locate them on the map.

Further, and most emphatically, **it does not consist of following the winding paths of rivers and railways**, in order to reach your destination. When visibility is normal, a well-trained pilot should be able to fly a direct route to his destination by map reading alone.

Map reading is a great deal more difficult at low altitudes than it is at higher ones, due to the relative nearness of the horizon and to the fact that the pilot cannot see enough of the terrain at one glance to enable him to relate what he sees to what is on his map. When first practising map reading and cross country navigation, flights should be conducted, at least, at 2 or 3 thousand feet above the ground.

When flying by map reading, it is advisable to orient the map to the direction in which the airplane is proceeding along its route, even though this may mean that the lettering on the map is sideways or upside down.

Ten Degree Drift Lines

For the pilot who has carefully prepared his charts and determined his compass heading, following a track over the ground between the point of departure and the destination would be a simple matter except for one thing, the wind. The wind has a perverse habit of blowing from the right or from the left and displacing the airplane from its track. In order to compensate for the wind, it is necessary to adjust the heading of the airplane somewhat into wind in order to maintain the desired track. In effect, the airplane is flying slightly sideways along the required track. This is known as **crab.**

Before taking off, the conscientious pilot gets a thorough weather briefing and learns from the briefer the expected direction and speed of the wind that will be will encountered. He then calculates the heading to fly in order to make good the required track. (How to do this is explained in **Navigation Problems** later in this chapter.) Unfortunately, even that does not always solve the problems caused by the wind.

Because winds are often different than what they were forecast to be, a pilot may find himself off track a short time after starting off on the flight. Determining what heading to take up in order to get back on the required track and what heading to fly once back on track in order to reach the destination is simplified if he has drawn 10 degree drift lines on either side of the route he has laid out on the chart.

These lines, opening out from the departure point (or the point at which he sets his heading) and closing down to the destination, enable a pilot to estimate track errors and to determine what heading changes to make. Each of the lines makes an angle of 10 degrees with the track line. If you have used a black pen to draw the track line, use a red pen (or some other colour) to draw the 10 degree drift lines so that, during the flight, there is no chance of confusing the track with the 10 degree drift lines,

First of all, here are some terms that apply to these calculations:

Required Track. The proposed path of the airplane over the ground.

Track Made Good. The actual path of the airplane over the ground.

Track Error. The angle between the required track and the track made good, measured in degrees either left or right of the required track.

Opening Angle. The angle between the required track and the track made good.

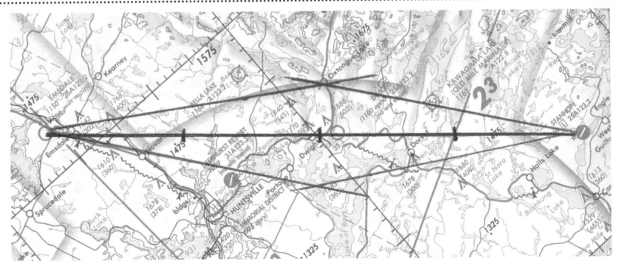

Fig. 39. Ten Degree Drift Lines.

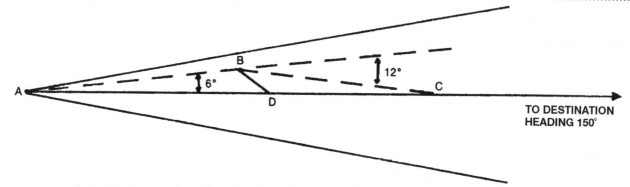

Fig. 40. Double Track Error Method and Visual Alteration Method of Regaining the Required Track.

Closing Angle. The angle between the old required track and the new required track necessary to arrive at the destination.

The required track and the ten degree drift lines are drawn on a chart as shown in Fig. 39. The airport at Emsdale is the point of departure and the destination is Stanhope. Ten nautical mile intervals have been marked on the track line to assist in distance calculations.

Having taken off from the airport of departure and flown for about 10 or 15 miles, you note that the airplane is not on the required track. You can change the heading either to return to the track or to fly directly to the destination on a new track. It is usually better, if you have not passed the half way point, to return to the required track since this is the route that has been studied and prepared on the chart. There are two methods by which you can return to the required track: (1) the double track error method or (2) the visual alteration method.

The Double Track Error Method

Fig. 40 depicts the double track error method of regaining the required track.

The heading necessary to fly the required track from point A to the destination is calculated to be 150 degrees. About 15 nautical miles from point A, you accurately pinpoint your position at B and estimate, using the ten degree drift lines as a guide, that the track error (or opening angle) of the track made good is about 6 degrees to the left of the required track. If you alter your heading by 6 degrees to the right and take up a heading of 156 degrees, you will be following a track that is parallel to the required track. In order to regain the required track, you need to double the track error; that is, you need to alter your heading by 12 degrees to the right and take up a new heading of 162 degrees (original heading of 150 degrees plus 12 degrees). You will regain the required track at C in approximately the same period of time as it took to reach position B from A. It is helpful if there are some terrain features along the track line to confirm that the required track has been regained. If there are none, you can assume that you have reached the required track at the calculated time. On regaining the track, you should subtract half of the correction applied to the original heading (162 degrees less half of 12 degrees) to establish the proper heading (156 degrees) to maintain the required track and reach your destination.

The Visual Alteration Method

If you can positively identify a landmark that is right on the required track, you may choose to regain the track by the visual alteration method. You establish your position at B (Fig. 40) and note that the track error is about 6 degrees and that a heading of 156 degrees (the original heading plus the track error) would produce a track parallel to the required track. You then fly visually to the positively identified landmark (D). On reaching that point, you take up a heading of 156 degrees which should keep you on the required track right to your destination.

Opening and Closing Angles Method

Sometimes more than half the flight has been completed before you are able to determine your exact position. Perhaps there have been no terrain features on which to make a fix. In this case, you may choose to regain the required track by the visual alteration method or you may choose to fly directly to the destination using the opening and closing angles method.

After flying for more than half the calculated flight time, you are able to pinpoint your position at C (Fig. 41). Using the 10 degree drift lines, you find your opening angle (a) to be about 4 degrees to the right of the required track. The closing angle (b) into the destination B appears to be about 8 degrees. In order to fly directly to B, add the opening angle and the closing angle and adjust your heading by this amount to the left. Your new heading would, therefore, be 073 degrees (085 degrees less 12 degrees).

The opening and closing angle method may be used at any distance along the track. It is not limited to use only after the half way point has been passed.

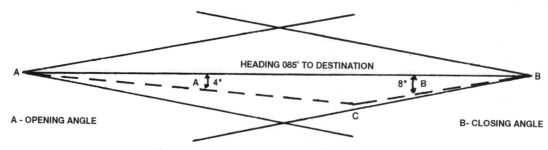

Fig. 41. Opening and Closing Angles Method.

"Two-Point" Method of Flying a Visual Range.

When it is not possible to obtain a wind before departure or time does not permit plotting it, a satisfactory method of determining the heading to steer is that known as flying a visual range.

Two points are selected along the track about 5 or 10 miles apart. The pilot, having gained the height it is intended to maintain throughout the flight, flies a heading by map reading that will pass over these two points. His compass will then indicate the heading to steer for the balance of the flight. Other points along the route may be selected subsequently and a series of ranges flown. This method is particularly useful if it is necessary to alter heading during a flight to a destination other than that originally intended.

Return to Point of Departure

In order to return to the point of departure, a pilot needs to calculate the **reciprocal** of his outbound heading to determine the heading to take up for the homebound flight.

By reciprocal is meant the reverse, or opposite, of a given direction such as a heading, course or bearing.

The reciprocal is found by adding or subtracting 180; **adding** if the direction is 180 degrees or less, **subtracting** if it is greater than 180 degrees.

Track .... 187°
Reciprocal .... 007°

However, determining the reciprocal of a heading is not so simple. Since wind drift allowance is the vital consideration in determining the heading to make good a particular track and since the wind will have quite a different effect in making good a reciprocal track, it can readily be seen that adding or subtracting 180 from the heading is not the method to calculate the reciprocal heading. The approximate heading to steer to make good the reciprocal track may be calculated by the following method:

Calculate the reciprocal of the heading out and double the wind correction angle the opposite way.

If the wind correction angle out has been 5° to port, to double it the opposite way, you **add** 10° (to starboard) to the reciprocal of the heading out to obtain the heading home. (Any correction applied to the right, or starboard, or clockwise, is **added.**)

If the wind correction angle out has been 5° to starboard, you **subtract** 10° (to port) from the reciprocal of the heading out to obtain the heading home. (Any correction applied to the left, or port, or anti-clockwise, is **subtracted.**)

Problem

You are flying a compass heading of 122°. You run into bad icing conditions and decide to turn for home.

A glance at your flight plan tells you that your true heading is 117° and your track (true) is 122°.

The reciprocal of your heading is 297° (117° + 180° = 297°).

You have been steering 5° to port of your track (to correct for a drift of 5° to starboard).

To double the wind correction angle the opposite way, you would now steer 10° to starboard of 297°.

The true heading for home would therefore be 297° + 10° = 307°.

Using the same variation you had on your heading out and the deviation applicable to the new heading, you can now easily determine the compass heading to steer for home.

Groundspeed Check

By noting the distance between check points along your route and keeping track of the time, the groundspeed may be found.

Fig. 42 shows a small section of a WAC chart on which is shown a portion of a flight track. At 10:24, the airplane passed over Amherst. At 10:43, it passed over Tatamagouche. The distance from Amherst to Tatamagouche is 39 nautical miles and the time taken to fly the distance was 19 minutes. The groundspeed, therefore, is 123 knots. Let us suppose that the destination is 82 nautical miles beyond Tatamagouche. At a groundspeed of 123 knots, the flight should take a further 40 minutes and the airplane can expect to arrive at its destination at 11:23.

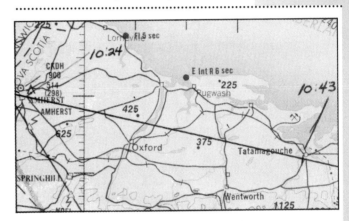

Fig. 42. Groundspeed Check

The "One-in-Sixty" Rule

An error in the track of one degree will cause an error in position of about one mile in a distance of 60 miles.

A pilot on a cross-country flight, who has got off track, will be able to estimate the distance he is off track quite easily, but it is very difficult to know how many degrees it is necessary to alter heading by compass to correct the error.

Suppose an airplane is 2 miles off its track after travelling 30 miles. The error in the track will be roughly 4°. Therefore the correction to the compass heading will be 4° to correct the error. This will put the airplane on a track parallel to the required track but 2 miles from it. Suppose the airplane is 60 miles from its destination. An additional 2° correction to heading will gradually close the track. Therefore a total correction of 6° will bring the airplane in to its destination.

The Pilot Navigator

Flying alone in the cockpit of an airplane, a pilot cannot really handle cumbersome plotting instruments. The basic preparation for a flight, marking the route and studying it, must be done in advance. However, not all contingencies can be foreseen and it is often necessary to make calculations and estimates during the flight.

The ability to make accurate estimates is a skill that is learned with practice. During any flight, you should practise making such estimates. Frequently select towns, lakes, airports, etc., that are off course, estimate the distance they are from the airplane, what heading would be required to fly to them and how long it would take to make the flight. A good knowledge of the principles of navigation will enable you with practice to make very accurate mental calculations en route.

With good average visibility, favourable weather, and a good map of the territory to be flown over, successful navigation depends on little more than the ability to read a map in general terms.

With low visibility or over comparatively featureless country, however, steering an accurate heading by compass becomes of primary importance. Accurate map reading is essential; the route and groundspeed must be checked frequently in order that the exact position of the airplane may be checked at frequent intervals.

Navigation Problems

Navigation Terms

WIND is air in motion, especially a mass of air having a common direction or motion. Wind moves horizontally. (A movement of air vertically is called a current.)

INDICATED AIRSPEED is the airplane's speed as indicated by the airspeed indicator.

TRUE AIRSPEED is the speed of the airplane relative to the air. It is calibrated airspeed corrected for the airspeed indicator error due to density and temperature.

GROUNDSPEED is the speed of the airplane relative to the ground. An airplane is affected by wind. If there is no wind at all, true airspeed and groundspeed will be the same. If, however, an airplane is flying in an air mass that is moving in the same direction, the airplane will have a tailwind that will help its progress over the ground, with the result that its groundspeed is in excess of the true airspeed. Conversely, a headwind will impede the progress of the airplane over the ground with the result that the groundspeed is slower than the true airspeed.

The HEADING of an airplane is the angle between the longitudinal axis of the airplane at any moment and a meridian. In other words, it is the direction the nose of the airplane is pointing, measured from an imaginary line running north and south. If the heading is measured from a true meridian, it is referred to as a **true heading,** if from a magnetic meridian, as a **magnetic heading.** If it is measured from the direction of a compass needle, it is referred to as a **compass heading.** The angle is measured clockwise through 360°.

The TRACK (intended) is the direction an airplane intends to travel over the ground. The intended track may be represented by a straight line drawn on a map. Its direction is the angle between this line and a meridian measured clockwise through 360°. As in the case of headings, tracks are named **true, magnetic or compass** with reference to the meridian from which they are measured.

The TRACK MADE GOOD is the actual path travelled by the airplane over the ground. Like the intended track, it may be represented by a line drawn on a map and (provided it is a reasonably straight line) its direction measured from a true or magnetic meridian or compass north.

DRIFT. A wind blowing from either the starboard or port side of an airplane will cause the airplane to drift away from its intended track. In order to maintain the intended track, it is necessary to turn the airplane slightly into wind to compensate for the force acting laterally upon it. Drift (or drift angle) is the angle between the heading being flown and the track made good over the ground. In other words, it is the angle at which the pilot heads the airplane across the track to keep the wind from blowing him off the track. It is expressed in degrees either port or starboard.

A TRUE MERIDIAN is a meridian on the surface of the earth joining the true north and south poles. On practically all maps used for air navigation purposes, the true meridians are shown.

A MAGNETIC MERIDIAN is the direction in which a compass needle will lie when influenced only by the earth's magnetic field. In actual practice, magnetic meridians are not shown on maps but are found by adding or subtracting the variation at any particular place to or from the true meridian. (Variation is indicated on maps by isogonic lines, which are lines joining all places of equal variation.)

COMPASS NORTH is the direction in which a particular compass needle will lie when influenced by both the earth's magnetic field and local magnetic influences (deviation) in the airplane. The actual reading on a compass at any time is the angle between compass north and the direction the airplane is heading.

AZIMUTH means direction measured as an angle clockwise from a meridian. It is the same as a bearing. The azimuth, or bearing, may be true, magnetic or compass.

The Composition of Velocities

The **velocity** of a body is the **rate of change of position** of a body in **a given direction;** hence it involves both speed and direction.

A velocity may be represented by a straight line. The direction may be shown by the direction of the line and the speed by the length of the line to scale.

The straight line referred to above which represents the velocity of a body may also be referred to as a vector. A **vector** may be defined as a **quantity having both magnitude and direction.** In aviation, vectors almost invariably represent speed and direction (or, in other words, velocity).

A body may be subjected to two or more velocities at the same time and these may not act in the same straight line.

Fig. 43. Composition of Velocities.

For example, an airplane may be flying on a heading due north at 80 knots with a wind blowing from the west at 20 knots. The airplane will have two velocities, one towards the north at 80 knots and the other towards the east at 20 knots.

These two velocities are equivalent to a single velocity which is called the **resultant**. The resultant, in this case, would represent the track and groundspeed of the airplane (Fig. 43).

When the components act in the same straight line, the resultant is obtained by addition or subtraction. When the components act in other than the same straight line, the resultant may be found by a **triangle of velocities**.

The Triangle of Velocities

If a moving point possesses, simultaneously, velocities representing speed and direction, these may be represented by two sides of a triangle (AB and BC in Fig. 44). They are, then, equivalent to a resultant velocity represented by the third side (AC).

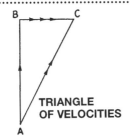

Fig. 44. Triangle of Velocities.

Note: Arrowheads should always be used to show the direction of the components clearly. A single arrowhead represents the heading of the airplane. Double arrowheads represent its track. Triple arrowheads indicate the wind.

The solution of navigational wind and drift problems is based on the principal of the triangle of velocities.

The **heading and true airspeed** of an airplane can be represented by one side of a triangle. The **wind and windspeed,** drawn to the same scale, can be represented by another, and the **track and groundspeed** by the third.

A knowledge of any four of these is sufficient to complete the triangle, from which the remaining two may be determined.

Note: In setting out a navigation problem involving a triangle of velocities, it is essential that all units of distance and speed be compatible. For example, all speed units must be in knots if the distance measurements are in nautical miles. Conversion tables for converting statute to nautical miles (or vice versa) appear on the final pages of this book.

The Flight Computer: Wind Side

Wind and drift problems may be solved on navigation computers in a fraction of the time it takes to plot them on paper. All pilots should be able to use a navigation computer and this section is written with the assumption that students have a computer available.

Such a navigation computer is commonly known as a dead reckoning computer. One side of the computer is a circular slide rule. The other side is a wind drift computer (shown in Fig. 45). The wind side of the computer is used to solve the wind and drift problems that follow.

The instrument has a **compass rose** which can be rotated. The dial inside the compass rose is a transparent window on which pencil marks can be made and erased. In the centre of the window is a dot with a ring around it, the **grommet,** which is a reference point used in plotting. Through the window can be seen a grid marked out in concentric arcs and radial lines. This is printed on a sliding plastic strip. The concentric arcs represent speed in mph or knots and will be referred to as **speed lines.** The radial lines represent degrees of drift to right or left and will be referred to as **drift lines.** At the top is a scale graduated in degrees of drift to right or left, the **wind correction angle (WCA) scale.** This may also be used as a variation scale to apply magnetic variation, east or west. In the centre is an arrow referred to as the **true index** (on some computers, it is called the true head).

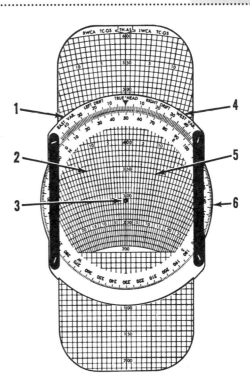

Fig. 45. Dead Reckoning Computer: Wind Drift Side.

1. Variation Scale.   4. Drift Scale.
2. Speed Lines.   5. Drift Lines.
3. Grommet.   6. Compass Rose.

Any navigation problem which can be plotted can be solved on the computer, including multiple drift, off-course, radius of action, interception, etc. Basic problems of primary interest to the pilot-navigator only are dealt with in the text which follows.

Wind and Drift Problems
Problem 1.

To find the heading to steer to make good a given track and to find the groundspeed when the airspeed and the wind and the windspeed are known.

A Skylark Deluxe is to proceed from Niagara Falls to Toronto Island airport. The intended track (true) is 332°. The true airspeed of the airplane is 85 knots. The wind is 280° (true) at 25 knots. (Note that the direction of the wind is the direction from which it blows.) What is the heading to steer and what is the groundspeed?

Solution by Plotting

First draw a vertical line to represent a true meridian on your plotting paper and mark the scale you are using in one corner. This applies to the solution of all wind and drift problems.

1. From A, draw A-B, 332° to represent the true track (Fig. 46).

2. From A, draw A-C at 280° to the meridian but downwind from A. The line A-C, therefore, is drawn at an angle of 100° to the true meridian. Draw the line A-C 25 n. miles to scale to represent the wind and windspeed.

3. With C as centre and radius 85 n. miles (the true airspeed), describe an arc cutting A-B at D. Join C-D.

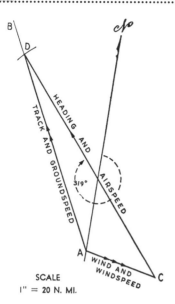

Fig. 46.

Therefore, in the triangle A-C-D, A-C represents the wind and windspeed, C-D the heading and true airspeed and A-D the track and groundspeed.

The heading to steer is the angle C-D makes with the meridian, which is 319°. This heading will be correct whatever the distance to be flown, so long as the wind and the airspeed of the aircraft remain constant.

The heading to steer by compass is found by applying the variation and deviation to the true heading.

The drift is the angle between the heading and the track. In this case, it is 13° to starboard.

The length of the line A-D represents the groundspeed, which is 67 kts.

The time taken is found by dividing the total distance by the groundspeed and multiplying by 60. Toronto Island is 39 n. miles from Niagara Falls. The estimated elapsed time would therefore be: (39 ÷ 67) x 60 = 35 minutes.

Solution by Computer

1. Set the wind direction (280°) on the compass rose opposite the true index.

2. Move the slide so that any convenient whole number shows under the grommet (centre of the disc).

3. Draw a line up from the grommet 25 units in length to represent the wind speed and draw an arrow at the end of the line (Fig. 47).

4. Set the true track (332°) at the true index.

5. Set the arrow head of the wind line on the 85 knot speed line (the true airspeed of the airplane) (Fig. 48).

6. Read the groundspeed at the grommet — 67 knots.

7. From the drift line passing through the arrow head of the wind line, read the **wind correction angle** (angle between the heading and the track). The heading is 13° to the **left** of track. The wind correction angle is 13° left.

8. Read the true heading on the wind correction angle (WCA) scale left at the 13° mark — 319° (Fig. 49).

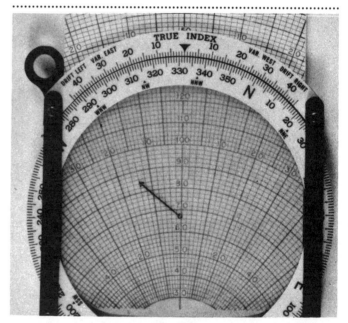

Fig. 48.

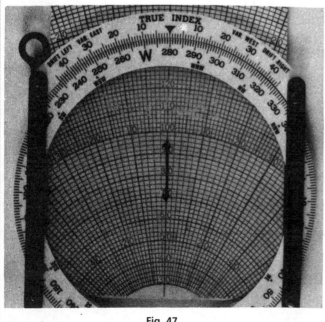

Fig. 47.

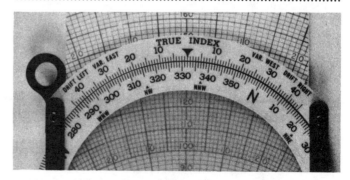

Fig. 49.

Problem 2

Suppose that it is desired to leave Niagara Falls and arrive at Toronto Island airport at a specified time, in other words, to make good a given groundspeed.

The true track is 332°. The wind is 280° (true) at 25 knots. The groundspeed required is 67 knots. What is the heading to steer and the true airspeed?

Solution by Plotting

1. Draw A-C (Fig. 46) to represent the wind and windspeed.

2. Draw A-D to represent the track and groundspeed.

3. Join C and D. C-D is the heading to steer (319°) and the true airspeed is 85 knots.

Solution by Computer

1. Set the wind direction (280°) under the true index.

2. Draw a wind vector up from the grommet to the proper length to represent 25 knots and draw an arrow at the end of the line.

3. Rotate the compass rose until the true track (332°) is set at the true index.

4. Move the slide until the grommet is on the speed line representing 67 knots.

5. Read the drift at the arrow head of the wind vector line — 13° left.

6. Read the true heading on the WCA scale left at the 13° mark — 319°.

7. Read the true airspeed at the speed line under the arrow head — 85 knots.

Problem 3.

To find the wind and windspeed, knowing the heading and true airspeed and the track and groundspeed.

A Nav Air Tutor is flying on a heading of 033° magnetic, at a cruising speed of 95 knots. It is making good a track of 048° magnetic and a groundspeed of 100 knots. What is the wind and windspeed?

Solution by Plotting

1. From A draw A-B, 033°, 95 kts. to scale, to represent the heading and true airspeed (Fig. 50).

2. From A draw A-C, 048°, 100 kts. to scale to represent the track and groundspeed.

3. Join B-C.

The angle B-C makes with the meridian (transferred) represents the wind and windspeed — from 299° magnetic at 26 kts.

The direction of the wind is always from the heading to the track.

Solution by Computer

1. Set the track (048° magnetic) opposite the true index.

2. Move the slide until the 100 knot speed line (representing the groundspeed) is under the grommet.

3. Mark a cross at the intersection of the 95 knot speed line and the appropriate drift line.

4. Compare the heading and the track. If the heading is to the left of the track, use the left drift lines. If the heading is to the right of the track, use the right drift lines. In this case, the heading is 15° to the left of the track. Therefore, mark a cross at the intersection of the 95 knot speed line and the 15° left drift line.

5. Rotate the compass rose until the cross is on the centre line above the grommet.

6. Read the wind direction (299° magnetic) opposite the true index and the windspeed (26 knots) on the centre line between the cross and the grommet.

The Flight Computer: Circular Slide Rule Side

Navigation calculations are much simplified by the use of a dead reckoning computer. The wind drift side of the computer has already been discussed. The other side is a circular slide rule.

The circular slide rule, illustrated in Fig. 51, is logarithmic in principle and can be used to solve any problem of multiplication, division or

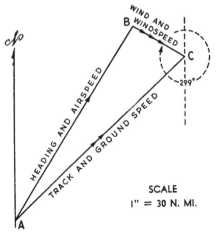

Fig. 50.

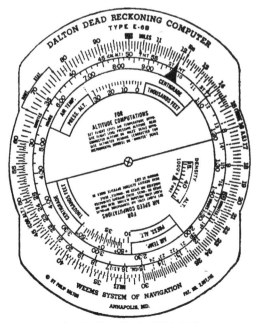

Fig. 51. The Circular Slide Rule

proportion. From a navigational point of view, however, its value lies in the rapid solution of time, speed, distance and conversion problems.

Note that the slide rule consists of three scales. The outer scale is fixed to the computer. The two inner scales are printed together on a disc that may be rotated to any position opposite the outer scale.

The outer scale represents miles, gallons, true airspeed and corrected altitude. The inner scale represents time in minutes, calibrated airspeed and calibrated altitude. The third scale represents time in hours and minutes.

The figures on a slide rule may represent any proportion or multiple of ten. 10 on the outer scale may therefore represent 1, 10, or 100. 45 on the inner scale may represent 4.5, 45, or 450. In the latter case, if 450 represented 450 minutes, the equivalent hours and minutes could be read directly on the third scale as 7 hrs. 30 min.

Solving Problems With the Circular Slide Rule
Time, Speed and Distance

For time, speed and distance problems, the outer scale represents miles, the inner scale minutes and the third scale hours and minutes. In such problems, there are three variables: time, distance and speed. Two must be known to solve a problem.

To find the distance when the groundspeed (140 kts) and the time (48 mins) are known, set 140 on the outer scale opposite 60 (the black arrow) on the inner scale. Read the distance on the outer (miles) scale opposite 48 on the inner (minutes) scale. The distance is 112 n. miles.

To find the elapsed time when the groundspeed (146 kts) and the distance (100 n. miles) are known, set 146 on the outer scale opposite 60 (the black arrow) on the inner scale. Read the time on the inner (minutes) scale opposite 100 (10) on the outer (miles) scale. The elapsed time is 41 minutes.

To find the groundspeed when the distance (30 n. miles) and the time (15.5 minutes) are known, set 30 on the outer scale opposite 15.5 on the inner scale. Read the groundspeed on the outer (miles) scale opposite 60 (the black arrow) on the inner (minutes) scale. The groundspeed is 116 knots.

Fuel Consumption

To solve problems relating to fuel consumption, the outer scale may be used to represent gallons and the inner scale to represent time.

To find endurance when the consumption in gallons per hour (16) and the quantity of fuel (72 gallons) are known, set 16 on the outer scale opposite 60 (the black arrow) on the inner scale. Opposite 72 on the outer scale, read the endurance on the inner scale. The answer is 270 minutes, or 4 hours and 30 minutes.

To find fuel consumption when the amount of fuel consumed (12 gallons) in a given time (2 hr. 21 min.) is known, set 12 on the outer scale opposite 2 hr. 21 min. on the third scale. Read the fuel consumption in gallons on the outer scale opposite 60 on the inner scale. The fuel consumption is 5.1 gallons per hour.

Conversion

Nautical to Statute Miles. The naut. and stat. mile indexes will be found on the outer scale (towards the left hand side in Fig. 51). Set any figure on the inner scale representing nautical miles opposite the nautical mile index and read the corresponding statute miles on the inner scale opposite the statute mile index. Or, to convert statute miles to nautical miles, set the figure representing statute miles opposite the statute miles index and read the corresponding nautical miles opposite the nautical mile index.

Imperial to U.S. Gallons. Mark 10 (for 1) on the inner scale as the imperial gallons index and 12 (for 1.2) as the U.S. gallons index. Set any figure on the outer scale (7.2 imp. gals. for example) opposite 10 on the inner scale. The figure on the outer scale that is opposite 12 will be the corresponding U.S. gals. Answer: 8.65 U.S. gals.

Litres to U.S. Gallons. Mark 10 (for 1) on the inner scale as the litre index and 26.4 (for .264) as the U.S. gallons index. Set any figure on the outer scale (89 litres for example) opposite 10 on the inner scale. The figure on the outer scale that is opposite 26.4 will be the corresponding U.S. gallons. In this case, 23.5 U.S. gallons.

Litres to Imperial Gallons. Mark 10 (for 1) on the inner scale as the litre index and 22 (for .22) as the imperial gallons index. Set any figure on the outer scale (75 litres for example) opposite 10 on the inner scale. The figure on the outer scale that is opposite 22 will be the corresponding imperial gallons. In this case, 16.5 imperial gallons.

Celsius to Fahrenheit. This conversion cannot be worked on the slide rule. Practically all navigation computers, however, have a Celsius to Fahrenheit conversion table printed on them. For the DR computer illustrated in Fig. 51, the table is printed on the reverse side.

Altitude Correction

The pressure altimeter used in airplanes is a relatively accurate instrument. However, it is calibrated to indicate the true altitude in ICAO Standard Atmosphere conditions (i.e. perfectly dry air with a MSL pressure of 29.92 inches of mercury and a MSL temperature of 15°C and standard lapse rate). With the subscale that is incorporated into the pressure altimeter, the pilot can correct for changes in atmospheric pressure but cannot correct for temperatures varying from that of standard atmosphere. In fact, the only time an altimeter correctly registers altitude is when the aircraft is sitting on the ground at an airport for which the current altimeter setting has been set on the subscale of the altimeter that is calibrated absolutely accurately. Whenever an aircraft is in flight, the altitude indication of an altimeter is in error.

The amount of error depends on the degree to which the average temperature of the column of air between the aircraft and the ground differs from the average temperature of the standard atmosphere of the same column of air.

To calculate true altitude using a computer, it is necessary to know the pressure altitude. Alter the subscale on the sensitive altimeter to read 29.92 and read the pressure altitude. Set the pressure altitude against the outside air temperature in the computer window marked "For Altitude Computations."

Calculate the difference between the published altitude of the airport for which you have set the current altimeter setting and the indicated altitude on the altimeter. Opposite that figure on the calibrated altitude scale (inner scale), read the true altitude on the corrected altitude scale (outer scale).

For example: Indicated altitude is 8500 feet. Pressure altitude is 9000 feet. The elevation of the airport for which the altimeter setting is set on the altimeter is 2600 feet. The outside air temperature at flight level is -25°C.

1. Set the pressure altitude of 9000 feet against the OAT, -25°C, in the altitude computations window.

2. The temperature error occurs only in the airspace between ground level (at 2600 feet) and the indicated altitude at flight level (at

From the Ground Up

8500 feet), that is, in 5900 feet of airspace.

3. Opposite 5900 feet (59) on the inner scale, read 5450 (54.5) on the outer scale.

4. Add the 2600 feet previously deducted as being errorless to the figure 5450 to find the true altitude of 8050 feet ASL (5450 + 2600) at flight level.

Airspeed Correction

The airspeed indicator is also subject to error as a result of temperature and the decreasing density of the air with altitude.

To find true airspeed, set the flight level air temperature opposite the flight level pressure altitude in the computer window marked "For Airspeed Computations". Pressure altitude is obtained by setting the reading 29.92 on the subscale of the sensitive altimeter. Convert the indicated airspeed (IAS) to calibrated airspeed (CAS) using the airspeed correction table in your Airplane Flight Manual. Read the true airspeed on the outer scale opposite the figure representing the calibrated airspeed on the inner scale.

For example: Pressure altitude is 16,000 feet. Temperature is -10°C. A calibrated airspeed of 200 knots is equivalent to a true airspeed of 260 knots.

Density Altitude

The performance data published in the Aircraft Flight Manual is related to standard atmosphere (29.92 inches of mercury at 15°C at sea level). Any increase in temperature or altitude means a decrease in the aircraft's optimum performance, since air density decreases with both altitude and temperature. Knowing the density altitude at the airport from which you are operating is therefore of special importance in determining the required take off and landing distances. If both elevation and temperature are high, the available runway may be insufficient for the performance capabilities of your airplane.

Density altitude is pressure altitude corrected for temperature.

To find density altitude, set the pressure altitude opposite the air temperature in the computer window marked "For Airspeed Computations". For example: pressure altitude is 6000 feet. Temperature is 20°C. The density altitude, read in the little window marked "Density Alt", is 8000 feet.

Navigation Problems
Fuel Hours

Many navigation problems are based on fuel hours combined with the wind drift problems which we have already studied. The fundamental factors required for the solution of all fuel hour problems are:

1. *The fuel consumption.*
2. *The quantity of fuel available in the tanks.*
3. *The groundspeed of the airplane.*

The **time** the airplane can fly on its available fuel (less reserve) can be found from 1 and 2.

The **distance** the airplane can fly in that time can be found from 3.

Problem 1.
A pilot starts to fly from Winnipeg, Manitoba to Regina, Saskatchewan. The track is 277° true and the distance is 288 nautical miles. The plane has a cruising speed of 120 knots. Fuel capacity is 45 gallons. Fuel consumption is 11 gallons per hour. Unable to get a forecast wind, the pilot assumes no wind. At the end of 28 minutes, the aircraft is over Portage La Prairie. With this wind, the pilot is anxious to know if there is sufficient fuel to reach Regina without drawing on the 45 min. reserve.

Solution: Fuel capacity is 45 gallons. Fuel consumption is 11 gal/hr. Using the circular slide rule, the fuel hours on board are calculated to be 246 minutes, or 4 hours 6 minutes. Deduct 45 minutes for reserve, and the safe fuel hours available are 3 hours 21 minutes.

Portage La Prairie is 42 n. miles from Winnipeg. The airplane has covered this distance in 28 minutes. Its groundspeed therefore is 90 knots.

At a groundspeed of 90 knots, the flight from Winnipeg to Regina (288 nautical miles) will take 3 hours 11 minutes which is sufficient to reach Regina without drawing on the airplane's 45 min. reserve.

Problem 2.
A pilot wishes to fly from Goshen, Indiana to Fargo, N.D., via Joliet, Illinois, along the airways. He wishes to know the farthest point he can reach along the route before refuelling. A schematic diagram of the route is shown in Fig. 52.

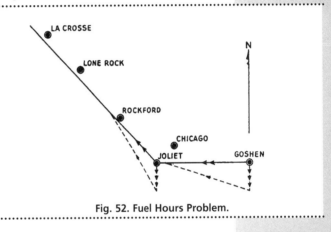

Fig. 52. Fuel Hours Problem.

The airplane has a cruising speed (true) of 100 knots. Its fuel capacity is 25 gals. and its consumption 5-1/2 gals/hr. The wind is from the north at 30 kts.

Solution: The airplane's endurance, found by using the circular slide rule, is 272 minutes. Deduct 45 minutes for reserve and the safe endurance is 227 minutes.

The track from Goshen to Joliet is 270°T and the distance is 107 nautical miles. By plotting or by wind drift computer, the groundspeed on the leg to Joliet is found to be 95 knots. At this groundspeed, the elapsed time to Joliet will be 73 minutes. Safe fuel hours remaining are 154 minutes (227 min. less 73 min.).

The track from Joliet towards Fargo is approximately 316°T. Using the same wind, by plotting or by wind drift computer, the groundspeed on this leg will be found to be 77 knots. At this groundspeed, in 154 minutes the plane will travel 202 n.m. The airport safely nearest this distance along the track is La Crosse (197 n.m.).

Radius of Action (R/A)

The radius of action (R/A) is the maximum ground distance an aircraft can fly outwards from a datum point before returning to the same or another datum point under given conditions of wind, true airspeed, fuel consumption and a given quantity of fuel.

The fundamental questions in determining the time to turn in order to reach its base at a given time are: when and where shall the turn be made and what will be the course back.

The radius of action problem presented here deals with an aircraft

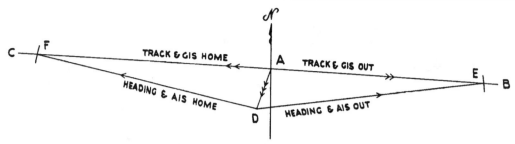

Fig. 53. Radius of Action.

that will return to its point of departure. Radius of action problems may also involve a return of the aircraft to a second base.

The problem is to find the R/A along a track, 093° true. Wind from 020°T. at 20 kts. Cruising speed 110 kts. Fuel endurance 5 hours. (i.e., 4 hrs. 15 min. plus reserve).

Solution by Plotting

1. From A (Fig. 53) draw A-B 93° (to represent the track out) and extend it back to some point C (to represent the track home).

2. From A, draw A-D at 020° to the meridian (the wind line is always drawn downwind), 20 knots to scale.

3. With centre D, and radius 110 knots (the true airspeed) describe arcs cutting A-B and A-C at E and F. Join D-E and D-F.

   A-E = Groundspeed Out = 102 kts.
   A-F = Groundspeed Home = 114 kts.

The safe fuel hours = 5 hours, less 45 min. reserve, which is 4 hrs. 15 min., or 4.25 hrs.

The Radius of Action is found by the following formula:
$F \times [(O \times H) \div (O + H)] = R/A$

F = Safe fuel hours.
O = Groundspeed out.
H = Groundspeed home.

Therefore, using the formula above:
R/A = 4.25 × [(102 × 114) ÷ (102 + 114)] =
4.25 × (11628 ÷ 216) = 229 n. miles

The heading to steer out is 083°T (D-E)
The heading to steer home is 283°T (D-F)

The time to turn is found by the formula: R/A ÷ O.

Therefore, the time to turn is 229 ÷ 102, or 2.24 hrs. or 2 hr. 14 min.

Solution by Computer
The solution of this type of problem, requiring the use of both sides of the computer, is broken down into two steps:

Wind-vector calculation, which computes the true heading and groundspeed outbound and inbound; and

Radius-of-action calculation, which computes time to turn and the radius-of-action distance.

1. Set up a wind triangle as in wind drift problem 1, using a wind velocity of 020° at 20 knots, true track of 093° and a true air speed of 110 knots, to determine heading and groundspeed.

2. The drift is 10° left.

3. Read true heading out (083°) at the 10° mark on the WCA Scale Left.

4. Read the groundspeed out (102 knots) at the grommet.

5. To determine the true heading for the return flight, rotate the compass rose until the reciprocal of the true track outbound (273°) is opposite the true index. To find the reciprocal of a true track, add or subtract 180 degrees. Therefore, the reciprocal of the true track 093° is 273° (093 + 180). Note: on a reciprocal heading, the drift will be opposite in direction.

6. Reposition the slide to place the arrow head of the wind vector line at the intersection of the 110 knot speed line (the true air speed) and the 10° right drift line.

7. Read the groundspeed (114 knots) at the grommet.

8. Read the true heading home (283°) at the 10° mark on the WCA Scale Right.

This completes the solution of groundspeed and true heading, outbound and inbound. The remainder of the problem is solved on the circular slide rule part of the computer.

9. Opposite the safe endurance time of the airplane (4 hrs. 15 mins.) on the inner scale (A in Fig. 54), set the sum of the groundspeed out and the groundspeed home (which is 216 knots) on the outer scale.

10. Opposite the groundspeed home (114 knots) on the outer scale, read the time until time-to-turn in minutes on the inner scale (134 mins. or 2 hrs. 14 mins.). See B on Fig .54.

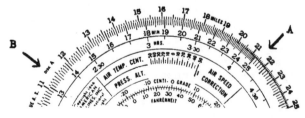

Fig. 54.

Therefore on the outbound flight, you can fly for 2 hrs. and 14 mins. before you must turn back.

11. To determine the distance flown in this time, place the groundspeed out (102 knots) on the outer scale opposite the pointer on the inner scale. Opposite the time (134 minutes) allowed for the outbound flight on the inner scale, read the distance on the outer scale, 229 nautical miles.

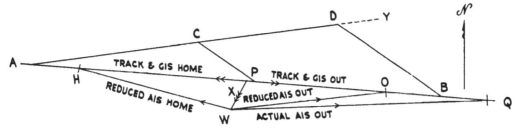

Fig. 55. Critical Point.

229 nautical miles is the radius of action.

Point of No Return
On long flights over oceans, jungles, or featureless country such as the Arctic, there is a point somewhere along the track beyond which, if trouble occurs, there is insufficient fuel remaining to return to base and it is mandatory to go on. This is called the point of no return. If the trouble is due to weather or some other cause which doesn't impair the performance of the airplane, this point is the time to turn which was determined in the radius of action problem above.

Critical Point
An airplane forced to turn back might be in difficulties (such as the loss of an engine in a multi-engine airplane) and be forced to fly at less than its normal cruising speed. In an emergency such as this, there is little time for calculation. The reduced airspeed of the airplane in distress should therefore be anticipated in advance. A good figure to assume is 60% of normal cruise speed. The **critical point** is a position somewhere along the route between two air bases from which it is equally **quick** to fly to either base. Based on a reduced airspeed, a predetermined critical point will enable a pilot in distress to instantly make up his mind whether it is advisable to turn back or to continue on.

An airplane cruising at a true airspeed of 250 kts. is flying from A to B along a track of 095° T, a distance of 1800 n. miles. The average wind over the route is estimated to be from 030°T at 30 kts. The reduced airspeed is assumed to be 60% of the normal cruise speed, or 150 kts.

Solution by Plotting
1. Draw A-B, the intended track, 095°, 1800 n. miles to any convenient scale (Fig. 55).

2. From some point X draw XW, 030°, 30 knots (the wind downwind). For convenience, a much larger scale can be used to plot this portion of the problem than that used to draw the line A-B.

3. With centre W and radius 150 n. miles (the reduced airspeed), describe arcs cutting X-B and X-A at O and H. Join W-O and W-H.
W-O is the heading to steer out, 085°T.
W-H is the heading to steer home, 285°T.
X-O is the reduced groundspeed out, 135 kts.
X-H is the reduced groundspeed home, 159 kts.

4. Draw any line A-Y inclined to A-B and mark off A-C equal to X-H, and C-D equal to X-O.

5. Join D to B. Through C draw a line parallel to D-B cutting A-B at P. P is the critical point. It is 973 n. miles from A.

6. With centre W and radius 250 kts. (the true airspeed), describe an arc cutting A-B (produced) at Q. Join W-Q.
W-Q is the actual groundspeed out, 236 kts.

The time to turn is the time to fly 973 n. miles at the actual groundspeed out, i.e.: 973 ÷ 236 = 4.12 hr. (4 hr. 7 min.)

Solution by Computer
1. Set up a wind triangle as in wind drift problem 1, using the true airspeed of 250 knots.

2. Read the groundspeed of 236 knots at the grommet.

3. Set up a wind triangle as in wind drift problem 1, using the reduced airspeed of 150 knots.

4. The drift is 10° left.

5. Read the true heading out (085°) at the 10° mark on the WCA scale left.

6. Read the reduced groundspeed out (135 knots) at the grommet.

7. To determine the inbound heading and groundspeed, rotate the compass rose to set the reciprocal track (095° + 180° = 275°) at the true index.

8. Reposition the slide to place the arrow head of the wind vector line at the intersection of the 150 knot speed line and the 10° right drift line.

9. Read the reduced groundspeed home (159 knots) at the grommet. Read the true heading home (285°) at the 10° mark on the WCA scale right.

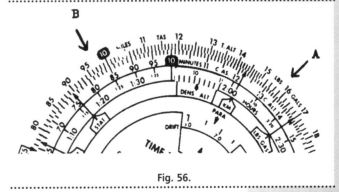

Fig. 56.

On the circular slide rule side of the computer:

10. Place the reduced groundspeed home on the outer scale opposite the reduced groundspeed out on the inner scale. See A on Fig. 56.

11. Search the slide rule for two numbers such that their sum will equal the total distance between the two points. One figure will be the distance from the base to the critical point, and the other

will be the distance from the critical point to the destination. As to which figure is the one to choose as the critical point, visualize the effect of the wind.

In this case, we find that 973 on the outer scale corresponds with 827 on the inner scale, their sum equalling 1800 nautical miles which is the total distance. The wind is a tail wind for the flight home and therefore the choice for the critical point is 973 nautical miles. See B on Fig. 56.

12. The time to the critical point is found by dividing the distance found in Step 11 (out to the critical point) by the normal groundspeed out (found in Step 2 to be 236 knots). On the computer, set the pointer opposite 236 and read the time opposite 973. The answer is 247 mins, or 4 hrs. 7 mins.

The critical point, therefore, is 973 nautical miles from base, and the time to turn is 4 hours and 7 minutes.

The critical point can also be determined by the following formula:
$P = (D \times H) \div (O + H)$

Therefore, $P = (1800 \times 159) \div (135 + 159) = 973$ n.m.

The time to turn is found by the formula $P \div G$

Therefore, the time to turn is $973 \div 236 = 4.12$ hrs. (4 hr. 7 min.).

In the above: P is the critical point.
D is the total distance to fly.
O is the reduced groundspeed out.
H is the reduced groundspeed home.
G is the actual groundspeed out.

Calculating the Climb or Let-Down

An airplane's airspeed is reduced during its climb and sometimes during a portion of its descent from cruising altitude. Its groundspeed is similarly affected. In planning a cross-country flight, the climb and descent vectors (that is, the portions of the flight required for climb and let-down) should each be calculated as separate time and distance problems, using the average wind and average airspeed temperature correction during the climb or let-down. The three separate segments of the flight should then be added to find the total time (time during climb + time at cruising altitude + time during let-down = total elapsed time). This may prove too time-consuming for the average pilot-navigator. A general rule that you may find helpful is: compute your flying time for the total distance using your true airspeed and the wind at cruising altitude. Then add 1 minute per 1000 ft. of cruising altitude above the elevation of the airport at point of take-off to allow for time lost in climb and letdown. For example: cruising altitude, 7000 ft. less elevation of airport, 1000 ft. = 6000 ft. Add 1 min/1000 ft. which is 6 minutes to your total time to allow for climb and letdown.

Problem
An airplane is flying towards its home airport where the elevation is 1267 ft. MSL (above mean sea level). It is flying at an altitude of 10,000 ft. MSL, at a groundspeed of 165 kts. The pilot wishes to descend at a rate of 400 ft. per min. and arrive over the airport at 1000 ft. above the ground, maintaining a constant groundspeed of 120 kts. How many minutes before his estimated time of arrival over the airport should he start his descent?

The airport has a field elevation of 1267 ft. MSL. Therefore 1000 ft. above ground will be 1267 + 1000 = 2267 ft. above sea level.

The pilot's descent will be 10,000 - 2267 = 7733 ft. At a rate of 400 ft./ min., the time required to let down will be 7733 ÷ 400 = 19.3 min.

Problem
After landing, the pilot referred to above decides to take off again and climb to a cruising altitude of 9,000 ft. MSL, at a rate of 300 ft./min. In the climb, he makes good a groundspeed of 165 kts. How far, in nautical miles, will he be from the airport when he reaches cruising altitude?

The airport has an elevation of 1267 ft. MSL. The climb to cruising altitude will therefore be 9,000 - 1267 = 7733 ft. At a rate of 300 ft./min., this will take him 7733 ÷ 300 = 25.7 min. At a groundspeed of 165 kts., he will travel 165 ÷ 60 x 25.7 = 70.6 n. miles.

The D/R computer can also be used for calculating letdown. Refer to the first problem above in which an airplane flying at 10,000 MSL is approaching an airport whose elevation is 1267 MSL. Descending at approximately 400 ft/min and maintaining a constant groundspeed of 120 knots, how many miles out should the pilot begin his descent in order to arrive over the airport at circuit height?

Find the number of feet to descend (10,000 - 2267 = 7733) in hundreds on the outer scale. Set under it the rate of descent in hundreds of feet on the hour scale (400 ft/min would be 4.00 hours) (See Fig. 57 "A"). On the minutes scale, find the groundspeed in tens of miles (120 knots would be 12) and read the number of miles directly on the miles scale (38.5 miles) (Fig. 57 "B").

The same setting can be used to determine what rate of descent will bring the airplane to circuit height above the airport, if the pilot begins his descent 38.5 miles out from the airport.

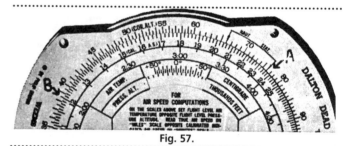

Fig. 57.

Calculating Climb Restriction Problems

At some airports, especially in congested areas, ATC may issue a departure clearance requiring a minimum climb rate per nautical mile. You will have to decide quickly if you are capable of this requirement. The following is a rapid method for solving this problem.

Problem
ATC has given a clearance. . . "This departure requires a minimum climb rate of 400 feet per mile until . . . . ." Your planned climb speed gives a 120 knot groundspeed. What climb rate in feet per minute must you meet to comply with the clearance?

Set your climb groundspeed of 120 knots on the miles scale adjacent to 60 minutes on the minute scale. Find the given climb requirement of 400 feet per nautical mile on the minutes scale. On the miles scale read 800. You will be required to maintain a climb rate of 800 feet per minute. If your airplane will do at least this considering its weight and the power available, then accept the clearance and go.

You must use nautical miles in computing this problem.

Finding the Wind Velocity in the Air

Of all the various methods devised for finding the wind velocity in the air, the only two which have any practical value from a pilot-navigator's point of view is (1) the track and groundspeed method and (2) the two drift method, since one does not involve any deviation of the airplane from its track and the other is so easily accomplished when flying over areas where the track and groundspeed method cannot be used.

Track and Groundspeed Method

An airplane on a navigation exercise passes checkpoint A at 10:10 and arrives at checkpoint B at 10:27. The cruising speed is 120 knots. The airplane is flying on a heading of 120°T.

A line joining A and B on the map represents the actual track made good (107°T).

The distance between A and B is 36 nautical miles. Therefore, the groundspeed is 127 knots.

Knowing the heading and airspeed (120°T — 120 knots) and the track and groundspeed (107°T — 127 knots), the wind velocity can now be found by plotting or computer as detailed in wind drift problem 3. The answer: wind from 218° at 29 knots.

Drift Method

A steady heading is flown at a constant airspeed and the drift noted. The heading is then altered, if possible by not less than 50°, and the drift again noted. If time permits, the heading should again be altered, either at least another 50° in the same direction, or back 100°, and a third drift noted.

It will be seen that very little time is lost by use of this method. If the heading alterations are made first to one side of the track and then to the other and, provided that approximately the same time intervals are flown on each leg of the flight to that point, the airplane will be very close to the original track when turned on to the original heading again.

This third drift angle is not essential, but it is desirable as a check. The wind velocity can then be found by combining this information on a mechanical computer or by plotting on the map or chart.

The Air Position

No pilot can ever become hopelessly lost if he knows his air position. The **air position** of an airplane at any time is its imaginary position, assuming that there has been no wind since it left the ground. It will be obvious then that, if the air position is known, the worst a pilot can be in error as regards **ground position** is a distance and direction equal to the speed and direction of the wind for the length of time he has been flying.

The air position is recorded by navigators on what is known as an **air plot.** This is the laying down on a map or chart of the **heading and airspeed** of the airplane during the entire time it is in the air.

While it is not possible for the pilot-navigator to keep an air plot and fly the airplane at the same time, it is of the **utmost importance** to keep a mental picture of air position. By jotting down in his log the exact time that each change in compass heading is made (should his whereabouts become uncertain in poor visibility) he can run back over his log and construct a rough diagram of his air position in his mind. Thus, when some landmark does appear in view below, he has a good general idea of the area on his map to search in order to identify it.

Note: The keeping of a log does not require recording endless volumes of figures on a comprehensive form. The few simple notations required by the pilot-navigator, such as time of departure, heading flown, times over points along the route, groundspeeds calculated, etc., can be jotted down on the border of the map if a note book doesn't happen to be handy.

Position Lines

Suppose you are standing on a street in Toronto down which you can see the Legislature rising above the end of the avenue of trees. You consult your street guide and identify the street as University Avenue. You know you are somewhere on University Avenue but you have no means of knowing exactly where. However, you note that you are standing at the crossing of a street named Dundas Street. Now you locate the intersection of University Avenue and Dundas Street on the map and you have fixed your exact position.

Position lines are obtained by observing the bearing of two or more objects and are plotted on a map or chart to fix the position of the observer in the same way that the intersection of University Avenue and Dundas Street was used to locate a position above.

Problem

A Twin Otter is patrolling out to sea off the Atlantic Coast. The port of St. Hilda is sighted and is found to bear 018° true. If the bearing of St. Hilda is 018° from the airplane, then the direction of the airplane is the reciprocal (opposite) of 018° or 198° from St. Hilda. A line is drawn on the chart from St. Hilda 198° (A-B in Fig. 58). This is a **position line** somewhere on which line it is known the airplane must be at a particular time. A position line is identified by arrowheads at either end and the time at which it was obtained is clearly marked, in this case 10.15 hrs.

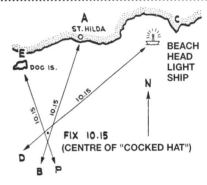

Fig. 58. Fix By "Cocked Hat", Three Position Lines.

Suppose that, at the same time, the Beach Head light ship was found to bear 052°. A second position line (C-D) may be drawn 232° from the light ship on the chart.

Where the two position lines intersect gives the position of the airplane at 10.15 hrs. This is called a **fix**.

It will be seen that in Fig. 58 a third position line (E-F) has been plotted from a bearing on Dog Island of 338° (reciprocal 158°). The purpose of including this third position line in the figure is to show that where three or more position lines are used, they may not always intersect at a common point. Note that E-F does not pass through the intersection of A-B and C-D, but forms a small triangle, or cocked hat. This is due to small errors in taking the bearings or in plotting. If the cocked hat is small, the fix is taken to be the dot (*) shown in the centre of the triangle. If the cocked hat is excessively large, no reliance can be placed on the fix.

When plotting a fix by two or more position lines, the angle between the position lines should not be less than 30° and preferably about 60°.

Upon obtaining a fix and marking it on a map or chart, it is possible to calculate the latitude and longitude of the airplane at the time the bearing was taken. This information is important in relaying position reports over water and uninhabited areas to air traffic control authorities.

There are many methods of obtaining position lines — by transit, by direction finding radio, by observations of celestial bodies, etc. It is of the **utmost importance** that the simple methods detailed above, explaining their use in fixing the position of an airplane,

Fig. 59. Electronic Flight Calculator.

should be thoroughly understood by the elementary student in order to enable him later on to tackle the more advanced problems he will encounter in radio and celestial navigation.

Electronic Flight Calculator

A wide selection of electronic flight calculators are available on the market. They are very versatile and very popular and may one day make wind drift computers and circular slide rules obsolete.

These calculators, similar to the one illustrated in Fig. 59, are pre-programmed to solve navigational problems, such as true heading, groundspeed, time enroute, distance, wind drift, fuel requirements, true airspeed, density altitude, rate of climb and descent, conversions (n. miles to s. miles, C° to F°, etc.), weight and balance, etc., in a matter of seconds.

Flight calculators made by different manufacturers will each operate slightly differently. Usually it is necessary first to select a function key that corresponds with the problem to be solved and then punch in the known information to solve the problem.

A careful reading of the instruction book that comes with the flight calculator you have purchased and some practice handling it will enable you to solve most of the navigation problems with which you will be confronted.

Transport Canada officials have approved several of these flight calculators for use in exams. They are listed in the **Personal Licensing Handbook** or can be determined by calling the testing centre prior to taking the examination. The manual on how to use the flight calculator **may not** be taken into the examination.

Navigation questions may be solved, at the examination, by plotting and formulae, by use of a flight computer and circular slide rule or by use of an electronic flight calculator.

From the Ground Up

Radio

Radio is the modern magic genie that creates invisible traffic arteries in the skies, whose voice reaches out to the airman in the overcast from the unseen world below — his guide to the weather that lies along his route, to the traffic pattern plan he must observe, to the vast amount of timely information he must receive to make flying as it is practised today a safe and practical undertaking.

Radio has been perfected to the point where it has come to be regarded in the present-day world of aviation as **indispensable.** While the vast airway radio networks of Canada and the United States offer a rapid and reliable means of communication and air navigation, it should never be taken for granted that radio equipment is **infallible.** It can, and does, on occasion fail. Complete dependence on radio has provided the prelude to all too many an airplane disaster. A pilot should by all means make full use of his radio equipment. He should never, however, forget the simple traffic rules and signals that apply when his radio fails to function, nor cease to practise elementary navigation as a precaution against the time when he may be caught with his radio dead.

Wave Length and Frequency

When a stone is dropped in water (Fig. 1), waves are set up. While the height, or strength, of the waves grows weaker as they travel away from the point where the stone hit the water, the length of the waves (W.L.) never varies.

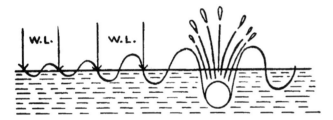

Fig. 1. Wave Length.

A radio transmitter sets up waves in the air in the same way that the stone sets up waves in water. The length of these waves remains constant, but the strength, or amplitude, decreases with distance from the transmitting station.

The actual linear measurement of the wave is known as the **wave length** and is referred to in meters. The period in which the wave vibrates (that is, rises and falls) between its crest and trough, is called a **cycle.** The number of cycles per second of time is called **frequency.**

Very low to high frequencies are expressed in **kilohertz (KHz),** that is, "thousands of cycles". For example, 3023.5 KHz stands for 3,023,500 hertz.

Very high frequencies (VHF) are expressed in **megahertz (MHz),** that is, "millions of cycles". For example 100 MHz stands for 100,000,000 hertz.

The relationship between wave length and frequency is as follows: the wave length in meters is equal to 300,000 divided by the frequency in kilohertz.

Conversely, the frequency in kilohertz is equal to 300,000 divided by the wave length in meters.

Note: The speed of radio energy or light is 300,000,000 meters a second. This figure is divided by 1000 to equal 300,000, the figure used in the formulae above.

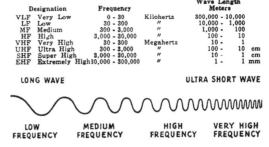

Fig. 2. Wave Length and Frequency.

Radio Bands

The airway radio communication facilities and radio navigation aids operate in the L/F (low frequency), M/F (medium frequency), H/F (high frequency), VHF (very high frequency) and UHF (ultra high frequency) bands.

Low and Medium Frequency

The non-directional beacons and marker beacons transmit navigation signals, and in some cases also voice transmissions, on the LF/MF bands of 200 to 415 KHz and 510 to 535 KHz.

Commercial broadcast stations transmit in the medium wave band on frequencies between 550 KHz and 1750 KHz which is therefore known as the "Broadcast Band". Broadcast stations, as well as providing news and entertainment, can be used for directional bearings or "homing" with automatic direction finding equipment.

High Frequency

Frequencies between 2,500 KHz and 30,000 KHz are known as high frequencies.

HF frequencies are allocated in 100 Hz steps so that more than 250,000 separate frequencies are available. Quite a number of these frequencies have been given to aviation use.

Because the range of HF signals is much greater than, for example, VHF signals, HF is an excellent air/ground communication facility in the northern remote areas of Canada and on long overwater flights.

HF radio is especially valuable as a communications facility for long range use because of a special characteristic of HF radio waves. They are reflected back to earth by the ionosphere. (See below **Characteristics of Radio Signals.**) HF radio is therefore the only way to maintain constant contact at ranges of 2500 miles or more and on transocean flights. Pilots use it regularly to relay the position reports that they are required to give at every 5 degrees of longitude change.

HF signals are, however, unpredictable. Since the height of the ionosphere varies from day to night and is affected by sunspots, auroras, etc., the angle at which the signals are reflected back to earth can be erratic.

HF stations in the upper range of the HF band have a greater reception distance during daylight hours, whereas stations in the lower range of the band have a greater reception distance during the night. Therefore, pilots should remember the rhyme

> Sun Up, Frequency Up
> Sun Down, Frequency Down

and select stations accordingly.

The frequency 5680 KHz is the designated HF air/ground communication channel in use in the remote areas of Canada. It is assigned to flight service stations in these northern regions to provide communication facilities for airplanes operating in the area. HF communication must be conducted on single sideband.

Single Sideband HF

Single sideband (S.S.B.) HF has replaced the old double sideband in all HF aeronautical communications in Canada. Because it permits transmissions to be carried over considerable distances (several thousands of miles in some cases), it is more efficient for use in areas where regular navigation aids and communication stations are non-existent.

S.S.B. is a method of compressing speech, or other intelligence, into a narrower band width. A fully amplitude modulated (A.M.) signal has two-thirds of its power in the carrier and only one-third in each of the sidebands. The sidebands carry the intelligence and the carrier serves only to demodulate the signal at the receiver. By eliminating the carrier and emitting only one sideband, available power is used to greater advantage. To recover the intelligence, the carrier must be re-inserted at the receiver. By this method, a gain of approximately 9 db over A.M. is obtained, equivalent to increasing the transmitter power 8 times.

In addition to the increase in power and range, S.S.B. communication conserves spectrum space. An S.S.B. voice signal requires less than 3 KHz of spectrum space, in contrast to the 7 or 8 KHz taken up by an A.M. signal.

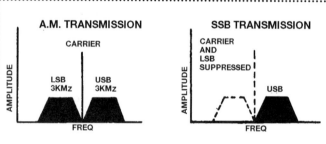

Fig. 3. Single Sideband.

Either the upper sideband (USB) or the lower sideband (LSB) may be employed. In more complex systems, both sidebands, with suppressed carrier, are used to convey two separate channels of intelligence.

The benefits of S.S.B. are greatest and most easily observed under poor propagating conditions. As a given transmission path deteriorates due to a combination of noise, severe selective fading and narrow band interference, the superiority of S.S.B. over A.M. is evident.

HF/SSB frequencies in the range of 2800 to 22,000 KHz (2.8 to 22 MHz) are allocated to aeronautical communication uses. An S.S.B. high frequency receiver is needed to receive the single sideband signal.

Very High Frequency

The VHF band lying between 30 MHz and 300 MHz is the most important from the pilot's point of view. Certain ranges of frequencies within this band have been allocated exclusively for aviation use for communications and for navigation aids.

The frequency range from 108.00 MHz through 117.95 MHz is used for navigation aids, such as the VHF omninavigation stations (VOR), and the instrument landing systems (ILS). Voice reception is also provided on these frequencies.

The frequency range from 118.00 MHz through 136.000 MHz is allocated to civilian aviation voice communications. Within the last number of years, 40 new channels between 136.00 and 136.975 MHz have been allocated to civilian aviation. These latter channels are mostly used by air carriers for en route communications.

VHF Frequency Utilization Plan

Time was when a pilot could travel anywhere in North America with a few simple transmitting and receiving frequencies, but today's multiplying traffic congestion has created vast and complex radio telephony problems. To call a radio facility today, a pilot must be able to select the right radio frequency — much like looking up the right number in a telephone directory. This information is available on aeronautical charts and in the **Canada Flight Supplement.**

For ATC purposes, channel allocation is based on 25 KHz spacing. Within the band of VHF frequencies used by aviation, there are 720 individual frequencies that can be allocated by utilizing 25 KHz spacing. (For example: 119.000, 119.025, 119.050, 119.075, 119.100, 119.125, etc.)

A 720 channel transceiver with 25 KHz spacing capability is required in all aircraft operating in the high level airspace, in all aircraft operating under instrument flight rules (IFR) and in all aircraft operating in the terminal airspace of the large, busy airports where frequency congestion is a problem.

For the present time, aircraft can still operate VFR in many areas of the country with radio equipment capable of only 50 KHz spacing but the time is fast approaching when the demand for discreet frequencies will be such that equipment with 720 channel capability will be universally required. Operators of equipment with only 50 KHz spacing may find their access to certain airspace restricted.

Equipment capable of only 50 KHz spacing can receive only 360 frequencies (for example: 119.05, 119.10, 119.15, 119.20, etc.) Owners of equipment with 360 channel capability need to be careful not to be confused by the way in which a frequency might be published. A 360 channel radio will receive 119.050 MHz and 119.100 MHz if the frequencies 119.05 MHz and 119.10 MHz are selected but it will not receive the intervening channels, 119.025 MHz and 119.075 MHz. Sometimes frequencies with 25 kilohertz spacing are published with only 2 digits to the right of the decimal point (the following 0 or 5 is omitted). A 360 channel receiver will receive 119.05 MHz and 119.10 MHz but will not receive 119.02 MHz or 119.07 MHz or 119.12 MHz.

The following are some of the most common VHF frequencies in use.

The universal VHF emergency frequency is 121.50 MHz. Aircraft equipped with dual communication systems should monitor 121.50 MHz continuously while in the air.

The frequency 122.20 MHz is used in Canada by flight service stations for both transmitting and receiving. In the United States, the frequency 122.20 MHz or 122.30 MHz is used by FAA flight service stations to reply to private aircraft at controlled airports and 123.60 MHz at non controlled airports.

Pilots operating in uncontrolled airspace are advised to remain continuously tuned to 126.70 MHz, to use this frequency when transmitting position reports and for general communications with flight service stations.

The frequency 122.90 MHz has been allocated for use by aircraft engaged in various private aeronautical activities (private multiple) which might include such things as parachute jumping, aerial crop spraying, formation flying, etc. and can be used both in ground to air and air to air transmissions.

The frequency 123.400 MHz is allocated for the use of soaring activites which include balloons, gliders, sail planes, ultralights and hang gliders for air to air and air to ground communications. It may be designated as the ATF at privately operated aerodromes used primarily for these activities.

UNICOM. Private advisory stations transmit and receive on a range of frequencies between 122.700 and 122.350 MHz. The most common frequency is 122.800 MHz. The word UNICOM is an acronym standing for "universal communication". A UNICOM facility is an air-to-ground communication facility operated by a private agency to provide **private advisory station (PAS)** service at uncontrolled airports. The facility may provide limited airport advisory service (when it is the designated mandatory frequency) and other information relating to fuel and service available. An **authorized approach UNICOM (AAU)** is also a privately owned air-to-ground communications facility. Its operators are trained to provide operational information to pilots conducting published instrument approaches. Approach limits are based on a local altimeter setting. At controlled airports, the private advisory station is used by aeronautical operators for company business. Aeronautical advisory stations in the U.S. transmit and receive on 122.80 MHz at airports where there is no control tower and on 123.00 MHz at airports that are served by a control tower.

Frequency Allocation. The following listing of the allocation of VHF channels to the various aeronautical facilities under the Frequency Utilization Plan, is relatively complete. However, changes in the plan are made from time to time and are published in Transport Canada Information Circulars.

In the case of air navigation aids (such as omniranges, ILS localizers, etc.), the frequencies are those on which the navigation signals are transmitted and are also the frequencies on which stations with simultaneous voice facilities will reply to a call.

- 108.05 through 117.95 MHz — Navigation aids.
- 108.1 through 111.9 MHz — ILS localizers with simultaneous voice channel, operating on odd tenth decimal frequencies.
- 110.1 MHz— ILS ramp check.
- 112.1 through 117.95 MH — Omniranges (VOR).
- 114.8 and 115.7 MHz — VOR ramp check.
- 118.00 through 119.65 and 119.75 through 121.40 MHz — Air traffic control. Two-way.
- 119.70 MHz — Aircraft to air traffic control. Below 12,000 ft.
- 121.50 MHz — Emergency. Air to ground.
- 121.6 MHz — Search and rescue.
- 121.6 through 121.95 MHz — All aircraft and air traffic control. Two-way. Airport utility (ground control).
- 122.00, 122.20, 122.30 and 122.50 MHz — All airplanes to flight service stations.
- 122.10 MHz — All airplanes to community aerodrome radio stations.
- 122.70, 122.80, 122.85 and 123.20 MHz — Private advisory stations. Two-way. Air to ground. Uncontrolled airports. UNICOM.
- 122.75 MHz — Private advisory. Private aerodromes. Air to air in Southern Domestic Airspace. General aviation.
- 122.90 MHz — Private multiple. The communication services normally encompassed by these stations would include forest fire fighting, aerial spraying, aerial advertising, parachute jumping, etc.
- 122.95 and 123.00 MHz — Private advisory stations. Controlled airports.
- 123.10 MHz — World wide search and rescue.
- 123.15 MHz — Flight service stations.
- 123.20 MHz — Private advisory. Uncontrolled airports. Air to ground. This frequency may also be used for communication with ground stations for the purpose of passing weather and other information to aircraft. Air to air to broadcast position and intention at an aerodrome for which there is no designated MF or ATF and no ground station.
- 123.30 and 123.50 MHz — Private advisory. Flight training. Flight test.
- 123.40 MHz — Soaring activity.
- 123.60 through 126.65 MHz — Air traffic control.
- 126.2 MHz — Military air traffic control. Two-way.
- 126.70 MHz — All civil airplanes to flight service stations.
- 126.75 and 126.80 MHz — Air traffic control communications. Below 24,000 ft.
- 128.825 through 132.025 MHz — Airplanes to company stations.
- 131.80 — Air to air in Northern Domestic Airspace.
- 132.050 through 134.95 MHz — Air traffic control. Above 24,000 ft.
- 135.00 through 136.00 MHz — Airplanes to company stations.
- 135.85 through 135.95 MHz — Flight inspection.
- 135.9 MHz — Military advisory.

Note: Because of the confusing number of VHF channels in use and the frequent changes that are continually taking place, it is advisable, when calling any airway communication station, control tower or other facility, to state the frequency on which you are calling. For example: **Gore Bay Radio. This is Golf Foxtrot India Bravo on 122.30.**

Ultra High Frequency
The UHF band includes the frequencies lying between 300 MHz and 3000 MHz. These frequencies are mostly allocated to special government use, except that distance measuring equipment (DME) and the glide slope portion of ILS operate on frequencies in the UHF Band.

Single and Double Channel Communication
The radio equipment installed in airplanes is capable of both receiving and transmitting. In some cases, a transmission from the airplane is carried on one frequency and reception from the ground station is carried on another. In some cases, the same frequency is used for both transmitting and receiving.

SINGLE CHANNEL SIMPLEX means communication in one direction only at a time — transmitting and receiving on the same radio frequency.

DOUBLE CHANNEL SIMPLEX means transmitting on one channel frequency and receiving on another frequency — but not simultaneously.

DOUBLE CHANNEL DUPLEX means transmitting on one channel frequency and receiving on another frequency simultaneously.

Characteristics of Radio Signals

Low, Medium and High Frequency
Waves emitted from low, medium and high frequency transmitting stations are of two types (Fig. 4).

Ground Waves
Ground waves follow the surface of the earth. The ground waves, by nature, travel in a straight line. They do, however, because of the phenomenon called **diffraction**, bend around obstacles in their path. As the earth's surface is full of large and small obstacles, the

repeated bending of the ground waves causes them to follow the earth's curvature. This bending is further enhanced by **surface attenuation**. As part of the ground wave comes in contact with the surface, it loses some of its energy to the surface and slows down. This slowing down causes a downward tilt in the wave that also helps the wave to follow the earth's curvature. The waves continue until they become undetectable due to surface attenuation.

The degree to which attenuation affects the ground waves depends on the nature of the surface. The conductivity of the earth's surface varies greatly. A radio wave will, for example, travel a greater distance over water than land. Conductivity is very poor over sand and ice but fairly high over rich soil.

Attenuation also depends on the frequency in use. The higher the frequency, the greater the attenuation. Very low frequency waves (VLF) are least affected by attenuation and bending is due to diffraction. Ranges of several thousand miles are common in VLF transmission. Attenuation has an increasing effect on the ground waves of low and medium frequencies, slowing them down and decreasing the range of their signals. High frequency is severely affected by attenuation so that the HF ground waves have a range of only about 100 nautical miles before they become undetectable. VHF and UHF waves (see below) are not affected by attenuation. Their radio waves do not bend.

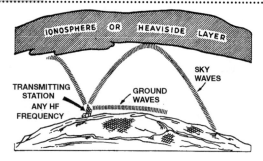

Fig. 4. Ground and Sky Waves.

Sky Waves

Sky waves travel up into the atmosphere and are reflected back to earth from the ionosphere. The behaviour of sky waves means that radio signals from the frequencies in the lower end of the radio spectrum (especially HF) can be heard at distances much beyond the range where the ground waves have become undetectable.

Between the point where the ground waves end and the reflected waves strike the earth, there is a skip zone, where very erratic signals or no signals at all are heard.

The skip characteristic of these signals accounts for the fact that sometimes you may hear a station, then lose it and later hear it again as you fly farther away from the station.

The behaviour of the sky waves is responsible for one of the chief advantages of HF radio — that of long range communication. It is possible on HF to communicate with a station 1000 kilometers away but at the same time be unable to reach a station 100 kilometers away.

Transmission capabilities of low, medium and high frequency radio waves may vary by night and day. At night, the sky waves travel at a flatter angle, causing a skip zone of greater extent, but a signal distance far greater than during the day.

Transmissions are also affected by such things as sunspot activity or electromagnetic disturbances which can upset the reflecting ability of the ionosphere. When this occurs, the radio waves are not reflected back to earth and a fade out is experienced.

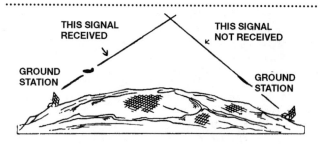

Fig. 5. Line of Sight Transmission.

Static is also a major problem in this frequency range.

Very High Frequency

Very high frequency radio waves have different properties entirely from the ground and sky waves described above. They do not bounce between the reflecting ionosphere and earth, but continue straight out into space. This means that they can only be received by an airplane on a **line of sight** position in relation to the station. VHF waves do not follow the curvature of the earth, nor bend around obstructions. For this reason, the higher the altitude at which the airplane is flying, the greater the distance at which it will be able to receive VHF signals (see Fig. 5 and Fig. 6).

Although line of sight transmission reduces the distance at which signals can be read at low altitude, this shortcoming is offset by the fact that stations below the horizon several hundred kilometers apart cannot interfere with one another.

VHF offers virtual freedom from atmospheric and precipitation static. Conversation is much like talking over an ordinary telephone. In addition to quiet, reliable communication, VHF equipment is smaller and lighter than corresponding L/MF equipment and, consequently, less power is required for normal communication.

Table of VHF Reception Distance

Ht. Above Station	Reception Dist.	Ht. Above Station	Reception Dist.
1000 ft.	39 n.m.	10,000 ft.	122 n.m.
3000 ft.	69 n.m.	15,000 ft.	152 n.m.
5000 ft.	87 n.m.	20,000 ft.	174 n.m.

Fig.6.

Precipitation Static

An aircraft flying in cloud or precipitation will accumulate electrical charges as a result of frictional contact with the liquid and solid particles in the atmosphere. These charges will accumulate until they are great enough to cause a discharge into the surrounding air. This discharge manifests itself as static on the aircraft radio receiver.

Precipitation static arises through contact with solid particles, such as dust and sand, ice crystals, rain, snow crystals, wet snow and freezing rain. It also arises through contact with an external electric field. Almost all clouds have electrical fields associated with them. The greater the turbulence in the cloud, the stronger the external field is likely to be. Strong electrical fields are associated with the precipitation areas below clouds and, of course, with thunderstorms. An airplane flying in the vicinity of an external electrical field will accumulate a strong charge that will cause considerable precipitation static in radio reception.

Static due to external fields is most common in summer when thunderstorms are most frequent. Static in the winter time is most commonly attributed to ice crystals which occur at ordinary flight levels at this time of year.

Data Link Communications

Data link extends the information superhighway to aviation, expanding the exchange of information through digital communications. The availability of information in the cockpit and the reliability and efficiency of communications between pilots and airspace managers are greatly enhanced through this space age telecommunication network. Data link communication allows messages, requests and air traffic clearances to be displayed both to pilots and air traffic controllers in textual form, eliminating the reliance on copying voice transmissions.

The data link communication infrastructure includes Mode S secondary surveillance radar, communication satellites and VHF links. Each of these links can be used to exchange data between aircraft and ground systems.

Mode S is a secondary surveillance radar system that provides two-way data communications capability for air traffic, flight information and surveillance services. In Mode S, each airplane is addressed individually. It gives identification, altitude and data. The opportunity to send data back and forth allows the controllers through their computer network to keep track of the aircraft's position and route. (See also **Transponder** in Chapter **Radio Navigation**.)

Satellite communications (SATCOM) are used for communications in the oceanic environment and remote areas.

VHF data link is widely used for aircraft operations and for limited flight information and air traffic services, such as pre-departure clearance, ATIS information, taxi instructions, take-off and landing requests and clearances, altimeter information, runway assignment, etc. Graphical weather service provides to the cockpit weather graphics information derived from ground based weather radars. This information is displayed on a cockpit control and display unit. Text weather products are also provided and displayed.

Satellite Communications/ VHF Data Link

The rapid progress in satellite technology has had a major impact on aviation communication, navigation and surveillance capabilities. New technology has increased the performance of satellite transmitter and receiver systems. They have already been proven in widespread use by the shipping industry. Geosynchronous satellites are also available for navigation use.

Satellite communications have the advantage of virtual global coverage, unlimited range, high reliability and good propagation conditions. High quality communication is available regardless of location, weather or ionospheric effects. The satellite communication network is capable of providing a data link capability via VHF radio. Aircraft can patch into the satellite system by means of a special antenna installed on the aircraft. A small antenna supports a data service, a larger one a voice service. Suitably equipped aircraft are also able to contact ATS on a satellite voice frequency. A unique telephone number is assigned to this facility in each flight information region. It may be used only by aircraft using the satellite network and is to be used only for non-routine flight safety calls.

Precise navigation service is also provided by the **global navigation satellite system (GNSS).** A complete GNSS includes satellites as well as ground based systems that are used to augment and monitor the satellites. There are two navigation satellite systems: the Russian GLONASS and the American GPS. The American global positioning system (GPS) operates with 25 low orbitting satellites (with a possible 28 planned) that claims accuracies in the region of 100 meters in three dimensions (latitude, longitude and altitude).

A satellite surveillance system is intended to compensate for the lack of radar coverage over ocean and remote areas. It implies some form of regular polling of an airplane's own navigation equipment (INS, Omega, VLF, etc.) so that ATC can monitor its position and ensure safe separation from other aircraft.

Satellites are not totally new to aviation. The signals of emergency locator transmitters (ELTs) have for some time been monitored by satellite and the ability of the system to accurately locate downed aircraft has been proven.

Communication Equipment

To be fully capable of two-way communication with ground stations or other airborne stations, the radio equipment installed in an airplane must consist of a transmitter, a receiver, an antenna, a microphone and a speaker or headset. When the transmitter and receiver are combined in one control unit, it is known as a transceiver.

In transmitting, the spoken words of the pilot are picked up by the microphone and relayed to the transmitter where they are converted to radio signals. The radio signals are carried by a wire to the antenna where they are broadcast in all directions to be received by radio receivers in various radio ground and airborne stations.

In receiving, the antenna picks up a radio signal from a transmitting station, carries it by a wire to the aircraft receiver where it is converted to a voice signal and heard through the speaker or earphones.

Proper transmitting technique is to hold the microphone in the palm of the hand with the thumb resting lightly on the "press-to-talk" switch. Hold the microphone within one or two inches of the lips and press the microphone button. Side tone permits you to hear the sound of your own voice in the earphones and to monitor your transmission.

Earphones usually come with a boom type of microphone attached. The boom holds the microphone at exactly the right distance from the lips. The microphone button is usually attached to the control wheel.

When finished speaking, you must release the button in order to receive. (The button automatically switches the set back to the receiver position.)

The VHF Transceiver

Most modern VHF communication equipment is incorporated in a unit that also includes the VHF navigation equipment. Such a radio unit is known as a 1-1/2 system and is usually called a NAV/COM.

In the radio illustrated in Fig. 7, the communication equipment occupies the left side of the unit and is marked COM. It has single channel simplex capability and 25 KHz spacing. It has crystal controlled tuning and an electronic display of the frequency selected.

Frequencies in the range from 118.000 to 135.975 MHz are tuned by the two large knobs. As is the case in many VHF transceivers, the frequency selector does not display the third digit after the decimal.

Fig. 7. VHF NAV/COM Equipment.

The 0 or 5 is automatically provided. The transceiver shown in Fig. 7 is tuned to 118.00 MHz. The frequency 136.97 MHz is on standby. The communication portion of this transceiver will receive 720 channels.

Volume is controlled by the small knob on the bottom right. The unit also includes a squelch control that eliminates undesirable noise.

To use the squelch control, turn the knob fully clockwise, then turn it back until the noise just ceases to be heard. Do not retard the squelch below this level at any time, as otherwise you may lose a weak signal which you desire to hear. In order to hear stations that are far off, it may be necessary to turn the squelch full on and put up with the noise to have the advantage of long distance communication.

Optimum operation of a 720 channel transceiver (i.e. strong reception and transmission over the full range of the channels) requires installation of a blade type antenna, rather than the whip type antenna commonly used with VHF communication equipment with less channel capability.

Operation of the NAV portion of the VHF transceiver is described in the Chapter **Radio Navigation.**

The HF Transceiver

The HF transceiver illustrated in Fig. 8 is typical of HF equipment capable of full single sideband HF communication. It is capable of frequency selection in .1 KHz steps in the band from 2 MHz to 26.999 MHz (2000 KHz to 26,999 KHz). It also has the ability to preset and store up to 24 channels. A single frequency synthesizer eliminates the need to change crystals when new channels are required (an inconvenience in old model HF radios). Other operating frequencies may be selected, however, without disturbing the presets. All frequencies are displayed in large gas discharge numerics.

Fig. 8. M/HF Transceiver.

Selection of upper sideband, lower sideband or AM operating modes is available and can also be preset. Either single simplex or semi duplex channels can be easily selected.

The system requires only a 10 foot wire antenna rather than the long trailing antenna normally associated with the old type HF radio. Antenna tuning is accomplished automatically. A squelch control helps to eliminate undesirable noise.

Care of Airplane Radio

Proper operation and care of airplane avionics is an important factor in maintaining its serviceability. **Avionics** is a term used to describe the electronic navigation and communication equipment in an airplane. It also includes any instruments and flight control equipment associated with radio.

Heat is the enemy of avionics. On most modern airplanes, all the avionics equipment is stacked in a central position on the control panel. The heat generated by each individual radio in the stack combines with that generated by the others to produce temperatures that are beyond the tolerance of the equipment. A radio cooling kit that draws cool air from outside the airplane and pumps it around the radios should be installed. A radio may be expected to be warm but it should not be uncomfortably hot to the touch.

Good radio reception, free of interference and static, is dependent on proper bonding of all metal parts of the airplane and on the use of **static discharge wicks.** The metal surfaces of an airplane are an essential part of the aircraft electrical ground system and for this reason it is important that all metal parts be connected together. If they are not, electrical fields may be generated around isolated metal components that not only may cause radio interference but may be conducive to formation of a spark that could cause a fire. Static discharge wicks are usually attached to the trailing edge of the wing to discharge static electricity into the atmosphere. Here are a few good practices to follow:

1. Turn off all radios when starting the engines of the airplane. Large voltage transients are created during the start procedure and it is wise to protect radio equipment from these.

2. Do not turn a radio on and off repeatedly. Once it is turned on, keep it on for the entire time that it may be needed. Then turn it off.

3. Allow communication equipment to warm up for a minute or so before beginning a transmission. This practice is especially important for the older tube type of equipment.

4. When an airplane is flown at high altitudes and the equipment is mounted outside the heated areas, all radio units should be operated during the descent in order to reduce moisture condensation in the boxes. This practice is especially recommended with radar units in which high voltages are generated.

5. Incorrect voltage affects the operation of radios and can shorten component life. For this reason, it is wise to monitor airplane generator voltage on a periodic basis.

6. Be sure the airplane battery is maintained in good condition. This is especially important with the new solid state radios and airplane alternator systems.

7. When a problem does occur in the radio equipment, accurate description of the symptoms will assist the service man's diagnosis of the trouble.

8. Have the microphone tested at an avionics shop. Carbon microphones age very quickly because the carbon has a tendency to dry up. Newly developed transistorized or dynamic microphones seldom deteriorate with age.

9. Keep antennas clean. Dirt build up decreases the efficiency of an antenna dramatically.

From the Ground Up

Radio Communication Facilities

Automatic Terminal Information Service (ATIS)

Non-control information relating to ceiling, visibility, wind, runway in use, NOTAMS affecting the airport, etc. is continuously broadcast on the automatic terminal information service. The information is recorded by tower personnel and is broadcast on the voice facility of the VOT (VOR Omnitest Facility) or on a specially assigned VHF/UHF channel.

Each recording is identified by a phonetic alphabet code letter, beginning each day with "Alfa". As airport conditions, such as wind, altimeter setting, etc. change, a new tape is prepared. Each succeeding message is given a new identifying letter (Bravo, Charlie, etc.) A typical ATIS message might be:

ATIS: **Vancouver International Airport. Information Bravo. Weather at 1400 Zulu. Three Thousand Scattered, Measured Ceiling Five Thousand Overcast. VisibilityTen. Haze. Temperature One Five. Dew Point Eight. Wind Two Five Zero (magnetic) at Ten (knots), Altimeter Two Niner Niner Eight. Runway Two Six. Inform Vancouver ATC on Initial Contact that you have received Information Bravo.**

The ATIS broadcast may also include special information for air traffic using precision instrument approach facilities: the type of instrument approach in use, information on the landing runway and the take off runway for both IFR and VFR traffic, information on parallel or simultaneous runway operations and stopping distance, any NOTAM affecting the airport, the serviceability of NAVAIDS and field conditions.

You should listen to the ATIS broadcast before contacting the tower or the terminal controller (when landing) or the ground or clearance delivery controller (when taking of) and inform the controller **on the initial contact** that you have the ATIS information, repeating the code word which identifies the message. The controller then does not have to issue this information in his message to you.

During periods of rapidly changing conditions which would create difficulties in keeping the ATIS message current, the following message will be recorded and broadcast:

ATIS: **Because of rapidly changing weather and airport conditions, contact ATC for current information.**

The word CAVOK (kav-oh-kay) is used in ATIS messages to denote a weather condition in which visibility is 6 statute miles (10 kilometers) or more, there are no clouds below 5000 feet (1500 meters) and no precipitation, thunderstorms, shallow fog or drifting snow.

At major airports in Canada, an operational information display system (OIDS) has been installed and displays for the controllers both the **now wind** and the **mean wind.** A computer samples the wind once every second. The now wind is averaged over the last five seconds; the mean wind, over the last two minutes. Mean wind is reported in ATIS broadcasts. Now wind (when it is in excess of 15 knots) is given by the tower controller for take-off and landing clearance and at any time on pilot request.

The ATIS service is in operation at most of the major international airports. Information on the frequency on which the ATIS is broadcast is printed on aeronautical charts and also in the **Canada Flight Supplement.**

Control Tower

Many airports have the service of a control tower to ensure the efficient and safe movement of air traffic. The air traffic controllers in the tower are responsible for the control of all traffic, taking off and landing, and for the control of all VFR traffic within the control zone surrounding the airport.

The radio frequency on which the tower operates can be found on aeronautical charts in the aerodrome information and in the **Canada Flight Supplement.** As well as the primary frequency on which the tower both transmits and receives, there are other frequencies on which the tower transmits and/or receives. These frequencies are listed in the **Canada Flight Supplement** in the COMM section of the information about a particular airport.

Ground Control

Most controlled airports also have ground control. The ground controller is responsible for the movement of all traffic on the ground at the airport, except aircraft landing and taking off.

The ground controller operates on a special frequency allocated for that purpose. At most airports, that frequency is 121.9 MHz, although other frequencies in the range 121.7 through 121.9 MHz are used if there is a possibility of interference with another ground controller at a nearby airport.

The ground control frequency in use at any particular airport is published in the **Canada Flight Supplement.**

Ground controllers will provide the pilot with precise taxi instructions and information about services and facilities.

Apron Advisory Service

At very busy airports where there is a lot of ground traffic in the vicinity of the terminal, apron advisory service is sometimes provided by ATS. This service normally provides gate assignment, push back instructions and advisories on other aircraft and vehicles on the apron. Aircraft entering the apron will be advised by the ground controller to contect the apron controller for explicit instructions as to where to park or what gate to use. Aircraft leaving the apron should contact the ground controller on the appropriate frequency to obtain taxi clearance before exiting the apron and entering the maneuvering area. Apron advisory service operates on its own assigned frequency.

Clearance Delivery

Airplanes intending to depart a major airport on a VFR flight plan may be required to call clearance delivery (if this facility is in operation) before starting to taxi. The clearance delivery controller will ask for the aircraft registration, type, destination and requested altitude. Although the main duty of the clearance delivery controller is to relay IFR clearances, with respect to VFR traffic he performs the clerical task of preparing the flight progress strips that are given to the tower controller, thus allowing the ground controller, who formerly performed this job, to concentrate more fully on his job of directing the ground movement of aircraft.

The clearance delivery frequency in use at any particular airport is published in the **Canada Flight Supplement.**

Terminal Arrival and Departure Control

Arrival control is provided to expedite the flow of IFR flights inbound within a terminal control area.

Departure control provides the same service for outbound IFR flights.

The frequencies for arrival and departure control for particular airports are listed in the **Canada Flight Supplement** and in the **Canada Air Pilot.**

Most arrival and departure control units are equipped with radar surveillance scopes, or scanning screens, which give the controller the range and azimuth of every airplane in the vicinity. Controllers are therefore able to issue any necessary instructions for the maintenance of safe separation and the avoidance of collision hazards. These facilities are purely traffic control services and should not be confused with a radar assisted approach (ASR and PAR), which provides instrument glide path landing guidance. They do, however, give position information and will furnish a lost pilot with the distance and the magnetic heading to the airport. They will also furnish the wind speed, the runway in use and pertinent traffic information.

Although arrival and departure control is principally in operation to handle IFR traffic, at some busy airports arriving and departing VFR traffic is also handled by this facility. In such a case, the pilot is instructed by an air traffic control unit or in the ATIS information to contact arrival or departure control on the assigned frequency.

Arriving VFR airplanes may contact arrival control on their own initiative and request **radar assistance** if they wish to do so. It is important to make the call at a reasonable distance from the facility and state your position as accurately as possible to assist the controller in locating you on the radar scope. You may be asked to make some identifying turns or, if you are transponder equipped, to operate the transponder on a specific code. Remain in communication with arrival control until advised to contact the tower.

Radar identification and radar vectoring by ATC do not relieve you, as pilot, of the responsiblity of collision avoidance and terrain and obstacle clearance. As a VFR flight, you are also responsible for maintaining VFR.

Departing VFR traffic may also contact departure control for **radar assistance.** The request should be made on initial contact with ground control or clearance delivery and information as to the direction of the flight should be given. The tower will advise when to contact departure control. Departure control takes over when the aircraft clears the runway on take-off and vectors it on radar until it is established on a course along the appropriate navigation facility.

Vectors are defined in the Chapter **Air Navigation***. In radar advisory service, the term has acquired a new meaning and is used to designate headings given to pilots to steer (for the purpose of avoiding other traffic or for detouring around thunderstorms). A radar traffic controller giving steering instructions to an aircraft is said to be vectoring the aircraft.*

VHF Direction Finding Service (VDF)

At selected airports across Canada, VHF direction finding equipment is now in operation to provide directional assistance to VFR airplanes. It is a service intended to provide aid in times of difficulty. This facility enables a flight service station or control tower operator to advise a pilot of his bearing from the airport or DF site. DF information is electronically derived from radio signals transmitted from the aircraft. VDF normally operates on six pre-selected frequencies in the 115.00 to 144.00 MHz range. Two factors limit the provision of the service. Since VHF transmissions are restricted to line-of-sight, altitude and location are a factor. Secondly, power of the transmitted signal will affect reception. Information may be obtained from either a speech transmission or a transmission in which there is no speech, only the mike button pressed.

DF service is provided when requested by the pilot or at the suggestion of ATC. The pilot requesting DF service should provide the facility with the following information: position of aircraft, if known, present heading and altitude.

Aircraft: **Fredericton Tower — This is Foxtrot Romeo Tango Mike — Request DF Homing — Approximately Two Zero Miles South of Fredericton — Heading Zero Four Five — Seven Thousand.**

The pilot will then be given the headings required to home to the airport at which the DF equipment is installed.

Tower: **Foxtrot Romeo Tango Mike — This is Fredericton Tower — For Homing to Fredericton Airport — Steer Heading Zero One Five.**

A pilot planning to use the heading indicator (directional gyro) as heading reference during a VHF DF homing should reset the heading indicator to the magnetic compass before calling the DF operator and then should not reset it without advising the DF operator.

The VHF DF Service is not a radar service and does not provide traffic separation. The pilot is responsible for avoiding other traffic, maintaining terrain clearance and remaining in VFR weather conditions.

Flight Service Stations (FSS)

Flight service stations have been established at some aerodromes across Canada. They are staffed by FSS specialists trained to provide an efficient flight safety service to pilots.

The majority of FSS provide service 24 hours a day to the airports where they are located and to any number of remote communication outlets (RCOs) assigned to them.

Flight service station communications with aircraft are conducted on a standard set of frequencies, including the emergency channel 121.50 MHz, the general purpose channel 126.7 MHz and the advisory channel 122.2 MHz. Most stations have 122.1 MHz (receive only), and a number of the more northerly stations have 5680 KHz.

During their hours of operation, through continuous monitoring of these channels, the flight service stations provide **en route flight information service.** Pilots are able to obtain and pass information or to report emergencies should the need arise. In addition, the FSS relays position reports and ATC clearances in areas where aircraft are beyond the communication range of ATC facilities.

Flight service stations, located at airports without control towers and at controlled airports when the control tower is not in operation, provide an **airport advisory service (AAS)** to arriving and departing aircraft. A landing or take off advisory may include wind direction and velocity, the favoured runway, visibility, altimeter setting, pertinent NOTAMs, airport conditions, such as snow on the runway, braking action, obstructions or any known hazards. The airport advisory service does not exercise actual traffic control but will advise all aircraft in the area of known and observed traffic, both in the air and on the ground in a manner that will convey the positions and intentions of conflicting traffic so that pilots can organize themselves into a safe traffic flow. In reporting aircraft and ground traffic that may affect the aircraft's safety, the word "traffic" is used to precede the phrase that summarizes the situation. If the FSS specialist is aware of a potential conflict, departing aircraft will be requested to hold short of the active runway until the conflicting aircraft or vehicle is off the runway.

At some uncontrolled airports where there is no FSS, advisory service is provided through remote communications outlets (RCOs). This service is called **remote aerodrome advisory service (RAAS)** and consists of weather reports which include wind and altimeter settings, the active or preferred runway (if known), field condition reports, NOTAMs, PIREPS and known aircraft and vehicle traffic. The RAAS is a remote service provided by an FSS not located at the airport. It must be remembered that the information about aircraft and vehicle traffic that the FSS personnel are able to provide is only the information that they have received by radio or landline; it is not based on personal knowledge of what they can see happening.

The FSS specialists control vehicles operating on the maneuvering area of controlled airports when the control tower is closed. This **vehicle control service (VCS)** is also provided at airports without a control tower. VCS is not available at airports served by RAAS and RCOs, although the FSS specialists will advise aircraft of the presence of vehicles on the maneuvering area about which they have knowledge.

At uncontrolled airports where **mandatory frequencies** have been designated, the FSS operate this air/ground facility and provide the airport advisory service and vehicle advisory service to aircraft operating at the airport and in the air within a designated distance from the airport.

The FSS provides a **flight planning service** to assist pilots in planning and conducting their flights in safety. Flight service specialists have available comprehensive and current weather information (**aviation weather information service — AWIS**) with which they are able to assist pilots at the pre-flight and en route stages of their flight to make decisions based on weather determinants. They also provide other information on airways, status of navaids, communication facilities, and airports, special regulations, NOTAMS, RSC/JBI reports and any other conditions a pilot may expect to encounter en route.

The FSS responds to declared emergencies, relays ATC clearances and other instructions, accepts flight plans for relay to ATC. They broadcast flight safety information such as unfavourable flight conditions and hazards along the route of flight to pilots concerned. They handle aircraft position reports and PIREPS.

Flight service stations are connected to ATC centres by direct phone lines in most cases, permitting quick access to centre facilities for those aircraft requiring in-flight instructions when flight paths do not permit radio contact with ATC units.

The **aviation weather briefing service (AWBS)** is provided by selected flight service stations in each region. It is a fully interpretive pre-flight and en route weather briefing service provided by briefers who are trained to adapt meteorological information to meet the needs of all aviation users and to provide advice on special weather problems. The sites have a full range of weather products including satellite and radar imagery.

VFR alerting service is provided for flights for which a VFR flight plan has been filed. If the flight plan is not closed within a specified period of time after the expected time of arrival, the FSS personnel will alert the search and rescue centre, will make a communication search of area airports to determine if the aircraft has landed safely and only forgotten to close the flight plan, and, if the aircraft is not located, will alert SAR to commence an air search.

The FSS provides **VDF and homing assistance** to aircraft in emergency or when requested by a pilot.

The FSS provide a **weather observing service** to provide weather information to AES for preparation of weather reports and forecasts. FSS personnel make regular hourly reports but also report any significant weather changes as they occur.

Flight service stations no longer broadcast weather on their navigation aid frequencies on a scheduled basis. However, pilots, who are proceeding into remote areas where public telephone is not available, may request that any FSS broadcast specified weather information over a voice capable navigation aid at prearranged times.

The FSS is responsible for the co-ordination and dissemination of NOTAM information and collects and distributes PIREPs. It provides customs notification service and serves as the focal point on an airport for the dissemination and collection of aeronautical information.

The call sign of an FSS is RADIO. In communicating with a flight service station while en route, call, for example, **Ottawa Radio** or **Saskatoon Radio**.

Remote Communication Outlets (RCO)

In order to provide extended communications capability, remote transmitting and receiving facilities have been established. These remote communication outlets (RCO) are operated from the nearest flight service station. Aircraft in the general area of the RCO can therefore communicate directly with an ATC centre which may be hundreds of kilometers away. The RCO is connected by land lines with the FSS. Not all of the FSS services are available through an RCO. It is primarily used to provide information to en route traffic.

Flight information service en route (FISE), available on the en route frequency, provides pilots with information on weather reports, forecasts, PIREPS, NOTAMs, altimeter settings and other operational information pertinent to the en route phase of flight. An RCO will also accept position reports and relay ATC clearances. **Remote aerodrome advisory service (RAAS)** is also provided (see above).

A **dial up remote communications outlet (DRCO)** is a standard RCO which has a direct connection to an ATS unit through a commercial telephone line which pilots can access by a dial up procedure. The pilot activates the system, via the aircraft radio transmitter, by keying the microphone four times on the published DRCO frequency. The frequencies for each DRCO are published in the **Canada Flight Supplement**. If the keying procedure is successful, the pilot will hear a dial tone, signalling pulses and a ringing signal. Once the communication link is established, the DRCO equipment will broadcast a pre-recorded voice message that confirms "Link Established". The pilot must then initiate the radio conversation using standard radiotelephony phraseology (for example, **Regina Radio. This is Piper Golf Papa Tango Echo).** If the keying procedure is unsuccessful in establishing the communication link, the aircraft may be out of radio range of the DRCO or the DRCO may in use by another aircraft.

Community Aerodrome Radio Stations (CARS)

Community aerodrome radio stations (CARS) are air/ground radio stations established at certain isolated airports and are operated by territorial or provincial governments or by air traffic services. Each CARS is interconnected with a flight service station.

CARS provide emergency communication advising the FSS of any distress, urgency calls or ELT signals. They provide airport information for landing and departing aircraft (surface wind, altimeter setting, aircraft and vehicle traffic and aerodrome conditions). They accept position reports and pass them to the FSS, accept and forward flight plans and flight itineraries and flight closures. They record surface weather data and relay it to Environment Canada. The CARS has available information relating to local conditions (weather reports and forecasts, NOTAM, runway conditions, PIREPS, SIGMETS). Requests for other than local data can be obtained by telephone from the FSS but may take time and may be available only through prior request.

Duats

The direct user access terminal system (DUATS) enables a pilot to access weather information and NOTAM information using a personal computer anywhere there is a telephone line or via terminals at airports. An **approved** DUATS service, like a flight service station, also will accept flight plans and flight itineraries and forward them to air traffic services. Transport Canada has developed standards for DUATS services which require, in order to be approved, that they provide current weather and NOTAM information for all of Canada and that no outdated information be disseminated to users. They must check flight plans and flight itineraries for accuracy and completeness before transmitting them to ATS. Some DUATS services also provide weather maps and other graphics such as radar and satellite images.

A DUATS service can be used as a sole source of pre-flight planning information. However, pilots are still responsible for ensuring that they obtain **all** the available information appropriate to the intended flight.

Private Advisory Stations

Aeronautical operators may establish their own radio stations at controlled airports for use in connection with company business, such as servicing of aircraft, etc. The private advisory station does not relay information relative to ATC, weather reports, condition of landing strips or other information normally provided by ATC units.

Commet

A commet is a ground communications station operated for the purpose of passing weather and other related information to aircraft.

Radar Service

The modernization and expansion of the radar system has greatly enhanced the smooth flow of air traffic and efficiency and safety in traffic movements. By reducing the separation between aircraft, radar has substantially increased airspace utilization.

Radar vectoring is used when necessary for separation purposes or for noise abatement procedures, when requested by the pilot or when vectoring will offer advantages to the pilot or controller. Vectoring instructions would take the form, for example: **Turn Right Heading 030 for Vectors to Victor 300.**

When requested by the pilot, VFR aircraft are provided with information on observed radar targets that may be of concern to the pilot. In advising of the existence of other traffic, the ATC unit uses the positions of the clock to indicate the whereabouts of these other aircraft. (For example: **Traffic at 3 O'Clock. Four Miles. Eastbound.**) The type of aircraft and altitude and the relative speed of the aircraft will be included if known.

Radar equipped ATC units can also provide information on the location and movement of areas of heavy precipitation and on severe weather conditions.

It is important that pilots respond only to those radar vectors directed to themselves. Following the radar instructions issued to another aircraft creates a potentially hazardous situation.

Radar Navigation Assistance

Radar navigation assistance is available from certain ATC units. As long as the aircraft is operating within areas of radar and communication coverage and can be radar identified, the ATC facility will provide assistance to navigation in the form of position information, vectors or track and groundspeed checks, information about traffic in proximity to the flight path of the aircraft and about weather phenomena such as thunderstorms.

VFR flights may be provided with the service at the request of the pilot, at the suggestion of the controller or in the interest of flight safety.

While being vectored by the radar service, the pilot is responsible for avoiding other traffic and for maintaining VFR. If the vector is leading the aircraft into IFR conditions, the pilot should request a vector that will allow the flight to remain VFR. The pilot is also responsible for maintaining adequate clearance above the terrain and any obstacles. If the vector is not providing adequate obstacle or terrain clearance, the pilot must inform the controller and request a heading that will allow adequate clearance, or climb to a more suitable altitude, or, if necessary, revert to navigation without radar assistance.

Table of Radar Coverage Area

Ht. Above Station	Radar Range	Ht. Above Station	Radar Range
1 ft.	1 n.m.	4,239 ft.	80 n.m.
17 ft.	5 n.m.	6,624 ft.	100 n.m.
66 ft.	10 n.m.	9.539 ft.	120 n.m.
414 ft.	25 n.m.	14,904 ft.	150 n.m.
1060 ft.	40 n.m.	26,496 ft.	200 n.m.
2385 ft.	60 n.m.	41,400 ft	250 n.m.

Fig. 9. Radar Coverage Area.

A pilot requesting radar assistance in an emergency situation should provide the following information to the ATC unit: the nature of the emergency and the type of assistance required, the position of the aircraft and the existing weather conditions, the type of aircraft, the altitude, whether the pilot has an IFR rating and whether the aircraft is IFR equipped.

Flight Information Service

Flight information service is provided by air traffic control units to provide pilots with information about hazardous flight conditions. Such data would include information about conditions and hazards which was not available at the time of take off or which may have developed along the route of flight.

Flight information service is provided whenever practicable to any aircraft in communication with an air traffic control unit either prior to flight or when in flight.

VFR flights are provided with information concerning severe weather conditions along the proposed route of flight, changes in serviceability of navigation aids, conditions of airports and associated facilities and other items considered pertinent to the safety of flight.

Flight information messages are intended as information only. If a specific action is suggested, the pilot is responsible for making the final decision concerning the suggestion.

Radiotelephone Procedure

Correct procedure on the part of operators of radiotelephone equipment is necessary for the efficient exchange of communications and is particularly important where lives and property are at stake. It is also essential for a fair sharing of "On the Air" time in the crowded radio spectrum.

Radiotelephone equipment installed in any civil aircraft may be operated only by persons holding an appropriate radio operator's licence — Restricted Radiotelephone Operator's Certificate. A reciprocal agreement between Canada and the U.S.A. permits citizens of either country to operate radio equipment installed in aircraft registered in either country if the operator holds a valid Canadian or U.S. pilot's licence and a valid Canadian or U.S. radio operator's licence.

The use of French is authorized in aircraft communication in or over the province of Quebec and at the Ottawa International Airport. Information on the bilingual services that are available at specific Quebec airports and at Ottawa can be found in the COMM section of the airport listings in the **Canada Flight Supplement.** ATIS is usually given in English on one frequency and in French on another. Other ATS services, such as tower, ground, arrival and departure, are bilingual. Elsewhere in Canada, English shall be used. For safety and operational efficiency, once the language has been determined, that language shall be used throughout the course of that communication. Pilots should make themselves thoroughly familiar with aeronautical phraseology in the language of their choice.

Phraseology

Standard phraseology is recommended in the interest of clarity and brevity. It is not compulsory for a pilot to use these standard phrases. If you wish to communicate with a radio facility in your own words, by all means do so. However, use of the standard phrases does provide a uniformity in transmission; it makes your transmissions more readily understood by the ground station operators and enables you to understand more easily transmissions to you.

Ground Stations

Ground radio facilities are identified by their names. For example:

AIRPORT RADIO	— *Community Aerodrome Radio Station*
	Remote Communication Outlet
	Mandatory Frequency
APPROACH	— *ATC Approach Control*
APRON	— *Apron Advisory Service*
ARRIVAL	— *ATC Arrival Control*
CENTRE	— *Area Control Centre*
CLEARANCE DELIVERY	— *Clearance Delivery*
COMMET	— *Commet*
DEPARTURE	— *ATC Departure Control*
GROUND	— *Airport Ground Control*
INFORMATION	— *Flight Information Centre*
METRO	— *Pilot to forecaster*
PRECISION RADAR	— *Precision Radar Approach Facility (PAR)*
RADAR	— *Enroute Radar Facility*
RADIO	— *Flight Service Station*
SURVEILLANCE RADAR	— *Surveillance Approach Radar Facility (ASR)*
TERMINAL	— *Terminal Control Unit*
TOWER	— *Airport Control Tower*
UNICOM	— *UNICOM or Aerodrome Traffic Frequency*

In calling a particular facility, a pilot would use the name of the station followed by the facility he is contacting.

For example: **Winnipeg Approach. This is Piper Foxtrot Tango Romeo Sierra.**
Vancouver Centre. This is . . .
Calgary Ground. This is . . .
Cleveland Precision Radar. This is . . .
Ottawa Radio. This is . . .
Quebec Tower. This is . . .
Fort Frances UNICOM. This is . . .

Numbers

Numbers are pronounced as follows:

0 ZEE-RO	5 FIFE	Decimal DAY-SEE-MAL
1 WUN	6 SIX	Thousand
2 TOO	7 SEV-en	TOU-SAND
3 TREE	8 AIT	
4 FOW-er	9 NIN-er	

When referring to numbers, except whole thousands, each digit is stated separately. For example:

Ten is spoken **One Zero.**
Nineteen is spoken **One Niner.**
Eight hundred is spoken **Eight Zero Zero.**
Eleven thousand is spoken **One One Thousand.**

Wind speed and cloud base numbers may be expressed in group form. For example:

Wind Two Seven Zero at Ten.
Three Thousand Four Hundred Broken.

Phonetic Alphabet

To avoid confusion due to the similarity in sound of some letters of the alphabet (such as B and C), a phonetic alphabet has been devised. It should be used for single letters or to spell out groups of letters as much as possible.

A — Alfa	N — November
B — Bravo "BRAH-voe"	O — Oscar
C — Charlie	P — Papa "POP-ah"
D — Delta	Q — Quebec "KAY-BECK"
E — Echo	R — Romeo
F — Foxtrot	S — Sierra
G — Golf	T — Tango
H — Hotel	U — Uniform
I — India	V — Victor
J — Juliet "Jool-ee-YET"	W — Whiskey
K — Kilo "KEE-loe"	X — X-Ray
L — Lima "LEE-mah"	Y — Yankee
M — Mike	Z — Zulu

Aircraft Call Signs

In radio communications, the call letters of your airplane must be expressed in phonetics at all times. On initial contact, you must give the manufacturer's name or the type of airplane, followed by the four letters of the registration. The words helicopter, glider or ultralight are acceptable substitutes for the type of aircraft.

For example: **Piper Tomahawk Golf Mike Oscar Tango** or **Ultralight India Bravo Charlie Hotel.**

In subsequent communications, the call letters may be abbreviated to the last three letters of the registration if the abbreviation is initiated by the ATC controller or FSS operator. For example, Piper Tomahawk Golf Mike Oscar Tango becomes **Mike Oscar Tango.**

In the case of foreign aircraft, on initial contact, the manufacturer's name or type of aircraft and the full aircraft registration must be given. The registration may be abbreviated, in subsequent communications, to the last three characters, if the abbreviation is initiated by ATC. For example: Cessna 172 N8723T becomes **Two Three Tango.**

A flight responding to a medical emergency for the transport of patients, organ donors, organs or other urgently needed medical material or services is called a medical evacuation flight (MEDEVAC). The word MEDEVAC is included after the four characters of the registration and is retained as part of the call sign even if the call sign is abbreviated to three letters. For example: **Whiskey Zulu India Medevac.**

Time

The 24 hour system is used in expressing time. It is expressed by four figures, the first two denoting the hour past midnight and the last two the minutes past the hour.

For example:
12:00 midnight is expressed	— 0000
12:30 a.m.	— 0030
2:15 a.m.	— 0215
5:45 a.m.	— 0545
12:00 noon	— 1200
3:30 p.m.	— 1530
10:50 p.m.	— 2250

Note: The day begins at 0000Z and ends at 2359Z. 2400Z is not used in flight operations.

Normally coordinated universal time (UTC) is used. In fact, all air traffic control facilities in North America operate on UTC. The letter Z after the hour indicates UTC.

Where operations are conducted solely within one time zone, standard zone time may be used, if care is taken to indicate clearly the time zone. Daylight Saving Time is not used. Standard time zones are indicated by letters as follows:

Newfoundland	— N
Atlantic	— A
Eastern	— E
Central	— C
Mountain	— M
Pacific	— P
Yukon	— Y

For example: 2:15 p.m. EST is expressed **1415E**
9:00 a.m. PST is expressed **0900P**
11:45 p.m. UTC is expressed **2345Z**

In air traffic control procedure, the hour is often omitted and the time referred to in minutes past the hour only. For example: 10:25 would be referred to as 25 and spoken **Two Five.**

However, if the hour is included in the time, it is spoken in four digits.

For example: 8:45 a.m. is spoken **Zero Eight Four Five.**
12:30 a.m. is spoken **Zero Zero Three Zero.**
12:45 p.m. is spoken **One Two Four Five.**

Airways and Air Routes

The phonetic alphabet is used in referring to airways and air routes. For example:

Airway V52 is spoken **Victor Five Two.**
Airway G1 (Green 1) is spoken **Golf one.**
Airway A2 (Amber 2) is spoken **Alfa Two.**
Airway R3 (Red 3) is spoken **Romeo Three.**
Airway B4 (Blue 4) is spoken **Bravo Four.**
Airway J500 (High Level 500) — **Juliet Five Hundred.**
Air Route GR1 (Green Route 1) — **Golf Romeo One.**
Air Route AR2 (Amber Route 2) — **Alfa Romeo Two.**
Air Route RR3 (Red Route 3) — **Romeo Romeo Three.**
Air Route BR4 (Blue Route 4) — **Bravo Romeo Four.**

Flight Altitudes and Headings

Flight altitudes are always given in thousands and hundreds of feet above sea level.

For example: **One Thousand.** (1000 feet ASL)
Two Thousand Five Hundred. (2500 feet ASL)
One Zero Thousand. (10,000 feet ASL)
One Six Thousand. (16,000 feet ASL)
Flight Level Two Four Zero. (FL 240)

Aircraft headings are given in groups of three digits, in degrees magnetic (if operating in the Southern Domestic Airspace) or in degrees true (if operating in the Northern Domestic Airspace).

For example: 060 is spoken **Heading Zero Six Zero.**
275 is spoken **Heading Two Seven Five.**

Aerodrome elevations are expressed in feet.

For example: **Field Elevation Six Seven Five.**

Radio Frequencies

Radio frequencies containing a decimal point are expressed with the decimal point in the appropriate place in the sequence:

For example: 121.5 MHz is spoken as **One Two One Decimal Five Megahertz.**

It is necessary that the pilot on the initial call state the frequency on which he is transmitting. This procedure is required in order to provide the ground station communicators, who guard many frequencies, with a positive indication of the correct transmitter to be selected for answering the call.

The ground station will normally reply on the frequency on which the call is received. Calls received on 122.2 MHz will be answered on the navigational aid frequency. Pilots requiring an exception to this procedure should state the frequency on which the reply is expected.

Aircraft: **Vancouver Radio This is Beechcraft Foxtrot Delta Lima Tango — On One Two Two Decimal Two — Over.**

The ground station will reply on the navigational aid frequency.

Aircraft: **Ottawa Radio This is Piper Foxtrot Alfa Golf Lima — On One Two Two Decimal**

Two — Reply on One Two Six Decimal Seven — Over.

The ground station will reply on 126.7 MHz.

Aircraft: **Montreal Radio This is Cessna Golf Bravo Romeo Yankee — On One Two Six Decimal Seven — Over.**

The ground station will reply on 126.7 MHz.

Transponder codes are given in numbers preceded by the word SQUAWK.

For example: Code 1200 is spoken **Squawk One Two Zero Zero.**

Runway Visual Range

Runway visual range information is provided by ATC, arrival control, PAR, control towers and flight service stations.

For example: **RVR Runway Two Five, Three Thousand Six Hundred Feet.**
RVR Runway One Niner, Variable from One Thousand Feet to Two Thousand Feet.

The Morse International Code

It is not necessary for a pilot to know morse code in order to use aeronautical radio facilities. However, the identifiers of VOR stations, ILS localizers and non directional beacons are broadcast in morse code and it is therefore advantageous to have some knowledge of the code.

```
A .—        N —.        1 .————
B —...      O ———       2 ..———
C —.—.      P .——.      3 ...——
D —..       Q ——.—      4 ....—
E .         R .—.       5 .....
F ..—.      S ...       6 —....
G ——.       T —         7 ——...
H ....      U ..—       8 ———..
I ..        V ...—      9 ————.
J .———      W .——       0 —————
K —.—       X —..—
L .—..      Y —.——
M ——        Z ——..
```

Standard Phrases

The following phrases and words should be used whenever applicable:

Word or Phrase	Meaning
Acknowledge	"Let me know you have received and understood this message."
Affirmative	"Yes,"
Break	"I hereby indicate the separation between portions of the message."
Confirm	"My version is... Is that correct?"
Correction	"I have made an error. The correct version is..."
Do you read?	"I have called you more than once. If you are receiving me, reply."
Go ahead	"Proceed with your message."
How do you read me?	Self explanatory
I say again	"I will repeat."
Negative	"No."
Out	"My transmission is ended. I do not expect a reply from you."
Over	"My transmission is ended. I expect a reply from you."
Read back	"Repeat this message back to me after I have given `Over'."
Roger	"Okay. I have received your message."
Say again	"Repeat." Never use the word "Repeat" — it is reserved for military purposes.
Speak slower	Self explanatory
Stand by	"I must pause for a few seconds." If the pause is to be longer than a few seconds, add "Out".
That is correct	Self explanatory
Verify	"Check with the originator."
Wilco	"Your instructions received, understood, and will be complied with."

Priority of Communication

In general, the following priorities are applied to radio communications by flight service stations.

1. *Emergency communications (Distress and Urgency)*
2. *Flight safety communications (ATC clearances, airport advisories, position reports, airfile flight plans, etc.)*
3. *Scheduled broadcasts*
4. *Unscheduled broadcasts*
5. *Other air-ground communications*

NOTAM, SIGMET or PIREP messages are generally handled as priority item 4.

Good Radio Technique

Optimum use of aircraft radio communication facilities depends on the good technique of the operator.

It is good procedure to listen briefly on the channel to be used before transmitting to ensure that you will not interrupt or cause harmful interference to stations already in communication.

Plan the content of your message before transmitting so that your call will be brief and clear. Use standard phraseology whenever practical. Do not, at any time, use CB terminology.

Radio telephony contact consists usually of four parts: the call-up, the reply, the message and the acknowledgement. For example:

Aircraft: **Buttonville Tower. This is Piper Cherokee Golf Alfa Victor Yankee. Over.**
Tower: **Golf Alfa Victor Yankee. (This is) Buttonville Tower.**
Aircraft: **Buttonville Tower. Golf Alfa Victor Yankee. One Five West. Two thousand five hundred feet. VFR. Landing Instructions.**
Tower: **Alfa Victor Yankee. Buttonville Tower. Runway One Five. Wind One Three Zero at Ten. Altimeter Two Niner Niner Two. Cleared to the Circuit.**
Aircraft: **Alfa Victor Yankee.**

"This is" and "Over" may be omitted.

1. Pronounce words clearly. Do not slur sounds or run words together.

2. Speak at a moderate rate, neither too fast nor too slow,

3. Keep the pitch of the voice constant. High pitched voices transmit better than low pitched voices.

4. Do not shout into the microphone.

5. Hold the microphone in the correct position — about 1 inch from the lips.

6. Know what you are going to say before starting the communication. "Ums" and "ahs" take up valuable air time.

7. Acknowledge receipt of all ATC messages including frequency changes directed to and received by you. The acknowledgement should take the form of a transmission of the aircraft call sign or a read back of the clearance or other appropriate message with the call sign. Clicking your microphone button as a form of acknowledgement is not proper radio procedure.

8. Profanity or offensive language is not permitted.

9. Do not request transportation, accommodation or other personal services from airways communication stations. They will not accept such messages.

The important thing is to prevent misunderstandings. Let the controller know who you are (on every transmission) and what you are doing. Repeat his message if necessary to avoid error.

Simple misunderstanding is one of the greatest problems in pilot/controller communication. In many cases, the pilot or the controller thinks he hears what he expects to hear. The practice of reading back to the controller any clearance received helps to avoid this kind of misunderstanding. The use of standard phraseology also helps to avoid misunderstandings and interjects professionalism into transmissions. If you are given a clearance that you do not understand or that does not make sense, ask for a clarification. Never "roger" anything that you do not fully understand.

Communication Checks

Aircraft operators may sometimes wish to check the serviceability of their communication equipment. Such a check may be made while the aircraft is airborne (signal check), while the aircraft is about to depart (preflight check), or by ground maintenance personnel (maintenance check).

The readability scale employed for communications checks has the following meaning: 1. Unreadable. 2. Readable now and then. 3. Readable with difficulty. 4. Readable. 5. Perfectly readable. The strength scale has the following meaning: 1. Bad. 2. Poor. 3. Fair. 4. Good. 5. Excellent.

Aircraft: **Churchill Radio This is Piper Foxtrot Romeo Yankee Tango — Signal Check on Five Six Eight Zero — Over.**

Tower: **Piper Romeo Yankee Tango This is Churchill Radio — Signal Check Reading You Strength Five — Over.**

Radio checks, when required, should whenever possible be requested on frequencies other than ATC frequencies since frequency congestion is a problem. Normally the establishment of two-way contact is sufficient proof that radios are serviceable.

Electronic Interference to Navigation and Communication Systems

Portable electronic devices produce radiation which may cause interference to aircraft navigation and communication systems. These systems are most vulnerable to this type of disruptive interference during the approach, landing and take-off phases of flight. During these stages, the aircraft is at low altitude and may also be subject to disruptive interference from ground based sources, the combined effect of which may cause unreliable indications by ILS, VOR and ADF equipment.

Citizen band (CB) radios, cellular telephones and transmitters that remotely control devices, such as toys, which intentionally radiate radio frequency signals are prohibited on board aircraft.

Audio or video recorders, audio or video playback devices, electronic entertainment devices, lap top computers, hand held calculators, FM radio receivers, TV receivers and electronic shavers may not be used during the critical phases of take-off, climb, approach and landing. They may be used during the cruise phase of flight if they have been demonstrated to be acceptable. Hand held electronic calulators, for example, have been found to cause interference to ADF equipment in the 200 to 450 KHz frequency range if the calculator is positioned within 5 feet of the loop or sense antenna. In a small general aviation airplane, it is probably wise to avoid using such equipment even during cruise if ADF, VOR and DME systems are being used to provide navigational direction.

Radio Telephone Procedures in Communication With Ground Stations

In the following examples of communications between aircraft and various ground stations, the initial call-up and reply have been omitted in order to conserve space.

Call-up and Taxi Authorization

Permission to taxi should be requested on the ground control frequency that is published in the **Canada Flight Supplement**. If no flight plan has been filed, you should inform the controller of the nature of your flight (e.g. local VFR). Do not leave the apron until authorized by the controller to do so. Ground control will reply, authorizing you to proceed to the runway in use and will give you the surface wind, the altimeter setting and the time.

At a number of major airports where clearance delivery is in operation, VFR flights are required to contact this controller before starting to taxi and before contacting the apron and/or ground controller.

Transponders should be adjusted to "standby" while taxiing and switched to "on" only immediately before takeoff.

Radio checks may be requested on the ground control frequency.

Aircraft: **Norwood Ground Control — This is Cessna Foxtrot Mike Kilo Alfa — Preflight Check — One Two One Decimal Niner (1) — Over.**

(1) Meaning you expect a preflight check on 121.9 MHz.

Ground: **Cessna Foxtrot Mike Kilo Alfa — This is Norwood Ground Control — Radio Checks Strength Five.**

Aircraft: **Norwood Ground Control — Cessna Foxtrot Mike Kilo Alfa on West Ramp — VFR to Forestville — Taxi Instructions — Over.**

Ground: **Mike Kilo Alfa (2) — Taxi to Runway Two Seven (3) — Wind Two Seven Zero at One Zero (4) — Altimeter Two Niner Seven Five (5) — Time Zero Five (6).**

(2) After communication has been established, the aircraft registration may be abbreviated to the last three letters, phonetically expressed.
*(3) Runway 27 lies approximately **270°** magnetic. Your compass or heading indicator should therefore read **270°** when you line up for take-off.*
(4) The wind is west at 10 kts. Winds given by control towers are magnetic whereas those given by meteorological stations are true.
(5) Set the barometric scale on your altimeter to 29.75
(6) Set your clock to 05 minutes past the hour.

With the increasing traffic around all airports, frequency congestion is becoming a problem and brevity in your transmissions is therefore desirable.

At many major airports, information about ceiling, visibility, runway in use, etc. is given over the automatic terminal information service. Listen to this information and when first calling ground control, state that you have the information broadcast on ATIS. Be sure to specify the phonetic identifier of the ATIS message (i.e. ATIS Information Bravo).

At an airport where there is no ATIS, listen on the ground control frequency to the controller talking to other airplanes and take note of the "numbers". When making your first contact, inform the controller:

Aircraft: **Norwood Ground Control — Cessna Foxtrot Mike Kilo Alfa with the Numbers — etc. etc.**

The controller then does not have to take up valuable air time repeating information that you already know.

If authorized to taxi to the runway in use, no further authorization is required to cross any inactive runway enroute. However, at no time, may a taxiing aircraft taxi on or across an active runway unless authorized to do so. If for any reason, the airport ground controller requires that you request further authorization before crossing or entering any runway en route, this requirement will be reflected in the taxi authorization.

To emphasize the protection of active runways and to prevent runway incursions, ATC will request a readback of a **"Hold/Hold Short"** of a runway or taxiway instruction. Examples of hold points that should be readback are:

Tower: **Hold** or **Hold On Runway** . . . . or **On Taxiway** . . . . **Hold East** (or **North**, etc.) **Of Runway** . . . . **Hold Short of Runway** . . . . or **Of Taxiway** . . . .

With the increased simultaneous use of more than one runway, instructions to enter, cross, backtrack or line-up on any runway should be acknowledged by a readback.

Authorization to taxi "to" the runway in use is not authorization to taxi "on" the assigned runway. You must hold short of the runway in the designated taxi holding position until receiving authorization to taxi on the runway to take-off position. If no taxi holding position is established, you should hold at sufficient distance from the edge of the active runway so as not to create a hazard to arriving or departing traffic (200 feet is the recommended distance).

Flight Plan

If you have not previously filed a flight plan, you may file one giving the information over the radio. However, filing a flight plan by radio is only permissible when it is impossible to file one in person or by telephone. A long transmission such as the filing of a flight plan aggravates the congestion of the air-ground communication channels and ties up controllers whose job is to expedite the safe and orderly movement of traffic.

Aircraft: **Summit Tower — This is Cessna Foxtrot Papa Tango Victor — Here is my Flight Plan — Cessna 172/SD/C — Airspeed One Two Five Knots — Departing Summit — Eight Thousand Five Hundred VFR (1) via Victor Seven — Destination Huntington — Departing at One Three — Elapsed Time One Hour Plus Four Minutes — Fuel on Board Three Hours Plus Ten Minutes — Garrett ELT — Standard VHF, ADF, DME and Transponder Equipped — Two Persons on Board — Pilot Ayer Worthy, 123 A Street, Anytown, Ontario — Amphibian — Aircraft Blue with White Trim — Arrival Report will be filed at Huntington FSS — Pilot Licence YZP 123456 — Over.**
Tower: **Papa Tango Victor — I have your Flight Plan.**
Aircraft: **Roger Papa Tango Victor.**

(1) VFR, Visual Flight Rules, do not require any stated altitude. However, air traffic control strongly recommends that you state your altitude and route. "Eight thousand five hundred" means 8500 feet altitude. "Victor seven" means that you will fly via VOR Airway V7.

Take Off Clearance

When you have run up your engine(s) and made your cockpit check, if you have not already done so, change to tower frequency and request a clearance for take-off:

Aircraft: **Moncton Tower — Stinson Foxtrot November No-vember Charlie — Ready for Take off on Runway Two Niner.**
Tower: **Stinson November November Charlie — Cleared for Take off on Runway Two Niner.**
Aircraft: **Roger November November Charlie.**

Having received and acknowledged take-off clearance, you should take off without delay. If, for some reason, you are unable to do so, inform the tower immediately.

Special Clearance

Should you wish to execute some special maneuver, such as a right-hand turn after taking off, the following phraseology would be appropriate:

Aircraft: **Bonanza Golf Alfa Tango Hotel — Ready for Take Off on Runway Zero Four — Request Right Turn Out — Over.**
Tower: **Bonanza Alfa Tango Hotel — Right Turn Approved — Cleared for Take off on Runway Zero Four.**
Aircraft: **Roger Alfa Tango Hotel.**

After take-off, remain tuned to the tower frequency. The tower will usually call you after take-off, give you any necessary traffic information and your time off. You should remain tuned to the tower frequency while in the control zone. (Airplanes doing local flying must remain tuned to the tower frequency at all times.) You do not need permission to change from the tower frequency once you are clear of the control zone. To make such a request would needlessly increase frequency congestion. Once outside the control zone, you should monitor 126.7 MHz.

VFR Position Reports

An airplane flying visual flight rules is not required to report its position to intermediate stations en route. However, it is a good idea to do so and there is a special sequence to follow in making this report.

After calling the flight service station and receiving their acknowledgement of your call, you should state your wish to give a report on your position. The sequence of this report is as follows:

1. *Identification*
2. *Position*
3. *Time over*
4. *Altitude*
5. *VFR Flight Plan*
6. *Destination*

Aircraft: **Ottawa Radio — This is Cessna Golf Delta Delta India — VFR Position Report — Over.**
Station: **Delta Delta India — This is Ottawa Radio — Go Ahead.**
Aircraft: **Ottawa Radio — Delta Delta India — By Ottawa at Five Eight — Six Thousand Five Hundred — VFR Flight Plan — Destination Muskoka.**

En Route Reports (Class B Airspace)

In Canada, VFR flights are subject to ATC control when flying in Class B airspace (at or above 12,500 feet on airways). Airplanes flying VFR in Class B airspace must obtain air traffic clearances and file en route position reports at all compulsory reporting points, or other reporting points which may be specified by ATC.

It is a good idea to keep a pencil and paper handy to copy down the clearances. Pilots in Class B airpsace shall read back the text of an ATC clearance when requested to do so by an ATC unit. (In fact, any VFR flight may be asked to read back the text of an ATC clearance.) The following shorthand symbols, used by ATC controllers, may prove useful to you:

Words and Phrases	Shorthand	Words and Phrases	Shorthand
ABOVE	ABV	AND	&
ADVISE	ADV	APPROACH	AP
AFTER (PASSING)	<	FINAL	F
AIRPORT	A	LOW FREQUENCY RANGE	R
ALTERNATE INSTRUCTIONS	( )	OMNI	O
ALTITUDE 6,000-17,000	60-170	PRECISION	PAR
STRAIGHT-IN	SI	FOR FURTHER CLEARANCE	FFC
SURVEILLANCE	ASR	FOR FURTHER HEADINGS	FFH
APPROACH CONTROL	APC	HEADING	HDG
AT (USUALLY OMITTED)		HOLD (DIRECTION)	H-W
(ATC) ADVISES	CA	INTERSECTION	XN
(ATC) CLEARS OR CLEARED	C	(ILS) LOCALIZER	L
(ATC) REQUESTS	CR	OMNI (RANGES)	O
BEARING	BEAR	OUTER COMPASS LOCATOR	LOM
BEFORE	>	OUTER MARKER	OM
BELOW	BLO	RADAR VECTOR	RV
BOUND	B	RADIAL	RAD
EASTBOUND, etc.	EB	RANGE (LF/MF)	R
OUTBOUND	OB	REMAIN WELL TO LEFT SIDE	LS
INBOUND	IB	REMAIN WELL TO RIGHT SIDE	RS
CLIMB (TO)	↑	REPORT DEPARTING	RD
CONTACT	CT	REPORT LEAVING	RL
COURSE	CRS	REPORT ON COURSE	R-CRS
CROSS	X	REPORT OVER	RO
CROSS CIVIL AIRWAYS	≠	REPORT PASSING	RP
CRUISE	→	REPORT REACHING	RR
DELAY INDEFINITE	DLI	REPORT STARTING	
DEPART	DEP	PROCEDURE TURN	RSPT
DESCEND (TO)	↓	REQUEST ALTITUDE	
DIRECT	DR	CHANGES ENROUTE	RACE
EACH	ea	REVERSE COURSE	RC
EXPECT APPROACH		RUNWAY	RY
CLEARANCE	EAC	STANDBY	STBY
EXPECT FURTHER		TAKEOFF (DIRECTION)	T→N
CLEARANCE	EFC	TOWER	Z
FAN MARKER	FM	TRACK	TR
		TURN LEFT	LT
		TURN RIGHT	RT
		UNTIL FURTHER ADVISED	UFA
		VICTOR	V

For example: C FTOM BLT/V5 ↑5 M5/UXB

Means: ATC clears Foxtrot Tango Oscar Mike to Bolton via Victor 5. Climb to 5000 feet immediately and maintain 5000 until reaching Uxbridge reporting point.

You have filed a flight plan from Yarmouth, Nova Scotia for a flight to St. Johns, (Torbay) Nfld., in Class B airspace, at 13,000 ft. via Victor airway 312. Compulsory reporting points along the way are: Halifax, Copper Lake, Sydney and Atlantic. Your airplane is fitted with a VHF transceiver and an omnirange receiver. (Maximum line-of-sight reception distance at 13,000' is 140 n.miles, so some of the latter portion of your flight will be by D/R navigation). You have taken off at 1915Z and are climbing to your intended altitude. You contact Yarmouth radio for a clearance (from Moncton Centre) to enter the Class B airspace.

Note: All clearance reports you make are to the air traffic control centre (ATC) but may be made through a communications facility such as a tower, flight service station or omnirange station, etc. The communications facility relays your message to ATC and issues the clearance to you. (When you report to a remote communication outlet (RCO), however, you are in effect in direct communication with the centre by remote control).

Aircraft: **Yarmouth Radio — This is Twin Comanche Golf Kilo Tango Mike — One Two Two Decimal Two** (listening on Omni Frequency) **— Over.**
Station: **Golf Kilo Tango Mike — This is Yarmouth Radio — Go Ahead.**
Aircraft: **Golf Kilo Tango Mike at One One Thousand Five Hundred — Two Zero Miles East on Victor Three One Two (1) — Requesting One Three Thousand Controlled VFR to Torbay via Victor Three One Two — Over.**

(1) Meaning the 050 radial of the Yarmouth omnirange which provides the outbound track for Victor Airway 312.

Station: **ATC Clears Kilo Tango Mike to Torbay via Victor Three One Two — Climb to and Maintain One Three Thousand VFR — Report Reaching One Three Thousand — Over.**
Aircraft: (Repeats clearance back) **Kilo Tango Mike is cleared to Torbay via Victor Three One Two — Climb to and Maintain One Three Thousand VFR — Report Reaching One Three Thousand — Over.**
Station: **Roger Yarmouth Radio Out.**

You report on reaching 13,000 feet as instructed.

Aircraft: **Yarmouth Radio — This is Golf Kilo Tango Mike at One Three Thousand at Two Two (2) — Over.**

(2) Meaning 22 minutes past the hour, i.e. 1922 Z.

Station: **Kilo Tango Mike — This is Yarmouth Radio — Check you at One Three Thousand at Two Two — Over.**
Aircraft: **Kilo Tango Mike.**

When you arrive over the Halifax omnirange (your first reporting point), you check the time, calculate your groundspeed and your estimated time of arrival over the next reporting point. You have contacted Halifax radio and the station has acknowledged your call.

| Aircraft: | Golf Kilo Tango Mike over Halifax at One Five — One Three Thousand — VFR — Estimating Copper Lake (3) at Two Zero Four Zero — Sydney (4) — Over. |

(3) Your next reporting point.
(4) The next succeeding reporting point.

| Station: | Check you over Halifax at One Five — One Three Thousand — VFR — Estimating Copper Lake at Four Zero — Over. |
| Aircraft: | Golf Kilo Tango Mike. |

You tune in the Charlottetown VOR on 114.9 MHz and select 162° on your omni bearing selector. This is the 162° radial of the Charlottetown omnirange. Where it intersects Victor airway 312 is the Copper Lake reporting point. When your needle centres, you are over Copper Lake.

| Aircraft: | Sydney Radio — This is Twin Comanche Golf Kilo Tango Mike — One Two Six Decimal Seven — Over. |

If the station does not reply immediately,

| Aircraft: | Sydney Radio — This is Twin Comanche Golf Kilo Tango Mike — One Two Six Decimal Seven — Do you read? — Listening on One One Four Decimal Niner — Over. |

If a station does not answer you after two or more calls have been made, you may issue a general call, requesting any station hearing you to contact the station you are calling. For example:

| Aircraft: | Any Station receiving me — Any Station receiving me — This is Golf Kilo Tango Mike — Advise Sydney Radio I am calling him on One Two Six Decimal Seven — Listening on One One Four Decimal Niner — Kilo Tango Mike Out. |

When contact with the station has been established,

Aircraft:	Over Copper Lake at Four Zero — One Three Thousand — VFR — Estimating Sydney at Two One One Five — Atlantic — Over.
Station:	Check you over Copper Lake at Four Zero — One Three Thousand — VFR — Estimating Sydney at Two One One Five — Atlantic — Sydney Altimeter Two Niner Seven Five (and any other information considered necessary, such as the weather ahead, etc.) — Over.
Aircraft:	Roger Kilo Tango Mike.

You arrive over Sydney at 2115Z and decide to request permission to descend below Class B airspace. You have contacted the station and he has acknowledged your call.

| Aircraft: | By Sydney at One Five — One Three Thousand — VFR — Estimating Atlantic at Two One Five Two — Requesting Descent Below One Two Thousand Five Hundred — Over. |

On account of traffic, or for some other reason, your request cannot be immediately approved.

| Station: | Golf Kilo Tango Mike Maintain One Three Thousand — Report Atlantic for Further Clearance — Call Five Minutes West of Atlantic — Over. |

Aircraft:	(Reads back clearance) Golf Kilo Tango Mike Maintaining One Three Thousand — Will Call Five Miles West of Atlantic on One Two Six Decimal Seven — Over.
Station:	Negative Five Miles — Call Five Minutes West of Atlantic — Over.
Aircraft:	Negative Five Miles — Call Five Minutes West of Atlantic — Over.
Station:	That is Correct — Sydney Radio Out.

You arrive five minutes west of Atlantic reporting point at 2147Z and have established contact with Sydney Radio:

| Aircraft: | Golf Kilo Tango Mike Five Minutes West of Atlantic at Four Seven — Requesting Further Clearance — Over. |
| Station: | ATC Clears Kilo Tango Mike to Descend Immediately — Report Leaving One Two Thousand Five Hundred — Traffic is Westbound Aero Commander Maintaining Eight Thousand Estimating Atlantic at Five Six (5) — Over. |

(5) Meaning there is an Aero Commander at 8000 ft. which will be over Atlantic at 2156Z.

| Aircraft: | (Reads back clearance) Kilo Tango Mike is Cleared to descend immediately — Report Leaving One Two Thousand Five Hundred — Check the Traffic as Westbound Aero Commander at Eight Thousand — Estimating Atlantic at Five Six — Over. |
| Station: | Roger Sydney Radio. |

You report as you pass through the 12,500 foot level on your descent and decide to check the weather at your destination:

Aircraft:	Leaving One Two Thousand Five Hundred at Five Zero — Requesting the Latest Torbay Weather and Altimeter.
Station:	Check you leaving One Two Thousand Five Hundred at Five Zero — Here is the Latest Torbay Weather — One Thousand Scattered... etc. etc. — Altimeter Two Niner Zero Three — Over.
Aircraft:	Weather Received — Kilo Tango Mike Over.

En Route Radar Surveillance

When operating in areas where radar coverage exists, VFR aircraft that are transponder equipped may request radar traffic information. ATC will provide this information, workload permitting.

| Aircraft: | Calgary Advisory — Cessna Foxtrot Bravo Charlie Delta — Two Zero Northwest at Nine Thousand Five Hundred — VFR — Squawking One Two Zero Zero — En route to Edmonton — Request Radar Surveillance. |

Arrival

When you are approximately 15 nautical miles from your destination but before entering the control zone, call the tower on the published frequency. When the tower replies, give your altitude, and position in miles and direction from the airport. The tower will reply, giving you the runway to use, wind direction and velocity, traffic, and any other information or instructions considered necessary.

If the airport has ATIS and you have listened to and noted the information, inform the tower on your initial contact that you have the ATIS information.

Or, if you have listened to communications between the tower and other airplanes and have noted the pertinent information about runway, wind, altimeter setting, etc., tell the controller that you have the "numbers".

Always check the **Canada Flight Supplement** for special procedures relating to a particular airport. If the airspace surrounding a control zone has been designated as classified airspace, it is required that you contact the appropriate air traffic control facility to secure a clearance before entering the airspace. The controller will give you instructions on how to join the circuit and will tell you when to switch to the tower frequency.

Aircraft:	**London Tower — This is Beechcraft Golf Sierra Tango Bravo.**
Tower:	**Beechcraft Golf Sierra Tango Bravo — This is London Tower — Go Ahead.**
Aircraft:	**Beech Golf Sierra Tango Bravo — One Five Miles Northwest — At Three Thousand — VFR — Landing London — Request Landing Instructions — Over.**
Tower:	**Beech Sierra Tango Bravo — Cleared to Enter Traffic Circuit (1) — Runway Three Two (2) — Wind Two Seven Zero Degrees at One Five — Altimeter Two Niner Seven Zero — Over.**
Aircraft:	**Roger Sierra Tango Bravo.**

(1) Should the traffic be right-handed instead of the conventional left hand circuit, this will be indicated as follows: **Cleared to the Right Traffic Circuit.**
(2) Should the field have parallel runways, the appropriate runway will be designated as follows: **Runway Three Two Left.**

Once established in the traffic circuit, you should advise the tower who will reply with further information.

Aircraft:	**London Tower — Golf Sierra Tango Bravo is Downwind.**
Tower:	**Sierra Tango Bravo — You are Number Two — Follow Piper Warrior on Base Leg.**
Aircraft:	**Sierra Tango Bravo.**

Final Approach

When you are on the base leg and about to turn in on final approach, request landing clearance. Normally, the tower controller will initiate landing clearance without the need for you to request it. If landing clearance is not received, you must pull up and make another circuit.

Aircraft:	**Winnipeg Tower — This is Mooney Foxtrot Echo Lima Sierra — Landing Clearance Runway Two Five — Over.**
Tower:	**Echo Lima Sierra — Cleared to Land on Runway Two Five.**
Aircraft:	**Echo Lima Sierra.**

or

Tower:	**Mooney Foxtrot Echo Lima Sierra — Widen your Approach — Cessna 310 Just Landing — You are Number Two.**
Aircraft:	**Wilco Foxtrot Echo Lima Sierra.**
Tower:	(When runway is clear) **Echo Lima Sierra — Cleared to Land on Runway Two Five.**
Aircraft:	**Echo Lima Sierra.**

If preceeding traffic has not cleared the runway, the tower may instruct you to pull up and make another circuit, even though landing clearance had been given.

Tower:	**Echo Lima Sierra — Pull Up and Go Around — Traffic Still on Runway.**
Aircraft:	**Echo Lima Sierra — Pulling Up to Go Around Again.**

At some high density airports, airplanes are simultaneously cleared to land on intersecting runways. In this situation, the tower may advise:

Tower:	**Yankee Sierra Uniform — You are Cleared to Land on Runway Three Three — Hold Short of Runway Two Four Right Which Has Landing Traffic.**
Aircraft:	(Reads back clearance) **Yankee Sierra Uniform — Cleared to Land on Runway Three Three — Hold Short of Runway Two Four Right.**

In accepting this landing clearance, you, as pilot, must ensure that your approach and landing are such that you will be able to stop before you reach the intersecting runway.

A landing aircraft may be "Cleared for the Option" which gives the pilot the option of making a touch-and-go, low approach, missed approach, stop-and-go or a full stop landing. This procedure will be used only during light traffic conditions.

Aircraft:	**Halifax Tower — This is Cessna Golf India Tango Juliet — Downwind Runway Three Four — Request The Option.**
Tower:	**India Tango Juliet — Cleared for the Option.**
Aircraft:	**India Tango Juliet.**

Straight in Approach

A pilot, after first reporting some distance from the field, may request a straight-in approach without entering the traffic circuit, if the traffic permits.

Aircraft:	**Regina Tower — This is Beaver Foxtrot Hotel India Oscar — Approaching Runway Three Zero on a Heading Three Zero Zero Degrees Magnetic — Requesting a Straight In Approach — Over.**
Tower:	**Foxtrot Hotel India Oscar — Cleared Straight In.**
Aircraft:	**Foxtrot Hotel India Oscar.**

Touch-and-Go Landings

Airplanes practising landings, who wish to touch down and take off immediately, may be given a touch-and-go clearance. This should be requested by the pilot at a point in the circuit, preferably the downwind leg, when the pilot is sure of his/her intentions.

Aircraft:	**Saint Stevens Tower — Whiskey Whiskey Delta (1) — Requesting a Touch-and-Go Landing — Over.**
Tower:	**Whiskey Whiskey Delta — Cleared Touch-and-Go.**

If it is not possible to approve the touch-and-go landing because of other traffic, the tower will reply:

Tower:	**Whiskey Whiskey Delta — Negative Touch-and-Go.**

(1) Since this would not be the initial communication, the call letters have been shorted to the last three of the four letter registration of the airplane

Taxi Clearance

After landing, taxi straight ahead and clear the runway as quickly as possible at the first available taxiway or turn off point. After leaving the runway, continue taxiing to a point at least 200 feet from the runway, or across the hold line, before coming to a stop.

Your time of landing will be given by the tower only if requested. When you are clear of the runway, taxi instructions will be given.

Aircraft: **Foxtrot Yankee Uniform Delta — Taxi Instructions — Over**
Tower: **Foxtrot Yankee Uniform Delta — On at One Five — Taxi to the Terminal via Taxiway Alfa.**
Aircraft: **Foxtrot Yankee Uniform Delta.**

At certain designated airports, ground control has been established to handle traffic on the ground and thus relieve traffic congestion on the tower frequencies. Ground control operates on specially designated frequencies, published in the **Canada Flight Supplement**.

In-bound airplanes, after landing and clearing the runway, should tune in the appropriate ground control frequency for taxi instructions and further communication with the tower.

Aircraft: **Richmond Ground — This is Aero Commander Golf Romeo Delta Tango — Taxi Instructions — Over.**

Mandatory Frequency (MF) and Aerodrome Traffic Frequency (ATF)

If you are intending to land at an uncontrolled airport for which a mandatory frequency has been designated, you should call the air/ground facility (FSS, CARS or RCO) on the published MF five minutes prior to entering the specified area associated with that airport.

Aircraft: **Timmins Radio — This is Piper Lance Golf Echo Echo Whiskey — Five Miles South at Three Thousand Five Hundred — VFR — Landing Timmins at One Six Four Five — Request Weather and Traffic Advisory — Over.**
Station: **Echo Echo Whiskey — Timmins Radio — Weather is . . . — Runway in Use is Zero Three — Traffic is . . .**

Having joined the traffic circuit, you report.

Aircraft: **Timmins Radio — Echo Echo Whiskey is Downwind for Runway Zero Three.**

Once established on the final approach, you report again.

Aircraft: **Timmins Radio — Echo Echo Whiskey is on Final for Runway Zero Three.**

Always prefix your transmission with the name of the airport, radio, FSS, etc. that you are calling so that there is no doubt about your intentions or the airport at which you are intending to land.

At other uncontrolled airports that are considered to be relatively active, aerodrome traffic frequencies have been designated instead of MFs. The ATF is usually the frequency of the ground station where one exists or 123.2 MHz if there is no ground station.

If there is an ATF and a ground station, you should call the facility on the published ATF immediately prior to entering the specified area and follow the same procedures in your transmissions as outlined for an airport with an MF.

At an uncontrolled airport for which neither a mandatory frequency nor an aerodrome traffic frequency has been designated, or during the hours when the ground station is not in operation, you should broadcast your reports blind on 123.2 MHz (or on the frequency of the inoperative ground station).

Aircraft: **Hinton Traffic — Piper Warrior Golf Hotel Alfa Bravo — Five Miles South — Landing Runway Zero Two.**

When established in the traffic circuit, you report.

Aircraft: **Hinton Traffic — Golf Hotel Alfa Bravo is Downwind for Runway Zero Two.**

After turning onto the final approach, you report again.

Aircraft: **Hinton Traffic — Hotel Alfa Bravo is on Final for Runway Zero Two.**

Distress

The first transmission of a distress call should be made on the air-ground frequency that is in use at the time. If you are unable to establish communication on that frequency, the distress call and message should be repeated on the general distress frequency (121.50 MHz or 3023.5 KHz) or on any other frequency that is available, in the effort to establish as quickly as possible communication with another station. However, before changing frequencies, you should transmit a message indicating the frequency to which you intend to change.

The distress message should be repeated at intervals until an answer is received, allowing sufficient time between transmissions for a receiving station to reply.

The distress call format is the word MAYDAY repeated three times, followed by your aircraft identification three times. If time permits, your message should give your estimated position, altitude, the type of aircraft, the nature of the emergency and your intended action (such as crash landing in timber, forced landing on water, etc.). Following the transmission of your message, hold the button on your microphone pressed down for 20 seconds to enable D/F bearings to be taken on you:

Aircraft: **Mayday — Mayday — Mayday — Foxtrot Oscar November Romeo — Foxtrot Oscar November Romeo — Foxtrot Oscar November Romeo — Five Zero Miles South of Grand Falls at One Seven Two Five Eastern — Four Thousand — Cessna 185 — Icing — Will Attempt Crash Landing on Ice — (Keep microphone button depressed 20 seconds) Foxtrot Oscar November Romeo — Over.**
Station: **Foxtrot Oscar November Romeo — Foxtrot Oscar November Romeo — Foxtrot Oscar November Romeo — This is Gander Radio — Gander Radio — Gander Radio — Roger Mayday — Gander Radio Out.**

The distress signal has priority over all other transmissions and must be sent only on the authority of the person in command of the aircraft. It indicates that the aircraft sending the signal is threatened by grave and imminent danger and requires immediate assistance. Any person who knowingly transmits a fraudulent distress signal is guilty of an offense and liable to imprisonment or fine.

All stations hearing a distress call must cease any transmission of their own so as not to interfere with the MAYDAY message. Any aircraft that hears a distress call that has not been immediately acknowledged should make every effort to attract the attention of a station that is in a position to render assistance. In relaying a distress call, always follow it with "from" and the call letters of your aircraft.

If and when the aircraft is no longer in distress, a message must be made on the same frequency (or frequencies) as the MAYDAY message was sent, cancelling the state of distress.

Aircraft: **Mayday All Stations — All Stations — All Stations — This is Foxtrot Oscar November Romeo — Foxtrot Oscar November Romeo — Foxtrot Oscar November Romeo — At One Seven Four Zero Eastern — Foxtrot Oscar November Romeo Distress Traffic Ended — Ice Cleared — Returning Grand Falls — Out.**

Urgency

Urgency signals are preceded by the word PAN repeated three times. They are given priority over all other communications except distress calls.

The urgency signal indicates that the station calling has an urgent message concerning the safety of a ship, or aircraft or of some person on board or within sight.

The urgency message should be addressed to a specific station and should be transmitted only on the authority of the pilot in command of the aircraft.

Aircraft: **Pan — Pan — Pan — Cleveland Tower — This is De Havilland Foxtrot Golf Foxtrot Echo — Advise Stinson Four Nine One November that his Undercarriage is Damaged — Over.**

Stations which hear the urgency signal should continue to listen for 3 minutes and then, if no further urgency message is heard, may continue normal operations.

Urgency messages may also be directed to all stations.

Station: **Pan — Pan — Pan — All Stations, All Stations, All Stations — This is Ottawa Radio, Ottawa Radio, Ottawa Radio — Emergency Descent at Ottawa Airport — Ottawa Tower Instructs All Aircraft Below Six Thousand Feet Within Range of One Zero Miles of Ottawa VOR Leave East and South Sectors Immediately — This is Ottawa Radio — Out.**

Urgency messages to all stations must be cancelled when the situation necessitating the message has been resolved or ended.

An urgency message may also be used by an aircraft experiencing difficulties that compel it to land but where immediate assistance is not required.

Aircraft: **Pan — Pan — Pan — This is Piper Golf Delta Bravo Kilo — Experiencing Partial Engine Failure — Will Land — Assistance Not Required.**

Safety

The safety signal is the word SECURITY repeated three times. The use of this signal indicates that a station is about to transmit a message concerning the safety of navigation or important meteorological warnings to aircraft in flight.

The safety signal has priority over all other communications except distress and urgency.

Radio Navigation

The use of radio in cross country flying has made navigation a relatively simple procedure. The VHF omnidirectional range (VOR) navigation system has made flight between VOR stations as easy as following a highway. Distance measuring equipment (DME) relieves the pilot of the time consuming chores of plotting, measuring and computing to determine groundspeed and distance from destination information. The L/MF NDBs and broadcast stations are beacons on which pilots can home or from which they can obtain a fix, if the airplane is equipped with automatic direction finding (ADF) equipment. Area navigation systems (such as Loran C, INS and the satellite based GPS system) guide the airplane to its destination effortlessly.

However, radio navigation aids should be considered a valuable aid, but not a substitute for, dead reckoning. Never neglect your compass. Keep an aeronautical chart handy and regularly check your position on it. Radios can, and do, fail and your insurance against becoming hopelessly lost is your conscientiousness in practising basic pilotage.

VHF Omnirange Navigation System (VOR)

The most widely used means of radio navigation is the very high frequency omnidirectional range (VOR), a ground based, short range navaid. Strategically placed throughout Canada and the United States are hundreds of VOR ground stations that transmit signals that, with the proper airborne receiving equipment, can be used for navigation purposes. It is the basis of the VHF airways system and is also used for VOR non-precision instrument approaches.

The omnirange system functions in the static free VHF frequency band between 108.10 and 117.95 MHz. Each station operates on its own assigned frequency.

The omnirange station puts out two signals. One of these signals is non-directional. It has a constant phase throughout its 360° of azimuth. It is transmitted at a rate of 30 times a second. It is called the **reference phase.**

The other signal rotates at a rate of 30 times per second. It is called the **variable phase.**

The reference signal is timed to transmit at the instant the variable signal passes magnetic north. In all other directions, the variable signal will occur sometime later than the reference signal.

The principle of the omnirange is based on the **phase difference** between the two transmitted signals (Fig. 1). By measuring the time it takes the rotating signal to sweep from north to where it reaches the receiving equipment in the airplane, it is possible to determine the bearing of the airplane from the station.

To simply illustrate this principle, let us use the case of an airport rotating beacon. The beam of light is rotating clockwise at, let us say, six rpm. This is one complete revolution every 10 seconds, or 36 degrees of azimuth per second. Suppose the green airport identification light was timed to flash each time the beam swept past magnetic north.

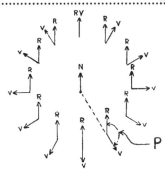

Fig.1. Phase Angle (P) and Azimuth.

Now if we had a stopwatch, we could determine our direction from the beacon. When we see the flash of the green light, we would start the watch. When the rotating beam sweeps past us, we would stop it. The number of seconds shown on the watch multiplied by 36 would be our magnetic bearing from the beacon.

Example: The time between the flash of green light and the sweep of the rotating beam past us is four seconds. 4 x 36 = 144. Then our bearing from the beacon is 144° magnetic (Fig. 2).

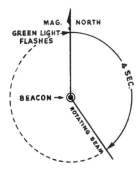

Fig.2. Principle of the Omnirange

This is exactly what the omnirange equipment does for a pilot, except that the timing between the reference and variable signals is done electronically, and the bearing obtained is any one of the infinite number of beams which radiate out like the spokes of a wheel from the omnirange station and which are referred to as **radials.**

Although, in theory, there are an infinite number of radials, only 360 are, in practice, usable. These radials are designated by numbers. Each radial is given a number which indicates its magnetic bearing **from** the station. For example: the radial 45 would be the radial you would follow if you wished to fly a magnetic track northeast away from the station. The radial is located 45° clockwise from the 0° radial which is always aligned with magnetic north.

There are, of course, omni bearings **towards** a station as well as away from it. If you were following the 45 radial to the station, you would be in a position southwest of the station. The 45 radial TO is a reciprocal of the 45 radial FROM.

Omni bearings are called radials regardless of whether they are to or from the station. However, some authorities refer to omni bearings as **courses.** Other authorities like to consider the radial as the magnetic bearing away from the station, and the course as the magnetic bearing towards the station. "45 FROM" would, therefore, be a radial away from the station, and "225 TO" would be a course towards the station.

A VOR station is depicted on aeronautical charts by a station azimuth (or compass rose) superimposed over the site of the VOR transmitting station, which is represented by a six sided symbol. Fig. 3 illustrates the azimuth for the Wainwright VOR station. The azimuth is marked off in increments of 5°. The 0° radial is aligned with magnetic north.

The name of the VOR is given in the information box along with the three letter identifier and the associated morse code identification. The frequency of the station is given in megahertz. The frequency for Wainwright is 114.5 MHz.

Several radials are drawn in (274° 254°, 034° and 087°). These radials are used to provide a flight path between two VOR stations. Such a specially designated flight path is called a Victor airway. The 274° radial is also Victor airway 350 (V350).

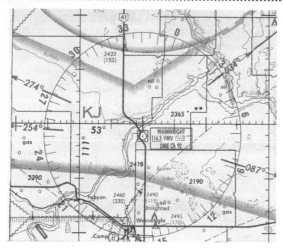

Fig. 3. How an Omnirange is Shown on a Map.

Victor Airways

Networks of VHF airways which make use of omnirange facilities crisscross the whole of the United States and most of Canada. These airways are referred to as Victor (or VOR) airways. They are flown by reference to the omni stations, which are located approximately 100 miles apart along the airways. The Victor airways are numbered like highways.

They are approximately 8 nautical miles wide. They are depicted on aeronautical charts by shaded lines the outer edge of which denotes the outer limits of the airway.

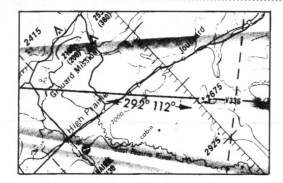

Fig. 4. Victor Airway (Controlled Airspace).

Note that the magnetic bearings of the airway illustrated are exact reciprocals: i.e. 292° and 112°. On some airways, these opposite bearings may not be exact reciprocals, due to differences in magnetic variation. For example, a certain Victor airway may be 077° eastbound but be 261° westbound. When flying Victor airways, you fly the omni radial which lies down the centre of the airway.

Under conditions of low visibility, it is extremely important to fly close to the centre line of the airway, particularly in the vicinity of terminal areas. An adjacent Victor airway may lie as close as 15° to the one you are using and traffic on that airway may be operating at the same altitude you are. Fig. 4 shows how a Victor airway appears on an aeronautical chart.

Advantages of VOR Navigation

The omnirange system provides a multiple number of courses towards and away from each station in the system. These courses are like invisible highways crisscrossing the continent on which the pilot can navigate to or away from any location.

The omniranges may also be used to determine a fix. A pilot is able to establish a position by rapidly tuning two omni stations and reading off their bearings. These will give two position lines. When the latter are drawn on a map, the point where they intersect will be the fix.

Because the omni is **position** sensitive, rather than **heading** sensitive, a pilot can fly a straight line and not worry about compass corrections, variation or deviation. Wind drift is also compensated for automatically. As a result, the airplane can be accurately kept on course.

The omni instruments in the airplane are designed to show the bearing or direction of the airplane from the omni station, **regardless of the direction the nose of the airplane may be pointing** (See Fig. 5). In other words, the omni bearing is a **magnetic bearing** between the airplane itself and the omni station. When you fly towards or away from an omni, you do so by maintaining a line of constant bearing. Obviously, to do this, you must keep kicking the nose around into wind to prevent the wind from drifting you off the bearing line you are endeavouring to maintain. In doing so, you are automatically correcting for wind drift with no effort or calculation of any kind on your part.

The accuracy of the course alignment of the omnirange is within plus or minus 3 degrees. Very precise navigation is therefore possible.

The VOR frequencies are free of static and interference and therefore give reliable indications. Reception is dependent, however, on distance from the station and altitude (See below).

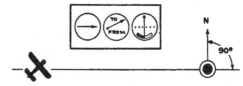

Fig. 5. Omnirange Bearing. The Bearing of the Station from the Airplane is 90 Degrees.

Disadvantages of VOR Navigation

The VOR signals, transmitted on VHF frequencies, are restricted to line of sight reception.

The curvature of the earth, therefore, has a pronounced effect on the distance at which a VOR signal can be received. At distances of about 50 nautical miles from the station, the signal will be received at an altitude of 1500 ft. AGL. The farther from the station that the airplane is, the higher it must be to receive the signal. (See Fig. 6. **Table of VHF Reception Distance** in Chapter **Radio.**)

At very high altitudes, it is possible to receive incorrect indications because of interference between two stations transmitting on the same frequency. Usually, the same frequency is assigned only to stations that are many hundreds of miles apart. However, on occasion, the areas may overlap and cause erroneous omni readings.

The omnirange signals are also affected by topographical features in the immediate vicinity of the omni station, such as fences, power lines, buildings, etc. This is called **site effect error** and causes erroneous readings and sometimes complete blanking out of the signal.

In the close vicinity of mountains, the signals are sometimes reflected or completely masked out in the shadow of the mountain. This is known as **terrain effect error.**

Errors are also occasioned by failure of parts of a particular receiver (**receiver error**) or by faults in the structure or equipment installed at a particular omnirange station (**ground station error**).

The vagaries of the VOR system are not of great magnitude.

Site and terrain errors are experienced for short periods and tend to average out before the airplane can respond. Receiver error can be eliminated by careful selection and proper maintenance of good reliable equipment. Ground station error, the most serious of those enumerated above, is seldom in excess of two degrees and in no case greater than plus or minus 3 degrees of error in displacement.

These vagaries of the omnirange appear in the form of minor irregularities such as course roughness, occasional brief flag alarm activity, course deflections and limited distance range. Over unfamiliar routes, be on the alert for these abnormalities.

The VOR Station

Each VOR station transmits on an assigned VHF frequency. As well as the omni signal, the station transmits a morse code three letter identification. For example: Toronto YYZ — . — — /— . — — /— — . .

Some omni stations are identified by a recorded voice identification, followed by three letters in code, e.g.: "Indianapolis VOR" (voice) . . /— . /— . . (IND in code). If the omnirange station is unattended and does not provide an air/ground communication facility, the word "unattended" will be included. For example: "Cooksville Unattended VOR".

Always listen for the identification signal to be sure you have the right station. During periods of maintenance, the coded identification is not broadcast. If you cannot hear the audio signal, take that as indication that the facility may be unreliable even though the VOR signal is being transmitted.

All Canadian VOR stations operate continuously. Most are equipped with an air/ground communications facility which allows pilots to contact flight service stations. The pilot transmits on 122.2 MHz or 126.7 MHz and receives on the VOR frequency. At selected VORs, taped broadcasts such as the transcribed weather broadcast (TWB) or automatic terminal information service (ATIS) are transmitted.

VOR stations can also be contacted to request direction finding (DF) assistance. The VHF DF equipment that is in operation at most flight service stations enables the FSS operator to advise a pilot of his bearing from the station. The VHF DF service is discussed more fully in the Chapter **Radio.**

The Omni Instruments

To use the omninavigation facilities, the airplane must be equipped with a VHF receiver, a VOR indicator and an antenna.

The Antenna

The VOR signals are received by the antenna which is normally located on the top of the fuselage or at the tail cone area, where it is free of interference from other electrical equipment. There are 3 types of antennas used with VOR equipment — the V-shaped, flexible steel antenna sometimes called a cat's whisker whip, the blade antenna or the towel bar antenna. The latter two types give somewhat better performance and are required if the VOR is being used in combination with RNAV.

The Receiver

The omnirange receiver is usually incorporated in a unit that has both a transceiver for transmitting and receiving on the communication channels and a receiver for receiving the navigation channels.

The VHF NAV/COM unit illustrated in Fig. 6 is a typical modern piece of radio equipment. The COM side of the unit has already been discussed in the Chapter **Radio.**

The NAV side of the unit is capable of receiving 200 channels in the frequency range from 108.00 MHz to 117.95 MHz with 50 KHz spacing.

The NAV receiver in Fig. 6 is tuned to 109.30 MHz. This particular unit allows the pilot to preselect a second VOR frequency (116.40 MHz) in the standby position.

Fig. 6. VHF NAV/COM Radio Equipment.

The NAV receiver also functions as a receiver for the glide path and runway localizer of the instrument landing system. ILS localizers operate on odd tenth decimals (i.e. 108.10, 108.30, 108.50, etc.) in the VHF frequency range from 108.10 to 111.90 MHz. When the set is tuned to an ILS frequency, the corresponding glide path frequency (which transmits in the UHF band) is tuned in automatically.

The VOR Indicator

The VOR indicator combines three functions in one unit: a bearing (or course) selector, a course deviation indicator and a TO/FROM indicator.

The VOR indicator illustrated in Fig. 7 incorporates these three functions. It also incorporates the cross pointer that is necessary to give a visual indication of the airplane's position when the NAV receiver is tuned to an ILS localizer.

The **omni bearing (or course) selector (OBS)** has an azimuth dial graduated in 360 degrees and numbered every 30° from 0 to 33. The numbers represent omni bearings (courses or radials). In Fig. 7, a

Fig. 7. Omni/ILS Indicator.

bearing (course or radial) of 000° (N) has been selected. The reciprocal bearing or course of 180° (S) is indicated at the tail end of the arrow.

The **TO/FROM indicator** is located on the right side of the instrument above and below the horizontal bar. Since the magnetic bearing of an omnirange radial can be ambiguous, the sense indicator is used to indicate whether the airplane is on a bearing **towards** or **away from** the omnirange station.

If the indicator shows TO, the bearing is towards the station from the airplane. If the indicator shows FROM, the bearing is away from the station towards the airplane.

The **course deviation indicator (CDI)** is a vertical needle that moves to the right or left of the row of vertical dots to indicate whether the airplane is to the right or left of the course or radial which the pilot has set up on the bearing selector. The course deviation indicator is often simply called the **left/right needle** or the **track bar indicator (TB)**.

When the NAV receiver is tuned to an ILS localizer, the vertical needle indicates deviation from the selected localizer course. The horizontal needle indicates deviation above or below the glide path. The ILS landing system is more fully discussed later in this Chapter.

VOR Receiver Checks

VHF equipment should be tested periodically to determine its accuracy. Low power VHF (VOT) omnitest transmitters have been installed at various locations throughout Canada and the United States and provide a means of checking a VOR receiver while the airplane is on the ground. These transmitters radiate a signal such that an airplane will receive an indication as if it were positioned on the magnetic north (000°) radial from the VOR facility. With the needle centered, the omni bearing selector should read 000° with the indicator in the FROM position or 180° with the indicator in the TO position. The bearing accuracy of the transmitted signal is maintained within a tolerance of 1°. An apparent error greater than 4° in the airplane VOR receiver is beyond acceptable tolerance. The equipment should be repaired or replaced. An apparent error of less than 4° is acceptable. Do not try, however, to make an allowance for the error and apply a correction factor when selecting a VOR radial. A makeshift practice such as this would only complicate VOR navigation procedures. Besides this, there is no guarantee that the error is uniform through 360 degrees.

VOR check points are established at some airports on the maneuvering area where there is a sufficiently strong VOR signal to facilitate a check of VOR equipment. A sign at the check point indicates the radial and the distance from the VOR transmitter. The radial indicated on the aircraft VOR indicator should be within 4° of the posted radial. The distance measurement is a check for DME equipment (see **Distance Measuring Equipment,** below).

The airports with VOT facilities are listed in the **Canada Flight Supplement.**

Where there are no special omnitest facilities, you may test your equipment for accuracy by the following method. Tune in the frequency of the nearest omni station and centre the needle. Then rotate the dial 180° and centre the needle again. The reciprocal of reading should be within 4 degrees of the original reading for permissable tolerance.

An alternate method of checking a VOR receiver while in flight is to select a VOR radial that lies along the centerline of a VOR airway. Select a prominent ground checkpoint on this radial about 20 miles from the VOR station. When directly over the ground checkpoint at reasonably low altitude, note the VOR bearing indicated by the receiver. A variation of 6° between the indicated bearing and the published radial is the maximum permissible.

In an aircraft equipped with dual VOR systems, one VOR receiver can be checked against the other when both are tuned to the same VOR facility.

VOR Navigation

To use the VOR navigation system, you must first tune in the proper station on the VOR receiver. The frequency of a particular VOR station is published in the information box for that station on the aeronautical chart. Always listen for the identification signal to be sure the proper station has been selected.

To select a radial, rotate the bearing selector. As you do so, the course deviation indicator (CDI) will begin to move towards the vertical row of dots. When the needle becomes centred, the reading on the bearing selector is the radial transmitted by the omni station on which the airplane is located. Regardless of the heading of the airplane, it is somewhere on the radial indicated by the bearing selector.

The CDI needle, when centred, indicates that the airplane is located on the selected radial but it does not indicate on which side of the station the airplane is located. The TO/FROM indicator resolves this ambiguity.

The To/From Indicator

Fig. 8 illustrates the sectors in which either a TO or a FROM signal is received when the bearing selector is indicating an omnirange radial. For the illustration on the left, the 160° radial of a particular omni station has been selected on the bearing selector. In the illustration on the right, the 45° radial has been selected. If the airplane is anywhere within the unshaded segment, regardless of the direction the airplane may be heading, the TO/FROM indicator will read FROM the station. If the airplane is anywhere within the shaded portion, the TO/FROM indicator will read TO

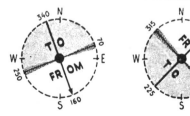

Fig. 8. Omnirange To/From Bearings.

the station, even though the airplane may actually be heading in a direction away from the station.

If the airplane happens to be in a position exactly 90° to the selected radial, the opposing signals that actuate the TO and FROM indicator cancel each other and produce an OFF situation.

In Fig. 9, the bearing selector in each airplane has been set to 0°. Airplanes A, B and H would show TO. Airplanes D, E and F would show FROM. Airplanes C and G would show OFF.

When the airplane passes over the omni station, the TO/FROM indicator will change from TO to FROM or vice versa to show station passage.

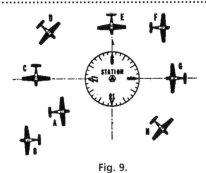

Fig. 9.

The Course Deviation Indicator

The course deviation indicator will remain centred as long as the airplane is actually on the radial indicated on the bearing selector. If the airplane is off course to the right of the radial, the CDI will swing to the left of centre. If the airplane is off course to the left of the radial, the CDI will swing to the right of centre. In other words, the CDI points in the direction in which the radial lies. To centre the needle, fly into the radial, that is, steer into the needle. When the needle centres, the airplane is back on course.

This is **normal needle sensing** and applies when the airplane is headed in the same direction as the radial selected. For example, if the selected radial is 90° and the airplane is headed in an easterly direction, normal needle sensing will apply.

If, however, the 90° radial has been selected, but the airplane is headed in a westerly direction, then **reverse needle sensing** will occur. To centre the needle, in this situation, it is necessary to fly away from the needle.

The easiest method of using the VOR system is to select a radial that is in the same general direction as the heading of the airplane and use normal needle sensing.

The CDI, as well as the TO/FROM indicator, gives an indication of station passage. As the airplane passes over the station, the needle swings to one side and then returns to its original position.

To Fly a Radial Inbound To an Omnirange Station

Suppose you wish to plan a flight from some place on the map to another place where an omnirange is located. Measure the magnetic track on the map from the place where you are now to the omnirange to which you wish to fly. (The magnetic compass rose printed around the omni station on the map is for your convenience in measuring.) The magnetic track you have just measured is the radial you need to fly to reach the station. First of all, tune and identify the station, and then

1. Set the bearing selector to the radial you obtained above.

2. Head the airplane in a direction calculated to make good the intended radial.

3. Observe the TO/FROM indicator. It should indicate TO.

4. The CDI will now point towards the direction the airplane must be steered to reach the selected radial. Steer into the needle until the vertical needle is centred. Then alter course as necessary to keep the needle centred.

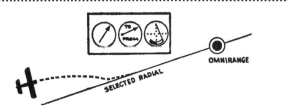

Fig.10. Flying a Radial Towards an Omnirange.

This procedure is known as **intercepting the radial** and the heading which was flown in order to make the interception is known as the **intercept heading**. The **angle of interception** is the angle the intercept heading makes with the selected radial. If you choose to intercept the radial at a large angle of interception (i.e. near 90°), you will reach the radial quickly but, unless you turn quickly, you may easily overshoot it. Selecting a shallower angle of interception (somewhere near 45° or less) will bring you to the selected radial not quite so quickly, but will allow you to make a smoother turn onto it. Of course, if you are quite close to the station, a shallow angle of interception may lead the airplane past the station before the interception can occur. You will have to take the factor of distance from the station into consideration when selecting the angle of interception.

How rapidly you are approaching the selected radial is indicated by the movement of the CDI. If it moves quickly from the side of the case towards the centre, a high rate of interception is indicated. Rate of interception is determined by groundspeed, angle of interception, distance from the station and whether you are flying towards or away from the station.

In order to effect a smooth transition from the intercept heading to the inbound radial, maintain the intercept heading until the CDI needle begins to move. As it nears the centre, begin to turn gradually onto the desired inbound heading. With practice, you will be able to achieve the desired heading for the inbound radial just as the needle centres.

When you pass over the station, the sense indicator will change from TO to FROM. This positively identifies your passage over the station. Suppose you should wish to return to the station. Execute a 180° turn. Be sure to turn the course selector 180°. If you do not do this, you will be flying a course towards a station on a FROM indication. The sensing of the needle will be reversed, and instead of turning towards the needle, you will have to turn away from it to centre it.

So far we have laid much emphasis on the need of doing no more than simply "flying the needle". Now let us offer a word of caution. Flying the needle can be overdone! Always remember that your omni equipment tells you where you are rather than how to get there. To get there, you must **head** the airplane in the right direction. Do not become "needle conscious" to the extent of neglecting your "heading consciousness".

Having intercepted the radial you intend to fly, always turn first to the heading of the radial. Probably after a short time, you will notice some drifting off course, since there is usually a cross wind component. To select a heading that will keep you on the radial, you must make a series of corrections to compensate for the wind drift. Suppose that you are tracking on the radial 030° inbound to station A. You note that on a heading of 030° you have drifted off course to the left. You, therefore, select a heading of 050° (an intercept angle of 20°) to intercept the radial and get back on course. Having intercepted the radial again, divide the intercept angle by 2 and take up a heading of 040°. You note that you are now drifting slightly off the radial to the right and therefore deduce that the correct heading to fly to stay on track is one somewhere between 030° which allows you to drift left of track and 040° which allows you to drift right of track. Intercept the radial again and fly a heading of approximately 035° which should allow you to remain on track directly to the station. This procedure of intercepting a radial and making corrections to remain on track is called **bracketing**.

In this way, you use the needle as a **guide** to correct the heading, rather than as a primary flight instrument, and you will avoid swinging wildly back and forth across the intended radial, trying to get the needle to stay in the centre.

Remember that the **omni bearing** and **airplane heading** will never be very far out of agreement, the difference being the wind correction angle. So centre the needle by making gentle corrections to your **heading** rather than concentrating on the needle itself and you will find that flying the omni is the easiest thing you ever tried to do.

To Fly a Radial Outbound From an Omnirange Station

The procedure is exactly the same as that described above, except that the airplane is headed in a direction calculated to make good the radial away from the station, and the sense indicator will read FROM.

To Obtain a Fix by Omnirange Bearings

In the same way that we were able to fix the position of an airplane by position lines obtained by compass bearings (See **Position Lines** in Chapter **Navigation**), we can also obtain a fix by bearings taken on two or more omni stations.

The pilot of the airplane in Fig. 11 is flying along a track, A-B. He wants to check his position. He tunes in omni station Y and rotates the bearing, or course, selector until the course deviation indicator is centred. As will be seen, the bearing selector will read 273° (FROM). He now tunes in omni station X and obtains a reading on the bearing selector of 185° (FROM). By drawing on the map the two position lines, 273° and 185° from the omni stations, the point where they intersect will be the position at the time the bearings were taken.

It is possible to determine groundspeed and to make time and distance calculations by taking, over a period of time, several fixes from the same two omni stations.

Dual Omni Installations

Many airplanes are equipped with two complete NAV/COM units. As well as providing the dependability factor that, if one radio fails, the second is available for both navigation and communication purposes, the two NAV units allow the pilot to select and fly one radial to or from an omni station and, at the same time, continuously take bearings on another off track station to determine position (i.e. to obtain fixes).

Tacan (Tactical Air Navigation)

TACAN operates in the UHF 960 to 1215 MHz band. Although it is primarily a military system, civil airplanes possessing the appropriate equipment may use it as a navigational aid. It combines the functions of the civil VOR (omnirange) and DME (distance measuring equipment) and therefore indicates to the pilot the azimuth, or direction, of the station to which the equipment is tuned and also its slant distance in nautical miles. Its use requires special TACAN equipment installed in the airplane.

An airplane equipped with distance measuring equipment (DME) can interrogate a TACAN for distance information but any apparent radial information obtained through a coupled VOR receiver is invalid and should be considered a false signal.

Vortac

In the past, TACAN was a separate navigation system providing azimuth and distance guidance to military airplanes. VOR/DME was a system designed to provide similar information to civil airplanes. It was found to be practical to combine them into a unified system by installing TACAN facilities at existing VOR stations, thereby bringing all air traffic, both military and civil, under one common control using coincidental facilities operating from the same geographical locations. The collocated facilities are called VORTACs.

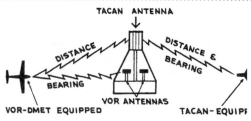

Fig. 12. VORTAC Station, showing facilities available for use with various types of airborne equipment.

VORTAC, therefore, gives the pilot the azimuth, or direction, to or from the station to which he is tuned on VHF and both the direction and distance on UHF.

A pilot with an omnirange receiver can tune in a VORTAC station for directional guidance.

A pilot with DME equipment can interrogate the VORTAC station for distance information. (DME is discussed more fully later in this Chapter.)

A pilot with TACAN airborne equipment can get both azimuth, and distance from the VORTAC facility.

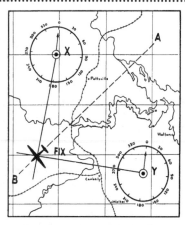

Fig. 11. Fix by Omnirange Bearings.

From the Ground Up

VORTAC stations are depicted on aeronautical charts in the same manner as are VOR stations, except that the symbol sited at the location of the VORTAC transmitter is a three sided figure as shown in Fig. 13. The TACAN channel, as well as the VHF frequency, is printed in the information box.

Fig.13. A VORTAC Station.

Horizontal Situation Indicator (HSI)

The horizontal situation indicator (HSI), also known as the navigation situation display (NSD) or the pictorial navigation indicator (PNI), is an instrument that combines the display of a heading indicator (or directional gyro) with the display of the omni indicator. It was developed to make the pilot's job simpler, eliminating the need to compare the displays of several instruments, to mentally calculate heading changes and to continuously scan and cross check. The instrument enables the pilot to visualize the airplane's position with reference to the selected radial by giving current heading, selected course, course deviation, station position and, when on an instrument approach, position relative to the glideslope and localizer.

The compass system of the HSI may be a simple gyroscopic heading indicator which must be set to correspond to the magnetic compass prior to take-off and must be adjusted during flight to correct for precession. The somewhat more expensive models of HSI incorporate a slaved compass system. A flux detector is mounted somewhere in the airplane where the magnetic influences are least (usually in a wingtip). It senses any movement of the airplane with reference to magnetic north and automatically corrects any precession errors in the HSI.

The **azimuth card** is controlled by the HSI's gyro system. It operates in exactly the same way as a standard heading indicator, moving as the airplane changes heading, so that the heading of the airplane is always at the 12 o'clock position under the lubber line. In Fig. 14, the heading of the airplane is about 285°. The nav display superimposed over the azimuth card moves with it, thus constantly showing the relationship between the current heading and the selected VOR radial. As a result, the HSI gives rather the same type of display as does an ADF. The arrow always points directly at the VOR station and indicates the approximate heading to take up to fly there.

The **azimuth card knob** (the outer portion of the knob at the lower right) is used to set the gyroscopically controlled azimuth card and to correct for precession. In a slaved system, the knob is used to set the system prior to take-off.

The **heading bug** can be set manually by the pilot (using the knob at the lower left of the instrument) as a reminder of the heading he wants to maintain or of the next heading to be flown. When the HSI system is coupled with an autopilot, the heading bug controls the heading mode of the autopilot. Pulling out the knob that controls the heading bug will cage the gyro.

The **course set knob** (the inner portion of the knob at the lower right) controls the position of the **course arrow.** The pilot sets this arrow manually to the desired radial of the VOR to which the omni receiver is tuned. In Fig. 14, the 240° radial has been selected. As the azimuth card rotates, the course arrow moves with the card to always indicate the difference between the airplane's heading and the selected course. The arrow consists of two parts. The front and tail end identify the selected radial. The centre portion of the course arrow acts as a **course deviation indicator (CDI).** By moving to either side of the course arrow, it indicates right or left deviation from the selected radial, just as the left/right needle of a standard omni indicator does. As with standard omni displays, fly into the needle to intercept the selected radial. The dots indicate the number of degrees the airplane is displaced from the radial. In VOR use, each dot represents 5 degrees. In Fig. 14, the CDI indicates that the airplane is off course 5 degrees to the left of the radial and must alter course to the right to intercept the 240° radial. When the course deviation indicator and the two parts of the course arrow are all lined up, the airplane is on the selected radial.

The **station indicator** consists of two white arrows aligned with the course arrow. They perform the same function as the TO/FROM indicator of a standard omni display. When the arrow points to the head of the course arrow (as in Fig. 14), it indicates that the selected radial is TO the station. When the arrow points to the tail of the course arrow, it indicates that the selected radial is FROM the station. On station passage, the arrow moves from the TO to the FROM position, or vice versa. When the selected radial is 90 degrees to the heading of the airplane, the white arrows will retract and not be visible. (Conventional TO/FROM indicators would show OFF in this condition.)

In the centre of the instrument, a fixed aircraft symbol shows the position and heading of the airplane with reference to the other indications of the instrument. Convenient 45 degree tic marks are painted on the outer edge of the display to help the pilot visualize procedure turns and reciprocals.

A **glideslope indicator** is also incorporated into the left side of the HSI. It consists of an arrowhead that moves up and down, showing the vertical position of the airplane with reference to the glideslope. The airplane must fly into the arrow to intercept the glideslope. In Fig. 14, the airplane is below the glideslope and must fly up to intercept it. When adequate glideslope signals are not being received, the indicator and scale are covered by a shutter, or OFF flag.

The HSI may be coupled with an area navigation system. When steering commands are generated by this system, the words RNAV appear in the lower right of the HSI display (See **Area Navigation** in this Chapter).

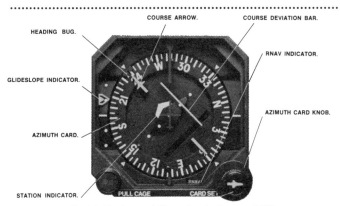

Fig.14. Horizontal Situation Indicator (HSI)

Radio Navigation

The HSI also incorporates a red **heading flag** that appears in the middle of the instrument to indicate loss of electrical power or failure of the azimuth card to track the gyro or slaved system. A **nav warning sign** appears at the lower right of the display whenever the navigation signal from the tuned VOR is not being received or is too weak for reliable navigation.

Radio Beacons

As additional aids to navigation, several types of radio beacons are in use along the airways.

Non-Directional Beacons (NDB)

Low powered non-directional beacons operate mostly in the 190-415 and 510-535 KHz band. They are installed at certain locations near airports to assist with instrument approach and holding pattern procedures.

Non-directional beacons may be installed on the localizer of the ILS. They operate, however, independently of the ILS on a continuous basis. They are installed at these locations to enable an airplane to home to the localizer and to track the approach course by means of ADF. For this reason, they are sometimes called **compass locators.** These NDBs transmit a single letter identifier: the NDB on the front course transmits the first letter of the two letter localizer identification and the NDB on the back course transmits the second. (For example: at Ottawa, the front course NDB would transmit O and the back course NDB would transmit W.)

Fig.15. Non Directional Beacon.

NDBs of higher power are also in use as en route facilities. They form the basis for the LF/MF airways/air routes system. ADF equipment is required in the aircraft to use these navigational facilities. They transmit a two-letter station identifier keyed in morse code. (For example: Flin Flon NDB transmits FO.) Voice transmissions are made from some NDBs.

NDBs are shown on aeronautical charts by shaded circles. Fig. 15 shows the NDB at Red Deer in Alberta. The name of the NDB, its two letter identification in letters and in morse code and the LF frequency on which it transmits are all given in the information box adjacent to the facility on the map.

Fan Marker Beacon

Fan marker beacons are installed to provide guidance on instrument approaches to some airports in mountain valleys.

Fan markers broadcast a signal that is elliptical in shape and has its major dimension at right angles to the localizer course. They are located on localizer facility courses to identify a designated position along the course. Fan markers are coded and emit an audible high pitched tone. They activate the white light on the marker beacon receiver in the aircraft.

Low powered fan markers are usually located on localizer facility courses where they are sited at right angles to the approach path of the aircraft and are designed to identify a designated position along the course. Fan markers are coded and their signal activates a light signal on the marker beacon receiver in the aircraft.

Marine Beacon

Marine radio beacons transmit in the low and medium frequency bands a signal consisting of several letters in code. Some marine radio beacons operate continuously during the navigation season and can be used by aircraft for navigational purposes. Only selected marine radio beacons are shown on enroute aeronautical charts.

Instrument Landing System (ILS)

The instrument landing system, or ILS, is a method of guiding a pilot through the overcast to a landing by means of visual reference to instruments in the cockpit. The system provides alignment and descent information about the approach path of the aircraft as it descends towards the runway.

The system consists of a localizer, a glide path transmitter and a non-directional beacon (NDB) along the approach path. A DME fix may replace the NDB in certain installations.

The **localizer** transmits a radio beam whose purpose is to keep the pilot lined up with the centre line of the runway. The localizer transmitter is located at the far end of the runway and operates in the VHF frequency band between 108.1 and 111.9 MHz (on odd frequencies). The transmitter sends out two signal patterns, one modulated at 90 hertz and the other at 150 hertz. The 90 hertz signal is to the left of the centre line and the 150 hertz signal is to the right of the centre line. When the aircraft is on an extension of the centre line of the runway, it receives the two signals equally and the localizer needle on the OMNI display is centered. The airplane will remain lined up with the runway if the pilot maintains a heading to keep the needle centered. The localizer course which is associated with the glide path and the outer marker beacon is called the **front course.** Many localizer installations also provide a **back course** which is the reciprocal of the front course and which enables a pilot to carry out a non-precision approach using the other end of the runway. However, when inbound on the back course, reverse needle sensing applies and it is necessary to correct opposite to the needle. There is no glide path information provided for a back course approach. Not all ILS localizers radiate a usable back course signal. Always check the published information about an ILS at any particular runway.

The **glide path** beam guides the aircraft along the proper downward approach path to enable it to touch down on the threshold of the runway. The glide path transmitter is situated about 1000 feet from the approach end of the runway and is offset approximately 400 feet from the runway centre line. It operates in the UHF frequency band between 329.3 and 335.0 MHz. Each glide path frequency is paired with a corresponding VHF localizer frequency. The centre of the glide path beam is aligned with the centre of the localizer beam (the area of equal strength between the 90 and the 150 hertz patterns) and extends about 1° above and below the centre. The glide path projection is normally set at an angle of 3° from the horizontal. There is no usable back course glide path. At some airports, an ILS is installed at each end of a runway to provide a front course approach to either end of the runway. The two systems are interlocked so that only one ILS can operate at any one time.

Identification for both the localizer and glide path is transmitted on the localizer frequency on the form of a two letter or letter number indicator preceded by the letter I: a 1.2 second space separates the I from the standard identifier. For example I OW — ILS Ottawa.

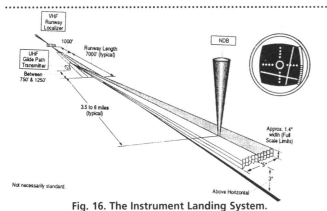

Fig. 16. The Instrument Landing System.

The insert shows the localizer and cross-pointer indicators. The instrument illustrated shows the pilot is off the localizer beam to the left and must steer right "into the needle" to centre the vertical needle and get back on the beam,. He is above the glide path and must fly down "into the needle" to centre the horizontal needle and get back on the glide path.

Note: the above needle indications apply when the pilot is inbound towards the field. When flying outbound away from the field, the needle sensing is reversed.

The instrument in the cockpit which enables the pilot to follow these two beams is the course deviation indicator and cross pointer indicator of the VOR indicator (See Fig. 7 and the insert in Fig. 16). The vertical (localizer) and the cross (glide path) pointers give the pilot directions. If the aircraft gets off to the right or left of the approach to the runway, the vertical needle indicates which side and how far. The horizontal needle indicates if the aircraft is above or below the proper glide path. A flag alarm signal appears when a weak, unreliable or null signal is being received.

A low powered NDB is located on the localizer (front and back courses) about 3.5 to 6 miles from the runway threshold to provide a fix to which the pilot can navigate for the transition to the ILS. In some cases, an en route NDB is located on the localizer to serve as a terminal as well as an en route facility. These NDBs transmit a single letter indicator: the NDB on the front course localizer transmits the first letter of the two letter localizer indicator and the NDB on the back course transmits the second.

A DME fix, which may be installed where it is not practicable to install an NDB provides distance information to define the initial approach fix and missed approach fix.

A further development of the system combines ILS with the automatic pilot, enabling the airplane to be flown down the two beams to a landing automatically.

The original ILS installations included several **marker beacons**. An outer and a middle marker were installed on the front course and a back marker was sometimes installed on the back course. Their purpose was to let the pilot know the distance of the aircraft from the threshold of the runway during an instrument landing approach. The marker beacons emitted an audible signal that was heard as a series of dots and/or dashes. They also illuminated signal lights on the marker beacon receiver in the aircraft. These marker beacons are, for the most part, being phased out and are being replaced by NDBs. They may be retained in a few locations where terrain difficulties have made the installation of an NDB impracticable.

The broadcast beam of the ILS localizer can be considered valid and reliable through 35° on either side of the front course centre line for a distance of 10 NM from the transmitter and through 10° for a distance of 18 NM from the transmitter, for both the front and back courses. Beyond these limits, false or erratic indications may be received.

Failure of certain elements of the multi-element localizer antenna systems can cause false courses or low clearances beyond 8° from the front or back course centre line. This could result in a premature cockpit indication of approaching or intercepting an on course centre line. A coupled approach should be initiated only when the aircraft is established on the localizer centre line. It is important also to confirm the localizer or course indication by reference to the heading indicator. False on course captures can occur anywhere from 8° to 35° but are most likely to occur in the vicinity of 8° to 12° azimuth from the published localizer course.

Electronic interference on ILS localizer system integrity is becoming significant especially in built up areas where power transmitter stations, industrial activity and broadcast transmitters generate interference with localizer receivers.

The interference may be transitory. Certain receivers are more susceptible than others. The interference is not likely to cause erroneous readings where the aircraft is being flown within the margins quoted above.

ILS Categories

Operational CAT I allows operation down to a minima of 200 feet decision height (DH) with a runway visual range of 2600 feet. If RVR information is not available, 1/2 statute mile ground visibility is substituted.

Interim CAT II allows operation to a minima as low as 150 feet DH with an RVR of 1600 feet.

Operational CAT II allows operation down to a minima as low as 100 feet DH with an RVR of 1200 feet.

Precision approach lighting systems are installed at airports that have been approved for CAT I and CAT II operations.

Localizer Approach

A localizer without glide path guidance is installed at some airports to provide track guidance during an approach. These aids may have a back course associated with them. Where the localizer is aligned within 3° of the runway, the approach procedure is called a **localizer (LOC) approach.**

Microwave Landing System (MLS)

The microwave landing system (MLS), like ILS, is designed to provide precise navigation guidance in the form of alignment and descent information to aircraft on approach to and landing on a runway. MLS is considered a more reliable and more accurate system than ILS and it can be installed at sites where, because of difficult terrain, excessive site preparation or shortage of frequencies, ILS is not practical.

Where ILS provides a single, fixed approach path, MLS transmits two scanning beams, one lateral and one vertical, to provide a variety of approach paths for different types of aircraft, from airliner jets to STOL aircraft to helicopters.

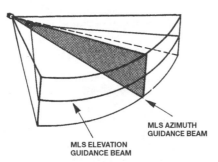

Fig. 17. Microwave Landing System.

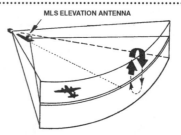

Fig. 19. MLS Elevation Antenna.

The **approach azimuth guidance** transmitter radiates a lateral scanning beam which sweeps from left to right of the extended centre line of the runway. The sweep is normally 40 degrees to either side but can be adjusted to suit local terrain or other requirements. The MLS receiver in the aircraft determines position by measuring the time between TO and FROM sweeps of the beam. In Fig. 18, a long interval (A) indicates that the aircraft is to the right of the centre line and a short interval (B) indicates that the aircraft is to the left. The range of the azimuth guidance beam is at least 20 n. miles and 20,000 feet.

Fig. 18. MLS Azimuth Antenna.

The **elevation guidance** transmitter radiates a fan shaped beam which sweeps up and down through the azimuth coverage area. In its up and down sweep, the beam covers an angle of 15° from the horizontal. The MLS receiver in the aircraft measures the time between sweeps of the beam to determine the descent angle of the aircraft.

The MLS receiver control unit in the aircraft permits the pilot to choose an approach path from 40° left to 40° right of the centre line of the runway and a descent angle from 1° to 15°. However, at every MLS installation, a minimum glide path angle is established that provides clearance of terrain and other approach obstacles. This number is transmitted as digital data and is the lowest glide path angle that can be selected. After setting the azimuth and elevation angles, the pilot will receive precise fly left/fly right, fly up/fly down guidance to the runway.

The MLS can be expanded to also incorporate a **back azimuth guidance** that transmits a beam to provide lateral guidance for missed approach and departure navigation. The lateral coverage is limited to 20° either side of the centre line. The range is about 5 n. miles up to about 5000 ft.

The MLS also includes as an integral part of the installation **range guidance.** Precise distance-to-go information is received in the aircraft from a **precision distance measuring equipment (PDME or DME/P)** transmitter. This equipment functions in much the same manner as regular DME installations, but with improved accuracy.

All MLS facilities transmit basic data, such as the station identification, DME/P channel and information associated directly with the operation of the landing system. Certain MLS may be expanded to broadcast auxiliary information, such as weather and runway conditions, etc.

The microwave landing system transmits in the C-band frequency range of 5031 to 5091 MHz. Two hundred individual channels are available for use.

MLS identification is a four letter designation starting with the letter M. The designation is transmitted in morse code.

Lateral and vertical guidance information can be displayed on conventional course deviation indicators. Range information can be displayed on conventional DME receivers. Lateral, vertical and range information may also be incorporated into multipurpose cockpit displays.

The great advantage of MLS is flexibility. It allows curved and segmented approaches, selectable glide path angles, accurate 3-D positioning of the aircraft in space, the establishment of boundaries to ensure clearance and departure phases and VTOL and STOL operations. The system has low susceptibility to interference from airport ground traffic and from weather conditions, such as heavy snowfall which adversely affect ILS.

Automatic Direction Finder (ADF)

The automatic direction finder is a very useful aid to aerial navigation. It can be used for homing to or finding direction from any station that broadcasts in the low and medium frequency radio bands.

Aviation installations that operate in the low frequency band include non-directional beacons and NDBs co-located with ILS localizers. The ADF is also able to receive commercial AM broadcasting stations and home on them.

An L/MF aviation radio installation (NDB, etc.) is readily recognizable by the morse code identifier that is broadcast continuously. In the case of a commercial broadcasting station, it is sometimes necessary to listen for some time for a station announcement before positive identification can be made. However, modern digital tuned receivers do make it easier to tune a station precisely.

The great advantage of ADF is that of range. Not being dependent on line of sight as is the VHF omnidirectional range (VOR) system, an ADF receiver is able to pick up stations at considerable distances.

ADF Equipment
The ADF system consists of an ADF receiver, a loop antenna, a sense antenna and a bearing indicator.

Antennas
Two antennas are required for ADF operation. One of these, known as the **sense antenna,** is a non-directional antenna that has the capability of providing directional information.

The other antenna is a **loop antenna** which senses magnetic bearing from the airplane to the station.

Back in the early days when ADF equipment was first introduced, the loop antenna was a complicated installation that had to be manually rotated to pick up maximum and minimum signals from a station. Modern ADF equipment is manufactured on the principle of solid state circuitry and has fewer moving parts and is therefore easier to operate and is considered more reliable.

The sense antenna usually is a long wire installed on the top of the airplane, stretching from an insulator near the top of the fuselage back to the stabilizer.

The loop antenna, a metal ring enclosing coils of insulated wires, is usually contained within a streamlined housing mounted well forward on the underside of the fuselage.

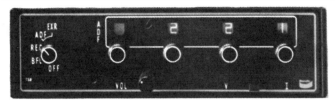

Fig. 20. Automatic Direction Finder.

Very recent innovations in ADF equipment incorporate both the sense and the loop antennas in a single blade antenna installed on the top of the fuselage.

The ADF Receiver

Modern ADF receivers are digital tuned providing rapid and precise tuning to any desired frequency. They receive stations throughout the low and medium frequency band.

The ADF receiver, like the ADF pictured in Fig. 20, receives information from the antenna and translates it into movement of the needle of the bearing indicator.

The ADF receiver incorporates a five position function selector switch.

The REC position (on other models of ADF equipment, this position is sometimes called ANT for antenna) selects only the non-directional or sense antenna. The loop antenna and therefore the bearing selector are disconnected from the system. With the switch in this position, the receiver operates as an ordinary L/MF receiver, receiving the voice facilities of the station to which it is tuned. This switch position is used for station identification or for listening only.

When operated in the ADF position, the receiver selects the automatic direction finding loop antenna which in combination with the signal received by the sense antenna gives the homing capability of the ADF. The audio feature of the REC position is retained but usually the pilot turns down the volume once the station has been positively identified

The EXR position is a feature of the particular ADF shown in Fig, 20. It provides extended range selection, permitting strong reception of stations at great distance.

The BFO (**beat frequency oscillator**) selection provides an aural aid to tuning. It should be used only in tuning unmodulated signals, such as those broadcast by stations in some foreign countries. It is not designed to be used with the modulated signal that is transmitted by commercial broadcast stations and aeronautical LF stations in Canada and the United States. The BFO function allows the underlying morse code identifier to be heard through the audio tone of the unmodulated signal. After the station is tuned and identified, the function selector switch should be repositioned at ADF.

The Bearing Indicator.

An ADF bearing indicator incorporates a bearing pointer and an azimuth dial which is graduated in degrees from 0 to 360. Some indicators have a fixed card, such as that in Fig. 21. Others have a movable card, such as that in Fig. 22.

With a bearing indicator with a fixed card, the nose of the airplane is always oriented to 0 on the bearing indicator, regardless of the actual magnetic heading of the airplane.

With a bearing indicator with a movable card, the pilot is able to rotate the azimuth card so that it corresponds with the heading of the airplane indicated by the heading indicator and the magnetic compass.

Regardless of which type of bearing indicator is used, the bearing pointer always points in the direction of the station to which the receiver is tuned.

Fig. 21. ADF Bearing Indicator (Fixed Card).

The loop antenna which senses the magnetic bearing of the station is positioned in relation to the longitudinal axis of the airplane. It measures, and the ADF pointer on the bearing indicator displays, the number of degrees, measured clockwise, between the longitudinal axis of the airplane and the station. This is known as a **relative bearing**.

The magnetic bearing of a station can, therefore, be determined from the displayed information. It is the sum of the magnetic heading of the airplane and the relative bearing.

> Magnetic Bearing = Magnetic Heading + Relative Bearing
> MB = MH + RB
> Or, conversely
> MH = MB - RB or RB = MB - MH

For example, the airplane in which the ADF in Fig. 21 is installed is flying on a magnetic heading of 090°. The ADF bearing indicator (fixed card) indicates that the station to which the ADF receiver is tuned has a relative bearing from the airplane of 290°. The magnetic bearing to the station (i.e. the magnetic heading to which the airplane would have to turn in order to fly directly to the station) would therefore be 020° (090° + 290°).

An ADF bearing indicator with a movable azimuth card automatically displays the magnetic bearing of the station. In Fig. 22, the azimuth card of the bearing indicator has been rotated to 090° to correspond to the magnetic heading of the airplane. The pointer indicates that the station to which the ADF in this case is tuned has a magnetic bearing of 120°(the relative bearing is 30°).

Fig.22. ADF Bearing Indicator (Movable Card).

ADF Navigation

First of all, let us review the behaviour of the bearing indicator in various flight situations.

Let's take an example of an airplane with a bearing indicator with a fixed card. If the airplane is heading directly towards the station, the bearing indicator pointer will point straight up to 000° (towards the nose of the airplane). If the airplane is heading directly away from the station, the pointer will point straight down to 180° (towards the tail of the airplane). If the airplane turns 90° to the signal from the station, so that the station is on its left, the pointer will point to 270° (towards the left wing). If the station is on the right, the pointer will point to 090° (towards the right wing).

Suppose that the bearing indicator has a movable card. The airplane in the above example is on a heading of 020° magnetic. The pilot has rotated the azimuth card of the bearing indicator to set 20 under the upper azimuth index. If the airplane is flying directly towards the station, the pointer will point straight up to 020°. If the airplane is heading directly away from the station, the pointer will point straight down to 200° (020° + 180°). If the station is directly off the left wing, the pointer will point to 290° (020°+ 270°). If the station is off the right wing, the pointer will point to 110°(020° + 090°).

Now, let us suppose that an airplane is flying a track that will take it directly over the station to which the ADF is tuned. On the heading in to the station, the ADF pointer will point to 000° (fixed card). As soon as the airplane passes over the station, the pointer will swing around and point to 180° as the airplane heads away from the station.

Let us suppose that an airplane is flying a track that will take it past a station that is off its right wing. As the airplane approaches and passes the station, the ADF pointer will gradually increase its reading from its initial indication, past 090°and on towards 180°. As long as the airplane maintains a constant heading, the bearing pointer, of course, will never point directly to 180°since the station will never be directly behind the airplane.

In the following discussion of how to use the ADF, we will assume that the airplane is equipped with an ADF bearing indicator with a fixed card.

Homing to a Station

In order to fly directly to the ADF facility to which the receiver is tuned, simply turn the nose of the airplane until the pointer of the bearing indicator points to 000°. Keeping the pointer on zero will result in a straight course to the station. On passage over the station, the pointer will swing around to 180° indicating that the station is now behind. The airplane is then able to track away from the station by keeping the pointer of the bearing indicator on 180°.

The above procedure is valid in a no-wind situation. The wind, however, tends to drift the airplane to one side or the other of a straight line to the station. A continuous process of correction of drift occurs as the ADF keeps heading the nose of the airplane towards the station.

The resultant track flown is a curved line (Fig. 23).

The progressive change in the heading of the airplane will be indicated to the pilot by the magnetic compass or heading indicator.

In a cross wind situation where the wind is causing the airplane to drift away from the straight line track to the station, it is possible to narrow the curvature of the track by the following method:

The pilot has tuned the station on which he wishes to home and has centred the ADF pointer to 000°. As the track that he intends to fly is 090° magnetic, the heading indicator will read 090° (Fig. 23. A).

After flying some minutes, the pilot notes the heading by H.I. has altered to 095°. He, therefore, realizes that the aircraft is drifting to port and that the wind is from the starboard. He heads the nose of the airplane, say, 3° more into the wind, steering 357° by ADF instead of 000° (098° by H.I.)(Fig. 23. B).

Note that when the heading is altered to starboard by the heading indicator (i.e. increases), the heading on the ADF alters to port (i.e. decreases).

If the heading continues to alter to starboard by H.I., the pilot should steer a few more degrees into wind and continue this process until he finally observes that his heading by H.I. remains constant. He has now counteracted wind drift and established a straight heading towards the station.

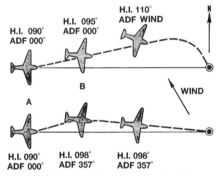

Fig. 23. Correction for Wind Drift.

Having reached and passed the station, a pilot must be especially careful to maintain the desired track when flying an outbound track from an ADF facility. If there has been a significant cross wind, keeping the ADF needle on the tail of the airplane (180°) and allowing the heading to change will result in a curved track leading away from the desired destination. This **reverse homing error** will take him far off course. Flying an outbound track requires special attention to wind drift, crab angle and heading.

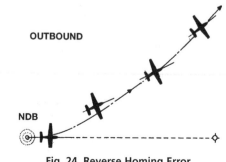

Fig. 24. Reverse Homing Error.

From the Ground Up

To Intercept an ADF Track

Sometimes a pilot may wish to fly a specific course into a station. This would happen when an ADF facility, such as an NDB, is located several miles from an airport. The extension of the magnetic track joining the NDB to the airport will be the track the pilot needs to intercept and turn to in order to arrive over the station on the proper heading to reach the airport.

After tuning in the station, the pilot should select a magnetic heading by H.I. that will intercept the required course to the ADF facility. By applying the formula given above (i.e., RB = MB - MH), it is relatively simple to determine the relative bearing shown on the ADF indicator that indicates interception of the required track.

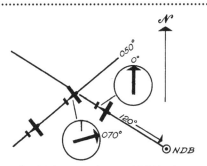

Fig. 25. Intercepting an ADF Track.

For example, the course to the NDB that the pilot wishes to intercept has a magnetic bearing of 120°. The magnetic heading selected by the pilot in order to make the interception is 050°. The ADF indicator should indicate a relative bearing of the station of 070° when the airplane intercepts the course (magnetic bearing of 120° minus magnetic heading of 050° equals relative bearing of 070°).

On intercepting the course, the pilot should turn the nose of the airplane until the bearing indicator pointer centres on 000° and then keep the pointer on 000° to arrive over the station.

In actual practice, some allowance should be made for the turn. The turn onto the desired course should be started a few degrees before the ADF pointer reaches the required relative bearing.

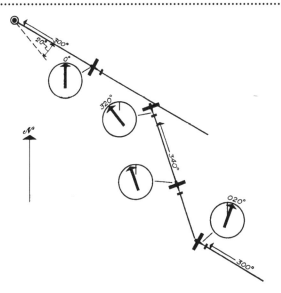

Fig. 26. Choosing the Proper Magnetic Heading in order to Intercept an ADF Course.

How does a pilot choose the proper magnetic heading to fly to intercept a particular ADF course? First of all, he should take up a heading which parallels the desired course, check the ADF indicator and note the number of degrees the needle is off the 0 index. The interception angle should be double this amount. For example, let us assume that you wish to intercept a course to an ADF station that has a magnetic bearing of 300° to the station. Take up a heading of 300° and you note on the ADF indicator that the station is to your right at a relative bearing of 020°. Make 40° your interception angle and fly a course of 340° magnetic. When the ADF bearing indicator indicates a relative bearing of 320°, you have intercepted the desired course. (Magnetic heading of 340° plus relative bearing of 320° equals magnetic bearing of 300° — 340 plus 320 less 360). Turn onto a magnetic heading of 300° and fly the ADF track into the station by keeping the ADF needle on 0.

Taking Bearings

Bearings can be taken on ADF facilities in the same way as they can be taken on VOR stations.

To take a bearing on an NDB or a broadcasting station, etc., that is off track, tune in to the frequency of the station. The pointer on the bearing indicator will indicate the direction of the station from the heading, or longitudinal axis, of the airplane. This is, as we have already learned, a relative bearing from the airplane to the station.

The magnetic bearing of the station is the sum of the magnetic heading of the airplane and the relative bearing.

To plot a position line on an aeronautical chart, the magnetic bearing of the station must be converted to a true bearing by applying deviation and variation.

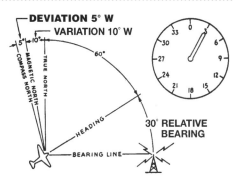

Fig. 27. Conversion of a Relative Bearing (shown on dial) to a Geographical Bearing

Fig. 27 graphically shows how to convert a relative bearing to a true, or geographical, bearing.

The relative bearing of the station from the airplane is 30°.

The compass heading of the airplane is 075°. Therefore, the compass bearing of the station is 105°.

By applying the deviation of 5°W and the variation of 10°W, the true bearing of 090° is determined.

The true bearing can be plotted as a position line on an aeronautical chart. Draw a line on the map away from the station, that is, using the reciprocal of the bearing. This is a position line, somewhere on which the airplane is known to have been at the time the bearing was taken.

When two or more bearings are taken nearly simultaneously on several different stations and the position lines plotted on a map, the point where the position lines intersect will be a fix (the position of the airplane at the time the bearings were taken).

When plotting a fix by two or more position lines, the angle between the position lines should not be less than 30°.

Dual ADF

Some airplanes are equipped with two automatic direction finders. With dual ADF, a pilot may take simultaneous bearings on two stations, or he may home on one station and at the same time obtain a continuous bearing on another station, giving him a running position fix.

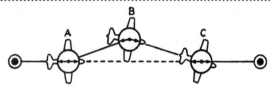

Fig. 28. Correction for Wind Drift with Dual ADF

A. Pilot tunes one ADF to station ahead and the other to station behind. Dual pointers are on reciprocal bearings. B. Wind drifts pilot to left of the course. Dual pointers no longer on reciprocal bearings. C. Pilot steers right until dual pointers are on reciprocal bearings again, indicating that the course has been regained. For remainder of flight pilot maintains heading which will keep dual pointers on reciprocal bearings.

Dual ADF may be used to correct wind drift when homing. This is done by tuning one receiver to a station ahead and the other to a station behind. The heading is adjusted to keep the two needles in line, that is, on reciprocal bearings (Fig. 28).

Dual ADF is particularly useful to a pilot on instruments executing a holding pattern procedure over an NDB. It can also be used to track the approach course to the runway.

Inaccuracies of the ADF System
Quadrantal Error

The ADF is affected by an error very similar to deviation in the magnetic compass. This is called **quadrantal error.** It is caused by refraction, or bending of the incoming radio waves by the metal structure of the airplane. Quadrantal error is minimal at the cardinal points (off the nose, tail and wing tips) of the airplane and maximum at bearings in between. Most ADF receivers have provisions for making a partial correction for this error. In this case, the loop antenna should be installed in a position on the airplane where the built-in quadrantal error compensation matches the error caused by the airplane. But this is difficult and time consuming and rarely done. The quadrantal error can be determined by swinging the airplane both on the ground and in the air. However, this too is difficult to do and therefore is rarely done. Usually the error is less than 10° and since the system is accurate off the nose and other cardinal points, the quadrantal error does not affect operation of the ADF significantly.

Oscillation of the Needle

Normally, a small degree of needle oscillation is a built-in feature of the automatic direction finder. It indicates to the pilot that the instrument is functioning. Wide fluctuations of the needle, however, may be due to the following causes:

1. Bad static conditions.

2. A weak signal. (The ADF is tuned to a station that is too far distant or not sufficiently powerful.)

3. The airplane is midway between two stations which are both transmitting on the frequency tuned.

When wide oscillation, or hunting, of the needle occurs, average the oscillation and fly the average heading.

Night Effect

The low and medium frequency radio bands produce both ground and sky waves. The ADF homes on the ground waves. At night, however, the sky waves are much stronger than in the day time since at night they are most reflected by the ionosphere. Since the sky waves may be arriving from any direction depending on how and where they have been reflected, they cause errors in the ADF bearing indication. The bearing indicator will swing rapidly as the sky wave changes. This is called **night effect.** The farther from the station, the more pronounced will be the night effect. Within some minimum distance of the station (varying from 20 to 40 nautical miles), the night effect usually disappears. Night effect is also more pronounced at the higher frequencies so that homing on a broadcast station will be more affected than on an NDB beacon. In order to minimize night effect, therefore, choose a station to home on, if possible, with a low frequency.

Terrain Effect
Coastal Effect

As radio waves pass from land to water, their direction of travel is changed. This is known as shoreline effect. Because of this, a bearing taken on an inland station from an airplane over water is inaccurate if it makes an angle of less than 30° with the shoreline. At greater angles, bending is negligible. When taking bearings over water, therefore, choose either stations right on the shore, such as marine stations, or stations so located that bearings on them make angles greater than 30° with the shoreline.

Mountain Effect

In mountain terrain, the signal is reflected off the side of mountains. The signal that arrives at the ADF loop antenna may not, therefore, be coming directly from the station. Erroneous and fluctuating bearings may be registered on the ADF bearing indicator.

Ore Deposits

Ore deposits will sometimes cause the needle of the ADF to deflect off course.

Ice and Sleet

Accumulation of ice and sleet on the antennae tends to produce erroneous readings of the ADF needle and also reduce signal strength.

Precipitation Static

When flying in heavy rain or snow, static may be encountered on the low and medium frequency bands. This static may be severe enough to interfere with the pointing capabilities of the bearing indicator needle. Reception of the audio signal may be so poor that identification of the station is impossible.

Precipitation static can be lessened by reducing the airspeed of the airplane.

Thunderstorms, containing large charges of static electricity, also interfere with the L/MF radio waves and cause erroneous readings of the ADF.

Radio Magnetic Indicator (RMI)

The radio magnetic indicator (RMI) is an instrument that combines the directional function of a heading indicator (or directional gyro) with an indicator which points to a specific selected navaid, in other words, a bearing indicator. The development of the radio magnetic indicator was a natural outgrowth of the introduction of ADF and is a valuable instrument because it tends to reduce pilot workload.

The rotating compass card of the RMI is controlled by a remotely mounted slaved compass so that the magnetic heading of the airplane relative to magnetic north is always presented under the index at the 12 o'clock position of the instrument. In a slaved system, there is no tendency to precession and there is, therefore, no requirement for periodic adjustment as is necessary in a gyroscopic heading indicator.

Fig.29. Radio Magnetic Indicator (RMI).

The bearing indicator needles of the RMI always point to the navigational facility and therefore display magnetic bearing to the station.

Most radio magnetic indicators use two indicator needles, one of which is tied to the ADF receiver while the other responds to signals from the VOR receiver. Each needle points to the station to which its complimentary nav receiver is tuned. Usually the ADF information is displayed on the double bar pointer and the VOR information on the single bar pointer. The arrowhead of the RMI VOR needle indicates the radial inbound to the station and the tail indicates the radial outbound from the station. (The RMI VOR needle does not represent the position of the airplane with reference to a desired radial regardless of the airplane's direction of flight, as does the course deviation indicator. It merely points to the station in exactly the same way as the RMI ADF needle points to a L/MF navaid.)

In the RMI depicted in Fig. 29, the airplane is on a magnetic heading of 324°. An NDB (or other L/MF) facility is off the port wing at a bearing of 290° magnetic. A VOR facility is off the starboard wing at a bearing of 012° magnetic. As long as the airplane continues on its present course, the ADF needle will continuously move counterclockwise as the airplane passes abeam of the station. The VOR needle will move clockwise as the airplane moves past that station. (In both cases, the needles will slow in their movement as they approach the 6 o'clock position and will settle short of it, since neither station will ever be directly behind, but always slightly to port and starboard respectively.) If the pilot wishes to intercept the 040° radial inbound to the VOR, he would continue on course (324°) until the arrowhead of the VOR needle points to 040° on the compass card and then turn to starboard (into the needle) to

that heading. The single bar needle will then be pointing straight up pointing to the VOR station directly ahead. The double bar needle will continue to respond to the ADF signal from the L/MF facility that is now astern of the airplane.

The latest generation of RMI incorporates a switching capability by which the pilot is able to select either VOR or ADF for each needle. The instrument can, at the pilot's option, work both needles on two different VORs or on two different L/MF navaids or on one VOR and one L/MF station. (The airplane must, of course, be equipped with dual VHF nav receivers and dual ADF receivers to utilize this feature to full capacity.)

The great advantage of the RMI is that it offers instant information on magnetic bearings to navaids on one instrument. The pilot is freed of the mental task of calculating relative bearings and magnetic bearings. In addition, the two needles provide simultaneous bearings on two different stations, giving the pilot position lines by which to continuously establish fixes on stations that are abeam of the airplane. The VOR feature of the instrument frees the pilot of the task of centering the needle of the OBS to determine which radial he is on. The RMI presents continuously the relative position of a VOR in the same way as it does an ADF station. The information can be used to identify a VOR intersection as has been described above. The RMI provides more warning of interception than does a conventional CDI. To fly directly to a VOR, it is only necessary to read the magnetic bearing to the station under the arrowhead of the VOR needle and turn to that heading.

Distance Measuring Equipment (DME)

DME (distance measuring equipment) is an electronic transmitter/receiver that gives the pilot a continuous reading of distance to a fixed ground station. It does this by measuring the time it takes a transmitted radio signal to travel to and return from the ground station. Since radio waves travel at a constant, known speed, it is easy to convert the radio signal travel time into distance. The ground installation from which DME can obtain distance information is a VORTAC or a TACAN station transmitting a radar beacon on UHF.

The airborne equipment transmits pulses to the ground station that, in effect, ask "How far am I from your station?". Each aircraft interrogation has a unique transmission rate or pattern which the ground station reproduces in its replies to that aircraft even though transmitting on the same frequency to all aircraft working the station. The airborne DME receiver selects the reply pulses designated for itself (i.e. those with the same rate relationship as used in its own original transmission), measures electronically the time interval between its own transmission and the received reply, converts the elapsed time into a mileage figure which is continuously displayed in the cockpit indicator as the **slant range** from the airplane to the station. The range increases as the plane flies away from a station and decreases as it approaches the station.

Slant range can be converted into ground distance by the following formula: ground distance is the square root of S^2 (slant range in nautical miles) minus A^2 (altitude in nautical miles). When the slant range is more than double the altitude, the difference between slant range and ground distance is negligible. However, when the aircraft is in the vicinity of a station and at high altitude, the difference between slant range and ground distance will be appreciable. In such a case, the pilot must employ the above formula to determine his actual ground distance from the station unless his DME includes a component solver that converts slant mileage to horizontal mileage automatically.

In practical application, the DME readings can be considered accurate if the airplane is over 5 miles from the station. Within 5 miles, slant range must be considered. Directly over a VORTAC installation, the DME will read altitude above the station in nautical miles.

DME units also have the capability to give groundspeed readings and a time to station reading.

The DME pictured in Fig. 30 is a typical airborne unit. It has 200 channel capability and a gaseous discharge display. It features two display modes. The first provides simultaneous distance to station, groundspeed and time to station information. The second displays distance to station and frequency. The frequency may be selected either internally or remotely from a VOR nav receiver.

All DME distance readings are in nautical miles and the groundspeed readings are in knots.

A DME groundspeed reading, however, is accurately displayed only when the airplane is flying on a course inbound or outbound from a VORTAC station. When the airplane is flying past a station, the DME groundspeed reading will be meaningless because of the principle on which the DME makes its computations. An inaccurate reading will also be displayed when the airplane is passing directly over the station. The groundspeed indicator will drop to 0 for the period that the airplane is over the VORTAC installation.

The groundspeed feature of the DME is of especial value to you for it allows you to choose the altitude where you will encounter the most favourable winds. By flying at several different altitudes and observing the groundspeed readout on the DME, you can determine at which altitude the groundspeed was highest (i.e. an indication of the most favourable winds).

Fig. 30. DME Receiver

The DME is also used in making orbiting or arc approaches to an airport. The **DME arc** is a course flown at a constant DME distance around a navaid such as a VORTAC which provides distance information. In a case where the VORTAC is located a few miles (for example, 5 miles) from the airport, by keeping that mileage reading constant on your indicator (5 miles in our example), you will be able to bring your airplane right over the airport. The exact position of the airport can be pinpointed by using the omni in conjunction with the DME. The airport lies on a particular omni radial from the VORTAC and will be located where the radial intersects the orbiting track.

DME operates in the 960 to 1215 MHz UHF range, transmitting in the range from 960 to 1024 MHz and from 1151 to 1215 MHz and receiving in the 1025 to 1150 MHz range.

DME is normally collocated with VOR installations (VOR/DME) It may also be collocated with an ILS or with localizers for LOC/LDA approaches. It is also sometimes collocated with en route NDBs to provide improved navigation capability. DME channels are paired with VOR, NDB or localizer channels so that, when you choose on

MHz	.0	.1	.2	.3	.4	.5	.6	.7	.8	.9
108	17	18	19	20	21	22	23	24	25	26
109	27	28	29	30	31	32	33	34	35	36
110	37	38	39	40	41	42	43	44	45	46
111	47	48	49	50	51	52	53	54	55	56
112	57	58	59	70	71	72	73	74	75	76
113	77	78	79	80	81	82	83	84	85	86
114	87	88	89	90	91	92	93	94	95	96
115	97	98	99	100	101	102	103	104	105	106
116	107	108	109	110	111	112	113	114	115	116
117	117	118	119	120	121	122	123	124	125	126

Fig.31. Chart of TACAN and VHF Channels. For example: TACAN channel 80 pairs with 113.3 MHz

your DME equipment a VOR channel, you automatically choose the UHF frequency as well. Channels on the DME channel selector are marked with the VOR frequencies even though it is controlling DME channels in the UHF MHz range. A DME frequency requiring only one decimal place (e.g. 110.3 MHz) is known as an X channel; one that requires 2 decimal places (e.g. 112.45 MHz) is known as a Y channel.

The TACAN military installations can also be used to derive distance measurement. The TACAN stations are designated on navigation charts by channel numbers 1 through 126. The chart in Fig. 31 shows how to translate a TACAN channel into megahertz in the VOR band.

Remember, however, that **only** the distance information derived from a TACAN station is valid. Any apparent radial information obtained through a coupled VOR receiver must be considered a false signal since the nature of the TACAN installation is such that only special TACAN equipment can derive azimuth information from it.

DME equipment is considered reliable up to 200 nautical miles from the station. Some equipment, however, limits its distance readout to a maximum of about 170 miles. It is considered accurate within about 2%. Since UHF reception is dependent on line of sight, altitude is important in the efficiency of the equipment. The higher the altitude, the greater the effective range.

Flight Director

A flight director is a highly sophisticated system that electronically collects the information provided by a number of instruments, feeds this information into a computer and presents it in two visual displays.

The two instruments that provide this visual display are a horizontal situation indicator (HSI) and an attitude director indicator (ADI). Usually the flight director installation is combined with an autopilot system. The entire package is then known as a **flight control system.** Different manufacturers give other names to the components of their systems. The HSI is sometimes called the **pictorial navigation indicator (PNI)** and the ADI is sometimes called the **flight command indicator.**

We have already discussed the HSI at some length. (See **Horizontal Situation Indicator** in this chapter.) The HSI pictured in Fig. 32 has two additional features to a conventional HSI. It displays a readout

Fig. 32. Flight Director System.
Top: Attitude Director Indicator (ADI).
Bottom: Horizontal Situation Indicator (HSI).

of the selected VOR radial and a DME readout of distance to the station. The ADI is somewhat similar to an attitude indicator but much more sophisticated. It incorporates a command display mounted over a horizon display. It is of prime value as an aid to flying instrument approaches with greater precision. An ADI may use either a cross pointer or a V bar as the command display. Fig. 33 shows a cross pointer display that indicates that the pilot is to the right of the localizer and below the glide path. He must correct to the left and fly up to bring the cross pointer back to the stationary airplane symbol. (The dotted line in Fig. 33 represents the horizon.)

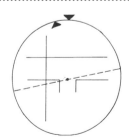

Fig. 33. Cross Pointer of ADI.

Because computers have the ability to accept and process an almost limitless amount of information, some manufacturers include as much information as possible on the ADI. Most of this additional information is displayed around the periphery of the instrument. It may include a ball to indicate slip or skid, an angle of attack indicator, a glideslope needle, a radio altimeter readout, a decision height indicator. The ADI pictured in Fig. 32 has several of these additional features.

When the flight director installation is combined with an autopilot system, information from the computer is fed directly to the autopilot which makes the necessary flight control corrections.

The flight director, as well as giving commands for an instrument approach, can intercept a radial inbound to a VOR or ILS, give pitch commands for a missed approach, for take-off and climb and for turns to heading, direct a pilot through a back course approach, give guidance on procedure turns and, when in the RNAV mode, capture a course to a waypoint. Altitudes, courses and approaches are selected by the push of a button to feed the flight director the information it needs to accomplish all its functions.

Autopilot

An autopilot system operates on the principle of gyroscopic rigidity in space. The system incorporates two gyroscopes to control flight in the longitudinal, lateral and vertical axes of the aircraft. If the aircraft deviates from the programmed attitude, the system computer detects the deviation and issues commands to the **servo motors** which move the primary flight control surfaces as required to return the aircraft to the required flight attitude. This is called **aerodynamic response.**

Different autopilot systems provide a wide variety of control. A very simple autopilot may control only the rudder to permit more accurate holding of flight headings. More complex systems offer a greater variety of options. Modes of operation in the roll channel include maintaining the wings level, following a selected heading, following a selected radio path. Modes of operation in the pitch channel include maintaining a certain pitch attitude, a certain vertical speed or following an ILS glide slope. Autopilot systems will react to centre of gravity changes to keep the aircraft in proper trim, will automatically shut down in the event of malfunction, will turn, climb or descend smoothly as directed.

Autopilot computers and flight director computers are interconnected in present day technology. The autopilot controls the airplane. The flight director monitors autopilot operation.

Ground Proximity Warning System (GPWS)

The purpose of the ground proximity warning system is to alert the flight crew to the existence of an unsafe condition due to terrain proximity.

It is activated when the aircraft is experiencing an excessive sink rate or when there is an excessive closure rate because of rising terrain. The system relays aural messages and activates warning lights on the instrument panel.

It also warns of terrain proximity during landing and take-off sequences. It is programmed to give warning of closure rates that are in excess of preset threshold values. It recognizes landing gear and flap positions and gives aural warning if they are not in the proper position for landing. It warns of excessive deviation below the glide slope on a front course ILS approach.

For proper operation, the GPWS requires input data relating to airspeed, barometric altitude, radio altitude and glide slope deviation.

Area Navigation

Area navigation systems, such as RNAV, INS, Loran C and GPS are invaluable in permitting flight to points that are not served by any kind of navigation aid, in finding airports in marginal weather. They permit straight line cross country flights over long distances. Area navigation greatly enhances air traffic control flexibility providing relief for some of the problems of air congestion and relieving some of the work load of air traffic controllers. It also contributes to air safety since it permits the pilot to choose a route that will bypass heavily congested areas and it reduces the funnelling of traffic over navigation aids. It permits the establishment of multiple side by side routes between main centres and permits segregation of traffic on the basis of speed. The direct point to point navigation reduces flight time by shortening flight paths.

RNAV

RNAV, by freeing the pilot from the need to fly directly to and away from radio navigation facilities, has provided greater lateral freedom for aircraft and, at the same time, allowed for more complete use of the airspace.

As a result, flight distances can be minimized, traffic congestion decreased and traffic flow at terminal areas can be expedited. In addition, instrument approaches, within certain limits, can be made at airports without local instrument landing systems.

With the capabilities of an RNAV system, precise off-airways, cross country navigation is possible using the present VORTAC facilities. The RNAV permits a pilot to fly directly from any spot in the service area of a VORTAC to any other spot in that service area, at all times knowing exact position and the distance to destination. Neither the destination point nor the starting point need be serviced by any kind of navigational aid to use this system of straight line navigation. The pilot no longer needs to track to a VORTAC installation, cross it and track out, doglegging his way to destination.

A typical RNAV airborne system is pictured in Fig. 34. This instrument is the computer of the system. In addition to it, the airplane must be equipped with a VHF NAV receiver, a DME and a Nav display unit, such as a conventional VOR indicator with RNAV mode or an HSI. The VHF NAV receiver provides the VOR information to the RNAV computer necessary for establishing a waypoint along the radial of a particular VORTAC. The DME provides range information so that the RNAV can compute the distance from the VORTAC to the waypoint. With this information, the RNAV computes the airplane's position relative to the waypoint. The HSI indicates the course to fly to reach the waypoint and the DME provides an updated display of distance to the waypoint.

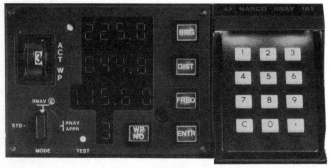

Fig. 34. RNAV System

In a typical RNAV system, it is possible for the pilot to pre-program the waypoints that he will use during a particular flight. To create a waypoint, it is only necessary to key into the RNAV computer the radial from the VORTAC and the distance from the VORTAC of the site chosen to be a waypoint. During the flight, the pilot selects the desired waypoint and all the information for it that has been stored in the computer's memory will appear on the digital readouts. The remotely channelled NAV receiver and the DME will automatically be channelled to the new VORTAC frequency. The HSI (or OBS) will indicate the airplane's position relative to the waypoint. The DME readout will show the distance of the airplane from the waypoint.

What the RNAV system does is create an imaginary VORTAC station over the waypoint. The HSI reacts to the radials radiating out from this imaginary VORTAC station. The DME computes distance to it.

When the HSI (or OBS) is working in the RNAV mode, the deflection of the CDI needle that indicates off course error is in nautical miles rather than in degrees. Each dot represents 1/2 mile off course.

Most manufacturers call their area navigation systems of this type RNAV. It is also known as a **course line computer** or **track line computer**.

Doppler

Doppler is a means of fixing position by methods somewhat similar to those used in D/R navigation, except that the plotting is automatically performed by electronic computers. Doppler provides information on groundspeed, drift angle, distance, latitude and longitude and deviation from track. The equipment is self-contained and does not depend upon radio or radar aids of any kind.

Groundspeed is measured by a phenomenon that is known as the doppler shift. Radar beams are directed from the transmitter towards a target. Due to the movement of the target, when these beamed signals are reflected back to the receiver they have undergone a shift, or change, in frequency. The difference in frequency is proportional to forward motion of the target and can therefore be used to accurately compute groundspeed.

To measure drift, several beams are directed to the right and left of the target. The doppler shift, or change of frequency experienced by one beam will be greater than that experienced by the other. The drift angle is thereby determined.

Doppler was once a favoured navigation system but has now been replaced by GPS, INS and other area navigation systems. The principle of the doppler shift has, however, gained recognition and is being used as a means to measure wind shear and to monitor the movement of thunderstorms and heavy rain areas.

Inertial Navigation System (INS)

The inertial navigation system is a completely self contained system, independent of any ground based navigation aids. After being supplied with initial position information, it is capable, by use of a computer, precision gyros and sensitive accelerometers, of continuously updating extremely accurate displays of position, groundspeed, attitude and heading. It also provides guidance and steering information for the autopilot and flight instruments. It can calculate the track to fly and the distance between two points and, in flight, the distance an airplane is off that track. The airborne equipment uses barometric pressure to resolve DME slant range.

Velocity and distance are computed from sensed acceleration. Two accelerometers, gyro stabilized, are mounted in a gimbal assembly, commonly called the platform. One measures acceleration in the north-south direction and the other in the east-west direction.

The gyro stabilized platform on which the two accelerometers are mounted is kept level with respect to the earth (rather than fixed in space) by a compensating mechanism that allows for the fact that the earth is rotating and that it is round. The computer, which knows the latitude and longitude of the take-off point, computes the information from the accelerometers into present position.

The accuracy of the INS is dependent on the accuracy of the initial position information. Therefore, system alignment, normally called gyrocompassing, at the beginning of the flight is very important. The accelerometers must be levelled and the system must be oriented to true north.

Some systems combine INS and full RNAV (VOR and DME) into one system. The system can therefore be used as a terminal area aid as well as a long distance en route navigation aid. A control display unit such as that shown in Fig. 35 displays navigation data generated by the system. Waypoint and station data is entered into the computer using optically encoded cards. Information on initial position, nine waypoint positions and 9 station positions, with their frequencies, elevation and magnetic variations can be entered into the computer at one time.

The latest generation of inertial navigation systems replaces the precision gyros with laser sensors which measure the aircraft's movement in all three axes (pitch, roll and yaw). While no more accurate than the earlier gyro systems, lasers have much higher MTBFs (e.g. 10,000 hours versus 2,500 hours) because of their lack of moving parts. Often, the laser system will provide its data direct to the onboard flight management system. In this application, the laser is called an **inertial reference system (IRS)**.

pulses transmitted by the master station, measured in microseconds. Each chain is designed a number determined by its GRI divided by ten. The Canadian West Coast chain, whose designation is 5990, transmits pulses from the master station every 59,900 microseconds.

A Loran C receiver measures the time interval that elapses between reception of the master signal and at least 2 of the secondary signals, plots hyperbolic lines of position and translates the data into position in terms of latitude and longitude. The airborne receiver will display groundspeed, range from and bearing to and estimated time of arrival at destination or any waypoint for which latitude and longitude data has been stored in its computer.

The Loran C receiver, pictured in Fig. 36, derives its database information from a small computer card which is inserted into the Loran unit by the pilot. The computer card can be regularly changed ensuring that the database information is as current as possible. Loran provides the pilot with information on hundreds of airports, TCA alerts, radio frequencies, minimum safe altitudes, nearest airport or VOR, VOR search and other flight facts. It will drive an external CDI, HSI and autopilot as well as compute and display the standard range of navigation information.

Fig. 36. Loran C Receiver.

Fig. 35. Inertial Navigation System.

Loran C

Loran C (an acronym for long range air navigation) is a low frequency, pulsed hyperbolic, radio navigation system that is widely used for air navigation. It operates in the LF range from 90 to 110 KHz.

Transmission stations are grouped into chains, each consisting of one master station and several secondary or slave stations. There are presently 21 Loran C chains (called GRI chains) in operation around the world.

The master station broadcasts an omnidirectional signal that is pulsed in a code unique to itself. A few microseconds later, each secondary station broadcasts its own LF signal in a unique code. These transmissions are repeated 10 to 20 times a second. Each chain is assigned a unique time sequence code called a **group repetition interval (GRI)**. The GRI is the time between the successive

Loran C uses low frequency, ground wave signals. One problem of Loran C is that wave propagation is not constant. It varies with the electrical properties of ground and water. Optimum reception and reliability is obtained when the signal has an uninterrupted path across salt water to the receiver. Range then may be as far as 2000 nautical miles. Signals that travel over land are subject to earth conductivity that causes slight changes in signal transit times. Over land, the range is also not so great, although stations sited as much as 800 nautical miles apart will give reasonably reliable signals.

The main factors affecting accuracy are distance from the transmitters, the geometry existing between the receiver and the transmitter and the type of terrain over which the signals have to travel to reach the receiver.

Wave propagation velocity is affected seasonally by such conditions as snow cover. Anomalies occur near metal ore deposits and reflecting surfaces.

The presence of sky waves, bouncing off the reflective ionosphere, will also disrupt time difference measurements. This error is usually only a problem in the outer portion of a coverage area where ground waves are weak.

The signals are subject to local interference from such sources as LF transmitters and high voltage power transmissions that employ signalling frequencies. The receiver system is also susceptible to precipitation static.

In spite of these disadvantages, Loran C has proved to be an attractive system for general aviation use.

The Loran C system was originally established as a navigation aid to marine navigation and, as a result, for a long time stations were commissioned only on the coastal areas of the continent and on the Great Lakes. Recently, the United States government established new chains in the centre of the continent that provide coverage throughout continental U.S. These new stations also provide some coverage of the continental gap that exists in Canada. Very little of northern Canada is within range of any Loran C chain.

Loran Phraseology

Acquisition — Signal reception is satisfactory for position computation.

Baseline extension — A line along which accurate position fixing is unlikely because of poor position lines.

Blink alert — A coded warning that the transmission by a secondary station is missing or in error.

Coverage area — The area surrounding a GRI chain in which reception of the Loran signal is satisfactory.

Database — Permanently stored and programmed data that makes possible access to waypoints and other nav information by identifiers.

GRI (group repetition interval) chain — A group of master and three or four slaved secondary stations.

Phase coding — The periodic shifting of the phase of transmission done by each station is used to distinguish ground waves from reflected sky wave contamination.

Triad — A master and two secondary stations that must be acquired to determine position lines for computation of a fix.

Global Navigation Satellite Systems (GNSS)

The International Civil Aviation Organization (ICAO) has recognized that a global navigation satellite system (GNSS) will satisfy navigation requirements for aviation into the 21st century. A complete GNSS may include satellites provided by various countries or commercial groups as well as ground systems to augment and monitor the satellites. At present, there are two satellite navigation systems: the U.S. GPS and the Russian GLONASS.

The Future Air Navigation Systems (FANS) committee of ICAO is tasked with monitoring and co-ordinating development of a global system to provide cost effective improvements in communications, navigation and surveillance (CNS) and air traffic management (ATM).

The **navigation** capability will be able to provide high integrity, highly accurate navigation service, suitable as a sole means of navigation for en route, terminal and non-precision approach landing operations.

FANS has developed a concept of **automatic dependent surveillance (ADS).** With ADS, an aircraft automatically transmits its position and other relevant data to the air traffic control centre via satellite data link. The aircraft position is displayed on a screen similar to present radar displays. This service is primarily intended for oceanic operations, remote land areas and areas where primary and secondary radar is impractical.

Global Positioning System (GPS)

The **global positioning system** was developed by the U.S. Department of Defence to provide a very precise, global navigation service. GPS is based on a constellation of 25 satellites orbiting the earth at an altitude of 11,000 n. miles. Each satellite propagates its signal to half of the earth. The system uses technology that can pinpoint position anywhere in the world, 24 hours a day, and claims accuracies in the region of 100 meters in three dimensions.

GPS is based on satellite ranging. Position on earth is determined by measuring distance from a group of satellites in space. The system works by timing how long it takes a radio signal from a satellite to reach the receiver. Since radio waves travel at a constant speed (the speed of light which is 186,000 miles per second), distance is calculated as a function of time.

Both the satellites and the receivers generate a very complicated set of digital codes known as "pseudo-random codes" at precisely the same time. An individual GPS receiver uses the ephemeris data sent by the satellite to calculate the time differential in receiving the satellite code and uses that time/distance information to establish latitude, longitude, altitude and to determine groundspeed.

To achieve the accuracy demanded by the system, very precise clocks are required. The satellites use atomic clocks, the most stable and accurate time reference as yet developed. Receiver clocks must also be very precise but can tolerate some imperfection. By taking readings on at least 4 satellites and using a method known as triangulation, receiver clock errors are cancelled out.

The GPS satellites are constantly monitored by the U.S. Department of Defence. The ground monitoring and control system evaluates the performance of the satellites which orbit the earth every twelve hours.

Ephemeris errors are determined and corrections are sent up to the satellites when necessary. These errors are usually very minor and are caused by gravitational pulls from the sun and moon and by pressure of solar radiation on the satellite. Each GPS satellite transmits not only its pseudo-random code but also a data message about these minor errors.

There are several sources of error in the GPS system. Water vapour in the atmosphere can affect the speed of the signals. The earth's ionosphere also affects the speed of the GPS radio signals. The electrically charged particles that make up the ionosphere slow down the signals at a rate inversely proportional to its frequency squared. The lower the frequency of the signal, the more it gets slowed down. Very sophisticated GPS receivers are able to eliminate this kind of error. However, even without a receiver based correction, these errors are negligible and will cause a position error of only a few meters.

Receivers sometimes make errors by rounding off a mathematical operation. Such receiver errors may incorporate a few feet of uncertainty into every measurement.

Another source of error is the **multipath error.** The signal transmitted from the satellite bounces around, taking a circuitous route before it reaches the receiver. Special antennas minimize this problem but, in severe cases, it can add some uncertainty to the measurements.

All of these inaccuracies, even if taken together, will affect the accuracy of position calculations by no more than 25 to 50 meters. Triangulation calculations can magnify or lessen these uncertainties depending on the relative angles of the satellites in the sky. The wider the angle between satellites, the better will be the measurement. A good GPS receiver will take into account this subtle principle of geometry that is called **geometric dilution of precision (GDOP),** will analyze the relative positions of all the satellites available and will choose the best four.

The satellites broadcast two separate forms of the pseudo-random code on the frequencies 1227.6 and 1575.42 MHz. The **P code,** which provides **precise position service (PPS),** is encrypted so that it can be used only by the military and provides the U.S. Department of Defence with some exclusivity to the system. The **C/A code,** which provides **standard position service (SPS),** is used by civilian receivers. The SPS signals are degraded through a technique called **selective** availability (SA) which artificially creates a significant clock error in order to deny high precision signals to unfriendly military forces. Even with SA, horizontal accuracy is within 100 meters and meets the requirements for en route and non-precision approach applications.

The accuracy of the system can be boosted using a technique called **differential GPS (DGPS).** A GPS receiver on the ground in a known location acts as a static reference. It measures distances to satellites, calculates corrections to match the GPS position to the actual position and transmits an error correction message to any other GPS receivers in the local area. They in turn use that error message to eliminate virtually all error in their own measurements. Differential GPS measurements are, therefore, much more accurate than standard GPS measurements.

Local area differential (LDGPS) is limited by line of sight communication from the differential stations and requires hundreds of ground stations. **Wide area differential (WDGPS)** has no line of sight limitations but has some degradation of accuracy due to transmission of the signal through the ionosphere. It requires fewer ground stations.

The integrity of the GPS system is compromised by the fact that the GPS satellite constellation does not warn users promptly if the system is providing faulty signals. Some receivers provide this integrity through **receiver autonomous integrity monitoring (RAIM)**. RAIM works by calculating aircraft position using different sets of satellites, comparing positions and warning the pilot if there is a discrepancy. It requires at least six satellites within the view of the GPS receiver to provide the receiver with enough information combinations to detect an unhealthy satellite.

Transport Canada requires that GPS equipment that is to be used in IFR application must meet certain standards (TSO C 129), one of which is RAIM capability.

Top of the line GPS receivers are able to track four or more satellites simultaneously and display instantaneous position and velocity information. They also can eliminate the GDOP problem by tracking all satellites in view. Single and double channel receivers with sequencing capability can achieve quite precise measurements but are slower in calculating position and velocity. A single channel receiver moves from one satellite to the next to gather data. The sequencing can interrupt positioning, limiting overall accuracy and limiting its capability to provide continuous positioning. A double channel receiver has increased capability, does not have to interrupt its navigation functions to receive satellite signals and provides more precise velocity measurements but does not have the capability to eliminate the GDOP problems.

GPS allows aircraft to make curved approaches to airports, fly more efficient routes and obtain positional information anywhere in the world. Its benefit is especially realized over oceanic areas.

Automatic Dependent Surveillance (ADS)

A further development of the GPS system permits ground surveillance of the position and progress of a GPS equipped aircraft. The airborne equipment has the capability of relaying to a ground based receiver the exact geographical co-ordinates of the aircraft as it proceeds along its track. This information is displayed on a surveillance screen at the air traffic control unit and is monitored by ATC controllers.

Known as **automatic dependent surveillance,** this system has the advantage of providing position monitoring of aircraft that are out of range of normal radar coverage, on transocean flights and in the remote areas of the continent. Flight control and flight separation can thus be provided for flights in these areas.

Automatic dependent surveillance is also possible with the Loran C navigation system (see above).

Moving Map Display

A moving map display based on GPS or Loran C (see above) uses the technology and capability of GPS or Loran C systems to depict the airplane's real-time flight path on a luminescent screen that also displays the applicable navigational chart. With position information obtained from the GPS or Loran C, the system displays the flight path with the airplane's position depicted either in the centre of the map as the map moves or with the map stationery while the airplane's position moves. Position relative to navigation aids and airports, TCAs, control zones and other special use airspace is continuously displayed.

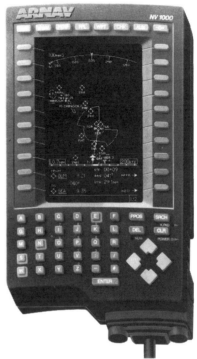

Fig. 37. Moving Map Display.

The database is supplied in a data card that is easily inserted into the system. All database information (such as airports and their runways and approaches, navaids, waypoints and intersections) can be called onto the screen. Flight plans can be created and keyed into the system. When coupled to an autopilot, the moving map display system will fly the airplane automatically on the flight plan, visually depicting its progress on the map on the screen. The map scale can be zoomed to display an area from 1000 miles down to 1 mile in length (the latter for approach mode). The system displays minimum en route safe altitude, CDI, groundspeed, ETA, ETE, bearing and distance to the next waypoint, nearest airport or VOR. etc.

VLF Navigation Systems

Very low frequency systems have been established on a world wide basis that are usable for very long range navigation. In these systems, radio waves are propagated at a constant velocity such that transmissions from two different locations can be compared by a receiver located at a third location. If the signals are transmitted at precisely the same known time, the difference in arrival time of the signals at the receiver can be used to establish a line of position. A position fix can therefore be established by using three or more stations.

There are two VLF systems now in existence; the U.S. Navy VLF and the omega systems. The VLF system has not been approved by Transport Canada as a prime air navigation aid. Omega, however, has been approved for use in the North Atlantic.

VLF – U.S Navy

The U.S. Navy has established seven VLF communication ground stations that are located worldwide in Australia, Japan, England, Hawaii, Maine, Washington and Maryland. They transmit at powers of 500 to 1000 kw. Although the Navy system is usable for navigation, it is not dedicated to navigation. Signal format and transmission are subject to change at the discretion of the U.S. Navy. The stations are also shut down for servicing for several hours each week. No NOTAM service is available regarding the service.

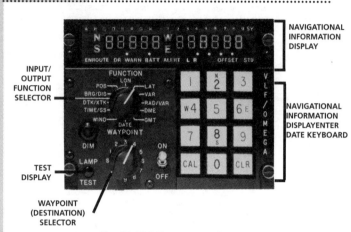

Fig. 38. VLF/Omega Receiver.

Omega

The omega navigation system, set up especially for navigation purposes, is a more reliable service. It is a very long range, very low frequency system that works on a network of only 8 stations that operate with 10 kw transmitters. The 8 stations are located in Norway, Liberia, Hawaii, North Dakota (U.S.A.), La Reunion, Argentina, Trinidad and Japan.

The stations transmit on the frequencies 10.2 KHz, 11.05 KHz 11.3 and 13.6 KHz. Each station transmits all these frequencies. However, each station transmits them in a sequenced format at a specified time and only for a very short interval. Consequently, no two stations are transmitting on a given frequency at the same time. Each station transmits at approximately 10 second intervals. In addition to the four basic navigation frequencies, each station transmits a unique frequency assigned only to itself.

Using the omega VLF system, an aircraft can fix its position to within about 1 mile during daylight or 2 miles at night. The reduced accuracy during darkness, characteristic of any low frequency radio navigation aid, results from diurnal changes in the ionosphere.

A typical VLF omega airborne system, such as that pictured in Fig. 38, is capable of receiving the 8 omega stations and 3 of the VLF Navy stations. It includes a digital computer and a cockpit display. The crew can preset up to 10 waypoints and the indicator can display at the choice of the crew a variety of navigation data, including longitude and latitude data, bearing to destination, cross track deviation, distance to go, groundspeed and estimated time to waypoint.

It is prey to the disadvantages of all very low frequency radio signals. It is subject to outage if the aircraft is flying through thunderstorms with heavy electrical activity or in weather conditions that cause precipitation static. It is also subject to disturbance during solar storms which cause ionospheric disturbances.

On the other hand, because there are no line of sight limitations with low frequency, it is just as accurate in mountain valleys as in open plains. The VLF signals propagate to great range so that omega will provide redundancy. In many locations, all eight stations will be detectable although perhaps only 5 or 6 will be usable.

For these reasons, the system is most useful if paired with an inertial navigation system. The accuracy of the omega system is useful to update such a self-contained system which is subject to error buildup during long endurance flights.

Electronic Flight Instrument System (EFIS)

The new technology in both avionics and computer generated displays is changing the look of the modern aircraft instrument panel. Popularly called the glass cockpit, the electronic flight instrument system (EFIS) is replacing the electromechanical instruments which traditionally have displayed information by means of pointers and dials. EFIS makes use of cathode ray tubes (CRTs), small high definition television style picture tubes, to display all kinds of information.

The display can be tied into any on board navigation system such as RNAV, Loran C, omega, etc. The adaptability of EFIS allows the flight crew to program the route they wish to travel and enter all particulars of the route into the computer's memory before the flight. The system can store and display alternate routes. In flight changes are computed instantly and the information and position displayed instantly. The system shows the position of the aircraft relative to any selected waypoint, displays heading, altitude and groundspeed information, creates new waypoints, projects the aircraft's true and intended courses. It can display the approach course to an ILS or MLS and at the same time depict the aircraft's position relative to that course. Information about weather ahead derived from the onboard weather radar, operational data relating to the aircraft (airspeed, rate of climb or descent, altitude, etc.) and the engines (engine temperatures, rpm or turbine speeds, etc.) all can be displayed. In fact, the scope of information that can be displayed by EFIS is virtually unlimited.

A recent modification of the EFIS system projects the displayed information onto the front window of the airplane, where it is directly in the pilot's field of vision outside the aircraft. The so

Fig. 39. Typical EFIS CRT Display
Left: Navigational Information Right: Operational Data.

called **heads up display system** permits the pilot to monitor both the instrumentation and the progress of the airplane, especially during the approach and landing.

One advantage of the electronic type of display is that information can be called up as it is required and returned to computer storage for automatic monitoring when it is not needed. New and customized features can be added without major changes or total replacement as is the case with the traditional electromechanical instruments.

The glass cockpit was first introduced in several of the new generation of jet transports and is now commonly found in a wide range of corporate aircraft. It will probably be some time before EFIS is adapted for use in small general aviation airplanes, although the innovative Beech Starship has a glass cockpit.

Cockpit Management System

Very advanced cockpit management systems are now available that control EFIS modes, navigation displays, ACAS (TCAS), weather radar, terrain and obstruction proximity warning systems, engine indications and crew advisory systems. All information is presented on a single multi-function display.

Aircraft position is displayed on a moving map display that incorporates a data base with thousands of waypoints, airports, runways, frequencies, airspace boundaries, VORs, NDBs, airways, etc. Geographical information such as elevation, hydrography, man made obstacles, highways, etc. are also presented.

The system monitors engine and airframe conditions, constantly scanning for out-of-range conditions. The system couples to existing aircraft systems, including analog instruments, engine instruments, encoding altimeters, fuel computers, Loran and GPS receivers.

The relationship between the aircraft and its environment is on constant display. Through the encoding altimeter and the Loran/GPS receivers, the system compares the aircraft altitude to the map data base. When the aircraft is below minimum safe altitude, the terrain/obstruction proximity warning system advises the pilot of ground proximity.

The system is designed to reduce crew workload and to enhance "eyes up" operation.

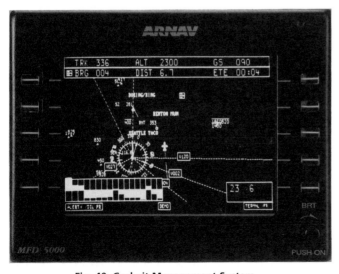

Fig. 40. Cockpit Management System..

Radar and Radar Facilities

The name radar is an abbreviation of "radio detection and ranging". To operate, radar requires a highly directional radio transmitter/antenna and a scope, or screen, to display the information received by the antenna.

The principal uses of radar in aviation are

1. Fixing positions of airplanes in flight.
2. Air traffic control.
3. Detecting thunderstorm activity.
4. Approach and landing guidance to airplanes

Radar operates on the 3,000 to 10,000 MHz frequency band. Extremely short bursts of super high frequency radio energy, called pulses, are fired into space along a **beam** from a highly directional radar antenna. These pulses strike airplanes, ships, buildings and other objects and are reflected back to the sender. The time it takes the pulse to travel to the reflecting object and return is timed electronically in millionths of a second to determine the distance of the object from the sender. This information is displayed on the face of a cathode ray tube called an oscilloscope or, more commonly, a scope or screen. The information appears in the form of a small blip of light that is called a **target.**

At the centre of the scope is the radar antenna. The scope is marked with concentric rings (called range markers) that measure distance from the antenna (Fig. 41). An azimuth scale is marked around the circumference of the scope (not shown in Fig. 41). The controller working the scope is thus able to determine the bearing and distance of a target representing an airplane.

The beam, transmitted by the radar transmitter and called in radar terminology a **trace,** will pick up only objects towards which it is pointed directly. In order to "see" objects around the entire horizon from the sender's position, the beam is rotated through 360° by a scanner at a constant rate of speed. A sweep line on the scope is synchronized with the rotation of the antenna. The target that appears on the scope fades out after the sweep line passes, but reappears again as the sweep line passes on its next rotation. As a result, the track of an airplane can be seen on the radar scope.

Radio waves travel at a rate of 186,000 miles per second. To measure them, the unit of time used is the **microsecond,** which is one millionth of a second.

Fig. 41. A Radar Scope.

Pulses of electromagnetic energy are beamed out to the target and reflected back (Fig. 42). The signal which is reflected back is called an echo. It takes 10.75 microseconds for a signal to travel one mile out to a target and for the **echo** to return. The round trip distance is known as a **radar mile.**

The use of radar in ATC procedure greatly increases utilization of the airspace and permits expansion of flight information services such as traffic and weather information and navigational assistance.

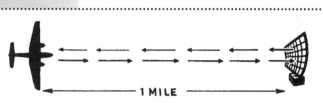
Fig. 42. Distance Measuring by Radar.

There are two basic types of radar systems.

A. **Primary Surveillance Radar (PSR):** Primary radar is based solely on the echo principle and requires no equipment in the airplane. Four types of primary radar are in use by ATC units.

1. Airport and Airways Surveillance Radar (AASR) — a medium range radar (150 to 200 nautical miles) designed for both air way and airport surveillance applications.

2. Airport Surveillance Radar (ASR) — a relatively short range radar intended primarily for surveillance of airport and terminal areas. (See below).

3. Precision Approach Radar (PAR) — a short range radar used as an approach aid. (See below)

4. Airport Surface Detection Equipment (ASDE) — radar surveillance of surface traffic that is used by control tower operators to monitor the position of aircraft and vehicles on the maneuvering areas of the airport, especially during conditions of reduced visibility.

RAMP. The ASR primary radar is scheduled for eventual replacement by RAMP units. The radar modernization project (RAMP) is designed to completely replace the existing ATC radar network with a system utilizing advanced computer technology.

B. **Secondary Surveillance Radar (SSR):** Secondary surveillance radar requires complimentary aircraft equipment.

1. Transponder — an airborne system designed to send a signal to the ground based radar receiver to reinforce the surveillance signal.

2. Radar Digitized Display (RDD-l) — a radar system that uses only transponder generated data on the ground radar display. Navigation assistance, therefore, is available only to aircraft with a functioning transponder.

Secondary Surveillance Radar

Although radar is a very valuable asset to the air traffic control system, it has several disadvantages.

In heavy traffic areas, all the targets on the radar scope look the same. There is no way a controller can distinguish between two targets in the same general area. He must ask one of the airplanes to take up a specific heading and then wait to see which target on the screen reacts to the instruction in order to make an identification.

The signal directed to low flying airplanes often experiences interference from terrain features, buildings, etc. As a result, the echo is weak. Precipitation also returns echoes.

To overcome these disadvantages of the primary radar system, a secondary surveillance radar system was developed. It requires airborne equipment that returns a strong signal to the primary radar transmitter. As a result, the target on the radar scope is stronger and more readily identified by the controller.

The airborne equipment is called a **transponder.**

Transponder

Designed to reinforce a surveillance radar signal, the transponder permits positive identification of an aircraft by the ground facilities of ATC. A typical airborne transponder system includes a control, a receiver/transmitter and a small L-bank antenna mounted on the underside of the aircraft. Ground equipment consists of a transmitter/receiver and a rotating directional antenna that usually is mounted on top of the regular surveillance radar antenna.

The transponder works like this. The ground radar equipment sends out a special interrogation signal asking all aircraft, "Who are you?". Only transponder equipped aircraft can answer. The airborne transponder picks up the signal and automatically sends back a strong pulsed signal in reply. The reply signal is computed into distance and direction (and into altitude on Mode C) by the ground radar station and is displayed with the aircraft's target on the traffic controller's scope as one or two slashes.

The interrogation signals from the ground radar are **moded.** Mode A asks for identification. Mode C requests altitude information. Pilots adjust their sets to the appropriate mode. On transponder controls used in general aviation, mode selection of A or A/C is provided. On Mode A, the transponder automatically replies to all interrogations. When the mode selector is placed in the A/C position, the transponder provides proper identification plus automatic altitude information.

Fig .43. Pulse Signals Between Airplanes and Ground Stations (either way) Moded and Coded for Signal Identification.

In order that altitude information can be encoded by the airborne transponder and relayed to ATC, an encoding altimeter must be installed in the airplane and linked into the transponder equipment.

In Canada, only aircraft equipped with transponders capable of automatic pressure altitude reporting (i.e. Mode C) may operate in Class A, B and C airspace.

In the transponder shown in Fig. 44, the Mode A selection is automatically selected when the set is activated. Mode C selection is made by selecting ALT.

Mode C should always be selected (if the transponder is capable of the altitude reporting function) unless directed by ATC otherwise. If the airplane is on a VFR flight plan, the appropriate VFR code (see below) **must** also be selected.

The reply signal sent by the airborne transponder is **coded.** Specific codes are assigned by an air traffic controller on the ground for various flight levels and conditions of flight. The pilot selects this code on the control in the cockpit.

From the Ground Up

Fig. 44. Transponder.

Most transponders have 4096 code capability. All transponder codes are, therefore, 4 digit codes. In Fig. 44, the code 0400 has been selected. The controller can identify the various code settings assigned to different flight requirements.

In assigning codes, for example, arrival control at a particular ATC centre may use 4700 while departure control will use 2400. One ATC centre in handing a flight over to another centre will ask the pilot to select the code of that centre. A pilot on a VFR flight plan may select 1200 (at or below 12,500 feet) or 1400 (above 12,500 feet) on the transponder control. This will automatically inform the ATC radar controllers that the aircraft is a VFR flight and they will be able to maintain positive separation of their IFR and VFR traffic.

When ATC issues instructions concerning transponder operation, the pilot is required to operate the transponder as directed until receiving further instructions or until the aircraft has landed. Only in the event of an emergency, communication failure or hijack should the transponder code be changed without authority from ATC.

In the event of an emergency and if unable to establish communication immediately with an ATC unit, a pilot wishing to alert ATC to the situation should select code 7700. Upon establishing communication, he should operate the transponder as directed by ATC.

In the event of a communications failure, the transponder should be adjusted to reply on Code 7600, in order to alert ATC to the problem. Code 7500 is used to alert ATC of unlawful interference (i.e. hijacking).

On an IFR flight plan, the loss of all radio communication ability necessitates termination of the flight at the first convenient airport. The versatility of the transponder can be seen in the situation in which the communications problem affects only the transmitting capacity. A flight, on an IFR clearance, can be continued by using the transponder as a transmitting device. The ATC controller is able to ask questions and give instructions in normal fashion; the aircraft acknowledges these transmissions by "squawking ident".

If the transponder or encoding altimeter fails during flight in airspace where Mode C is required, the aircraft may be operated to the next airport of intended landing and, thereafter, to complete its itinerary or to a repair base, if authorized by ATC.

ATC may, upon written request, authorize an aircraft not equipped with a Mode C transponder to operate in airspace where its use is mandatory. Such authorization is given only if the safety of air traffic within that airspace is not compromised.

If the controller decides to verify a "target" when more than one aircraft are on the same code, he will call the pilot and ask him to "squawk ident". When the pilot presses the ident button on the transponder control, a special signal is sent out that causes the target to "blossom" on the controller's scope. The target will blossom for a few seconds and then return to its regular pattern. The ident feature should be operated only when directed by ATC.

Most controls have a function tester or small reply light which blinks every time the transponder replies to a ground interrogation signal, letting the pilot know that the transponder is operating.

ATC radar units are equipped with an alarm system that is triggered when an aircraft within its radar coverage has selected the emergency, communication failure or unlawful interference transponder codes. It is possible for those codes to be momentarily and unintentionally selected while the aircrew is changing from one code to another. To avoid unnecessary activation of the alarm system at ATC, pilots should be very careful not to select a 7 in the left hand position on the transponder head while changing codes. However, do not select "standby" while changing codes as this action will cause the target to be lost on the radar screen at ATC.

Transponders should be adjusted to "standby" while taxiing for take-off, to "on " as late as practicable before take-off, and to "standby" or "off' as soon as practicable after landing.

Mode S. A recent development in transponder technology is the implementation of Mode S, transponders with a data link capability. With 16 million separate codes, Mode S gives each aircraft a unique signature of its own. It is fully compatible with the ATC radar beacon system. An SSR with Mode S capability will transmit the conventional ATCRBS interrogation, followed by a series of additional signals. The Mode S transponder will reply to the conventional ATCRBS interrogation but also will send back the additional requested data. The information sent through the Mode S system will probably include clearance delivery, take-off clearance confirmation and other information relating to clearances and routes. It could also include weather information and even weather maps. The information would be displayed on a cockpit CRT display and the crew would acknowledge receipt of the information by pushing a button. Mode S will have the advantage of helping to reduce the load on the heavily used VHF channels. Mode S transponders are an integral component of all ACAS (TCAS) installations (see below).

Airborne Collision Avoidance System (ACAS)
With the substantial increase in air traffic and the concern about mid air collisions and near misses, it was inevitable that an automatic collision avoidance system would be developed. ACAS is based on transponder technology and requires Mode S capability in the airborne equipment. ACAS equipment designed and manufactured in the United States is called **Traffic Alert and Collision Avoidance System TCAS).**

An ACAS (TCAS) equipped aircraft constantly sends out interrogations to which any aircraft equipped with at least a Mode C transponder will automatically respond, with no pilot action required.

ACAS (TCAS) is designed to operate independently of ATC. Depending on the type of equipment, ACAS (TCAS) will provide **traffic alerts (TA)** and/or **resolution advisories (RA).** TAs provide information on nearby traffic and are intended to help the crew in visual acquisition of conflicting traffic. RAs may be **preventative advisories** which instruct the pilot to maintain or avoid certain vertical speeds or **corrective advisories** which instruct the pilot to take preventative action by deviating from the current flight path.

If two ACAS II equipped aircraft are on potential collision courses, their computers will communicate using the Mode S transponder data link which has the capability of providing co-ordinated and complimentary RAs (e.g. one being advised to descend, the other to climb).

There are currently three versions of ACAS (TCAS).

ACAS I is the simplest version. It provides a proximity alert function that gives range, altitude and a rough indication of relative bearing, accurate to about 14 degrees. Its range is about four nautical miles. ACAS I provides a traffic alert only, advising of a potential collision. No guidance (RA) is given. The pilot must visually acquire the traffic and take what avoidance maneuver is necessary.

ACAS II has greater range (about 14 to 15 n. miles) and bearing information is accurate to about 9 degrees. It consists of a computer, controls, wiring and antennas which provide both TAs and vertical plane RAs. ACAS II has sophisticated collision avoidance logic built in. It will compute a suitable avoidance maneuver and display it on a separate display screen, weather radar or EFIS cockpit display. If two ACAS II equipped aircraft are on potential collision courses, the avoidance maneuvers are co-ordinated. The system transmits the avoidance maneuver of one aircraft to the other and the second aircraft takes a complimentary maneuver.

ACAS III is a more advanced system. It has even greater accuracy and range and will provide for both vertical and horizontal plane maneuver advisories (that is, turns as well as climbs and descents) instead of the vertical advisory only that is given by ACAS II.

ACAS equipped aircraft require two antennas, instead of the one only that is usually installed on the belly of the aircraft. Since the unit is meant to interact with aircraft both above and below, an antenna on the upper surface of the fuselage is also required.

ACAS (TCAS) does not diminish a pilot's responsibility to ensure safe flight. ACAS responds only to aircraft which are transponder equipped whose transponders are functioning. Aircraft without transponders or with only Mode A are invisible to ACAS (TCAS) equipped aircraft. Alerts and advisories are not provided.

Air Regulations permit pilots to deviate from an ATC instruction or clearance in order to follow ACAS/TCAS resolution advisories. When responding to a resolution advisory, the pilot is required to advise the appropriate ATC unit of the deviation in flight path as soon as possible. When clear of the conflicting traffic, he should expeditiously return to the last ATC clearance or instruction received and accepted. Aircraft maneuvers conducted during a resolution advisory should be kept to the minimum necessary to avoid the conflicting traffic.

Transponder Phraseology
Air traffic controllers use certain phraseology when referring to the operation of transponders.

Squawk (code) — Operate the transponder on the specified code.

Squawk ident — Push the IDENT button.

Squawk standby — Switch transponder to "standby" position.

Stop squawk — Switch off transponder.

Squawk Low/Normal — Operate the transponder on low or normal sensitivity as specified. The transponder is operated on normal unless low is requested by ATC.

Squawk Altitude — Activate Mode C with automatic altitude reporting.

Encoding Altimeter
Altitude information is relayed by the airborne transponder equipment to the ATC radar display for the information of controllers. The instrument that feeds the altitude information into the transponder is known as an encoding altimeter.

There are two basic types of encoding altimeters — the pneumatic or mechanical type and the servo type. The pneumatic type is essentially a conventional barometric altimeter with an encoding device that converts altitude into digitally coded electrical signals in 100 foot increments. These signals are wired to the transponder which, when set to Mode C, are transmitted as replies to the "What's your altitude?" interrogation of ground radars. It uses evacuated capsules called aneroids to sense altitude, by expanding and contracting with static pressure. The movement of the aneroid walls, via gearing, drives the altimeter pointers for visual display. At the same time, the pressure of the moving aneroids drives a small glass disc providing the electrical signal to activate the encoding device.

The servo type of encoding altimeter also uses the aneroid system to sense altitude but the power to drive the encoding and display mechanism is derived from the airplane's electrical system. The servo type, having an outside source of power, can drive pointers and encoders and at the same time provide ancillary outputs for an altitude alerter, rate of climb indicator, vertical nav system and auto-pilot. The servo type encoding altimeter is very accurate and very reliable. Its disadvantage is its complete dependence on electrical power to operate.

Fig. 45. Encoding Altimeter.

Like a conventional altimeter, the encoding altimeter is fitted with a barometric scale by which the pilot is able to set the current altimeter setting on the instrument. The pointers on the face of the instrument respond to any change in the altimeter setting in the same way as does a conventional altimeter. However, the altitude information that is relayed through the encoding altimeter and the transponder to the radar receivers at the ATC unit is based on pressure altitude (a barometric reading of 29.92" Hg). The ATC computers correct the received pressure altitude signals to the local altimeter setting in order to have a uniform standard for altitude information for all the airplanes within radar range of any one particular ATC unit.

Primary Surveillance Radar
ASR (Airport Surveillance Radar)
ASR is a relatively short range radar that gives surveillance of the immediate area surrounding the airport. The ASR scope provides range and azimuth information only, unless the aircraft is equipped with a transponder with altitude reporting capability. The ASR controller will direct the aircraft to a selected runway. If the controller is receiving altitude information, glide path assistance can be given. If no altitude information is available, the pilot must judge his own descent, based on the distance information furnished by the controller.

Par (Precision Approach Radar)
PAR guides the airplane to final approach on the designated instrument

runway. In addition to azimuth and range information, it furnishes the pilot with glide path assistance down to the point of touchdown on the runway.

Radar Assisted Approach

The capabilities of the radar systems provide a method of guiding a pilot on instruments down to the runway for a landing by means of verbal instructions from a controller working on a primary surveillance radar system such as ASR or PAR.

A radar assisted approach can be provided when no alternative method of approach is available. It requires no special airborne equipment other than a radio receiver and normal flight instruments. Radar assistance can be requested by a pilot who may have become lost or accidentally caught in instrument conditions. It can also be used in an emergency situation.

Fig. 46. Radar Scanning Screens.

Right: The SURVEILLANCE or ASR SCOPE. Circular lines are 5 miles apart. The blips are airplanes. The radar surveillance antenna, rotating through 360°, scans a radius 30 miles out and 10,000 feet up. The position of every airplane within a 2,800 square mile area is continuously shown on the surveillance scope.

Left: The PAR (PRECISION APPROACH RADAR) SCREEN. Exact position of the aircraft is constantly shown on the screen in three dimensions: altitude, azimuth and range. (The airplane silhouettes have been sketched in to make the principle more easily understood. In actual practice, the airplanes appear as blips, as they do on the scope on the right.) The PAR operator has two such scopes, one covering a 10 mile area and the other a 3 mile area. The latter is scaled up to provide greater precision on final approach. The equipment is accurate to within plus or minus ten feet.

The radar operator on the ground can see the plane on the radar scanning screens (Fig. 46) which he has in a panel before him. Lines and circles on the scanning screens provide the radar operator with azimuth, range and altitude data which enables him to guide the pilot down to the runway of the airport.

To Use Radar Assisted Approach

To request radar assistance on an approach if you are flying VFR, call the tower. State your approximate position as accurately as you can and your altitude. The tower will assign a transmitting and receiving frequency for communication with the radar controller. You will then be requested to execute a few simple turn maneuvers or, if you are transponder equipped, you will be asked to "Squawk Ident". This is for positive identification of your airplane on the scanning screen. At this point the controller may give you lost communications instructions as follows:

Controller: **If you lose Radio Contact for any 5 Second Interval, Maintain VFR and Contact the Tower on — MHz** (Tower frequency) **for Secondary Instructions.**

If you are lost under VFR conditions, the surveillance approach (ASR) controller will give you a heading to steer to bring you to the runway and will continue to vector your approach in case you happen to stray from the required track.

If the visibility is marginal and it is necessary to make a landing under instrument conditions, you will be given precision radar navigation assistance to guide you at a safe altitude, into the traffic pattern and on to the base leg. At this point you will be requested to check your gyro compass and not to reset it during the remainder of the approach. In case your gyro is not working, advise the controller immediately. He will give you a no gyro approach and will probably brief you as follows:

Controller: **This will be a No Gyro Precision Approach. Runway One Three. Make Standard Rate Turns. Execute Turn Instructions on Receipt.**

Obey the instructions implicitly, using your attitude indicator as your turn reference. Do not attempt to make corrections on your own.

On base leg, as you are about to intercept the final approach leg, the PAR controller will take over and turn you on to the correct heading to line you up for final approach to the selected runway. As you intercept the glide path, you will be given the correct rate of descent to maintain. Instructions will be simple.

Controller: **You are now on Final Approach for Runway One Three. Six Miles from Touchdown.**

Once the descent has started, if the approach is normal, the controller will merely give you distance, steering and glide path checks. For example:

Controller: **Three Miles from Touch Down. You are on Course. On the Glide Path.**

Should you get off the line of approach or glide path, he will advise you of the deviation and give you the necessary correction. For example:

Controller: **Four Miles from Touch Down. 200 Feet to the Left. Turn Right. Heading One Three Six.** Followed by, **You are on Course.** Or **You are 50 Feet Above the Glide Path. Ease It Down.** Followed by, **You are on the Glide Path. Adjust Your Rate of Descent** As you near the threshold of the runway, he will caution you, **Wheels Down and Locked.** When you reach it, he will advise you, **Over the End of the Runway. Take it Visually.** You are then cleared to take over and land.

Emergency Locator Transmitter (ELT)

An emergency locator transmitter is a battery operated radio transmitter that sends out a distinctive distress signal on the international emergency frequency 121.5 MHz and/or 243.0 MHz.

A new generation of ELTs transmits a signal on 406 MHz. A 406 MHz beacon is uniquely coded and therefore identifiable to the aircraft in which it is installed. A 406 MHz beacon transmits for about half a second every 55 seconds. A 121.5 MHz beacon transmits continuously.

The ELTs in use in general aviation aircraft contain a **crash activation sensor (G-switch)** which is designed to detect the abnormal deceleration characteristics of a crash and automatically activate the transmitter. They are constructed to survive that crash and to transmit a signal for at least 48 hours at a wide range of ambient temperatures. The ELT is activated if subjected to a force of 5 to 7 gs for 11 milliseconds.

Most Canadian aircraft are required to have an approved ELT installed. Those exempted are aircraft used in training operations that are proceeding only out to a defined training area, multi-engine and jet airplanes that operate only in controlled airspace, gliders, balloons and ultra-light airplanes.

The ELT has been proven to reduce significantly the time necessary to locate a downed aircraft, thus substantially improving the chances of survival of injured crew and passengers. Do not assume that the automatic activation feature has worked. Ensure that the ELT is indeed sending a signal. Turn the switch to ON. To improve your chances of being located as soon as possible, do not delay turning on the ELT until your flight planned time expires. Such a delay will only delay rescue. Do not cycle the ELT through OFF and ON positions in an attempt to preserve battery life, as irregular operation reduces localization accuracy and will hamper homing efforts. Once the ELT is turned on, leave it on until you have been positively located and have been directed to turn it off by the SAR forces.

A system of orbiting satellites (SARSAT) is used to detect and locate ELT signals. The SARSAT system is very efficient. ELT signals will be detected normally within 90 minutes. In many cases, a downed aircraft has been located before it was reported missing. The SARSAT satellite can pick up the signal anytime and relays the information to the search and rescue (SAR) aircraft. The SAR aircraft home on the signal to find you.

ELT signals are effective only in line-of-sight. For best range, the ELT transmitter should be placed as high as possible on a level surface with no obstructions between it and the horizon. The antenna should be vertical.

As well as the type of ELT that is installed in the airplane and is activated automatically by the impact of a crash landing, there is also a portable hand held type of ELT that is operated manually. Installation of the former type is mandatory but many pilots feel that having the portable ELT aboard as well is an extra precaution.

The ELT will only do its job, however, if it has been "armed". The transmitter has a three position switch — ON, OFF and ARMED. Arming the ELT should be part of the preflight check. Turning it off should be part of post flight shut down procedures.

If you have landed because of bad weather or for some other non-emergency reason and no emergency exists, do not activate your ELT. If, however, the delay will extend more than one hour beyond your flight plan or more than 24 hours past your flight itinerary or beyond your specified SAR time, you will be reported overdue and a search will be initiated. To avoid an unnecessary search, you should notify the nearest ATS unit of your changed flight plan. If you cannot contact an ATS unit, attempt to contact another aircraft on one of the following frequencies in order to have that aircraft relay the information to ATS: 126.7 MHz, a local common frequency, 121.5.MHZ or 5680 KHz.

There are five categories of ELT.

1. **Type A or AD — Automatic ejectable or automatic deployable** automatically eject from the aircraft and are set in operation by inertia sensors when the aircraft is subjected to a crash deceleration.
2. **Type F or AF — Fixed (not ejectable) or automatic fixed** are automatically set in operation by an inertia switch when the aircraft is subjected to crash deceleration forces. The transmitter can be manually activated or deactivated. Provision is also made for recharging the batteries. An additional antenna may be provided for portable use of the ELT. Most general aviation aircraft use this type of ELT.
3. **Type AP — Automatic portable** is similar to type F except that the antenna is integral to the unit for portable operation.
4. **Type P — Personnel** has no fixed mounting and does not transmit automatically. A manual switch is used to start or stop the transmitter.
5. **Type W or S — Water activated or survival** transmits automatically when immersed in water. It is waterproof, floats and operates on the surface of the water. It has no fixed mounting. It should be tethered to survivors or life rafts.

Inadvertent activation of ELTs has been a problem and it is recommended that all pilots operating aircraft with this equipment installed listen on the ELT frequency (121.5 MHz) immediately after arming the ELT and again immediately prior to shut down to ensure that it is not transmitting. The battery must be disconnected whenever the ELT is removed from the airplane for servicing. Any accidental activation of the ELT should be reported immediately to the nearest ATS unit.

The ELT is a battery operated transmitter. Pilots must bear in mind that the battery must be replaced at or about the expiration date marked on the battery. Only approved batteries may be used. ELTs also require an annual recertification to assure that, in a crash situation, they will transmit a signal which will be picked up by the COSPAS/SARSAT satellite. An ELT signal, which is off frequency or low power, could be missed by COSPAS/SARSAT. The satellite requires a coherent signal to develop the position of a crashed aircraft.

Testing of an ELT may be carried out in the first 5 minutes of any hour UTC. A maximum 3 audio sweeps taking about 2 seconds is all that is necessary to determine that the ELT is in good operating condition. In no case should the testing take longer than 5 seconds. Longer testing can result in false alarms being registered with ATC.

Any accidental activation of the ELT should be reported to the nearest ATC unit giving the location of the transmitter and the time and duration of the accidental transmission, in order to forestall an unnecessary search activity.

It is the pilot's responsibility to ensure that all passengers and crew know the location of the ELT and how to operate it. There must be a placard in the aircraft cabin giving this information.

Airmanship

Airmanship is proficiency in the handling and operating of airplanes on the ground and in the air. It involves understanding of the necessity and care in the execution of keeping an airplane in airworthy condition. A proficient pilot is fully conversant with the capabilities of his airplane and understands and regularly refers to the flight performance charts in the Airplane Flight Manual. He understands the principle of weight and balance and knows how to load an airplane properly. He knows how to plan and carry out a cross country flight proficiently, how to avoid wake turbulence, how to make good landings, how to deal with emergencies, etc. This chapter is designed to introduce a pilot to the concepts involved in good airmanship and whet his appetite to seek further information and experience in broadening his proficiency as a pilot.

Care of the Airplane

The wise pilot carefully and systematically supervises the inspection, maintenance, and servicing of his airplane. "Air mechanics are, on the whole," he observes, "more precise, painstaking and conscientious about their work than is the average pilot. They are, however, only human and as such are not infallible. They could forget to clean the filter or fit a split-pin or ground the airframe while refuelling. It pays a pilot to check every detail himself before taking an airplane off the line."

Airplane Cleanliness

Dirt covers defects, making them more difficult to detect. Dirt also increases skin friction, detracting from the performance of the airplane.

Dead insects commonly found on the leading edge of wings and propeller blades cause drag and should be removed to assure maximum aerodynamic efficiency. This is a sensitive area and should be extremely smooth at all times if the manufacturer's performance figures are to be realized. (The manufacturer obtained his performance calculations in a clean airplane). Therefore, keeping an airplane clean is very important.

External Airframe

For washing airplanes, use mild detergent soaps and water. Linseed soap is an approved soap for fabric covered airplanes and, when properly used (after the surface is dry, rub down with a soft dry cloth as if to shine), it has the advantage of leaving a thin film which makes the surface slippery and thereby helps to reduce drag. Remove mud with water. Never use gasoline as a cleaning fluid.

Be sure that the drain grommets located at the lowest points in the wings, fuselage and control surfaces are not clogged with dirt or foreign matter. If these are clogged, water is prevented from draining away; moisture left sitting in the interior of the airframe tends to start corrosion and rot. Ventilation, especially important in wood and fabric structures, will also be hampered if the drain grommets are clogged.

Varsol is the approved cleaning fluid for removing oil and grease from engines, fire walls, cowlings and all external metal surfaces including the underside of the fuselage which is most susceptible to collecting grime. Aircraft constructed of composite materials should be kept clean using solutions recommended by the manufacturer.

Metal airplanes are susceptible to corrosion and must be kept scrupulously clean to prevent corrosion. Cleaning agents which do not contain abrasive compounds should be used. Wax applied to metal surfaces prevents oxidation and reduces the amount of cleaning and polishing required.

Visibility is much better through a clean windshield than through a dirty one. The windshield should always first be washed with clean water using the bare hands to loosen accumulated dirt. Then polish with special non-abrasive liquid cleaners made especially for this purpose. Clean the inside of the windshield as well.

Make certain that the cool air inlets are open and free of foreign objects. In the spring of the year especially, it may be necessary to cover these openings while the airplane is parked outdoors to prevent birds from building nests in the engine.

Interior

Dirt and dust in the fuselage increase weight and add to the fire hazard.

Keep the interior of the cabin and fuselage free of dirt and foreign material. Use a vacuum cleaner on floor areas. Be extremely careful when cleaning upholstery as it may have been treated with fire resistant chemicals and this protection should not be removed; follow the manufacturer's recommendations.

Spilled acids, chemicals, fire extinguisher fluids, etc. should be reported as they could cause corrosion and weakness of the structure.

Dirt falling beneath the floor where control pulleys and connections are located can cause wear of the bearing surfaces and lead to incorrect operation. Dirt and dust may also interfere with proper operation of electrical switches.

Propellers

Propellers should be kept clean. As these are rotating airfoil sections, what applies to wing cleanliness applies also to them. They should be kept clean of dead insects, grass and dirt. A clean and waxed propeller will actually produce a higher rpm at full power settings than one that is dirty. A propeller constructed of composite material will require different cleaning methods than those used on metal propellers. Always refer to the Owner's Manual for the manufacturer's recommendations.

Carburetor Air Filter

Inclined to be overlooked by pilots between scheduled maintenance inspections is the condition of the carburetor air filter. It should be kept clean of bugs, plant seeds, dust and dirt at all times. One of the most frequent causes of premature engine wear is a dirty air filter. Constant attention to the cleanliness of this part will assure many additional hours of running before an engine overhaul becomes necessary.

Exhaust Heaters

Cabin heaters which obtain heat from the exhaust system should be inspected at frequent intervals for leaks which might allow carbon monoxide fumes to enter the cabin.

Ground Handling of the Airplane

Tricycle geared airplanes are most easily and safely maneuvered, during ground handling, by the use of a tow-bar attached to the nose wheel; therefore, it is recommended that one be used whenever possible. When moving tricycle geared airplanes by hand when no tow-bar is available, push down at the point where the stabilizer attaches to the fuselage in order to raise the nose wheel. The airplane can then be turned by pivoting it about the main gear.

On both tail wheel and tricycle geared airplanes, when moving a high wing airplane forward or backward by hand, push at the wing strut root fitting or at the main landing gear strut. Low wing airplanes are most easily pushed backwards by hand by pushing against the leading edge of the wing; the wing tips may be used to help maneuver the airplane.

Under no circumstances should an airplane be moved by pulling or pushing on the propeller blades. This action imposes unnecessary

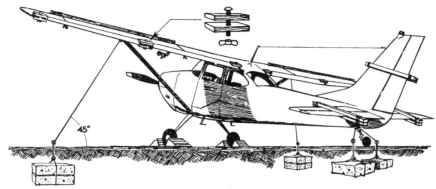

Fig.1. Picketing an Airplane in the Open.

stress on the propeller blade, but, more importantly, there is always the danger of serious injury if the propeller is alive.

Picketing an Airplane Outdoors

1. Park the airplane in a sheltered location, such as behind a row of trees or buildings. (However, parking beside a building is not recommended in winter since ice falling from the roof could damage the airplane.)

2. Head the airplane into the prevailing wind.

3. Use external control locks to hold the ailerons, flaps, elevators and rudder in the neutral position to prevent damage occurring to control surfaces. (Cockpit control surface locks are not recommended for continuous outdoor storage as gusts can still exert pressure on the various control surfaces and stretching of the control wires will result.)

4. Place chocks behind and in front of all wheels.

5. Fasten ropes to appropriate places on the wings and fasten them to weights buried in the ground. Stakes, pegs or corkscrews are not recommended. If the ground becomes saturated with rain, these will easily pull out.

6. Cover the engine and cabin with a tarpaulin, polyethylene, or canvas, etc. to prevent dust from entering the engine where it might cause excessive wear and to prevent birds from building nests, etc. The covering over the cabin will prevent rain from entering and causing corrosion and general deterioration. Cover the pilot pressure source (pitot tube).

7. In very cold weather, drain the oil into a clean container while it is still warm.

8. Place a spoiler spanwise over the top of the wing about 20% back from the leading edge. Such a spoiler will prevent lift from being developed. A 2 x 4 piece of lumber held in place by a light rope looped around the wing chordwise is an effective spoiler.

Refuelling

Always ground the airplane and the hose nozzle when refuelling, owing to the danger of fire from static electricity. Static electricity builds up in aircraft simply as a consequence of moving through the air. After landing, the charge cannot dissipate to ground because of the insulating effect of the rubber tires. In order to dissipate this electrical charge and prevent the generation of a spark that could ignite fuel vapour into a fireball, it is required procedure to properly ground the aircraft before refuelling begins. The grounding wire should be attached to some part of the aircraft; usually the landing gear is selected.

The hose nozzle should make metal to metal contact with the filler neck of the fuel tank that is being filled. Although grounding of the aircraft is important at any time, it is especially important in very cold temperatures which are more favourable to the generation of static electricity.

Static free clothing should be worn at all times when working around airplanes in order to reduce the possibility of fire. Nylon and other synthetics are very susceptible to the generation of static electricity.

Plastic containers, funnels or other devices of non-conductive material should not be used for the transfer of fuel to an airplane.

Be sure that the proper type of fuel is used in refuelling your airplane. Only fuel of the type and octane rating specified by the engine manufacturer should be used. Be especially conscientious in checking the fuel supply if you use jet fuel which is colourless. Two grades of fuel, each with its own distinctive colour, will become colourless when mixed together. Water also is colourless.

Tanks should be filled immediately after flying. Water vapour in the air in an empty or partially filled tank will condense and cause water contamination of the fuel.

All aviation fuels absorb moisture from the air and contain water in both suspended and liquid form. The water in liquid form, being heavier than the fuel, will settle to the bottom of the tank. This water **must** be removed from the tank. During the inspection prior to flight, a reasonable quantity of fuel should be drawn from the lowest point of the fuel system into a glass container. Any water in the sample will settle to the bottom where it can be readily detected. Uncontaminated fuel has an inherent brilliance and will sparkle in the presence of light. Water or fine dirt particles will make the fuel appear cloudy or hazy.

If there is any suspicion of serious water contamination of the airplane fuel system, the entire system should be thoroughly checked.

The water that exists in the fuel in suspended form does not ordinarily cause problems. It passes harmlessly through the system and evaporates in the heat of the combustion process. However, in very cold weather, any water in the fuel, including the minute particles in suspension, may freeze into ice crystals that can block fuel lines and filters. A de-icing additive usually prevents this problem but it is important to use only an additive that is recommended by the engine manufacturer. A drop in engine power or in fuel consumption or rough running of the engine are signs of fuel icing.

Water is not the only contaminant that might be present in fuel. Sand, dust, rust and microorganisms may also find their way into fuel supplies. Sand and dust can be blown into tanks during refuelling or introduced through unclean refuelling equipment. A fuel nozzle, for example, that is dragged across the ground will likely pick up dirt. Rust may develop in storage tanks and fuel trucks and rust particles could be introduced into the aircraft fuel tanks if the filtering devices of the refuelling system are poorly maintained. Microorganisms tend to occur in unleaded fuels especially and may multiply sufficiently rapidly to cause corrosion in the fuel tank and clog filters and screens. Careful maintenance, proper filtration during refuelling and thorough fuel inspections will prevent the problems caused by these foreign bodies.

Fuel additives to prevent icing and fouling should be used only if they are approved and only in strict compliance with the manufacturer's instructions.

Do not completely fill the tanks with cold fuel if the airplane is to be stored in a warm hangar. Aviation gas will expand about 1/2% in volume for a 5°C rise in temperature.

Because of the danger of static electricity and possible fire, refuelling from drums and cans is not advisable. Unfortunately, in some situations, there is no alternative. Proper bonding and grounding connections are very important and tank caps should not be removed until grounding has been done. To be as safe as possible, first ground the aircraft, then attach the funnel to the aircraft by a grounding wire and finally attach the hand held refuelling container to the funnel by a grounding wire.

When refuelling from a drum, it is important to use a proper filter/ water separator with a portable pump bonded to the drum. Never pump fuel from the bottom of a drum. The pump intake pipe should be made shorter than the depth of the drum by approximately one inch. Do not pump fuel from a drum which has just been moved. Let the drum stand for some time to allow water or dirt to settle to the bottom. A chamois lined funnel should be used only in emergency, since the passage of fuel through the chamois increases the possibility of generation of a charge of electricity. Chamois fibres may clog fuel system filters and nozzles. If you are using a chamois as a filtering device, use a chamois of good quality only. Any reduction in the flow through the chamois should be reason to suspect water contamination. Never wring a chamois free of water as this action nullifies the water separation properties of the chamois. After the flight during which the aircraft was refuelled using a chamois or felt lined funnel, the aircraft should be thoroughly checked for a contaminated fuel system.

All fuel filters, sumps, bowls and screens should be regularly inspected and cleaned. During the preflight inspection, check all drain cocks and tank caps to be sure they are properly closed and seated. The condition of fuel tank caps should be especially noted. Deterioration of the cap seal or corrosion of the cap itself or the cap seating area will permit water to enter the fuel tank.

Only oil of the S.A.E. number or grade specified by the engine manufacturer should be used. The manufacturer usually recommends different oils for varying climatic conditions.

Oil should be changed only in accordance with the recommendation of the engine manufacturer.

The oil tank should never be filled to the top. Leave room for the expansion of the oil when it gets hot.

Battery. Be alert for possible battery overcharging which may occur due to faulty current/voltage regulator malfunction. Overcharging produces heat that causes plates within the battery to warp and allows destructive acid to leak out and corrode adjacent parts.

Winter Operations

Flying in the wintertime necessitates extra precautions on the part of the pilot. Snow, ice and cold weather create problems that the pilot who flies only in the summer never has to worry about.

Aircraft tied down outside require special attention because of the hazards posed by the build up of snow and ice on wings and tail surfaces. They are not designed for the stress of heavy loads of snow and ice; they should be removed even if the aircraft is not going to be flown.

When the aircraft is parked outdoors, blowing snow can enter pitot and static pressure sources, fuel tank vents, airscoops, wheel wells and any other uncovered openings and freeze solid causing malfunctions of the parts affected. Engine covers and pitot pressure source covers should be used in winter.

If the drain holes in the airframe are clogged, water that gets inside the structure will be trapped and will freeze. Ice can jam the controls, throw the weight and balance off, crack the airframe and interfere with proper venting. Inspect the drain holes regularly to be sure they are open.

Fuel Lines. Water in the fuel lines will at any time of year affect engine performance but, during the winter, there is the added risk of the water freezing, blocking the fuel lines and causing engine stoppage. Refuel after each flight to help prevent condensation and, in addition, drain a substantial amount of fuel from the lowest point of the fuel system before each day's flight and after each refuelling. Special care needs to be taken when the temperature hovers around the freezing level. Moisture in the fuel may be present as suspended ice crystals. They will not show up when a contamination drain check is done. If the temperature rises above freezing, these ice crystals may turn to water and, if they exist in sufficient quantity, may cause engine failure.

Cabin Heaters. Since cabin heaters will be in regular use during winter operations, it is essential to ensure that the cabin heater system is free of cracks or holes. In light aircraft, heat to the cabin is produced by engine exhaust. Any failure in this system may result in carbon monoxide entering the cabin. Even small amounts of this dangerous gas affect judgment and flying ability. Large amounts can be lethal. It is essential to carry out a thorough visual inspection of the external parts of the exhaust system as part of the pre-flight inspection and on a regular basis perform preventative maintenance. It is also good practice to carry out periodic tests for the presence of carbon monoxide.

Carburetor Heat. Regularly check the carburetor heat control to ensure that it is operating properly throughout its entire range. Be sure also that the carburetor temperature gauge is working properly and is giving true indications. It is your ice warning device. During cold weather operations, when checking proper operation of the carburetor hot air control as part of the normal pre-take-off check, allow the engine to run for a few seconds with the heat control in the full ON position to assure that any ice that may have formed in the carburetor venturi area during taxiing and warm-up will be eliminated.

Control Cables. Control cables should be checked for proper winter tension. Colder temperatures can cause cables to contract, may lower the cable's tension and make the flight controls less effective.

Landing Gear. Retractable landing gear and wheels equipped with wheel pants can become clogged by the mixture of ice, slush and mud that is thrown up during landing and taxiing. Be sure the landing gear is clean and capable of spinning freely. Check strut extensions. Oleos should be serviced with nitrogen for winter operations to prevent the formation of ice crystals in the hydraulic fluid. Struts should be wiped with clean hydraulic fluid to remove snow, ice and dirt. Check tires for proper inflation. Extreme cold temperatures may require that they be reserviced. It is wise, in the winter, to remove the streamline wheel covers on fixed gear aircraft to prevent an accumulation of slush that may freeze and lock the wheels or brakes. On retractable gear aircraft, the condition of shields, boots and curtains used to protect activating devices and switches must be carefully maintained. During take-off, more slush and mud will be thrown on the landing gear and may freeze solid after the gear is retracted. To prevent this, delay retracting the gear for a moment or so to allow the slipstream to blow off most of the slush.

Taxiing. Taxiways and runways are often icy and slippery in the wintertime. Taxi very slowly and keep lots of distance between you and other aircraft taxiing ahead. Avoid taxiing through puddles. The water that splashes up on the brakes may later freeze. Avoid the use of brakes as much as possible. When the brakes are used during taxi, heat is generated at the wheel hubs, melting any snow or slush that has accumulated there. As the wheels cool while the airplane is stopped (awaiting take-off clearance, for example, or to do the take-off check), the moisture in the wheels is likely to freeze and lock the wheels. Even worse is the situation of moisture in the wheels freezing after the airplane takes off. At the next landing, the wheels will be frozen and the landing roll may be unusually short.

Preflight Inspection. Above all, do not hurry your preflight inspection because of the discomforts of the cold weather. Be even more thorough than you would normally be. Dress warmly so that you are comfortable enough to take your time.

Be sure that you have carefully studied the handbooks pertaining to your airplane, understand all the systems of the airplane and know the recommended winter operation procedures.

It is important to conduct your preflight planning with care, so that your flight can be carried out safely within the limitations of both the weather and your abilities as a pilot. Winter operations mean fewer hours of daylight, snow showers that completely and suddenly obscure the terrain, whiteout conditions, icy runways, etc. Do not fly in icing conditions if you have no de-icing equipment. Even light freezing drizzle will quickly spread a cover of ice over the windshield and side windows, producing instrument conditions inside the cockpit. In planning any winter flight, it is wise to be prepared for the worst conditions that winter can bring.

Winter cold is just as hard on the human body as it is on the airplane. Proper warm winter clothing, including warm boots, gloves and hats, should be worn by all crew members and passengers. Several layers of loose clothing provide more warmth than one bulky layer. Clothing should be kept as clean and dry as possible to be most effective. The clothing worn for an airplane flight in the winter should be adequate to keep you warm in the event of a forced landing, not just to keep you comfortable during the flight. The survival kit should be designed for the worst combination of terrain, temperature and precipitation that may be encountered. THINK SURVIVAL! You want to be able to survive the environment after surviving the crash.

Engine. Special procedures for the operation of the engine in cold weather have been discussed in the Chapter **Aero Engines**.

Runways. Directional control is difficult on an icy runway during take-off, especially in a cross wind situation. Stopping distances, in the event of an aborted take-off, will increase. Acceleration is reduced on snow and slush covered runways. There are limits to the depth of slush that can be tolerated. Consult your flight manual for limits for your aircraft.

During landing, it is important to establish a good cross wind correction and stabilize your approach so that you are properly lined up with the runway centre line. Landing with crab or off centre will aggravate control problems on an icy runway. Touchdown should be made at low speed so that deceleration can be accomplished without use of brakes and within the available runway. Reliable information about the current conditions of the runway at your destination (plowing, ice cover, slush) should be obtained before even contemplating the flight.

Critical Surface Contamination

An accumulation of frost, snow or ice on the wings or other horizontal surfaces will substantially alter the lifting characteristics of the airfoil. Even a very light layer of frost will sufficiently alter the lifting characteristics of the airfoil that the airplane's take-off capabilities will be substantially affected. The roughness of the frost covered surface spoils the smooth flow of air over the airfoil, separates the vital boundary layer and causes an increase in stall speed and a decrease in the stall angle of attack. Tests have shown that frost on the wings of a conventional light airplane has increased the stalling speed by at least 5 to 10% and even as much as 30% and the takeoff distance by as much as 100%. In one test, a ski-equipped STOL airplane with frost on the wings could not reach a speed at which it could take off even though a large lake was available for the takeoff run.

If frost can cause so much trouble, ice is worse. It is possible for the stalling speed of an ice contaminated airplane to be increased to such a degree that the airplane cannot reach sufficient speed to achieve take-off or, if having achieved take-off, to maintain flight. The airplane is very likely to stall as soon as it flies out of ground effect. Ice contamination severely affects the flight characteristics of all airplanes. In addition to adding significantly to the weight of the airplane, it changes horizontal stabilizer forces, upsetting the aerodynamic balance and centre of gravity location. The turbulent airflow over the ice contaminated wing means that the airplane will stall at a higher airspeed and at a lower angle of attack. Lateral control of the aircraft deteriorates requiring larger and larger control wheel deflections to maintain stable flight. Ice and snow on the airfoils can also cause flutter that may result in deformation of wing or aileron structures.

It is, therefore, absolutely critical that all frost, snow and ice be removed from critical surfaces before take-off. Critical surfaces are defined as wings, control surfaces, rotors, propellers, horizontal stabilizers, vertical stabilizers or any other stabilizing surfaces of an aircraft and, in the case of an aircraft with rear mounted engines, the upper surface of the fuselage.

Never assume that snow will blow off. Snow can actually change to ice during the take-off run. Always clean off any accumulation and be sure the wings are clean before getting underway. Pay special attention to the hinge areas of flaps and control surfaces. Small pieces of ice can restrict their movement and prevent full travel. Wings covers are a good investment for airplanes that operate regularly during the winter.

Under the right conditions, ice can form very rapidly, completely coating the airplane with a thick covering in a matter of minutes.

Clear ice may form on the upper surface of the wing over the fuel tank after prolonged flight at below freezing temperatures. Rain striking the wing over the cold fuel tank will turn to a sheet of clear ice that is so transparent that it is hard to detect. The refuelling process, if the fuel is warm, can melt snow or ice on the wings that will later refreeze or, if the fuel is cold, can cause moisture to freeze.

In icing conditions, you cannot rely on the stall warning systems to give a warning of an impending stall. These devices are calibrated to perform under clean wing conditions and do not recognize the degraded performance of a contaminated wing.

If the aircraft does not have any de-icing or anti-icing capability, flight through known icing conditions must be scrupulously avoided.

There are a number of major factors that contribute to critical surface contamination and a knowledgeable pilot will recognize them as indicators of an icing condition.

Ambient temperature provides a good indication of the potential for icing conditions.

Aircraft surface temperature indicates the susceptibility of the aircraft to icing. Aircraft surface temperature is affected by solar radiation. An aircraft will have a warmer surface temperature on a sunny day than on an overcast day with identical ambient temperatures. When the fuel in a wing fuel tank is very cold, the cold fuel in the tanks can so chill the aluminum wing surface that moisture in humid air or rain will turn to frost or ice over the fuel tank (**cold soaking**).

Precipitation type and rate. Dry snow tends not to stick to aircraft surfaces whereas wet snow will adhere and build up rapidly. Freezing rain combined with snow creates slush. Heavy precipitation is potentially more critical than precipitation falling at a light rate.

Relative humidity. Very moist air in combination with cold air temperatures and cold aircraft surface temperatures can give icing conditions even on a clear sunny day.

Wind direction and velocity. North winds are cold and dry but south winds are usually moist. High winds will blow snow onto a parked aircraft.

Operation in close proximity to other aircraft. The exhaust or prop wash from other aircraft can blow snow or slush onto your aircraft or can cause snow on your aircraft to melt and then later freeze as ice. Don't follow too closely behind another aircraft when taxiing to take-off position.

Operation on snowy, slushy or wet surfaces increases the possibility that contaminants will be splashed onto the wings, flaps and control surfaces. It is important to remember that contaminants on the underside of wings are as degrading to performance as contaminants on the upper surface. Keep taxi speeds slow to reduce the possibility of splashing snow and slush onto your aircraft.

Aircraft configuration and surface roughness. Aircraft collect snow, slush, freezing rain and ice on any exposed surface or angle. Smooth surfaces don't collect as much contamination as do rougher surfaces. However, a smooth surface sometimes gives the appearance of being wet when it, in fact, is covered in clear ice.

Be alert to the conditions that cause icing even before going out to your aircraft. Get a thorough weather briefing and the most up-to-date forecast so that you are aware of temperatures and precipitation at your stops and en route.

Ice and snow not only affect the lifting surfaces. They can clog and close off vital openings, such as the pitot and static pressure sources of the pitot-static system, rendering the airspeed, altimeter and vertical speed useless. Clogged fuel breather vents will prevent the free flow of fuel. Gyroscopic instruments powered by a venturi would be affected by ice building up on the venturi throat. Ice on antennas impedes radio reception.

See also **Snow, Frost and Ice** in Section **Stall** in Chapter **Theory of Flight** and **Icing** in Chapter **Meteorology**.

De-Icing

Examine your aircraft very carefully prior to flight. Use your eyes and hands to examine the surfaces to ensure that your aircraft is "clean" before departing on a flight. Have the aircraft de-iced by ground crews if there is any contamination.

The de-icing process is intended to restore the aircraft to a clean configuration so that there is no degradation of the aerodynamic characteristics nor mechanical interference with moving parts, antennas, sensors, etc.

There are several suitable methods to de-ice an aircraft. It may be parked in a heated hangar until all the contamination has melted. Wing covers and other temporary shelters will reduce the amount of contamination and the time required to de-ice the aircraft. Light, dry snow can be removed with a broom. Light frost can be rubbed off using a rope sawed across the contaminated area.

De-icing is most commonly accomplished, especially on larger aircraft, using heated solutions of water and FPD (**freezing point depressant**) fluids. These fluids lower the freezing point of water in the liquid or crystal (ice) phase. De-icing is often followed by anti-icing using cold, rich solutions with a very low freezing point. The anti-icing solution is relatively thick and forms a protective layer over the aircraft that impedes the formation of ice, snow and frost. De-icing and anti-icing fluids should not be used unless approved by the aircraft manufacturer.

The FPD fluids used to de-ice aircraft in North America are usually composed of ethylene glycol or propylene glycol combined with water and other fluids. They should not be used in an undiluted state. Ethylene glycol has a much higher freezing point in its pure state than when diluted with water and may freeze when exposed to slight temperature decreases associated with cold-soaked fuels in wing tanks, reduction of solar radiation by clouds obscuring the sun, wind effects and lowered temperature during development of wing lift. Pure propylene glycol, being quite viscous, can cause lift reductions.

FPD fluids in concentrated forms should not be applied to pitot and static pressure sources, angle of attack sensors, control surface cavities, cockpit windows, the nose of the fuselage, the lower side of the radome, air inlets or engines.

SAE and ISO Type II fluids contain no less than 50% glycols and have a minimum freeze point of -32°C. They are thickened and thus have the property of adhering to aircraft surfaces until the time of take-off. These fluids are used for de-icing, when heated, and for anti-icing, when used unheated. They are effective anti-icers because of their high viscosity and pseudo-plastic properties. They remain on the wings during ground operations but flow off the wings readily during take-off. SAE and ISO Type II fluids should not be used on aircraft with a rotation speed (V_R) of less than 100 knots. Below this speed, some of the fluid may not completely flow away and may cause some degradation of performance.

SAE and ISO Type I fluids are unthickened. They are used for de-icing but provide very limited anti-icing protection.

SAE and ISO Type III is a thickened fluid with properties between Types I and II. It has a longer hold over time than Type I but less than Type II. It is suitable for use on aircraft with a shorter ground roll to rotation and a rotation speed of less than 100 knots.

Temperature buffer is the temperature difference between the freezing point of the fluid being applied and the ambient temperature. Hold over time is increased with an expansion of the temperature buffer. (It should not be less than 10° for Type I fluids or 3° for Type II fluids, depending on the ambient temperatures.) The maximum buffer is the best choice. However, greater buffers require the use of more glycol which is costly and increase the burden for collection and processing of FPD spillage and run-off. Nevertheless, the maximum buffer provides the greatest margin of safety and reduces the possibility of refreezing of the FPD fluid during take-off, climb and at altitude.

If your aircraft is fitted with anti-icing or de-icing systems, be sure that this equipment is operational before taking off on a flight during which it may be required. Be sure that you know how and when to use it.

Weight and Balance

You as pilot are responsible for the safe loading of your airplane and must ensure that it is not overloaded. The performance of an airplane is influenced by its weight and overloading it will cause serious problems. The take-off run necessary to become airborne will be longer. In some cases, the required take-off run may be greater than the available runway. The angle of climb and the rate of climb will be reduced. Maximum ceiling will be lowered and range shortened. Landing speed will be higher and the landing roll longer. In addition, the additional weight may cause structural stresses during maneuvers and turbulence that could lead to damage.

The total gross weight authorized for any particular type of airplane must therefore never be exceeded. A pilot must be capable of estimating the proper ratio of fuel, oil and payload permissible for a flight of any given duration. The weight limitations of some general aviation airplanes do not allow for all seats to be filled, for the baggage compartment to be filled to capacity and for a full load of fuel as well. It is necessary, in this case, to choose between passengers, baggage and full fuel tanks.

The distribution of weight is also of vital importance since the position of the centre of gravity affects the stability of the airplane. In loading an airplane, the C.G. must be within the permissible range and remain so during the flight to ensure the stability and maneuverability of the airplane during flight.

Airplane manufacturers publish weight and balance limits for their airplanes. This information can be found in two sources:

1. *The Aircraft Weight and Balance Report.*
2. *The Airplane Flight Manual.*

The information in the Airplane Flight Manual is general for the particular model of airplane.

The information in the Aircraft Weight and Balance Report is particular to a specific airplane. The airplane with all equipment installed is weighed and the C.G. limits calculated and this information is tabulated on the report which accompanies the airplane log books. If alterations or modifications are made or additional equipment added to the airplane, the weight and balance must be recalculated and a new report prepared.

Weight

Various terms are used in the discussion of the weight of an airplane. They are as follows:

Standard Weight Empty: The weight of the airframe and engine with all standard equipment installed. It also includes the unusable fuel and oil.

Optional or Extra Equipment: Any and all additional instruments, radio equipment, etc. installed but not included as standard equipment, the weight of which is added to the standard weight empty to get the basic empty weight. It also includes fixed ballast, full engine coolant, hydraulic and de-icing fluid.

Basic Weight Empty: The weight of the airplane with all optional equipment included. In most modern airplanes, the manufacturer includes full oil in the basic empty weight.

Useful load (or Disposable load): The difference between gross take-off weight and basic weight empty. It is, in other words, all the load which is removable, which is not permanently part of the airplane. It includes the usable fuel, the pilot, crew, passengers, baggage, freight, etc.

Payload: The load available as passengers, baggage, freight, etc., after the weight of pilot, crew, usable fuel have been deducted from the useful load.

Operational Weight Empty: The basic empty weight of the airplane plus the weight of the pilot. It excludes payload and usable fuel.

Usable Fuel: Fuel available for flight planning.

Unusable Fuel: Fuel remaining in the tanks after a runout test has been completed in accordance with government regulations.

Operational Gross Weight: The weight of the airplane loaded for take-off. It includes the basic weight empty plus the useful load.

Maximum Gross Weight: The maximum permissible weight of the airplane.

Maximum Take-Off Weight: The maximum weight approved for the start of the take-off run.

Maximum Ramp Weight: The maximum weight approved for ground maneuvering. It includes the weight of fuel used for start, taxi and run up.

Passenger Weights: Actual passenger weights must be used in computing the weight of an airplane with limited seating capacity. Allowance must be made for heavy winter clothing when such is worn. Winter clothing may add as much as 14 lbs to a person's basic weight; summer clothing would add about 8 lbs.

On larger airplanes with quite a number of passenger seats and for which actual passenger weights would not be available, the following average passenger weights may be used. The specified weights for males and females include an allowance for 8 lbs of carry-on baggage.

	Summer	Winter
Males (12 yrs & up)	182 lbs	188 lbs
Females (12 yrs & up)	135 lbs	141 lbs
Children (2-11 yrs)	75 lbs	75 lbs
Infants (0-up to 2 yrs)	30 lbs	30 lbs

Fuel and Oil: The Airplane Flight Manuals for airplanes of U.S. manufacture give fuel and oil quantities in U.S. gallons. Canadian manufactured airplanes of older vintage may have manuals that give fuel and oil quantities in Imperial gallons. Some recently printed manuals may give fuel and oil quantities in litres. At most airports in Canada, fuel is now dispensed in litres. It is therefore necessary to convert from litres to U.S. or Imperial gallons as required for your particular airplane. To convert litres to U.S. gallons, multiply by .264178. To convert litres to Imperial gallons, multiply by .219975.

The following weights are for average density at the standard air temperature of 15°C. At colder temperatures, the weights increase slightly. For example, at -40°C, one litre of aviation gasoline weighs 1.69 lbs.

	Litre	U.S. Gallon	Imp. Gallon
Av. Gas	1.59 lb	6.0 lb	7.20 lb
J.P.4	1.77 lb	6.6 lb	8.01 lb
Kerosene1	.85 lb	7.0 lb	8.39 lb
Oil (65)	1.95 lb	7.5 lb	8.85 lb

Maximum Landing Weight: The maximum weight approved for landing touchdown. Most multi-engine airplanes which operate over long stage lengths consume a considerable weight of fuel. As a result, their weight is appreciably less on landing than at take-off. Designers take advantage of this condition to stress the airplane for lighter landing loads, thus saving structural weight. If the flight has been of short duration, fuel or payload may have to be jettisoned to reduce the gross weight to maximum landing weight.

Maximum Weight — Zero Fuel: The weight of the airplane exclusive of usable fuel, but including useful load. Some manufacturers designate this weight for structural reasons. Transport planes carry fuel in their wings, the weight of which relieves the bending moments imposed on the wings by the lift. The maximum weight — zero fuel limits the load which may be carried in the fuselage. Any increase in weight in the form of load carried in the fuselage must be counterbalanced by adding weight in the form of fuel in the wings.

Float Buoyancy: The maximum permissible gross weight of a seaplane is governed by the buoyancy of the floats. The buoyancy of a seaplane float is equal to the weight of water displaced by the immersed part of the float. This is equal to the weight the float will support without sinking beyond a predetermined level (draught line).

The buoyancy of a seaplane float is designated by its model number. A 4580 float has a buoyancy of 4580 lb. A seaplane fitted with a pair of 4580 floats has a buoyancy of 9160 lbs.

Regulations require an 80% reserve float buoyancy. The floats must, therefore, have a buoyancy equal to 180% of the weight of the airplane.

To find the maximum gross weight of a seaplane fitted with, say 7170 model floats, multiply the float buoyancy by 2 and divide by 1.8. (7170 x 2) ÷ 1.8 = 7966 lb.

Computing the Load
A typical light airplane has a basic weight of 1008 lb. and an authorized maximum gross weight of 1600 lb. An acceptable loading of this airplane would be as follows:

Basic Empty Weight	1008 lb
Consisting of Weight Empty	973 lb.
Oil	15 lb.
Extra Equipment	20 lb.
Useful Load	592 lb
Consisting of Pilot	150 lb.
Fuel	146 lb.
Payload: Passenger.	175 lb
Baggage	121 lb.

Problem
To find the maximum payload that can be transported a given distance and the amount of fuel required.

A seaplane on contract with a mining company is required to transport a maximum load of freight a distance of 300 nautical miles to a bush operation. The estimated groundspeed is 110 knots. The useful load for this airplane is 1836 pounds. Fuel capacity is 86 U.S. gallons. Fuel consumption is 20 gallons per hour or 120 lb of fuel per hour.

The time to fly 300 nautical miles is 164 minutes (300 ÷ 110 x 60). Add to that the 45 minutes required for reserve and the amount of fuel required must be sufficient for 209 minutes of flying time.

The amount of fuel required at 20 gallons per hour is 69.7 U.S. gallons (20 ÷ 60 x 209). That quantity of fuel weighs 418 lb (69.7 x 6lb.).

The fuel calculations can also be computed by using the weight of fuel consumed per hour. The weight of fuel necessary for the flight is 418 lb. (120 ÷ 60 x 209).

The useful load is 1836 lb. The weight of the pilot (170 lb.) and fuel (418 lb.) is 588 lb. Therefore, the maximum payload permissible is 1248 lb.

What quantity of fuel in litres will be required? One U.S. gallon equals 3.785332 litres. The quantity of fuel required is, therefore, 263.8 litres (69.7 x 3.785332). (See **Conversion Tables**.)

Balance Limits
The position of the centre of gravity along its longitudinal axis affects the stability of the airplane. There are forward and aft limits established by the aircraft design engineers beyond which the C.G. should not be located for flight. These limits are set to assure that sufficient elevator deflection is available for all phases of flight.

If the C.G. is too far forward, the airplane will be nose heavy, if too far aft, tail heavy. An airplane whose centre of gravity is too far aft may be dangerously unstable and will possess abnormal stall and spin characteristics. Recovery may be difficult if not impossible because the pilot is running out of elevator control. It is, therefore, the pilot's responsibility when loading an airplane to see that the C.G. lies within the recommended limits.

Usually the Airplane Owner's Manual lists a separate weight limitation for the baggage compartment in addition to the gross weight limitation of the whole airplane. This is a factor to which the pilot must pay close attention, for overloading the baggage compartment (even if the plane itself is not overloaded) may move the C.G. too far aft and affect longitudinal control.

The Airplane Owner's Manual may also specify such things as the seat to be occupied in solo flight (in a tandem seating arrangement) or which fuel tank is to be emptied first. Such instructions should be carefully complied with.

As the flight of the airplane progresses and fuel is consumed, the weight of the airplane decreases. Its distribution of weight also changes and hence the C.G. changes. The pilot must take into account this situation and calculate the weight and balance not only for the beginning of the flight but also for the end of it.

Definitions
The **centre of gravity (C.G.)** is the point through which the weights of all the various parts of an airplane pass. It is, in effect, the imaginary point from which the airplane could be suspended

and remain balanced. The C.G. can move within certain limits without upsetting the balance of the airplane. The distance between the forward and aft C.G. limits is called the centre of gravity range.

The **balance datum line** is a suitable line selected arbitrarily by the manufacturer from which horizontal distances are measured for balance purposes. It may be the nose of the airplane, the fire wall or any other convenient point (see Fig. 2).

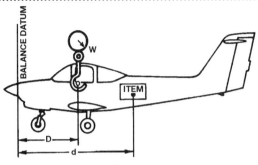

Fig. 2. Balance.

The **moment arm** (D in Fig. 2) is the horizontal distance in inches from the balance datum line to the C.G. (0 in Fig. 2). The distance from the balance datum line to any item, such as a passenger, cargo, fuel tank, etc. is the arm of that item (d in Fig.2).

The **balance moment** of the airplane is determined by multiplying the weight of the airplane by the moment arm of the airplane. It is expressed in inch pounds. The balance moment of any item is the weight of that item multiplied by its distance from the balance datum line. It is, therefore, obvious that a heavy object loaded in a rearward position will have a much greater balance moment than the same object loaded in a position nearer to the balance datum line.

The **moment index** is the balance moment of any item or of the total airplane divided by a constant such as 100, 1000, or 10,000. It is used to simplify computations of weight and balance especially on large airplanes where heavy items and long arms result in large unmanageable numbers.

If loads are forward of the balance datum line their moment arms are usually considered negative (-). Loads behind the balance datum line are considered positive (+)★. The total balance moment is the algebraic sum of the balance moments of the airplane and each item composing the disposable load.

★*In many cases the positive (+) sign is omitted, but the negative (-) sign is always shown. To simplify matters, both are included in our example.*

The C.G. is found by dividing the total balance moment (in inch-pounds) by the total weight (in lb.) and is expressed in inches forward (-) or aft (+) of the balance datum line.

The **centre of gravity range** is usually expressed in inches from the balance datum line (i.e. +39.5" to +45.8"). In some airplanes, it may be expressed as a percentage of the mean aerodynamic chord (25% to 35%). The MAC is the mean aerodynamic chord of the wing.

To calculate the position of the C.G. in percent of MAC. Let us assume that the weight and balance calculations have found the C.G. to be 66 inches aft of the balance datum line and the leading edge of the MAC to be 55 inches aft of the same reference (Fig. 3). The C.G. will, therefore, lie 11 inches aft of the leading edge of the MAC. If the MAC is 40 inches in length, the position of the C.G. will

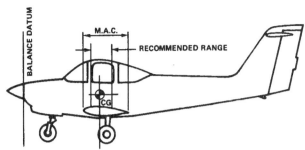

Fig. 3. Mean Aerodynamic Chord.

be at a position (11 ÷ 40) 27% of the MAC. If the calculated C.G. position is within the recommended range (for example, 25% to 35%), the airplane is properly loaded.

There are several methods by which weight and balance calculations may be made for any loading situation.

Finding Balance by Computation Method

For this example, an airplane with a basic weight of 1575 lb. and an authorized gross weight of 2600 lb. has been selected. The balance datum line for the airplane, selected by the manufacturer, is the fire wall. The recommended C.G. limits are 35.5" to 44.8".

List in table form the airplane (basic weight), pilot, passengers, fuel, oil, baggage, cargo, etc., their respective weights and arms. Calculate the balance moment of each. Total the weights. Total the balance moments. Divide the total balance moment by the total weight to find the moment arm (i.e. the position of the C.G.).

(Note: In this example, the oil is listed as a separate item and the balance datum line is the fire wall in order to give an example of a negative moment arm.)

The moment arm for this loading of the airplane is 42.52" (110,270 ÷ 2593). The total weight (2593 lb.) of the loaded airplane is less than the authorized gross weight (2600 lb.). The moment arm falls within the C.G. range (35.5" to 44.8"). The airplane is, therefore, properly loaded.

Item	Weight Lb.	Moment Arm Inches	Balance Arm Inch-Lb.
Basic Airplane	1575	+36	+56,700
Pilot	165	+37	+ 6,105
Passenger (front seat)	143	+37	+ 5,291
Passenger (rear seat)	165	+72	+11,880
Child (rear seat)	77	+72	+ 5,544
Baggage	90	+98	+ 8,820
Fuel (60 U.S.Gal. @ 6 lb.)	360	+45	+16,200
Oil (2.4 U.S.Gal. @ 7.5 Lb.)	18	-15	-270
	2593		110,270

The above example examines the situation of an airplane almost at gross weight with the C.G. in a rearward position but within the C.G. range. If this calculation had resulted in a C.G. position that was aft of the C.G. limits, even though the total weight of the airplane was under the authorized gross weight, it would be necessary either to lighten the load or to shift the load by, for example, having the passengers change seats.

A lightly loaded airplane at the end of a flight when the fuel is almost all consumed may experience the situation that the C.G. moves forward beyond the permissible C.G. range. In some airplanes, when flying with only the pilot on board and no passengers or baggage, it is necessary to carry some suitable type of ballast to compensate for a too far forward C.G. Every pilot should, therefore, calculate the moment arm for the lightest possible loading of his airplane to determine if it is acceptable.

Finding Balance by Graph Method

Most Airplane Flight Manuals include tables and graphs for calculating weight and balance. The charts in Fig. 4 to 6 are typical of those found in such manuals. They are very easy to use and eliminate the time consuming mathematical steps of the computation method.

Fig. 4.

Using the charts in Fig. 4, 5 and 6, let us work out a weight and balance problem.

1. Take the licensed empty weight and moment/1000 from the Weight and Balance Data Sheet carried in the airplane and enter them in the columns of the chart in Fig. 4. For the airplane used in our problem, these figures are 1507 lbs. and a moment of 56.6.

2. Enter the weight and moment/1000 for the oil in the proper columns. The sample airplane in this problem uses 10 U.S. quarts (19 lbs. and a moment of -0.4). You usually have a full load of oil for a trip. Therefore these figures can be considered non-variables.

3. Compute the weight of the pilot and the front seat passenger. Refer to the Loading Graph (Fig. 5) and find the moment/1000 on the scale. Enter these figures on the chart — 340 lbs. and a moment of 12.0.

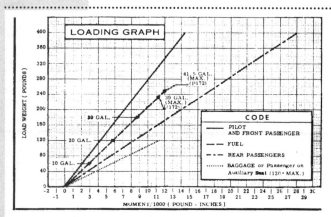

Fig. 5.

4. Determine the weight of fuel on board. (In this case, 41.5 U.S. gallons at 6 lbs. per gallon). Find the moment/1000 from the Loading Graph and enter the figures on the chart ¤ 249 lbs. and a moment of 12.0.

5. Calculate the weight of passengers in the rear seat. Using the proper line on the Loading Graph, find the moment/1000 and enter these figures on the chart — 340 lbs. and a moment of 23.8.

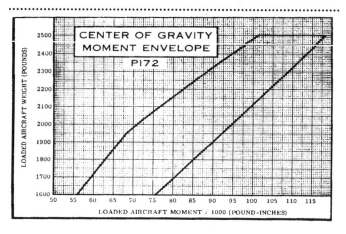

Fig. 6.

6. Weigh the baggage and read the moment/1000 on the proper line on the Loading Graph. Enter these figures on the chart — 45 lbs. and a moment of 4.3.

7. Total the weight column. The total must be less than the gross weight allowed for your airplane. The gross weight of the airplane used in this problem is 2500 lbs. If the total in the weight column exceeds the gross weight, you must lighten the airplane's load.

Total the moment column. Remember to subtract the oil moment.

8. Refer to the Centre of Gravity Moment Envelope (Fig. 6). Locate the intersection of the lines connecting the total weight with the total moment/1000. If this point is within the envelope, your airplane is loaded within proper limits. If this point of intersection is outside the envelope, the load must be adjusted.

Weight Shift

Sometimes in weight and balance calculations, the gross weight of the airplane or the position of the C.G. turns out to be beyond acceptable limits. If the total weight is too much, some part of the load has to be removed. If, however, the gross weight is not exceeded but the centre of gravity is outside the C.G. envelope, the problem may be solved by shifting some portion of the load from a rearward position to a more forward position (if the C.G. is aft of limits) or vice versa (if the C.G. is forward of limits).

Suppose that the loaded airplane weighs 3000 pounds which is within limits but the C.G. is located in a position that is too far aft. Some baggage or cargo stowed in the rear cargo hold could be moved forward. How much weight must be moved forward to bring the C.G. into acceptable limits?

Apply the following formula:

$$\frac{\text{Weight to move}}{\text{Weight of the Airplane}} \text{ equals } \frac{\text{Distance C.G. Must Move}}{\text{Distance Between Arms}}$$

From the Ground Up

The moment arm of the rear baggage compartment is 125. A more forward position for a piece of baggage has an arm of 40. The aft C.G. limit is 46 but the unacceptable loading has produced a C.G. of 48.

X	equals	2 (48 less 46)	X equals 70.5 lbs.
3000		85 (125 less 40)	

This formula can also be used to determine how far forward a piece of cargo, equipment, etc. must be moved in order to achieve an acceptable C.G. position. After the piece of cargo/luggage has been moved, recalculate the weight and balance to ensure that the desired result has been achieved.

Weight and Balance and Flight Performance

The flight characteristics of an airplane at gross weight with the C.G. very near its most aft limits are very different from those of the same airplane lightly loaded.

For lift and weight to be in equilibrium in order to maintain any desired attitude of flight (see **Theory of Flight**), more lift must be produced to balance the heavy weight. To achieve this, the airplane must be flown at an increased angle of attack. As a result, the wing will stall sooner (i.e. at a higher airspeed) when the airplane is fully loaded than when it is light. Stalling speed in turns (that is, at increased load factors) will also be higher. In fact, everything connected with lift will be affected. Take-off runs will be longer, angle of climb and rate of climb will be reduced and, because of the increased drag generated by the higher angle of attack, fuel consumption will be higher than normal for any given airspeed. Severe g forces are more likely to cause stress to the airframe supporting a heavy payload.

An aft C.G. makes the airplane less stable, making recovery from maneuvers more difficult. The airplane is more easily upset by gusts. However, with an aft C.G., the airplane stalls at a slightly lower airspeed. To counteract the tail heaviness of the aft C.G., the elevator must be trimmed for an up load. The horizontal stabilizer, as a result, produces extra lift and the wings, correspondingly, hold a slightly lower angle of attack.

An aft C.G. means that the airplane will have a better range in cruise because of the lower angle of attack and the reduced downward force on the horizontal tail surfaces. Under no circumstances, however, should the airplane purposely be loaded with the C.G. outside the envelope in order to achieve better range.

An airplane with a forward centre of gravity, being nose heavy, is more stable but more pressure on the elevator controls will be necessary to raise the nose — a fact to remember on the landing flare. The forward C.G. means a somewhat higher stalling speed — another fact to remember during take-offs and landings.

Every pilot should be aware of these general characteristics, shared by most airplanes, when they are loaded to their weight and balance limits. The important thing to remember is that these characteristics are more pronounced as the limits are approached and may become dangerous if they are exceeded. Overloading, as well as the immediate degradation of performance, subjects the airplane to unseen stresses and precipitates component fatigue.

Airplane Performance

Effect of Temperature and Altitude on Airplane Performance

The figures published in the Flight Manual for the performance capabilities of a certain model of airplane are always related to standard atmosphere (29.92 inches of mercury at 15°C at sea level). However, only rarely will the airplane actually operate under conditions that approximate standard atmosphere. Any increase in temperature or altitude means a decrease in the aircraft's optimum performance.

Air density decreases with altitude. At 10,000 feet, the pressure exerted by a column of air is considerably less than at sea level. As a result, at high elevation airports, an airplane requires more runway to take off. Its rate of climb will be less, its approach will be faster (because the TAS will be faster than the IAS) and the landing roll will be longer.

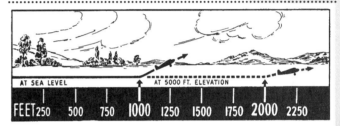

Fig. 7. Effect of Altitude and Temperature on Take-off & Climb.

Air density also decreases with temperature. Warm air is less dense than cold air because there are fewer air molecules in a given volume of warm air than in the same volume of cooler air. As a result, on a hot day, an airplane will require more runway to take off, will have a poor rate of climb and a faster approach and will experience a longer landing roll.

In combination, high and hot, a situation exists that can well be disastrous for an unsuspecting, or more accurately, an uninformed pilot. The combination of high temperature and high elevation produces a situation that aerodynamically reduces drastically the performance of the airplane. The horse power out-put of the engine is decreased because its fuel-air mixture is reduced. The propeller develops less thrust because the blades, as airfoils, are less efficient in the thin air. The wings develop less lift because the thin air exerts less force on the airfoils. As a result, the take-off distance is substantially increased, climb performance is substantially reduced and may, in extreme situations, be non-existent.

Humidity also plays a part in this scenario. Although it is not a major factor in computing density altitude, high humidity has an effect on engine power. The high level of water vapour in the air reduces the amount of air available for combustion and results in an enriched mixture and reduced power. In situations of high humidity and high density, it would be wise to add 10% to the computed take-off distance and to expect a reduced rate of climb.

Mountain airports are particularly treacherous when temperatures are high, especially for a low performance airplane. The actual elevation of the airport may be near the operational ceiling of the airplane without the disadvantage of density altitude. Under some conditions, the airplane may not be able to lift out of ground effect or to maintain a rate of climb necessary to clear obstacles or surrounding terrain.

Density altitude is pressure altitude corrected for temperature. It is, in layman terms, the altitude at which the airplane thinks it is flying based on the density of the surrounding air mass.

Too often, pilots associate density altitude only with high elevation airports. Certainly, the effects of density altitude on airplane performance are increasingly dramatic in operations from such airports, especially when the temperature is also hot. But it is important to remember that density altitude also has a negative effect on performance at low elevation airports when the temperature goes above the standard air value of 15°C at sea level. Remember also that the standard air temperature value decreases with altitude.

Standard Temperature	Elevation	Density Altitude at			
		27°C	32°C	38°C	43°C
15°	Sea Level	1200'	1900'	2500'	3200'
11°	2000'	3800'	4400'	5000'	5600'
7°	4000'	6300'	6900'	7500'	8100'
3°	6000'	8600'	9200'	9800'	10,400'
-1°	8000'	11,100'	11,700'	12,300'	12,800'

The accompanying chart gives some examples of how high temperatures increase the density altitude even at low elevation fields:

Fig. 7 illustrates the effect of density altitude. The take-off run at sea level with a temperature of 15°C (standard air) is 1000 feet. At a 5000 foot elevation, with a temperature of 20°C, the take-off run is doubled and the rate of climb diminished 55%.

Airplane Flight Manuals publish tables or charts (See Fig. 11, 13 and 20) that give specific figures for performance at varying density altitudes. These figures should always be referred to when operating in high density altitude situations.

In order to compute the density altitude at a particular location, it is necessary to know the pressure altitude. To determine the latter, set the barometric scale of the altimeter to 29.92" Hg and read the altitude. (See also **Density Altitude** in the Chapter **Theory of Flight**.)

Density altitude can be calculated for any given combination of pressure altitude and temperature on the circular slide rule portion of a flight computer. The procedure for doing so has already been discussed in the Chapter **Navigation**.

Density Altitude Koch Chart

The **Koch Chart** (Fig. 8) shows in representative form the kind of decrease in performance a pilot can expect when operating in situations of high altitude and/or higher than normal temperatures. It can be used to estimate the increase in an airplane's take-off roll, and the decrease in its rate-of-climb over its known standard-air, sea-level values. For example, assume a pressure altitude of 6000 feet and an airport temperature of 37°C. Locate 6000 feet on the airport pressure altitude index on the right edge of the chart. Locate 37°C on the temperature index on the left edge of the chart. Join these two points and note where the line cuts the percentage index. The airplane's normal take-off distance would be lengthened by 230% and its initial rate of climb would fall off 76%.

If you have neither your Airplane Flight Manual or a Koch Chart, take-off performance can be calculated by a relatively simple rule of thumb. Remember, first, that an **increase** in density altitude will cause a corresponding **decrease** in take-off performance. Add 10% to the normal take-off ground roll and to the distance to clear a 50 foot obstacle for every 1000 foot increase in the density altitude

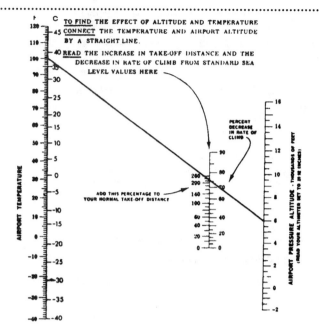

Fig. 8. Koch Chart.

up to a density altitude of 3000 feet. Above 3000 feet, add 20% per 1000 foot increase in the density altitude.

For example: At an airport at sea level, the altimeter setting is 29.92" Hg and the altimeter indicates that the pressure altitude is 0 feet. The temperature is 15°C which is standard air. The take-off run under these conditions would be 1000 feet. Suppose that the temperature were to rise to 25°C and the pressure altitude remain constant. The density altitude of the airport at sea level would now be 1000 feet. For the increase of 1000 feet in density altitude, add 10% to the take-off run which would now be 1100 feet.

Denalt Performance Computer

A small circular computer called the Denalt Performance Computer may also be used to calculate expected take-off distance and rate of climb in high density altitude situations. Rotate the computer until the outside air temperature appears in the appropriate window. The take-off factor can then be read opposite the pressure altitude in a second window. (Pressure altitudes from sea level through

Fig. 9. Denalt Performance Computer.

14,000 feet, in increments of 2000 feet, are given.) Multiply this take-off factor by the take-off distance that is normally required for your airplane at sea level in standard conditions and at the same gross weight. The percent rate of climb (ROC) can be read opposite the pressure altitude in a third window. Multiply the percent ROC by the normal sea level rate of climb for your airplane in standard conditions and at the same gross weight.

Take off Performance Charts

The Airplane Flight Manual publishes information, usually in chart form, on the take-off performance of a specific model of airplane. As a pilot you should familiarize yourself with these charts to be able to predict how your airplane will perform under varying conditions and should refer to these charts whenever there is any doubt that the take-off conditions may not be sufficient for the performance capabilities of the airplane. In addition, it is important to remember that the charts for any particular airplane were compiled from performance figures of factory new equipment in optimum conditions. Any typical general aviation airplane, with considerable time on both airframe and engine, will have a poorer performance potential than that predicted by the charts. In addition, underinflated tires, dragging brakes, dirt on the wings, etc. will also affect performance negatively.

TAKEOFF DISTANCE
MAXIMUM WEIGHT 2400 LBS

WEIGHT LBS	TAKEOFF SPEED KIAS		PRESS ALT FT	0°C		10°C		20°C		30°C		40°C	
	LIFT OFF	AT 50 FT		GRND ROLL FT	TOTAL FT TO CLEAR 50 FT OBS	GRND ROLL FT	TOTAL FT TO CLEAR 50 FT OBS	GRND ROLL FT	TOTAL FT TO CLEAR 50 FT OBS	GRND ROLL FT	TOTAL FT TO CLEAR 50 FT OBS	GRND ROLL FT	TOTAL FT TO CLEAR 50 FT OBS
2400	51	56	S.L.	795	1460	860	1570	925	1685	995	1810	1065	1945
			1000	875	1605	940	1725	1015	1860	1090	2000	1170	2155
			2000	960	1770	1035	1910	1115	2060	1200	2220	1290	2395
			3000	1055	1960	1140	2120	1230	2295	1325	2480	1425	2685
			4000	1165	2185	1260	2365	1355	2570	1465	2790	1575	3030
			5000	1285	2445	1390	2660	1500	2895	1620	3160	1745	3455
			6000	1425	2755	1540	3015	1665	3300	1800	3620	1940	3990
			7000	1580	3140	1710	3450	1850	3805	2000	4220	---	---
			8000	1755	3615	1905	4015	2060	4480	---	---	---	---
2200	49	54	S.L.	650	1195	700	1280	750	1375	805	1470	865	1575
			1000	710	1310	765	1405	825	1510	885	1615	950	1735
			2000	780	1440	840	1545	905	1660	975	1785	1045	1915
			3000	855	1585	925	1705	995	1835	1070	1975	1150	2130
			4000	945	1750	1020	1890	1100	2040	1180	2200	1270	2375
			5000	1040	1945	1125	2105	1210	2275	1305	2465	1405	2665
			6000	1150	2170	1240	2355	1340	2555	1445	2775	1555	3020
			7000	1270	2440	1375	2655	1485	2890	1605	3155	1730	3450
			8000	1410	2760	1525	3015	1650	3305	1785	3630	1925	4005
2000	46	51	S.L.	525	970	565	1035	605	1110	650	1185	695	1265
			1000	570	1060	615	1135	665	1215	710	1295	765	1385
			2000	625	1160	675	1240	725	1330	780	1425	840	1525
			3000	690	1270	740	1365	800	1465	860	1570	920	1685
			4000	755	1400	815	1500	880	1615	945	1735	1015	1865
			5000	830	1545	900	1660	970	1790	1040	1925	1120	2070
			6000	920	1710	990	1845	1070	1990	1150	2145	1235	2315
			7000	1015	1900	1095	2055	1180	2225	1275	2405	1370	2605
			8000	1125	2125	1215	2305	1310	2500	1410	2715	1520	2950

CONDITIONS:
Flaps 10°   Paved, Level, Dry Runway
Full Throttle Prior to Brake Release   Zero Wind

NOTES:
1. Short field technique as specified in Section 4.
2. Prior to takeoff from fields above 3000 feet elevation, the mixture should be leaned to give maximum RPM in a full throttle, static runup.
3. Decrease distances 10% for each 9 knots headwind. For operation with tailwinds up to 10 knots, increase distances by 10% for each 2 knots.
4. For operation on a dry, grass runway, increase distances by 15% of the "ground roll" figure.

Fig. 10. Take-Off Data Chart.

Fig. 10 depicts a Take-Off Data Chart which tabulates the expected ground run and the expected distance over the ground needed to clear a 50 foot obstacle. The figures are given for a gross weight of 2400 lb. and for lighter loadings of 2200 lb. and 2000 lb. The table includes take-off performance data at sea level and at elevations in 1000 foot increments up to 8000 feet and at temperatures varying from 0°C to 40°C.

The performance figures in Fig. 10 are given for take-off from a hard surface, level, dry runway with no wind and with 10° of flap. For operation on a dry, grass runway, the distance required for the ground roll should be increased by 15%. Less favourable conditions such as long grass, sand, mud, slush, standing water, glassy water (in the case of seaplanes) and soft snow (in the case of skiplanes) can easily double the take-off. A 10% reduction in distances may be applied for every 9 knots of headwind; a 10% increase in distances for every 2 knots of tailwind.

What would be the expected take-off performance of the airplane to which the table in Fig. 10 applies at an airport with an elevation of 4000 ft., with a dry, grass runway, an 18 kt headwind and an outside air temperature of 10°C? At an operational gross weight of 2400 lbs., the expected ground roll with zero wind on a paved dry runway could be 1260 ft. Eighteen knots of headwind would decrease the ground roll by 20% to 1008 ft. However, the dry grass runway would increase the roll by 15%. The expected ground roll would be approximately 1160 ft. The distance over the ground to clear a 50 foot obstacle would be approximately 2044 ft.

For some airplanes, the takeoff data is given only for gross weight. A useful rule of thumb to remember is: A 10% reduction in weight will result in a 10% reduction in take-off roll. A 10% increase in weight will result in a 20% increase in take-off roll.

Some Airplane Flight Manuals publish a graph that relates take-off distance to density altitude. Fig. 11 illustrates such a graph. To use this graph, the pilot must first calculate the density altitude of the airport of operation.

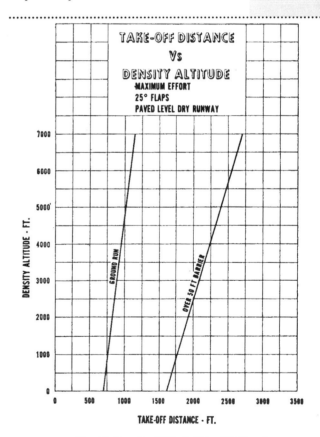

Fig. 11. Density Altitude Take-Off Graph.

For example, let us assume that the density altitude is 5000 feet. Locate the figure 5000 on the left edge of the graph. Locate the intersection of the 5000 foot line with the "ground run" line and the "over 50 ft. barrier" line. The ground run to achieve take-off will be approximately 1000 feet and the distance required to clear a 50 foot obstacle will be approximately 2335 feet.

If after calculating density altitude and checking the tables, it appears that the take-off run will require more runway than is available, you, as pilot-in-command, have several alternatives. You can lighten the load, if possible, or you can wait until the temperature decreases. Generally, the most critical time for flight operations when the temperature is very hot is from mid-morning through mid-afternoon. This is especially true at high elevation airports, but even at lower elevations, aircraft performance may be marginal. Aircraft operations should, therefore, be planned for early morning or late evening hours.

It is important to remember that in taking off from airfields that are at high elevation, you should use as a reference the same indicated airspeed that you would use during take-off from an airfield at sea level. It is the true airspeed that is affected by the increase in elevation and temperature.

In a high density altitude situation (above 5000 ft.), a normally aspirated engine should be leaned to achieve maximum power on take-off. The excessively rich mixture that is produced at full rich would further degrade the overall performance. (See also **When to Lean the Engine** in Chapter **Aero Engines**.)

Climb Performance Charts

The Airplane Flight Manual also publishes data for climb performance. Fig. 12 illustrates a typical table of figures for maximum rate of climb.

MAXIMUM RATE OF CLIMB

WEIGHT LBS	PRESS ALT FT	CLIMB SPEED KIAS	RATE OF CLIMB - FPM			
			-20°C	0°C	20°C	40°C
2400	S.L.	76	805	745	685	625
	2000	75	695	640	580	525
	4000	74	590	535	480	420
	6000	73	485	430	375	320
	8000	72	380	330	275	220
	10,000	71	275	225	175	---
	12,000	70	175	125	---	---

CONDITIONS:
Flaps Up
Full Throttle

NOTE:
Mixture leaned above 3000 feet for maximum RPM.

Fig. 12. Maximum Rate of Climb Data Chart.

The maximum, or best rate of climb, it will be remembered, is the rate of climb which will gain the most altitude in the least time and is used to climb after take off until ready to leave the traffic circuit. The table in Fig. 12 lists the rate of climb in feet per minute at 2400 lb. gross weight. The chart includes climb performance data at sea level and at pressure altitudes up to 12,000 feet and at temperatures varying from -20°C to 40°C. Note how the rate of climb decreases with altitude. At gross weight, at sea level, with an ambient temperature of 20°C, the rate of climb is 685 feet per minute, at 10,000 feet 175 ft/min, at 12,000 feet 0 ft/min. At 12,000 feet, therefore, (at an outside air temperature of 20°C) the airplane has reached its service ceiling.

These climb performance figures are calculated with flaps up, full throttle and the mixture leaned for maximum rpm above 3000 feet.

Some Airplane Manuals also publish a graph that relates climb performance to density altitude. The graph in Fig. 13 is such a chart. It indicates the climb performance that can be expected at varying density altitudes. At a density altitude of 10,000 feet, for example, the airplane for which this chart is applicable will climb at about 250 feet per minute. At a density altitude of 16,000 feet, it has reached its ceiling and will not climb any higher.

Many Airplane Manuals also publish charts for cruise climb (Fig. 14). Cruise climb, or normal climb, is the climb airspeed used for a prolonged climb. The chart indicates the fuel used, time required to reach altitude, and still air distance covered in order to reach various altitudes when climbing at a certain indicated airspeed with various power settings.

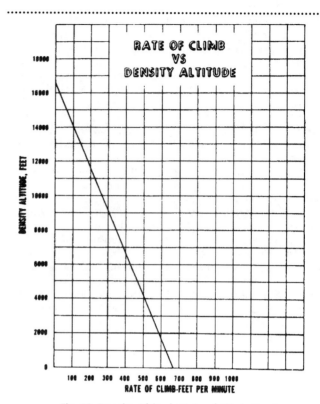

Fig. 13. Density Altitude Rate of Climb Graph.

CRUISE CLIMB

POWER SETTING		Climb IAS MPH	Fuel Flow Gl. Hr Per Eng	5000 FT. 41°F FROM SEA LEVEL			10000 FT. 23°F FROM SEA LEVEL			15000 FT. 5°F FROM SEA LEVEL		
RPM	M.P.			Dist. Miles	Time Min.	Fuel Used Gal.	Dist. Miles	Time Min.	Fuel Used Gal.	Dist. Miles	Time Min.	Fuel Used Gal.
2600	F.T.	150	Full Rich	9	3	7	18	7	10	30	11	13
2450	29	130	15.5	10	5	7	22	10	9	36	15	12
2450	29	150	15.5	15	6	7	32	12	10	52	13	13
2300	28	150	12.6	20	8	7	43	16	11	70	25	14

NOTE: WARM-UP AND TAKE-OFF ALLOWANCE 4 GALLONS AT SEA LEVEL.
MIXTURE AT RECOMMENDED FUEL FLOW, FLAPS AND GEAR UP.

Fig. 14. Cruise Climb Chart.

Cruise Performance Charts.

Performance figures for cruise at gross weight are also given in most Airplane Flight Manuals. These charts show the fuel consumption, true airspeed, endurance and range that may be expected **when cruising** at a certain altitude with the engine being operated at normal lean mixture, at various combinations of rpm and MP settings (to give a required % of power).

The figures in the chart in Fig. 15 are for a typical general aviation airplane. They are tabulated for cruise at various altitudes. 2500 feet and 5000 feet are illustrated. The Manual includes figures for 7500 feet, 10,000 feet, etc. up to the airplane's service ceiling.

It must be noted, however, that these figures are calculated on standard air conditions and zero wind. Wind direction and velocity will not affect fuel consumption or endurance but will certainly affect range. Interpolation of the figures must also be made for variations as a result of density altitude.

Fig. 15. Cruise Performance Chart.

CRUISE PERFORMANCE

LEAN MIXTURE

Standard Conditions — Zero Wind — Gross Weight - 2800 Pounds

RPM	MP	% BHP	GAL/HOUR	TAS MPH	60 GAL (NO RESERVE) ENDR. HOURS	60 GAL (NO RESERVE) RANGE MILES	79 GAL (NO RESERVE) ENDR. HOURS	79 GAL (NO RESERVE) RANGE MILES
2500 FEET								
2450	23	76	14.2	158	4.2	670	5.6	885
	22	72	13.4	154	4.5	690	5.9	910
	21	68	12.7	151	4.7	715	6.2	940
	20	63	12.0	148	5.0	730	6.6	965
2300	23	71	13.1	154	4.6	700	6.0	925
	22	67	12.2	149	4.9	740	6.5	970
	21	62	11.5	145	5.2	760	6.9	1005
	20	59	11.0	142	5.5	775	7.2	1020
2200	23	67	12.1	149	5.0	745	6.5	980
	22	63	11.4	146	5.3	770	6.9	1010
	21	59	10.8	142	5.6	790	7.3	1040
	20	55	10.2	138	5.9	810	7.7	1065
2000 MAXIMUM RANGE SETTINGS	20	47	8.7	126	6.9	865	9.1	1135
	19	43	8.2	121	7.3	890	9.6	1170
	18	39	7.5	113	8.0	900	10.5	1185
	17	35	7.0	105	8.6	905	11.3	1190
5000 FEET								
2450	23	78	14.5	163	4.1	670	5.4	885
	22	73	13.6	159	4.4	700	5.8	925
	21	70	13.0	156	4.6	720	6.1	950
	20	65	12.2	151	4.9	750	6.5	985
2300	23	73	13.4	158	4.5	710	5.9	930
	22	69	12.6	155	4.7	730	6.3	965
	21	64	11.9	151	5.0	760	6.6	1005
	20	60	11.2	146	5.4	785	7.1	1035
2200	23	68	12.4	155	4.8	750	6.4	985
	22	64	11.7	151	5.1	775	6.8	1020
	21	60	11.0	146	5.5	800	7.2	1050
	20	57	10.5	143	5.7	815	7.5	1075
2000 MAXIMUM RANGE SETTINGS	19	45	8.5	126	7.1	895	9.3	1175
	18	41	7.9	118	7.6	905	10.0	1190
	17	37	7.3	111	8.2	910	10.8	1200
	16	34	6.8	103	8.8	905	11.6	1190

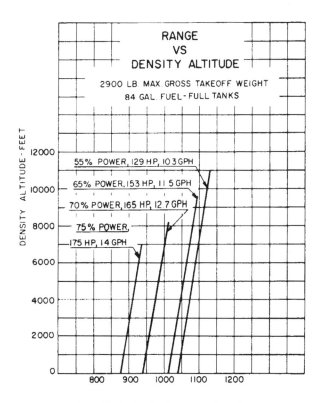

Fig. 16. Cruise Performance Graph.

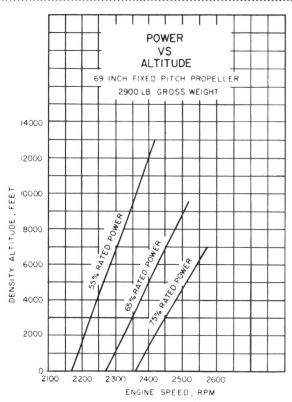

Fig. 17. Power Graph.

Some manuals publish their cruise performance data in graph form such as that shown in Fig. 16. By reference to the graph, a pilot can determine the expected range for any combination of density altitude and percentage of power. For example, at 5000 feet density altitude and 65% power, the airplane for which this graph applies, has a range of approximately 1050 miles.

To use the graph in Fig. 16, the pilot must also refer to a graph (Fig. 17) relating power to altitude to determine what rpm will give the percentage of power required. For example, to achieve 65% power at 5000 feet density altitude, it is necessary to operate at a power setting of 2400 rpm.

Landing Performance Charts

Perfect landings are usually preceded by deliberately planned and well executed approaches. Correct approach speeds are important. The Airplane Flight Manual for any particular model of airplane recommends the speeds to use on approach with various flap settings. These airspeeds should always be used.

If information on approach speeds is not available, a useful rule of thumb to determine final approach speed for the average conventional design airplane may be found by applying the formula presented in Fig. 18.

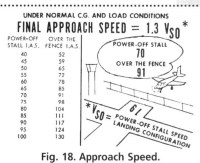

Fig. 18. Approach Speed.

The 1.3 V$_{SO}$ formula should be calculated using calibrated airspeed. Calibrated airspeed is indicated airspeed corrected for position and instrument error. CAS values are published in the Owner's Manual. For some airplanes, there is a significant discrepancy between CAS and IAS, especially at the lower end of the speed range near the stall.

The factor of weight is important in determining landing speed. All airplanes stall at slower airspeeds when they are light. A lightly loaded airplane, landing at the same airspeed that is used when it is heavily loaded, will float before touchdown to dissipate the excess energy, thus extending the landing distance. The 1.3 V$_{SO}$ formula should be calculated using the stalling speed for the **actual** weight of the aircraft, not the maximum landing weight. If the Owner's Manual does not publish a table of approach speeds as a function of reduced weight, a rule of thumb is to reduce the calibrated approach airspeed for the maximum weight of your airplane by one-half of the percentage of the weight decrease. If, for example, the airplane weight is 20% below maximum, the calibrated approach airspeed would be decreased by half of that, or by 10%.

The 1.3 V$_{SO}$ airspeed should be used only after all maneuvering is completed, that is, on **short final only.**

On some airplanes, the manufacturer may require a particular approach speed for **all** weights because, during certification flight testing, it was found that for stability and control reasons, or for go-around safety, a fixed airspeed is required. Always comply with the manufacturer's recommendations.

Most Airplane Flight Manuals publish information on landing performance in a table such as that in Fig. 19. The table assumes 30° of flaps, power off, no wind and maximum braking on a paved, dry, level runway. It lists the expected ground roll and distance to clear a 50 foot obstacle when landing at airports of varying elevations and at varying ambient air temperatures. A probable 10% reduction in landing distance is indicated for every 9 knots of headwind; a 10% increase for every 2 knots of tailwind. If, for example, the velocity of the headwind at an airport at 3000 feet (temperature 20°C) is 9 knots, the ground roll can be expected to be 554 feet [615 - (615 x 10%)].

Since there is some loss in the quality of braking action on the grass of a sod runway, the ground roll after landing can be expected to be longer than it would be on a hard surface runway. In the chart in Fig. 19, a 45% increase is suggested as probable for the ground roll when landing on a sod runway when the grass is dry. In any less favourable conditions, such as wet grass, slush, or standing water on either a sod or a hard surface runway, braking action will be degraded even further and the landing roll will be increased substantially.

Density altitude affects the landing performance of an airplane as greatly as it affects take-off performance. High temperature and

LANDING DISTANCE

WEIGHT LBS	SPEED AT 50 FT KIAS	PRESS ALT FT	0°C GRND ROLL FT	0°C TOTAL FT TO CLEAR 50 FT OBS	10°C GRND ROLL FT	10°C TOTAL FT TO CLEAR 50 FT OBS	20°C GRND ROLL FT	20°C TOTAL FT TO CLEAR 50 FT OBS	30°C GRND ROLL FT	30°C TOTAL FT TO CLEAR 50 FT OBS	40°C GRND ROLL FT	40°C TOTAL FT TO CLEAR 50 FT OBS
2400	61	S.L.	510	1235	530	1265	550	1295	570	1325	585	1350
		1000	530	1265	550	1295	570	1325	590	1360	610	1390
		2000	550	1295	570	1330	590	1360	610	1390	630	1425
		3000	570	1330	590	1360	615	1395	635	1430	655	1460
		4000	595	1365	615	1400	635	1430	660	1470	680	1500
		5000	615	1400	640	1435	660	1470	685	1510	705	1540
		6000	640	1435	660	1470	685	1510	710	1550	730	1580
		7000	665	1475	690	1515	710	1550	735	1590	760	1630
		8000	690	1515	715	1555	740	1595	765	1635	790	1675

CONDITIONS:
Flaps 30°
Power Off
Maximum Braking
Paved, Level, Dry Runway
Zero Wind

NOTES:
1. Short field technique as specified in Section 4. For operation with tailwinds up to 10 knots, increase distances by 10%
2. Decrease distances 10% for each 9 knots headwind. For operation with tailwinds up to 10 knots, increase distances by 10% for each 2 knots.
3. For operation on a dry, grass runway, increase distances by 45% of the "ground roll" figure.
4. If a landing with flaps up is necessary, increase the approach speed by 7 KIAS and allow for 35% longer distances.

Fig. 19. Landing Distance Chart.

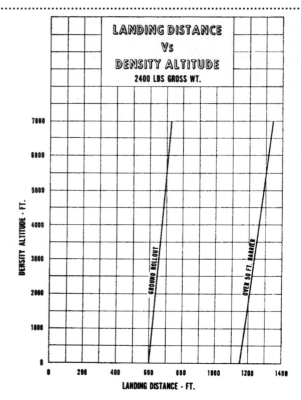

Fig. 20. Density Altitude Rate of Climb Graph.

high elevation will cause an increase in the landing roll because the true airspeed is higher than the indicated airspeed. Therefore, even though using the same indicated airspeed for approach and landing that is appropriate for sea level operations, the true airspeed is faster, resulting in a faster groundspeed (with a given wind condition). The increase in groundspeed naturally makes the landing distance longer and should be carefully considered when landing at a high elevation field, particularly if the field is short.

Some Airplane Flight Manuals contain performance charts and tables such as the one in Fig. 20 which relate landing distance to density altitude. Pilots should develop the habit of referring to these charts in order to anticipate the distance that will be required to land their airplane under various conditions of flight.

Landings and Take-Offs at Unimproved Airstrips

The data published in the Airplane Flight Manual generally applies to dry paved runways. It is almost impossible to produce accurate figures applicable to turf runways because of the number of variables involved. The conditions change depending on whether the turf is wet, dry, frozen, snow or ice covered and change still further depending on how wet, how much snow, etc. The weight of the airplane is a factor since a heavier airplane tends to sink into the surface. The size of tires, tire pressure and tire tread all have an effect both on take-off roll and on braking action. New tires and new brakes can produce greater braking action. In addition, take-off and landing performance vary with the condition of the airplane and the powerplant.

For the most part, a pilot has to learn by experience what his airplane will do by way of performance at unimproved airstrips.

There are, however, some rules of thumb that will supplement the approved take-off and landing performance charts and increase the safety of operations at unimproved airstrips. They should not be used at variance with data published by the manufacturer.

Density Altitude: For every 1000 foot increase in the density altitude, add 10% to the ground roll and to the distance to clear a 50 foot obstacle up to 3000 feet density altitude and add 20% above 3000 feet density altitude.

Weight: For a 10% decrease in the weight of an airplane, there is a 10% decrease in the take-off ground roll and in the distance to clear a 50 foot obstacle. For a 10% increase in the weight, there is a 20% increase in the ground roll/distance (a ratio of 1:2).

Surface: On a firm turf runway, add 7% to the ground roll/ distance.

On a rough or rocky runway or on short grass (up to 4 inches), add 10%.

On long grass (more than 4 inches), add 30%.

On a soft surface, such as mud or snow, add 75%. In some situations, it may be impossible to achieve take-off, if the conditions are bad enough.

Headwind: Use the following formula to determine the effect of a headwind. The formula should not be used for a headwind of less than 10 knots.

90% minus the headwind component divided by the rotation speed equals % of ground roll/distance.

For example: If the headwind is 10 knots and the rotation speed is 50 knots.

90% - (10 knots ÷ 50 knots) = 90% - 20% = 70% of ground roll/distance.

Tailwind: Use the following formula to determine the effect of a tailwind. The formula should not be used for tailwinds greater than 5 knots. Take-offs with tailwinds greater than 10 knots are not recommended.

110% plus the tailwind component divided by the rotation speed equals % of ground roll/distance.

Sloped Runway: For every 1° of upward slope in the runway, add 10% to the take-off roll. An upslope of 2° will add quite significantly to take-off distances. Sloping runways provide a different visual perspective than a level surface and landings and take-offs demand more planning and closer attention to flying techniques. When attempting to take off from a runway with an upslope, it is important to know if the airplane can climb safely away from the rising terrain. Uphill take-offs with tailwinds should be avoided.

A 1° downward slope in the runway reduces the take-off roll by only 5%. If a take-off is commenced downhill with a tailwind, the airplane will accelerate more rapidly. Avoid the temptation to lift off prematurely because the groundspeed seems high. Wait for a safe indicated airspeed.

Combined Effect. The combined effect of a change in aircraft weight, runway surface, runway slope and/or density altitude, in one or more combinations, can dramatically alter take-off distances.

Aborting a Take-Off: At 25% of the ground roll to take-off, the airplane should have achieved 50% of its lift-off speed.

At 50% of the ground roll, it should have achieved 70% of its lift-off speed.

At 80% of the ground roll, it should have achieved 90% of its lift-off speed.

Lift-off speed should be reached within the first 75% of the usable runway. If lift-off has not been achieved in this distance, the take-off should be aborted. For example, with a ground roll to take-off of 1200 feet, the minimum runway length must be 1600 feet.

Tailwind Landings

On occasion, because of dangerous terrain near an airport, it is necessary to operate from a runway with a tailwind component. Safe operation in this situation requires special training and firm guidelines. The tailwind component should be limited to less than 10 knots. Density altitude and aircraft capabilities must be considered.

Tailwinds contribute to an illusion of a "hot" landing (or take-off). The pilot must pay special attention to the airspeed indicator and ignore the illusion of excessive speed. A headwind over the vertical stabilizer contributes to directional control. A tailwind does not, resulting in some degradation of directional control. A tailwind will lengthen the take-off roll and decrease the angle of climb after take-off. During landing, a tailwind will require the pilot to increase the rate of descent to maintain a given glideslope.

The best advice on deciding on whether or not to land or take-off with a tailwind is DON'T. If you have an option, change runways or go to another airport.

Cross Wind Landings and Take-Offs

The amount of headwind is an important factor in determining the take-off and landing distance that will be required in any given situation. In Fig. 19, for example, it was noted that a 10% reduction in landing distance could be expected for every 9 knots of headwind for the airplane for which that chart was applicable. Similar reductions in take-off distance were noted in Fig. 10. The wind, however, is rarely a direct headwind, especially at single strip airports. It is

Flight Path

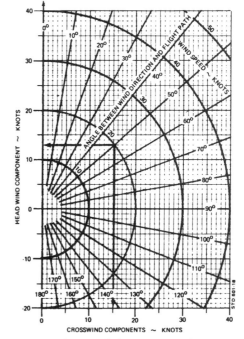

Fig. 21. Wind Component Chart.

more likely to be blowing from some angle off the runway and will, therefore, be a headwind component.

Wind component charts similar to that in Fig. 21 are published in Airplane Flight Manuals for use in determining headwind components in take-off and landing situations. Let us assume that the wind as reported by the tower is from 080° at 30 knots. The runway in use is 05 and the angle between the wind and the runway is, therefore, 30 degrees. By locating the intersection on the chart of the 30° line representing "angle between wind direction and flight path" and the wind curve, the headwind component is found to be 26 knots and the cross wind component 15 knots. The pilot would, therefore, base his take-off performance on a 26 knot headwind component.

If you don't have a cross wind component chart handy, here are some rules of thumb. If the cross wind is 20 degrees off the runway, the cross wind component is 25% of the wind velocity. If the wind is 40 degrees off, the cross wind component is 50%. If the wind if 60 degrees or more off, the cross wind component is 75% of the wind velocity. If the wind is 90 degrees to the landing path, the cross wind component is 100%.

The headwind component can also be quickly calculated using the wind side of a flight computer that has a squared graph section. Using the above wind and runway situation, rotate the compass rose to set the wind direction (080°) under the true index. With the grommet on the straight 0 line, draw a wind vector arrow 30 units long to represent the wind speed (Fig. 22). Rotate the compass rose to set the runway (050) under the true index. The wind vector arrow indicates a headwind component of 26 knots and a cross wind component of 15 knots (Fig. 23).

Strong cross winds that are blowing at angles approaching 90 degrees to the runway cause a problem in directional control. Any pilot who has attempted to land a light airplane at a small airport with a single strip that happened that day to be directly 90° to a heavy wind knows the value of good cross wind technique.

There are essentially two kinds of contrary wind conditions that affect light planes on landing. There is the fairly steady wind at a pronounced angle to the runway which can be determined. There is also the gusty changeable wind. This second cross wind condition requires instinctively

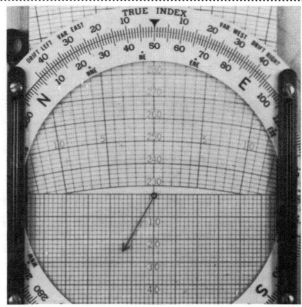

Fig. 23.

correct responses from the pilot, who must use aileron for keeping the airplane over the runway and opposite rudder to line it up with the runway and, at the same time, be prepared to deal with sudden changes in airspeed and altitude due to the gusty conditions.

When the wind is gusting, a shift in almost any direction, even to 180°, can occur. The pilot is well advised to pay special attention to the behaviour of trees, bushes and dust and not just the windsock.

How much contrary wind can an airplane handle? This information is usually given in the Owner's Manual. If the manual has been conscientiously written, it will list windspeed maximums for every conceivable angle. It may, however, give only the maximum cross wind angle and the maximum wind velocity.

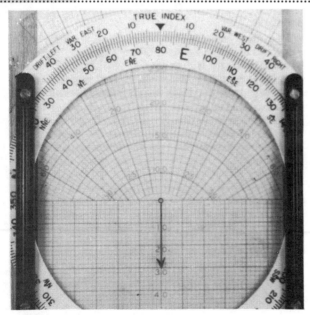

Fig. 22.

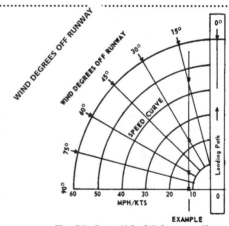

Fig. 24. Cross Wind Tolerance Chart.

A cross wind tolerance chart (Fig. 24) can be used to determine the maximum permissible windspeeds for varying cross winds. All airplanes of North American manufacture are designed to handle cross winds of 90° with a velocity of 20% of their stall speed. An airplane with a stall speed of 60 knots, for example, can tolerate a cross wind of 12 knots (60 x 20%). Draw a line on the graph parallel to the landing path from the point 12 knots on the index at the bottom of the chart. The point at which this line intersects each of the lines representing wind degrees off runway and the speed curves indicates the permissible wind speeds for the varying cross wind conditions. For a cross wind of 60 degrees off the runway, the maximum permissible wind speed is 14 knots; at 30 degrees, 24 knots; at 15 degrees, 45 knots.

Cross wind considerations are of considerable importance for take-off as well as for landing. In take-off, the effect of torque must be remembered. Since torque will tend to push the airplane's nose to the left, a cross wind from the right will not be too difficult to handle. In a cross wind from the left, however, it could be possible to run out of right rudder. If the cross wind is near maximum permissible speed, a left-hand cross wind calls for a decision to stay at home, unless the airplane is a twin with counter-rotating propellers.

In a stiff cross wind with full upwind aileron held, it is possible to end up making the final segment of the take-off on the upwind wheel only. This makes the airplane difficult to handle but the pilot should avoid the temptation to haul the airplane off prematurely. It is advisable to break ground at a slightly higher than normal speed, even if it is necessary to force the airplane to remain on the runway. Once in the air, the pilot wants to have the positive control responses that come with good flying speed.

On the other hand, it is inadvisable to carry too much airspeed when landing in a cross wind. The greater the angle of the cross wind, the less head wind component there is, so that in a 90° cross wind there is no headwind component at all. Although in extremely gusty conditions, some extra speed is advisable, too much may complicate the job of bringing the airplane to a stop before the end of the runway.

Taxiing

"Don't ever stop flying an airplane until you have parked it, shut off the engine and tied it down."

Aircraft are just that: *air*craft. They are not necessarily very good *ground* craft. The effect of wind on an aircraft during its roll out and while taxiing demand that the pilot be alert to wind direction. A strong wind that was right down the runway on landing will be a strong cross wind or tailwind while taxiing and will expose the aircraft to directional control problems. The Owner's Manual for your aircraft will give directions on how to taxi in high winds and how to keep the aircraft from tipping over.

Cruising Speeds

Between an airplane's top speed and its stalling speed, there is a wide range of speeds that have often been referred to as the working speeds. Each is the answer to a particular problem of practical flying.

Best Gliding Speed or Normal Glide. The speed at which an airplane with power off will glide the farthest distance. This is the speed at which the lift/drag ratio is best and therefore might be called the "speed of least drag".

Speed of Slowest Descent. The speed at which the airplane loses altitude at the least rate. In a nose high attitude, near the stalling speed, the airplane sinks most slowly. This speed is made constant use of by glider pilots.

Maximum Range Speed. The speed at which the airplane travels the most miles per gallon of fuel. This speed is about 5 to 10% faster than the best gliding speed.

Maximum Endurance Speed. The speed at which the fuel consumption is the least per hour. To achieve this condition of flight, the airplane is flown quite slowly, with minimum power, very nose-high, with the wings at a high angle of attack. The airplane is not covering much distance for it is too slow but its endurance is substantially extended.

Maneuvering Speed. The maximum speed at which the airplane should be flown in rough air. It is the speed least likely to permit structural damage to the airplane in rough air or during aerobatic maneuvers. Maneuvering speed for most airplanes is set at about twice the flaps-up, power-off stall speed.

Optimum Cruise Speed. The speed at which the best balance between use of fuel and maximum range is achieved.

Best Rate of Climb Speed. The speed at which the airplane will gain the most altitude in the least time. This speed is rather fast; for most airplanes it is approximately the same speed as that of best glide.

Best Angle of Climb Speed. The speed at which the airplane climbs most steeply, gaining the most altitude for distance covered over the ground. In routine flying, it is seldom necessary to use best angle of climb speed but it is of importance when it is necessary to climb out over a high obstacle at the end of a runway. It is, however, not a climbing procedure that should be used except in emergency. Firstly, it is a speed that is not much above the stall and the pilot must be very vigilant that he does not permit the airplane to stall. Secondly, at such a slow speed, with full power, insufficient air is passing over the engine to cool it properly. The engine is therefore likely to overheat and the resulting engine wear is undesirable.

Air Endurance. The maximum time an airplane can continue to fly under given conditions ★ with a given quantity of fuel.

Range. The maximumn **distance** an airplane can fly under given conditions ★ with a given quantity of fuel.

★Given Conditions. Both airspeed and altitude affect the range and endurance. The best airspeed for maximum endurance is generally less than that for maximum range. Generally speaking, a propeller driven airplane may be flown at a relatively low altitude for maximum endurance. For maximum range, a propeller driven airplane should be flown at that height at which, at full throttle, the indicated airspeed is about 5 to 10% better than the best glide speed. This is realized at a relatively high altitude. A jet, on the other hand, would fly at high altitude for both range and endurance.

Range and endurance, when stated in the airplane specifications, are usually based on still air, standard atmosphere conditions. However, they may be computed for any assumed wind direction and wind speed.

Airplane performance is always a compromise between such variable factors as range vs payload, endurance vs airspeed, etc.

Use of Performance Charts

The performance information that is available to you, as a pilot, in the form of the tables and graphs that have been illustrated in the preceding sections is very valuable. You should thoroughly familiarize yourself with the tables and graphs published in the Airplane Flight Manual for your airplane and should practise using them at every opportunity in order to gain confidence in their use and to determine for yourself that your calculated or anticipated performance equals the actual or realized performance on any particular flight.

The charts are particularly useful in planning a cross country flight and in solving the navigational problems of range, fuel consumption, etc. that are part of proper pre-flight planning.

Wake Turbulence

The theory of wing tip vortices was introduced in the Chapter **Theory of Flight** in the discussion of induced drag. Lift is generated by the creation of a pressure differential over the wings: low pressure over the upper wing surface and higher pressure under the wing. As the lift producing airfoil passes through the air, the air rolls up and back about each wing tip producing two distinct counter rotating vortices, one trailing each wing tip (Fig. 25). The intensity of the turbulence within these vortices is directly proportional to the weight and inversely proportional to the wing span and the speed of the airplane. The heavier and slower the airplane, therefore, the greater is the intensity of the air circulation in the vortex cores. The most violent vortices would be generated by airplanes during take-off and landing and at near gross weights when the aircraft are flying at high angles of attack.

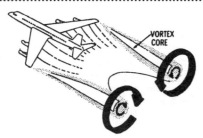

Fig. 25. Wing Tip Vortices.

Air density influences vortex strength. In cold air, the vortices can be expected to be more severe. Vortex strength is also affected by the configuration of the airplane producing the vortices. The position of flaps and undercarriage, as well as the location of the engines and the tail configuration, modify vortex pattern and persistence. For example, the high tail of the DC9 increases the persistence of the vortices.

The greatest vortex strength occurs under conditions of clean configuration, high weight and slow speed.

Similar vortices are generated by rotary wing aircraft. These vortices have the same internal air circulation as those generated by fixed wing airplanes, but are potentially more dangerous because the helicopter's lower operating speeds produce more concentrated wakes.

Aircraft which fly directly into the core of a vortex will tend to roll with the vortex. The effectiveness of control response and the wing span of the airplane determine its capability to counteract the roll. If the wing span and the ailerons extend beyond the vortex, counter-roll control is usually effective. Aircraft with a short wing span, even though they may be high performance types, will have great difficulty counteracting the roll.

The vertical gusts encountered when crossing laterally through the vortex can impose structural loads as high as 10g on a small airplane flying at a high angle of attack. The combined effect of an upgust immediately followed by a downgust has been estimated as high as 80-feet-per-second. Most small planes are designed to withstand vertical gusts of 30-feet-per-second at their normal operating speeds.

Aircraft in the ultra-light category are especially vulnerable to the devastating effects of being caught in wake turbulence, even the wake turbulence generated by relatively small transport aircraft.

Fig. 26. Hazards of Wing Tip Vortices.

There is a distinct possibility of structural failure when an airplane crosses a pair of vortices at a large angle (about 90°). The severe up-down-down-up forces and the pilot's attempt to counteract them could result in airframe design limits being exceeded. Loss of control is another result of encountering these unseen monsters. Vortex cores can produce a roll rate of 80 degrees per second, a situation with which the small light airplane is structurally unable to cope.

Vortex generation starts with rotation (the raising of the nose from the runway) and reaches its peak intensity at lift off when the full weight of the airplane is sustained by the wings and when the airspeed is low. It ends when the airplane touches down. The vortices of large, heavy airplanes settle below and behind the airplane at about 300 to 500 feet per minute for about 30 seconds. The vortices of smaller, lighter airplanes settle more slowly. In all cases, within about 2 minutes, they level off about 1000 feet below the aircraft's flight path. They may trail the generating aircraft by 10 to 16 miles depending on its speed and, in still air, decay slowly.

Headwind less than 10 knots.

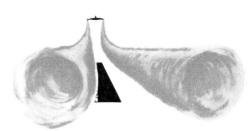

Light cross wind — less than 5 knots.

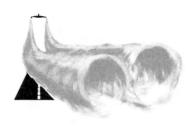

Cross wind — over 5 knots.

Fig. 27. Wake Turbulence in the Runway Area.

The strength of the vortex remains relatively constant during the first 2 minutes after the passage of the generating aircraft. After that, dissipation of the vortices occurs at varying rates, first in one vortex and then the other. Vortices have been known to last as long as 5 minutes. Atmospheric turbulence helps to break up the vortices; the greater the turbulence, the more rapid the dissipation of the vortices.

The vortex flow field covers an area about 2 wing spans in width and one wing span in depth. The rotating vortex may, therefore, have a diameter as great as 200 feet behind a very large airplane.

When the vortices sink to the ground, they tend to move laterally outward over the ground at a speed of about 5 knots. This characteristic means that the laterally moving vortices may position themselves over a parallel runway and constitute a hazard to airplanes operating there. In a cross wind condition, the lateral movement of the upwind vortex will be decreased while the movement of the downwind vortex will be increased. As a result, the upwind vortex may remain stationary in the touchdown zone. A tailwind condition can move the vortices of the preceding airplane forward into the touchdown zone of the airplane following even though the latter takes the precaution of landing beyond the touchdown point of that preceding airplane.

The advent of the huge jumbo jets, in the class of the 747, 757 and C-5, makes the problem of wake turbulence even more acute. These aircraft generate roll velocities exceeding the roll control capability of some of the larger light airplanes and far in excess of the capability of small airplanes.

The phenomenon of wake turbulence is of immense importance to a pilot. In fact, it can be a matter of life and death. It is of prime importance, therefore, that you, for your own safety, learn to envision the location of the vortex wake generated by large aircraft so that you can avoid encountering the very serious hazards that are associated with this situation.

Wake Turbulence Avoidance

Remember that the vortices, in still air, have a downward and outward movement but that the ambient wind will alter their normal pattern of movement. Pilots should be particularly alert to wind conditions and try to picture mentally how those wind conditions will affect the vortices. Will they remain in the touchdown zone, or drift onto a nearby runway, or sink into the landing path from a crossing runway, etc.?

During flight. Although wake turbulence is most likely to be encountered during the arrival and departure procedures, it can be a hazard during cruising flight as well. Avoid crossing behind and less than 1000 feet below the flight path of a large, heavy airplane or of a helicopter, especially at low altitude when even a momentary wake turbulence encounter could be hazardous. Alter course or climb to be above the expected vortices.

During taxi. Stay well behind large airplanes that are taxiing or maneuvering on the ground. Do not cross behind a large airplane that is doing an engine run-up. Avoid taxiing below a hovering helicopter. The downwash from its rotors is significant and hazardous.

During take-off. On the same or a parallel runway. Start the take-off roll at the beginning of the runway and plan to be airborne before the rotation point of the previous airplane. Stay upwind of the large airplane. This action and a normal climb should keep you above the descending vortices of the preceding airplane.

Do not make intersection take-offs when large aircraft are using the same runway.

On an intersecting runway. Plan to be airborne before you cross the intersection, remembering to keep your flight path well above that of the aircraft that departed on the other runway.

When following an aircraft that has just landed, plan to become airborne beyond the point of touchdown of the preceding aircraft.

During landing. When following a heavy aircraft that has just taken off, plan to touch down before the rotation point of the preceding airplane.

When following an aircraft that has just landed, plan to touch down beyond the point where the preceding aircraft touched down, remembering to keep your approach path above that of that aircraft.

When landing on an intersecting runway behind a large airplane that has taken off, note the airplane's rotation point and, if it was past the intersection, continue your approach and land prior to the intersection. If the rotation point of the large airplane was prior to the intersection, it might be best to abandon the landing unless you can assure a landing well before the intersection.

Departing or landing after a large airplane that has executed a missed approach or a touch-and-go landing. The vortex hazard may exist along the entire runway. Ensure than an interval of at least 2 minutes has elapsed before making your take-off or landing.

Above all, avoid a long dragged-in approach. The largest number of dangerous encounters with wake turbulence has been in the last half mile of the final approach.

In an effort to minimize the hazards of wake turbulence, ATC has established separation minima that are to be applied to landing and departing aircraft. Controllers generally apply a 6 mile radar separation between a heavy jet and a light airplane (one that weighs less than 12,500 lbs.) that is following at the same altitude or at less than 1000 feet below. A light airplane is kept at least 4 miles behind a medium sized airplane.

In non radar departures, a minimum of two minutes is required between the departure of a heavy airplane and any lighter airplane. This will be extended to 3 minutes if the second airplane will use more runway than the first or if it is a intersection take-off.

However, ATC cannot guarantee that wake turbulence will not be encountered. When the tower controller advises, "**Caution Wake Turbulence**," he is warning you of the possible existence of this phenomenon.

In some situations, the full wake turbulence separation may not be required. Take-off clearance will usually be issued to a pilot who, on his own initiative, has requested a waiver of separation. The waiver may, however, be refused to a light aircraft taking off behind a heavy aircraft if the take-off is started from an intersection or a point significantly farther along the runway or if the preceding heavy aircraft made a low or missed approach along the same runway. A request for a waiver indicates to the controller that the pilot accepts responsibility for wake turbulence separation and avoidance. The controller will still issue a wake turbulence caution with the take-off clearance.

Even when the controller advises of the possibility of wake turbulence, it is solely the pilot's responsibility to avoid encountering it. Don't hesitate to ask for further information if you believe it will

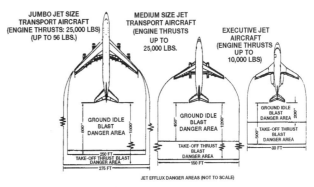

Fig. 28. Jet Blast Danger Areas

assist you in analyzing the situation and deciding on a course of action. Even though you have received a clearance to land or take off, if you believe it safer to wait, or to use a different runway, or in some other way to alter your operation, ask the controller for a revised clearance. The controller is interested in the prevention of accidents too and will assist you in any way he can while accomplishing his job of expediting traffic.

Jet Blast Hazard

The area immediately behind jet airplanes can be particularly dangerous to small airplanes maneuvering on the ground or about to take off or land. Pilots should exercise caution when operating near active runways or taxiways. With the increasing use of intersecting runways, there is always the possibility of jet blast or propeller wash affecting other aircraft.

Light weight airplanes with high wings and narrow track undercarriages are particularly susceptible to jet blast accidents. Stay at least 600 feet away from a jumbo jet when its engines are idling and 1600 feet away from it when it is at full throttle for take-off. Fig. 28 illustrates the danger areas.

Airmanship

Good airmanship involves careful attention to the many facets of preparation for and execution of a flight whether it is a short hop around the circuit or a long cross country flight.

Preparation for a Flight

Before embarking on any cross country flight, careful preparations and planning should be done.

1. Up-to-date aeronautical charts appropriate for the route of the flight should be obtained and carefully studied to select a route that will take into consideration the nature of the country to be flown over, refuelling points, emergency landing areas, prohibited areas, etc., all of which are factors which influence the selection of a good route. Avoid areas where there is no chance of making a safe forced landing.

If the flight is to be conducted in accordance with visual flight rules (VFR) by contact navigation (i.e. navigating with reference to landmarks), the track selected should be drawn on the map and its direction measured by a navigation plotter. Distance between check points should be measured, using the mileage scale at the bottom of the chart. The track should then be converted from true to magnetic (applying the variation from the map) and its direction indicated on the map by figures above and below the line. Arrows should be used to distinguish between the track out and home. If the flight is to be a long one, the figures should be spotted in at convenient intervals, allowing for changes in variation and for convergency.

A distance scale should be marked along the track to assist in making groundspeed checks at, say, 50 or 100 mile intervals. Some pilots prefer to mark prominent check points (such as communities, omnirange stations, reporting points, etc.).

Draw 10 degree drift lines on either side of the track opening out from the departure point and closing down into the destination.

The character of the country should be carefully studied, particularly as regards high ground and dangerous obstructions. Outstanding landmarks should be noted.

2. Visit the flight service station or the weather office at the airport to obtain the meteorological information available or access weather information through transcribed weather broadcasts. Weather information is also available via computer terminals by accessing DUATS or the Environment Canada Bulletin Board Service. Study the weather map for wind circulation, frontal activity, and the weather over the whole general area in which you intend to fly. (This knowledge will be valuable in case you have to alter your flight plan to an alternate destination.) Check the aerodrome, area and upper winds forecasts, hourly weather reports along your route.

3. Select the height at which you wish to conduct the flight, taking into consideration the winds aloft, high ground and obstructions en route and in the vicinity of landing fields, weather minima for VFR flight, cruising altitude rules.

4. Calculate the true heading to steer and apply the variation to obtain the magnetic heading. Calculate the groundspeed for each leg of the route.

5. Compute the safe and maximum range of your airplane. Compute the fuel required, based on the estimated times en route, using

FLIGHT PLANNING FORM													
FROM	TO	ALT	I.A.S.	T.A.S.	TRACK (T)	W/V	HDG (T)	VAR	HDG (M)	G/S	DIST	TIME	FUEL

Fig. 29. Flight Planning Form.

RADIO FACILITES					
AIRPORT	NAVAID	FREQ	TWR	MF ATF	ATIS

Fig. 30. Radio Facilities Planning Form.

the cruise consumption chart in your Flight Manual. Select refuelling points.

6. Check suitability of airports, runway patterns and field conditions.

7. Check radio facilities which are available en route and the frequencies on which they operate.

8. Note the time of sunrise and sunset (VFR). Review the latest NOTAMS (Notices to Airmen). Glance over the airport bulletin board.

9. Prepare a flight log entering in appropriate sections all the information about the flight, altitude to be flown, track, wind velocity and direction, heading, calculated groundspeed, checkpoints, mileage between checkpoints, radio frequencies for navaids, information about the destination airport and refuelling stops, tower or mandatory frequencies, weather forecast for destination, etc. Forms, such as those shown in Fig. 29 and Fig. 30, can be used in preparing your flight log.

10. You are required by Air Regulations to be familiar with pertinent flight information such as navaids — good operating practices — VFR procedures — air traffic control procedures — light gun signals — radio telephone phraseology and techniques — VOR check points — VHF/DF direction finding data and procedures — emergency, search and rescue, SCATANA rules — weather information, how and where obtainable.

11. Note the location of classified airspace (terminal control areas, control zones, etc.). Note the location of specified areas that surround aerodromes at which mandatory frequencies are established. Note Class F airspace (restricted and advisory areas) and air defence identification zones (ADIZ).

12. If the flight is to be conducted VFR at night, check the elevation of the land over which the airplane will fly during the climb after take-off and on the approach to landing. Check also for obstacles on these flight paths. Plan the flight path to clear all terrain and obstacles.

13. If the flight is to be conducted in accordance with visual flight rules (VFR) but by using only air navigation radio aids, such as omni, ADF, etc., enroute low altitude charts and/or the appropriate IFR Charts must be used. These charts contain all the distance and en route information necessary for flying from point to point by using radio equipment only.

14. Fold maps correctly and place them in the proper sequence in the cockpit. Be sure to carry other maps for areas adjacent to your line of flight. You may need them if you have to fly around bad weather or if you become temporarily lost.

15. Be sure that you are properly licensed and qualified for the particular type of airplane you are flying and for the nature of the flight you are about to undertake.

16. See that the airplane is properly certified. Ensure that it is serviceable and fit for the flight. Be sure that the mandatory manuals and certificates are on board, along with the licences and/or permits for all members of the crew. See also **Pilot's Inspection Prior to Flight**.

17. Be sure that fuel of the proper grade is on board and that it is sufficient for the flight plus 45 minutes reserve.

18. Check the load (all-up weight must not exceed the authorized limit) and the distribution of the load (to insure that the centre of gravity is within safe limits). Make sure the load is properly secured and does not block the emergency exits.

19. Check the temperature and the elevation of the field. Be sure that runway lengths are adequate if high temperature or high elevation conditions prevail.

20. Review the flight load factors and airspeed limitations as outlined in your Airplane Flight Manual.

21. Prepare a table, such as that shown in Fig. 31, showing all the information relating to your airplane's performance.

22. File a **flight plan** or **flight itinerary** with ATC or leave a **flight itinerary** with a responsible person.

Pilot's Inspection Prior to Flight

A careful inspection of the airplane should be carried out prior to flight.

This inspection guide is applicable to almost any single-engine or light twin-engine airplane, **provided** it is modified to suit the airplane type and the manufacturer's recommendations.

This checklist and all others in this manual are representative only.

AIRSPEEDS
Normal Takeoff ___
Normal Climb ___
Best Rate of Climb ___
Beast Angle of Climb ___
Max Maneuvering ___
Best Glide ___
Max Flap Extented ___
Max Gear Down ___
Clean Stall ___
Stall with gear and flaps down ___
Never exceed ___
Normal Approach ___

PERFORMANCE
Takeoff distance to clear 50'
Sea Level 15°C ___
5000' ASL 25°C ___

WEIGHT AND BALANCE
Center of Gravity Limits
Forward ___ Aft ___
Empty Weight ___
Max Takeoff Weight ___
Max Baggage Compartment Weight ___

FUEL AND OIL
Fuel Grade ___
Usable gallons of fuel ___
Consumption per hour 65% at 5000 feet (ISA) ___
Oil Grade ___
Quantity ___

Fig. 31. Aircraft Performance Chart.

Always refer to and follow the recommendations of the manufacturer in carrying out any inspections and procedures. Individual models of airplane may have special procedures and inspection guidelines that will vary from those published in this manual.

The circled numbers in Fig. 32. correspond to the numbers indicated on the list below. By following the numerically indicated route, a systematic inspection of the airplane can be accomplished.

Stand off and observe the general overall appearance of the airplane for obvious defects.

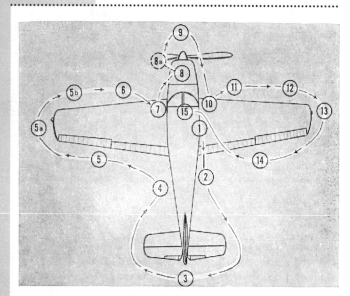

Fig. 32. Preflight Inspection.

1. **Cockpit/Cabin.**
   - Battery and ignition switches — OFF.
   - Control locks — REMOVE.
   - Landing gear switch — gear DOWN position.

2. **Fuselage.**
   - Baggage compartment — contents properly secured and within the safe C.G. limits, that is, not too far forward or too far back to upset the trim.
   - Airspeed static source — free from obstructions.
   - Condition of covering — missing or loose rivets, cracks, tears in fabric, etc.
   - Anticollision and navigation lights — condition and security.
   - Avionics antennae (nav com, VOR, transponder, marker beacon, glideslope, ADF, ELT) — cracks, oil or dirt, proper mounting and damage.

3. **Empennage.**
   - Deicer boots — condition and security.
   - Control surface locks — REMOVE.
   - Fixed and movable control surfaces — dents, cracks, excess play, hinge pins and bolts for security and condition.
   - Tailwheel — spring, steering arms and chains, tire inflation, and condition.
   - Lights — navigation and anticollision lights for condition and security.

4. **Fuselage.**
   - Same as Item 2.

5. **Wing.**
   - Control surface locks — REMOVE.
   - Control surfaces, including flaps — dents, cracks, excess play, hinge pins and bolts for security and condition.
   - General condition of wings and covering — torn fabric, bulges or wrinkles, ripples, loose or missing rivets, etc.
   - 5a. Wing tip and navigation light — security and damage.
   - 5b. Deicer boots — general condition and security.
   - Landing light — condition, cleanliness, and security.
   - Stall warning vane — freedom of movement. Prior to inspection, turn the master switch ON so that the stall warning signal can be checked when the vane is deflected.

6. **Landing gear.**
   - Wheels and brakes — condition and security, indications of fluid leakage at fittings, fluid lines and adjacent area.
   - Tires — cuts, bruises, excessive wear, and proper inflation.
   - Oleos and shock struts — cleanliness and proper inflation.
   - Shock cords — general condition.
   - Wheel fairing — general condition and security. On streamlined wheel fairings, look inside for an accumulation of mud, ice, etc.
   - Limit and position switches — security, cleanliness, and condition.
   - Ground safety locks — REMOVE.
   - Seaplanes: floats — remove float covers and inspect for water. Pump out with bilge pump if necessary.

7. **Fuel tank.**
   - Fuel quantity in tank — visually check the tanks to verify the amount and the correct grade or type of fuel. The fuel quantity must be sufficient to allow the aircraft to fly for a period of 30 minutes beyond the destination airport (in VFR flight by day) and for 45 minutes beyond the destination airport (in VFR flight at night).
   - Fuel tank filler cap and fairing covers — secure.
   - Fuel tank vents — obstructions. If the fuel tank is equipped with a quick or snap-type drain valve, drain a sufficient amount of fuel into a container to check for the presence of water and sediment.
   - Drain cocks — make sure they open and close properly. No drips.

8. **Engine.**
   - Engine oil quantity — secure filler cap and dipstick. General condition and evidence of fuel and oil leaks.
   - Cowling, access doors, and cowl flaps — condition and security.
   - Carburetor air filter — cleanliness and security.
   - Exhaust stacks — check for cracks and studs for tightness.
   - Spark plugs — check terminals for security and cleanliness.
   - Engine mount — for cracks and security. Drain a sufficient quantity of fuel from the main fuel strainer (often referred to as the filter bowl or gascalator) to determine that there is no water or sediment remaining in the system.
   - When closing the cowling, be sure to inspect cowling and baffle seals to assure that they are snug and in place. This is important to assure proper cooling of the engine.

8a. **Nose landing gear.**
   - Wheel and tire — cuts, bruises, excessive wear, and proper inflation.
   - Oleo and shock strut — proper inflation and cleanliness.
   - Wheel well and fairing — general condition and security.
   - Limit and position switches — cleanliness and security.
   - Ground safety lock — REMOVE.

9. **Propeller.**
   - Propeller and spinner — security, oil leakage, and condi-

tion. Be particularly observant for deep nicks and scratches.
- Assure that the ground area under the propeller is free of loose stones, cinders, etc.

10. Fuel tank.
  - Same as Item 7.

11. Landing gear.
  - Same as Item 6.

12. Pitot.
  - Pitot cover — REMOVE.
  - Pitot and static ports — remove obstructions.
  - General condition and alignment.

13. Wing.
  - Same at Item 5, 5a, and 5b.

14. Cockpit.
  - Cleanliness — see that there are no loose articles which might foul controls, or cause distracting noises.
  - Windshield and windows — obvious defects and cleanliness.
  - Safety belt and shoulder harness — condition and security. See that there is a safety belt for every passenger. Secure belts in unoccupied seats.
  - Fire fighting equipment — check cabin fire extinguisher for capacity and ease of release from its holder and for availability to each crew member. A hand held fire extinguisher in the cockpit should be of a type suitable for extinguishing any fire that is likely to occur and that is designed to minimize toxic gas concentrations.
  - Emergency exits — not obstructed by disposable load.
  - First aid kit (mandatory) — on board and accessible.
  - Timepiece that is readily available to every crew member.
  - Flashlight, if the flight is to be conducted at night.
  - ELT — secure in mounting, connections tight, general condition and security (no corrosion), antenna secure, annual recertification completed and current, battery not time-expired, ELT switch in ARMED position.
  - Appropriate charts on board and readily accessible. Where the aircraft will be operated VFR Over The Top, in night VFR flight or IFR flight, all necessary charts and publications covering the route of the proposed flight and any possible diversionary route must be on board.
  - Cockpit checklists or placards that enable the aircraft to be operated in normal, abnormal or emergency conditions in accordance with the limitations specified in the Aircraft Operating Handbook. These include checklists for pre-start, pre-take-off, post-take-off, pre-landing and emergency procedures. The emergency checklist shall include emergency operation of fuel, hydraulic, electrical and mechanical systems, engine inoperative procedures and any other procedures necessary to ensure the safety of the flight and passengers.
  - Mandatory documents on board — certificate of registration, certificate of airworthiness (or flight permit relating to the flight), journey log book, the Aircraft Operating Manual, radio equipment licence, licence of each crew member, radio operator's licence, certificate of insurance. The Canadian Aviation Regulations require that a copy of the procedures to be followed in the event of interception be carried on board. Also have on board any special authorization for the flight (e.g. authorization for an employee to collect expenses for flying on business). Check that the C of A is valid and that the airplane has been signed out within the required period.
  - Survival equipment — on board. It shall include survival equipment sufficient for the survival on the ground of each person on board, given the geographical area, the season of the year and the anticipated climatic variations. The survival equipment shall include the means of starting a fire, providing shelter, purifying water and visually signalling distress. (See also **Survival Equipment** in Chapter **Aeronautical Rules and Facilities.**
  - Adjust rudder pedals so that full rudder travel may be assured.
  - Parking brake — SET.
  - Check all instruments for proper reading and, where applicable, fluid levels (turn and bank and compass).
  - Landing gear and flap switches or levers in proper position.
  - Check all switches and controls.
  - Check fuel selector valve to be sure it is moving freely.
  - Trim tabs — SET.
  - Pilot's seat — LOCKED.

15. Cabin Briefing.
  - Ensure that all passengers know the following vital information:
  - Operation of seat belts — how to secure and how to release.
  - Securing of seat backs and chair tables, where applicable.
  - Stowage of carry-on baggage.
  - Smoking limitations — when smoking is permitted, if at all.
  - Location of normal and emergency exits — how to operate door handles.
  - In the case of an over-water flight, the location and use of personal flotation devices and life preservers
  - In the case of a flight that will require the use of oxygen, the location and the use of oxygen equipment.
  - Crash position, location of fire extinguishers, first aid kit, ELT — how to operate.
  - Emergency procedures should be outlined on cards which passengers should be advised to read so they can act knowledgeably and positively if an emergency situation should arise.

It is extremely important to carry out a thorough pre-flight inspection before every flight. Small clues indicating a malfunctioning or damaged component may easily be missed in a hurried pre-flight check. The extra effort involved in not just looking at but really inspecting every part of the airplane may make the difference between a safe flight and one that ends in an accident. The time to be even more vigilant is after a maintenance, painting or modification job has been performed on the airplane. It is, unfortunately, not uncommon for components to be reinstalled incorrectly.

Cockpit Check Prior to Flight

A systematic and careful cockpit check should also be carried out prior to flight.

1. Do a complete runup of the engine as described in the Chapter **Aero Engines.**

During the runup of the engine and especially when checking the engine for correct and normal operation, if there is any indication that the engine(s) is malfunctioning, under no circumstances should you attempt to take off. There is only one course of action in such a circumstance and that is to return to the ramp and conduct a thorough investigation as to the possible cause of the malfunction. The trouble may be serious, such as wrong fuel in the tanks or air leaks in the fuel line plumbing. If you attempt to take off knowing that the engine(s) is not operating smoothly, you are inviting disaster. Power failure is very likely to occur at the most critical phase of the take-off or during the initial climb.

2. Check all instruments systematically (usually from left to right) and adjust each as they are checked if adjustment is necessary. For example: Set the altimeter setting on the altimeter and check that the height indicated is the elevation of the airport (plus or minus 50 to 75 feet). If the altimeter setting is not known, set the airport elevation under the indicator needle. Wind the clock and set the correct time. Etc.

3. Check hydraulics to assure the proper pressure reading on the gauge.

4. Set trim tabs of both elevator and rudder to take-off position according to the particular load and C.G. of the airplane.

5. Mixture — Full rich (unless high elevation of the airport requires slight leaning).

6. Carburetor heat — COLD (unless atmospheric conditions necessitate heat).

7. Pitch — Propeller in full fine for take-off.

8. Fuel — Check the fuel gauge(s) for the proper quantity of fuel in each tank and adjust the fuel selector to the proper tank for take-off. Adjust cross feed and booster pumps. Primer locked. Be sure that you have enough fuel for the flight you are planning plus the reserve fuel that is required by the Canadian Aviation Regulations.

9. Flaps — Adjust to take-off position.

10. Switches — Magneto ON. Generator ON. Anti-collision beacon ON. Pitot heat, navigation lights etc. as required.

11. Gyros — Adjust the heading indicator to the runway heading. Adjust the attitude indicator if necessary. Allow 5 minutes after engine start for gyros of vacuum operated instruments to reach normal operating speed. Allow 3 minutes if the gyros are electrically driven. If the gyro instruments are venturi driven, only after the airplane is well established in flight will the instruments give reliable indications.

12. Gills — Adjust cowl gills to take-off position.

13. Arm the ELT if you have not already done so. Listen on 121.5 MHz to make sure that the ELT is not transmitting.

14. Safety belts of all passengers and crew fastened. No smoking.

15. Parking brake off. Tail wheel lock adjusted. Water rudders up (seaplanes).

16. Doors and windows or canopy top — Closed and secure.

17. Check freedom of all controls — ailerons, elevators and rudders. While moving the control column and rudder pedals, check that the control surfaces are responding in the proper direction of travel. This check is especially important if the airplane has undergone maintenance during which the control connections have been adjusted or removed and reinstalled. It is not unheard of for the controls to be reinstalled in reverse.

The cockpit check should be made **deliberately without haste from a written check list.** A definite sequence should be followed, moving clockwise around the cockpit. Touch each control with your hand and name it aloud. For example: "Trim Tabs — Flaps — Fine Pitch", etc. With larger and more sophisticated types of airplane, the cockpit check becomes more and more involved. On some large transport aircraft, the cockpit check requires several typewritten pages to list. Always work from a written, not a memorized, check list no matter how small your airplane. You may be interrupted while doing your chores by a radio call, forget where you left off and overlook some vitally important procedure.

There are many check lists relating to the various phases in the operation of an airplane. There are check lists for preflight, before starting engines, engine starts, before taxiing, engine runup, before take-off, take-off and climb, cruise, descent, before landing, aborted landing, after landing and after shut down, as well as check lists relating to emergency situations. Small, single engine airplanes may use only a portion of these. Larger, multiengine airplanes may use them all. Whatever check lists are prepared for your airplane, make a habit of always using them during the phases of operation to which they apply.

Take-Off Procedure

Taxiing an airplane is an elementary operation, the first thing a student pilot is taught. Nevertheless, no pilot should become complacent about the procedure. It is always necessary to pay close attention to the control of the airplane. Taxi slowly enough that the airplane will stop instantly when the brakes are applied, slowly enough that it will stop on its own when the throttle is closed. Always check the brakes before moving more than a few meters. Keep a sharp lookout outside of the cockpit and have someone on the ground guide you through any area where the clearance seems marginal. Avoid taxiing too closely behind large turbine powered airplanes. Be very careful taxiing in conditions of high winds and gusts.

You are responsible for ensuring that there is no likelihood of collision with another aircraft or vehicle, both during taxi and during take-off, and that the aerodrome is suitable for your intended operation.

Uncontrolled Airports

At an uncontrolled airport at which a mandatory frequency has been designated, always follow the procedures for reporting your intentions to the FSS or CARS operating the MF. If an aerodrome traffic frequency has been designated, report your intentions to the operator of the ATF or broadcast your intentions blind. See **Mandatory Frequency** and **Aerodrome Traffic Frequency** in the Chapter **Aeronautical Rules and Facilities.**

Always taxi downwind to the **extreme end** of the field or runway to take off. By doing this, the full length of the runway is available for the take-off run if it is needed. This practice is a good habit to cultivate. One day, in a short field on a hot day, you will be glad to have every foot of the runway to use. In a seaplane, it is wise to allow at least twice the distance that is really required for the take-off run.

Make a final check of the instruments and everything in the cockpit again. Make a conscientious visual check of the approach path to the runway to see that no airplanes are approaching to land. Report your departure intentions on the MF or ATF.

Trim the stabilizer properly for take-off and turn and take off into the wind.

Controlled Airports
With Radio

If your airplane is radio equipped, taxi instructions will be given to you verbally by the tower. If you have been authorized to taxi to the runway in use, taxi to the indicated holding position on the taxi strip. If there is no indicated holding position marked, taxi to a position approximately 200 feet from the boundary of the runway in use to do your engine run-up and cockpit check. (You may taxi across non-active runways without further authorization from the tower to reach the runway in use unless otherwise instructed.) Turn at an angle of about 45° to the taxi strip to do your engine run-up in case there may be another airplane behind you. When you have made your cockpit check, request a take-off clearance from the tower. (Radio traffic procedures are described in the Chapter **Radio).**

If requesting take-off clearance from an intersection, it is your responsibility as pilot to ensure that the portion of runway available to you is sufficient for the requirements of your airplane. The tower controller bases his authorization on the existing traffic, noise abatement rules, etc. and on not the capabilities of your airplane.

If you have approached a runway at an intersection but wish to back track on that runway to achieve enough runway length for take-off, you must indicate your intentions to the tower and obtain a clearance for the maneuver prior to entering the runway.

Sometimes, to expedite the movement of traffic, the controller in giving take-off clearance includes the word "immediate". On acceptance of the clearance, you must taxi onto the runway and take off in a continuous movement. If, however, in your opinion, this action would affect your operation, you should refuse the clearance and request a static take-off (i.e. a full stop in position prior to starting the take-off roll). There is documented evidence that the loss of an engine on take-off has been caused by an air gap in the fuel system occasioned by the centrifugal force of the tight turn onto the runway leaving the fuel tank line momentarily exposed.

Always turn left after take-off, unless otherwise authorized by the tower.

Nordo

If your airplane is not fitted with radio, you must inform the control tower of your intentions and make arrangements for visual signals. Do your engine runup and cockpit check on the apron before taxiing to the runway. (The apron is a surfaced area in front of the hangars, which oldtimers refer to affectionately as the tarmac.) When ready for take-off, taxi to position on the taxi strip approximately 200 feet from the boundary line of the runway in use. Turn your airplane toward the tower to attract the attention of the controller. Wait for the steady green light signal to proceed onto the runway and take off.

Steady Green Light — Clear to take-off.

Flashing Green Light — Clear to taxi, but do not take off.

Steady Red Light (or Red Flare) — Stop. Do not taxi.

Flashing Red Light — Taxi clear of the landing area in use.

Flashing White Light — Return to the ramp or hangar.

Flashing Red and Green Light (U.S.) — Danger. Be on alert.

Blinking Runway Lights — Vehicles and pedestrians are to vacate the runway immediately.

Acknowledge reception of visual signals by the full movement of the rudder or ailerons of your airplane at least 3 times in succession or by taxiing the airplane to the authorized position.

If your airplane is equipped with a receiver only (RONLY), before embarking on the flight, you must advise the tower, preferably by telephone, of this fact. The tower will then transmit clearances and instructions to you on the tower frequency rather than use light signals. Otherwise, the procedures that apply to NORDO airplanes apply also to RONLY airplanes.

Take-Off Considerations

Be sure, before beginning your take-off run, that you have given full consideration to the effect on your take-off of the following: gross weight and centre of gravity, density altitude, wind direction and velocity, runway conditions, ground effect and emergency procedures in the event of power failure.

It is difficult for controllers to track the many aircraft and vehicles moving on the maneuvering areas of the airport, especially at large airports. Practise ground safety awareness by proceeding with caution when approaching any other taxiway or runway. Always communicate clearly and listen to the other traffic on your frequency. If you have received a clearance that you do not understand, ask for clarification. Have your checks completed and be ready for take-off before requesting it. When cleared for take-off, move onto the active runway decisively and take off, so as not to spoil the timing of the controller and other pilots who are anticipating your action.

After take-off, maintain a listening watch on the appropriate frequency for aerodrome control communications and keep a watchful eye on other aerodrome traffic to avoid potential collision situations.

En Route Procedure

1. Note the time of take-off in the journey log. Conform to the traffic circuit procedures for departing the circuit. When well clear of the circuit traffic, turn on to the desired heading. Climb to the selected altitude, adjusting the power and the mixture in accordance with the manufacturer's operating instructions. During the climb, check the instruments regularly. On reaching the desired altitude, level off and adjust the power and mixture as necessary.

2. Check distance, time and groundspeed by D.R. navigation as detailed in the Chapter **Air Navigation** or by radio aids as detailed in the Chapter **Radio Navigation.** Make course corrections and revise your ETA as necessary. Keep a log of the progress of the flight and record the time over the various checkpoints.

3. Fly straight and level when checking headings on the magnetic compass. Reset the heading indicator frequently.

4. Avoid thunderstorms. Take appropriate action if this is unavoidable. Avoid turbulent air if possible. If unavoidable, slow the airplane to the recommended **maneuvering speed.** Do not get stranded over a cloud layer. Avoid the wake of large airplanes.

5. Check the weather, as you go, by radio or observation. Listen for the weather reports from stations en route. Maintain a continuous listening watch for en route advisories, SIGMETS and AIRMETS. Flight information service does provide information regarding unfavourable weather developing along your route, thunderstorms, icing, unserviceability of radio aids, airports or other hazards to safety providing you contact an ATC unit prior to take-off or en route.

6. Apply carburetor heat, windscreen defroster, wing de-icers when necessary.

7. Observe air traffic rules, especially those dealing with distance from cloud (both vertically and horizontally), visibility, cruising altitudes both on and off airways, weather minima in control zones and other controlled airspace.

8. Keep a constant lookout for other aircraft.

9. Keep an accurate check on fuel consumption, constantly.

10. Observe wind and weather changes.

11. Report to communications stations en route any unusual weather conditions encountered, or observed, as an aid to other pilots. Reports so transmitted by pilots en route are referred to as PIREPs.

12. Fly around bad weather if possible. If this is impracticable, turn back — repeat — turn back.

Airmanship

13. Check your flight and engine instruments frequently. Be sure you understand altimeter setting and altimeter errors thoroughly. Compute true airspeed based on indicated airspeed, outside air temperature and pressure altitude.

14. Report time over intermediate stations en route. This is good practice if you intend to acquire an instrument rating later on: it also provides the search and rescue service with a clue as to your whereabouts if you fail to arrive at your destination. Stay away from any area classified as Class F Airspace.

15. If the flight is being conducted at a high altitude, calculate the point at which to begin a gradual descent that will result in arrival at the destination airport at approximately circuit height. A gradual descent avoids rapid power off let downs that contribute to carburetor icing. During such a gradual descent, the airplane is able to pick up a bit of airspeed and consequently a better groundspeed which in turn gives a better en route time.

Landing Procedure

On approach to destination, contact the ATC facility that is appropriate for that airport (i.e. terminal area controller, approach control, control tower) several miles outside the boundaries of the classified airspace. If an ATIS broadcast is available, listen to it first and note the information. Follow the instructions of the control facility.

If the airport has no control facilities but does have a designated mandatory frequency or aerodrome traffic frequency, report on the MF or ATF 5 minutes before you reach the specified area and again when you have joined the traffic circuit.

If the airport is uncontrolled and has no published MF or ATF, keep a careful watch for other traffic and monitor 123.2 MHz and report your own intentions on this frequency. Observe the wind T or windsock and note the direction of traffic. Conform to the standard procedure for traffic circuits at uncontrolled airports. See also **Aerodrome Traffic Procedures** in Chapter **Aeronautical Rules and Facilities.**

When operating in the vicinity of an aerodrome, you must conform to the traffic pattern that is published for that aerodrome or that is formed by other aircraft in operation there. You must be observant of other traffic for the purpose of avoiding collision.

Pre-Landing Check

1. Fuel. Check gauges as to contents of fuel tanks. Pressures. Tank selectors, crossfeed, booster pump(s) as required.

2. Parking brake off.

3. Gear down. (Amphibians, as applicable)

4. Water rudders up. (Seaplanes)

5. Mixture rich.

6. Fine pitch.

7. Carburetor heat. Check. Set as required.

8. Flaps. As recommended.

9. Safety belts fastened. No smoking.

Be sure, before beginning your descent towards a landing, that you have given full consideration to conditions which might affect the landing: proper approach speed, gross weight and centre of gravity, density altitude, wind direction and velocity and gustiness, runway conditions, runway slope.

Landing and Post-Landing Procedures

1. Keep a sharp look out for other aircraft. Be ready to pull up and go around if necessary.

2. Adhere strictly to approach speeds and flap settings recommended for the airplane.

3. In rain and in winter, be prepared for adverse braking conditions. Many airports prepare James Brake Index reports on the condition of the runway and report these to pilots on request.

4. Touch down as near the threshold as possible to have as much runway as possible for the ground run, especially on a slippery runway or when the density altitude is high.

5. Be thoroughly familiar with traffic control light signals as they apply to traffic both on the ground and in the air and be prepared to comply with any signal directed to you.

6. Clear the active runway as quickly as possible.

7. Taxi with the utmost caution to the parking area.

8. Monitor 121.5 MHz just before shutting down your engine to be sure that your ELT has not been inadvertently activated.

9. Record your flight time and air time.

10. Close your flight plan by filing an arrival report with ATC. If air traffic control fails to receive an **arrival report** within a reasonable time, you will be assumed lost and a costly search and rescue operation begun.

11. Refuel the tanks and service your airplane.

12. Tie the airplane down securely if hangar storage is not available. Put wing covers and engine covers on if the airplane is to be stored in the open in a cold climate.

Post-Flight Inspection

A post-flight inspection is a recommended practice for small airplanes as well as for large ones. Mechanical deficiencies which arise during a flight can best be identified and attended to at the completion of the flight. It is an ideal time to check the condition of the airplane, noting signs of strain or wear.

Wrinkled skin — could indicate internal structural damage as a result of exposure to severe turbulence or excessive airspeeds.

Metal damage — from stones or other debris. Propellers are especially vulnerable.

Mud or ice — that clogs small openings such as the pitot pressure source or vent holes.

Tire surfaces — scuffing and tearing of tires may occur as a result of poor runway conditions.

Uneven landing gear extension — could be caused by the loss of tire pressure, improper hydraulic pressure in struts.

Fuel stains — or other signs of leakage of fuel, oil or hydraulic fluid.

Your experience during the flight may indicate other potential trouble areas. Excessive fuel consumption may indicate a problem with the seating of the fuel caps, with the fuel drains or line fittings. Look for oil drips if the oil consumption was high. Check out any other condition that was abnormal during the flight.

Braking Technique

Excessive braking shortens the life of the brakes and the aircraft tires. Therefore, airplanes should not be taxied at excessive speeds with unnecessarily high throttle settings that require use of brakes to maintain control.

Excessive landing speeds also result in accelerated tire wear and increase the possibility that excessive braking may be required to bring the airplane to a stop. If you have landed with too much speed and the tower then calls for a short turn off or if the runway is short or if a vehicle or airplane unexpectedly enters the runway, heavy braking will be required to comply with the situation.

Don't begin to brake as soon as you touch down. Right after touchdown, the airplane is still producing lift and the application of brakes has no effect on slowing the airplane. Aerodynamic drag is the major factor in slowing the airplane in the first quarter of its speed decay. Brakes become increasingly effective as airspeed and lift decrease. Once the airplane has slowed to at least 75% of its touchdown speed, rolling friction and the use of brakes will bring the airplane to a stop.

Form the habit of landing at the proper recommended speed and touching down at the end of the runway so that excessive braking will not be required.

In nosewheel airplanes, lower the nosewheel gently to the runway as quickly as possible after touchdown to reduce wing lift. Be sure to fully close the throttle so that there is no extra power counteracting the stopping process. Apply the brakes smoothly and firmly. For the brakes to be most effective, it is necessary to keep as much weight as possible on the main gear wheels. Therefore, while braking, apply some back pressure on the control wheel. Under conditions of deceleration, braking and rolling friction, there is a nosedown pitching force that transfers the weight of the airplane to the nosewheel rather than keeping it on the main gear where it is most effective. Do not, however, apply so much back pressure that the nosewheel is lifted off the ground. You need it to steer the airplane.

Maintaining positive brake pressure until the aircraft has come to a stop is preferable to pumping the brakes. The former technique results in a shorter landing distance. More significant, however, is the fact that the constant brake pressure allows the brake system to absorb more heat before reaching the point at which the brake fluid would boil.

If brake discs are subjected to excessive heat during a landing, the brake discs may cone (become distorted). As the disc starts to cone, the clearance between the disc and the linings decreases, causing non-uniform pressure distribution and poor braking action. Excessive heat energy is generated if the brakes do not have sufficient time to cool between successive applications (as in two landings within a short time span of each other). This residual energy is stored in the disc and is added to the heat energy created during the next brake application. The resulting high temperature may exceed the brake insulation threshold and cause the brake fluid behind the brake cylinder pistons to boil. Continued brake application will produce excessive pedal travel and poor brake response. Brake fluid will sustain a higher temperature without boiling when it is under pressure. If it is close to its boiling point when under pressure, when the pressure is removed, the brake fluid will boil. It is for this reason that pumping the brakes is an inadvisable technique.

Braking is reduced by sliding wheels. In addition, directional control is jeopardized. During braking, take care not to lock the wheels. Braking effectiveness drops significantly in a skid. To make matters worse, tires can be easily damaged in a skid.

Braking action is virtually non-existent on wet sod or wet runways or on frosted or icy runways. Landing rolls will be substantially increased under such conditions when brakes can be expected to be ineffective.

Braking problems can best be avoided by the use of proper speeds during taxi, landing and take-off and by proper maintenance. Mostly, airplane brakes are trouble free. However, if the airplane flies less than 200 hours per year and if it is exposed to unusual amounts of moisture, salt or industrial chemicals, corrosion and rust will contaminate the discs. The prime component of the brake system is the hydraulic fluid. It transmits pressure and energy but also lubricates and cools the moving parts. Obviously, the hydraulic fluid must always be at full capacity. But it is also important to use the exact type of brake fluid specified by the manufacturer. Mixing fluids may render the system useless. Some brake fluids may actually dissolve the rubber seals of incompatible systems. In addition, the fluid must be kept clean. Particles of dirt can render the system inoperative.

Tire Pressure

Tire pressure has a direct bearing on the life span and general condition of tires during their useful life. Both overinflation and underinflation result in external signs of misuse and may cause structural damage that leads to blow-outs.

Because they sustain a considerable impact on landing, aircraft tires are designed to have much more flex than automobile tires. Under inflation, however, imposes additional flexing that builds up internal heat that may cause material damage to the inner lining and sidewalls.

Low pressure can also allow a tube-type tire to slip on the rim, resulting in valve shear. Underinflation makes tires more susceptible to hydroplaning and loss of control on wet surfaces.

Ambient temperature has a direct bearing on tire pressure. There is about one psi (lb. per square inch) of pressure loss for every two to three degrees (C) drop in temperature. In a country like Canada where there can be wide ranges of temperature during one 24 hour period, a very significant loss of tire pressure can take place overnight on a particularly cold night.

Aircraft tires are designed for an optimum pressure and this level should be maintained as closely as possible. Be alert to possible tire pressure loss as a result of seasonal temperature changes, slow leakage developing after a hard landing or terrain damage during taxiing. Acquire a tire gauge and check your tire pressure regularly.

Landing Errors
Wheelbarrowing

Low wing airplanes with steerable nose gear are most susceptible to the pilot error of wheelbarrowing, although any tricycle gear airplane can be put into this unfavourable attitude.

It occurs when the pilot has inadvertently thrust too much weight onto the nose wheel. Loss of directional or braking control is the usual result.

Wheelbarrowing usually occurs when the pilot uses excess speed while making an approach in a full flap configuration. He may, in this situation, touch down with little or no rotation and then may try to hold the airplane on the ground with forward pressure on the control column. As a result, the main wheels are carrying insufficient weight for normal braking response. In winter, when snow or ice make the runway pavement more slippery than usual, the problem is compounded.

Some wheelbarrowing accidents have occurred, under strong cross wind conditions, when the pilot is using the slip method of drift correction.

If there is sufficient runway remaining with no obstructions, the best corrective action is probably a go-around. If this is not possible, close the throttle and ease the column to aft of the normal position. This action may lighten the load on the nose gear and put sufficient weight on the main gear to achieve normal braking response.

The chief cause of wheelbarrowing on take-off is the tendency to hold the airplane on the ground with forward control pressure in order to build up a faster than normal groundspeed before rotating.

Balloons and Bounces

Sometimes in the final stages of the approach, the pilot has the sensation that the ground is coming up faster than it in fact is. He increases the pitch attitude too rapidly and causes the airplane to climb instead of descend. This climbing during round out is known as ballooning. It is also caused when the flaps are lowered too late in the landing sequence.

A bounce will almost certainly occur if the round out is made too slowly or too late so that the nose gear (in the case of tricycle geared airplanes) or the main gear (in the case of tail wheel airplanes) touches first. A bounce will also occur if the round out is completed too high and the airplane is dropped onto the runway.

A bounce or a balloon each produces a critical situation. Not only is the airplane gaining height above the runway but it may also be approaching a stall. As well, the cross wind correction may have been lost and the pitch attitude of the airplane may be in excess of that of the normal landing attitude.

The best corrective action for a severe balloon or bounce is to apply power and execute a go around. If the balloon or bounce is low in height and the pitch change is not extreme and there is sufficient runway left, a landing may be made by establishing directional control, applying power to cushion the landing and adjusting the pitch attitude to that landing attitude proper for touchdown.

However, during a bounce or a balloon, the cross wind correction will almost invariably be lost and the airplane will start to drift. A landing should not be attempted unless the longitudinal axis of the airplane is straight with the runway and all drift has stopped. If the situation is not right, go around again.

If the pitch attitude has exceeded the landing attitude, the nose high situation means a rapid decrease in airspeed and diminishing response to the controls as the stall is neared. Lowering the nose is essential but lowering it too much may result in a hard landing that could cause structural damage. It is best to apply full power, release some of the back pressure to achieve level flight and then initiate the procedure for a go around.

Porpoising

Porpoising is the name given to the condition in which the airplane bounces back and forth between the nose wheel (or tail wheel) and the main gear after touchdown. It occurs most frequently as a result of an incorrect landing attitude and excessive airspeed that results in the nose wheel coming in contact with the runway before the main gear.

The porpoise may become progressively worse resulting in violent unstable oscillation of the airplane about the lateral axis that can damage the landing gear and the airplane structure.

The best corrective action is to smoothly use the controls to establish the normal landing attitude and add power to get the airplane airborne again. If there is sufficient runway left, land. Otherwise, go around again.

Hydroplaning

Hydroplaning is a condition that can develop whenever a tire is moving on a wet surface. The tire squeezes water from under the tread generating water pressures which can lift portions of the tire off the runway and reduce the amount of friction the tire can develop.

On a runway contaminated by rain or wet snow, it can be impossible for an airplane to accelerate to take-off speed and then to stop on the remaining runway in an aborted take-off. During landing, deceleration and stopping an airplane can be similarly compromised.

There are three types of hydroplaning.

Viscous hydroplaning occurs when there is a thin film of water and relatively low tire speeds. The water lubricates the surface and decreases traction. A water film of only a tiny fraction of a centimetre will drastically reduce the friction between the tire and the pavement and double the stopping distance.

Dynamic hydroplaning requires deeper water and results in complete loss of tire contact with the pavement. The tire lifts off the runway and rides on a wedge of water.

Reverted-rubber hydroplaning can occur when a locked tire skids on a wet or icy runway. Frictional heating raises the tire temperature causing rubber particles to shred off the tread. These particles accumulate behind the tire forming a dam that blocks the escape of water. The trapped water heats and turns to steam. The steam pressure lifts the tire from the surface.

The point at which dynamic hydroplaning begins is determined by water depth, runway texture, tire tread depth and tire pressure. The depth of water on the runway must exceed the depth of the tire's tread and the depth of the runway surface texture. Nosewheel tires will begin hydroplaning earlier than main wheel tires because nose wheel tire pressures are normally lower than those of the main wheels. (It is, therefore, important to lower the nose as soon as possible after touchdown in order to quickly get a vertical load on the nose wheel tire and get it tracking effectively.) The minimum speed at which dynamic hydroplaning can occur can be calculated by multiplying the square root of the tire pressure by nine. Once established, hydroplaning can continue at speeds well below this figure.

Tests have shown that one half an inch of slush on the runway increases take-off distance by about 15%. One and a quarter inches of slush increases safe runway distance by 100%. With two inches of slush, the test airplanes could not accelerate to take-off speed.

One of the best ways to minimize the chances of hydroplaning is to ensure that your tires have good treads and are properly inflated and that your brakes are effective. Tires with less than one-sixteenth of an inch of tread and brakes worn to 90 percent of limits compound the problem. When landing on a dry runway, the required landing distance should not exceed 60% of the available runway. If the runway is wet, the required landing distance is 115% of the dry runway requirement. Ensure that the runway is long enough for your needs before attempting a landing. Remember that runways contaminated by water following or during a rain shower can be as slick as wet ice.

Go Around / Overshoot

A go around, or an overshoot (the two terms means substantially the same thing), is a basic flight maneuver to be used when it becomes inadvisable to continue a landing approach. It should be the pilot's obvious choice of action when there is interfering traffic or when the landing procedures (flare, cross wind correction, touchdown, etc.) do not come together properly to ensure a safe landing and roll out. At a controlled airport, the tower may, because of a potentially unsafe situation on the runway, advise the pilot of an airplane on final approach to pull up and go around again.

A go around can become a very risky flight procedure if the pilot does not decide soon enough that a go around is the best choice and delays making a decision until the situation has become critical.

Do not be indecisive about executing a go around. If there is any suspicion that a landing is turning out poorly, start the go around procedure. The sooner you make the decision, the easier and safer the procedure will be. A go around should be viewed as an opportunity to correct what was an unacceptable landing approach and to make a good and safe landing on the next try.

A go around initiated at relatively high altitude is hardly more than a fly through. Initiated at low altitude, it demands attention to proper procedure. Be mentally prepared for the go around on every landing. Complete the landing checklist early. Be sure that rich mixture and fine pitch have been selected. They are needed to achieve full power if a go around is necessary.

Make the go around decision early. Once having made it, don't change your mind. First, smoothly apply full power, select carburetor heat to the cold position and re-trim the airplane for full power operation. Establish and maintain the proper airspeed for level flight and then retract the flaps. When a safe airspeed is attained and there is no chance of the airplane settling onto the runway, raise the landing gear. Continue to maintain straight and level flight until you reach climb speed.

Climb at the airspeed for best angle of climb, using the maximum allowable power, until all obstacles have been cleared. Then, assume the airspeed for best rate of climb and reduce the power to that recommended in the Pilot's Operating Handbook.

If the go around was initiated because of other traffic on the runway, make a slight turn to parallel the runway when a safe altitude and airspeed have been obtained. Keep the other aircraft in sight until the possibility of a collision no longer exists.

Make a radio call to advise the tower (or on the mandatory frequency at an uncontrolled airport) of your action at the same time as you are stabilizing the climb at the proper power, speed and trim.

Wind Shear

One of the basic principles of flight is that airspeed is not affected by the movement of the air body itself. This is true except in the situation of wind shear which involves an abrupt or sudden change in wind velocity. Since the inertia of the airplane is far greater than that of the surrounding body of air, there is an inevitable lag in the airplane's response to the sudden increase or decrease of wind, resulting in a temporary gain or reduction in airflow over the wings and therefore in airspeed. The change in airspeed may not last more than a few seconds. In cruising flight, there would be no serious problem but, under some landing or take-off conditions, it could be critical, since it can cause stalls, undershoots or overshoots.

If, for example, the wind changes suddenly from a no wind condition to a 20 knot tailwind or if there is a sudden 20 knot decrease in the headwind (negative wind shear), inertia causes a several second delay before the airplane reacts to the change in wind during which the airspeed will fall by almost 20 knots. If the airplane is approaching to land at an airspeed near the stall, the approach path could steepen or a stall could occur since any loss in airspeed means a reduction in lift.

Conversely, if a positive wind shear (increase in the velocity of the headwind or a decrease in the velocity of the tailwind) of 20 knots takes place, again inertia causes a delay in the reaction of the airplane and the airspeed briefly builds up by about 20 knots. During this brief lag, there is an increase in lift as a result of the increased airspeed, the rate of descent decreases and there is a tendency to overshoot.

Cross wind shears and vertical shears also have an adverse effect on the airplane. An abrupt cross wind shear will make it weathercock into the new wind. An abrupt downdraft causes a brief decrease in the angle of attack of the wing with a resultant loss of lift. An abrupt updraft causes an increase in the angle of attack that, if the airplane is already near the stall speed, may push the angle of attack beyond the stall angle.

The important parameter is the rate of change of the wind's velocity with respect to time. This determines the pilot's and the airplane's ability to cope with the wind shear. If the airplane could be instantaneously accelerated or decelerated to respond to changes in wind speed, there would be no problem. However, there is an inevitable lag in pilot reaction time and in airplane response. Large airplanes with larger inertia factors adjust less quickly than do smaller general aviation airplanes. The outcome of a critical wind shear encounter depends, therefore, upon quick and correct action on the part of the pilot. Even a slight delay in initiating corrective action can result in the loss of crucial performance capability needed for an airplane to recover from a severe wind shear encounter at low altitude.

If a pilot is aware of a wind shear condition, he can compensate for the expected value of the shear by varying normal approach airspeeds. This is a solution only up to a point. Wind shear is complex and unpredictable. Wind changes can be gradual or abrupt. Strong headwinds or tailwinds can suddenly become weak or vice versa. There may be turbulence. There may be a lateral component that introduces drift and heading problems.

There are certain clues that, by alerting a pilot to its possible presence, can help him to avoid a wind shear encounter. They include pilot reports (PIREPS) from other pilots who have encountered it, low level wind shear alerting system warnings, the presence of thunderstorms and virga.

During any landing or take-off, a pilot should be especially alert to any fluctuation in airspeed, vertical speed and attitude. Any excessive variation should be taken as an indication of wind shear and corrective action instantly initiated.

See also **Wind Shear** in the Chapter **Meteorology.**

Ground Effect

Every pilot has encountered the term ground effect. What exactly is it?

The total drag of an airplane is divided into two components, parasite drag and induced drag. Induced drag is the result of the wing's work in sustaining the airplane in flight. The wing lifts the airplane simply by accelerating a mass of air downward. (It is perfectly true that reduced pressure on top of an airfoil is essential to lift; that, however, is only one of the factors that contributes to the generation of lift.) The amount of downwash is directly related to the work of the wing in pushing the mass of air down and therefore to the amount of induced drag produced. At high angles of attack, induced drag is high. As this corresponds to lower airspeeds in actual flight, it can be said that induced drag predominates at low speed. (See also **Drag** in Chapter **Theory of Flight.**)

When a wing is flown very near the ground, there is a substantial reduction in the induced drag. Downwash is significantly reduced;

Fig. 33. Ground Effect.

the air flowing from the trailing edge of the wing is forced to parallel the ground. The wing tip vortices that also contribute to induced drag are substantially reduced; the ground interferes with the formation of a large vortex (Fig. 33).

Many pilots think that ground effect is caused by air being compressed between the wing and the ground. This is not so. Ground effect is caused by the reduction of induced drag when an airplane is flown at slow speed very near the surface.

Ground effect exerts an influence only when the airplane is flown at an altitude no greater than its wing span, which for most light airplanes is fairly low. A typical light airplane has a wing span of perhaps 35 feet and will experience the effect of ground effect only when it is flown at or below 35 feet above the surface (ground or water).

A low wing airplane is generally more affected by ground effect than a high wing airplane because the wing is closer to the ground. High wing airplanes are, however, also influenced by this phenomenon.

At the moment of take-off, when the wing is only 3 or 4 feet above the ground, a low wing airplane will experience a reduction in induced drag of about 48% as compared to the induced drag generated at flight altitude. At 18 feet altitude, the reduction in induced drag drops off to about 8%. When the airplane reaches an altitude equal to its wing span, the influence of ground effect disappears.

Pilots get into trouble because of ground effect when they precipitate take-off before the airplane has reached flying speed. Take the scenario of a pilot trying a take-off from a poor field. He uses full power and holds the airplane in a nose high position. Ground effect reduces induced drag and the airplane is able to reach a speed where it can stagger off. As altitude is gained, induced drag increases as the effect of the ground effect diminishes. Twenty or thirty feet up, ground effect vanishes, the wing encounters the full effect of induced drag and the struggling airplane which got off the ground on the ragged edge of a stall becomes fully stalled and drops to earth.

A mixture of short runways, rough ground, grass or snow, high airport elevation, high air temperature, a weak engine and a heavy load, in any of many combinations, sets the scene for this type of accident.

When you do find yourself in a marginal take-off situation, know your airplane's take-off speed for the conditions prevailing, the distance required to accelerate to that speed and then allow a generous margin of safety by picking up as much speed as possible just off the ground before trying to climb. An airplane accelerates more rapidly in ground effect than above it. Therefore, just a few moments in level flight just a few feet off the ground achieves the goal of attaining a safe climb speed more quickly.

Ground effect is also influential in landing. As the airplane flies down from free air into ground effect, the reduction of induced drag as it nears the runway comes into effect to make the airplane float past the point of intended touchdown. In the common case of an airplane coming in with excessive speed, the usable portion of the runway may slip by with the airplane refusing to settle down to land. A go around will probably be necessary. On short fields, approach as slowly as is consistent with safety.

An airplane also tends to be more longitudinally stable in ground effect. It is slightly nose heavy. The downwash from the wing normally passes over the tail at an angle that produces a download on the tail. Ground effect deflects the path of the downwash and causes it to pass over the tailplane at a decreased angle. The tailplane produces more lift than usual and the nose of the airplane tends to drop. To counteract this tendency, more up elevator is required near the ground. During take-off as the airplane climbs out of ground effect, the download on the tailplane increases and the nose tends to pitch up.

Gust Conditions

A gust or bump increases the load on the wings. The speed of the airplane should therefore be reduced when flying in gusty air.

In approaching to land, on the other hand, a little higher speed should be maintained to assure positive control.

A tail wheel airplane making a landing in gusty wind conditions should make a **wheel landing.** With this type of landing, the airplane makes contact with the ground while still maintaining flying speed. There is no critical period between positive air and positive ground control. If the tail is held high until all forward speed has been lost, there is no tendency for a gust to lift the airplane back into the air.

Low Flying

Remember the one about the dear old lady who cautioned her son to "Always fly low and slow"? There is one occasion on which her counsel turns out to be good sound airmanship after all. That is when a pilot is forced to fly low in poor weather. Under these conditions the following precautions should be observed:

Reduce speed because of the limited visibility but keep a safe margin in case it is necessary to turn quickly to avoid a collision. Always keep one hand on the throttle, ready for any sudden emergency.

Keep more than a customary sharp look-out both ahead and on either side.

Remember that it is easy to overestimate actual airspeed when flying low downwind, because of the apparent high groundspeed, and a pilot has therefore a tendency to stall after turning downwind.

It should be noted that the tendency to stall when flying downwind is due to an **optical illusion on the part of the pilot.** The actual stalling speed of an airplane is the same whether flying upwind or downwind.

When forced to fly low, detour around any towns or cities that lie along your route.

Flying low in poor weather should be resorted to, of course, only in an emergency or when an air traffic controller authorizes you to proceed in conditions that are below the VFR minima.

Visibility in Rain

Rain falling on the windshield of an airplane causes a distortion that will make the terrain contours appear lower than they actually are. A hilltop half a mile ahead may appear to be 200 feet lower than it actually is. This distortion is the result of **refraction.** Light beams are refracted (change direction) as they pass from one medium to another, as from air to water. Water slows up the passage of light and causes it to bend. As a result, objects, terrain, lights appear lower than their actual elevation relative to the airplane. Refraction error also causes the eye to see a horizon below the true horizon.

Diffusion also causes a distortion. Lights seen through moisture tend to spread apart and appear less intense and therefore farther away. However, on landing approaches, diffusion has a different effect. Approach lights appear larger and therefore nearer. The degree of distortion will vary according to weather and terrain conditions so no rule of thumb applies. Pilots simply must be aware of the phenomenon and when flying in the rain be alert to the problem. It is essential to ensure proper terrain clearance when flying en route and on final approach to landing.

Rain repellant used on the windshield may help reduce the refraction error. Effective windshield wipers also help to reduce the problem.

Flight in Volcanic Ash

Flight in volcanic ash has the potential for serious and hazardous complications to aircraft components. Aircraft surfaces and windshields can be damaged. Aircraft heat and vent systems, hydraulic and electronic systems may be contaminated. Powerplant failures are a common result of the ingestion of volcanic ash. In addition, volcanic ash is heavy and accumulations of it on the wings and tail surfaces can adversely affect weight and balance.

Ash from volcanic eruptions can rapidly rise to heights in excess of 60,000 feet and be blown downwind of the source for considerable distances. Encounters have been reported 2400 n.m. from the source and up to 72 hours later.

If ash is visible, **do not enter it.** The risk of entering ash in IMC conditions or at night is particularly dangerous because there is no visual warning. Unfortunately, aviation radar cannot detect volcanic ash. St. Elmo's Fire is a tell tale sign of a night encounter, although engine problems may be the first indication. Exit the cloud as quickly as possible.

In remote areas, pilots are often the first to be aware of a volcanic eruption. If an eruption or ash is observed, an urgent PIREP should be filed with the nearest ATS unit.

Flying the Laminar Flow Airfoil

Laminar flow airfoils, introduced in **Theory of Flight**, have certain flight characteristics that the pilot flying them must learn to appreciate. The laminar airfoil takes less energy to slide through the air since it has the advantage of producing less drag.

There are, however, also disadvantages to this type of airfoil. The principle disadvantage is that anything that mars the absolutely smooth surface of the wing (mud splashes, dust and dirt, grime or any foreign material) will reduce the efficiency of the wing much more than would be the case with the same amount of foreign material on a wing of conventional airfoil design.

Pilots flying airplanes with laminar flow wings must be more precise in their technique. Abrupt changes in speed and angle of attack can suddenly cause large areas of the wing to change from laminar flow to turbulent flow. This can cause large variations in the drag of the wing resulting in hunting and porpoising especially during low speeds.

An airplane with a laminar flow wing is subject to more abrupt stalls should the angle of attack be increased suddenly. If the stall is approached gradually, the airplane will perform normally. But an abrupt increase in the angle of attack or in g load can result in a violent stall. For this reason, pilots flying this type of airfoil must be especially careful during landing not to approach with too high an angle of attack near the stalling speed. A sudden disturbance can cause instant stall.

Mid Air Collision Hazard

With the progressive increase in both speeds and traffic density, risk of collision becomes an increasingly serious hazard. For example, a pilot in a high speed executive transport airplane who observes a jet 1 1/2 miles distant on a 90° converging track **has only approximately 7 seconds to take evasive action.** However, it takes a minimum of 10 seconds for a pilot to spot traffic, identify it, realize that there is a collision hazard and take evasive action.

Surprisingly, however, a very small percentage of mid air collisions occur between aircraft converging at a head-on angle. Most occur as a result of a faster airplane overtaking and hitting a slower airplane. The closing speed in this case would be relatively slow, giving the pilot ample time to react and take evasive action.

Nearly all mid air collisions occur during daylight hours, in VFR conditions, within five miles of an airport, usually in the traffic circuit and primarily on final approach.

Given these facts, it would appear that pilots are not practising the principle of "see and avoid". Seeing is a full time job for every pilot regardless of the type of airplane he flies. A pilot must visually scan in all directions constantly.

Scanning

The eye is a miraculous organ but it does have certain limitations. It is vulnerable to dust, fatigue, age, optical illusions, emotion, germs. In flight, vision is affected by atmospheric conditions, hypoxia, acceleration, glare, aircraft design, windshield distortion, etc.

More than this, however, the eye only sees what the mind lets it see. A day dreaming pilot sees nothing.

The eye is subject to focusing problems. It takes **time to adjust** the focus from near to far objects. In hazy conditions with no distinct horizon, there is nothing to focus on and a pilot will experience **empty field myopia**, not seeing even opposing traffic when it does enter his field of vision. Another problem is that of **narrow field of vision or tunnel vision.** The eyes are limited to a relatively narrow field of vision in which they can actually focus and classify an object. The mind cannot identify targets in the periphery.

Glare on a sunny day makes objects hard to see, especially during flight directly into the sun. Contrast creates another problem. An airplane over a cluttered landscape blends into the background and can be almost impossible to see.

Since visual perception is affected by many factors, it follows that pilots must learn to use their eyes in the most efficient and effective manner in an external scan.

Learn how to scan properly, knowing how to concentrate on the areas most critical at any given time. In normal flight, the critical area is about 60° to the left and to the right of the centre of your visual area and about 10° up and down from your flight path. The slower your airplane, the greater your vulnerability and therefore the greater the scan area required.

The most effective scan pattern is the block system. Traffic detection can only be made through a series of eye fixations at different points in space. The viewing area (the windshield) should be divided into segments (blocks), each about 10° to 15° wide. The pilot should methodically scan for traffic in each block of airspace in sequential order. In other words, stop and look in block A, move to block B and stop and look, move to block C and stop and look, etc. The sequence may be from left to right, from right to left, or front to left side, front to right side. There is no one scan sequence that is best. You should develop a scan that is comfortable and workable for you. The scan must, however, include a series of fixations in the block areas. When the head is in motion, vision is blurred and the mind will not register targets. You will not see a stationary target, yet it is the one which constitutes a potential mid air collision hazard.

External scanning must be shared with internal (instrument panel) scanning. Generally, the external scan will take about 3 times as long as the internal scan. The internal scan should start with the attitude indicator, move to the heading indicator, the altimeter, the airspeed indicator, the rate of climb, the turn and bank and back to the attitude indicator. Navigation and engine instruments might be included in every third or fourth scan.

Collision Avoidance

Collision avoidance involves more than proper scanning techniques. There are other important factors in the see and avoid principle.

Plan your flight ahead of time. Have charts folded and in sequence. Prepare a flight log with all the information that might be required during the flight so that you need to spend as little time as possible with your head down in the cockpit.

To do a competent scan, the windshield and windows must be clean and free of obstructions such as solid sun visors and window curtains.

Encourage your passengers to look for other aircraft and bring aircraft sightings to your attention.

All airplanes have blind spots because of their inherent design: a window frame, the wing or wing strut, the forward fuselage, etc. These blind spots are inevitable but can be compensated for by the pilot.

Never let down, turn or climb into a blind spot. When letting down, turning, or climbing, it is advisable to make a slight S turn to have a look before initiating the maneuver. During prolonged climbs or descents, outside positive control areas, execute gentle left and right banks every few thousand feet in order to broaden your field of vision and also to increase the likelihood of being seen as a result of motion and light reflection.

Sustained periods of straight and level flight outside positive controlled airspace should be broken at intervals by gently banking the airplane in each direction in order to broaden your field of vision.

Be especially mindful of the fact that pilots of high wing and low wing airplanes can be in each other's blind spot. Collisions of this type happen most frequently at uncontrolled airports, when the low wing airplane descends on top of the high wing airplane, especially on final approach or just before touchdown, although it can happen anywhere in the circuit.

With the introduction of mandatory communication procedures at uncontrolled airports, the chance of a collision of this type is substantially reduced but not eliminated. There have been incidents of pilots who were talking to each other, but unable to see each other, colliding. They were in each other's blind spot. In such a situation, it would be advisable for the pilot, caught in a situation in which he knows another airplane is in the same landing pattern as he but is unable to see him, to assume that the other airplane is in his blind spot and to make a gentle level turn to the left. At the same time he should report his intentions on the mandatory frequency. Any turn, no matter how slight, will increase the separation between the two airplanes and decrease the risk of collision.

While flying at circuit height, your airplane will cast a shadow on the ground on a sunny day. So will other airplanes. Glance at your shadow occasionally and scan a wide circle around it. Two converging shadows could foretell a collision.

When another aircraft is approaching and it has movement, left, right, up or down, there is no danger of colliding. The rate of movement governs the margin of separation. If, however, the other airplane is approaching your track and there is no apparent change in the relative position at which you first saw it, you are on a collision course and should take immediate evasive action. Any turn, climb or descent will provide a margin of separation.

An important thing to remember is that **you cannot hit anything that has moved out of the spot from which it was first noticed.** If, however, it becomes stationary at any point, a collision is imminent.

The most common error is turning the wrong way. The rule of turning towards the target and keeping it in sight as long as possible is the rule for visual collision avoidance. By keeping the other aircraft in sight, you remain in control of the situation.

When meeting another aircraft head on, alter course to the **right** in order to avoid any chance of collision.

Observe rigidly the traffic rules governing flight altitude when flying both on and off airways. Update your altimeter setting as often as practicable.

Be especially alert when flying in areas where traffic is likely to be heaviest — near busy airports especially at lower altitudes and at any height over navigation facilities. When below 3000 feet and within 10 miles of an airport, reduce speed to the minimum airspeed which permits safe control for any necessary maneuver.

When flying cross country, avoid high density areas unless landing.

The use of landing lights greatly enhances the probability of an airplane being seen and thus is an excellent technique to avoid mid-air collisions. Therefore, turn on your landing lights both during day and night, while landing or taking-off, when flying below 2000 feet AGL within terminal areas and in aerodrome traffic circuits, while operating under Special VFR conditions or in conditions of reduced visibility such as haze or at dusk.

High intensity strobe and anti-collision lights should be on at any time the airplane is in the air. However, they should not be used on the ground as they are distracting to pilots taxiing, awaiting take-off or on final approach to landing. Strobe lights should be activated only immediately prior to take-off and extinguished after landing.

A Mode C transponder will permit ATC to use radar to see you, know your altitude and provide you and other aircraft with timely traffic advisories.

If visibility drops below VFR minima and you are not qualified to file an IFR Flight Plan, do not proceed. Land or turn back.

From the Ground Up

Under an overcast with low ceilings (particularly around 1000 ft.), be careful to avoid approach corridors in the close vicinity of an airport. IFR traffic will be dropping out of the clouds. Approach corridors are shown on instrument approach charts. If you don't happen to have one, you can roughly estimate the location of an approach corridor by visualizing a line from the radio navigation facility to the runway, extending the line 10 miles outbound from the facility and allowing room for the procedure turn which an airplane doing an instrument approach will make.

Be careful flying on top of cloud. Don't skim the tops. Jets may come barrelling up out of the clouds as fast as 6000 fpm. VFR over the top is now legal in Canada. Pilots flying VFR OTT must be especially vigilant and maintain a flight path that is at least 1000 feet above the overcast. It none too ample a margin.

At night, the see and avoid concept depends on aircraft lights. Navigation and anti-collision lights need to be in good working order.

At night, the eyes see in a different way than they do during the day. Peripheral vision is better than direct staring because the cones which are concentrated in the centre of the lens need a lot of light to function properly. The rods around the edge of the lens, however, do not require so much light and are more efficient at night. Using peripheral vision is, therefore, better than direct staring. Objects will be more apparent if you look 10 to 15 degrees to one side of the object. (See also **Vision** in Chapter **Human Factors**.)

Aviation Safety Reports have established that air collisions most always happen in ideal weather. When you are flying VFR — **look where you are going**. Do not argue the right-of-way. You may be dead right but, if the other pilot doesn't realize it, you may be just dead, period.

Good Airmanship – Some Tips

1. Whenever possible, schedule your flight during a time when traffic is least heavy.

2. Be so familiar with the layout of your airplane that you can find and work every control without looking. Practise this blindfolded on the ground until you can do it perfectly.

3. Keep the windows of your airplane unobstructed. Don't put maps and computers in the windows.

4. Keep the windows of your airplane clean.

5. Wear sunglasses on bright days. Do not use opaque sun visors.

6. Be sure you are flying at the correct altitude for your direction of flight.

7. Always look around carefully before starting a turn, while in the turn and after resuming straight and level flight.

8. When climbing or descending, do several clearing turns to check the path of your flight for other airplanes.

9. Keep up a regular routine of scanning the sky around you for other airplanes. Don't forget to do this while concentrating on one airplane. There may be others around also.

10. Keep a watch for older or large airplanes and stay out of their way. Visibility out of these types is usually poor.

11. Give your passengers the job of looking for and pointing out other airplanes in the air.

12. Pay attention to the radio as other pilots call in and report their positions.

13. Be precise in reporting over the geographic location requested by the controller. Call at that point and not two or three miles beyond.

14. Once you have accepted an ATC clearance, follow it. If an ATC clearance is not acceptable because of the operational capabilities of your airplane, so inform ATC and ask for other instructions. Ask to "Say Again" if you don't understand. Never acknowledge a transmission if you have not understood.

15. If you fly a high performance airplane, try to slow it down to around 100 knots in the circuit so that your speed matches that of other circuit traffic.

16. Be especially alert when joining the circuit and when turning on final.

17. Be sure to remain VFR and well clear of cloud unless you are cleared IFR.

18. If you are IFR and in good visibility, remember that VFR traffic could be at your altitude and track.

Emergency Procedures

Aircraft today are very safe and, especially if they are well maintained, rarely experience malfunction or emergency situations. You should not, however, become complacent but should always be prepared to deal with such a situation should it occur. Being prepared by having a pre-determined plan of action which you periodically review and practise will help you make a good decision when the worst happens.

In almost all situations, the aircraft will be capable of controlled flight. Flying the airplane is, therefore, your primary task while you assess and deal with the problem.

No one can predict when an in-flight emergency will occur. When it does, there is not much time to react. It is critical that the correct procedures be followed. This can happen only if you have taken training in emergency procedures and regularly review and practise them. Only by regularly rehearsing emergency procedures will you be confident and ready to react correctly when faced with the real thing. Your emergency checklist and Aircraft Operating Manual should always be readily available in an accessible place in case you need to refer to it. Make sure also that the manual you have is for the make, model and year of manufacture of the aircraft you are flying and is current with amendments from the manufacturer.

Some emergency responses will be a combination of memory and non-memory items with vital actions being completed first (your instinctive response thanks to training) and the remaining items completed later on (by referring to your manual or checklist).

Engine Failure on Take-Off

If an engine fails immediately after take-off, land straight ahead. The worst possible action is trying to turn back to the airport. An airplane needs both time and altitude to make a 180° turn, neither of which it has if the engine fails at low altitude during climbout. At anything less than 700 feet AGL, the only option is to land straight ahead. Of course, this does not mean ploughing into a brick wall head-on. Some maneuvering within a 60° arc right or left to select an open area that presents an acceptable landing field is a reasonable and practicable action.

It is always best to climb out after take-off at the highest possible airspeed that is consistent with the principles of good airmanship and with the local situation. At the higher airspeed, not only is the engine obtaining better cooling but the flight controls are more effective. In the event of a complete and sudden power failure during the climb out, the margin between the higher airspeed and the stall speed and the responsive controls give the pilot time to transition safely from climbing attitude to normal glide. If the engine fails when the airplane is climbing steeply at a low airspeed, there is little margin above the stall speed, the speed decays rapidly and there is poor response from the flight controls. The airplane is sluggish in responding to the pilot's attempt to transition to a glide attitude and will probably stall.

Engine Failure During Take-Off
1. Close the throttle.
2. Brake firmly.
3. Maintain runway heading.
4. Turn off the fuel, switch off the magnetos and pull the mixture into idle cut-off.

Engine Failure After Take-Off
1. Lower the nose and assume the speed for best glide.
2. Select the best available landing area in an area of about 60° left or right of the airplane heading.
3. Turn off the fuel and the magnetos. Pull the mixture to idle cut-off.
4. Make gentle turns to avoid obstacles.
5. When sure of reaching the chosen landing area, lower the flaps. Do not allow the airspeed to increase.
6. On short final, turn off the master switch. Unlatch the cabin doors (to guard against them becoming jammed.)
7. Resist the temptation to turn back to the airport.

Unlatched Door in Flight
If it has not been latched properly, the cabin door may come unlatched during flight. The door will trail in a position a few inches open and will cause a lot of noise and wind in the cockpit. The flight characteristics of the airplane will not be affected except that the rate of climb will be reduced. The important thing is to **fly the airplane** and not be distracted by the noise and wind.

Slow the airplane down because higher speeds will increase the suction forces to pull the door farther open. Don't try to close the door during flight. If the door opened during take-off, return to the airport in a normal manner. If it opened during flight, proceed to the nearest airport. If practicable, during the landing flare out, have a passenger hold the door to prevent it from swinging open.

Action in the Event of Fire
The following check lists for procedures to follow in the event of fire is representative only. Always use the procedure recommended by the manufacturer of the airplane you are flying.

Engine Fire on the Ground
Engine fires on the ground usually are the result of improper starting techniques, most specifically, overpriming. Overpriming may cause a backfire and a carburetor fire if the engine does not start. In most cases, any fire resulting from a backfire can be quenched by continuing to crank the engine until it starts. If an engine fire has started and is not put out by getting the engine running, the fuel should be shut off, the primer locked and a fire extinguisher used to put out the fire.

The procedures for dealing with an engine fire will be detailed in your **Aircraft Flight Manual.**

Do not fly the airplane after the fire has been extinguished until it is inspected by an aircraft maintenance engineer in case there has been internal engine damage. (See also **Procedure for Backfire During Starting** in Chapter **Aero Engines.**)

Engine Fire During Flight
Single Engine Aircraft – Engine Fire
1. ★ Mixture Control — IDLE CUT-OFF.
2. ★ Propeller pitch control — COARSE PITCH.
3. Fuel and oil cocks — CLOSED.
4. Operate engine fire extinguisher.
5. Ignition switch — OFF.
6. Throttle — CLOSED.
7. Reduce airspeed.
8. Radio MAYDAY. State your position.
9. Make a dead engine landing in the most suitable area.

Twin Engine Aircraft – Engine Fire
1. Identify the problem engine.
2. ★ Mixture Control — IDLE CUT-OFF.
3. ★ Propeller pitch control — FEATHERED.
4. Fuel and oil emergency shut-off — CLOSED.
5. Operate engine fire extinguisher.
6. Ignition switch — OFF.
7. Throttle CLOSED.
8. Fuel selector switch, booster pump, generator switch, etc. (if fitted) — OFF.
9. Radio MAYDAY and advise your point of intended emergency landing.

★ *If fitted. Otherwise begin at 3. (single) or 4 (twin).*

Electrical Fire
Modern aircraft have elaborate electrical systems that may short circuit and cause a fire. Your **Aircraft Operating Manual** will detail the procedures for dealing with this type of emergency.

The first step, in most procedures, is to turn off the master switch to stop all power to the electrical systems and to turn off all electrical components in order to prevent further shorting while you attempt to identify and isolate the faulty system or unit. While attempting to locate the cause of the fire, turn on each unit (with the master switch back on) one at a time and wait a significant time between each. With electrical fires, it often takes a while for the faulty electrical component to heat up again and start smouldering. If a circuit breaker has popped or a fuse blown, it is likely related to the fire. Do not reset a circuit breaker or replace a fuse before consulting your **Aircraft Operating Manual.**

Any fire causes anxiety and stress. Do not get preoccupied with locating the source of the fire and forget to fly the airplane

Fuselage Fire
The modern airplane is not a potential torch as were the early wood and fabric airplanes. Nevertheless fires do sometimes start, usually through carelessness or from electrical short circuits. A means of controlling a fire should be near at hand.

Portable fire extinguishers suitable for combating cabin fires and approved for use are of 5 types: carbon dioxide, dry chemical, water, Halon 1211 and Halon 1301 extinguishers. Generally, any fire extinguishing agent is a compromise between the hazards of the fire, smoke and fumes and the toxicity of the extinguishing agent. Certain types of extinguishers should not be installed in aircraft because of their high toxicity rating; these include those using carbon tetrachloride, Halon 1001, Halon 1011 and Halon 1202. Check carefully before installing a particular type of fire extinguisher that it is of an approved

From the Ground Up

type and be sure that you know how to use it and the possible dangers involved.

There are three classifications of fires:

Class A — in ordinary combustible material. On these, water or solutions containing large percentages of water are most effective.

Class B — in flammable liquids, greases, etc. On these, a blanketing effect is essential.

Class C — in electrical equipment. On these, the use of a non-conducting extinguishing agent is of first importance.

An extinguisher using Halon 1211 is effective against all 3 classes of fire and has proven to be the most effective on gasoline based upholstery fires. A Halon 1211 extinguisher is, therefore, probably the most useful type to install in a light aircraft. Exposure to high levels of Halon 1211 produces dizziness and impaired co-ordination but these effects disappear quickly when fresh air is introduced. Most nonpressurized aircraft cabins are well enough ventilated to clear out the Halon fumes before they become toxic. Halon 1301 is less toxic than Halon 1211 but is not effective against Class A fires.

Fire extinguishers require periodic inspection. About once a year, they should be discharged, inspected, new seals installed and refilled. Extinguishers should be secured in an accessible position with a quick release bracket.

In the event of fire:

1. Select all electrical switches (except ignition) — OFF.
2. All windows, ventilators and cabin air extractors — CLOSED.
3. Apply portable fire extinguishers to source of fire.
4. After fire subsides, leave all switches, except ignition, OFF. Land immediately.

Distress Signals

Radio procedures for airplanes experiencing emergencies in flight have been discussed in the Chapter **Radio**.

An emergency condition is classified in accordance with the degree of danger or hazard being experienced.

Distress presumes a condition of being threatened by serious and/or imminent danger requiring immediate assistance.

Urgency presumes a condition concerning the safety of an aircraft or other vehicle or of some person on board or within sight which does not require immediate assistance.

Forced Landing

Engine failure and a forced landing are circumstances every pilot hopes never to have to face. However, with proper preparedness, they need not be life threatening. Most frequently, engine failure is a result of fuel starvation or carburetor icing. Apply carburetor heat. Select another fuel tank. Flip on an electric fuel pump if there is one. Select a full rich mixture.

While running through these procedures, prepare the airplane for a forced landing. Immediately establish the airplane in the airspeed for maximum distance glide (best glide speed). Trim the airplane. If it is not properly trimmed, gliding distance will be reduced.

If nothing has worked to get the engine going again, accept the inevitable and start thinking about the emergency landing. Select a suitable field.

A seasoned pilot will seldom get caught with engine failure without having a forced landing area already selected in his mind. He forms the habit of picking suitable fields as he goes and of constantly checking the wind. As a result, if the engine unexpectedly fails, he immediately knows of a suitable field and is free to concentrate entirely on the approach.

Preflight planning, taking into account the possibility of a forced landing, is important. Try to select a route that keeps away from high rough terrain where there would be a poor selection of emergency landing sites. Plan also to fly at fairly high altitudes since excess height can be converted into distance when the engine fails. It also can be converted into time, time to think about what to do.

Requirements for a Good Forced Landing Field

The requirements of a good forced landing ground are:

1. Firm surface, reasonably smooth, with sufficient space to effect a landing into wind. Pasture land is satisfactory because the short grass cannot hide holes, ditches, boulders or other hazards. Avoid ploughed fields.

2. Since landing should be made into the wind, select a field, if possible, that is into wind. If your altitude is not sufficient to maneuver into wind or a suitable into wind field is not available, this may not be possible. Be prepared for a cross wind landing.

3. The approach should be clear of obstacles such as trees, telegraph wires, high tension lines, houses, etc. Avoid fields immediately alongside highways or railway lines because of wires.

4. Clear take-off run (for when the trouble is corrected).

5. Try to choose a field that is near houses or a road so that help is readily available.

In winter, if on wheels, it is better to attempt a forced landing on a highway comparatively free of traffic than a deep snow covered field.

If forced to fly over very rough bush country on wheels, it is wise to keep within gliding distance of a highway as a possible emergency landing place. In choosing to land on a highway, special care must be taken as there may be unseen hazards such as power lines and signs as well as vehicle traffic. If no such highway exists, by all means pancake into a lake close to a shoreline, or into muskeg, rather than risk crashing through timber.

Flying seaplanes over land was at one time considered a highly hazardous undertaking. Experimental landings, however, have proved conclusively that a seaplane can be landed in a restricted field with less risk of damage than can the average landplane. A pilot should have no hesitation about flying a seaplane overland when necessary, provided good level country lies below, such as would be considered reasonably safe for landplanes.

Forced Landing Sequence

The most important principle of coping with a forced landing is acceptance of the inevitable. Time is critical. Don't waste time trying over and over again to get the engine started. Turn your attention to landing safely. Once you have made up your mind to execute a forced landing, stick to your decision even if the engine should start again as you near the ground. Should the engine start up again on the line of approach and should you decide to carry on, there is no guarantee that the engine may not fail again a few moments later when you are no longer in a position to reach the field.

Airmanship

Practise forced landings and have the procedures down pat so that you act instinctively. Indecisiveness and lack of co-ordination are your worst enemies when faced with the real situation.

Immediately the engine fails:

1. Reduce speed to gliding speed and set trim. Any excess speed can be used to gain height. However, be sure to maintain flying speed especially in turns to avoid a stall and spin. Remember also that you will lose a fair amount of height in your turns.

2. Select a suitable field. Select a key position, preferably on the downwind side of the field, some physical feature on the ground that will provide continuing orientation to the selected field. Check the direction and glide to the downwind side of field. If sufficient altitude is available, conform to the familiar circuit pattern. It may, however, be necessary to fly directly to the key position or to fly a straight in approach.

3. Check for the possible cause of the engine failure: fuel selector on the correct tank; fuel selector ON. Switch to another tank that has fuel. Switch on the fuel pump in case the failure is in the regular fuel feed system. Check that the primer is in and locked; that the mixture control is in the full rich position. If the cause of the engine failure cannot be found and rectified, close the throttle, turn off the fuel, place the mixture control in the idle cut off position, turn off the magnetos and turn off the alternator and generator switches. Do not turn off the master switch until the flaps and gear are set for landing.

4. Transmit a MAYDAY radio call. Make this call before you have descended below the radio range of ground stations. Select code 7700 on your transponder. Be sure that your ELT is ARMED (that should have been part of your pre-flight check).

5. At the downwind side of the field, glide cross wind and look for any obstacles in the field. Turn onto base leg using the key position as a guide. Base leg must be well within the into wind gliding range of the aircraft. Note the drift, which will indicate the strength of the wind. A strong wind will necessitate a base leg closer to the threshold of the chosen field. Look for any obstacles in the field. Decide on your exact landing path.

If you are much too high over the key position, keep the proper distance from the landing site and lose height by flying between two selected key positions, always turning towards the field so that you can make the final approach at any time. Do not lose too much altitude. It is better to be slightly high than too low.

6. At about 500 feet AGL, turn onto the final approach. When certain of clearing all obstacles on the downwind boundary of the field, sideslip off the surplus height or lower the flaps. Do not lower the flaps or gear until absolutely necessary as they increase the drag and steepen the glide.

7. Tend to overshoot. Aim to land well into the field. It is better to overshoot and hit the far fence at low speed than to undershoot and hit the near fence at flying speed. Reduce your speed as much as possible and keep your angle as shallow as possible. Set the airplane down nice and easy. Hard landings as a result of too high a sink rate produce as serious injuries as do excessive deceleration rates.

If the field is hopelessly small or rough for a safe landing, retract the undercarriage and make a belly landing.

Procedure After Forced Landing

If the forced landing is due to engine failure, examine the tanks to see if there is gas and oil. Check the high tension leads to the plugs to see that they are intact. Examine the filters for dirt. Do not attempt to repair the trouble, if you have located it, unless you are experienced and qualified to do so.

If the failure is repaired, a thorough test of engine performance must be made before attempting to take off.

If no self-starter is fitted and it is necessary to get an inexperienced person to assist in swinging the prop, see that he is made familiar with propeller swinging routine before attempting to start the engine. Use wheel chocks.

Never, under any circumstances, have an inexperienced person handle the throttle or switches. The pilot only must operate these when starting an engine.

If the trouble cannot be repaired and it is necessary to obtain assistance, be sure to have the following information ready to transmit over the long distance telephone or radio:

1. Type of airplane and engine model number.

2. Exact location of the field in which the forced landing was made and the nearest village or town.

3. Reason for the forced landing and cause, if it can be ascertained.

4. Whether gas or oil is required for the return trip.

5. Any parts, special tools, or materials required to make the necessary repairs.

6. Whether the forced landing field is suitable for take-off.

7. The type of airplane that may safely land and take off from the field.

8. The telephone number at which you may be reached.

Precautionary Landing

A precautionary landing is one made with the engine and aircraft functioning normally but when landing is made compulsory due to shortage of fuel, aircraft malfunction, partial loss of engine power, being lost, bad weather, darkness, illness on board, etc.

If you become lost in smoke, haze, rain, fog or snow, do not fly around aimlessly exhausting your fuel supply. Land immediately and wait until conditions improve sufficiently to enable you to take off again and pick up your bearings.

In selecting an appropriate field for a precautionary landing, the criteria are: suitability of the land area, wind velocity and direction, inspection of the landing area, type of approach.

Landing Area: The requirements for a landing area for a precautionary landing are the same as those for a forced landing. It should be sufficiently long for both landing and for a later take-off. It should be into wind. It should be smooth, firm, level and free of obstructions on the approaches. The approach and overshoot areas should be free of high obstacles. It should be close to transportation or communication.

Wind Velocity and Direction: It is possible to judge wind velocity and direction by nearby smoke trails. Grass and grain fields ripple in the direction of the wind. Dust is blown with the wind. Water is calm on the leeward side of lakes.

Field Inspection: Having chosen the field, fly over it at a low altitude (500 to 1000 feet) at least once, preferably to one side of the proposed landing path, to check the size, surface and gradient of the field and the suitability of the approaches, to identify landmarks and reference points relative to the field which can be used as turning points in the landing circuit. Fly over the field a second time to look for hidden obstacles and rough or swampy ground. During the inspection. a reduced airspeed is preferable. Be alert to illusions created by drift.

Type of Approach: The type of approach will be determined by the information you learned during the inspection. If there is any doubt and no alternate field is available, consider the field short and the surface soft and plan your approach accordingly. The final circuit should be done at circuit height or (if weather does not permit) at the highest altitude allowable under the circumstances. Get into position for a normal engine-assisted approach and landing. Do not neglect the pre-landing cockpit check. Lower the flaps to the landing position (if applicable). Approach with as low a forward speed as possible but with a safe margin above the stalling speed. Aim to touch down as close to the near boundary as is practicable. The aim is to reduce the landing run to the minimum. If the field is very small, switch off the engine after touching down.

Should a landing be made solely as a precautionary measure that will result in the flight being delayed beyond the ETA specified in your flight plan, you should make every effort to notify an FSS or ATC unit that no emergency exists. Search and rescue activity is initiated one hour after the ETA you specified in your flight plan. If you cannot contact anyone, switch on your ELT at the appropriate time and leave it on until search crews locate you. Once located, use your aircraft radio to advise the SAR crew of your condition and intentions.

Take-Off: When you are ready to take off again, the precautions you take are as important as those you took on landing. Inspect the take-off path carefully. This you can do on foot. If the landing path proved satisfactory, use that ground run path for the take-off.

The procedure for a precautionary landing should also be used **when landing at any unfamiliar aerodrome.** The procedure gives you an opportunity to study the field before attempting a landing. However, in the case of an aerodrome, the pass over the airstrip to study it must, of course, be made well above circuit height. The recommended landing procedure ensures adequate ground control no matter what the surface and length of the runway once the landing is made.

Emergency Landing Precautions

If you have landed with trouble sufficiently serious to preclude all hope of getting away, REMAIN WITH YOUR AIRPLANE. An airplane is a comparatively easy object for search and rescue planes to spot but a human being in a forest is almost impossible to sight. Anything you can do to attract the attention of the spotters on the search airplane will enhance your chances of being found promptly. This is especially important if your emergency landing has turned out to be a crash landing and you and your aircraft are in the middle of a bush. Even when guided by an ELT signal, SAR spotters may have trouble finding the aircraft.

Remain with your airplane, reserve your energy and conserve your food supply. Start and maintain a smudge fire. Your smoke signal will assist search planes in locating you.

If your radio is serviceable, use it sparingly. (Short messages save batteries.) State your position as accurately as you can. You may transmit distress signals on any frequency, but almost all aeronautical communication stations listen on 121.5 MHz and 243.0 MHz. Bear in mind that reception is better at night and there is less traffic on the communication channels.

If you have an HF radio, it will be more effective than VHF or UHF. The HF range is much greater than the line of sight capability of VHF and UHF. The HF channel to use is 5680 KHz. It is monitored by many FSS, especially those in the remote areas of Canada. The recommended time for distress signals is 15 and 45 minutes past the hour for 3 minutes duration. Canada maintains DF facilities that can pinpoint the source of HF transmissions made anywhere in the country.

Be sure that your ELT has been activated and is sending a signal. It is your best assurance that you will be located promptly. Do not delay activation of your ELT until your flight planned time expires. That only delays rescue. Once you have turned on your ELT, **do not turn it off** until you have been positively located and directed by SAR forces to turn it off.

It is beyond the scope of this text to discuss in detail the art of survival in the bush after a forced landing. Every airplane should be equipped with a survival kit packed with the recommended equipment and supplies that will enable you to make a camp and exist for some period of time until help comes. There are excellent texts available that discuss the subject of **Survival** and we recommend that every pilot secure and study them carefully.

Ground-Air Emergency Signals

1. Make a smudge fire by pouring oil on rags or by using aircraft tires, green grass, brush wood or anything that will produce heavy smoke to attract attention during the day. At night, fire is more effective, the brighter the better. However, you must be careful not to allow the fire to get out of control and start a forest fire or grass fire. In wooded areas, fire, smoke or pyrotechnics must be used with discretion.

2. Make a large SOS on open ground. In snow, outline the SOS with boughs or moss.

3. Make trails in virgin snow. These can be readily seen from search airplanes.

4. Lay your cowlings out so they shine in the sun. One of the best conspicuity items now available is a cloth panel of brilliant fluorescent colour. It is a highly effective ground signal when laid out on the ground during the day. It can also be used as a lean-to shelter or as a warm blanket.

5. Keep your aircraft clear of snow or brush.

6. When using your radio, call at appropriate times when others are most apt to be listening. Save your battery as much as possible.

7. Point the flashlight at approaching airplanes and flash SOS. At night, a flashlight beam can be spotted from the air from a considerable distance.

Radar Assistance

Radar Assistance is available on a 24 hour basis to all aircraft within the limits of the identification zones. This assistance is available to any airplane that is in distress or in an emergency situation. Navigational assistance will be rendered when and wherever possible in the form of position information, vectors or track and groundspeed checks, weather advice, etc. Flights requesting this assistance must be operating within areas of radar and communication coverage and be radar identified. To request radar assistance in an ADIZ, call **"Radar Assistance"** on 126.7 MHz. When circumstances warrant a MAYDAY call, use 121.5 MHz.

If advisory service cannot be furnished because of air defence priority, the station will reply "unable".

When requesting service, climb to the best possible altitude. This will improve your chances of being picked up by the radar station.

A pilot, who is lost or in distress and unable to make radio contact, should attempt to alert available radar systems by the following procedure:

Squawk the emergency transponder code 7700. Monitor the emergency frequencies. Fly a triangular pattern as outlined below.

If the radio receiver of the airplane is working but the transmitter is inoperative, it is possible to attract the attention of radar assistance by flying a triangular pattern such as that shown in Fig. 34. Fly the pattern **to the right.** Hold each heading for 2 minutes. (Jet airplanes will hold each heading for 1 minute.) Repeat the triangular pattern twice. Resume the original heading and repeat the triangular pattern at 20 minute intervals.

If the radio is not operating at all or if the airplane is NORDO, fly the triangular pattern **to the left.**

An airplane may be seen as a blip on a radar scope but cannot be individually identified. By flying the prearranged pattern, the radar facility will be able to identify the blip that is the airplane in need of help.

A pilot lost or in need of assistance whose radio receiver is working will receive instructions by radio to guide him to the nearest landing place. An airplane whose radio is inoperative will be intercepted by a search and rescue plane and led to the nearest landing field.

Never comply with radar vector instructions intended for another airplane.

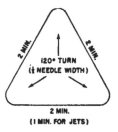

Fig. 34. Triangular Pattern for Lost Aircraft.

Triangular pattern for aircraft lost or requiring assistance. If the radio receiver is operating, fly the pattern to the right or clockwise. If no radio is operating, fly the pattern to the left or anti-clockwise.

Bush Sense

The vast hinterlands of Canada and Alaska offer a challenge to the spirit of youth which few other lands the world over can rival. To the sportsman pilot, they spell romance, the thrill of the great outdoors, the lure of unfished waters, the zest of virgin forest hunting areas. To the commercial pilot, their mining, forest, fur and oil industries offer rare and profitable opportunity.

Flying in the trackless wilderness can be just as safe and practicable as flying the organized airways. Plain, ordinary airmanship, a knowledge of the basic laws and proper provision for essential needs are the secrets of the bush pilot's much publicized sixth sense. Bush sense is, after all, just reasonable proficiency in the handling of seaplanes and skiplanes, plus some fundamental knowledge of the woods.

Flying single-engine airplanes in the area, commonly called the sparsely settled area of Canada, is an undertaking that calls for some special precautions. In this area, radio aids to navigation, weather information, fuel supplies, aircraft servicing facilities, accommodation and food are limited or non-existent. Careful preflight planning is therefore of utmost importance.

Airplanes must be capable of 2-way radio communication with a ground station from any point along the route. Capability on 121.5 MHz and 126.7 MHz is required. Any airplane operating in the sparsely settled area must continuously monitor 121.5 MHz. HF capability on 5680 KHz is required for any airplane operating in the Arctic Archipelago, as is an emergency transmitter capable of operating independently of the aircraft battery.

Navigation equipment must include a good heading indicator or gyrosyn compass and a means of checking heading using the sun or other celestial bodies. All headings and bearings are measured in degrees true because of the unreliability of the magnetic compass in the areas near the north magnetic pole. A manual entitled **Finding the Sun's True Bearing** can be used to keep the heading indicator aligned. GPS navigational capability is of great benefit.

A flight plan must be filed when proceeding northbound, when entering the Arctic Archipelago and when returning southbound. Adequate emergency equipment must be on board. Within the Arctic Archipelago, fuel must be sufficient for 500 miles range, plus 45 minutes reserve.

Seaplanes

A seaplane, or floatplane, behaves in the air much like a landplane, being capable of the same normal maneuvers. On the water, the techniques for handling and maneuvering a seaplane are very different from those needed for handling an airplane on land. The pilot needs to understand seamanship. A seaplane has no brakes and is affected by both wind and currents. A seaplane always tends to turn into wind. Maneuvering is therefore difficult especially in a strong wind.

Because of the weight of the floats, the useful load of a seaplane is normally less than that of the same airplane on wheels. One might assume that the increased drag produced by the floats also means a reduced rate of climb and cruising speed. In fact, seaplanes are usually fitted with fine pitch propellers or with constant speed propellers with extra fine pitch stops that produce the extra power needed to take off from a water surface. The high rpm of the fine pitch propeller gives an increased rate of climb overcoming the drag of the floats. Except in a few odd cases, the drag of the floats does results in a reduced cruising speed.

Fig. 35. Seaplane.

From the Ground Up

The basic techniques of flying a seaplane, in terms of take-off, landing, taxiing, sailing and general seamanship, mooring and docking, are beyond the scope of this manual. A few pertinent points only will be mentioned.

The pre-flight inspection of a seaplane involves the same items as a landplane with one important addition. The float compartments must be inspected for water and pumped with a bilge pump, if necessary. Each float has 6 to 10 compartments. Each one must be inspected to be sure it is dry. Be sure the hose in each compartment is connected. Float fittings, struts, attachments and cables and the water rudders must be inspected. Water rudders are essential to operation on the water. Be sure they operate through their full range of travel.

In cold weather when air temperatures are below freezing, the water clinging to the water rudders may freeze after lift off. It is a good precaution to play with the rudders for a little while after lift off until the water is blown off and the chance of the water rudders freezing is minimized.

Turbulence. Temperature variations between large areas of rock and water in the north country cause turbulence more pronounced than in more settled areas. Shore-line bumps can be sudden and violent. A seaplane approach over a shore-line should be made in a straight line and with a margin of surplus speed.

Shore-line bumps also present hazards to take-off. In a strong wind, a very severe down-current of air may be encountered in taking off towards a steep shore-line. Always allow yourself plenty of room when taking off towards a steep rising shore.

Glassy Water. One of the most difficult conditions a bush pilot has to contend with is glassy water. When water is glassy, its surface is practically invisible from above.

1. Try to land alongside a weed bed if one exists within reasonable distance of the landing area. If necessary, land in the weeds. Wild rice is not heavy enough to turn a seaplane over.

2. Land close to a shore-line, using the shore-line as a guide to the height of the surface of the water. Before doing so, fly over the shore-line at low altitude to look for reefs and shoals.

3. Make a power-on approach using the elevators to hold the nose above the horizon and to maintain a constant airspeed. Adjust the throttle to establish a shallow rate of descent. Fly the airplane onto the water in this nose high attitude. On making contact with the water, gradually close the throttle and hold the control column right back in order to slow the aircraft down as soon as possible.

This method of approach and landing on glassy water should be practised under normal landing conditions at every possible opportunity. If caught with a dead engine over glassy water and beyond gliding distance of a shore-line or weed bed, consider throwing out loose cushions or other items that will float on the surface and provide a visual clue as to the height of the aircraft above the water.

Rough Water. Do not attempt step landings on rough water. Land slow. If gusty weather has necessitated coming in fast, hold off until all surplus speed has been lost. Waves, like sheep, travel in flocks. In very rough water, comparative calm patches can sometimes be found by looking ahead.

Entering a Strange Area. When landing in a strange area, circle the vicinity at low altitude several times to look for rocks, reefs or floating timber. If you are not sure of their location when about to take off, taxi slowly downwind over the area you intend to use on take-off.

The Wind Direction for landing on water can be observed by noting the calm space which always exists on the lee side of the shore line over which the wind is blowing. Boats lying at anchor point into the wind. Smoke and flags are also good wind indicators.

Taxiing. Taxi either very slowly or very fast (up on the step). It is important to produce as little spray as possible. Any speed faster than a very slow taxi will produce spray that will strike the propeller, causing erosion of the blades and the possibility of serious damage. There is also the possibility of overheating the engine as there is not much forward speed to provide engine cooling. Apply throttle positively and quickly to get the seaplane on the step. Once on the step, the spray line is behind the propeller.

Do not approach a dock or rocky shore downwind in a high wind head-on. Allow the airplane to weathercock and sail it in tail first, using bursts of engine if necessary to steer it.

Taking Off. Special techniques are sometimes necessary when taking off from a very restricted area. One method is to get the plane up on the step when taxiing downwind and then to skid around into wind in a wide half circle. This maneuver is more applicable to flying boats than seaplanes because of their lower centre of gravity. This procedure must be carried out with caution, however. If you turn too abruptly, the combination of centrifugal force in the turn and the wind striking the airplane broadside may be sufficient to overturn the airplane.

A take-off from glassy water can be assisted by taxiing over the area to make ripples or by having a boat do so.

Immediately after taking off in very cold weather, move the controls to prevent spray from freezing on the control hinges.

Skiplanes

Ski equipped airplanes offer an almost limitless potential in the wintertime in a country like Canada. Any flat open area, any frozen lake or river is a landing place. Although skis are aerodynamically dirty, they do not affect the performance of the airplane significantly. Some care must be taken to avoid bare spots while taxiing, landing and taking off to prevent wearing the bottom coating of the skis.

Maneuvering a skiplane on the ground requires special precautions. A skiplane has no brakes. Rolling friction alone brings it to a stop after landing. Taxiing turns can be a problem especially in high winds. In some cases, help from persons outside the airplane may be required to keep directional control. Steerable tail skis are usually fitted but, in some wind and snow conditions, their steering effectiveness is considerably diminished.

Fig. 36. Skiplane.

Snow conditions affect greatly the performance of a skiplane. Even a slight rise or fall in the ambient temperature changes the texture of the snow, making the difference between a surface over which the skis slide freely and a surface that clogs up the bottom of the ski and impedes its progress.

Skiplanes should land and take off directly into wind whenever possible since the ski assembly cannot cope with the side loading in a cross wind drift situation.

White-out. An unbroken snow surface with an overcast condition is sometimes as difficult to judge as glassy water. This condition is known as white-out. Although visibility may not be markedly diminished, depth perception and distance judgment are lost due to loss of horizon and gradient. A condition in which there is blowing snow for a height of three or four feet above the surface produces another kind of white-out that is similar to ground fog. In this case, visibility is drastically reduced. Continued flight into white-out conditions is possible and safe only if the pilot recognizes the phenomenon and changes from visual to instrument flight. When landing in white-out conditions and if in doubt as to your approach height above the surface, fly the airplane on, using the same method as detailed above for glassy water.

Soft Snow. If snow conditions are soft, sweep a runway, or make one by taxiing up and down with the airplane lightly loaded before attempting to take off with a full load.

When taxiing in loose snow, make good wide turns to avoid too much torsional strain on the undercarriage.

If the snow is sticky, oil the skis before taking off. One simple method of doing this is to freeze a couple of bags to the ice. Saturate these with coal oil and taxi over them when about to take off.

Ice on Wings. Make sure that all surfaces are free from moisture, frost, ice or snow before attempting to take off. Even a light coating of frost can destroy the lift of the wings sufficiently to prevent a take-off.

Picketing in the Open. Skiplanes can be picketed by freezing the ends of the wing and tail mooring lines into the ice. (Dig a channel in the ice through which the lines can be passed and frozen in.)

When picketing overnight, block the skis up off the ice or snow, lay green garbage bags under the skis or place a mat of evergreen boughs under them. Skis become warm from taxiing, may melt the ice or snow underneath and, by morning, become frozen in.

If the ice has become soft and honeycombed, it may be necessary to move the airplane several times during the night. The weight of the airplane generates heat and it may start to thaw its way through the soft ice.

Landing on Glare Ice. Unless the airplane is fitted with a steerable tail ski or some other form of anti ground-looping device, allow yourself as much room as possible in all directions when landing on glare ice.

Wet and Slushy Snow is similar to mud. With this condition, therefore, always land with the tail well down. In the spring when there is some doubt as to ice conditions on a lake, land out in the middle. The ice is always thickest there (unless there is a current).

Equipment

The following minimum equipment should be carried by aircraft operating over remote areas in bush country:

Emergency survival equipment is outlined in Section **Sparsely Settled Areas** in Chapter **Aeronautical Rules and Facilities.**

All crew and passengers should be dressed in or have available warm clothing to enable them to survive in the temperatures to which they would be exposed in the event of an emergency landing.

Seaplanes

1 small lightweight folding anchor.

2 wing lines (about 30' long).

1 line (about 50' long) for anchoring, or for tying the tail to a tree or rock when beached on a shore.

1 paddle.

1 bilge pump (for pumping out the floats or hull).

1 engine cover.

1 float repair kit.

Life jackets for all on board. Some seaplane pilots are lured into a false sense of security because they have equipped themselves and their passengers with life jackets. They reason that if an accident should happen during take-off or landing, everyone in the airplane would be able to swim to shore with the aid of the life jackets. However, it must be remembered that the water in thousands of the Canadian lakes that are frequently used by float planes is so cold that human beings could not survive in it for longer than a few minutes before succumbing to death by exposure.

A gallon pail is not a necessity but comes in handy. It can be used to store some of the emergency kit, for carrying water, as a drogue (sea anchor) or as an anchor (by filling it with stones).

Skiplanes

1 snow shovel, ice chisel, snow knife.

2 wing lines (about 15' is ample).

1 line (about 50' long).

1 engine heating tent. This fits completely over the nose and reaches to the ground. It is designed to house a stove for preheating the engine in sub zero weather which is too severe for oil dilution to be effective (below -30°C.).

1 blowpot, or some other form of stove, for preheating the engine and lubricating oil.

1 scraper approved for use on airplanes. Avoid using a scraper that will scratch the plexiglass windscreens.

1 set of wing covers.

A tent. (If engine or wing covers are carried, they may be used to improvise a tent.)

Small hydraulic jack (for jacking up skis).

All Bush Aircraft

1 metal fuel funnel fitted with a water separating filter.

1 oil can.

Engine and airframe tools and spares sufficient to effect at least minor repairs.

Most bush pilots carry a small portable semi-rotary pump for refuelling out of gas drums.

Ultra-Lights

In many ways, the ultra-light movement today is very reminiscent of the original flying machines of the Wright Brothers and of the famed Silver Dart designed by Alexander Graham Bell and first flown in Canada in 1909. Like those first aircraft, the first generation of ultra-light airplanes had single surface wings with exposed tubing and bracing cables that held everything together. The pilot sat in an unenclosed seat, feeling the rush of wind around him. There are still some ultra-lights of this design manufactured and flying today but the new generation of airplanes in the ultra-light category has advanced to utilize wings of conventional airfoil section and enclosed cabins. In fact, many of them look very much like the early models in the lines of famous airplane manufacturing firms.

The origins of the ultra-light airplane can be traced directly to the hang glider movement. Someone, experimenting to achieve greater flexibility and control, attached an engine to a hang glider and gave birth to the first ultra-light airplane.

Fig. 37. Ultra-light.

The simplest and earliest models of this new breed of flying machine were controlled by pilot weight shift. They had no landing gear.

The pilot launched his machine by running over the ground fast enough to achieve flying speed. He landed it the same way.

Ultra-lights, for the most part, no longer employ weight shift as a means of control and all have advanced their design to include shock absorbing wheel landing gear, either in the tricycle configuration or a main gear with a tail wheel. They feature independent three axis aerodynamic controls, that is, ailerons, elevators and rudder, that are operated by a conventional stick control and rudder pedals.

The joy of the ultra-light airplane is the slow speed at which it flies. Many ultra-lights have stall speeds as low as 15 to 18 knots. As a result, landing speeds are low and the runway needed for take-off and landing is quite short. They also cruise at relatively slow speeds. Because ultra-light airplanes have the capability for sustained flight and would therefore be apt to stray into controlled airspace, certain regulations as to their operation have been formulated by Transport Canada.

An ultra-light single place airplane is defined as a power-driven heavier-than-air aircraft, designed to carry not more than one person, having a launch weight not exceeding 165 kg (363 lbs) and a wing area, expressed in square metres, of not less than the launch weight minus 15 divided by 10 and in no case less than 10 square metres.

An ultra-light two place instructional airplane is defined as a power-driven heavier-than-air aircraft, designed to carry not more than two persons, having a launch weight not exceeding 195 kg (429 lbs) and a wing area, expressed in square metres, of not less than 10 square metres and a wing loading of not greater than 25 kg/m2 calculated using the launch weight plus the occupant weight of 80 kg per person.

The **launch weight** of an ultra-light airplane is the total weight when it is ready for flight including any equipment, instruments and the maximum quantity of fuel and oil that it is designed to carry, but not including any float equipment to a maximum weight of 34 kg, the weight of the occupant(s) and the weight of any ballistic parachute installation.

The **wing loading** is the total of the launch weight plus the weight of the occupant(s) divided by the total wing area.

At the time of publication of this edition, there was a consultative process taking place at Transport Canada that was intended to bring about some changes in the definitions of launch weight and gross weight of ultra-lights and advanced ultra-lights. The new regulations are intended to streamline the whole spectrum of ultra-light operation. (Refer to the **Ultra-light Aeroplane and Hang Glider Information Manual** for the latest information on these aspects of ultra-light operation.)

Ultra-light airplanes must be registered, must be issued with a certificate of registration and must have their registration marks painted or attached in some permanent fashion on the airplane. The registration marks, as with all Canadian aircraft, begin with the letter C to denote nationality. A four letter group beginning with the letter I follows. The nationality and registration, the name of the manufacturer, the date and place of manufacture, the model and serial number of the airplane must all be inscribed on a fireproof identification plate that is securely affixed in a prominent position on the airplane.

Anyone piloting an ultra-light airplane must be the holder of a private or commercial pilot licence or have been issued a private or commercial ultra-light pilot licence. An applicant for a Private Pilot Licence — Ultra-light Category will be issued with a licence only on satisfactory completion of an ultra-light ground school course and successful completion of a Department of Transport written examination. The exam requires knowledge of the Canadian Aviation Regulations, air traffic rules and procedures, Information Circulars and A.I.P. Canada Supplements relating to ultra-lights, as well as basic knowledge of weight and balance requirements. Flight training shall be conducted under the direction and supervision of the holder of a Commercial Pilot Licence — Ultra-light Category or a Flight Instructor Rating — Aeroplane Category. The applicant must demonstrate proficiency in piloting an ultra-light airplane. There are two categories of ultra-light pilot licence, the private pilot licence and the commercial pilot licence. In order to take instruction in piloting an ultra-light airplane, an individual must first be issued with a student pilot permit for ultra-light airplanes.

Passengers may not be carried in an ultra-light, other than for the purpose of providing dual instruction.

Ultra-lights may be flown only in accordance with visual flight rules and only during daylight hours.

They may not be operated within 5 nautical miles from the centre of any airport or within the control zone associated with any uncontrolled airport, unless prior permission has been obtained in writing or by two-way radio from the airport operator. To operate within the control zone associated with a controlled airport, it is necessary to obtain an air traffic control clearance by two-way radio voice communication from the air traffic control unit in the tower at that airport.

Human Factors

The preceding chapters of this manual have been devoted to the technical aspects of flying. The reader should now understand how lift is produced by an airfoil to make an airplane fly, the basic construction of an airframe, all about the operation and care of an aero engine, how to use aircraft communication and navigation radio equipment, how to navigate from A to B, the vagaries of weather, etc. etc.

When the first editions of this book were published and, indeed until not too many years ago, it was generally believed that if an individual had a good understanding of all these technical aspects of pilotage, he or she had acquired the basic prerequisites to be a successful, efficient and safe pilot.

In the last few years, however, it has been learned that a thorough grasp of these subjects, though essential, is not enough. Human factors are a very important part of flight crew training. Human aspects, such as cockpit organization, crew co-ordination, fitness and health, sensory illusions and decision making are as vital to safety in the air as are flying techniques. The relationship of people with machines, the environment and other people is part of the human factors equation.

There is much to understand about the pilot himself and his physical and involuntary reactions to the unnatural environmental conditions of flying. During the Second World War, it was first realized that some airplane losses were due to pilot incapacitation rather than to enemy action. The challenge of explaining these unusual occurrences was taken up and since that time much research has been conducted into such subjects as hypoxia, spatial disorientation, hyperventilation, the bends, impairment due to drugs and alcohol, and mental stress. Startling and sobering information is now available.

Man is essentially a terrestrial creature. His body is equipped to operate at greatest efficiency within relatively narrow limits of atmospheric pressure and, through years of habit, has adapted itself to movement on the ground.

In his quest for adventure and his desire for progress, man has ventured into a foreign environment, the air high above the ground. But these lofty heights are not natural to man. As altitude increases, the body becomes less and less efficient to a point, at sufficient altitude, of incapacitation and unconsciousness. Completely deprived of oxygen, the body dies in 8 minutes. Without ground reference, the senses can play tricks, sometimes fatal tricks.

Airplane accidents are an occurrence that every conscientious pilot is concerned with preventing. Most aircraft accidents are highly preventable. Many of them have one factor in common. They are precipitated by some human failing rather than by a mechanical malfunction. In fact, statistics indicate that human factors are involved in 85% of aircraft accidents. Many of these have been the result of disorientation, physical incapacitation and even the death of the pilot during the flight. Others are the result of poor management of cockpit resources.

It is the intention of this chapter to explain briefly some of these human factors to help pilots understand and appreciate the capacities and limitations of their own bodies so that flying might never be a frightening or dangerous undertaking but instead the enjoyable and safe and efficient experience all lovers of airplanes and the airways have always believed it to be.

General Health
Since flying an airplane demands that the pilot be alert and in full command of his abilities and reasoning, it is only common sense to expect that an individual will ensure that he is free of any conditions that would be detrimental to his alertness, his ability to make correct decisions, and his rapid reaction times before seating himself behind the wheel of an airplane.

Certain physical conditions such as serious heart trouble, epilepsy, uncontrolled diabetes and other medical problems that might cause sudden incapacitation and serious forms of psychiatric illness associated with loss of insight or contact with reality may preclude an individual from being judged medically fit to apply for a licence.

Other problems such as acute infections are temporarily disqualifying and will not affect the status of a pilot's licence. But they will affect his immediate ability to fly and he should seek his doctor's advice before returning to the cockpit of his airplane.

In fact, any general discomfort, whether due to colds, indigestion, nausea, worry, lack of sleep or any other bodily weakness, is not conducive to safe flying. Excessive fatigue is perhaps the most insidious of these conditions, resulting in inattentiveness, slow reactions and confused mental processes. Excessive fatigue should be considered a reason for cancelling or postponing a flight.

Hypoxia
The advance in aeronautical engineering in recent years has produced more versatile airplanes capable of flying at very high altitudes. At such high altitudes, man is susceptible to one of the most insidious physiological problems, **hypoxia.** Because hypoxia comes on without warning of any kind, supplementary oxygen must be available in any aircraft that will be flown above 10,000 feet. The general rule of oxygen above 10,000' ASL by day and above 5000' ASL by night is one the wise pilot will practise to avoid the hazard of this debilitating condition. Hypoxia can be defined as a lack of sufficient oxygen in the body cells or tissues.

The greatest concentration of air molecules is near to the earth's surface. There is progressively less air and therefore less oxygen (per unit volume) as you ascend to higher altitudes. Therefore each breath of air that you breathe at, for example, 15,000 feet ASL has about half the amount of oxygen of a breath taken at sea level.

The most important fact to remember about hypoxia is that the individual is unaware that he is exhibiting symptoms of this condition. The brain centre that would warn him of decreasing efficiency is the first to be affected and the pilot enjoys a misguided sense of well-being. Neither is there any pain nor any other warning signs that tell him that his alertness is deteriorating. The effects of hypoxia progress from euphoria (feeling of well-being) to reduced vision, confusion, inability to concentrate, impaired judgment and slowed reflexes to eventual loss of consciousness.

There are four types of hypoxia: **Hypoxic hypoxia** is a normal effect of altitude and is avoided by the use of on board oxygen systems. **Anaemic hypoxia** is caused by an over abundance of carbon monoxide in the haemoglobin (see **Carbon Monoxide** below). **Stagnant hypoxia** is a condition in which the brain is deprived of an adequate blood supply. **Histotoxic hypoxia** is caused by chemical poisoning and by high blood alcohol (See **Alcohol** below).

Effects on Vision at 5000 Feet
The retina of the eye is actually an outcropping of the brain and as such is more dependent on an adequate supply of oxygen than any other part of the body. For this reason, the first evidence of hypoxia occurs at 5000 feet in the form of diminished night vision. Instruments and maps are misread; dimly lit ground features are misinterpreted.

Above 10,000 Feet
It is true that general physical fitness has some bearing on the exact

altitude at which the effects of hypoxia will first affect a particular individual. Age, drinking habits, use of drugs, lack of rest, etc., all increase the susceptibility of the body to this condition. However, the average has been determined at 10,000 feet.

At 10,000 feet, there is a definite but undetectable hypoxia. This altitude is the highest level at which a pilot should consider himself efficient in judgment and ability. However, continuous operation even at this altitude for periods of more than, say, four hours can produce fatigue because of the reduced oxygen supply and a pilot should expect deterioration in concentration, problem solving and efficiency.

At 14,000 feet, lassitude and indifference are appreciable. There is dimming of vision, tremor of hands, clouding of thought and memory and errors in judgment. Cyanosis (blue discolouring of the fingernails) is first noticed.

At 16,000 feet, a pilot becomes disoriented, is belligerent or euphoric and completely lacking in rational judgment. Control of the airplane can be easily lost.

At 18,000 feet, primary shock sets in and the individual loses consciousness within minutes.

At higher altitudes, death may result after a prolonged period.

The Canadian Aviation Regulations rule that an aircraft should not be operated for more than 30 minutes between 10,000 feet and 13,000 feet or at all above 13,000 feet unless oxygen is readily available for each crew member.

Prevention of Hypoxic Hypoxia

The only way to prevent hypoxia is to take steps against it before its onset. Remember the rule: Oxygen above 10,000 feet by day and above 5,000 feet at night.

Stagnant Hypoxia

Stagnant hypoxia is a condition in which there is a temporary displacement of blood in the head. It occurs as a result of positive g forces and can be attributed to the fact that the circulatory system is unable to keep blood pumped to the head.

G is the symbol for the rate of change of velocity and it represents both a force and a direction. Positive g involves a force that acts from the feet to the head; negative g from the head to the feet.

G tolerance is lowered by ill health, low blood pressure, obesity, fatigue, smoking, hypoxic hypoxia, hangovers. It is affected by the peak value of the force imposed, the duration and the rate of onset.

Positive g is experienced in an abrupt pull-out from a high speed dive, during an inside loop and in a co-ordinated turn.

The first symptom of stagnant hypoxia is deterioration in vision. The pilot experiences **grey-out,** a condition in which the vision becomes dim and colourless. As the g force increases, **black-out** (temporary loss of vision) occurs as the blood supply to the eyes is completely interrupted. As the g forces are further increased, consciousness is lost.

Negative g forces increase the blood pressure in the eyes and cause **red-out.** Negative g in the excess of -5g can cause rupture of the small blood vessels in the eyes. Negative g is experienced in pushovers and outside loops.

G tolerance is affected by diet, requiring adequate hydration and normal blood sugar. Good physical condition also is important: weight bearing programs are more effective than aerobic training.

Ozone Sickness

Another problem associated with flight at very high altitudes is ozone sickness. Although it has been evident only with flights operating at altitudes of 30,000 feet or more, the advent of general aviation airplanes that operate at subsonic speeds at such levels makes this a problem of which even the private pilot should be aware.

Ozone is a bluish gas that exists in relatively high concentrations in the upper levels of the atmosphere, especially in the tropopause. Because the tropopause fluctuates in its average altitude from season to season, any flight operating above 35,000 feet is likely to come into contact with ozone at some time.

Although ozone does have a distinctive colour and odour, passengers and flight crew who have experienced ozone sickness have been unaware of the apparently high concentrations of ozone prior to the onset of the symptoms.

The symptoms of ozone sickness are hacking cough, poor night vision, shortness of breath, headache, burning eyes, mouth and nose, mild chest pains, leg cramps, fatigue, drowsiness, nose bleed, nausea and vomiting. The symptoms become more severe with continued exposure and with physical activity but do diminish rapidly when the airplane descends below 30,000 feet.

Some relief from the symptoms can be achieved by breathing through a warm, moist towel. Limiting physical activity to a minimum and breathing pure oxygen are also effective in alleviating the symptoms.

Carbon Monoxide

Oxygen is transported throughout the body by combining with the haemoglobin in the blood. However, this vital transportation agent, haemoglobin, has 210 times the affinity for carbon monoxide that it has for oxygen. Therefore, even the smallest amounts of carbon monoxide can seriously interfere with the distribution of oxygen and produce a type of hypoxia, known as **anaemic hypoxia.**

Carbon monoxide is colourless, odourless and tasteless. It is a product of fuel combustion and is found in varying amounts in the exhaust from airplane engines. A defect, crack or hole in the cabin heating system may allow this gas to enter the cockpit of the airplane.

Susceptibility to carbon monoxide increases with altitude. At higher altitudes, the body has difficulty getting enough oxygen because of decreased pressure. The additional problem of carbon monoxide could make the situation critical.

Early symptoms of CO poisoning are feelings of sluggishness and warmness. Intense headache, throbbing in the temples, ringing in the ears, dizziness and dimming of vision follow as exposure increases. Eventually vomiting, convulsions, coma and death result.

Although CO poisoning is a type of hypoxia, it is unlike altitude hypoxia in that it is not immediately remedied by the use of oxygen or by descent to lower altitudes.

If you notice exhaust fumes or experience any of the symptoms associated with CO poisoning, you should shut off the cabin heater, open a fresh air source immediately, avoid smoking, use 100% oxygen if it is available and land at the first opportunity and ensure that all effects of CO are gone before continuing the flight. It may take several days to rid the body of carbon monoxide. In some cases, it may be wise to consult a doctor.

Cigarettes

Cigarette smoke contains a minute amount of carbon monoxide. It has been estimated that a heavy smoker will lower his ceiling by more than 4000 feet. Just 3 cigarettes smoked at sea level will raise the physiological altitude to 8000 feet. Haemoglobin has great affinity to the carbon monoxide in the cigarette smoke and absorbs it readily. Its oxygen absorbing qualities are reduced to about the same degree as they would be reduced by the decrease in atmospheric pressure at 8000 feet ASL.

The carbon monoxide from cigarettes has detrimental effects not only on the smoker but on the non-smoker as well. After prolonged exposure to an increased level of carbon monoxide in a confined area such as a cockpit, crew or passengers will experience symptoms such as respiratory discomfort, headaches, eye irritation from the "second hand" smoke.

Cigarette smoking has also been declared as hazardous to health, contributing to hypertension and chronic lung disorders such as bronchitis and emphysema. It has been linked to lung cancer and coronary heart disease.

Hyperventilation

Hyperventilation, or overbreathing, is an increase in respiration that upsets the natural balance of oxygen and carbon dioxide in the system, usually as a result of emotional tension or anxiety. Under conditions of emotional stress, fright or pain, a person may unconsciously increase his rate of breathing, thus expelling more carbon dioxide than is being produced by muscular activity. The result is a deficiency of carbon dioxide in the blood.

The most common symptoms are dizziness, tingling of the toes and fingers, hot and cold sensations, nausea and sleepiness. Unconsciousness may result if the breathing rate is not regulated.

The remedy for hyperventilation is a conscious effort to slow down the rate of breathing and to hold the breath intermittently to allow the carbon dioxide to build up to a normal level. Sometimes, the proper balance of carbon dioxide can be more quickly restored by breathing into a paper bag, that is, by re-breathing the expelled carbon dioxide.

The early symptoms of hyperventilation and hypoxia are similar and may be confused. In fact, both conditions can occur at the same time. A pilot, flying at high altitude, may think that he can counteract the effects of hypoxia by taking more rapid breaths. Hyperventilation does not help you get more oxygen. It only increases the emission of carbon dioxide. Hypoxia is unlikely to occur below 8000 feet ASL. Above 8000 feet, if oxygen is available, take three or four deep breaths of 100% oxygen. If the symptoms persist, the problem is hyperventilation and should be treated as such.

Decompression Sickness
Trapped Gases

Because of the change in barometric pressure during ascent and descent, gases trapped in certain body cavities expand or contract.

The inability to pass this gas may cause abdominal pain, toothache or pain in ears or sinuses cavities. In some cases, the pain may be so severe as to lead to incapacitation. The conditions caused by changing barometric pressure are known as **dysbarisms**: any physical damage that results is called a **barotrauma.**

Ear Block

The ear is composed of three sections. The outer ear is the auditory canal and ends at the eardrum. The middle ear is a cavity surrounded by bones of the skull. It houses the organs of hearing and is filled with air. The eustachian tube connects the middle ear to the throat.

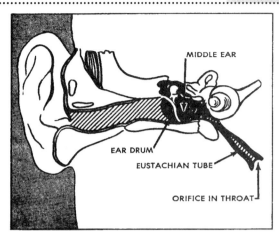

Fig.1. Ear Structure and Eustachian Tube.

The inner ear controls certain equilibrium senses and contains the cochlea, a small organ that analyzes sound vibrations.

During ascents and descents, air must escape or be replenished through the eustachian tube to equalize the pressure in the middle ear cavity with that of the atmosphere. If air is trapped in the middle ear, the eardrum stretches to absorb the higher pressure. The result is pain and sometimes temporary deafness. Eardrum rupture is even possible.

During climbs, there is little problem since excess air escapes through the tube easily. However, during descents, when pressure in the middle ear must be increased, the eustachian tubes do not open readily. The situation is aggravated if the individual has a cold, an allergy or an infected throat. Pilot and passengers must consciously make an effort to swallow or yawn to stimulate the muscular action of the tubes. Sometimes it is advisable to use the **valsalva technique,** that is, to close the mouth, hold the nose and blow gently. This action forces air up the eustachian tubes. Children may suffer severe pain because of ear blocks during descents. They should be repeatedly reminded to swallow or yawn. Small babies are incapable of voluntarily adjusting the pressure in the middle ear and should be given a bottle to suck during descents.

Painful ear block generally occurs as a result of too rapid descent. If the pilot or the passengers are unable to relieve the pain of ear block by the methods described, it may be necessary to climb to altitude again and make the descent more gradually.

After a flight in which 100 per cent oxygen has been used, the valsalva procedure should be used several times to ventilate the middle ear and thus reduce the possibility of pain occurring later in the day.

Sinus Block

The sinuses are air filled, bony cavities connected with the nose by means of one or more small openings. If these openings are obstructed by swelling of the mucous membrane lining of the sinuses (as during a cold), equalization of the pressure is difficult. Pain, in the cheekbones on either side of the nose, in the upper jaw or above the eyes, will result. The valsalva procedure will relieve sinus pain.

For both ear and sinus block, the prudent use of nasal inhalants may be helpful. A nasal inhalant containing antihistamine, however, should not be used for the reasons stated in the section on drugs below.

Toothaches

Toothaches may occur at altitude due to abscesses, imperfect fillings, inadequately filled root canals. Anyone who suffers from toothache at altitude should see a dentist. However, the pain caused by a sinus block can be mistaken for toothache.

Gastrointestinal Pain

Gas pains are caused by the expansion of gas within the digestive tract during ascent into the reduced pressure at altitude. Relief from pain may be accomplished by descent from altitude.

Certain foods, such as beans, spicy food, carbonated beverages, are known gas producers and should be avoided by individuals who know from personal experience that they can cause a problem.

The Common Cold

Don't fly with a cold. A cold that is a mere discomfort on the ground can become a serious menace to pilots and passengers in the air.

Tiredness, irritability, drowsiness and pain are all symptoms of a cold and work together to make a pilot unsafe in the air. More insidious, however, is the effect a cold may have on the sinuses and on the middle and inner ear. Swollen lymph tissue and mucous membranes may block the sinuses causing disabling pain and pressure vertigo during descent from altitude. Infection of the inner ear, that is a common symptom of a cold, can also produce severe vertigo. The tissue around the nasal end of the eustachian tube will quite likely be swollen and middle ear problems associated, under normal conditions (see above), with descent from altitude will be severely aggravated. A perforated eardrum is a possible result. Although a perforated eardrum usually heals quickly, in some cases there is permanent hearing impairment or prolonged infection of the middle ear.

Cold remedies do not prevent symptoms. They usually only bring on other problems, drowsiness being the most common.

Evolved Gases

Nitrogen, always present in body fluids, comes out of solution and forms bubbles as barometric pressure decreases in ascents to higher altitudes. If the bubbles lodge in small blood vessels, they can cause circulation blockage, pain and tissue damage. Certain conditions predispose to the development of **decompression sickness.** It is more common in the elderly and in females and in people who are active. Obese, overweight persons are more susceptible to it as fatty tissue contains more nitrogen.

Bends or cramps are characterized by pain in and around the joints and can become progressively worse during ascent to higher altitudes. The symptoms occur most commonly in large joints, such as the knee or shoulder, but may also occur in joints that have been previously injured. Exercise or movement will make the pain worse. The symptoms can usually be reduced by descent to a lower altitude.

Chokes are pains in the chest caused by blocking of the smaller pulmonary blood vessels by innumerable small bubbles. In severe cases, there is a sensation of suffocation. This is a serious condition and an immediate descent to lower altitudes must be made.

Paraesthesia or Creeps is another decompression sickness caused by tiny gas bubbles under the skin. Symptoms include tingling, itching, cold and warm sensations.

Central nervous system disturbances include visual disturbances, headache and, more rarely, paralysis and sensory disturbances.

Decompression sickness is unpredictable. One of the outcomes may be shock, characterized by faintness, dizziness, nausea, pallor, sweating and even loss of consciousness. Collapse is possible if larger bubbles interfere with the blood supply to the brain or the spinal cord. Usually the symptoms disappear when a return to the ground is made. However, the symptoms may continue and special treatment (**recompression**) may be needed.

Decompression sickness, caused by evolved gas, is rare below 20,000 feet. The best defence against this painful problem is a pressurized cabin. Some protection against it can be achieved by breathing 100% oxygen for an hour before ascending to altitudes above 20,000 feet. This action washes the nitrogen out of the blood. Oxygen does not come out of solution or form bubbles. Refrain also from drinking carbonated beverages or eating gas producing foods.

Scuba Diving and Flying

A person that flies in an airplane immediately after engaging in the sport of scuba diving risks severe decompression sickness at much lower altitudes than this problem would normally be expected. The breather tanks, used by a scuba diver to counteract the greater pressure of the water on the body, deliver compressed air that supersaturates the body tissues with nitrogen even in a shallow dive. At a depth of 30 feet, the body absorbs twice as much nitrogen as it would on the ground. Ascending to 8000 feet ASL could bring on incapacitating bends.

After non-decompression dives, flights up to 8000 feet ASL should be avoided for 12 hours. On dives where decompression stops were required during the ascent to the surface, the interval should be 24 hours. Regardless of the type of dive, flights above 8000 feet ASL should avoided for at least than 24 hours even in pressurized aircraft since even pressurized aircraft may lose pressurization. A good rule, if you have dived to a depth below 30 feet, is not to fly for 24 hours to permit the nitrogen content of the body to return to normal.

Vision

Good vision is of primary importance in flying, in judgment of distance, depth perception, reading of maps and instruments and should, therefore, be scrupulously protected.

Pilots are exposed to higher light levels than is the average person. Very high light levels prevail at altitude because the atmosphere is less dense. In addition, light is reflected back at the pilot by cloud tops. This light contains more of the damaging blue and ultra-violet wavelengths than are encountered on the surface of the earth. Prolonged exposure can cause damage to the eye and especially to the lens. Sunglasses should, therefore, be worn to provide protection against these dangers and to prevent eyestrain.

Instrument panels should be dull grey or black to harmonize with the black instruments, so that the eye does not have to adjust its lens opening constantly as the line of vision moves from the dark instruments to a light coloured panel.

When flying into the sun, the eyes are so dazzled by the brightness that they cannot adjust quickly to the shaded instrument panel. This situation causes eyestrain and is fatiguing to the pilot. Good quality sunglasses, with high definition, coated against glare and ultraviolet light help to minimize the problem.

Atmospheric obscuring phenomena such as haze, smoke and fog have an effect on the distance the normal eye can see. The ability of the eye to maintain a distance focus is weakened. Distant objects are not outlined sharply against the horizon and, after a short lapse of time, the eye, having no distance point to fix on, has difficulty maintaining a focus at a distance of more than a mile or two (a condition known as **empty field myopia**). As a result, scanning for other aircraft becomes difficult and requires special effort on the part of the pilot. With the pilot's focal range reduced, the span of time in which to perceive the danger and take evasive action is considerably shortened. Pilots must learn to recognize the limitations of the human eye under varying weather conditions and realize that the see and avoid maxim has limitations under some atmospheric conditions.

A small percentage of the general population is colour blind. For the NORDO pilot who expects clearances by light signals, colour vision (especially red/green) is essential. Because modern cathode ray tube (CRT) instruments, weather radar and collision avoidance displays are introducing more complex colour displays, colour vision is important for pilots flying aircraft with this sophisticated equipment.

Depth Perception

Clues for accurate depth perception are often absent in the air. Clouds are of varying size and there is no way to estimate their distance. Landings on glassy water or on wet runways are a problem as is the condition known as white out that occurs in blowing snow and other winter situations.

Night Vision

At night, the pilot's vision is greatly impaired. The cones that are concentrated in the centre of the lens need a lot of light to function properly. As a result, there is a blind spot in the centre of the eye at night. This blind spot is sufficiently large to block out the view of another airplane some distance away if the pilot is looking directly at it.

At night, it is necessary to develop the technique of using peripheral vision. One sees at night by means of the rods that are concentrated on the edges of the lens and are responsible for peripheral vision. It takes the rods about 30 minutes to adjust fully to darkness. Even a small amount of white light will destroy the dark adaptation.

You should wear sunglasses during the day but remove them after sunset. Avoid looking at bright lights when you propose to undertake a night flight. Do everything possible to adapt your eyes to the darkness prior to your flight. Wearing red goggles for 30 minutes prior to a night flight helps the eyes adapt to darkness.

Force your eyes to view off centre. To see at your best, you must look off centre at about 15 to 20 degrees to be in the area of maximum rod density for both eyes. With good off-centre vision, faint distant objects, like the dim lights on a landing strip or another aircraft, will be easily seen.

Don't use yellow or pink highlighting on flight charts. These colours do not show up at night with the red cockpit lighting on. Use blue-high lighting instead. Always carry a quality flashlight with a red night vision filter and a spare set of batteries and a spare bulb. Ensure that they are within easy and immediate reach.

Night vision is also sensitive to hypoxia. Supplementary oxygen should be used above 5000 feet to avoid depriving the eye of oxygen. Dirt and reflection on the windshield cause confusion at night. A very clean windshield is important.

Night Lighting of Instruments

Lighting of instruments is a problem in that the instruments must be well enough lit to be readable without the light destroying the pilot's dark adaptation.

Ultraviolet flood lighting of fluorescent instrument marking is probably the least satisfactory. The instruments are marked with fluorescent paint that shows up under fluorescent lighting as a bluish green colour. The disadvantages are that the instruments can't be kept in focus, dark adaptation may be lost, eyes are irritated, vision becomes foggy.

Red lights. Lighting of instruments by indirect individual red lights is unsatisfactory because uniform light distribution over all parts of the instrument cannot be achieved. There is no illumination of knobs and switches. Red flood lighting of the whole instrument panel is more satisfactory. However, the ability to distinguish colours one from another is lost. Coloration of maps is indecipherable and information printed in red becomes unreadable.

White lights. Low density white light is considered the best cockpit lighting system. The instruments can be clearly read and colours recognized. Because the low density white light can be regulated, dark adaptation is not destroyed although it is somewhat impaired.

Thunderstorms

It is not advisable to fly an airplane through or near thunderstorms. The blinding flashes destroy night adaptation. Turn the cockpit lights full bright if you are in the vicinity of lightning activity in order to prevent lightning blindness.

Anti Collision Lights

When flying in the clouds, strobe lights and rotating beacons should be turned off as the reflection off the cloud of the blinking light is irritating to the eye.

Design Eye Reference Point (DERP)

In every airplane, there is a position at which the pilot, when seated, is assured of an optimum vision zone. The **design eye reference point (DERP)** is a certain height above the cockpit floor and a certain distance from a datum line. External visibility is affected by windshield size, posts, glare shield and the fore/aft position of the seat, the latter being the only variable in this list.

Any change in seat position will alter the external field of view. If, for example, a pilot has the seat positioned so that his eyes are at the DERP and then moves the seat rearward and downward, he will reduce his view both above and below the nose of the aircraft. His vision lines inside the cockpit will also be altered. Any change, therefore, from the DERP is a compromise between external and internal vision and one's ability to manipulate the controls.

Aircraft are designed to meet certain cockpit visibility requirements established by the aviation authorities in the country of manufacture. Some manufacturers provide reference points that the pilot is to use in making seat adjustments. These reference points may be something as simple as two balls affixed to the glare shield which the pilot must line up visually. However, many manufacturers do not provide guidance to pilots as to the DERP for their aircraft. As a result, through ignorance, pilots risk restricting their external field of view and jeopardize the safe operation of their aircraft. Determining the DERP for the aircraft you fly is a safety precaution it is wise to make. Write to the manufacturer and request the information.

The following guidelines should be considered in locating the correct seat placement (height as well as fore/aft position): (1) all flight controls must be capable of full travel and free of restriction; (2) flight instruments and warning lights must be visible to the pilot; (3) forward visibility over the nose should not be restricted; and (4) the seat position should be comfortable for the pilot.

Noise, Vibration and Temperature
Noise

Noise is both inconvenient and annoying. It produces headaches, visual and auditory fatigue, air sickness and general discomfort with an accompanying loss of efficiency. Even at levels which are not uncomfortable, noise has a fatiguing effect, especially when the pilot is exposed for a long period as on a lengthy cross country flight. To arrive at destination suffering from noise induced fatigue and have to make a landing under minimum conditions is clearly an undesirable situation.

Sound is measured in decibels (dB). The zero level is defined as the weakest sound that can be heard by a person with good hearing in a quiet location. The loudest sound that most people can bear is 140 dB. High levels cause pain or nausea. Noise levels approaching these levels should not be experienced without ear protection. In fact, ear protection should be used for continuous noise levels above 80 dB.

Little has been done to reduce and control noise in aircraft cockpits. Tests have measured the sound level in modern aircraft at 90 to 100 decibels. Noise levels in jets can approach 140 decibels.

With noise levels of this magnitude, hearing damage is a distinct problem unless some sort of hearing protection is used. Many pilots report temporary loss of hearing sensitivity after flights. Still others have reported an inability to understand radio transmissions from the ground, especially during take-off and climb when the engine is operating at full power. In fact, there is documented evidence to show that continued exposure to high levels of aircraft noise will result over the years in loss of hearing ability.

The detrimental effect of noise is not a sudden thing but builds up progressively over years of exposure. Pilots of helicopters and aerial application aircraft are particularly susceptible because of the relatively high levels of noise experienced in these cockpits and the long durations of exposure. But even pilots, who put in only three or four hours a week in their airplanes, have been found to have slightly impaired hearing after several years.

Everyone experiences some hearing deterioration as the process of growing old. Add this to a level of deafness caused by exposure to noise and it becomes obvious that a pilot reaching middle age could have a serious hearing deficiency.

Protective devices against noise are therefore important, first of all, in helping to reduce fatigue during individual flights and, secondly, in helping to minimize the possibility of hearing loss or deterioration in later years.

The best protection is a pair of properly fitting earplugs. They lower noise levels by as much as 20 to 30 decibels. The use of ear covering devices, such as earphones, can also help if they are tight fitting. If they fit poorly, they can be worse than nothing in that they give the wearer a false sense of security. The use of earplugs as well as earphones is recommended.

The wearing of earplugs does not impair ability to hear. In fact, speech intelligibility is improved because the earplugs filter out the very noises that interfere with voice transmissions.

The regular wearing of earplugs, especially by pilots but also by passengers, is therefore a good precautionary measure to ensure continued good hearing throughout a pilot's lifetime.

Vibration

The power plant of the airplane is the principal source of vibration. At subsonic speeds, this vibration is responsible for fatigue and irritability and can even cause chest and abdominal pains, backache, headaches, eyestrain and muscular tension. If the vibration happens to occur in the frequency of about 40 cycles per second, blurring of the eyes may occur. It is even possible to become hypnotized as a result of rhythmic and monotonous vibrations.

Vibration can rarely be completely controlled but it can be dampened by placing a barrier between the pilot and the source to reduce the effect. Dampening can also be accomplished by reducing the source of the vibration or by modifying the transmission pathway.

Temperature

At temperatures over 30°C, discomfort, irritability and loss of efficiency are pronounced. High temperatures also reduce the pilot's tolerance to mental and physical stresses, such as acceleration and hypoxia.

At cold temperatures, the immediate danger is frostbite. Continued exposure will result in reduced efficiency to the point where safe operation of the airplane is impossible.

Hypothermia

The most serious result of extended exposure to extremely cold temperatures is a condition known as hypothermia. Hypothermia is a lowering of the temperature of the body's inner core. It occurs when the amount of heat produced by the body is less than the amount being lost to the body's surroundings. As it progresses, vital organs and bodily systems begin to lose their ability to function. It is a condition that can develop quickly and may be fatal.

In the early stages, the skin becomes pale and waxy, fatigue and signs of weakness begin. As the body temperature drops farther, uncontrollable intense shivering and clumsiness occur. Mental confusion and apathy, drowsiness, slurred speech, slow and shallow breathing are the next stage. Unconsciousness and death follow rapidly.

Hypothermia certainly can attack a pilot in the airplane cockpit if there is no cabin heating system and if he is not adequately dressed to protect against very cold ambient temperatures. Usually, however, hypothermia is considered to be a danger to the pilot who has been forced down and is exposed to the elements. Cold, wetness, wind and inadequate preparation are the conditions which cause it. Wet clothing, caused by weather, immersion in water or condensed perspiration, acts like a wick and extracts body heat at a rate many times faster than would be the case with dry clothing. Immersion in cold water greatly accelerates the progress into hypothermia.

The best protection against this condition is adequate clothing, shelter, emergency rations and, above all, knowledge of the danger. Every wintertime flier should have a survival kit that includes a lightweight tent, plastic sheet, survival blanket, etc., that can be used to construct a shelter. Always wear (or take along as extras) proper clothing for the worst conditions you might encounter. Several layers of clothing are more effective than one bulky layer. Protect high heat loss areas, such as the head, neck, underarms, sides of the chest. Carry effective rain gear and put it on before you get wet. High energy foods that produce heat and energy should be included in the survival kit. Hot fluids help to keep body heat up. Guard against becoming tired and exhausted. A tired person, exposed to a cold, wet and windy environment, is a prime candidate for hypothermia.

Hyperthermia

Hyperthermia (or heat stroke) occurs when the body is unable to dissipate heat either through radiation or by sweating. If the body core temperature rises above 41°C, thermostatic control is mostly lost. People with heat stroke should be treated in hospital as soon as possible.

Sensory Illusions

Under normal conditions, the **kinaesthetic sensors** (skeletal muscles, bones and joints), **vision** (eyes) and the **vestibular (labyrinth) organs** (inner ears) provide the brain with information about the position of the body in relation to the ground. In flying, however, conditions are sometimes encountered which fool the senses.

The eyes are the prime orienting organs but are dependent on reference points in providing reliable spatial information. Objects seen from the air often look quite different than they do when seen from the ground. If the horizon is not visible, a pilot might choose some other line as reference, such as a sloping cloud bank. Fog and haze greatly affect judgment of distance. Lights on the ground at night are commonly confused for airplanes. Even stars can be confused with ground lights.

The tension of various muscles in the body assists in a small way in determining position. The body is accustomed to the pull of one g force acting in only one direction. In an airplane, if a second force is introduced as in acceleration, deceleration and turns and, if there is no outside visual reference, illusions may result. For example, in a bank, centrifugal force directed outward and the normal downward pull of gravity combine to give an illusion of level flight. Acceleration gives an illusion of climbing and deceleration of diving.

The three semicircular canals of the inner ear are primarily associated with equilibrium. They are filled with fluid and operate on the principle of the inertia of fluids. Each canal has tiny hair like sensors that relate to the brain the motion of the fluid. Rotation of the body tends to move the fluid, causing the displacement of the sensors which then transmit to the brain the message of the direction of their displacement. However, if the turn is a prolonged and constant one, the motion of the fluid catches up with the canal walls, the sensors are no longer bent and the brain receives the incorrect message that the turning has stopped. If the turn does then indeed stop, the movement of the fluid and the displacement of the sensors will indicate a turn in the opposite direction. Under instrument conditions or at night when visual references are at a minimum, incorrect information given by the inner ear can produce **vestibular illusions** that can be dangerous.

The following factors contribute to **visual illusions:** optical characteristics of windshields; rain on the windshield; effects of fog, haze, dust, etc. on depth perception; the angle of the glide slope makes a runway appear nearer or farther as does a very wide or very narrow runway; variations in runway lighting systems; runway slope and terrain slope; an approach over water to the runway; the apparent motion of a fixed light at night **(autokinetic phenomenon).** The visual cues by which a pilot makes judgments about the landing approach are largely removed if the approach is over water, over snow or other such featureless terrain or carried out at night. A particularly hazardous situation is created if circumstances prevent him from appreciating ground proximity before touchdown.

The following factors contribute to **kinaesthetic illusions:** change in acceleration or deceleration; low level flight over water; frequent transfer from instrument to visual flight conditions (choose either VFR or IFR and stick with the choice); unperceived changes in flight altitude.

There is just one way to beat false interpretation of motion. Put your faith in your instruments and not in your senses. **Refer to the attitude instruments constantly when flying at night or in reduced visibility conditions.** Always trust the attitude instruments no matter what your senses tell you.

Spatial Disorientation

Spatial disorientation means loss of bearings or confusion concerning one's sense of position or movement in relation to the surface of the earth. Disorientation rarely occurs without reduced visual references in such situations as fog, cloud, snow, rain, darkness, etc.

A type of spatial disorientation is caused in some individuals by flickering shadows. When, for example, letting down for a landing into the setting sun in a single engine airplane, the idling propeller can induce reactions that range from nausea to confusion and, in rare cases, complete unconsciousness. Other causes of this sensation are helicopter rotor blade shadows, the flashing illumination caused by anti-collision lights when flying in cloud, and runway approach strobe lights when viewed through the propeller at night.

The term **vertigo** is sometimes used in relation to spatial disorientation. Vertigo is a sensation of rotation or spinning, an hallucination of movement of either the individual himself or of the external world.

Coriolis Effect

Coriolis effect is probably the most dangerous type of disorientation. The three semicircular canals of the inner ear are interconnected. If movement is occasioned in two of them, a sympathetic but more violent movement is induced in the third. This is known as tumbling and causes extreme confusion, nausea, and even rolling of the eyeballs that prevents the pilot from reading correctly the airplane instruments. This situation can occur if, when the airplane is in a turn, the pilot suddenly turns his head in another direction. The rule should always be to avoid head movements, especially quick ones, when flying under instrument conditions.

Somatogravic – False Climb Illusion

The **otolith** is a small organ which forms part of the inner ear and vestibular apparatus. Its function is to sense and signal to the other organs the position of the head relative to the vertical. This signal has a profound influence on the balance and orientation of the body.

The otolith, simply described, is an erect hair with a small weight or mass at its tip. The base of the hair is embedded in a sensory cell which conveys to the brain information about the angle of the hair.

When the head is tilted backward, the small mass bends the hair and the message relayed to the brain indicates a backward tilt. If the head is held vertical but is subjected to acceleration, the hair bends owing to the inertia of the mass at the tip of the hair. Both tilt and acceleration, therefore, produce the same response by the otolith. If there are no visual cues to compliment the information from the otolith, the brain is unable to differentiate between tilt and acceleration. If tilt and acceleration are experienced simultaneously, the interpretation is that of a much steeper tilt. This is known as the **somatogravic (false climb) illusion**.

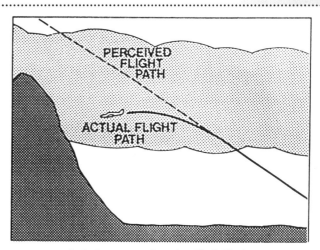

Fig.2. Somatogravic- False Climb Illusion.

In such a situation, a pilot is tempted to lower the nose of the airplane. This increases the forward acceleration component and increases the illusion of climbing steeply. Owing to lag in the altimeter and vertical speed indicator, the loss of height may go unnoticed.

There are three situations in which the false climb illusion may occur: (1) take-off at night or in IFR conditions, (2) an overshoot in reduced visibility or in IFR conditions, and (3) a climb from VFR into IFR conditions. During the latter situation, the illusion can be compounded by turbulence, in a turn, or by reliance on an artificial horizon that is not quite erect.

With deceleration, a similar illusion occurs. The activation of speed brakes or the lowering of flaps, which both cause sudden deceleration, swing the otoliths forward and produce an illusion of increased descent. This illusion most commonly happens on final approach at slow speed. The pilot's reaction of pulling back on the controls may cause a stall.

All pilots irrespective of experience or skill are susceptible to the illusion. Pilots must learn to anticipate the illusion and ignore it, to establish a positive climb or descent attitude and to rely on the aircraft instruments for confirmation of attitude.

Black Hole Illusion

Darkness, absence of visual cues and few ground lights combine to induce a false perception of altitude and/or attitude that is known as the **black hole illusion.** Under certain circumstances, during night and in clear weather, when an aircraft is on approach to land over dark terrain with no lights either below or to the sides of the approach path, with only the distant airport runway lights to provide visual stimuli, even the most experienced pilots may visually overestimate altitude and fly too low with the risk of landing short of the runway.

The most relevant source of visual information is the vertical angle, the angle subtended at the eye by the nearest and farthest lights. If, during a descent, a pilot maintains this angle at a constant value, the approach path follows the arc of a circle centred above the pattern of lights towards which the aircraft is descending, with the result that the aircraft would be flown into the ground short of the runway threshold.

The situation can be aggravated by a long, straight in approach to an airport located on the near side of a small city and by substandard runway and approach lighting.

A back up reference, such as an altitude alerter, a ground proximity warning system or a radio altimeter will help pilots to avoid being caught by this dangerous illusion.

Alcohol

Alcohol, taken even in small amounts, produces a dulling of judgment, comprehension and attention, lessened sense of responsibility, a slowing of reflexes and reduced coordination, decreases in eye efficiency, increased frequency of errors, decrease of memory and reasoning ability, and fatigue.

When a pilot undertakes a flight along a given course from one airport to some landing place, hundreds of decisions must be made relating to the operation of the airplane and the navigational aspects of the flight. Proper procedures must be accomplished to effect the safe completion of the flight and to ensure that no hazard is created to other airplanes in nearby airspace. Obviously, anything that impairs the pilot's ability to make decisions will increase the accident potential.

Alcohol is absorbed very rapidly into the blood and tissues of the body. Its effects on the physiology are apparent quite soon after ingestion and wear off very slowly. In fact, it takes about 3 hours for the effects of 1 ounce of alcohol to wear off. Nothing can speed up this process. Neither coffee nor hard exercises nor sleep will minimize the effects of alcohol.

Scientists have recently discovered that alcohol is absorbed into the fluid of the inner ear and stays there after it has gone from the blood and brain. Since the inner ear monitors balance, the presence of alcohol in the middle ear can be responsible for incorrect balance information and possibly spatial disorientation.

The presence of alcohol in the blood interferes with the normal use of oxygen by the tissues **(histotoxic hypoxia).** Because of reduced pressure at high altitudes and the reduced ability of the haemoglobin to absorb oxygen, the effect of alcohol in the blood, during flight at high altitudes, is much more pronounced than at sea level. The effects of one drink are magnified 2 to 3 times over the effects the same drink would have at sea level.

A pilot should never carry a passenger that is under the influence of alcohol. Such a person's judgment is impaired. His reactions during ascent to higher altitudes are unpredictable. He may become belligerent and unmanageable and a serious hazard to the safety of the flight.

The rule for both pilot and passengers in relation to alcohol quite simply should be **"No alcohol in the system when you fly".** The Canadian Aviation Regulations require that a pilot allow at least 8 hours between the consumption of alcohol and piloting an airplane. In fact, more time is probably necessary. An excellent rule is to allow 24 hours between the last drink and take-off time and at least 48 hours after excessive drinking. The after effects (hangover) of alcohol consumption also affect performance capability, causing headache and impairing emotional stability and judgment.

Drugs

Drugs, as well as the conditions for which they are taken, can interfere with the efficiency of the pilot and can be extremely dangerous. Even over the counter drugs such as aspirin, antihistamines, cold tablets, nasal decongestants, cough mixtures, laxatives, tranquilizers and appetite suppressors impair the judgment and co-ordination. They are responsible for drowsiness, dizziness, blurred vision, confusion, vertigo and mental depression. The effects of some drugs are even more pronounced at higher altitudes than on the ground. Some over the counter drugs taken in combination will react with each other resulting in a larger effect than even the sum of their individual effects. Some prescription drugs, such as antibiotics, are equally or more dangerous. Usually, however, a person sick enough to be on antibiotics is too sick to be flying.

Any use of illicit drugs is incompatible with air safety. Some common over the counter drugs contain prohibited drugs. For example, some headache remedies contain codeine, cold symptom depressants may contain drugs that are on the prohibited list. Check the labels carefully. Professional commercial pilots are subject to random drug testing and the presence of an illicit drug, even if taken unknowingly, may result in immediate grounding, loss of job and possible branding as a drug abuser.

The so-called soft drugs should be particularly avoided as they affect performance, mood and health.

Sedatives and tranquilizers that are prescribed for insomnia, anxiety and depression are unacceptable for pilots.

Antihistamines. (For allergic disorders). Cause sedation with varying degrees of drowsiness, decreased reaction time, disturbances of equilibrium. Cold cures usually contain an antihistamine. Do not pilot an airplane within 24 hours of taking an antihistamine.

Sulpha Drugs. Cause visual disturbances, dizziness, impaired reaction time, depression. Remain off flying for 48 hours.

Tranquilizers. Affect reaction time, concentration and attention span.

Aspirin. Toxic effects are relatively rare and are almost always associated with large doses. If you take aspirin in small dosage and have had no reactions in the past, it is probably safe to take it and fly.

Motion Sickness Remedies. Cause drowsiness and depress brain function. They cause temporary deterioration of judgment making skills. Do not take either prescribed or over the counter motion sickness remedies. If suffering from airsickness while piloting an aircraft, open up the air vents, loosen the clothing, use supplemental oxygen if available and keep the eyes on a point outside the airplane. Avoid unnecessary head movements.

Reducing Drugs. Amphetamines and other appetite suppressing drugs cause feelings of well being that affect good judgment.

Barbiturates. Produce a marked depression of mental alertness. Similar symptoms are attributable to nerve tonics and nerve pills.

Anaesthetics. With spinal or general anaesthetics, you should not fly until your doctor says it is safe. With local anaesthetics used in minor surgery or dental work, it is wise to wait 24 hours before flying.

Blood Donations

Because it takes several weeks for the blood circulation to return to normal after a blood donation, it is recommended that pilots who are actively flying refrain from volunteering as blood donors. If a blood donation has been made, you should consult your doctor before flying again. It is recommended that you should wait at least 48 hours.

Fatigue

Fatigue is one of the most common physiological problems for air crew members and will adversely affect individuals who are otherwise in good health. It has repeatedly been cited as the causal factor in airplane accidents. Fatigue degrades performance. A tired pilot cannot carry out tasks as reliably and accurately as he should. He is irritable and less alert, willing to accept lower standards of accuracy and performance.

Fatigue begins when the pilot begins a flight and increases with each hour in the air. As a result, at the time of landing when reflexes and judgment should be keenest, the pilot is most affected by the cumulative effects of fatigue.

The biggest danger of fatigue is that an individual may not recognize its effects.

The onset of fatigue is accompanied by numerous symptoms: deterioration in timing of movements, irritability and lack of patience, a tendency to lock the attention on individual instruments rather than to see the instrument panel as a whole, a tendency to become forgetful and ignorant of relevant cues, a tendency to overcontrol the airplane, an awareness of physical discomforts, a loss of "seat of the pants" flying ability, a tendency to accept a wider margin of error than normal.

Fatigue is caused by many things: lack of sleep, poor nutrition, stress, prolonged and repeated flights, aircraft noise, eye strain, vibration, wide variations in temperature and humidity, heavy workload and uncomfortable working conditions, boredom, monotony, night flights, frustrations from work and family.

Acute fatigue is easily treated by a meal and a good sleep. Chronic fatigue is more serious and is caused by difficult or stressful work with inadequate rest and is often aggravated by disturbed Circadian rhythms.

A sound physical condition, a healthy mental attitude, proper diet and adequate rest are a pilot's best weapons in fighting fatigue. On very long flights where the pilot and co-pilot share the piloting duties, the advantage of taking a 15 to 20 minute pre-planned power nap shows beneficial results in improved performance. Conversation and discussion help to keep us awake. Caffeine in moderation helps. Frequent small meals to keep up the blood sugar relieve fatigue.

Circadian Rhythm

Serious problems are associated with disturbed biological rhythms and related sleep disturbance and deprivation. The body operates on a **circadian, or 24 hour, rhythm** which is related to the earth's rotation time. It is maintained principally by the cycles of light and darkness, but also by meals and physical and social activities. Safety, efficiency and well being are affected by the disturbed pattern of biological rhythms occasioned by long range flight, irregular schedules and late night flights. Long distance transmeridian air travel, especially, is responsible for sleep disturbance, disruption of eating and elimination habits that result in lassitude, anxiety, irritability and depression, all symptoms of what is commonly called **jet lag**. Wide differences are found amongst individuals in their ability to sleep out of phase with their biological rhythms. The use of drugs or tranquilizers to induce sleep is not recommended as they have a lasting adverse effect on later performance. The use of alcohol is also not recommended since it is a drug, a depressant and, while it does induce sleep, it interferes with deep sleep.

Pregnancy

Providing the pregnancy is normal and without complications, pilots may continue to fly up to 30 weeks into the pregnancy. There are certain physiological changes that may affect flight safety. In the first trimester, nausea is common and may be worsened by motion, engine fumes and g forces. In the second trimester, anaemia is common and may affect the pilot's susceptibility to hypoxia. At 12 to 14 weeks, the fetus may be subject to seat belt injury. The fetus may be exposed to potentially hazardous conditions, especially cosmic radiation. Cosmic radiation increases with altitude, is greater at the poles than at the equator.

Because 6 to 10% of normal pregnancies deliver preterm, even pilots with a normal pregnancy are considered temporarily unfit after the 30th week of the pregnancy.

Eating

The stresses of flying, or indeed of any activity, consume energy. This energy is derived from oxygen and from blood sugar. The pilot is unwise to fly for too long without eating. His blood sugar will be low (**hypoglycaemia**); that is, his energy reserve will be low. Reactions will be sluggish and efficiency will be impaired. It is a good precaution to carry a nutritious snack on long flights.

Overeating is equally as unwise as not eating. Drowsiness and excessive gas formation are the result of over indulgence at the dinner table just before a flight.

In fact, we should all eat three nutritious meals a day, starting with a good breakfast. A doughnut and a cup of coffee is not a substitute for a breakfast or lunch containing complex carbohydrates and some protein.

At altitudes above 5000 feet ASL, the body experiences a higher loss of water through the surface area of the lungs than it does at sea level. This loss occurs because the percentage of water vapour in a given volume of air decreases with altitude. Because this water loss is not accompanied by a loss of salt, as occurs with perspiration, there is no accompanying sensation of thirst. Especially on long flights at higher altitudes, it is advisable therefore to have a drink of water every hour or so to replace the lost body fluid.

A common cause of incapacitation in the air is gastrointestinal problems caused by stomach cramps, nausea, vomiting and diarrhoea. It is important to avoid foods that may be contaminated. In multicrew aircraft, pilots should not eat at the same time nor eat the same food.

Stress

Flying fitness is not just a physical condition. It has a definite meaning in the psychological sense as well. It involves the ability of the pilot to perceive, think and act to the best of his ability without the hindering effects of anger, worry and anxiety.

Studies have shown that emotional factors, mental upsets and psychological maladjustments are repeatedly present in airplane accidents. The ability to think clearly and act decisively is greatly influenced by the feelings and emotions. In fact, every individual will panic earlier than normal if he is suffering from fatigue, illness, worry or anger. But, even well away from the panic threshold, good judgment is seriously impaired under stress.

There are many factors that contribute to stress in the cockpit. They are generally classed into three categories: physical, physiological or psychological.

Physical stressors include extreme temperature and humidity, noise, vibration, lack of oxygen.

Physiological stressors include fatigue, poor physical condition, hunger, disease.

Psychological stressors relate to emotional factors such as a death or illness in the family, business worries, poor interpersonal relationships with family or boss, financial worries, etc.

It is essential that a pilot be able to recognize when stress levels are getting too high. If you are suffering from domestic stress, if you are undergoing divorce or separation, if you have suffered bereavement, if an argument with your spouse or your boss is still rankling, if worries are building up to an unbearable load, if you have been despondent and moody, the cockpit of your airplane is probably no place for you.

Nevertheless, stress levels do naturally build up in the airplane cockpit, when there are a multitude of decisions to make and tasks to perform. Stress is, in effect, generated by the task itself and is not always negative. The sympathetic nervous system responds to stress and provides us with the resources to cope with the new sudden demands. However, the stress load may easily become unmanageable and a pilot needs to take measures to manage the stress load so that it does not become so. He needs to learn how to reduce or prevent in advance those stressors over which he has control.

Signs of chronic stress are many and varied: loss of a sense of humour, frequent anger, disturbed sleep, weight change, forgetfulness, excessive drinking, repeated mistakes, distraction, backaches, tense stomach, headaches, fatigue, suppressed immune system that leaves you prey to disease. Stress may make you a nag, critical, argumentative. It may erode your self image and leave you full of doubt, cynical, unconfident, directionless and pessimistic.

Stress can, however be managed. The physiological stressors can be controlled by maintaining good physical fitness and bodily function, by engaging in a program of regular physical exercise, by getting enough sleep to prevent fatigue, by eating a well balanced diet, by learning and practising relaxation techniques. The physical stressors can be reduced by making the cockpit environment as stress free as possible. A conscious effort to avoid stressful situations and encounters helps to minimize the psychological stressors. Support from family, friends, colleagues, organizations and, if necessary, professional counselling help restore psychological equanimity. Relaxation exercises, meditation, recreation renew and recharge the psyche.

Panic

There are many things that can happen in the air that cause fear and anxiety. These are normal reactions to a predicament that is out of the ordinary. What is to be avoided is allowing that normal anxiety to progress to panic.

Panic is a complete disregard for reason and learned responses, a feeling of extreme helplessness. A pilot in the grip of panic will freeze at the controls, will make a totally wrong response or succumb to completely irrational action.

Fatigue, hangover, emotional stress, chronic worry, illness, all substantially reduce the amount of anxiety an individual can withstand before he succumbs to panic.

The best way to prevent panic is through training and frequent rehearsal of emergency techniques. A pilot who knows his emergency routines so well that they are automatic will be less likely to panic when faced with a real emergency situation.

Lack of self confidence is, in itself, self defeating and an open door to panic. Not that a pilot should be fearless, for the fearless pilot has suspended reality-testing. He refuses to admit that there is any situation into which he is not competent to venture. Self confidence is quite another thing. The self confident pilot can assess the reality of a situation, can call on his reserves of training and knowledge to cope with the situation and does not permit emotion to cloud his reason.

Physical Fitness

The purpose of this book has been to instruct the pilot in what he should know to be a competent aviator. What he should do is, however, of equal importance. The most competent, knowledgeable and experienced pilot is in business only so long as his medical is valid. Maintaining physical fitness is therefore of prime importance.

Throughout the flying fraternity, there are thousands of pilots in their senior years who are still enjoying the privileges of their licence and using their airplane for pleasure, business and travel. If you want to be flying when you are eligible for the old age pension, now is the time to start looking after your health and maintaining your physical fitness.

The person who is physically active, participating in a regular routine of exercise or sports, will most likely have a healthy heart, lungs and not be overweight. Diet is important, not only to keep weight at an acceptable level, but also in the control of heart disease. The case against smoking as a contributor to lung disease and heart disease is heavily documented. Protection of hearing by wearing earplugs has already been mentioned as has the need to protect the eyes from undue eyestrain.

Pilot Decision Making

Aeronautical knowledge, skill and judgment have been considered the three essential faculties that pilots must possess to be professional in the execution of their duties. The knowledge and skill have been taught in ground school and flight training programs but decision making skills have usually been considered a trait that pilots innately possess or that are acquired through experience. In fact, good decision making skills can also be taught.

Training in decision making skills is now a part of the pilot training program. Pilots can learn good judgment just as thoroughly as they learn the mechanical concepts and basic skills of flying. But what is good judgment? It is the ability to make an instant decision which assures the safest possible continuation of the flight.

"Pilot judgment is the process of recognizing and analyzing all available information about oneself, the aircraft and the flying environment, followed by the rational evaluation of alternatives to implement a timely decision which maximizes safety. Pilot judgment thus involves one's attitudes toward risk-taking and one's ability to evaluate risks and make decisions based upon one's knowledge, skills and experience. A judgment decision always involves a problem or choice, an unknown element, usually a time constraint, and stress." (Transport Canada: Judgment Training Manual).

The causal factor in about 80% to 85% of civil aviation accidents is the human element, in other words, pilot error, a poor decision or a series of poor decisions made by the pilot-in-command. This concept is known as the **poor judgment chain.** One poor decision increases the probability of another and as the poor judgment chain grows, the probability of a safe flight decreases. The judgment training program teaches techniques for breaking the chain by teaching the pilot to recognize the combination of events that result in an accident and to deal with the situation correctly in time to prevent the accident from occurring.

The Process
Pilot decision making is a process. There must, first of all, be **situation awareness,** which means you need to determine everything you can about a flying situation and assess whether the information is important. This kind of awareness is acquired only through training, experience, practice and study.

The next step is to establish various courses of action. Give yourself as many **options** as possible and consider the likely outcome of each one.

Now you have to **choose** what action to take and do it before time runs out. Whatever you choose, be sure to leave yourself an "out".

Then you must **act.** Carry out your decision with all your skill.

Monitor the results of your action to be sure you are getting the desired outcome. The **evaluation** of the results is in effect re-evaluating the whole situation and beginning the process again.

Factors
Many factors influence your ability to make good decisions. Clear thinking and careful reasoning enhance the decision making process. Cloudy, confused thinking hampers it.

Knowledge. Every step of the decision making process requires knowledge. Knowledge is the result of what you have learned in the classroom and in the briefing you received prior to the flight. It is also the result of your pre-flight planning and preparation and the details you notice from the start to the finish of your flight.

Situational awareness. Situational awareness is knowing what is going on around you. Decision making can't even begin unless you are aware of what is happening. Assuming that everything is normal when it is not can be very dangerous. A false assumption, once made, can be difficult to change and can generate a whole chain of incorrect decisions — the poor judgment chain. It is very easy to shape reality to fit your expectation. You see what you expect to see, hear what you expect to hear. You focus your attention on one item and something more important goes unnoticed (known as **target fixation**). Sometimes your subconscious prefers to ignore the bad news, changing incoming information to provide the message you want to hear.

Skill. Skill is an important part of the process. Skills should be practised until they become automatic responses. Emergencies are not planned. They happen when you least expect them. Be prepared.

Stress. Stress is very much a part of our daily lives. Poorly managed, it can effectively impair our ability to make good decisions. Stress may be considered in terms of one's ability to cope. Trouble arises when the demands of the task exceed the pilot's capacity to deal with them, whether because the demands are overwhelming or because capacity is reduced for some reason. The body reacts to this stress overload by increased heart beat, respiration, blood pressure and perspiration. Stress management is, therefore, an important factor in good decision making (see also **Stress** above**).**

Risk management. Any human activity involves some level of risk. There are, of course, risk elements in flying. Assessing that risk is part of the decision making process.

Pilots need to do a pre-flight inspection of themselves. Are you competent to undertake a flight if you are tired, on medication, not current, under stress? If the level of risk involved in any one of such personal elements is marginal, perhaps the flight should not be undertaken.

Critical decisions about the aircraft must be made on the ground during pre-flight planning: weight and balance, take off and landing performance, cross wind limits and cruise performance, condition of the aircraft. Throughout the flight, assessment of the operating systems must also be continually made.

The risks inherent in the environment are an important area to consider. Weather is, of course, the most obvious environmental concern. There is also potential risk in the selection of airports to use. Runway length and width, runway surface, obstacles and landing aids are components of the aviation environment.

The type of operation also introduces a risk element. The risk factor to complete a critical air ambulance mission under adverse conditions would perhaps be more acceptable than the same circumstances for a sight seeing or pleasure flight.

Assessment of risk is assessment of probability; estimation of the "odds". Always leave yourself an "out"; have an escape plan if the situation turns too bad. This is the most important tool of risk management.

An accurate assessment of the risks associated with the risk elements (pilot, aircraft, environment and operation) enable a pilot to arrive at decisions that ensure a safe conclusion to a flight. Maybe the good decision is not to take off at all.

Attitude
Your ability to make good decisions depends to a great extent on your attitude. Attitudes are learned. They can be developed through training into a mental framework that encourages good pilot judgment. The pilot decision making training program is based on recognition of five hazardous attitudes.

Anti-authority. This attitude is common in those who do not like anyone telling them what to do.

Resignation. Some people do not see themselves as making a great deal of difference in what happens to them and will go along with anything that happens.

Impulsivity. Some people need to do something, anything, immediately without stopping to think about what is the best action to take.

Invulnerability. Many people feel that accidents happen to other people but never to themselves. Pilots who think like this are more likely to take unwise risks.

Macho. And then there are the people who need to always prove that they are better than anyone else and take risks to prove themselves and impress others.

Pilots who learn to recognize these hazardous attitudes in themselves can also learn how to counteract them, can learn to control their first instinctive response and can learn to make a rational judgment based on good common sense.

DECIDE. The DECIDE acronym was developed to assist a pilot in the decision making process.
    D — detect change.
    E — estimate the significance of the change.
    C — choose the outcome objective.
    I — identify plausible action options.
    D — do the best action.
    E — evaluate the progress.

Using the DECIDE process requires the pilot to contemplate the outcome of the action taken. The successful outcome should be the action that will result in no damage to the aircraft or injury to the occupants.

When a pilot receives a licence to fly, he or she is being given the privilege to use public airspace and air navigation facilities. He is expected to adhere to the rules and to operate an aircraft safely and carefully. He is expected to use good judgment and act responsibly. Decision making is a continuous adjustive process that starts before take-off and does not stop until after the final landing is made safely, the engine is shut down and the aircraft is parked. Positive attitudes toward flying, learned judgment skills, will improve a pilot's chances of having a long and safe flying career.

Human Factors

The human factor is the most flexible, adaptable and valuable part of the aviation system, but it is also the most vulnerable to influences which can adversely affect its performance. Optimizing the role of people in the aviation environment involves all aspects of human performance and behaviour: decision making, the design of displays and controls and the cabin layout, and even the design of aircraft operating manuals, checklists and computer software.

Human factors is about people in their living and working situations, about their relationships with machines, with procedures, with the environment about them and with other people.

In most cases, accidents result from performance errors made by healthy and properly certificated individuals. The sources of some of these errors may be traced to poor equipment or procedure design or to inadequate training or operating instructions. Reduced levels of human performance capability and limitations in human behaviour result in less than optimum performance.

There would appear to be a direct relationship between workload and performance. At low levels of workload, such as during the cruise phase of long haul flights, performance is poor and the ability to react in an emergency is potentially negatively affected. The standard of performance increases as workload increases up to an optimum level of workload and performance. At extremely high levels of workload (overload), performance is again jeopardized. In the aviation industry, the concept of workload is of primary importance to ensure that the demands of the task never exceed the capabilities of the pilot.

Recognition of human factors is based on the effectiveness, the safety and the efficiency of the system and on the well being of crew members.

The central figure in the human factors equation is the pilot, or other crew member, who is the most critical but also the most flexible component of the system. However, people have limitations and are subject to considerable variations in performance.

Design of cockpit space is important to pilot performance. Comfortable seats designed to fit the human body, instrument displays designed to match the sensory and information processing characteristics of the user, controls with standardized movement, coding and location, are recognized as important factors in providing a compatible and comfortable working environment. All too often, pilot error can be attributed to knobs and levers that are poorly located, that operate differently from one airplane to another, that are improperly coded.

An essential part of flying safely is proper pilot and cockpit organization before a flight. Before you take off on a flight, it is important that you have all the necessary documents, charts and equipment on board, properly stowed and accessible. Nothing should be stowed where it would interfere with the safe operation of the airplane. Everything should be organized and neatly arranged so that it is readily available when needed. Loose items should be secured so they will not be thrown around the cockpit during sudden maneuvers.

The non physical aspects, such as procedures, manuals and checklists, symbology and computer programs, are responsible for delays and errors if these are confusing, misleading or excessively cluttered in their presentation and documentation.

The effect of environmental factors, such as noise, heat, lighting and vibration, are recognized as causal factors in human error, as are fatigue, distorted circadian rhythm, stress.

Cockpit Resource Management

Traditionally, crew members have been trained individually and it was assumed that individually proficient crew members would be proficient and effective members of a crew team. However, flight crews function as groups and group influences play a role in determining behaviour and performance. Leadership, crew co-operation, teamwork and personality interaction are vital factors in cockpit resource management. Training programs aimed at increasing the co-operation and communication between crew members are vital in ensuring efficient and safe airplane operation. Cockpit resource management training focuses on the functioning of the flight crew as an intact team and provides opportunities for crew members to practise their skills together. The program teaches crew members how to use their own personal and leadership styles in ways to foster crew effectiveness. The program focuses on the importance of proper communication, division of responsibilities, leadership and teamwork and has established five interrelated concepts to enhance crew skills.

Attention management includes understanding how distractions and error chains can be avoided.

Crew management teaches the importance of proper communications, division of responsibilities, leadership and teamwork.

Stress management focuses on understanding the effects of life stress events and provides stress coping strategies.

Attitude management focuses on methods of recognizing and controlling certain hazardous attitudes and behavioural styles.

Risk management addresses the evaluation of operational hazards.

Air Safety

What do we mean by air safety? In essence, air safety is the putting into practice of all that knowledge on aviation and technical matters that a pilot has acquired in the course of his aeronautical training and flying career. It is, in other words, perpetual and continued **concern** for the well being of and **respect** for the personal property and endowed rights of our fellow men as well as of ourselves. It is based on sound judgment and a thorough knowledge of the subjects which have been dealt with between the covers of this manual. However, the ultimate in flight safety is more than just the possession of **knowledge.** It is also the exercise of **common sense** and, perhaps what is more important, of **self discipline,** that is, the self discipline to apply the knowledge and the common sense in a positive attempt to do the right thing or make the right decision at the right time regardless of other distracting and prejudicial influences. The safety conscious pilot bases his decisions on good judgment rather than on emotional or impulsive whims.

For a lifetime of pleasant and safe flying, not only must a pilots have an appreciation of the foundation on which "the rules of flight" were built, but they must also practise these rules at all times.

Government aviation agencies and flight safety experts all agree that the biggest challenge they face is that of making pilots aware of their own limitations based on their knowledge and experience or, to be more explicit, based on their lack of knowledge and inexperience. The conclusions reached by accident investigators tend very substantially to support the widely held opinion that almost every pilot, who has found himself in trouble while at the controls of an airplane, got into his predicament either because he had insufficient appreciation of the degree of competence and skill required during a particular flight or because he over-estimated the limitations of his own ability to cope with the conditions he knew he would encounter.

It is important, therefore, for pilots to realize not only the value of securing adequate training but also how much retraining they need and how frequently they need such retraining in order to execute their pilot-in-command responsibilities adequately, efficiently and with the maximum amount of safety during flight. It is important for every pilot to review, review and review again those basic aeronautical fundamentals to be sure that the essentials which are important for safe flying are not forgotten. He should never give up trying to learn more about aviation and flight safety. No one really ever graduates from a course in flying. The acquisition of knowledge goes on forever. The wise pilot knows that "When you think you've learned all that there is to know about flying, that's a sign that you'd better quit flying." The pilot who flies an airplane as a hobby still requires the skill of the professional pilot and must approach his flying with equal earnestness.

It is important that every pilot should feel a sense of responsibility both to himself and to all his fellow pilots. All pilots share the same airspace and an error in judgment by one can have serious detrimental effects on others. Don't be considered a hazard to yourself and to others.

The Proper State of Mind

Possessing the proper mental attitude towards piloting an airplane is the prime factor governing safe flying where the pilot is concerned. It governs all of his actions. With the proper state of mind, a pilot is able to make correct judgments as to what are and what are not the important considerations at hand, where emphasis should be placed, what are the pros and cons of various courses of action. Pilots possessing the proper state of mind are positive while being open-minded; they are confident in their judgment because they are aware of their own shortcomings, limitations of knowledge and experience, and because they respect their reasonable doubts. They are self reliant but are not dogmatic in believing in their inability to be wrong. Confidence without doubt is over-confidence. Freedom from inhibitions, complexes and worries is the pilot's ultimate goal in achieving the ambition of being a safe pilot. The problems of one's personal world do not belong in the cockpit of the airplane. If safety in flight depends on the pilot's actions in his cockpit, then it is easy to realize the importance of directing his whole attention and concentration, free of extraneous problems and thoughts, to the tasks of flying the airplane. A thinking and responsible pilot cares about mistakes and avoids repeating them. A pilot possessing the proper mental outlook will not hesitate to turn back when encountering bad weather. He has no wish to run the risk of being a statistic in the next morning's newspaper.

The rules and maxims listed below are the product of the wisdom of experienced pilots and experts in aviation safety. They are not new or startling but are rules that wise pilots have memorized and practised for years. If followed, they will give you too thousands of safe flying hours and many years to enjoy those things in life that you value and enjoy most of all.

Checklist for Safe VFR Flight

Transport Canada has prepared the following checklist as a guide to good flight planning procedures.

1. Before flying.
  Current on type? If not, get a checkout.

  Review aircraft operating manual: speeds, fuel consumption/capacities, systems operation, take-off and landing distances, emergency procedures.

  Weather briefing: VFR for entire flight? Forecast: 1,000 foot en route cloud clearance above ground? Plan bad weather procedures.

  Prepared for forced landing? Rations, survival gear, life jackets. Wear or carry clothing for outside conditions.

2. Planning your flight.
  Calculate all-up weight. C of G in limits?

  Are VFR charts and Supplement current?

  NOTAMS: destination airport, snow clearing, airports.

  Study VFR map for route, select alternate airports.

  Prepare map and flight log: safety heights, tracks, headings, groundspeeds, times, approach and departure procedures.

  Sufficient daylight on arrival?

  File flight plan.

3. Preflight.
  Brief passengers: door, seatbelts, prop hazard, ELT.

  Check you have: aircraft operating manual, log, maps, computer, pencils, survival equipment, clothing and sun glasses.

4. Walkaround.
  Drain fuel: examine for water.

  Dipstick fuel: enough for flight plus 45 minutes.

  All surfaces completely free of frost/ice? (Even a little could be deadly.)

  Fuel caps secured? Oil sufficient and cap secure.

5. Take-off.
Avoid intersection take-offs: go to the end of the runway.

Mental review: liftoff speed and engine failure procedures. Plan to abort if not airborne in first half of runway.

Engine/prop at full power: temperatures and pressures in the green.

6. Inflight.
Map read. Watch chart to ground.

Visual contact at all times. Anticipate landmarks ahead.

Fuel consumption per plan? When 3/4 fuel gone, plan to refuel.

Groundspeed per plan? Revise ETAs.

Radio FSS on flight progress and weather.

7. Arrival.
20 minutes before: review **Canada Flight Supplement,** reconfirm airport information (elevation, runway lengths and heading and circuit height).

Review approach speeds, cross wind limit, flaps selection.

10 minutes before: broadcast intentions. Get traffic.

5 to 10 minutes before: altimeter setting, surface wind, traffic.

On long descents, warm engine periodically.

Check cross wind: if over limit, land elsewhere.

Landing lights on.

Fly a circuit (avoid a straight in approach). Report joining. Check windsock.

Report and keep sharp lookout on base and final.

8. Landing.
Recheck windsock on final. Review overshoot procedure.

Plan touchdown 200 feet after threshold.

Have weight on wheels by first one third of hard runway or one quarter if grass.

Undershooting or turbulence? Add power early.

9. Shutdown.
Passengers remain seated until prop(s) stopped.

Flight plan closed.   Refuel.

ELT off.   Tie down, wheel chocks, etc.

10. Emergencies.
Engine failure on take-off: attain glide speed, close throttle, land straight ahead, gentle turns.

Engine failure in cruise: attain glide speed, close throttle, check mixture and fuel controls, attempt restart.

Rough engine: carb heat full on, check mixture, fuel.

Low cloud or low visibility ahead: carburetor heat full on, turn around, follow bad weather plan.

Uncertain of position: check time flown from last landmark for probable position, go to prominent landmark, map read ground to chart.

Lost: slow down, radio for assistance, D/F or radar steer, the 4 Cs (climb, confess, consult, comply). If no radio assistance: alert radar, fly 2 minute triangles until help comes, prepare for forced landing before fuel is gone.

Make These Resolutions

1. I shall look upon the licence to fly as a privilege to which I shall be entitled only as long as I execute this privilege strictly within the limits of my ability and experience.

2. I shall conduct a thorough inspection of my airplane prior to every flight and I shall refrain from flying in any airplane with a known operational defect.

3. I shall check the weather prior to every cross-country flight and, when filing a flight plan, shall remember to close it upon arrival at my destination.

4. I shall carry passengers only in an airplane with which I am familiar and in which I have had adequate recent experience.

5. I shall not plan any flight beyond the limits which can be reached under prevailing wind conditions with 3/4 of the fuel on board and I shall land at the nearest available airport when my fuel supply indicates one hour remaining.

6. I shall fly at legal altitudes at all times and refrain from making unauthorized low passes, hedge-hopping, buzzing and similar childish stunts which constitute a danger to myself and others.

7. I shall report unusual weather and atmospheric conditions as soon as possible to the nearest radio facility.

8. I shall refrain from flying close to or entering into weather conditions considered marginal with reference to my capabilities, experience and the equipment aboard the airplane.

9. I shall offer to yield landing priority to jet airliners and other high-speed airplanes when in the traffic circuit of an airport serving scheduled airlines.

10. I shall personally supervise refuelling, tie-down and other ground servicing activity which may affect the safety of persons and equipment.

11. I shall refrain from entering an airplane when recently having partaken of drugs, alcohol or other debilitating substances, when not having had any sleep during the preceding 16 hours or when physically or mentally impaired for any reason whatever.

12. I shall conduct myself in the process of operating an airplane and at all other times in a manner giving a favorable impression of aviation and the flying community.

Obey These 12 Important Rules

1. CHECK OUT — Plan ahead so that you never attempt to exercise the privilege of your pilot certificate in any airplane unless checked out by a well-qualified instructor and having successfully completed a minimum of

(a) One hour ground familiarization with controls/systems and airplane operating limitations.

(b) 8 regular take-offs and landings (day or night).

(c) 2 short field take-offs and landings.

(d) 2 cross wind take-offs and landings.

2. PRE-FLIGHT — Plan ahead so that you never start the engine until

(a) You have checked weight and balance data.

(b) You have made sure any objects carried in the passenger cabin are properly secured and free of the controls.

(c) Prescribed walk around and preflight inspection completed.

(d) Fuel quantity double checked for proposed flight plan plus 15 minutes for take-off plus 45 minutes reserve.

(e) Fuel contamination checked into transparent container from under engine and wing tank quick drains.

3. VIGILANCE — Plan ahead so you never occupy any area on the ground or in the air without "double checking" for possible existing or potential hazards.

4. CONTROLS AND SYSTEMS — Plan ahead so that you

(a) Never operate an airplane unless you are familiar with the operation and correct use of all controls and systems.

(b) Never start engine, start take-off, start landing, start cruising or start let-down until all prescribed procedures are accomplished from a **written** check list.

(c) Never attempt to operate an airplane with a known malfunction. If malfunctioning occurs in flight, head for nearest airport.

(d) Are always alert for the formation of carburetor ice. Use full carburetor heat at the first indication of carburetor icing.

(e) Never raise flaps after landing a retractable gear airplane until well clear of the active runway and only after double-checking the control you are activating.

5. WEATHER — Plan ahead.

(a) Pre-flight — Study the weather establishing en route forecast, en route conditions and escape route to good weather.

(b) In-flight — Never get even close to losing good ground reference control. When encountering clouds bases of 1000 feet or visibility of less than 5 miles, plan to retreat to a good alternate airport. Execute a retreat on encountering clouds bases of 800 feet or under 3 miles visibility. (Unless current and qualified for IFR)

(c) Night — Never, unless assured of 2000 foot ceiling and 5 miles visibility and assured that no frontal or ground fog or storm conditions will be encountered. (Unless current and qualified IFR)

6. SPEED/STALL CONTROL — Plan ahead so that you never abruptly change attitude of airplane nor allow airspeed to drop below

(a) At least 160% of stall speed when maneuvering below 1000 feet.

(b) At least 140% of stall speed during straight approach or climb out.

(c) At least 120% of stall speed over threshold and ready for touchdown.

7. NAVIGATION — Plan ahead so that you

(a) Reach destination one hour before sunset unless qualified and prepared for night flight.

(b) Never operate at an altitude less than 500 feet above the highest obstruction (2000 feet in mountain area) except on straight climb from take-off or straight in approach to landing.

(c) Predetermine ETA over all check points. If lost, never deviate from original course until orientated. Always hold chart so plotted course coincides with flight path.

(d) Divert to nearest airport if periodic fuel check indicates you won't have 45 minutes reserve at destination.

8. TAKE-OFF OR LANDING AREA — Plan ahead so that you

(a) Never take off or land unless on designated airports with known, current runway maintenance.

(b) Restrict operations to runway length equal to airplane manufacturer's published take-off or landing distance, plus 80% safety margin if hard surface, double manual distance if sod, and triple manual distance if wet grass (about same traction as ice.)

(c) Night — Never operate except on well-lighted, night operated airport, and then using steeper approach attitude to clear unlighted obstacles.

9. TAKE-OFF OR LANDING LIMITS — Plan ahead so that you

(a) Always plan touchdown within the first 1/3 of a runway in order to roll to a stop within the second 1/3, having the final 1/3 for spare.

(b) Abort take-off if not solidly airborne in first 1/2 of runway.

(c) Abort landing if not solidly on in first 1/2 of runway. (First 1/4 if wet grass.)

(d) Never relax control until the wheels have ceased to roll.

10. WIND LIMITS — Plan ahead so that you

(a) Never attempt taxiing in cross winds or gusts exceeding 50% of stall speed unless outside assistance is available and used. Taxi very slowly when winds exceed 30% of stall speed.

(b) Never attempt take-off or landing when 90° cross surface winds exceed 20% of stall speed or 45° surface winds exceed 30% of stall speed.

(c) Never taxi closer than 1000 ft. from the blast end of powerful airplanes, and then only when headed into remaining blast effect.

(d) Never get close to powerful airplanes on take-off, in air, on landing without allowing time for wake turbulence to subside.

11. PHYSICAL CONDITION — Plan ahead so that you never pilot an airplane when

Air Safety

a) Having less than 8 hours from bottle to throttle (24 hours from big bottle).

(b) Extremely fatigued.

(c) Taking tranquilizing or sleep inducing drugs.

(d) Hypoxic from oversmoking or operating above 10,000 feet (without oxygen).

(e) When emotionally upset.

12. STARTING ENGINE — Plan ahead so that you never, never attempt to hand start an airplane unless a qualified person is at the controls or, in an emergency, unless all wheels are securely blocked and strong, tight, and secure tiedowns are affixed to both wings and tail.

Do's and Don'ts for Safe Flight

DO get a weather briefing before taking off on any cross country flight. DON'T assume that the weather will be as it usually is or recently was, or as you wish it to be.

DO seek the advice of others who are better informed about the area's weather flying. Frequently area and route phenomena are known to local pilots or other individuals who may be junior to you in years and overall flying experience. DON'T be too proud to heed this advice.

DO cancel the flight that may be hazardous. DON'T become a pusher or you might become a statistic.

DO have an alternate plan of action if adverse weather is encountered. DO use it while still VFR. DON'T forget the time honoured 180° method of survival.

DO remember that sucker holes are always present to tempt the inexperienced pilot. DON'T be lured into pressing on just a little farther.

DO reduce airspeed in turbulence. DON'T operate your airplane or your skills at continuous maximum performance.

DO follow regularly traversed routes through mountains or over unfavorable terrain.

DON'T fail to allow yourself a margin for error. Should you be forced to land, be where you can land safely.

DO obtain sufficient altitude before crossing a mountain range. DON'T be caught in the turbulence or downdrafts of ridge crossing.

DO allow sufficient clearance between yourself and the bases of the clouds. DON'T expect all clouds to have flat bases. Often they are quite ragged and irregular with hills and other high ground penetrating their low-hanging portions.

DO have the utmost respect for thunderstorms. They are dangerous! DON'T fly into or near them.

DO fly as a VFR pilot if you are a VFR pilot. DO exercise caution when engaging in over-the-top flights. The cloud deck may extend farther than you anticipate. Cloud tops may reach altitudes that you and your airplane will not be able to maintain resulting in your becoming involved in instrument flight with the consequent violation of safety and the Aviation Regulations. Also remember you are dependent upon an adequate oxygen intake. DO take the proper training or obtain an instrument rating and fly an IFR equipped airplane if you want, on a regular basis, to fly above cloud without reference to the ground or horizon.

DO proceed with your alternate plan if you find your destination weather poor. DON'T hold over an airport waiting for ground fog to dissipate. Go to the nearest airport where weather is suitable.

DO remember that poor weather today will probably be good weather tomorrow or the next day.

DO learn to be relaxed yet alert in the cockpit. If your mind and body are tense, fatigue will set in before the flight is terminated and your ability may become impaired due to such fatigue.

DON'T let anything influence your good judgment.

DON'T get into bad flying habits.

Principles of Safe Landing

1. Never land over traffic already lined up on the runway.

2. On approach, line up with the white center line of the runway and stay lined up during approach, touchdown and rollout.

3. In a cross wind landing, the slip method of alignment with the center line of the runway will allow you to determine if the cross wind is too strong. If you can't stay lined up by slipping, you cannot land safely.

4. Never land on a runway if the cross wind component exceeds your self-imposed maximum, or the maximum for the airplane, or the maximum for runway conditions.

5. Always fix a land/no land point on the runway. When you determine you are overshooting, go around.

6. Never land without completing your landing checklist. The item you forgot may be your gear.

7. Never land at an airport if runway conditions are unknown.

8. Never land at a **closed** field.

9. Clear for traffic constantly (the final approach is a high potential area for mid-air collision).

10. Be prepared for an aborted landing under emergency circumstances. **Know the procedure.** Review it briefly while still en route.

11. Learn to use all the airplane equipment provided for the landing operation. **Know how and when to use flaps.**

12. In a properly executed approach, you will not need full flaps until on final approach.

13. Never get into the habit of dragging in low on final approach. Learn a method of gauging the right approach angle. Conversely, don't develop the habit of diving on the airport. Pilots who do this may end up overshooting.

14. Review approach/landing type accidents. Learn from the mistakes of others. Learning about accidents first-hand is too costly.

15. Recurrent dual instruction and practice provide the only known ways of staying proficient. **Stay proficient.**

16. When runway length is marginal and runway surface conditions and temperature are factors and when landing at high elevations, compute density altitude.

17. At a controlled field, make initial contact 15 miles out.

18. When landing at a field lacking communication, transmit your position blind. Another pilot may be arriving or departing at the same field.

19. When landing VFR at a field with one or more instrument approaches, keep your approach path well clear of the instrument approach radio aids. These aids act as funnels for all landing traffic and have a high mid-air collision potential.

20. Use the procedures recommended for a precautionary landing when landing at a small or unfamiliar airstrip.

Principles of Safe Take-Off

1. Always use a checklist and always complete the checklist prior to takeoff.

2. Review the emergency checklist prior to takeoff. You may need the emergency procedure sooner than you think!

3. Always clear area for traffic. Don't take off if you must hurry to avoid traffic.

4. Never take off until you have fully advised the tower, unicom, MF, ATF, etc. of your proposed operation. Don't keep it a secret. If no communication facility is bases at the field, transmit blind your departure runway, turn after take-off, and direction of flight.

5. Always have an abort point on runway selected and use it if all systems are not GO at that point.

6. Don't take off at a closed field.

7. Don't attempt to take off if you don't know runway conditions.

8. Don't take off when the cross wind component exceeds the maximum component you have imposed on yourself or if it exceeds the maximum for runway conditions.

9. Always use the centre line of the runway.

10. Know the critical V speeds for your aircraft: stall speed (dirty and clean), best angle and best rate of climb. Review them prior to take-off.

11. Always compute density altitude when runway length, temperature, field elevation or obstructions ahead create doubt in your mind. Compute it often enough to remember how to do it!

12. For multi-engine aircraft, all of the above apply as well as the V speeds and performance limits of single engine operation.

13. On cross wind take-offs, start the take off roll with ailerons fully deflected in proper direction and then decrease the deflection as speed builds.

14. Never become airborne, in a cross wind especially, until the aileron controls are effective.

15. Never take off over gross weight or out of C.G.

16. Never take off with an unresolved problem, either with yourself or the aircraft.

17. During take-off, devote full attention to the take-off.

18. With retractable gear, always have a positive climb indication before retracting gear.

19. Even when VFR , don't take off until your radio gear is set up. It is much easier to do this on the ground than in the air, especially if an emergency comes along.

20. Even when VFR, if you are at a field with an omnitest and you are going cross-country, make a VOR accuracy check. If there is no omnitest on the field, make the accuracy check in the air as soon as possible.

21. Never take off downwind unless special conditions dictate.

22. Review take-off accidents for your type of aircraft.

23. Periodic practice and recurrent dual instruction are the ONLY ways to make sure you are **current, proficient and safe.**

24. Never exceed the maximum recommended flap setting for take-off for your aircraft.

Shoulder Harnesses

Experiments carried out by government testing agencies and by airplane manufacturers have proved conclusively that shoulder harnesses provide much more protection than do seat belts. In fact, it is conjectured that, had shoulder harnesses been worn, quite a number of victims of airplane accidents would not have been killed.

When seat belts only are worn, the hips become a pivot point. When the airplane decelerates suddenly, the upper body swings like a pendulum from the hips. The head strikes the control panel or windshield suffering serious and often fatal injury. Properly installed shoulder harness will restrain the whole upper torso and prevent this type of head injury.

The pilot is responsible for showing passengers how to fasten and unfasten seat belts and shoulder harnesses and should insist that they be secured during take-offs and landings and, preferably, throughout the flight. Severe turbulence can be encountered unexpectedly at any time, even in clear air. Serious injury can be prevented if the seat belt is fastened at all times.

Seat belts should be worn as low as possible across the hips. Single strap shoulder harnesses should be snug, but not tight, across the chest.

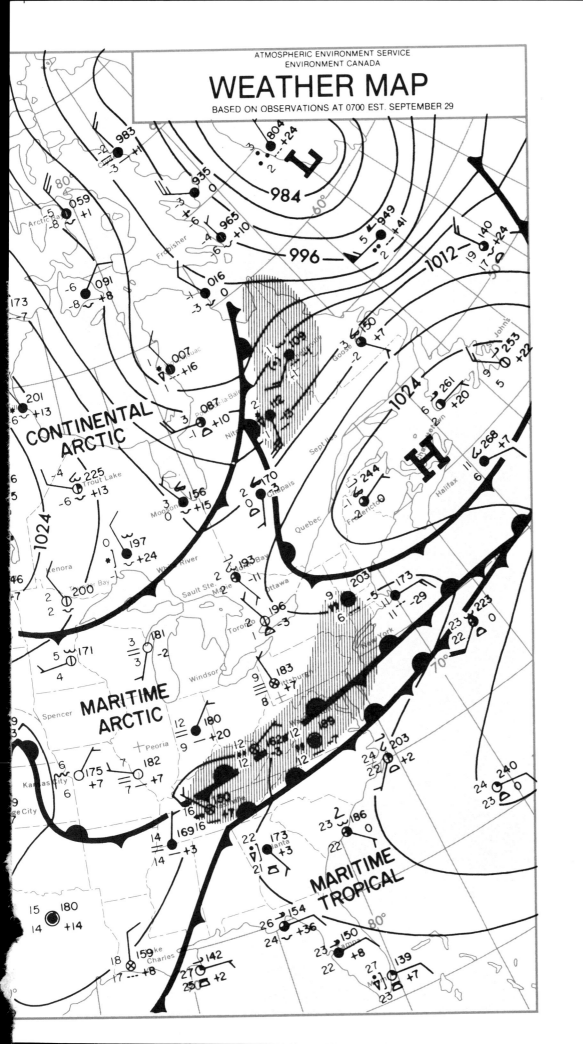

STATION MODEL

Weather stations are indicated on weather maps by a ball or circle, called the Station Circle. This circle is also used to indicate the cloudiness of the sky — black being completely overcast, white, clear, and the percentage of cloudiness between these two extremes being represented by a corresponding percentage of black and white.

The letter "M" in the station circle means that the weather report for the station is missing that day.

The direction of the wind is indicated by a line radiating from the ball, and the force of the wind is indicated by "barbs" or "feathers".

Around the Station Circle numbers and symbols are arranged to indicate all the various elements of the weather which have been observed at the station at map making time. The complete diagram is known as a STATION MODEL.

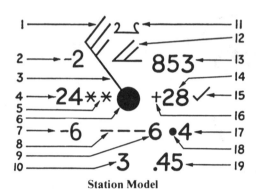

Station Model

1. Wind Force (23 to 27 kts.). 2. Temperature in degrees Celsius. 3. Direction of Wind (320°). 4. Visibility in code (24 = 1½ miles). 5. Present State of Weather (Continuous light snow). 6. Cloudiness of sky (completely overcast). 7. Dewpoint, in degrees Celsius. 8. Low Clouds (Fractocumulus)*. 9. Amount of Cloud, in code (6 = 7 or 8 tenths). 10. Height of Lower Cloud, in code (3 = 600 - 999 feet). 11. High Cloud (Cirrostratus)*. 12. Middle Cloud (Altostratus)*. 13. Barometric Pressure MSL (985.3 mb.) See "Sea Level Pressure". 14. Amount of Barometric Change in past 3 hrs., in 10ths of millibars (2.8 mb.). 15. Barometric Tendency in past 3 hrs. (Falling or steady, then rising — or, rising, then rising more quickly)*. 16. Plus or Minus indicates Pressure higher or lower than 3 hrs. ago. 17. Time precipitation began or ended, in code (4 = 3 to 4 hrs. ago)*. 18. Past Weather (Rain). 19. Precipitation in hundredths of inches (.45 inches).

* Omitted when not observed or recorded.

WEATHER MAP SYMBOLS

The various symbols and code figures which are grouped around the Station Model are shown in the following tables:

SKY COVER

0	○	Absolutely no clouds	5	◐	5/8
1	⊙	1/8 or less	6	◐	6/8
2	◔	2/8	7	◕	7/8
3	◕	3/8	8	●	8/8 (overcast)
4	◑	4/8	9	⊗	sky obscured

Wind Force

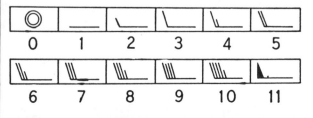

0: Calm. 1: 1-2 kts. 2: 3-7 kts. 3: 8-12 kts. 4: 13-17 kts. 5: 18-22 kts. 6: 23-27 kts. 7: 28-32 kts. 8: 33-37 kts. 9: 38-42 kts. 10: 43-47 kts. 11: 48-52 kts. 12: 53-57 kts. 13: 58-62 kts. 14: 63-67 kts. 15: 68-72 kts. 16: 73-77 kts. 17: 103-107 kts.

FRONT SYMBOLS

On Printed Charts		On Weather Map
▲▲▲▲	Cold Front	Continuous blue line
△△△△	Cold Front (aloft)	Broken blue line
●●●●	Warm Front	Continuous red line
○○○○	Warm Front (aloft)	Broken red line
▲●▲●	Stationary Front	Alternate red and blue line
△○△○	Stationary Front (aloft)	Alternate red and blue line (broken)
▲●▲●	Occluded Front	Continuous purple line
△○△○	Occluded Front (aloft)	Broken purple line

Frontogenesis
Symbols

On Printed Charts		On Weather Map	
▲ ▲ ▲ ▲	Cold Front	● ● ● ●	Blue
● ● ● ●	Warm Front	○ ○ ○ ○	Red
● ▲ ● ▲	Stationary Front	● ○ ● ○	Blue and Red

Frontolysis
Symbols

On Printed Charts		On Weather Map	
▲ – ▲ – ▲ – ▲	Cold Front	/////////	Blue
● – ● – ● – ●	Warm Front	/////////	Red
● – ▲ – ● – ▲	Occluded Front	/////////	Purple
● – ▲ – ● – ▲	Stationary Front	/////////	Blue and Red

Squall Line

— . . — . . — . . — . . —
Purple

Trowal/Trough
Symbols indicate the position of a trough of warm air aloft.

— — — — —
Purple

Ridge

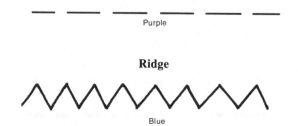

Blue

Centre of Pressure Movements
Symbols indicate the position of high or low centres of pressure — generally at 12 hr. intervals.

Air Masses

cA—Continental Arctic   mP—Maritime Polar
mA—Marine Arctic   mT—Maritime Tropical

Precipitation and Obstructions to Vision on Surface Weather Maps

Because surface weather maps provide a complete picture of actual surface conditions, they must, in addition to indicating pressure patterns, air masses, and frontal systems, show areas and types of precipitation and obstructions to vision. These features, like those already described, are depicted symbolically.

To fully interpret surface weather maps, you must be able to recognize and decode the symbols shown on the table below. Be sure to STUDY THE TABLE CAREFULLY and learn each symbol.

SYMBOL	DESCRIPTION OF SYMBOL	TRANSLATION
1. (dotted shading)	green shading	continuous rain
2. (hatched shading)	green hatching	intermittent rain
3. ,	green comma	drizzle
4. ▲	green triangle with green period in centre	ice pellets
5. ✶	green star	snow
6. ∼	green curved line (used with comma or period)	freezing rain (with period) freezing drizzle (with comma)
7. •/▽	green period over green inverted triangle	rain shower
8. ✶/▽	green star over green inverted triangle	snow shower
9. ↳	green T with green lightning symbol	thunderstorm
10. (yellow shading)	yellow shading	fog

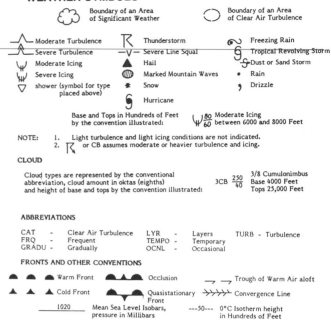

SIGNIFICANT WEATHER PROGNOSTIC CHARTS — WEATHER SYMBOLS

- Boundary of an Area of Significant Weather
- Boundary of an Area of Clear Air Turbulence
- Moderate Turbulence
- Thunderstorm
- Freezing Rain
- Severe Turbulence
- Severe Line Squall
- Tropical Revolving Storm
- Moderate Icing
- Hail
- Dust or Sand Storm
- Severe Icing
- Marked Mountain Waves
- Rain
- shower (symbol for type placed above)
- Snow
- Drizzle
- Hurricane

Base and Tops in Hundreds of Feet by the convention illustrated: Moderate Icing 80/60 between 6000 and 8000 Feet

NOTE:
1. Light turbulence and light icing conditions are not indicated.
2. ↳ or CB assumes moderate or heavier turbulence and icing.

CLOUD

Cloud types are represented by the conventional abbreviation, cloud amount in oktas (eighths) and height of base and tops by the convention illustrated: 3CB 250/40   3/8 Cumulonimbus Base 4000 Feet Tops 25,000 Feet

ABBREVIATIONS

- CAT - Clear Air Turbulence
- LYR - Layers
- TURB - Turbulence
- FRQ - Frequent
- TEMPO - Temporary
- GRADU - Gradually
- OCNL - Occasional

FRONTS AND OTHER CONVENTIONS

- Warm Front
- Occlusion
- Trough of Warm Air aloft
- Cold Front
- Quasistationary Front
- Convergence Line
- 1020 Mean Sea Level Isobars, pressure in Millibars
- ---50--- 0°C Isotherm height in Hundreds of Feet
- H 1020   L×996   Centres of Low and High Pressure respectively Pressure in Millibars

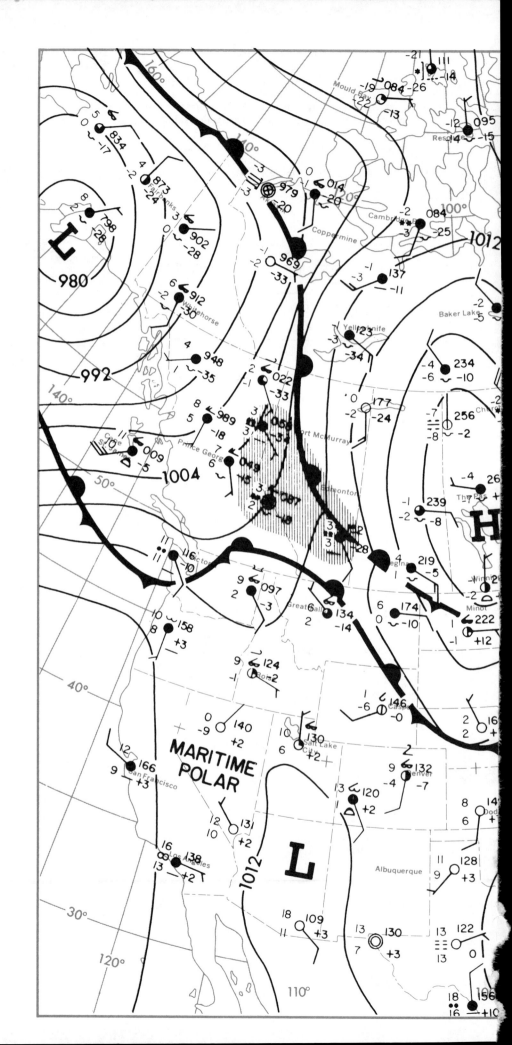

Examination Guide

The specimen examination questions which appear in this chapter are based on both the Private and Commercial Pilot Examination Guides. They are designed to assist you in preparing for your own examination.

It cannot be too strongly emphasized that they are **sample questions only,** not the **actual questions** which you will be called upon to answer — so do not attempt to memorize the answers. The questions are offered merely as a guide to the subject matter on which they are based. Study that subject matter until you know it sufficiently well to be prepared to answer any question of a similar nature which could possibly be constructed on an examination paper.

Always read the statement first. Be sure you understand what it means. Decide the correct answer in your own mind (or plot it if it is a problem) — **before you look at the multiple choice answers.** Then select the answer which most nearly coincides with your own reasoning.

The correct one of the four multiple choice answers will be found in "Answers To The Questions" following the questions. Mark your selection of the correct answer before you consult the table.

Do not be discouraged if your answer does not exactly agree with the one which appears in this manual. Every effort is made to keep this publication completely up-to-date by constant revision but changes occur so fast that it is difficult to keep pace with them all.

Canadian Aviation Regulations (CARS) are also subject to endless change. No attempt has been made in this manual to reproduce these regulations in their entirety although some of the regulations have been outlined in the Section **Rules of the Air** and in the Chapter **Aeronautical Rules and Facilities.** The Canadian Aviation Regulations (to be published in 1996) are a consolidation of the Air Regulations, the Air Navigation Orders and the Air Regulations Series. They are divided into eight Parts. Each Part will incorporate all regulations relating to a particular area of aviation, as follows: Part I — General Provisions; Part II — Identification, Registration and Leasing of Aircraft; Part III — Aerodromes and Airports; Part IV — Personnel Licensing and Training; Part V — Airworthiness; Part VI — General Operating and Flight Rules; Part VII — Commercial Air Services; Part VIII — Air Navigation Services. It is important for a student pilot to study the regulations, with particular reference to the sections which apply to the type of licence towards which he/she is working (i.e. private, commercial, recreational, ultra-light, VFR, IFR, etc.).

A.I.P. Canada (Aeronautical Information Publication) — A Transport Canada publication in which all information of interest to pilots and navigators is consolidated. Amendments are published and distributed every 56 days. The publication contains information on aerodromes, communications, meteorology, rules of the air and air traffic services, facilitation, search and rescue, aeronautical charts, licensing, registration and airworthiness and airmanship. The publication is intended to consolidate in one publication all preflight reference information required by pilots in planning flights. It incorporates in plain language a description of the Canadian Aviation Regulations.

NOTAM — An acronym for Notices to Airmen, NOTAMs are used to advertise information concerning the establishment, condition or any change in any aeronautical facility, service or procedure, the knowledge of which is essential to pilots.

Canadian NOTAMs are distributed by an Automated Data Interchange System (ADIS) to all flight service stations and are available to pilots on a continuing basis They may also be broadcast as a Voice Advisory on the FSS frequency in the locality for which the NOTAM has been issued. NOTAMs are issued to disseminate vital information about the condition of aeronautical facilities such as navigation aids, airports, runways, approach systems, communication frequencies, changes in designated airspace or air traffic procedures, hazards, military exercises or maneuvers and special events.

A.I.P. Canada Supplements are published and distributed by mail to advise pilots of changes to procedures and to disseminate information of a general nature, to advise of amendments, temporary changes of operational significance. A.I.P. Supplements replace Class II NOTAMS.

Information Circulars and Aviation Notices. Information Circulars provide advance information of major changes to legislation, regulations or procedures. Aviation Notices are used to disseminate information which is mainly of interest to specific regions or when the information is of a one-time nature and the notice will have served its purpose on first reading.

Canadian Examinations

Private Pilot — Aeroplane. The written examination for a Private Pilot's Licence in Canada consists of a single paper of 100 questions made up of 20 questions on air law, 20 questions on navigation (which includes airways and cross country procedures), 30 questions on meteorology and 30 questions on general knowledge (which includes theory of flight, engines, airframes and general airmanship).

Commercial Pilot — Aeroplane. The Commercial Pilot's examination consists also of one paper of 100 questions on the following subjects: Canadian Aviation Regulations, air traffic rules and procedures, navigation, flight planning, radio aids to navigation, meteorology, theory of flight, aircraft operating procedures, airframes and aero engines, airmanship.

Recreational Pilot Permit (RPP). The RPP is a new permit that was introduced in 1995. It requires that the applicant complete not less than 25 hours of flight training and pass an examination composed of 80 questions, 20 on each of the following subjects: air law, navigation, meteorology and general knowledge (which includes airframes, engines and systems, theory of flight, flight instruments, flight operations and human factors). The holder of a Recreational Pilot Permit is authorized to fly single engine, piston powered, non high performance aircraft with a maximum of four seats; is authorized to fly landplanes or seaplanes, or both, as endorsed, day VFR within Canada; may carry not more than one passenger.

Student Pilot Permit or Private Pilot Licence for Foreign and Military Applicants, Air Regulation Examination (PSTAR). All student pilots (with the exception of ultra-light and glider pilot applicants) must write the PSTAR examination before being issued their student pilot permits. The exam consists of 50 questions on the following subject areas: Canadian Aviation Regulations, air traffic control clearances and instructions, air traffic control procedures as they apply to the control of VFR traffic, air traffic procedures at uncontrolled airports and aerodromes, special VFR regulations, Information Circulars and NOTAMs. The pass mark is 90%.

Restricted Radio Operator's Licence. In addition to the knowledge of radio aids required to obtain either a Private or Commercial Pilot's Licence, a Restricted Radio Operator's Licence is required in Canada to operate an aircraft radio transmitter. The examination for the licence consists of a few simple questions to ensure that you understand how to operate receiving and transmitting equipment, that you know the regulations applicable to radiotelephone communications and procedures, and that you know the radio regulations relating to distress, urgency and safety. You are not required to know morse code or to understand electronic circuits.

Study Material

In addition to your text, A.I.P. Canada, the Canadian Aviation

Regulations, there are several booklets that are very helpful in preparing for the written pilot examination.

Sample Examination for Private Pilot Licence — Aeroplane (TP 2894). A small booklet containing questions typical of those found in current Transport Canada examinations. Because this booklet is prepared by the authorities who set the civil pilot examinations, it is a good guide to a candidate preparing to sit for an examination.

Study and Reference Guide for the PSTAR Examination (TP 11919E) contains over 200 questions typical of those found in the PSTAR examination.

Study and Reference Guide for Private Pilot and Commercial Pilot Licences (TP 5717). A complete syllabus of Canadian pilot examination requirements. Outlines the subject matter and degree of knowledge required to pass the Transport Canada examinations in the following subjects: aviation regulations, airframes and engines, theory of flight, flight instruments, meteorology, navigation and radio aids, and flight operations.

These publications are available from Transport Canada, Ottawa, Canada.

The following publication, while not a "must" for examination requirements, contains extremely valuable information for both VFR and IFR operations, and should be in the possession of all Commercial and Airline Transport Pilots in Canada.

Canada Air Pilot. Published in seven volumes, one for each of the seven Transport Canada regions in Canada. Contains complete and detailed data on all aerodromes, seaplane bases, customs airports, radio aids (VOR and LF/MF), airways, radio stations, beacons, broadcasting stations, radio frequency bands, instrument approach and landing procedures, D/F procedures, ADIZ, and mountain regions. Aerodrome charts show airport runways, radio aids, instrument landing procedures, field data, ground facilities, etc.

The above publication is obtainable from the Department of Energy, Mines and Resources, Ottawa, Ontario.

Sample Questions

1. It is the pilot's responsibility to ascertain that certain certificates and documents are aboard the airplane and appropriately displayed. These required documents include
   1. registration certificate, Flight Manual and current airworthiness certificate.
   2. current airworthiness certificate, operations limitations (Form 309) and Airplane Flight Manual.
   3. certificate of registration, certificate of airworthiness, journey log books, radio equipment licence or permit for each crew member.
   4. airplane operating limitations set forth in a Flight Manual, on placards, listings or marking, current airworthiness certificate, registration certificate.

2. When operating in accordance with VFR, aircraft shall be flown
   1. outside designated airways.
   2. outside control zones.
   3. with visual reference to the ground or water.
   4. only in Class D Airspace.

3. A private pilot may conduct a flight without visual reference to the ground or water if
   1. the flight is conducted in Class B airspace.
   2. outside of controlled airspace.
   3. cleared for Special VFR.
   4. endorsed for instrument flight.

4. Except for the purpose of taking off or landing, an airplane may not be flown at a height of less than_____over an aerodrome, except as otherwise directed by the ATC unit.
   1. 1000 feet.   3. 3000 feet.
   2. 2000 feet.   4. 500 feet.

5. You are flying at night, off airways, at 3000 feet and observe the white light of an airplane at approximately your altitude and in your immediate vicinity. If there is a possibility of collision, the Canadian Aviation Regulations require that
   1. the other pilot alter course to the right.
   2. the other pilot alter course to the left.
   3. you alter course to the right.
   4. you alter course to the left.

6. When on final approach for landing, the right-of-way is normally given to
   1. faster airplanes.   3. airplanes at higher altitude.
   2. slower airplanes.   4. airplanes at lower altitude.

7. All facts relating to maintenance, repairs, new installations and modifications must be recorded in the
   1. Aircraft Journey Log.
   2. Aircraft Flight Manual.
   3. Aircraft Technical Log.
   4. Certificate of Airworthiness.

8. A tear in the skin of a monocoque fuselage
   1. will not affect the stress capability of the structure.
   2. might affect the stress capability of the structure.
   3. will cause flutter.
   4. increases the load factor.

9. A device fitted to a control surface to relieve control pressure on that control surface is called a
   1. rudder.   3. mass balance.
   2. adjustable stabilizer.   4. trim tab.

10. The ratio of the actual load acting on the wings to the gross weight of the airplane is called
    1. the aspect ratio.   3. the load factor.
    2. the power load.   4. compression.

11. V_a, the design maneuvering speed, is not indicated on the air speed indicator. It can normally be found in the Airplane Flight Manual or on a placard in the cockpit. This airspeed is important

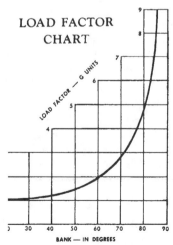

Fig. 12. Load Factors Produced at Varying Degrees of Bank at Constant Altitude.

because it is the recommended airspeed to be used when
1. flying the traffic circuit.
2. maneuvering with both the landing gear and wing flaps extended.
3. flying in extremely rough air or severe turbulence.
4. maximum range (miles per pound of fuel) is desired.

12. Referring to the Load Factor Chart in Fig. 12, if the maximum flight load factor is 3.8, what is the maximum bank at constant altitude and at maximum gross weight which could be made without exceeding this load factor?
1. about 67°.
2. about 74°.
3. less than 50°.
4. not over 76 gs.

13. A torsion stress is caused by
1. bending.
2. stretching.
3. corrosion.
4. twisting.

14. The five principal factors which affect the magnitude of lift and drag of an airplane wing are
1. angle of attack, gross weight, lift coefficient, shape of the airfoil, wind velocity.
2. angle of attack, shape of the airfoil, wing area, airspeed, air density.
3. angle of attack, shape of the airfoil, aspect ratio, coefficient of lift, gross weight.
4. angle of attack, thrust, gross weight, airspeed, air density.

15. When an aircraft is flying in normal straight and level flight, at a constant airspeed,
1. weight equals drag and thrust equals lift.
2. drag equals lift and weight equals thrust.
3. lift equals weight and drag equals thrust.
4. weight and thrust equal lift.

16. The relative wind has an effect on the
1. angle of incidence.
2. angle of attack.
3. coefficient of drag.
4. asymmetric thrust.

17. When the angle of attack of the wing is increased to the point where the wing stalls, the centre of pressure will
1. move rearward and then forward.
2. move forward.
3. move forward and then rearward.
4. remain stationary.

18. What is profile drag?
1. drag produced by the generation of lift.
2. drag produced by a positive angle of attack.
3. form drag plus skin friction.
4. wing-tip vortices.

19. The best lift/drag ratio is achieved
1. gliding for maximum range.
2. gliding for maximum endurance.
3. climbing at best angle of climb airspeed.
4. climbing at best rate of climb airspeed.

20. When an airplane is flying very near the ground and influenced by ground effect,
1. parasite drag is reduced.
2. skin friction is reduced.
3. the wing tip vortices are strengthened.
4. induced drag is reduced.

21. On certain airplanes, the manufacturer recommends the use of some degree of flap during take-off. Flaps, under these circumstances,
1. permit a higher take-off speed.
2. permit a better angle of climb.
3. increase the upper camber of the wing and produce more lift.
4. change the angle of incidence of the wing.

22. The motion of an airplane about its normal axis is known as
1. yawing.
2. looping.
3. rolling.
4. pitching.

23. A mass of streamlined shape fitted in front of the hinge of a control surface is incorporated to
1. countcract flutter.
2. counteract adverse yaw.
3. increase the centre of gravity range.
4. improve directional stability.

24. An aircraft that exhibits positive stability on the longitudinal axis and neutral stability on the lateral axis would correct
1. a pitch displacement but would require positive correction of a roll.
2. a yaw displacement but would require positive correction of a pitch.
3. a pitch displacement but would require positive correction of a yaw.
4. a roll displacement but would require positive correction of a pitch.

25. Lateral stability is improved by
1. balanced controls.
2. offsetting the vertical fin.
3. a high aspect ratio.
4. dihedral.

26. In comparison with a take-off and climb under calm conditions, in climbing an airplane into wind you should recognize that one of the following conditions exists.
1. a lower angle of attack is normally employed.
2. the rate of climb is increased.
3. the angle of climb is increased.
4. increased power is normally used.

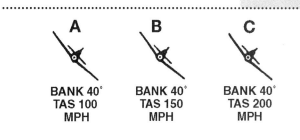

FIG. 27.

27. Referring to Fig. 27 and assuming that all the airplanes are making co-ordinated turns, which of the following statements is correct?
1. Airplane B's rate of turn and radius of turn is the same as airplanes A and C.
2. Airplanes A, B and C have the same rate of turn, but airplane C has the largest radius of turn.
3. Airplane A has the greatest rate and smallest radius of turn.
4. Airplane C has the greatest rate and largest radius of turn.

28. An aircraft which has a normal stalling speed of 70 knots is in a steep 60° turn. What is the stalling speed in the steep turn?
1. 99 knots.
2. 70 knots.
3. 140 knots.
4. 75 knots.

COMMENTS: In a 60° bank turn with a load factor of 2, multiply the normal stall speed by the square root of the load factor being imposed. Refer to Section **Stall** in the Chapter **Theory of Flight**.

29. If a thin coating of frost or light snow has formed on the wings of an airplane, the take-off should not be attempted until it has been removed because

1. the coating will disturb the airflow over the wings and destroy some of the lifting capability.
2. the added weight overloads the airplane.
3. the covering is cold and brittle.
4. the cold air has contracted the covering of the wings, thus changing the airfoil section.

30. An aircraft stalls at an indicated airspeed of 60 knots at sea level. At 10,000 feet ASL, the aircraft, at the same weight, stalls at
   1. an indicated airspeed of 70 knots.
   2. a true airspeed of 60 knots.
   3. an indicated airspeed of 60 knots.
   4. a true airspeed lower than the indicated airspeed.

31. During autorotation, the downgoing wing has
   1. the same angle of attack as the upgoing wing.
   2. a greater angle of attack than the upgoing wing.
   3. a lesser angle of attack than the upgoing wing.
   4. is more affected by the angle of incidence.

32. An airplane flying at 10,000 feet ASL in the Altimeter Setting Region should have its altimeter set to
   1. the altimeter setting of the aerodrome of departure.
   2. the altimeter setting of the aerodrome of intended landing.
   3. 29.92 inches mercury.
   4. the altimeter setting of the nearest aerodrome.

33. If the pitot pressure source becomes blocked by dirt, water or ice, inaccurate readings will be caused in the
   1. altimeter.   3. airspeed indicator.
   2. attitude indicator. 4. vertical speed indicator.

34. Of the four altimeters illustrated in Fig. 34, select the one that indicates an altitude of 880 feet.
   1. W.   3. Y.
   2. X.   4. Z.

W   X   Y   Z

Fig. 34.

Listed below are eight calibrated airspeed limitations and ranges depicted by the airspeed indicator illustrated in Fig. 35.
   V_{NE} — Never exceed speed.
   V_{SO} — Power-off stall with gear and flaps "landing" (down)
   V_{SL} — Power-off stall with gear and flaps "clean" (up)
   V_{NO} — The maximum permitted speed for normal operations.
   V_{FE} — Max. flaps down speed,
   Flap operating range.
   Normal operating range.
   Caution range.

35. By reference to the airspeed indicator illustrated in Fig. 35, you determine that V_{NO} is
   1. the upper limit of the white arc.
   2. 38 knots less than V_{NE}.
   3. the upper limit of the yellow arc.
   4. 140 knots (the mid point of the yellow arc).

36. In a reciprocating engine operating on a four stroke cycle, during the compression stroke,
   1. the intake valve is closed and the exhaust valve is open.

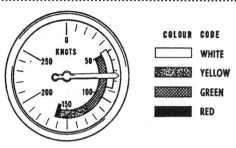

Fig. 35.

   2. both valves are open.
   3. both valves are closed.
   4. the exhaust valve is closed and the intake valve is open.

37. In a piston engine, the camshaft
   1. rotates at the same speed as the crankshaft.
   2. rotates at half the speed of the crankshaft.
   3. rotates at twice the speed of the crankshaft.
   4. is independent of the crankshaft.

38. The volume of the cylinder at the bottom of the compression stroke compared to the volume at the top of the stroke is called
   1. the mean effective pressure.
   2. the volumetric efficiency.
   3. the piston displacement.
   4. the compression ratio.

39. When a check of the dip stick indicates the need for two litres of oil, the attendant informs you that he has only detergent type oil available. This means that
   1. it should never be used in engines over 125 hp.
   2. it can be used only in engines with suitable filters and flexible oil lines.
   3. it is the latest type of oil and should always be used when available.
   4. the oil contains additives and should be added only to the same type of oil.

40. After oil dilution has been applied to an engine,
   1. the sump should be topped up with engine oil.
   2. the oil should be drained into a container and kept in a warm place.
   3. the oil should be vented.
   4. the engine should be run for an appropriate time at a sufficiently high temperature to boil off the fuel before departing on the next flight.

41. You make certain that no less than the specified octane gas is used in filling your tanks. Use of gasoline with a lower octane rating may cause
   1. rough operation and lower manifold pressure.
   2. detonation.
   3. pre-ignition and increased output.
   4. the spark plugs to foul.

42. The fuel is drawn from the carburetor float chamber into the engine by means of
   1. a fuel pump.   3. a venturi effect.
   2. an acceleration pump. 4. a turbocharger.

43. At cruise power, above 5000 feet, it is generally advisable to lean the engine to "lean best power"
   1. to achieve economy of fuel consumption.
   2. to prevent detonation.
   3. to prevent carburetor icing.

4. to cool the engine.

44. The purpose of an exhaust gas temperature (EGT) analyzer is to indicate the
   1. fuel flow to the engine.
   2. temperature of the air entering the air intake.
   3. temperature of the exhaust gas in the exhaust manifold to accurately adjust mixture setting.
   4. percentage of fuel in the cylinders.

45. Carburetor ice may be detected by
   1. an increase in rpm.
   2. a decrease in exhaust gas temperature.
   3. an increase in cylinder head temperatures.
   4. a decrease in manifold pressure.

46. If carburetor ice is present, the restriction to airflow frequently causes a richer mixture and some loss of power. Application of carburetor heat will likely cause a further immediate loss of power. In this case you should
   1. leave full heat applied until all ice is dissipated.
   2. turn carburetor heat off and adjust the mixture control to obtain maximum rpm.
   3. decrease the amount of carburetor heat until the rpm increases.
   4. turn carburetor heat off and open the throttle to obtain desired power.

47. One of the principal advantages of the fuel injection system is
   1. economy of fuel consumption.
   2. improbability of carburetor icing.
   3. higher horsepower.
   4. better general performance.

48. A device that uses the energy of the hot exhaust gases to supply the engine with dense air at high altitude is called a
   1. supercharger.   3. turbocharger.
   2. augmentor system. 4. carburetor.

49. When the magneto switch is in the OFF position,
   1. the primary circuit of the magneto is grounded to the airframe.
   2. the contact breaker points are closed.
   3. the electrical components installed in the airplane will not operate.
   4. the high tension current in the secondary circuit is directed to the battery.

50. If during a flight at normal cruising speed, one magneto of a dual ignition system failed completely, it would normally cause
   1. a loss of approximately 75 rpm.
   2. the engine to overheat.
   3. excessive vibration of the engine.
   4. considerable extra load on the other magneto.

51. Coarse pitch, in reference to a propeller,
   1. is used for take-off.
   2. also is known as "increase rpm".
   3. is used for cruise.
   4. should be selected for landing.

52. In an airplane fitted with a constant speed propeller, any adjustment of the throttle
   1. is registered on the manifold pressure gauge.
   2. is registered on the tachometer.
   3. adjusts the speed of the propeller.
   4. adjusts the pitch of the propeller.

53. Before engaging the starter to warm up the engine, the propeller
   1. should be pulled through several times in order to build up adequate compression for starting.
   2. need not be pulled through. This is only necessary on an airplane without a starter.
   3. should be pulled through several times to loosen congealed oil and to partially prime the engine.
   4. should not be pulled through because of the danger of a kickback.

54. If your airplane has a maximum angle of climb at a speed of 82 knots but the manufacturer recommends climbing at 87 knots, the most probable reason for the higher recommended airspeed would be
   1. to obtain better visibility with the nose lowered.
   2. to obtain better cooling of the engine.
   3. to maintain a higher head temperature at the higher airspeed.
   4. danger of stalling in a climb at the lower airspeed.

55. What is the most commonly recommended practice for preventing condensation in the fuel tanks?
   1. drain a pint of fuel from the tank sumps each night.
   2. strain all fuel as it is put into the tanks.
   3. fill each fuel tank after every flight.
   4. install a quick drain gascolator.

56. A layer of dark grey cloud from which continuous rain or snow falls is called
   1. cirrostratus.   3. nimbostratus.
   2. altostratus.   4. cumulonimbus.

57. When flying through a layer of nimbostratus, you encounter sudden severe turbulence. The most probable cause is
   1. embedded towering cumulus or cumulonimbus.
   2. embedded altocumulus.
   3. embedded stratocumulus.
   4. embedded cirrocumulus.

58. The tropopause is lower
   1. in summer than in winter.
   2. over the equator than over the north pole.
   3. over the north pole than over the equator.
   4. south of the jet stream than north of it.

59. The clouds which appear, in sequence, in advance of an approaching warm front are
   1. cirrus, cirrostratus, altostratus and nimbostratus.
   2. cumulus, cumulonimbus, and nimbostratus.
   3. cirrus, cumulonimbus and nimbostratus.
   4. altostratus, cumulus, cumulonimbus and nimbostratus.

60. A low pressure area is
   1. also called an anti-cyclone.
   2. a region of relatively low pressure with the highest pressure at the centre.
   3. a region of relatively low pressure with the lowest pressure at the centre.
   4. a region in which the winds blow clockwise.

61. A steep pressure gradient indicates
   1. calm air.   3. strong winds.
   2. light winds.   4. rising pressure.

62. An aircraft is flying at 10,000 feet ASL on a track of 200°. In order to maintain this track, the pilot is holding a heading of 185°. An area of low pressure exists
   1. to the right of track.   3. ahead of the airplane.
   2. to the left of the track. 4. behind the airplane.

63. At night as the sides of the hills or mountains cool, the air in contact with them tends to become denser and blows down the slope into the valley. This wind is called a
   1. anabatic wind.   3. mountain wave.

2. katabatic wind.   4. land breeze.

64. If you were flying on a westerly heading and encountered a range of mountains lying in a north south line and if you attempted to fly through a saddle on a west heading with the wind from the west, you would expect
    1. to lose altitude rapidly on the west side of the saddle.
    2. to lose altitude rapidly on the east side of the saddle.
    3. to lose altitude rapidly while in the saddle.
    4. to gain altitude rapidly on the east side of the saddle.

65. Lenticular clouds are usually associated with
    1. a warm front.   3. a mountain wave.
    2. a cold front.   4. inversions.

66. A sudden violent change in wind speed or direction that can impose severe penalties on an airplane's performance is called
    1. wind tip vortices.   3. diurnal variation.
    2. wind shear.   4. clear air turbulence.

67. When unsaturated air is forced to rise, the expansion of the rising air causes it to cool. This cooling is called
    1. the normal lapse rate and is 1.98°C per 1000 feet.
    2. advection cooling and is 2°C per 1000 feet.
    3. radiation cooling and is 3°C per 1000 feet.
    4. the dry adiabatic lapse rate and is 3°C per 1000 feet.

68. The amount of water that a given volume of air can contain at a given pressure is governed by
    1. the temperature.   3. the stability.
    2. the relative humidity. 4. the lapse rate.

69. In certain circumstances, air at higher altitudes may be warmer than the air below it. This is called
    1. convection.   3. radiation.
    2. inversion.   4. lapse rate reversal.

70. Select the correct statement from the following.
    1. A shallow lapse rate indicates unstable air.
    2. An isothermal layer favours vertical motion.
    3. Vertical currents develop readily in unstable air.
    4. Visibility is always good in stable air.

71. An air mass may be defined as a large section of the____with uniform properties of ____and ____in the horizontal. The missing words are
    1. stratosphere, temperature, pressure.
    2. tropopause, stability, pressure.
    3. atmosphere, stability, moisture.
    4. troposphere, temperature, moisture.

72. A front is
    1. a narrow transition zone between a cyclone and an anticyclone.
    2. a line of thunderstorms.
    3. a narrow transition zone between two air masses.
    4. a mass of layer cloud which is very thick and which covers a wide area.

73. The cloud and precipitation that often develop at a cold front are caused by
    1. cold air climbing over the warm air.
    2. cold air being heated as it moves over the warm ground.
    3. warm air expanding as it is lifted by the advancing cold air.
    4. convergence.

74. In order for clouds to form in the atmosphere, relative humidity must be____, there must be____of the air and____must be present. The missing words are
    1. low, cooling, coalescence.
    2. high, cooling, condensation nuclei.
    3. high, heating, condensation nuclei.
    4. low, heating, coalescence.

75. Fog that forms on clear nights with light winds is known as
    1. advection fog.   3. upslope fog.
    2. steam fog.   4. radiation fog.

76. The dissipation of fog is assumed to result from
    1. an increase in the wind velocity that blows the fog away.
    2. the release of heat through the latent heat of vaporization.
    3. terrestrial radiation as sunlight filters down through the fog or stratus layer.
    4. subsidence.

77. Hoar frost forms on an aircraft as a result of
    1. supercooled water droplets freezing on impact.
    2. freezing rain striking the aircraft.
    3. water vapour turning directly into ice crystals on the aircraft.
    4. water vapour condensing on the aircraft and then freezing.

78. As a thunderstorm matures, strong downdrafts develop and cold air rushing down out of the cloud spreads along the surface well in advance of the storm itself. This is called
    1. a microburst.   3. virga.
    2. the gust front.   4. thunderstorm turbulence.

79. Mechanical turbulence is the result of
    1. convection.
    2. friction between the air and the ground.
    3. diurnal variation of wind.
    4. orographic lift.

80. The surface weather map (see insert) indicated that at map time
    1. Maritime Tropical Air is responsible for a Bermuda High over the Maritimes.
    2. the cold air over the prairies will move north and not affect central Ontario.
    3. central Alberta is enjoying clear skies and warm temperatures.
    4. a stationary front over the Eastern Seaboard is bringing precipitation to that area.

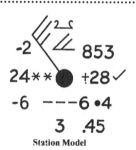

Station Model

Fig. 81.

81. Referring to the station model in Fig. 81, the temperature of the station is____, the dewpoint is____, the wind is from ____ and the barometric pressure is____.
    1. -2, -6, NW, 985.3 hPa.
    2. 24, 6, NE, 853.0 hPa.
    3. -6, 3, SE, 1028.0" Hg.
    4. -6, 02, W, 985.3 hPa.

Weather Data

TAF Aerodrome Forecasts: The following aerodrome forecasts were available at the Calgary Weather Office. They were issued at 1040Z and are valid from 1100Z on the date of issue to 1100Z on the following day.

YYC (Calgary) 02015KT P6SM SKC

YQL (Lethbridge) 36025KT P6SM OVC 030 FM 1600Z 31010KT P6SM SCT 200 BKN 600

YXH (Medicine Hat) 36020KT 5SM OVC 030 FM 1700Z 31015G25KT 5SM SCT 040 BKN 800

YQR (Regina) 31025KT 4SM -TSRA OVC 015 FM 1800Z 36015G25KT 5SM SCT 3000 BKN 900

METAR Aviation Routine Weather Reports: The following sequences were available at the Calgary Weather Office. They are the weather reports for 1300Z.

YYC 36020KT 15SM SKC 09/02 A3012

YQL 36010KT 8SM OVC 250 09/02 A3003

YXH 02025G30KT 10SM OVC 250 10/06 A2992

YQR 31020G25 KT 4SM -RA BKN 015 OVC 850 09/08 A2976

Upper Winds Forecast: The available forecast of winds at Calgary is valid from 1000Z to 2200Z.

	6000	9000	12000	18000
YYC	0215	3520	2930	2840

82. In the METAR aviation routine weather reports, cloud heights are reported in hundreds of feet above
    1. the highest terrain within a radius of 25 miles of the station of observation.
    2. the highest obstruction within 5 miles of the perimeter of the airport.
    3. the surface at the station of observation.
    4. MSL (mean sea level) at the station of observation.

83. At Calgary (YYC), the reported weather is
    1. wind is 360° at 20 kts — visibility 15 statute miles — sky clear — temperature 9°C — dewpoint 2°C — altimeter setting 30.12 inches.
    2. wind is 360° at 20 kts — visibility 15 statute miles — sky condition is overcast at 2000 feet — temperature 9°C — dewpoint 2°C — altimeter setting 30.12 inches.
    3. wind is 180° at 20 kts — visibility 15 nautical miles — sky clear — temperature 9°C — dewpoint 2°C — altimeter setting 30.12 inches.
    4. wind is 360° at 20 mph — visibility 15 statute miles — scattered clouds at 2500 feet — dewpoint 9°C — temperature 2°C — altimeter setting 3012 hPa.

84. The report of weather phenomena at Regina (YQR) is
    1. moderate rain.   3. very light freezing rain.
    2. light rain showers. 4. light snow showers.

85. Upper winds forecasts are available for many stations. These forecasts often prove very helpful in trip planning. Referring to the upper winds forecast, we find that the wind direction and speed at YYC between 1000Z and 2200Z at 7500 feet (by interpolation) would be approximately
    1. 350° at 20 knots. 3. 030° at 15 knots.
    2. 005° at 18 knots. 4. 040° at 32 knots.

86. The TAF aerodrome forecast for Medicine Hat is
    1. surface winds from 360° at 20 knots — sky condition overcast at 3000 feet.
    2. surface winds from 036° at 20 knots — sky condition overcast at 300 feet.
    3. surface winds from 310° at 25 knots — sky condition overcast at 3000 feet.

    4. surface winds from 200° at 36 knots — sky condition overcast at 3000 feet.

87. The geographical co-ordinates of Swift Current are (refer to the 1: 1,000,000 chart at the back of the book)
    1. 50°18'N, 107°41'W   3. 50°18'N, 107°41'E
    2. 51°42'N, 108°19'E   4. 50°36'N, 107°82'W

88. 1045 Mountain Daylight Time is
    1. 1745Z.   3. 0245Z.
    2. 1645Z.   4. 0345Z.

89. A curved line on the surface of the earth that cuts all meridians at the same angle is called a
    1. rhumb line.   3. azimuth line.
    2. great circle.   4. grivation.

90. Magnetic variation is defined as
    1. errors in the compass caused by the magnetic field associated with metal in the airplane frame and engine.
    2. the difference between the magnetic track and the magnetic heading of the airplane.
    3. the angle between true north and magnetic north at any given point on the earth.
    4. the difference between true track and magnetic track.

91. In the northern hemisphere, if your airplane is on a northerly heading and if a turn is made towards the east or the west, the compass reading
    1. will lag or indicate a turn in the opposite direction.
    2. will lead or indicate a turn in the opposite direction.
    3. is inversely proportional to the magnetic variation for the particular area.
    4. will be correct, providing the turn has been perfectly co-ordinated.

92. In an airplane in straight and level flight, the magnetic compass will indicate a turn to the left when
    1. accelerating on a heading of 270°.
    2. decelerating on a heading of 090°.
    3. decelerating on a heading of 270°.
    4. accelerating on a heading of 000°.

93. Associate the following terms with the appropriate definitions.
    A. Isogonic Line   C. Agonic Line.
    B. Isobar.   D. Isotherm.
    W. A line connecting points of equal pressure on a constant altitude chart.
    X. A line on a chart joining points of zero variation.
    Y. A line on a chart joining points of equal variation.
    Z. A line on a chart joining places or points having equal temperatures.

    1. A-W, B-X, C-Z, D-Y.   3. A-Y, B-W, C-X, D-Z.
    2. A-X, B-W, C-X, D-Z.   4. A-Y, B-Z, C-W, D-X.

94. In the vicinity of the magnetic poles there is an area
    1. in which all airplanes must be under positive control.
    2. in which the compass is unreliable.
    3. in which you must reset the heading indicator every 3 minutes.
    4. in which the northerly turning error of the compass is more pronounced than usual.

95. VFR Navigation Charts (VNC Series) are based on a Lambert Conformal Conic Projection. The following list contains 3 properties characteristic of this projection and 3 other properties characteristic of a Mercator Projection.
    A. Meridians are straight, parallel lines, intersecting the parallels (which are also straight, parallel lines) at right angles.
    B. Meridians are converging lines, intersecting the parallels

(which are concentric arcs of circles) at right angles.
C. A straight line represents a great circle.
D. A straight line represents a rhumb line.
E. Scale errors are small, hence the scale may be considered constant over a single sheet.
F. Scale varies with latitude and is only constant at the equator, hence distances must be measured on the latitude scale adjacent to the area.

Which of the following includes the 3 items which are characteristic of the Lambert Conformal Conic Projection?
1. A — C — F. 3. A — D — E.
2. B — D — F. 4. B — C — E.

96. What does a map scale expressed as 1: 1,000,000 mean?
1. 1 inch = 8 miles.
2. 1 inch on the map = one millionth of a mile.
3. 1 mile equals = 1,000,000 inches.
4. 1 inch on the map = 1,000,000 inches on the ground.

97. Complete familiarity with the compass rose is essential to navigation. Which of these statements is true?
1. The reciprocal of 267° is 117°.
2. The direction NE is 090°.
3. To turn right 90° from a heading of 145° would be a turn to 235°.
4. The reciprocal of 115° is 225°.

98. You plan to conduct a flight from Lindsay (N44° 22' W78° 47') to Collingwood (N44° 27' W80° 09') at an airspeed of 120 kts. (Refer to the l: 500,000 chart of the map insert at the back of the book.) The wind is forecast to be from 300° at 20 knots. What is the magnetic heading to steer and what is the anticipated groundspeed?
1. 279°M and 139 knots. 3. 271°M and 101 knots.
2. 289°M and 101 knots. 4. 281°M and 139 knots.

99. What is the distance from Lindsay to Collingwood?
1. 68 n. miles. 3. 59 n. miles.
2. 59 s. miles. 4. 118 n. miles.

100. Based on your anticipated groundspeed, how long should the flight take?
1. 35 minutes. 3. 41 minutes.
2. 30 minutes. 4. 70 minutes.

101. You are on a flight between Peterborough (N44° 13' W78° 21') and Wiarton (N44° 44' W81° 06'). At 1015 you cross the eastern shore of Lake Simcoe on track. At 1033 you see immediately below you the eastern shore of Georgian Bay at Wasaga Beach. Assuming that you are on track, what groundspeed are you making?
1. 90 mph. 3. 140 knots.
2. 120 mph. 4. 120 knots.

102. Based on that groundspeed, when will you arrive over Wiarton?
1. 1050. 3. 1045.
2. 1057. 4. 1110.

103. You are planning a flight from Calgary (N51° 06' W114° 01') to Swift Current (N50° 17' W107° 41') at an estimated groundspeed of 140 knots. (Refer to the map insert at the back of the book. Use the 1: 1,000,000 chart.) Your airplane's average fuel consumption is 12 gals U.S. per hour. Usable fuel capacity is 48 gallons U.S. Allowing for a 45 minute fuel reserve, what are the safe fuel hours still available when you land at Swift Current?
1. 1 hr. 15 min. 3. 1 hr. 30 min.
2. 1 hr. 4. 45 minutes.

104. On a flight between Midland (N44° 41' W79° 55') and Oshawa (N43° 55' W78° 53'), some useful landmarks would be
1. Highway 400 — Victor Airway V 37 — the east shore of Lake Simcoe — the town of Uxbridge.
2. Highway 400 — the east shore of Lake Simcoe — Alert Area CYA 503 — the town of Uxbridge.
3. Highway 400 — Victor Airway V 216 — the town of Uxbridge — Highway 401.
4. Highway 400 — the west shore of Lake Simcoe — the rail road running north from Toronto to Beaverton — the town of Uxbridge.

105. An airport is
1. a registered aerodrome.
2. an aerodrome with a control tower.
3. an aerodrome in respect of which a certificate is in force.
4. an aerodrome with paved runways.

106. The west end of a runway oriented east and west is numbered
1. 09. 3. 27.
2. 90. 4. 270.

107. The runway is lighted at night by two parallel lines of ___ lights; taxiways by two parallel lines ___ of lights. The missing colours are
1. white, green. 3. white, blue.
2. white, red. 4. green, white.

108. Where taxiway holding positions have not been established, aircraft waiting to enter an active runway should normally hold
1. 100 feet from the edge of the runway.
2. at the edge of the runway.
3. at least 200 feet from the edge of the runway.
4. behind the threshold line.

109. You are well advanced on final approach and the runway is clear of traffic. You have not received landing clearance and are unable to establish contact with the tower. You should
1. continue and land as the runway is clear.
2. continue and land as you assume the tower controller is busy with other duties for the moment.
3. continue and land because with a runway clear of traffic, landing clearance is just a formality.
4. pull up and make another circuit.

110. Your destination airport is one for which a Special Procedures NOTAM has been published. It states that the circuit height is 1500 feet ASL. The airport elevation is 400 feet ASL. The ceiling is 1000 feet overcast and the visibility is 3 miles. Under these circumstances, your circuit altitude should be
1. 1100 feet above the airport elevation.
2. 1000 feet above the airport elevation.
3. 500 feet below the cloud base.
4. as high as possible without entering the cloud.

111. An aircraft is "cleared to the circuit" where a left hand circuit is in effect. Without further approval from ATC, a right turn may be made to
1. enter the circuit from the upwind side, crosswind or on the downwind leg.
2. enter the base leg.
3. enter the final leg.
4. descend on downwind leg.

112. You have been authorized to taxi to the runway in use. To get there you must cross two taxiways and one runway. Your authorization allows you to
1. taxi to position on the runway in use.
2. taxi to hold short of the runway in use but get clearance to cross the other runway.

3. taxi to hold short of the runway in use, but get clearance to cross each taxiway and runway en route.
4. taxi to, but hold short of the runway in use.

113. Accepting a clearance for an "immediate take-off" allows the pilot to
   1. back track on the live runway to use the maximum available length for take-off.
   2. taxi to a full stop in position on the runway and take off without further clearance.
   3. taxi onto the runway and take off in one continuous movement.
   4. complete the run-up and pre-take-off check on the runway.

114. A departing VFR aircraft will normally clear tower frequency when
   1. an altitude of 2000 feet above the ground has been reached.
   2. the flight is 15 miles from the airport.
   3. the flight is 10 minutes flying time from the airport.
   4. the flight is clear of the control zone.

115. The specified area associated with a mandatory frequency and within which a pilot is required to contact the ground station and report his intentions normally is
   1. 5 n.m. in radius and extends up to 3000 ft. AGL.
   2. 10 n.m. in radius and extends up to 3000 ft. AGL.
   3. 5 n.m. in radius and extends up to 5000 ft. AGL.
   4. 10 n.m. in radius and extends up to 5000 ft. AGL.

116. Pilots flying VFR and intending to land at an aerodrome where no mandatory frequency, aerodrome traffic frequency or ground station exists, should broadcast their intentions on
   1. 123.2 MHz.   3. 126.7 MHz.
   2. 122.8 MHz.   4. 121.5 MHz.

117. A series of red flashes directed to an airplane in flight means
   1. airport unsafe, do not land.
   2. give way to other aircraft and continue circling.
   3. do not land for the time being.
   4. you are in a prohibited area, alter course.

118. A flashing white light from the control tower to an airplane taxiing or about to take off means
   1. clear the runway.
   2. delay your take-off.
   3. return to the ramp or hangar.
   4. be on the alert for hazardous conditions.

119. When in VFR flight in the Northern Domestic Airspace, the altimeter should be set to_____ and the direction of flight should be determined by_____ track calculations.
   1. the current altimeter setting of the nearest station along the route of flight — true.
   2. 29.92' Hg — magnetic.
   3. the current altimeter setting of the nearest station along the route of flight — magnetic.
   4. 29.92' Hg — true.

120. To transit VFR through any part of Class C Airspace, pilots must
   1. have a special Class C endorsement to their licences.
   2. monitor 126.7 MHz.
   3. advise ATC of their intentions and obtain a clearance.
   4. obtain prior permission in writing from the ATC unit.

121. To conduct a VFR flight in Class B Airspace, a pilot must
   1. have a Class B licence endorsement.
   2. be flying an airplane equipped with a full IFR instrument panel.
   3. advise ATC of his intentions and obtain a clearance.
   4. squawk code 1200 on the transponder.

122. The primary function of Class C Airspace to VFR flights is to
   1. provide headings to the pilot that will "home" the aircraft to the airport.
   2. provide assistance to a pilot flying above an overcast cloud condition.
   3. more efficiently integrate the flow of VFR and IFR flights within the classified airspace.
   4. permit "Special VFR" flights to be conducted within the control zone.

123. An airspace classified as Class F and indicated on an aeronautical chart by the code CYA(S) denotes
   1. a danger area with artillery activity.
   2. a restricted area with surveillance activity.
   3. an advisory area with soaring activity.
   4. an altitude reservation.

124. Which of the following extends upwards from the surface of the earth?
   1. Terminal Control Area.   3. Low Level Airway.
   2. Control Zone.   4. Control Area Extension.

125. An area in which air traffic control extends upwards from a specified height above the surface is called
   1. a control zone.   3. a flight information area.
   2. a restricted area.   4. a control area.

126. A flight plan must be filed for all VFR flights
   1. along designated airways.
   2. at night.
   3. above 3000 feet AGL.
   4. to or from a military aerodrome.

127. When a VFR flight plan has been filed with the appropriate ATC unit, the pilot in command shall report his arrival within
   1. 30 minutes after landing.
   2. 1 hour after landing.
   3. 12 hours after landing.
   4. 24 hours after landing.

128. A VFR flight plan must be closed by the pilot
   1. by advising ATC either directly or via a communications base of his arrival time.
   2. except at airports served by a control tower in which case the tower will close the flight plan.
   3. by parking his aircraft in close proximity to the tower.
   4. by filing a return flight plan.

129. Using the following data, which time should be entered in the "Elapsed Time" section of the flight plan form?
   Estimated elapsed time A to B — 1 hour 10 minutes
   Estimated stopover time at B — 45 minutes.
   Estimated elapsed time B to C — 1 hour 30 minutes.
   1. 3 hours 25 minutes.
   2. 2 hours 40 minutes.
   3. 3 hours 25 minutes plus 45 minutes for reserve.
   4. 1 hour 55 minutes.

130. An Air Traffic Control "Instruction"
   1. is the same as an Air Traffic Control "Clearance".
   2. must be complied with when received by the pilot to whom it is directed unless he considers it unsafe to do so.
   3. must be "read back" in full to the controller and confirmed before becoming effective.
   4. is in effect advice provided by ATC and does not require acceptance or formal acknowledgement by the pilot concerned.

131. A pilot accepts a clearance and subsequently finds that all or part of the clearance cannot be complied with. He should
   1. comply as best he can under the circumstances to carry out the clearance and say nothing to ATC.

313

Examination Guide

2. comply as best he can under the circumstances and advise ATC as soon as possible.
3. disregard the clearance.
4. comply with only the part that suits him.

132. You are flying an airplane outside of a control area on a cross country flight at an altitude of more than 3000 feet but below 18,000 feet. Your magnetic track is 155°. Flight visibility is 2 1/2 miles. Your altitude should be
1. odd thousands plus 500 feet.
2. even thousands plus 500 feet.
3. odd thousands.
4. assigned to you by ATC.

133. What is the minimum height AGL above which aircraft shall comply with the Cruising Altitude Order?
1. 700 feet.   3. 3000 feet.
2. 2200 feet.   4. 3500 feet.

134. In Canada, flights conducted VFR over the top
1. may not be conducted in Class B Airspace.
2. may be conducted if the weather at ground level is IFR.
3. are not permitted.
4. may be conducted if the departure point and the destination are forecast to have VFR weather.

135. The minimum flight visibility for airplanes under VFR outside controlled airspace is ____; within control zones, the minimum flight visibility is ____. The missing figures are
1. 500 feet, 1 mile.   3. 1 mile, 1 mile.
2. 1 mile, 3 miles.   4. 3 miles, 3 miles.

136. The pilot of a fixed wing airplane taking off on a VFR flight must have sufficient fuel to complete the flight with forecast weather conditions, plus reserve fuel for
1. climb to cruising altitude.
2. 1 hour at normal cruise speed.
3. 45 minutes at normal cruise speed.
4. 100 miles diversion to an alternate airport.

137. Special VFR may be authorized by the appropriate ATC unit to an airplane with a functioning two way radio when flight and ground visibility are not less than
1. 5 miles.   3. l/2 miles.
2. 2 miles.   4. 1 mile.

138. An aircraft flying in accordance with Special VFR would be flying within a
1. control zone.   3. terminal control area.
2. aerodrome traffic zone. 4. the speed limit area.

139. A pilot on a Special VFR flight has been cleared to the circuit. Ahead and below his flight path there is an overcast layer of stratus cloud. The responsibility of remaining clear of cloud is
1. the responsibility of the pilot.
2. the responsibility of the tower controller as it is within a control zone.
3. the responsibility of ATC because the weather is below VFR.
4. the responsibility is shared equally by the pilot and ATC.

140. ADIZ rules apply
1. only to aircraft flying at a true airspeed of 180 or more.
2. only to aircraft flying above Flight Level 180.
3. to all aircraft.
4. to all southbound aircraft.

141. In radio transmission, ground waves follow the surface of the earth. As they come in contact with the surface, the ground waves slow down. This slowing down is called
1. diffraction.   3. surface attenuation.
2. skip.   4. terrain effect.

142. The long range capabilities of HF radio are attributable to the behaviour of
1. ground waves.   3. line of sight transmission.
2. sky waves.   4. phase differential.

143. Two of the following four statements are characteristic of VHF equipment. Select the two which are correct.
A. One of the disadvantages of VHF is its susceptibility to atmospheric and precipitation static.
B. VHF is relatively free from atmospheric and precipitation static.
C. Reception distance for usable service is generally based on radio "line of sight".
D. Reception distance remains constant regardless of the altitude.
1. A and D.   3. A and C.
2. B and D.   4. B and C.

144. ATIS information for Calgary airport is broadcast on the frequency____. ATIS gives information relating to_____ (Refer to the 1: 1,000,000 chart.)
1. 114.8 MHz — ceiling, visibility, wind, altimeter setting, runway in use.
2. 125.0 MHz — ceiling, wind, altimeter setting, runway in use, traffic density.
3. 114.8 MHz — ceiling, visibility, wind, runway in use, landing clearance.
4. 125.0 MHz — ceiling, wind, altimeter setting, runway in use.

145. Your initial radio contact with the control tower shall always be
1. immediately after entering the control zone.
2. upon entering the aerodrome traffic zone.
3. prior to entering the control zone.
4. immediately prior to joining the circuit.

146. The minimum content of an initial radio call up to Winnipeg tower by the pilot of C-GULT is "**Winnipeg Tower, This is Piper Warrior_____**".
1. Golf, Uniform, Lima, Tango. Over.
2. Lima, Tango. Requesting Landing Information. Over.
3. Uniform, Lima, Tango. Over.
4. Charlie, Golf, Uniform, Lima, Tango. Over.

147. In subsequent transmissions, if initiated by ATC, the following may be omitted:
1. the first two letters of the registration.
2. the first three letters of the registration.
3. the phonetic equivalents.
4. the whole registration.

148. The tower answers your transmission with the following instruction:
"**Golf Uniform Lima Tango — Runway Two Five — Wind Three Six Zero at One Zero — Altimeter Three Zero Zero Three — Report Five Miles North — Over.**"
From the preceding message, you know that
1. you are cleared to land.
2. the altimeter setting is three zero one two.
3. your downwind magnetic heading will be approximately 070°.
4. the wind is 10 mph.

149. The tower frequency at Lethbridge is
1. 248 MHz.   3. 120.7 MHz.
2. 118.3 MHz.   4. 115.7 MHz.

150. VHF Direction Finding equipment provides directional

assistance. The DF information is electronically derived from
1. automatic direction finding equipment.
2. the VHF Omnidirectional Range transmitter.
3. radar surveillance radar.
4. radio signals transmitted from the aircraft.

151. Pilots using a heading indicator as a reference during VHF/DF homing should
1. reset the indicator every 15 minutes.
2. not reset the indicator without advising the DF operator.
3. not reset the indicator unless the DF operator recommends such action.
4. mentally compensate for gyroscopic precession every 5 minutes.

152. A communication station that provides airport and vehicle advisory service, assists pilots in planning their flights by providing weather and other information and accepts and relays flight plans and provides other flight safety services is called
1. a flight information office.
2. a control tower.
3. a flight service station.
4. a commet.

153. A pilot on a VFR flight has requested radar assistance and is flying a radar vector, provided by ATC, to the nearest airport. Ahead and below is a solid overcast cloud condition. The pilot should
1. maintain heading and altitude as it is a clearance issued by ATC.
2. maintain heading and altitude because ATC knows of the cloud and will issue further instructions.
3. climb above the cloud and fly 1000 feet on top.
4. alter course as necessary to remain VFR and advise ATC.

154. The prefix of a call to the London flight service station should be
1. London Radio, this is.
2. London Flight Service Station, this is.
3. London Advisory, this is.
4. London Centre, this is.

155. A radio equipped aircraft has been cleared to land at a controlled airport. The pilot should acknowledge the clearance by
1. clicking the microphone button.
2. transmitting the appropriate aircraft call sign.
3. replying "Roger".
4. replying "Wilco".

156. Pilots flying VFR in uncontrolled airspace should monitor
1. 121.5 MHz   3. 122.2 MHz.
2. 126.7 MHz.   4. 123.2 MHz.

157. The emergency VHF frequency is
1. 122.1 MHz.   3. 121.5 MHz.
2. 118.3 MHz.   4. 5280 MHz.

158. An airplane has an engine on fire over a remote area. What radiotelephony call-up signal would the pilot transmit three times?
1. Mayday.   3. Urgent.
2. Pan.   4. S.O.S.

159. You are instructed to fly the 260° radial outbound from the VOR. You would set
1. 260 on the OBS and expect a FROM reading with the needle centred.
2. 260 on the OBS and expect a TO reading with the needle centred.
3. 080 on the OBS and expect a FROM reading with the needle centred.
4. 026 on the OBS and expect a TO reading with the needle centred.

160. You are flying inbound to a VOR and the bearing selector is set to 090. With a magnetic heading of 090°, you note that the CDI needle starts to move to the left. You would expect that
1. the wind is from the right and would turn to 060° to intercept the radial.
2. the wind is from the left and would turn to 060° to intercept the radial.
3. the wind is from the right and would turn to 120° to intercept the radial.
4. the wind is from the left and would turn to 120° to intercept the radial.

161. With the bearing selector set to 270, list all the airplane positions (Fig. 161) that would give a TO reading on the TO/FROM indicator.
1. G and B.   3. F, G and H.
2. A, D and G.   4. E, F, G and H.

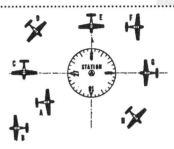

FIG. 161.

162. What is one of the indications you observe when you fly directly over an omnirange station on a preselected course?
1. The needle will swing through 180°.
2. The bearing selector will return to 0°.
3. You will penetrate a cone of silence.
4. The TO/FROM indication will reverse.

163. Which of the following is not a VOR frequency?
1. 115.2 MHz   3. 126.7 MHz.
2. 113.4 MHz.   4. 117.5 MHz.

164. A pilot is able to derive both azimuth or directional information and distance information from a VORTAC transmitter, if the aircraft is equipped with both
1. a VOR receiver and a DME.
2. a VOR receiver and an ADF.
3. an ADF and a DME.
4. a VOR receiver and a marker beacon receiver.

165. The navigational aid located at Midland (N44° 41' W79° 55') (Refer to the 1: 500,000 chart) is a
1. VOR and transmits on 112.8 MHz.
2. NDB and transmits on Channel 75.
3. VORTAC and transmits on 122.8 MHz.
4. ILS and transmits on 112.8 MHz.

166. A DME displays the distance between the airplane and the ground transmitting station as
1. altitude plus ground distance.
2. ground distance.
3. slant range.
4. a DME arc.

167. The ILS indicator in Fig. 167 shows the airplane to be off the localizer beam and off the glide path. The pilot must fly_____ into the glide path and correct to the to intercept_____the localizer.
1. down, left.   3. down, right.

315

Examination Guide

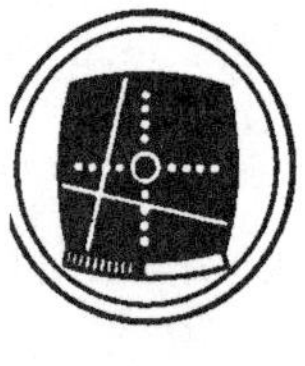

FIG. 167.

   2. up, left.   4. up, right.

168. The glide path projection of an ILS normally
   1. is set at an angle of 8°.
   2. is set at an angle of 3°.
   3. is also usable on the back course.
   4. activates the localizer indicator of the course deviation indicator.

169. The GPS navigation system
   1. determines position from signals propagated by satellites and ground stations.
   2. broadcasts three separate forms of the pseudo random code.
   3. is subject to errors caused by precipitation static.
   4. determines position and altitude from signals propagated from geosynchronous satellites.

170. What does the needle of the ADF normally indicate?
   1. the loop position of maximum signal strength.
   2. the relative bearing of the station.
   3. the relative heading of the airplane.
   4. the magnetic bearing of the radial towards the station.

171. How do you determine that drift is present when homing by ADF?
   1. by variation of the ADF when maintaining a constant heading by heading indicator or magnetic compass.
   2. by variation on the heading indicator when maintaining a constant heading by ADF.
   3. if the reading on the heading indicator is higher than the ADF, the drift is to starboard.
   4. when the reading on the magnetic compass and ADF do not agree.

172. With a magnetic heading of 350° and an ADF reading of 300°, the magnetic bearing to the station is
   1. 050°.   3. 110°.
   2. 290°.   4. 300°.

173. Your heading is 180° and the ADF reading is 090°. If you turn to a magnetic heading of 150°, the ADF will read
   1. 240°.   3. 120°.
   2. 160°.   4. 030°.

174. A transponder transmitting on Mode C
   1. automatically reports altitude to the radar operator.
   2. causes the target representing the aircraft on the radar screen to blossom.
   3. does not need to be equipped with an encoding altimeter because the pressure altimeter can be tied into the system.
   4. identifies the aircraft as being a VFR flight below 10,000 feet ASL.

175. For VFR flight below 10,000 feet ASL, the proper transponder code to select is
   1. 0600.   3. 7700.
   2. 1400.   4. 1200.

176. An aircraft suffering communication failure may alert ATC to the problem, by selecting code
   1. 7700.   3. 4700.
   2. 7600.   4. 1400.

177. To minimize the danger of static electricity starting a fire during refuelling operations, the aircraft should be
   1. bonded.
   2. fitted with static discharge wicks.
   3. grounded.
   4. refuelled immediately after landing.

178. The maximum gross weight of an airplane is the
   1. basic empty weight plus the payload.
   2. operational weight empty plus the payload.
   3. maximum weight approved for operation of the airplane.
   4. basic empty weight plus the useful load.

179. If the airplane is improperly loaded and the aft C.G. limit is exceeded to any appreciable extent, it will be
   1. difficult to put into a stall and spin.
   2. nose heavy.
   3. difficult to get off the ground.
   4. easy to stall and spin and recovery would be more difficult than under conditions of proper loading.

180. Your Airplane Flight Manual specifies a basic empty weight of 1640 lbs. and a maximum gross weight of 2550 lbs. You are planning a flight with 2 friends who weigh 172 lbs. and 195 lbs. respectively. Your weight is 179 lbs. You have three pieces of baggage weighing 30, 24 and 17 lbs. each. You have full fuel tanks giving you 48 U.S. Gallons of usable fuel. The take-off weight of the airplane would be
   1. the same as the maximum gross weight.
   2. slightly more than the maximum gross weight.
   3. 5 lbs less than the maximum gross weight.
   4. 56 lbs less than the maximum gross weight.

181. You are planning a trip with three friends. The front seat passenger weighs 200 lbs. Your weight is 180 lbs. The back seat passengers weigh 145 lbs and 125 lbs respectively. You have baggage that weighs 100 lbs. Use the charts in Fig. 4, 5 and 6 in the Chapter **Airmanship** to do the calculations for this problem. Assume a basic empty weight of 1507 lbs. for the airplane and a full fuel load of 41.5 gallons U.S. Maximum gross weight for this airplane is 2500 lbs.
   1. The total weight is 25 lbs. less than the maximum gross weight but the C.G. is outside the moment envelope.
   2. The total weight is 25 lbs. over the maximum gross weight and the C.G. is outside the moment envelope.
   3. The total weight is equal to the maximum gross weight and the C.G. is within the moment envelope.
   4. The total weight is 25 lbs. less than the maximum gross weight and the C.G. is within the moment envelope.

182. Pressure altitude is 8000 feet. The outside air temperature is 10°C. Density altitude is____. (Use the circular slide rule side of a flight computer to do the calculation.)
   1. 7000 feet and well within safe limits for any general aviation airplane manufactured today.
   2. 9000 feet and could be above the safe operating limitations for your airplane, particularly if the airport has a short runway.
   3. 8000 feet and well within safe limits for any general aviation airplane manufactured today.
   4. 10,000 feet and could be above the safe operating limitations for your airplane, particularly if the airport has a short runway.

183. Two of the most important factors which affect the performance of an airplane are temperature and altitude. Assume that you will take off under the following conditions:
Airport — 5000 feet above sea level.
Temperature — 28°   Airspeed — 70 knots.

TAKE-OFF DISTANCE

Flaps Up

★ T/O Distance (Feet) to clear 50 ft. obstacle at 80 mph (70 kts) (Ground run 40% of distance shown).

Altitude	Outside Air Temperature			
	4°C	16°C	28°C	40°C
	Distance			
Sea Level	1590	1700	1800	1920
2000 Feet	1920	1950	2190	2340
4000 Feet	2290	2450	2610	2800
6000 Feet	2740	2940	3140	3380
7000 Feet	3050	3280	3510	3770

★ Landing and take-off correction: Reduce distances by approximately 10 percent for each 5 knots windspeed.

FIG. 183.

You will take off directly into a surface wind of 10 knots. With flaps up, your take-off to clear a 50 foot obstacle will be approximately ___. (Refer to Fig. 183)
1. 2875 feet
2. 2610 feet
3. 2300 feet
4. 3130 feet

184. In the example given in the previous question, your ground run will be approximately
1. 920 feet.
2. 1150 feet.
3. 1044 feet.
4. 1380 feet.

185. Since the surface wind is gusting, causing turbulence as you approach to land, it is important to
1. always use full flaps.
2. land as slowly as possible.
3. maintain the same airspeed you would in smooth air.
4. carry a little excess airspeed to assure positive control.

186. You have been cleared to land in a gusty wind condition. You are concerned at the wind angle as indicated by the windsock at the approach end of the runway, which you feel now favours another runway. You should
1. continue to land as you must obey the landing clearance.
2. use full flaps and approach at a reduced approach speed.
3. alter heading and land on another runway which is more into wind.
4. overshoot and request an into wind runway.

187. The wind is blowing at 90° to the runway. The airplane you are flying has a stalling speed of 50 knots. What is the maximum wind velocity that the airplane can handle? (See **Cross Winds** in Chapter **Airmanship**.)
1. 10 knots.
2. 15 knots.
3. 5 knots.
4. 20 knots.

188. "Daylight" in Canada, at any place where the sun rises and sets daily, is defined as that period of time
1. between sunrise and sunset.
2. commencing one hour before sunrise and ending one hour after sunset.
3. commencing one half hour before sunrise and ending one half hour after sunset.
4. when the centre of the sun's disc is not more than 12° below the horizon when viewed from the surface.

189. An airplane flying at night must display on its right, or starboard, wing tip the following light

1. a red light.
2. a white light.
3. a flashing green light.
4. a green light.

190. Which statement concerning wing tip vortices is false?
1. They are caused directly by jet wash.
2. These vortices normally settle below and behind the aircraft.
3. With a light cross wind, one vortex can remain stationary over the ground, for some time.
4. Lateral movement of vortices, even in a no wind condition, may place a vortex core over a parallel runway.

191. The pilot of a light aircraft approaching to land immediately following a large aircraft should plan his approach (other circumstances permitting) to remain
1. above the approach path of the large aircraft and land beyond its touch down point.
2. on the upwind side of the large aircraft's approach path and land on the upwind side of the runway.
3. below the approach path of the large aircraft and land just prior to its touchdown point.
4. on the normal approach path for his airplane but at higher speed, as there is only a slight chance of encountering hazardous turbulence.

192. A helicopter in forward flight produces
1. rotary blade vortices which rise above the helicopter.
2. trailing vortices similar to wing tip vortices.
3. turbulence which remains at the same level as the helicopter.
4. turbulence ahead of the helicopter.

193. Wing tip vortices generated by a departing airplane are most severe
1. before rotation.
2. at lift-off.
3. immediately following the application of take-off power.
4. during cruise climb configuration.

194. An isolated thunderstorm is in close proximity to an aerodrome. A pilot wishing to land should
1. not be concerned about wind shear on final approach because the thunderstorm is off to the west and not over the approach path.
2. hold over a known point clear of the thunderstorm until it is well past the aerodrome.
3. land as quickly as possible to be on the ground before the thunderstorm reaches the aerodrome.
4. add one half the wind gust factor to the recommended landing speed.

195. At high altitude, a pilot is susceptible to hypoxia. Hypoxia is
1. excess carbon monoxide in the haemoglobin.
2. imbalance of oxygen and carbon dioxide in the body system.
3. spatial disorientation.
4. lack of sufficient oxygen in the body cells.

196. To prevent hypoxia, a pilot should
1. use oxygen above 5000 feet ASL during daytime.
2. breathe into a paper bag.
3. use oxygen above 5000 feet ASL at night.
4. open the windows and air vents on the airplane.

197. Decompression sickness
1. is caused when nitrogen comes out of solution and forms bubbles.
2. is caused by breathing ozone when flying at high altitudes.
3. causes severe equilibrium problems.
4. is not aggravated by a severe obesity problem.

198. One alcoholic drink taken at 5000 feet as compared to the same drink taken as sea level

1. has a lesser effect.
2. has a greater effect.
3. has no appreciable difference in effect.
4. increases the chance of coriolis effect.

199. Which of the following statements pertaining to the use of alcohol is true?
   1. Relatively small amounts of alcohol significantly decrease a pilot's tolerance to hypoxia.
   2. Small amounts of alcohol will not affect a pilot's judgement.
   3. The effects of alcohol are constant regardless of altitude.
   4. Coffee accelerates the body's ability to recover from the effects of alcohol.

200. Select the statement that best describes the effects of fatigue.
   1. Financial or family problems do not influence fatigue.
   2. Fatigue slows reaction time and contributes to errors.
   3. A fatigued person recuperates more quickly as altitude is gained.
   4. A fatigued person must have food immediately before and during a flight.

201. After any underwater activity where compressed air is used for respiration and decompression stops are required during the ascent to the surface, it is recommended that within 24 hours following such activity a pilot should
   1. not fly.
   2. restrict flight to altitudes of less than 7000 feet.
   3. restrict rates of climb and descent to less than 300 feet per minute.
   4. restrict flight to passenger status only.

202. Many common drugs such as cold tablets, cough mixtures, antihistamines and other over the counter remedies may seriously impair the judgment and co-ordination needed while flying. The safest rule is to
   1. read the manufacturer's warning to ensure that you are aware of possible reactions to such drugs.
   2. allow at least 8 hours between taking any medicine or drug and flying.
   3. allow at least 48 hours after commencing medication
   4. restrict flight to passenger status only.

203. A condition in which there is a lowering of the temperature of the body's core is called
   1. hyperventilation.   3. coriolis effect.
   2. hypoglycaemia.   4. hypothermia.

Answers to the Questions

1.3	27.3	53.3	79.2	105.3	131.2	157.3	183.3
2.3	28.1	54.2	80.4	106.1	132.1	158.1	184.1
3.4	29.1	55.3	81.1	107.3	133.3	159.1	85.4
4.2	30.3	56.3	82.3	108.3	134.4	160.2	186.4
5.3	31.2	57.1	83.1	109.4	135.2	161.3	187.1
6.4	32.4	58.3	84.2	110.3	136.3	162.4	188.3
7.3	33.3	59.1	85.2	111.1	137.4	163.3	189.4
8.2	34.1	60.3	86.1	112.4	138.1	164.1	190.1
9.4	35.2	61.3	87.1	113.3	139.1	165.1	191.1
10.3	36.3	62.3	88.2	114.4	140.3	166.3	192.2
11.3	37.2	63.2	89.1	115.1	141.3	167.1	193.2
12.2	38.4	64.2	90.3	116.1	142.2	168.2	194.2
13.4	39.4	65.3	91.1	117.1	143.4	169.4	195.4
14.2	40.4	66.2	92.3	118.3	144.1	170.2	196.3
15.3	41.2	67.4	93.3	119.4	145.3	171.2	197.1
16.2	42.3	68.1	94.2	120.3	146.1	172.2	198.2
17.3	43.1	69.2	95.4	121.3	147.1	173.3	199.1
18.3	44.3	70.3	96.4	122.3	148.3	174.1	200.2
19.1	45.4	71.4	97.3	123.3	149.2	175.4	201.1
20.4	46.1	72.3	98.2	124.2	150.4	176.2	202.4
21.3	47.2	73.3	99.3	125.4	151.2	177.3	203.4
22.1	48.3	74.2	100.1	126.4	152.3	178.3	
23.1	49.1	75.4	101.4	127.2	153.4	179.4	
24.1	50.1	76.3	102.2	128.1	154.1	180.3	
25.4	51.3	77.3	103.3	129.1	155.2	181.2	
26.3	52.1	78.2	104.4	130.2	156.2	182.2	

Glossary

Absolute Altitude. Actual height above the earth's surface.

Active Runway. Any runway or runways currently being used for take-off or landing. When multiple runways are used, they are all considered active runways. When an aircraft is landing or taking off on an air surface other than a runway, the direc- tion of flight will determine the active runway.

Advisory Airspace. Class F airspace of defined dimensions within which an activity such as pilot training, soaring, etc. is carried out and of which non-participating pilots should be aware. Non-participating pilots may enter advisory airspace at their own discretion but extra vigilance is essential.

Advisory Service. Advice and information provided by a facility to assist pilots in the safe conduct of flight and aircraft movement.

Aerodrome. Any area of land or water (including the frozen surface) or other supporting surface used or designed, prepared, equipped or set apart for the arrival and departure, movement or servicing of aircraft. It includes any buildings, installations and equipment in connection therewith.

Aerodrome Elevation (or Field Elevation). Elevation of the highest point of the landing area.

Aerodrome Traffic. All traffic on the maneuvering area of an aerodrome and all traffic flying in, entering or leaving an aerodrome circuit.

Aerodrome Traffic Frequency (ATF). A VHF frequency designated for use of radio equipped aircraft operating on the surface or in the vicinity of certain specified uncontrolled airports.

Aerodrome Traffic Zone. Class G Airspace of defined dimensions extending upwards from the surface of the earth.

Aerodynamic Coefficients Nondimensional coefficients for aerodynamic forces and moments (i.e. drag and lift coefficients).

Airborne Collision Avoidance System (ACAS). An aircraft system based on secondary surveillance radar (SSR) transponder signals which operates independently of ground-based equipment to provide advice to the pilot on potential conflicting aircraft that are equipped with SSR transponders. ACAS equipment designed and manufactured in the U.S. is called Traffic Alert and Collision Avoidance System (TCAS).

Air Carrier. A person who operates a commercial air service.

Aircraft. Any machine capable of deriving support in the atmosphere from the reactions of the air.

Aircraft Engine. An engine used for propelling an aircraft. It includes turbochargers, appurtenances and accessories necessary for its functioning, but does not include propellers.

Air Defence Identification Zone. Airspace of defined dimensions within which the ready identification, location and control of aircraft is required.

Airframe. Includes fuselage, booms, nacelles, cowlings, fairings, airfoil surfaces and landing gear of an aircraft and their accessories and controls.

Airmet (Airmen's Meteorological Advisory). A weather advisory issued to aircraft having limited capability. It advises of weather significant to the safety of the aircraft.

Airplane. Power driven heavier than air aircraft, deriving its lift in flight from aerodynamic reactions on surfaces that remain fixed under given conditions of flight.

Airport. An aerodrome in respect of which a certificate is in force.

Airport and Airways Surveillance Radar (AASR). A medium range radar designed for both airway and airport surveillance applications.

Airport Surface Detection Equipment. Radar equipment designed to detect all principal features of an airport including all aircraft and vehicle traffic.

Airport Surveillance Radar (ASR). Relatively short range radar intended for surveillance of airport and terminal areas.

Airport Traffic. All traffic on the maneuvering area of an airport and all aircraft flying in the vicinity of an airport

Airspeed Indicator. An instrument that indicates the speed that an airplane is travelling through the air.

Air Time. The period of time commencing when the airplane leaves the supporting surface and terminating when it touches the supporting surface at the next point of landing.

Air Traffic. All aircraft in flight and all aircraft operating on the maneuvering area of an aerodrome.

Air Traffic Control Clearance. Authorization by an air traffic control unit for an air craft to proceed within controlled airspace under specified conditions.

Air Traffic Control Instruction. A directive issued by an air traffic control unit for traffic control purposes.

Air Traffic Control Service. A service provided for the purpose of preventing collisions between aircraft and, on the maneuvering area, between aircraft and obstructions and of expediting and maintaining an orderly flow of air traffic.

Air Traffic Control Unit. An area control centre established to provide air traffic control service to IFR flights and to VFR flights that are subject to control; a terminal control unit established to provide air traffic control service to IFR flights and to VFR flights that are subject to control when operating within a terminal control area; an airport control tower unit established to provide air traffic control service to airport traffic.

Airway A control area or portion thereof established in the form of a corridor equipped with radio navigation aids.

Airworthy. In respect to an aeronautical product, in a fit and safe state for flight and in conformity with the applicable standards of airworthiness.

Alerting Service. Service provided by ATS units to notify appropriate organizations regarding aircraft in need of search and rescue aid.

Alternate Airport. An aerodrome specified in a flight plan to which a flight may proceed when landing at the intended destination becomes inadvisable.

Altimeter. An instrument that, by measuring the pressure of the atmosphere, displays altitude information.

Altimeter Setting. The barometric pressure reading used to adjust a pressure altimeter for variations in existing atmospheric pressure or to the standard altimeter setting (29.92" Hg).

Altitude. The height of a level, point or object measured in feet above ground level (AGL) or from mean sea level (MSL).

Altitude Reservation. Airspace of defined dimensions in controlled airspace reserved for the use of a specific agency during a specified time.

Angle of Attack Indicator. An instrument that indicates a continuous readout of the margin above the stall.

Approach/Arrival Control. An ATC service provided to expedite the flow of IFR flights inbound within a terminal control area.

Apron. That part of an aerodrome, other than the maneuvering area, intended to accommodate the loading and unloading of passengers and cargo; the refuelling, servicing, maintenance and parking of aircraft; the movement of aircraft, vehicles and pedestrians necessary for such purposes.

Area Control Centre (ACC). A unit located within a control area which provides supervision of IFR en route air traffic within the area. In the U.S., it is called an air route traffic control centre (ARTC).

Area Navigation (RNAV). A navigation system that permits aircraft operation on any desired course within an area serviced by ground based navigation signals or within the capabilities of a self contained airborne system.

Arrival Report. A report containing the aircraft registration, the aerodrome of departure, the time of arrival and the aerodrome of arrival.

Attitude Indicator. An instrument designed to provide an artificial horizon as a means of reference for judging attitude of an airplane.

Automated Data Interchange System. An aeronautical fixed telecommunications network to ensure the safety of air navigation and the regular, efficient and economical operation of air services.

Automatic Altitude Reporting. The function of a transponder which responds to Mode

319

Glossary

C interrogations by transmitting the aircraft's altitude in 100 foot increments.

Automatic Direction Finder (ADF). An aircraft radio navigation system which senses and indicates the direction to a L/MF directional radio beacon ground transmitter.

Automatic Terminal Information Service (ATIS). The continuous broadcast of recorded non-control information in selected terminal areas.

Backtrack. The taxiing of an aircraft on an active runway in a direction opposite to the landing or take-off direction.

Balanced Field Length. The field length where the distance to accelerate and stop is equal to the take-off distance of an airplane experiencing an engine failure at the critical engine failure recognition speed (V_1).

Bearing. The horizontal direction to or from any point, usually measured clockwise from true north, magnetic north, or some other reference point, through 360 degrees.

Below Minimums. Weather conditions below the minimums prescribed by regulation.

Blind Spot (or Blind Zone). An area from which radio transmissions and/or radar echoes cannot be received.

Blind Transmission. A transmission from one station to another in circumstances where two-way communication cannot be established but where it is believed that the called station is able to receive the transmission.

Brake Horsepower (BHP). The power delivered at the propeller shaft of an aircraft engine.

Broadcast. A radio transmission that originates from an aircraft and that is not directed to any particular receiving station.

Calibrated Airspeed (CAS). Indicated air speed corrected for position and instrument error.

Call Up. Initial voice contact between a facility and an aircraft using the identification of the unit being called and the unit initiating the call.

Canadian Domestic Airspace. All navigable airspace of Canada.

Canadian Minimum Navigation Performance Specifications. Specifications relative to the navigation performance capability of aircraft operating in a specified portion of the Canadian Domestic Airspace.

Canard. A control surface incorporating a horizontal stabilizer and elevator attached to the forward part of the airplane.

CAVOK. A term used to indicate no cloud below 5000 feet, visibility 6 miles or more and no precipitation or thunder storm activity or fog or drifting snow.

Ceiling. The lowest height at which a broken or overcast condition exists, or the vertical visibility when an obscured condition such as snow, smoke or fog exists.

Celestial Navigation. The determination of geographical position by reference to celestial bodies.

Centre. A unit located within a control area which provides supervision of IFR en route traffic within the area.

Certificate of Airworthiness. A conditional certificate of fitness for flight issued in respect of a particular aircraft.

Certificate of Registration. A certificate issued to the owner of an aircraft with respect to the registration and the registration markings for that aircraft.

Classification of Airspace. A, B, C, D, E, F and G.

Clear Air Turbulence. Turbulence encountered in air where no clouds are present.

Clearance Limit. The fix, point or location to which an aircraft is cleared when issued an air traffic control clearance.

Climb (or Climb-out). The portion of flight operation between take-off and the initial cruising altitude.

Climb, Best Angle of. The angle which will gain the most altitude in a given distance.

Climb, Best Rate of. The angle which will gain the most altitude in the least time.

Codes/Transponder Codes. The number assigned to a particular multiple pulse reply signal transmitted by a transponder.

Commercial Air Service. Any use of aircraft for hire or reward.

Commet. A ground communication facility associated with a weather station and authorized to pass weather information to arriving and departing aircraft.

Community Aerodrome Radio Station (CARS). An air/ground radio station operated by the Territorial Government at airports in the northern areas of Canada.

Compass Rose. A circle graduated in degrees printed on some charts as a reference to true or magnetic direction.

Composite Flight Plan. A flight plan which specifies VFR operation for one portion of flight and IFR for another portion.

Compulsory Reporting Point. A reporting point over which an aircraft must report to ATC.

Contact Approach. An approach wherein an aircraft on an IFR flight plan, having an ATC authorization, operating clear of cloud and with at least 1 mile flight visibility and a reasonable expectation of continuing to the destination airport in those conditions, may deviate from the instrument approach procedure and proceed to the destination airport by visual reference to the surface of the earth.

Control Area. A controlled airspace extending upwards from a specified height above the surface of the earth in which air traffic control is provided.

Control Area Extension. Controlled airspace of defined dimensions within the low level airspace extending upwards from 2200 feet above the surface of the earth.

Controlled Airport. An airport at which an ATC unit is provided.

Controlled Airspace. An airspace of defined dimensions within which ATC service is provided.

Control Tower. A unit established to provide air traffic control service to aerodrome traffic.

Control Zone. Controlled airspace of defined dimensions extending upwards from the surface of the earth up to 3000 feet above airport evaluation unless otherwise specified.

Co-ordinates. Intersections of lines of reference usually expressed in degrees, minutes and seconds of latitude and longitude used to determine position.

Course. The intended direction of flight in the horizontal plane measured in degrees from north. (U.S. term. In Canada, called track) Also, an omni bearing towards a VOR, VORTAC or TACAN navigational facility.

Course Deviation Indicator. A vertical needle incorporated into an OMNI/ILS indicator to show deviation from the selected VOR course or radial.

Critical Altitude. The maximum altitude at which it is possible to maintain a rated power or specified manifold pressure.

Critical Engine. The engine which, should it fail, will have the most adverse effect on the aircraft flight characteristics.

Critical Field Length. The balanced field length as it applies to a particular airplane. The critical field length is calculated for each individual airplane to meet the balanced field length criteria.

Cross Wind. A wind not parallel to the runway or to the path of an aircraft.

Cruising Altitude/Level. An altitude or flight level, as shown by a constant altimeter indication, maintained during level flight.

Customs Notification Service (ADCUS). A service provided, by ATC units, for advance notification to Customs officials for transborder flights at specified Ports of Entry.

Danger Area. Class F Restricted Airspace of defined dimensions established over international waters.

Daylight. The period of time in any day when the centre of the sun's disc is less than 6 degrees below the horizon, and, in any place where the sun rises and sets daily, may be considered to be the period commencing one half hour before sunrise and ending one half hour after sunset.

Day VFR. A flight conducted in accordance with VFR during the hours of daylight.

Dead Reckoning. Navigation by use of pre determined vectors of wind and true airspeed, precalculated heading, groundspeed and estimated time of arrival.

Decision Height (DH). A specified height at which a missed approach must be initiated during a precision approach if the required visual reference to continue the approach to land has not been established.

Defence VFR (DVFR) Flight. A flight conducted in accordance with visual flight rules within an Air Defence Identification Zone.

Density Altitude. Pressure altitude corrected

for temperature.

Departure Control. An ATC service for departing IFR, and under certain conditions, VFR aircraft.

Designated Intersection. A point on the surface of the earth over which two or more designated position lines intersect. The position lines may be magnetic bearings from NDBs, radials from VHF/UHF aids, centre lines of designated airways, air routes, localizers and DME distances.

Deviation. The angle through which the compass needle is deflected from magnetic north due to the influence of magnetic fields in the airplane.

Direct User Terminal System (DUATS). A computer based system provided by a vendor to pilots or other operational personnel. It supplies aviation weather and NOTAM information necessary for pre-flight planning via computer terminals or personal computers owned by the vendor or users. DUATS services approved by Transport Canada provide flight plan filing capability.

Discrete Frequency. A separate radio frequency for use in direct pilot/controller communications in ATC which reduces frequency congestion by controlling the number of aircraft operating on the frequency at any one time.

Displaced Threshold. A threshold that is located at a point on the runway other than the designated beginning of the runway.

Distance Measuring Equipment (DME). Equipment used to measure in nautical miles the slant range distance of an aircraft from the DME navigation aid.

Emergency Locator Transmitter (ELT). A radio transmitter which operates from its own power source on 121.5 MHz, or on 121.5 MHz and 243.0 MHz, and is activated automatically by the g forces experienced in a crash landing.

Empennage. The tail section of an airplane (vertical stabilizer, or fin, horizontal stabilizer or tail plane, rudder, elevators, and all trimming and control devices).

Equivalent Airspeed (EAS). Calibrated airspeed corrected for compressibility factor.

Expected Approach Time. The time at which ATC expects that an arriving aircraft, following a delay, will leave the holding fix to complete its approach for a landing.

Expected Further Clearance Time. The time at which it is expected that further clearance will be issued to an aircraft.

Expedite. An expression used by ATC when prompt compliance is required to avoid the development of an imminent situation.

Fan Marker Beacon. A type of radio beacon, the emissions of which radiate in a vertical fan-shaped pattern.

Feathered Propeller. A propeller whose blades have been rotated so that the leading and trailing edges are nearly parallel with the aircraft flight path to minimize drag and engine rotation.

Final. Commonly used to mean an aircraft is on the final approach course or is aligned with a landing area.

Final Approach. That segment of an instrument approach between the final approach fix or point and the runway, airport or missed approach point, whichever is encountered last, wherein alignment and descent for landing are accomplished.

Final Approach Area. That area within which the final approach portion of an instrument approach procedure is carried out.

Fix. A geographical position determined by visual reference to the surface, or by reference to one or more radio navaids, or by celestial plotting, or by another navigational device.

Flag/Flag Alarm. A warning device incorporated in certain airborne navigation and flight instruments to indicate that the instrument is inoperative or that signal strength is below acceptable values.

Flaps Down Speed. The maximum speed at which the airplane may be flown with the flaps lowered.

Flameout. Unintended loss of combustion in turbine engines resulting in the loss of engine power.

Flight Information Region (FIR). An air space of defined dimensions extending upwards from the surface of the earth within which flight information service and alerting service is provided.

Flight Itinerary. Specified information relating to the intended flight of an aircraft, that is filed with either ATC or a responsible person.

Flight Level (FL). An altitude expressed in hundreds of feet indicated on an altimeter set to 29.92" Hg or 1013.2 hectopascals.

Flight Path. A line, course or track along which an airplane is flying or intended to be flown.

Flight Permit. A permit issued to an amateur built aircraft or to a private aircraft, that does not qualify for a certificate of airworthiness, for purposes of experiment, test, demonstration or other special flight.

Flight Plan/Notification. Specified information, relating to the intended flight of an aircraft, that is filed with ATC.

Flight Service Station (FSS). An aeronautical facility providing mobile and fixed communications, flight information, search and rescue alerting and weather and flight planning services to pilots and other users.

Flight Time. The time from the moment the aircraft first moves under its own power for the purpose of flight until the moment it comes to rest at the next point of landing.

Flight Visibility. The average range of visibility at any given time forward from the cockpit of an aircraft in flight.

Flight Watch. An en route flight advisory service provided by Flight Service Stations in the U.S. on frequency 122.0 MHz.

Fuel Remaining. The amount of fuel remaining on board until actual fuel exhaustion. It should be expressed in the approximate number of minutes the flight can continue with the fuel remaining. All reserve fuel should be included in the time stated, as should an allowance for established fuel gauge system error.

Full Throttle Altitude. The maximum altitude at which maximum engine power output is available. Above that altitude, the horsepower of the engine will be less than the rated horsepower.

General Aviation. That portion of civil aviation which encompasses all facets of aviation except scheduled air services and non-scheduled air transport operations for remuneration or hire and military aviation.

Glide Path/Glide Slope. A descent profile determined for vertical guidance during a final approach.

Glide Speed for Endurance. The airspeed that gives, with no power, minimum sink.

Glide Speed for Range. The airspeed that results in an angle of attack that gives the maximum lift/drag ratio.

Global Positioning System (GPS). A precise navigation system based on satellites orbiting the earth at very high altitude. Position in 3 dimensions is calculated by triangulation.

Grid Navigation. A navigation system devised for use in areas in close proximity to the north pole.

Gross Weight. The maximum permissible weight of the airplane.

Ground Control. The operating position in the control tower that provides clearances and instructions for the movement of airport traffic and the pertinent information to all traffic within the airport perimeter.

Groundspeed. The speed of an aircraft relative to the surface of the ground.

Ground Visibility. The visibility at an aerodrome as reported by an ATC unit, an FSS, a CARS, a Commet station or a ground based radio station operated by an air carrier.

Hang Glider. A motorless heavier-than-air aircraft deriving its lift from surfaces that remain fixed in flight, designed to carry not more than two persons and having a launch weight of 45 kg or less.

Heading. The direction in which the longitudinal axis of an aircraft is pointed, usually expressed in degrees from north (true, magnetic or compass).

Heading Indicator. An instrument designed to indicate the heading of an airplane.

Height Above Aerodrome (HAA). The height in feet of the minimum descent altitude above the published aerodrome elevation. HAA is published for all circling minima.

Height Above Touchdown Zone Elevation (HAT). The height in feet of the decision height or the minimum descent altitude above the touchdown zone elevation.

Hertz (Hz). The standard radio equivalent of frequency in cycles per second of an elec tromagnetic wave. Kilohertz (KHz) is a

Glossary

frequency of 1000 cycles per second. Megahertz (MHz) is a frequency of one million cycles per second.

High Frequency (HF). The frequency band between 3 and 30 MHz.

High Level Air Route. In the High Level Airspace, a prescribed track between specified radio aids to navigation along which ATC service is not provided.

High Level Airspace. All airspace that is within the Canadian Domestic Airspace at or above 18,000 feet ASL.

High Level Airway. In the controlled high level airspace, a prescribed track between specified radio aids to navigation and designated "J", along which ATC service is provided.

Holding Fix/Holding Point. A specified location, identified by visual or other means in the vicinity of which the position of an aircraft in flight is maintained in accordance with an ATC clearance.

Hold Short. Instructions to hold at least 200 feet from the edge of a runway while awaiting permission to cross or proceed onto that runway.

Homing. Flight toward a navaid, without correcting for wind, by adjusting the aircraft heading to maintain a relative bearing of zero degrees.

Horizontal Situation Indicator (HSI). A navigational instrument that combines the display of a heading indicator with the display of an omni indicator.

Identification Zone. An airspace of defined dimensions extending upwards from the surface of the earth within which certain rules apply for the security control of air traffic.

IFR Aircraft/Flight. An aircraft conducting flight in accordance with instrument flight rules.

IFR Conditions. Weather conditions below the minimum for flight under visual flight rules.

Indicated Airspeed (IAS). The uncorrected speed of the aircraft read from the airspeed indicator. It is the measurement of the difference between impact and static pressure.

Indicated Altitude. The reading on the altimeter when it is set to the current barometric pressure.

Inertial Navigation System (INS). A self contained long range navigation system which by means of computers, precision gyros and sensitive accelerometers provides guidance and steering information and accurately displays position, groundspeed and heading.

Initial Approach. That segment of an instrument approach between the initial approach fix or point and the intermediate fix or point wherein the aircraft departs the en route phase of the flight and maneuvers to enter the intermediate segment.

Inner Marker. A marker beacon used with an ILS, located between the middle marker and the threshold of the ILS runway.

Instrument Flight Rules (IFR). Rules governing the conduct of flight under instrument meteorological conditions.

Instrument Landing System (ILS). A precision instrument approach system consisting of a localizer, glide path, outer and middle markers and approach lights.

Instrument Meteorological Conditions (IMC). Meteorological conditions expressed in terms of visibility, distance from cloud and ceiling less than the minima specified for visual meteorological conditions.

Instrument Runway. A runway equipped with electronic and visual navigation aids for which a precision or nonprecision approach procedure has been approved.

Intermediate Approach. That segment of an instrument approach between the intermediate fix or point and the final approach fix or point wherein the aircraft configuration, speed and positioning adjustments are made in preparation for the final approach.

International Civil Aviation Organization (ICAO). A specialized agency of the United Nations whose objective is to develop the principles and techniques of international air navigation and to foster planning and development of international civil air transport.

Intersection. (1) A point defined by any combination of courses, radials or bearings of two or more navigation aids. (2) The point where two runways cross or meet within their length.

Jet Stream. A migrating stream of high speed winds present at high altitude.

Joint En Route Terminal System (JETS). An automated ATC system that displays digitally on a radar console altitude information on airplanes with Mode C transponder capability.

Known Aircraft. Aircraft of whose movements ATS has been informed.

Landing Gear Extended Speed. The maximum speed at which an aircraft can be safely flown with the landing gear extended.

Landing Gear Operating Speed. The maximum speed at which the landing gear can be safely extended or retracted.

Landing Run/Roll. The distance from the point of touchdown to the point where the aircraft can be brought to a stop or exit the runway.

Landing Sequence. The order in which aircraft are positioned for landing.

Launch Weight. The total weight of a hang glider or an ultra-light airplane when it is ready for flight. It includes any equipment, instruments and the maximum quantity of fuel and oil that it is designed to carry but does not include the weight of any float equipment (to a maximum of 34 kg), the weight of the occupant or the weight of any parachute installation.

Load Factor. The ratio of a specified load to the total weight of the aircraft.

Local Traffic. Aircraft operating in the traffic circuit or within sight of the tower.

Localizer. The component of an ILS which provides course guidance to the runway.

Loran (Long Range Navigation). An electronic navigation system by which hyperbolic lines of position are determined by measuring the difference in the time of reception of synchronized pulse signals from two fixed transmitters.

Low Approach. An approach over an airport or runway following an instrument or VFR approach, including the go-around maneuver, where the pilot intentionally does not intend to land.

Low Frequency (LF). The frequency band between 30 and 300 KHZ.

Low Level Air Route. In the low level airspace, a route extending upwards from the surface of the earth and within which ATC service is not provided.

Low Level Airspace. Ail airspace within the Canadian Domestic Airspace below 18,000 feet ASL.

Low Level Airway. In the low level airspace, a prescribed track between specified radio aids to navigation along which ATC service is provided. It extends upwards from 2,200 feet above the surface up to, but not including 18,000 feet ASL.

Mach Indicator. An instrument that provides a continuous indication of the ratio of airspeed to the local speed of sound.

Mach Number. The ratio of true airspeed to the speed of sound.

Mandatory Frequency (MF). A VHF frequency designated for use of radio equipped aircraft operating on the surface or in the vicinity of certain specified uncontrolled airports.

Maneuvering Area. That part of an aerodrome intended for the taking off and landing of aircraft and for the movement of aircraft associated with take-off and landing. It excludes the aprons.

Maneuvering Speed. The maximum speed at which the flight controls can be fully deflected without damage to the airplane structure.

Manifold Pressure. Absolute pressure measured at the appropriate point in the induction system, expressed in inches of mercury.

Manifold Pressure Gauge. An instrument that indicates in inches of mercury the pressure of the fuel/air mixture in the engine intake manifold.

Marker Beacon. An electronic navigation facility transmitting a 75 MHz vertical fan or bone shaped radiation pattern.

Mayday. The international radiotelephony distress signal.

Mean Wind. Wind direction and speed as determined from a sample reading every second over the last two minutes.

MEDEVAC. A term used to request ATS priority handling for a medical evacuation flight based on a medical emergency in the transport of patients, organ donors, organs or other urgently needed life-saving medical material.

Medium Frequency (MF). The frequency

band between 300 and 3000 KHz.

Microwave Landing System. A precision instrument approach system operating in the microwave spectrum.

Middle Marker. A marker beacon that defines a point along the glide path of an ILS, normally located at or near decision height.

Military Operations Area. An airspace of defined dimensions established to segregate certain military activities from IFR traffic and to identify for VFR traffic where these activities are conducted.

Military Terminal Control Area (MTCA). Controlled airspace of defined dimensions designed to serve arriving, departing and en route aircraft and within which special procedures and exemptions exist for military aircraft.

Minimum Crossing Altitude (MCA). The lowest altitude at certain fixes at which an aircraft must cross.

Minimum Descent Altitude (MDA). A specified height referenced to sea level for a nonprecision approach below which descent must not be made until the required visual reference to continue the approach to land has been established.

Minimum En Route Altitude (MEA). The published altitude above sea level between specified fixes on airways or air routes which assures acceptable navigational signal coverage and which meets the IFR obstruction clearance requirements.

Minimum Fuel. An aircraft's fuel supply has reached a state where, upon reaching the destination, it can accept little or no delay.

Minimum Holding Altitude. The lowest altitude prescribed for a holding pattern which assures navigational signal coverage, communications and meets obstacle clearance requirements.

Minimum Obstruction Clearance Altitude (MOCA). The altitude above sea level in effect between radio fixes on a VHF/UHF low level airway which meets the IFR obstruction clearance requirements for the route segment.

Minimum Reception Altitude (MRA). At a specific VHF/UHF intersection, the lowest altitude above sea level at which acceptable navigation signal coverage is received to determine the intersection.

Minimum Safe Altitude. Altitudes depicted on approach charts which provide at least 1000 feet of obstacle clearance for use within a specified distance from the navigation facility upon which a procedure is predicated.

Minimum Sector Altitude. The lowest altitude which will provide a minimum clearance of 1000 feet above all objects located in an area contained within a sector of a circle of 25 NM radius centred on a radio aid to navigation.

Minimum Sink Speed. The airspeed that gives, with no power, minimum sink.

Minimums/Minima. Weather condition requirements established for a particular operation or type of operation.

Missed Approach. A maneuver conducted by a pilot when an instrument approach cannot be completed to a landing.

Mode. The letter or number assigned to a specific pulse spacing of radio signals transmitted and received by ground interrogator and airborne transponders. Mode C is used for altitude reporting.

Mountainous Area/Region. An area of defined lateral dimensions above which special rules concerning minimum en route altitudes apply.

Movement Area. That part of an aerodrome intended for the surface movement of aircraft, including maneuvering areas and aprons.

Navigational Aid/Navaid. Any visual or electronic device airborne or on the surface which provides point to point guidance information or position data to aircraft in flight.

Never Exceed Speed. The maximum speed at which the airplane can be operated in smooth air.

Night. The period of time when the centre of the sun's disc is more than 6 degrees below the horizon or the period of time commencing one half hour after sunset and ending one half hour before sunrise, in any place where the sun rises and sets daily.

Night VFR. A flight conducted in accordance with VFR during hours of night.

Nondirectional Beacon (NDB). An L/MF radio beacon transmitting nondirectional signals whereby the pilot of an aircraft equipped with ADF can determine his bearing to or from the radio beacon and "home" on or track to or from the station.

Nonprecision Approach Procedure. A standard instrument approach procedure in which no electronic guide path is provided.

NORDO. Aircraft in which no radio equipment is installed.

Normal Operating Limit Speed. The maximum safe speed at which the airplane should be operated; the cruise speed for which it was designed.

Northern Domestic Airspace (NDA). All airspace within the Canadian Domestic Airspace that lies north of a line that, specifically defined by regulation, begins at the Alaska/Canada border on the Arctic Ocean and, more or less, extends southward through Yellowknife to Churchill and thence northeast to Frobisher and the Atlantic Ocean.

NOTAM. Notices to airmen are transmitted on the telecommunication networks and are also published in printed form and convey information of special import regarding the facilities in their respective areas.

Now Wind. Wind direction and speed as determined from a sample reading every second and averaged over the last five seconds.

Obstacle Clearance Altitude/Height. The lowest altitude above the elevation of the relevant runway threshold or above the aerodrome elevation used in establishing compliance with appropriate obstacle clearance criteria.

Omega Navigation System. A long range navigation system based on very low frequency navigation signals transmitted by 8 world wide stations.

Omni. The commonly used shortened name for the very high frequency omnidirectional navigation system.

Outer Marker. A marker beacon at or near the glide path intercept altitude of an ILS approach.

Overshoot. A phase of flight wherein a landing approach of an aircraft is not continued to touchdown.

Pilotage. Navigation by visual reference to landmarks.

Pan. The international radiotelephony urgency signal.

Pilot-In-command. The pilot responsible for the operation and safety of an aircraft during flight time.

Pilots' Automatic Telephone Weather Answering Service. A continuous telephone recording of meteorological and aeronautical information.

PIREP (Pilot Weather Report). A report of meteorological phenomena encountered by aircraft in flight.

Pitch Setting. The propeller blade setting as determined by the blade angle.

Position Report. A report over a known location as transmitted by an aircraft to ATC.

Positive Control. Control of all air traffic within designated airspace by air traffic control.

Precipitation. Any or all forms of water particles (rain, sleet, hail or snow) that fall from the atmosphere and reach the surface.

Precision Approach Radar (PAR). A high definition, short range radar approach aid. It provides altitude, azimuth and range information.

Preferred Runway. When there is no active runway, the preferred runway is considered to be the most suitable operational runway taking into account such factors as: the runway most nearly aligned with the wind; noise abatement or other restrictions which prohibit the use of certain runways; ground traffic and runway conditions.

Pressure Altitude. The reading on the altimeter when it is set to standard barometric pressure (29.92" Hg).

Prevailing Visibility. The maximum visibility value common to sectors comprising one-half or more of the horizon circle.

Primary Frequency. The radiotelephony frequency assigned to an aircraft as a first choice for air-ground communication.

Procedure Turn. A maneuver in which a turn is made away from a designated track followed by a turn in the opposite direction, both turns being executed so as to permit the aircraft to intercept and proceed along the reciprocal of the designated track.

Propeller. A device for propelling an airplane by its action on the air. It comprises two or more blades on an engine driven shaft for the purpose of producing thrust approximately perpendicular to its plane of rotation.

RADAR (Radio Detection and Ranging). A device which, by measuring the time interval between transmission and reception of radio pulses and correlating the angular orientation of the radiated antenna beam, provides information on range, azimuth and/or elevation of objects in the path of the transmitted pulses.

Radar Altimeter/Radio Altimeter. Airborne equipment which makes use of the reflection of radio waves from the ground to determine the height of that aircraft above the ground.

Radial. A magnetic bearing extending from a VOR, VORTAC or TACAN navigation facility, except for facilities in the Northern Domestic Airspace which may be oriented on true north.

Radio. (1) A device used for communication. (2) Used to refer to a flight service station.

Radio Magnetic Indicator (RMI). A navigational instrument coupled with a gyro compass that indicates the direction to a selected navaid and indicates bearing with respect to the heading of the aircraft.

Rate-One-Turn. The turn rate of three degrees per second normally used by aircraft operating at less than 250 kt.

Rating. A statement that, as a part of a pilot license, sets forth special conditions, privileges or limitations.

Readback. Procedure whereby the receiving station repeats a received message back to the transmitting station so as to obtain confirmation of correct reception.

Remote Communications Outlet (RCO). An unmanned air/ground communications station remotely controlled by an FSS or CARS.

Reporting Point. A geographical location in relation to which the position of an aircraft is reported.

Required Visual Reference. In respect of an aircraft on an approach to a runway, that section of the approach area of the runway or those visual aids that, when viewed by the pilot of the aircraft, enable the pilot to make an assessment of the aircraft position and the rate of change of position relative to the nominal flight path.

Responsible Person. An individual who has agreed with the person who intends to commence a flight that he will notify ATC of the non-arrival of the flight at the ETA specified in the flight itinerary.

Restricted Area. Class F Airspace of defined dimensions above the land areas or territorial waters within which the flight of aircraft is restricted in accordance with certain specified conditions.

RONLY. An airplane equipped with a radio receiver only.

Rotate. A word used to indicate one of several take-off sequences. The term stems from "rotate about the lateral axis of an airplane" and applies to a change of attitude during take-off.

Runway. A defined rectangular area, on a land aerodrome, prepared for the landing and take-off run of aircraft along its length.

Runway Heading. The magnetic or true direction that corresponds with the runway centre line.

Runway Visual Range (RVR). For a particular runway, the maximum distance in the direction of take-off or landing a pilot will be able to see the lights or other delineating markers along the runway from a specified point above the centre line that corresponds to the average eye level at the moment of touchdown.

Secondary Surveillance Radar (SSR). A radar system that requires complementary aircraft equipment (transponder).

Separation. In ATC, the spacing of aircraft to achieve their safe and orderly movement in flight and while landing and taking off.

Sigmet (Significant Meteorological Information). A weather advisory issued concerning weather significant to the safety of aircraft (e.g. tornadoes, thunderstorms, severe turbulence, icing, etc.)

Southern Domestic Airspace (SDA). All airspace within the Canadian Domestic Airspace that lies south of a line, that, specifically defined by regulation, begins at the Alaska/Canada border on the Arctic Ocean and, more or less, extends southward through Yellowknife to Churchill and thence northeast to Frobisher and the Atlantic Ocean.

Special VFR Conditions. Weather conditions in a control zone which are less than basic VFR and in which some aircraft are permitted flight under visual flight rules.

Specified Area. An area in the vicinity of an uncontrolled aerodrome for which a mandatory frequency or an aerodrome traffic frequency has been designated and within which MF or ATF procedures apply. Usually a circle with a 5 nautical mile radius extending up to 3000 feet above aerodrome elevation.

Speed Brake. Moveable aerodynamic devices on aircraft that reduce airspeed.

Squawk. Activate specific mode, codes or function on an aircraft transponder.

Standard Atmosphere. A standard unit of atmospheric pressure, 29.92 inches of mercury at 15°C at sea level, as defined by ICAO.

Standard Rate Turn. A turn of three degrees per second.

Stop and Go. A procedure in which an aircraft lands, makes a complete stop on the runway and then commences a take-off from that point.

Stopover Flight Plan. A flight plan which includes two or more separate en route flight segments with an intermediate stop at one or more airports.

Straight-In Approach. Entry into the traffic circuit by interception of the final approach leg without executing any other portion of the traffic circuit.

TACAN (Tactical Air Navigation). An ultrahigh frequency air navigation aid which provides suitably equipped aircraft a continuous indication of bearing and distance to a Tacan station.

Tachometer. An instrument that indicates the speed at which the engine crankshaft is turning.

Target. The indication shown on a radar display resulting from a primary radar return on a radar beacon reply.

Terminal Control Area (TCA). An airspace of defined dimensions extending upwards from a defined base AGL within which a terminal control unit provides ATC services designed to serve arriving, departing and en route aircraft.

Tetrahedron. A device located on the airport which is used as a landing/wind direction indicator.

Threshold. The beginning of that portion of the runway usable for landing.

Threshold Crossing Height (TCH). The height of the glide slope above the runway threshold.

Touch and Go. An operation by an aircraft that lands and departs on a runway without stopping or exiting the runway.

Touchdown. (1) The point at which an aircraft first makes contact with the landing surface. (2) In ILS, the point where the glide path intercepts the landing surface.

Touchdown Zone (TDZ). The first 3000 feet of runway or the first third of the runway, whichever is less, from the threshold in the direction of landing.

Touchdown Zone Elevation (TDZE). The highest runway centreline elevation in the touchdown zone.

Tower. A terminal facility which through use of air/ground communications, visual signalling, provides ATC services to airborne aircraft operating in the vicinity of the airport and to aircraft operating on the movement area of the airport.

Track. The projection on the earth's surface of the path of an aircraft, the direction of which at any point is usually expressed in degrees from north (true, magnetic or grid).

Traffic Circuit/Pattern. The traffic flow that is prescribed for aircraft landing at, taxiing on or taking off from an airport.

Transcribed Weather Broadcast (TWB). A continuous recording of meteorological and aeronautical information that is broadcast on L/MF and/or VOR facilities.

Transponder. A receiver/transmitter which will generate a reply signal upon proper interrogation (SSR).

True Altitude. Exact height above mean sea level.

True Airspeed. Calibrated airspeed (or equivalent airspeed) corrected for air density error.

Turn and Slip Indicator/Turn CoOrdinator. An instrument designed to indicate the direction and rate of turn of an aircraft and the amount of slipping or skidding in the turn.

Ultrahigh Frequency. The frequency band

between 300 and 3000 MHz.
Ultra-Light Aircraft. A power driven heavier-than-air aircraft designed to carry not more than one person (in the case of a single place) and not more than two persons (in the case of a two place) with restrictions as to launch weight.
Uncontrolled Aerodrome/Airport. An aerodrome or airport without an air traffic control tower in operation.
Uncontrolled Airspace. Airspace within which ATC has neither the authority nor the responsibility to exercise control over air traffic.
UNICOM. Universal Communications. An air/ground communication facility operated by a private agency to provide private advisory station service at uncontrolled aerodromes and airports with no ATS air/ground communications.
Variation. The angle between the true meridian and the magnetic meridian.
Vector. A heading issued to an aircraft to provide navigational guidance by radar.
Vertical Speed Indicator. An instrument that indicates the rate, in feet per minute, at which an airplane is ascending or descending.
Very High Frequency (VHF). The frequency band between 30 and 300 MHz.
Very Low Frequency (VLF). The frequency band between 2 and 30 KHz.
VFR Aircraft/Flight. An aircraft conducting flight in accordance with visual flight rules.
VFR Over The Top. Authorization for flight above an en route cloud layer providing certain conditions exist.
VHF Direction Finding Service (VHF/DF). A facility designed to provide bearing information to airplanes.
Visibility. The ability, as determined by atmospheric conditions and expressed in units of distance, to see and identify prominent unlighted objects by day and prominent lighted objects by night.
Visual Approach. An approach wherein an aircraft on an IFR flight plan, operating in VFR weather conditions under the control of an ATC facility and having an ATC authorization, may proceed to the airport of destination in VFR weather conditions.
Visual Flight Rules. Rules that govern the procedures for conducting flight under visual conditions.
Visual Meteorological Conditions (VMC). Meteorological conditions expressed in terms of visibility, distance from cloud and ceiling equal to or better than specified minima.
VOR (Very High Frequency Omnidirectional Range) Station. A ground based navigation aid transmitting VHF navigation signals 360° in azimuth oriented from magnetic north.
VORTAC (VOR/TACAN). A navigation aid providing VOR azimuth, TACAN azimuth and TACAN DME at one site.
Vortices/Wing Tip Vortices. Circular patterns of air created by the movement of an airfoil through the air when generating flight.
VOT (VOR Test Signal). A ground facility which emits a test signal to check VOR receiver accuracy.
Wake Turbulence. The turbulence associated with the pair of counter-rotating vortices trailing from the wing tips of an airplane in flight.
Waypoint (RNAV). A predetermined geographical position used for route or progress reporting purposes that is defined as relative to a VORTAC station position.
Wind Direction Indicator. A device which visually indicates the wind direction for the purpose of determining the direction in which landings and take-offs should be made.
Wind Shear. A change, either vertically or horizontally, in wind speed and/or direction in a short distance resulting in a tearing or shearing effect.

Abbreviations

A or AST	— Atlantic Standard Time	
AAE	— Above Aerodrome Elevation	
AAS	— Airport Advisory Service	
AASR	— Airport and Airways Surveillance Radar	
AAU	— Authorized Approach UNICOM	
A/C	— Aircraft	
ACA	— Arctic Control Area	
ACAS	— Airborne Collision Avoidance System	
ACC	— Area Control Centre	
A/D	— Aerodrome	
AD	— Airworthiness Directive	
ADCUS	— Advise Customs	
ADF	— Automatic Direction Finder	
ADI	— Attitude Direction Indicator	
ADIS	— Automatic Data Interchange System	
ADIZ	— Air Defence Identification Zone	
ADS	— Automatic Dependent Surveillance	
AFTN	— Aeronautical Fixed ` Communication Network	
AES	— Atmospheric Environment Service	
A/G	— Air/Ground	
AGL	— Above Ground Level	
AI	— Attitude Indicator	
AIC	— Aeronautical Information Circular	
AIP	— Aeronautical Information Publication Canada	
AIREP	— Meteorological Report	
Airmet	— Airmen's Meteorological Advisory	
Alt	— Altitude	
Altn	— Alternate	
AM	— Amplitude Modulation	
AME	— Aircraft Maintenance Engineer	
ANAL	— Analyzed Charts (Weather)	
ANO	— Air Navigation Order	
AOE	— Airport of Entry	
ARCAL	— Aircraft Radio Control of Aerodrome Lighting	
ARTC	— Air Route Traffic Control (U.S.)	
ARU	— Altitude Reservation Unit	
ARV	— Air Recreational Vehicle	
ASDA	— Accelerate Stop Distance Available	
ASDE	— Airport Surface Detection Equipment	
ASL	— Above Sea Level	
ASR	— Airport Surveillance Radar	
ATC	— Air Traffic Control	
ATCRBS	— ATC Radar Beacon System	
ATF	— Aerodrome Traffic Frequency	
ATIS	— Automatic Terminal Information Service	
ATM	— Air Traffic Management	
ATS	— Air Traffic Services	
ATZ	— Aerodrome Traffic Zone	
AUW	— All-up-weight	
AWBS	— Aviation Weather Briefing Service	
AWIS	— Aviation Weather Information Service	
AWOS	— Automated Weather Observation System	
BBS	— Bulletin Board Service	
BC	— Back Course	
BDC	— Bottom Dead Centre (Engine)	
BFO	— Beat Frequency Oscillator	
BHP	— Brake Horse Power	
BM	— Back Marker	
BMEP	— Brake Mean Effective Pressure	
BRG	— Bearing	
C or CST	— Central Standard Time	
C	— Celsius	
CA	— Conflict Alert (ACAS)	
cA	— Continental Arctic (Air Mass)	
CAE	— Control Area Extension	
Can.	— Canada	
CAP	— Canada Air Pilot	
CAR	— Canadian Aviation Regulations	
CARS	— Community Aerodrome Radio Station	
CAS	— Calibrated Airspeed	
CASRP	— Confidential Aviation Safety Reporting Program	
CAT	— Civil Aviation Tribunal	
CAT	— Clear Air Turbulence	
CAT I	— Category I	
CAT II	— Category II	
CAVOK	— Ceiling and Visibility OK	
CAVU	— Ceiling and Visibility Unlimited	
CCI	— Condition and Conformity Inspection	
CDA	— Canadian Domestic Airspace	
CDI	— Course Deviation Indicator	
CFB	— Canadian Forces Base	
CFS	— Canada Flight Supplement	
CG	— Centre of Gravity	
CMNPS	— Canadian Minimum Navigation Performance Specifications	
CNS	— Communication, Navigation and Surveillance	
C of A	— Certificate of Airworthiness	
C of R	— Certificate of Registration	
COM	— Communication	
cP	— Continental Polar (Air Mass)	
CP	— Centre of Pressure	
CRT	— Cathode Ray Tube	
CTA	— Control Area	
CVFR	— Controlled VFR	
CW	— Continuous Wave (Wireless Key)	
CZ	— Control Zone	
DADS	— Digital Altimeter Display System	
DAH	— Designated Airspace Handbook	
DCD	— Double Channel Duplex	
DCS	— Double Channel Simplex	
dB	— Decibel	

DERP	— Design Eye Reference Point	HAA	— Height Above Aerodrome	METAR	— Aviation Routine Weather Report
DEW	— Distant Early Warning	HAT	— Height Above Touchdown Zone Elevation	METO	— Maximum Except Take-Off Power
DF	— Direction Finding			MF	— Mandatory Frequency
DG	— Directional Gyro	HE	— High Level Enroute Chart	MF	— Medium Frequency
DGPS	— Differential GPS	HF	— High Frequency	MHA	— Minimum Holding Altitude
DH	— Decision Height	Hg.	— Inches of Mercury	MHz	— Megahertz
DME	— Distance Measuring Equipment	HI	— Heading Indicator	Min.	— Minutes
DND	— Department of National Defence	HIAL	— High Intensity Approach Lighting	MLS	— Microwave Landing System
DOC	— Department of Communications	HLA	— High Level Airspace	MM	— Middle Marker
DOT	— Department of Transport	HP	— Horse Power	MNPS	— Minimum Navigation Performance Specifications
D/R	— Dead Reckoning	hPa	— Hectopascal		
DRCO	— Dial Up Remote Communication Outlet	HR	— High Level Air Route	MOA	— Military Operations Area
		Hrs.	— Hours	MOCA	— Minimum Obstruction Clearance Altitude
DT	— Daylight Time	HSI	— Horizontal Situation Indicator		
DUATS	— Direct User Access Terminal System	Hz	— Hertz	mP	— Maritime Polar (Air Mass)
		H24	— Continuous Operation	MP	— Manifold Pressure
DVFR	— Defence Visual Flight Rules			Mph	— Miles per Hour
		IAF	— Initial Approach Fix	MPS	— Meters Per Second
E or EST	— Eastern Standard Time	IAS	— Indicated Airspeed	MRA	— Minimum Reception Altitude
EAS	— Equivalent Airspeed	ICAO	— International Civil Aviation Organization	MSA	— Minimum Sector Altitude
EAT	— Expected Approach Time			MSL	— Mean Sea Level
EFIS	— Electronic Flight Instrument System	IF	— Intermediate Fix	mT	— Maritime Tropic (Air Mass)
		IFR	— Instrument Flight Rules	MTCA	— Military Terminal Control Area
EGT	— Exhaust Gas Temperature	ILS	— Instrument Landing System	MVFR	— Marginal Visual Flight Rules
EHF	— Extremely High Frequency	IMC	— Instrument Meteorological Conditions		
ELT	— Emergency Locator Transmitter			N. miles	— Nautical Miles
EPR	— Engine Pressure Ratio	INS	— Inertial Navigation System	Nav.	— Navigation
ETA	— Estimated Time of Arrival	IRS	— Inertial Reference System	Navaid	— Navigation Aid
ETD	— Expected Time of Departure	ISA	— International Standard Atmosphere	NAVSTAR	— Global Positioning System
ETE	— Expected Time En Route			NCA	— Northern Control Area
		J or JET	— High Level Airway	NDA	— Northern Domestic Airspace
F	— Fahrenheit	JBI	— James Brake Indicator	NDB	— Non Directional Beacon
FA	— Aviation Area Forecast	JETS	— Joint En Route Terminal System	NM	— Nautical Mile
FAA	— Federal Aviation Administration (U.S.)			NORDO	— No Radio
		KHz	— Kilohertz	NOTAM	— Notices to Airmen
FACN	— Area Forecast	km	— Kilometer	NT Error	— Northerly Turning Error
FAF	— Final Approach Fix	KMH	— Kilometers Per Hour	NWS	— North Warning System
FANS	— Future Air Navigation System	kPa	— Kilopascal		
FAR	— Federal Air Regulations (U.S.)	kts	— knots	OAT	— Outside Air Temperature
FDCN	— Upper Winds and Temperature Forecast			OBS	— Omni Bearing Selector
		L	— Litre	OCL	— Obstacle Clearance Limit
FIR	— Flight Information Region	LAHS	— Land and Hold Short Operation	OIDS	— Operational Information Display System
FISE	— Flight Information Service En Route	Lat.	— Latitude		
		LDA	— Landing Distance Available	OKTAS	— Eighths
FL	— Flight Level	LDA	— Localizer Type Directional Aid	OM	— Outer Marker
FLT PLN	— Flight Plan	LHA	— Local Hour Angle	ONS	— Omega Navigation System
FM	— Frequency Modulation	LIAL	— Low Intensity Approach Lighting	O/R	— On Request
FMS	— Flight Management Systems	LLA	— Low Level Airspace	O/T	— On Time
FP	— Flight Plan	L/MF	— Low/Medium Frequency		
FPM	— Feet Per Minute	LO	— Low Level Enroute Chart	QNH	— Altimeter Setting
FPD	— Freezing Point Depressant	LOC	— Localizer		
FSS	— Flight Service Station	Long.	— Longitude	P or PST	— Pacific Standard Time
FT	— Aerodrome Forecast	LORAN	— Long Range Air Navigation	PAL	— Peripheral Station
FVFR	— VFR Flight Following Service	LSB	— Lower Side Band	PAPI	— Precision Approach Path Indicator
		LVL	— Level	PAR	— Precision Approach Radar
GA	— General Aviation			PAS	— Private Advisory Station
GASA	— Geographic Area Safe Altitude	M	— Mach Number	PATWAS	— Pilot Automatic Telephone Weather Answering Service
GCA	— Ground Controlled Approach	M or Mag	— Magnetic		
GDOP	— Geometric Dilution of Precision (GPS)	M or MST	— Mountain Standard Time	PCZ	— Positive Control Zone
		mA	— Maritime Arctic (Air Mass)	PIREP	— Pilot Report
GHA	— Greenwich Hour Angle	MAC	— Mean Aerodynamic Chord	PNI	— Pictorial Navigation Indicator
GMT	— Greenwich Mean Time	Mag.	— Magnetic	PPS	— Precise Position Service (GPS)
GNSS	— Global Navigation Satellite System	MANOT	— Missing Aircraft Notice	PROG	— Prognostic Chart (Weather)
		MAP	— Missed Approach Point	psi	— Pounds Per Square Inch
GP	— Glide Path or Glide Slope	Mb	— Millibar	PSID	— Design Limited Maximum Allowable Pressure Differential
GPS	— Global Positioning System	MCA	— Minimum Crossing Altitude		
GS	— Glide Slope	MDA	— Minimum Descent Altitude	PSR	— Primary Surveillance Radar
GPWS	— Ground Proximity Warning System	MEA	— Minimum En Route Altitude	PSTAR	— Student Pilot Permit Exam
		MEDEVAC	— Medical Evacuation Flight		
GRI	— Group Repetition Interval (Loran)	MEP	— Mean Effective Pressure	RA	— Resolution Advisory (ACAS)
		Met.	— Meteorology	R/A	— Radius of Action

RAAS	—	Remote Aerodrome Advisory Service
RADAR	—	Radio Detection and Ranging
RAIM	—	Receiver Autonomous Integrity Monitoring (GPS)
RAMP	—	Radar Modernization Project
RAPCON	—	Radar Approach Control
RAREPS	—	Weather Radar Reports
RASO	—	Regional Aviation Safety Officer
RAT	—	Rectified Air Temperature
RCO	—	Remote Communications Outlet
RCR	—	Runway Condition Report
RDD-1	—	Radar Digitized Display
RFSS	—	Remote Flight Service Station
RMI	—	Radio Magnetic Indicator
RNAV	—	Area Navigation
RONLY	—	Receiver Only
RPM	—	Rotations per Minute
RSC	—	Runway Surface Condition
RTF	—	Radiotelephony Frequencies
RVCS	—	Remote Vehicle Control Service
RVR	—	Runway Visual Range
Rwy	—	Runway
S.miles	—	Statute Miles
SA	—	Aviation Weather Report
SAR	—	Search and Rescue
SARSAT	—	Search and Rescue Satellite
SATCOM	—	Satellite Communications
SAWR	—	Supplemental Aviation Weather Report
SCA	—	Southern Control Area
SCATANA	—	Security Control of Air Traffic and Air Navigation
SCS	—	Single Channel Simplex
SD	—	Radar Report
SDA	—	Southern Domestic Airspace
SHA	—	Sidereal Hour Angle
SID	—	Standard Instrument Departure
Sigmet	—	Significant Meteorological Information
SIRO	—	Simultaneous Intersecting Runway Operations
SM	—	Statute Mile
SPS	—	Standard Position Service (GPS)
SSB	—	Single Side Band (HF Radio)
SSR	—	Secondary Surveillance Radar
SST	—	Supersonic Transport
STAR	—	Standard Terminal Arrival Route
STOL	—	Short Take-Off and Landing
SVFR	—	Special VFR
T	—	True
TA	—	Traffic Alert (ACAS)
TACAN	—	Tactical Air Navigation
TAF	—	Aerodrome Forecast
TAS	—	True Airspeed
TAT	—	True Air Temperature
TB	—	Track Bar Indicator
TC	—	Transport Canada
TCA	—	Terminal Control Area
TCAS	—	Traffic Alert and Collision Avoidance System (U.S.)
TCH	—	Threshold Crossing Height
TCU	—	Terminal Control Unit
TDC	—	Top Dead Centre (Engine)
TDZ	—	Touchdown Zone
TDZE	—	Touchdown Zone Elevation
T/O	—	Take-Off
TSB	—	Transportation Safety Board
TWB	—	Transcribed Weather Broadcast
Twr	—	Tower
UDF	—	UHF Direction Finder
UHF	—	Ultra High Frequency
UNICOM	—	Universal Communication, Private Advisory Station
US	—	United States
USB	—	Upper Side Band (HF Radio)
UTC	—	Co-ordinated Universal Time
V	—	Victor Airway (VOR)
VASIS	—	Visual Approach Slope Indicator System
VCS	—	Vehicle Control Service
VDF or VHF/DF	—	VHF Direction Finding Service
VFR	—	Visual Flight Rules
VFR OTT	—	VFR Over The Top
VHF	—	Very High Frequency
VLF	—	Very Low Frequency
VMC	—	Visual Meteorological Conditions
VNC	—	VFR Navigation Chart
VOLMET	—	Inflight Meteorological Information
VOR	—	VHF Omni-directional Range
VORTAC	—	VOR/TACAN
VOT	—	VOR Receiver Test Facility
VSI	—	Vertical Speed Indicator
VTA	—	VFR Terminal Area Chart
VTOL	—	Vertical Take-off and Landing
VTPC	—	VFR Terminal Procedure Chart
WAC	—	World Aeronautical Chart
WAS	—	Water Aerodrome Supplement
WL	—	Wave Length
WP	—	Waypoint
WSCN	—	Airmet Message
Wx	—	Weather
XC	—	Cross Country
Z or ZULU	—	Co-Ordinated Universal Time
Z	—	VHF Station Location Marker

The V Speeds

V_A	—	design maneuvering speed.
V_B	—	design speed for maximum gust intensity.
V_C	—	design cruising speed.
V_D	—	design diving speed.
$V_{DF/MDF}$	—	demonstrated flight diving speed.
V_F	—	design flap speed.
$V_{FC/MFC}$	—	maximum speed for stability characteristics.
V_{FE}	—	maximum flap extended speed.
V_H	—	maximum speed in level flight with rated rpm and power.
V_{LE}	—	maximum landing gear extended speed.
V_{LLO}	—	landing light operation speed.
V_{LO}	—	maximum landing gear operating speed.
V_{LOF}	—	lift-off speed.
V_{MC}	—	minimum control speed. The minimum speed at which it is possible to control the airplane in the air, with the critical engine inoperative.
V_{MCA}	—	minimum control speed in the air in a take-off configuration.
V_{MCG}	—	minimum control speed on the ground. The minimum speed at which it is possible to suffer an engine failure on takeoff and maintain control of the airplane.
V_{MCL}	—	minimum control speed in the air in an approach or landing configuration with an engine inoperative.
$V_{MO/MMO}$	—	maximum operating speed. The maximum permitted speed for all operations.
V_{MU}	—	minimum unstick speed. The minimum speed at which the airplane can be lifted off the runway without displaying any hazardous flight characteristics.
V_{NE}	—	never exceed speed.
V_{NO}	—	maximum permitted speed for normal operations.
V_{RA}	—	rough air speed. The recommended speed for flight in turbulence.
V_R	—	rotation speed.
V_{REF}	—	cross the threshold speed
V_S	—	stall speed. The minimum steady flight speed at which the airplane is controllable.
V_{SL}	—	stall speed obtained in a specified configuration. Power-off stall speed ("Clean").
V_{SO}	—	stall speed in landing configuration.
V_X	—	speed for best angle of climb.
V_Y	—	speed for best rate of climb.
V_1	—	critical engine failure speed. Take-off decision speed. The speed above which the take-off is continued, and below which the take-off is abandoned in the event of an engine failure.
V_2	—	take-off safety speed. Take-off and climb speed.
V_{2min}	—	minimum take-off safety speed.
V_3	—	flap retraction speed.

Turbine Technology

Pilots of turbine airplanes should be familiar with the following terms:
N numbers refer to speeds of various components.
Ng — gas generator speed.
Np — power turbine speed.
Nh — high rotor speed.
N — Compressor speed (rpm or percent) for a single compressor engine.

N_1 — Speed of the low pressure compressor of a dual compressor engine, or the speed of a single compressor engine equipped with a free turbine.
N_2 — Speed of the high pressure compressor of a dual compressor engine, or the free turbine speed of a single compressor engine equipped with a free turbine.
N_3 — Free turbine speed of a dual compressor engine equipped with a free turbine.

Beta Range — the ground range of operation of the propeller in a turboprop configuration. For ground operation, controllable thrust can be obtained by scheduling and co-ordinating fuel flow and blade angle according to the dictates of the power lever. The beta range refers to the angle of travel of the propeller from fine pitch up to the point of reverse propeller angle

CONVERSION TABLE

litres/imperial gallons

L	IMP	L	IMP	L	IMP
1	.22	41	9.02	81	17.82
2	.44	42	9.24	82	18.04
3	.66	43	9.46	83	18.26
4	.88	44	9.68	84	18.48
5	1.10	45	9.90	85	18.70
6	1.32	46	10.12	86	18.92
7	1.54	47	10.34	87	19.14
8	1.76	48	10.56	88	19.36
9	1.98	49	10.78	89	19.58
10	2.20	50	11.00	90	19.80
11	2.42	51	11.22	91	20.02
12	2.64	52	11.44	92	20.24
13	2.86	53	11.66	93	20.46
14	3.08	54	11.88	94	20.68
15	3.30	55	12.10	95	20.90
16	3.52	56	12.32	96	21.12
17	3.74	57	12.54	97	21.34
18	3.96	58	12.76	98	21.56
19	4.18	59	12.98	99	21.78
20	4.40	60	13.20	100	22.00
21	4.62	61	13.42	200	44.00
22	4.84	62	13.64	300	66.00
23	5.05	63	13.86	400	88.00
24	5.28	64	14.08	500	110.00
25	5.50	65	14.30	600	132.00
26	5.72	66	14.52	700	154.00
27	5.94	67	14.74	800	176.00
28	6.16	68	14.96	900	198.00
29	6.38	69	15.18	1000	220.00
30	6.60	70	15.40	2000	440.00
31	6.82	71	15.62	3000	660.00
32	7.04	72	15.84	4000	880.00
33	7.26	73	16.06	5000	1100.00
34	7.48	74	16.28	6000	1320.00
35	7.70	75	16.50	7000	1540.00
36	7.92	76	16.72	8000	1760.00
37	8.14	77	16.94	9000	1980.00
38	8.36	78	17.16	10000	2200.00
39	8.58	79	17.38		
40	8.80	80	17.60		

litres/U.S. gallons

L	U.S.	L	U.S.	L	U.S.
1	.26	41	10.83	81	21.40
2	.53	42	11.10	82	21.66
3	.79	43	11.36	83	21.93
4	1.06	44	11.63	84	22.19
5	1.32	45	11.89	85	22.46
6	1.59	46	12.15	86	22.72
7	1.85	47	12.42	87	22.99
8	2.11	48	12.68	88	23.23
9	2.38	49	12.95	89	23.51
10	2.64	50	13.21	90	23.78
11	2.91	51	13.47	91	24.04
12	3.17	52	13.74	92	24.31
13	3.44	53	14.00	93	24.57
14	3.70	54	14.27	94	24.84
15	3.95	55	14.53	95	25.10
16	4.23	56	14.80	96	25.36
17	4.49	57	15.06	97	25.63
18	4.76	58	15.32	98	25.89
19	5.02	59	15.59	99	26.16
20	5.28	60	15.85	100	26.42
21	5.55	61	16.12	200	52.84
22	5.81	62	16.38	300	79.26
23	6.08	63	16.65	400	105.68
24	6.34	64	16.91	500	132.10
25	6.61	65	17.17	600	158.52
26	6.87	66	17.44	700	184.94
27	7.13	67	17.70	800	211.36
28	7.40	68	17.97	900	237.78
29	7.66	69	18.23	1000	264.0
30	7.93	70	18.49	2000	528.4
31	8.19	71	18.76	3000	792.6
32	8.45	72	19.02	4000	1056.8
33	8.72	73	19.29	5000	1321.0
34	8.98	74	19.55	6000	1585.2
35	9.25	75	19.82	7000	1848.4
36	9.51	76	20.08	8000	2113.6
37	9.78	77	20.34	9000	2377.8
38	10.04	78	20.59	10000	2642.0
39	10.30	79	20.87		
40	10.57	80	21.14		

UNITS OF MEASUREMENT

Dimensions	Canada
Distance	Nautical Miles and Tenths
Altitudes, Heights, Elevations and Dimensions on Aerodromes and Short Distances	Feet
Horizontal Speed	Knots
Vertical Speed	Feet Per Minute
Wind Speed	Knots
Wind Direction for Landing and Taking Off	Degrees Magnetic in Southern Domestic Airspace Degress True in Northern Domestic Airspace
Wind Direction for All Other Purposes	Degrees True
Cloud Altitude and Height	Feet
Visibility	Statute Miles
Altimeter Setting	Inches of Mercury and Millibars
Temperature Surface Upper Air	Celsius
Weight	Pounds
Time	Hours and Minutes the Day of 24 Hours Beginning at Midnight Greenwich Mean Time

CONVERSION TABLES and FACTORS

TEMPERATURE

C°	F°	C°	F°
235	455	90	194
230	446	85	185
225	437	80	176
220	428	75	167
215	419	70	158
210	410	65	149
205	401	60	140
200	392	55	131
195	383	50	122
190	374	45	113
185	365	40	104
180	356	35	95
175	347	30	86
170	338	25	77
165	329	20	68
160	320	15	59
155	311	10	50
150	302	05	41
145	293	00	32
140	284	-05	23
135	275	-10	14
130	266	-15	-05
125	257	-20	-04
120	248	-25	-13
115	239	-30	-22
110	230	-35	-31
105	221	-40	-40
100	212	-45	-49
95	203	-50	-58

MULTIPLY	BY	TO OBTAIN
Celsius	9/5 then add 32	Fahrenheit
Centimeters	0.3937	Inches
Fahrenheit	Subtract 32 then x 5/9	Celsius
Fathoms	6	Feet
Feet	0.30481	Meters
Gallons (Imp.)	1.20095	Gallons (U.S.)
" "	4.54597	Litres
" (U.S.)	0.83268	Gallons (Imp.)
" "	3.785332	Litres
Inches of Hg.	33.86395	Millibars
" " "	0.491174	Lbs. sq. inch
Inches	25.4	Millimeters
Kilograms	2.20462	Pounds
Kilometers	0.62137	Miles (Statute)
"	0.539553	" (Nautical)
Lbs. sq. inch	2.036	Inches of Hg.
Litres	0.219975	Gallons (Imp.)
"	0.264178	" (U.S.)
Meters	3.28083	Feet
Miles (Nautical)	76/66	Miles (Statute)
" "	1.853249	Kilometers
" (Statute)	66/76	Miles (Nautical)
" "	1.609347	Kilometers
Millibars	0.029531	Inches of Hg.
"	0.7501	Millimeters of Hg.
Millimeters	0.03937	Inches
" of Hg.	1.33315	Millibars
Pounds	0.453592	Kilograms

AVIATION GASOLINE

Conversion table
litres/weight in pounds at 15°C

L	lbs	L	lbs
10	15.8	60	94.8
20	31.6	65	102.7
25	39.5	70	110.6
30	47.4	75	118.5
35	55.3	80	126.4
40	63.2	85	134.3
45	71.1	90	142.2
50	79.0	95	150.1
55	86.9	100	158.0

EQUIVALENTS

MINUTES	HOURS
5	.1
10	.2
15	.3
20	.3
25	.4
30	.5
35	.6
40	.7
45	.8
50	.8
55	.9

Index

Abbreviations 325
Abnormally Cold Temperature 40
Abnormally High Pressure 39
Absolute Altitude 41
Absolute Humidity 126
Absolute Temperature 127
Acceleration 58, 171
Acceleration Error (Compass) 171
Acceleration Pump 58
Accidents 102
Additives (Fuel) 57
Additives (Oil) 53
ADF 224-9
ADIZ 95, 110-1
Adiabatic Lapse Rate 127-8
Advection 126, 127
Adverse Yaw 28
Advisory Airspace (Class F) 99
Aerobatic Category 37
Aerobatics 102
Aerodromes 86-94, 175, 176, 178
Aerodrome Forecasts 156-8
Aerodrome Traffic Frequency(ATF) 91, 213
Aerodrome Traffic Zone 100, 108
Aero Engines 10, 49
Aeronautical Advisory Station 197
Aeronautical Charts 173-84
Aeronautical Rules & Facilities 86
Aeronautical Information (Charts) 178, 179
Agonic Lines 167
Aileron Drag 23
Ailerons 9, 13, 23
A.I.P. Canada/Supplements 100, 305
Air Almanac 172
Airborne Collision Avoidance
  System (ACAS) 239-40
Aircraft Call Signs 205
Aircraft Occurrences 102
Aircraft Radio Control of Aerodrome
  Lighting 88
Air Defence Identification
  Zone 95, 110-1
Air Endurance 261
Airflow 20
Airfoils 19, 24
Airframes 8, 244
Airmanship 244, 264-77
Air Masses 129-30
Airmets 160
Air Navigation 164
Air Navigation Orders 100
Airplane, The 8
Airplane, Care of the 244
Airplane Performance 253-61
Air Plot, The 193
Airport 86
Airport Advisory Service 202
Airport Runways 86-7
Airport Surveillance Radar 238, 240
Air Position 193
Air Regulations 100
Air Route 96, 99, 206
Air Safety 299
Airside Guidance Signs 87
Airspace System, Canadian 94-100
Airspeed 41-2, 184, 261
Airspeed Correction 189
Airspeed Definitions 42
Airspeed Indicator 41-2
Airspeed Indicator Errors 41-2
Airspeed Limitations 36

Air Temperature Gauge 73
Air Time 17
Air Traffic Control 103
Air Traffic Control Clearances
  and Instructions 103
Air Traffic Rules and
  Procedures 103-12
Air Traffic Services 103
Airways/Airway Systems
  ..... 95, 96, 206, 216
Alcohol 294
Alerting Service 203
Alpha Range 71
Alternator 68
Altimeter 38
Altimeter, Encoding 240
Altimeter Errors 39, 121
Altimeter, Radar/Radio 43
Altimeter Setting and Altimeter
  Setting Region 39, 94, 117-8, 153
Altitude Correction 188
Altitude Definitions 41
Altitude, Effect on Performance 253
Altitude Reservation 99
Ammeter 69
Ammeter 69
Amplitude 195
Anabatic Winds 121
Analyzed Charts 149
Angle of Attack 20
Angle of Attack Indicator 47
Angle of Incidence 25
Anhedral 30
Annual Airworthiness Information
  Report 100
Annual Change 167
Antenna 199, 217, 224-5
Anti-Cyclone 118
Anti-Icing 145
Apparent Solar Day 165
Approach or Arrival Control 202
Approach Lights 88
Apron 86
Apron Advisory Service 201
Arctic Control Area 95
Area Forecast 155-6
Area Navigation 231-6
Arrival Control (Radio) 202
Arrival Procedure 211-2
Arrival Report 160, 270
Artificial Horizon 45
Aspect Ratio 25
ASR 238, 240
Astro Compass 171-2
Asymmetric Thrust 30
Atmosphere, The 114
Atmospheric Pressure 117
Attenuation (Radio) 198
Attenuation (Weather Radar) 143
Attitude Director 231
Attitude Indicator 45-6
Augmentors 53, 66
Authorized Approach UNICOM 197
Automated Weather Observation
  System (AWOS) 150
Automatic Dependent
  Surveillance (ADS) 234, 235
Automatic Direction
  Finder (ADF) 224-9
Automatic Mixture Control 160
Automatic Terminal Information

Service (ATIS) 201
Autopilot 231
Autorotation 35
Auxiliary Drives 51
Aviation Notice 305
Aviation Weather Briefing
  Service (AWBS) 160, 203
Aviation Weather Forecasts 155-9
Aviation Weather Information
  Service (AWIS) 159, 203
Aviation Weather Reports 150-5
Avionics 200
AWOS Reports 150
Axes of an Airplane 27
Azimuth 166, 171, 184
Azimuth Card 221

B

Backfire 78
Backing (Wind) 122
Baffles (Engine) 53
Balance (Controls) 28
Balance Datum Line 251
Balance Limits 250
Balance Moment 251
Balance, Weight and 249-53
Balloons 272
Barometer, The 117
Barometric Scale 39
Basic Weight 249
Battery 68, 80, 246
Beacons, Radio 222
Bearings 166, 216, 225, 227
Bearing Indicator (ADF) 225, 229
Bearing Selector (Omni) 217
Beat Frequency Oscillator (BFO) 225
Bends 290
Bernoulli's Principle 20
Best Angle/Rate of Climb 31, 261
Best Glide Speed for
  Range/Endurance 32, 261
Beta Range 71
Black Hole Illusion 294
Blood Donations 295
B.M.E. 49
Boil-off (Oil Dilution) 55
Bonding (Radio) 200
Boost 64
Border Crossing Flights 106
Bounces 272
Boundary Layer 23
Bracketing 220
Brakes 12
Braking Technique 271
Briefing (Weather) 159, 160, 203
Broadcast 90
Bucket Target Thrust Reverser 84
Bulletin Board Service (Weather) 160
Bus Bar 68
Bush Sense 282-5
Buys Ballot's Law 119
By-Pass Engine 83

C

Cabin Briefing 267
Cabin Pressurization 14
Calibrated Airspeed 42
Camber 19
Canada Flight Supplement 86, 175-6
Canadian ADIZ 95, 110-1
Canadian Airspace System 94-100

Canadian Aviation Regulations
(CARS) 100, 305
Canard 10
Canopy Static 148
CANPASS-Private Aircaft 107
Cap Cloud 121
Carbon Monoxide 288
Carburetor 57-65, 80
Carburetor Air Filter 62, 244
Carburetor Air Temperature
Gauge 62, 72
Carburetor Icing 61-4
Categories: Normal, Utility 37
CAVOK 1 .... 153, 161, 201
Ceiling 153
Celestial Navigation 164, 171-2
Celsius Scale 127
Central Nervous System
Disturbances 290
Centre of Gravity 20, 29, 34, 250
Centre of Pressure 20, 29
Centrifugal Force 32, 119, 170
Centripetal Force 32. 170
Certificate of Airworthiness 100
Change of Lat. and Long. 165
Chart Legends 178
Charts (Maps) 149-50. 173-84
Charts (Performance) 255-60
Chokes 290
Chord 1 0
Cigarettes 289
Circadian Rhythm 295
Circuit (Traffic) 89, 91
Circuit Breakers 68
Circular Slide Rule 187
Clamshell Type Thrust Reverser 84
Class A, B, etc, Airspace 97-100
Class B Airspace 97, 109, 210
Class F Airspace 100, 179
Clean Aircraft Concept 34
Clear Air Turbulence 124, 161
Clearance (Air Traffic) 103
Clearance Delivery 201
Clearance Limit 110
Clear Ice 144
Climb Calculation 192
Climbing 31
Climb Performance 256
Climb Restriction Problems 192
Closing Angle 182
Clouds 115-7. 126, 135-6, 145
Coalescence 136,145
Coastal Effect (ADF) 228
Cockpit Check Prior to Take-Off ... 267
Cockpit Management System 237
Cockpit Resource Management 298
Code (Radar) 238-9
Col 118
Cold Front 132, 134
Cold Soaking 35
Cold Stream Thrust Reverser 84
Cold Weather Operation 79-80, 246-9
Collision Hazard/Avoidance 275-7
Commercial Pilot Exam/Licence 305
Commet 204
Common Cold 290
Communication Checks 208
Communication Equipment 199-200
Communication Facilities 201
Community Aerodrome Radio
Station (CARS) 203-4

Compass/Compass Errors 168-71
Compass Locators 222
Compass North 168, 184
Compass Rose 179, 216
Compass Swinging 169
Composite 15
Composite Flight Plan 105
Composition of Velocities 184
Compression (Atmospheric Heating) 126
Compression Ratio 51
Computers 185, 187
Condensation 125, 135, 145
Condensation Nuclei 125, 135
Confidential Aviation Safety Report 103
Conformal (Charts) 173
Conic Projection 173
Construction of an Engine 50
Construction Materials 15
Contouring (Weather Radar) 142
Contour Lines (Weather Charts) 149
Contours 177
Contrails 129
Control Area/Control Area
Extension 96
Control Locks (Pilot's Inspection
Prior to Flight) 266
Control Systems 13
Control Tower 201
Control Zone 96, 108
Controlled Airports, Traffic
Procedures 91-3, 268
Controlled Airspace 95, 96, 108
Controls (Balanced) 28
Convection 126, 128, 136
Convergence 119. 129, 136
Convergency 173
Conversion Problems 173, 188-9
Conversion Tables 328
Cooling (Engines) 53
Co-ordinated Universal Time 165, 206
Co-ordinates 164
Coriolis Effect 293
Coriolis Force 119
Corridors (Weather Radar) 142
Corrosion 15
Couples 24
Course (Omni) 215
Course Deviation
Indicator 219, 221, 223
Course Line Computer 232
Cowl Flaps 53
Cowling 10
Crab 181
Creeps 290
Critical Point 191
Critical Surface Contamination ... 247-9
Cross-Country Procedure 209-11, 269
Cross Wind Landings/Take-Offs
.... 259-60, 264-70
Cruise Performance 256-7
Cruising Altitudes 39, 107-8
Cruising Speeds 108, 261
Customs Notification Service
(ADCUS) 103, 106
Customs Regulations 106-7
Cycle 195
Cyclone 118
Cylinder Head Temp. Gauge 72

D

Danger Area 99

Dangerous Goods 103
Data Link Communications 199
Dead Reckoning 164
Dead Reckoning Computer 185
Deceleration 171
Deceleration Error (Compass) 171
Declination 167, 172
Decompression Sickness 289
Definitions 315
De-Icing 145, 248
Denalt Performance Computer 254
Density (Airspeed Error) 41
Density Altitude 40, 41, 189, 253-8
Density and Temperature 127
Departure Control (Radio) 202
Depressions 118, 131
Depth Perception 291
Design Eye Reference Point 291
Detonation 57
Deviation (Compass) 168
Deviation Indicator 219, 221, 223
Dew 126
Dewpoint 126, 153
Dial Up Remote Communication
Outlet 203
Differential Ailerons 23
Differential GPS 235
Diffraction 197
Dihedra 129
Dimensional Units 328
Dip (Compass) 166, 169
Directional Gyro 44
Directional Stability 30
Direction Finding Service (VHF) 202
Direct User Access Terminal
System (DUATS) 105, 160, 204
Discharge Wicks (Radio) 200
Disorientation 293
Distance Measuring Equipment
(DME) 220, 229-30, 232
Distress Signals 213, 279
Diurnal Variation (Temperature) 126
Diurnal Variation (Wind) 121-2
Divergence 119
Dive Speed Max. Permissible 36
DME 220, 229-30
DME Arc 230
Domestic Airspace 94
Doppler 154, 232
Double Channel Duplex and Simplex 197
Double Track Error Method 182
Downburst 123, 140
Downwash 19, 34
Drag 19, 20-4
Drag and Anti-Drag Wires 9
Drag Coefficient 22
Drift/Drift Angle 184
Drift Correction (ADF) 226
Drift Method of Finding Wind
in the Air 193
Drizzle 137, 145
Droop Wing Tip 25
Drugs 294
Dry Sump Lubrication 54
Dry Vertical Card Compass 171
Dual ADF 228
Dual Ignition 67
Dual Instruction 101
Dual Omni 220
DUATS 105, 160, 204
Dust Devils 122

DVFR Flight Plan	110
Dynamic Balance	28

E

Ear Block	289
Ear Plugs	292
Earth's Magnetism	166
Eating	295
Echo (Radar)	43, 237
Eddy Motion	122, 136
Electrical Fire	278
Electrical System	68-9
Electronic Flight Instrument System (EFIS)	236-7
Electronic Interference (Radio)	208
Elevators	10, 13
Emergency Equipment	111-2
Emergency Locator Transmitter (ELT)	111, 241-2
Emergency Procedures	110, 277-82
Emergency Signals	213-4, 279, 281
Empennage	10
Encoding Altimeter	240
Endurance	261
Engines, Aero	10, 49
Engine Failure	277-8
Engine Fault Finding Table	80
Engine Fire	278
Engine Instruments	72-5
Engine Maintenance	76
Engine Mountings	10
Engine, Operation	49, 75-81
Engine, Running Up	78
Engine, Starting	77
Engine, Winter Operations	79-80
Enroute Charts	175
En Route Flight Information Service	202
En Route Procedures	269-70
En Route Radar Surveillance	211
En Route Reports	210
Ephemeris Errors (GPS)	234
Equator	164, 166
Equilibrium	19
Equipment, Emergency	111
Equipment, Night	101
Equipment (Seaplanes and Skiplanes)	284-5
Equipment, Survival	111-2
Equivalent Airspeed	42
Evolved Gases	290
Examination Guide (Sample Questions)	305
Exhaust Gas Temperature Gauge	60
Exhaust Heaters	244
Exhaust System	65-6
Exhaust Trails	129
Exosphere	115
Explosives	103

F

Fahrenheit Scale	127
False Climb Illusion	293
Fan Marker	222
Fatigue	295
Feathering Propeller	71
Fennel's Law	119
Field Level Pressure	39
Fin	10, 30
Finding the Sun's True Bearing	172
Finding the Wind in the Air	192
Fire, Action in the Event of	278-9
Fire Extinguisher	279
Fire Wall	10
First Point of Aries	172
Fix	193, 220, 228
Flaps	9, 26-7, 34
Flight Altitude	109, 206
Flight Calculator	194
Flight Computer	185, 187
Flight Control System	230
Flight Director	230-1
Flight Information Region	100
Flight Information Service	103, 204
Flight Information Service En Route (FISE)	203
Flight Instruments	37-47
Flight Level	39
Flight Performance	30, 253-61
Flight Planning	264-5
Flight Planning Service (FSS)	203
Flight Plans/Flight Notifications/Flight Itineraries	104-6, 209
Flight Rules, VFR & IFR	104
Flight Service Station	159, 160, 202-3
Flight Time	17
Flight Visibility	138
Float Buoyancy	250
Flutter	16, 28
Fog: Radiation, Advection, Upslope, Steam, Precipitation Induced, Ice	135, 137-8
Forced Induction	64
Forced Landing Procedure	279-80
Force Feed Lubrication	54
Forces Acting on an Airplane	19
Forecasts (Weather)	155-9
Forest Fires	99
Form Drag	21
Four-Stroke Cycle	51
Freezing Point Depressant Fluids	248
Freezing Rain, Drizzle	135, 137, 145
French Language, Use of	205
Frequency	195, 206
Frequency Utilization Plan	196, 197
Frise Ailerons	23
Frontal Clouds	132-3
Frontal Lift	129, 136
Frontal Surface	131, 133
Frontal Turbulence	146
Frontogenesis and Frontolysis	134
Fronts	130-5
Frost/Frost (Icing)	34, 126, 144, 247
Fuel Consumption Problems	188
Fuel Hours	189
Fuel Icing	80
Fuel Injection	64
Fuel Management	77
Fuel Management System	59
Fuel Pump Fuel System	55
Fuel System	55-7
Fuel Tank Installations	55-6
Fuel Vapourization Ice	62
Fuels	56-7, 80, 246
Fuselage, The	8
Fuselage Fire	278-9

G

G or g (Load Factor)	16
Gases, Trapped and Evolved	289-90
Gasoline (See Fuels)	
Gastrointestinal Pain	290
Gas Turbine (Jet) Engines	81-4
Gear (Landing)	10-3
Generator	68
Generator Warning Light	69
Geographical Co-ordinates	164
Geometric Dilution of Precision (GPS)	234
Glare Ice	284
Glassy Water	283
Glaze Ice	144
Glide Path (ILS)	222
Glideslope Indicator	221, 223
Gliding	31-2
Global Navigation Satellite System (GNSS)	199, 234
Global Positioning System (GPS)	199, 234-5
Glossary	319
Go-Around	272-3
GPS	199, 234-5
Graticule	164, 177
Gravity Feed Fuel System	55
Great Circle	166, 173
Greenwich Hour Angle (GHA)	172
Greenwich Mean Time (GMT)	165
Gross Weight	16, 249
Ground-Air Emergency Signals	281
Ground Control	201, 213
Ground Effect	22, 273-4
Ground Handling of the Airplane	244-5
Ground Proximity Warning System (GPWS)	231
Groundspeed	173, 183, 184, 185
Groundspeed Check	183
Ground Visibility	138
Ground Waves (Radio)	197
Group Repetition Interval	233
Gust Conditions	121, 140, 151, 274
Gust Front	123, 140
Gust Load	17
Gyro Horizon	45
Gyro Instruments	43-7
Gyroscope	43
Gyroscopic Precession/Inertia	44
Gyrosyn Compass	47

H

Hail	127, 140, 141
Haze	138
Heading	166, 184, 185, 206
Heading Indicator	44-5
Heads Up Display System	237
Health	287
Heavy Rain (Stall)	35
Hectopascal	117
Hemispheric Prevailing Winds	120
HF Radio	195-6, 197, 200
HF Radio Transceiver	200
High Altitude Charts	175
High Level Airspace	95
High Level Weather	147-8
High Pressure Areas	118
Holding Pattern, Standard	110
Hold Short	209
Homing (ADF)	226
Horizontal Stabilizer	29
Horizontally Opposed Engine	49-50
Horizontal Situation Indicator (HSI)	221, 231
Horse Power	49
Human Factors	287, 298
Humidity (Atmospheric)	125-6
Humidity and Power	75
Hydroplaning	272
Hyperventilation	289

Hypoglycaemia 295
Hypothermia/Hypothermia 292
Hypoxia 287
Hypsometric Tints 177

I

ICAO Standard Atmosphere 115
Ice on Wings (Skiplanes) 284
Ice Pellets, Prisms 137
Icing 34, 80, 121, 140 141
..... 143-6, 148, 161, 228, 247, 284
Icing (Airspeed Error) 42
Icing (Carburetor) 61-4
Identification Zone 95, 110-1
Idle Cut-Off 59
Idling (Engine) 58
IFR and VFR 104
Ignition System 66-8
ILS 222-3
IMC and VMC 139
Impact Ice 62
Impulse Coupling (Magneto) 67
Inches of Mercury 117
Indicated Airspeed 41, 42, 184
Indicated Altitude 41
Induced Drag 21
Inertia, Gyroscopic 44
Inertial Navigation System (INS) 164, 232-3
Information Circulars 100, 305
In-Line Engine 50
Inspection, Pilot's, Prior to Flight 265-7
Inspection (Airworthiness) 17
Instruction (Air Traffic) 103
Instrument Flight Rules 101, 104
Instrument Landing System (ILS) 222-3
Instrument Lighting 291
Instrument Meteorological Conditions (IMC)) 139
Instruments, Engine 72-5
Instruments, Flight 37-47
Instruments, Plotting 179
Instruments, Radio Navigation 217-37
Intercept Heading 219, 227
Interception of Civil Aircraft 111
Interference Drag 21
International Date Line 164
Inversion 123, 128
Ionosphere 115, 198
Isobars 118, 134
Isogonic Lines/Isogonals 167, 177
Isothermal Layer 128
Isotherms 127

J

James Brake Indicator 155
JBI Reports 155
Jet Blast Hazard 263-4
Jet Lag 295
Jet Propulsion 81-4
Jet Stream 123, 124-5
Judgment 296, 297

K

Katabatic Winds 121
Keel Effect 30
Kilohertz 195
Kilometer 173
Kilopascal 117
Knot 173
Koch Chart 254

L

Lag 42, 43
Lambert Conformal Projection 173-4
Laminar Airfoils 23, 24, 275
Laminar Layer 23
Land and Hold Short Operation (LAHSO) 92
Land Breeze 120
Landing Errors 271-2
Landing Gear 10-13, 246
Landing Performance 257-61
Landing Procedure 89-93, 212, 262, 263
Landmarks 180
Lapse Rate 127-8
Latent Heat Of Vapourization, Fusion 125
Lateral Axis 28
Lateral Stability 29
Latitude 126, 164, 177
Launch Weight (Ultra-lights) 285
Layer Tinting (Charts) 177
Leading Edge Flaps 26
Leaning the Engine 60-1
Left/Right Needle 218
Lenticular Cloud 121
Let-Down Calculations 192
LF/MF Radio 195, 197
Licences 100
Lift, Lift Coefficient, Lift-Drag Ratio/Curves 19, 22
Lifting Agents (Air) 128
Lightning 140, 141
Lightning Detection Equipment 142
Light Signals, In-Flight 93
Light Signals, Taxiing 93, 269
Lights, Aerodromes 88-9
Lights, Instruments 291
Lights, Navigation 102
Lights, Night (Airplane) 102
Line of Sight Transmission 198
Line Squalls 134
Load Factor Chart 33
Load Factors 16, 33
Load Factors in Turns 33
Loads 16
Local Hour Angle 172
Localizer (ILS) 222
Localizer Approach 223
Local Mean Time (LMT) 165
Log Books 17, 100, 101
Longerons 8
Longitude 164, 177
Longitude and Time 165
Longitudinal Axis 28
Longitudinal Stability 29
Loop Antenna (ADF) 225
Loran C 233-4
Low Altitude Charts 170
Low Flying 274
Low Level Airspace 96
Low Level Nocturnal Jet Stream 123
Low Level Wind Shear 123, 157
Low Pressure Areas 118, 131
Lubber Line 167, 168
Lubrication 53-5

M

Mach Indicator 47
Mach Number 37
Macroburst 123, 140
Magnetism, Earth's 166
Magneto, The 66-7
Mandatory Frequency (MF) 90, 203, 213
Maneuvering Area 86
Maneuvering Speed 17, 36, 261
Manifold Pressure and Power 74
Manifold Pressure Gauge 73
Map Reading 181
Maps 149-50, 173-84
Maps, Weather 149-50
Marine Beacon 222
Marker Beacon (ILS) 223
Mass Balance 28
Materials, Construction 15
Max. Speeds (Flaps Down, Cruise, Normal, Gust, Diving) 36
Max. Speeds (Range, Endurance) 261
Max. Weight (Landing, Zero Fuel) 249-50
Mayday 213
Mean Aerodynamic Chord 10, 25, 251
Mean Sea Level (MSL) Pressure 117
Mean Solar Day 165
Mean Time 165
Mean Wind 150, 201
Mechanical Turbulence 121, 129, 136, 146
Megahertz 195
M.E.P. 49
Mercator Projection 174
Mercury Barometer 117
Meridians of Longitude 164, 184
Meridians, Magnetic 166, 184
Mesosphere and Mesopause 115
Metar Weather Reports 151
Meteorology 114
METO Power 49
Microburst 123, 140
Microsecond (Radar) 237
Microwave Landing System (MLS) 223-4
Mid-Air Collision Hazard 275-7
Military Terminal Control Area 97
Millibar 117
Min. Cruising Altitudes (MEA, MOCA) 108
Min. VFR Flight Altitudes 109
Mixture Control 58-60
Mode (Transponder) 238
Mode S 199, 239
MOGAS 57
Moment Arm/Index 251
Monocoque Fuselage 8
Morse International Code 207
Mountain Effect (ADF) 228
Mountain Effect (Altimeter) 40
Mountain Wave/Winds 40, 120-1
Movement Area 86
Moving Map Display 235
Multipath Error (GPS) 234

N

Nautical Mile 172
Navigation 164, 215
Navigation Lights 102
Navigation Plotter 179
Navigation Problems 184-94
Navigation Terms 184
Never Exceed Speed 36
Newton's Laws of Motion 19, 43
Night 101
Night Effect (ADF) 228

Night Vision	287, 291
Noise	291-2
Non-Directional Beacons	222, 223
NORDO	90, 92-3, 98, 99, 269
Normal Axis	28
Normal Category	37
Normal Climb	31
Normal Glide Speed	261
Normal Operating Speed	36
Normally Aspirated Engine	64
Northerly Turning Error (Compass)	170-1
Northern Domestic Airspace & Control Area	94, 95
NOTAMS	100, 155, 305
Now Wind	150, 201
NWS	111

O

Obscuration	152
Obstructions to Vision	152
Occluded Fronts/Occlusions	133
Occurrences, Aircraft	102
Octane Rating	56
Oil Change	76, 79, 246
Oil Dilution	54-5
Oil, Lubrication	53-5
Oil Pressure Gauge/ Temperature Gauge	72
Oil, Requirements	53
Oil Temperature	54
Okta	152
Oleo, The	13
Omega Navigation System	236
Omni Navigation Instruments	217-8
Omnirange Navigation System	215-22
One-in-Sixty Rule	183
Opening and Closing Angles Method (Navigation)	182
Opening Angle	181
Operational Information Display System (OIDS)	150, 201
Optimum Cruise Speed	261
Orographic Lift	129, 136
Oscillation (ADF)	228
Otolith	293
Outside Air Temperature Gauge	73
Overshoot	272-3
Over Water Flights	102
Oxidation Stability (Oil)	53
Oxygen above 10,000 ft	287-8
Ozone Sickness	288

P

Pan	214
Panic	296
PAR	238, 240-1
Parallels of Latitude	164
Parasite Drag	21
Paresthesia (Creeps)	290
Parts of an Airplane	8
Performance Charts	255-60
Performance (Flight)	30, 253-61
Performance Number (Fuel)	56
P Factor	30
Phonetic Alphabet	205
Phraseology, Radio	205, 234, 240
Physical Fitness	296
Picketing Aircraft	245, 284
Pilotage	164
Pilot Decision Making	296-8
Pilot-Navigator, The	183

Pilot Reports (PIREPS)	154, 160-1
Pilot's Automatic Telephone Weather Answering Service	160
Pilot's Inspection Prior to Flight	247, 265-7
Pilot's Licence	100
PIREPS	154, 160-1
Pitch (Axis)	28
Pitch (Propeller)	69
Pitot Pressure Source	38
Pitot-Static Instruments	37-43
Planform	25
Plotting Instruments	179
Plotting Model	149
Plotting a Track	180
Point of No Return	191
Polar Front, The	130
Polarity of a Magnet	66
Porpoising	272
Position Error (Airspeed)	42
Position Lines	193, 220, 228
Position Reports (Radio)	103-4, 209
Post Landing Check/Inspection	270
Power	49
Power Approach	32
Power Loading	16
Precautionary Landing	280-1
Precession (Gyro)	44, 45
Precession (Performance)	30
Precipitation	124, 125, 135, 136-7, 145, 152
Precipitation Static	141, 198, 228
Precision Approach Path Indicator (PAPI)	89
Precision Approach Radar	238, 240-1
Pre-Flight Inspection	247, 265-7
Pregnancy	295
Pre-Ignition	57
Pre-landing Check	270
Preparations for a Cross-Country Flight	180, 264-8
Pressure (Altimeter Error)	39
Pressure Altitude	39, 41
Pressure, Atmospheric	117-9, 134, 135, 140, 141
Pressure Gradient	119
Pressure Ratio	82, 83
Pressure Tendency	119
Pressurization (Cabin)	14
Prevailing Visibility	138, 152, 157
Primary Surveillance Radar	238, 240-1
Priority of Communications	207
Private Advisory Stations	197, 204
Private Pilot Exam/Licence	305
Prognostic (Weather) Charts	149
Projections (Charts)	173-4
Propeller, The	69-71, 80
Propeller Maintenance	77, 244
Propeller Reversing	71
Propulsion System	10
Protractor	179
Pseudo Random Codes (GPS)	234
PSTAR Exam	305

Q

Quadrantal Error (ADF)	228

R

Radar	237-41
Radar Altimeter	43
Radar Assistance	111, 202, 204, 211, 241, 282
Radar Mile	238
Radar, Weather	142-3
Radar Weather Reports	154, 160
Radial (Omni)	215, 219
Radial Engine	50
Radial Spread (Lightning Detection Equipment)	142
Radiation (Temperature)	126, 127
Radio, Radio Bands	195
Radio Altimeter	43
Radio Beacons	222
Radio, Care of	200
Radio Magnetic Indicator (RMI)	228-9
Radio Navigation	164, 215
Radio Operator's Licence	100, 305
Radio Receiver	199, 217, 225, 230
Radio Technique	207
Radiotelephone Procedure	205-14
Radio Transmitter/Receiver	199
Radius of Action	189
Radome	143
Rain	137, 142, 145, 274-5
Rain Gradient (Weather Radar)	142
Ram Jet	81
RAMP	238
Range	261
Rate of Climb Indicator	42
Receiver Autonomous Integrity Monitoring (GPS)	235
Reciprocals	183
Reciprocating Engine, Construction of	50
Recreational Pilot Permit/Exam	305
Reduction Gears	51
Refuelling	245
Relative Airflow/Wind	20, 36
Relative Bearing	225
Relative Humidity	126
Relief (Charts)	177
Remote Airport Advisory Service	203
Remote Communication Outlet	203
Reporting Point	104
Representative Fraction	176
Restricted Airspace (Class F)	99
Return to Point of Departure	183
Rhumb Line	166, 174
Ribs	9
Ridge of High Pressure	119
Right of Way	101
Rigidity	16
Rigidity in Space	30, 44
Rime Ice	144
RNAV	232
Roll (Axis)	28
RONLY (Radio Receiver Only)	94, 269
Rotor Cloud	121
Rough Water (Seaplanes)	283
RPM Indicator	73
RSC Reports	155
Rudder	10, 13
Rules of the Air (Right of Way)	100-3
Running up the Engine	78
Runways, Airport	86-7, 247
Runway Surface Condition Reports	155
Runway Visual Range/Visibility	138, 152, 207

S

S.A.E. Number	54

333

Safety, Air	299	
Safety Signals	214, 281	
SARSAT	242	
Satellite Communications	199	
Saybolt Universal Viscosity	54	
Scale (Charts)	176	
Scanning	275-6	
SCATANA	111	
Scuba Diving and Flying	290	
Sea Breeze	120	
Sea Level Pressure	117	
Seaplanes	282-3, 284	
Seat Belts	303	
Secondary Low	118	
Secondary Surveillance Radar	238-40	
Security (Radio Signal)	214	
Sense Antenna (ADF)	224	
Sensory Illusions	292-4	
Separation Point	33	
Sequential Operations	92	
Servo Tab	14	
Shear Zones	140	
Shielding	67	
Shock Absorbers	12	
Shoulder Harnesses	303	
Sidereal Hour Angle (SHA)	172	
Sigmets	160	
Signals, Distress/Urgency	213-4, 281	
Signals, Ground Control	93, 269	
Signals, Tower	93, 269	
Simultaneous Operations	92	
Single Channel Simplex	197	
Single Sideband HF	196	
Sinus Block	289	
Site Effect Error (Omni)	217	
Skin Friction	21	
Skiplanes	83-4	
Sky Condition	117, 138, 152, 157	
Sky Waves (Radio)	198	
Slant Range (DME)	229	
Slant Range Visibility	138	
Slats	26	
Slipstream	30	
Slots	26	
Slowest Descent Speed	261	
Snow Conditions (Skiplanes)	284	
Snow, Snow Grains, Snow Pellets, Soft Hall	137	
Snow (Icing)	34, 145	
Solar Day	165	
Solar Radiation	126	
Solid State Ignition	67	
Somatogravic Illusion	293-4	
Southern Domestic Airspace and Control Area	94, 95	
Space	115	
Span	10	
Span Loading	16	
Spark Plugs	51, 80	
Spars	9	
Sparsely Settled Area	95, 111	
Spatial Disorientation	293	
Special VFR	108	
Specified Area	90	
Speed Brakes	26	
Speed of Slowest Descent	261	
Speed of Sound	47	
Spinning	35	
Spiral Dive	36	
Spoilers	26	
Spot Heights	177	
Squall/Squall Line	121, 134	
Stabilator	10, 13	
Stability (Air)	125, 128-9	
Stability (Airplanes)	28-30	
Stabilizer	10, 14, 29	
Stall	33-6	
Standard Atmosphere	115	
Standard Holding Pattern	110	
Standard Pressure Region	39, 94	
Starter Motor	68	
Starting The Engine	77	
Static	141, 148, 198, 228	
Static Balance	28	
Static Discharge Wicks	200	
Static Pressure Source	38	
Stationary Front	132	
Station Model	149	
Station Pressure	117	
Statute Mile	172	
St. Elmo's Fire	141	
Stopway	87	
Strain	15	
Stratosphere and Stratopause	115, 147	
Streamlining	23	
Stress (Airplane)	15	
Stress Analysis	16	
Stress, Emotional	296	
Stressed Skin Structure	8	
Struts	9	
Student Pilot Permit/PSTAR Exam	305	
Study Material (Study and Reference Guide)	305-6	
Sublimation	125, 135	
Subsidence	126, 136	
Sub Stellar Point	172	
Suction Method, Boundary Layer Control	123	
Sunrise and Sunset	101	
Sun's True Bearing, Finding The	72	
Supercharging	64-5	
Supercooled Warm Rain Process	145	
Supercooled Water Droplets	125, 135, 137, 143, 145	
Surface Attenuation (Radio)	198	
Surface Friction	119	
Surface of Discontinuity	131	
Surface Weather Charts	149	
Surface Winds	120, 134	
Survival	111-2	
Sweepback	30	
Swinging the Compass	169	
Symbols, Air Traffic Clearances	210	
Symbols, Weather Map (See Weather Map Insert)		
Symbols, Weather Report	151-3	

T

Table of VHF Reception Distance	198	
TACAN	220, 229, 230	
Tachometer	73	
TAF Aerodrome Forecast	156	
Tail Plane	10, 29	
Tail Plane Stall	144	
Tail Section	10	
Tailwind Landings	259	
Take-Off Performance	255-6, 258, 259	
Take-Off Procedure	89-93, 209, 263, 268	
Tank Installations	55	
Taxi Procedure	208, 212, 247, 261	
Taxiways	87	
TCAS	239	
Temperature	125, 126-8, 134, 135, 153, 291, 292	
Temperature (Altimeter Error)	40	
Temperature, Effect on Performance	253	
Temperature Forecasts	158	
Temperature Scales	127	
Ten Degree Drift Lines	181-2	
Terminal Area Charts	175	
Terminal Control (Radio)	202	
Terminal Control Area	96	
Terrain Characteristic Tints	177	
Terrain Effect (ADF)	228	
Terrain Effect Error (Omni)	217	
Terrestial Radiation	126	
Tetrahedron	87	
Theory of Flight	19	
Thermal Turbulence	146	
Thermosphere	115	
Threshold	86	
Threshold Lights	88-9	
Throttle Ice	62	
Thrust	19, 20, 69, 82, 84	
Thrust and Drag	22	
Thrust Reverser	84	
Thunderstorms	123, 139-43, 291	
Time and Longitude	165	
Time, Speed and Distance Conversions and Problems	173, 188	
Timing (Valves, Ignition)	51-2, 68	
Tire Pressure	271	
To/From Indicator	218, 221	
Toothaches	289-90	
Topographical Symbols	178	
Topsails	25	
Torque	30, 69	
Tornado	118, 122, 141	
Track	182, 184, 185	
Track and Groundspeed Method of Finding Wind in Air	193	
Track Bar Indicator	218	
Track Line Computer	232	
Traffic Alert and Collision Avoidance System (TCAS)	239	
Traffic Circuit (Pattern)	89, 91	
Traffic Procedures, Aerodromes	89-94	
Transborder Flights	106	
Transceiver Radio	199-200	
Transcribed Weather Broadcasts	160	
Transition Point	23	
Transition Area	96	
Transmission of Loads	9	
Transponder	238-9	
Transportation Safety Board	102	
Transverse Mercator Projection	174	
Trapped Gases	289	
Triangle Pattern, Lost Aircraft	282	
Triangle of Velocities	185	
Trim Systems	10, 14	
Troposphere and Tropopause	114, 124, 147	
Trough of Low Pressure	118	
Trowal	133, 135	
True Airspeed	42, 184, 185	
True Altitude	40, 41	
Truss Fuselage	8	
Turbine Terminology	327	
Turbocharging	65, 80	
Turbojet, Turboshaft, Turboprop Turbofan Engines	82-4	

Turbulence 34, 121, 122, 126, 129
.... 134, 135, 136, 140,146-7,148,283
Turbulent Layer 23
Turn and Slip (or Turn and
   Bank) Indicator 46
Turnaround Bay 87
Turn Co-ordinator 47
Turning Moment 24, 69, 171
Turns 32-3, 34
Turns, Load Factors in 33
Turns (Stalling) 34
Turns to Compass Headings 171
"Two Point" Method of Flying
   a Visual Range1 83
Two Stroke Cycle 52

U

UHF Radio 197
Ultra-Light Aircraft 285-6
Uncontrolled Airports, Traffic
   Procedures 89, 268
Uncontrolled Airspace 99, 108, 112
Undercarriage 10-3
UNICOM 197
Unimproved Airstrips 258-9
Units of Distance and Speed 172-3
Universal Co-ordinated Time 165, 206
Unlatched Door in Flight 278
Upper Front 133, 135
Upper Level Weather Charts 149
Upper Level Winds 120
Upper Winds and Temperature
   Forecasts 158
Urgency Signals 214, 281
Utility Category 37

V

V Speeds 327
Vacuum Systems (Gyros) 44
Valley Breeze 121
Valves 51-2
Vapourization Ice 62
Vapour Lock 57
Vapour Trails (Contrails) 129
Variable Pitch 70
Variation 167
VDF and Homing Assistance 203
Vector 184, 202
Veering (Wind) 122
Vehicle Advisory Service 203
Velocity 184
Venturi 44
Vernal Equinox 172
Vertical Axis 28
Vertical Card Compass 171
Vertical Speed (Velocity) Indicator 42-3
Vertigo 293
VFR Alerting Service 203
VFR and IFR 104
VFR in Class B Airspace 109
VFR Navigation Charts 174-5
VFR Over The Top 109
VFR Position Reports 209
VFR Terminal Area Charts 175
VFR Visual Holding Pattern 110
VHF Data Link 199
VHF Direction Finding Service 202, 203
VHF Omnirange Navigation System....215-22
VHF Radio 196, 199
VHF Radio Transceiver 199, 217
VHF Reception Distance 198

Vibration 291, 292
Victor Airways 96, 216
Virga 140
Viscosity (Oil) 53
Visibility 134, 135, 138, 148, 152
Visibility in Rain 274-5
Vision 287, 290-1
Visual Alteration Method 182
Visual Approach Slope Indicator
   System (VASIS) 88
Visual Flight Rules 104
Visual Meteorological Conditions 139
VLF Navigation Systems 235-6
VMC and IMC 139
Volcanic Ash 161,275
Volmet Broadcasts 160
Voltage Regulator 68
Voltmeter 69
VOR Station 217
VOR Navigation System 215-22
VORTAC 220, 229, 230, 232
Vortex Generators 23

W

Wake Turbulence 261-4
Warm Fronts 132, 134-5
Wash Out/Wash In 25
Water Aerodrome Supplement 86, 176
Wave Length and Frequency 195
Weather Briefing 159, 160
Weather Broadcasts 160
Weather Charts 149-50
Weather Forecasts 155-9
Weather Information
   Sources 149, 159-61
Weather Minima for VFR Flight 108
Weather Observing Systems 150
Weather Radar 142-3
Weather Radar Network/Maps 154, 160
Weather Reports 150-5
"Weather Sense" 159
Weather Signs 148-9
Weight 19, 20, 34, 249
Weight and Balance 249-53
Weight Shift 252-3
Wet Sump Lubrication 54
Wheelbarrowing 271
Wheel Landing 274
White Out 284
Wind 119-25, 134, 141, 147
.... 151, 157, 184, 185, 192
Wind and Drift Problems 185-7, 188-94
Wind Indicators 87, 283
Wind, Mean and Now 201
Wind Shear 121, 123, 146, 147
.... 153, 157, 161, 273
Windsock/Wind Cone 87
Wing, The 9, 24-7
Wing Fence 25
Winglet 9, 22, 25
Wing Loading 16, 285
Wings, Slotted 26
Wing Tip Design 25
Wing-Tip Trails 129
Wing-Tip Vortices 21, 129, 261-4
Winter Kit (Engine) 79
Winter Operations 79-80, 246-9, 283-5
World Aeronautical Charts 175

Y

Yaw (Axis) 13, 28

Z

"Z" Time (UTC) 165, 206
Zone (Time) 165

Index

335